Birkhäuser

More information about this series at http://www.springer.com/series/4927

Frontiers in Studies of Earthquakes and Faults

Edited by
Yehuda Ben-Zion
Xiaofei Chen
Haijiang Zhang

Previously published in *Pure and Applied Geophysics* (PAGEOPH),
Volume 176, No. 3, 2019

Birkhäuser

Editors
Yehuda Ben-Zion
Department of Earth Sciences
University of Southern California
Los Angeles, CA, USA

Xiaofei Chen
Department of Earth and Space Sciences
Southern University of Science and Technology
Shenzhen, Guangdong, China

Haijiang Zhang
School of Earth and Space Sciences
University of Science and Technology of China
Hefei, Anhui, China

ISSN 2504-3625
Pageoph Topical Volumes
ISBN 978-3-030-24520-7

Cover illustration: Figure 1 taken from Structural Properties of the San Jacinto Fault Zone at Blackburn Saddle from Seismic Data of a Dense Linear Array by Pieter-Ewald Share, Amir A. Allam, Yehuda Ben-Zion, Fan-Chi Lin and Frank L. Vernon, *Pure and Applied Geophysics*, 176, 1169–1192, doi: 10.1007/s00024-018-1988-5

This book is published under the imprint Birkhäuser, www.birkhauser-science.com by the registered company Springer Nature Switzerland AG
The registered company address is: Gewerbestrasse 11, 6330 Cham, Switzerland

Contents

Pure Appl. Geophys. 176 (2019), 979–982

https://doi.org/10.1007/s00024-019-02139-x

Pure and Applied Geophysics

Frontiers in Studies of Earthquakes and Faults: Introduction

Yehuda Ben-Zion,[1] Xiaofei Chen,[2] and Haijiang Zhang[3]

The knowledge on properties and dynamics of earthquakes and faults remains highly incomplete despite considerable research spanning over 100 years. The main difficulties stem from the limited resolution of the available observations, strongly nonlinear behavior and other complexities in the governing physics, along with existence of multi-scale multi-phase (solid, granular, fluid) crustal heterogeneities. Recent theoretical and technique developments, novel laboratory experiments, dense seismic arrays, and other high quality data sets provide new opportunities for advancing the understanding of earthquakes and fault zones. These and related topics were discussed in a workshop titled "Frontiers in Studies of Earthquakes and Faults", conducted during Nov 27–Dec 1, 2017 at the Southern University of Science and Technology, Shenzhen, China. The papers in this topical issue are based on presentations given at the workshop and additional contributions, and they describe the state-of-the-art in several key frontiers. Topics covered in this volume include analysis of earthquake source properties, models of dynamic ruptures and slow slip events, imaging fault zones and the crust, detection of small earthquakes, high-resolution results from laboratory fracturing experiments, quantifying temporal changes of seismic properties, inversions of focal mechanisms to stress and more.

[1] Department of Earth Sciences, University of Southern California, Los Angeles, CA 90089-0740, USA. E-mail: benzion@usc.edu

[2] Department of Earth and Space Science, Southern University of Science and Technology, Shenzhen, China.

[3] School of Earth and Space Sciences, Laboratory of Seismology and Physics of Earth'S Interior, University of Science and Technology of China, Hefei 230026, Anhui, China.

Madariaga et al. (2019) discuss spectra of the Mw 8.4, 2014, Iquique earthquake observed in Chile at various distances from the source. They show that many displacement spectra have ω^{-1} low frequency decay associated with near-intermediate fields and the finite static displacement, and that the standard ω^{-2} decay is observed only at large distances compared to the rupture zone size. Not recognizing the near and intermediate field terms when present can lead to various erroneous interpretations of the data. Kurzon et al. (2019) present simulation results of dynamic ruptures and seismic radiation in a model governed by a continuum damage–breakage brittle rheology. Dynamic changes of elastic moduli during the rupture process produce expansion–compaction around the rupture front and isotropic source term. Significant shear heating is generated only at some distance behind the rupture front with sufficient accumulation of slip. The S/P energy partitioning and far field radiation patterns of the waves are significantly different from those in standard models without brittle damage, especially for S waves.

Ide and Yabe (2019) use a 2D probabilistic cellular automaton model to simulate slow slip events. Each cell can have at any time a state of either slip or stuck, which is updated as a Markov process that depends on the state of the cell and its neighbors. The simulation results mimic various observational properties of slow earthquakes. Separating the effects of loading and dissipation can explain variations in episodic slow slip activity. Bergen and Beroza (2019) compare four methods for producing fingerprints for a detection approach based on waveform similarity without templates from time–frequency data. They synthesize a dataset of earthquake seismograms from real earthquake and noise data recorded in northern California and use this dataset to systematically analyze the performance of each algorithm. They quantify the performance of different methods in the context of blind similarity-based detection and show that the FAST fingerprinting method generally performs best in terms of discriminative ability.

Hauksson and Meier (2019) use earthquake depth distributions in 16 distinct litholotectonic regions in southern California and constraints on strain rate and temperature to infer the crustal yield strength envelope (assuming Byerlee's friction and non-linear dislocation creep law) in each region. Five different crustal lithologies and the inferred thermo-mechanical properties agree well with expectations from past tectonics in the area. The differential yield stress is generally smaller and peaks at a shallower depth in hotter and more quartz rich regions, and is larger and peaks at greater depths for colder and predominantly diorite crust. Renard et al. (2019) present results from triaxial compression deformation experiments of low porosity sandstone under confining pressure of 10–20 MPa and room temperature, accompanied by X-ray tomography imaging at various stages of the experiments. The images allow separating the rock volume into host rock matrix, pore space and fractures, and digital volume correlations provide time series of three-dimensional incremental strain tensor fields throughout the experiments. The samples deform after the yielding level first by dilatant opening and propagation of microfractures, and then by shear sliding via grain rotation and strain localization producing system-size faults.

Share et al. (2019b) derive high-resolution V_P and V_S models for the southern California plate boundary from a triple-difference tomography incorporating both event-pair and station-pair differential times. The results include low velocities along major fault segments, V_P/V_S anomalies, and clear across-fault velocity contrasts. The geometries of contrasts in seismic properties across fault sections imply a near vertical San Andreas fault northwest of San Gorgonio Pass, and northeast dipping San Andreas fault southeast of that area. Abolfathian et al. (2019) invert earthquake focal mechanisms around the San Jacinto fault zone in southern California for the background stress field, and compare the results with coseismic strain parameters derived from earthquake potencies and *b*-values of the seismicity. The results indicate clockwise rotation of maximum horizontal compression direction with increasing depth, along with changes in the principal stress and strain plunges. Significant stress changes are not observed across the time of the 2010 Mw 7.2 El Mayor–Cucapah earthquake $\sim$ 150 km to the southeast of the study area.

Share et al. (2019a) use earthquake waveforms recorded by a dense linear array to image the internal structure of the San Jacinto fault zone at Blackburn Saddle. Analysis of fault zone head waves reveal a deep bimaterial interface with significant velocity contrast that extends at least 50 km to the southeast. Analyses of body wave delay times and trapped waves indicate a damage zone northeast of the fault that is several hundred m wide and includes an inner trapping structure with width of $\sim$ 150 m and depth extent of $\sim$ 2 km. The results are consistent with preferred propagation direction of earthquake ruptures in the area to the northwest. Gu et al. (2019) image the region around the Tanlu fault zone in eastern China using ambient noise data recorded by a local seismic array. Fundamental-mode Rayleigh waves with periods in the range 0.2–5 s are constructed from the vertical component of the noise data and are inverted to 3D shear wave velocity model using a direct surface-wave tomographic method. The fault zone is shown to be associated with shallow high-velocity anomaly corresponding to metamorphic rocks. Low-velocity anomalies to the west of the fault are associated with thick sedimentary rocks in the Hefei Basin.

Huang and Yao (2019) use an improved compressive-sensing back-projection method to study seismic radiation by the 2008 Mw 7.9 Wenchuan earthquake, China, in multiple frequency bands. The radiation from the earthquake rupture is shown to be frequency-dependent and related to geometrical fault properties. Low frequency waves are radiated from smooth fault segments with large coseismic slip. High frequencies are radiated primarily from fault complexities that produce strong changes of rupture velocity. Jiang et al. (2019) relocate earthquakes around the time of the 2016 Yunlong Ms5.0 earthquake in China and use these as templates to detect additional events. The derived catalog has about three times more events than those listed in the Yunnan Seismic Networks catalog. The results suggest three stages of seismic activity during the Yunlong earthquake sequence. The event locations and refined focal mechanisms for a dozen $M \geq 3$ events provide information on the fault strike and dip angles.

Obermann et al. (2019) use 3-D sensitivity kernels of multiply scattered body and surface waves in a heterogeneous solid to provide constraints on the depth of changes of seismic velocities inferred from analysis of coda waves. The accuracy of the estimated location depends on the sources-receivers geometry and proper characterization of scattering at the subsurface. The method is applied to estimate the depth of velocity changes obtained from correlations of ambient noise in the region around the 2008 Wenchuan earthquake. Yang et al. (2019) use synthetic calculations in simple 1-D layered models with low velocities and attenuation coefficients in the top 2 km to examine effects of shallow and deep changes of seismic properties on Rayleigh waves. Large structural changes limited to the top 1–3 km are found to produce significant phase-velocity drops in the period range 5–20s. The results indicate that analyses using different frequency ranges and calculations of delay times over multiple period bands are essential for resolving the depth range of temporal changes of properties.

Wollin et al. (2019) invert earthquake focal mechanisms for the background stress field in the Sea of Marmara region in Turkey. The results show an overall combined strike-slip and normal faulting consistent with the transtensional setting along with internal variations. The eastern Sea of Marmara has primarily a strike-slip regime in the north and normal faulting in the south. The central Marmara region is dominated by normal faulting, while the western Sea of Marmara has primarily strike-slip regime. The results suggest that future large events can be associated with either strike-slip or normal faulting. Kaduri et al. (2019) assess aseismic deformation along sections of the North Anatolian fault in Turkey based on microstructural and chemical analyses of rock samples collected along locked and creeping sections of the fault. The relation between finite strain and volume change is used to quantify the evolution of the penetrative pressure solution cleavage mechanism of creep. The cumulative aseismic displacement is estimated with the aid of geodetic and geologic data to be in the range 9–49% of the total displacement in the creeping sections and negligible in locked sections.

Li (2019) performs stress inversions of focal mechanisms in the focal region of the Tohoku-Oki mainshock for six time intervals between 2006 and 2017, and uses a 2D model to estimate from the inversion results corresponding changes in the deviatoric stress field. The results suggest that the stress accumulation may have had an acceleration phase several years before the Tohoku-Oki earthquake. This was followed by a significant release of the deviatoric stress field during the Tohoku-Oki event and rapid post mainshock reloading. Yoshida et al. (2019) analyze earthquake focal mechanisms and stress inversion results to examine triggering mechanisms of seismicity in inland Tohoku by the 2011 M9 Tohoku-Oki mainshock. Earthquake clusters triggered by the Tohoku-Oki earthquake tend to occur in regions having similar orientations of the background stress field to that associated with the static stress change by the Tohoku-Oki mainshock. Swarm sequences in central Tohoku in regions associated with stress reduction by the Tohoku-Oki earthquake may have been triggered by fluid movements along major geological boundaries.

Acknowledgements

We are grateful to Ying Chen for valuable help with the organization of the 2017 workshop on Frontiers in Studies of Earthquakes and Faults. The workshop was funded by the Southern University of Science and Technology, Shenzhen, China. We thank the authors of the papers for their contributions and the referees for critical reviews that improved the scientific quality of the volume. The referees include Yinshuang Ai, Greg Beroza, Peter Bird, Nicolas Brantut, Marine Denolle, Georg Dresen, Ahmed Elbanna, William Frank, Jeanne Hardebeck, Gregor Hillers, Renée Heilbronner, Chris Johnson, Jianshe Lei, Zefeng Li, Yajing Liu, Patricia Martínez-Garzón, Diane Moore, Fenglin Niu, Anne Obermann, Zhigang Peng, Zacchary Ross, Antonio Rovelli, Wayne Thatcher, Takahiko Uchide, Ersheng Wang, Jianping

Wu, Qingju Wu, Hongfeng Yang, Yingjie Yang, Keisuke Yoshida, Ilya Zaliapin, Miao Zhang.

References

Abolfathian, N., Martínez-Garzón P, & Ben-Zion, Y. (2019). Spatio-temporal variations of stress and strain parameters in the San Jacinto fault zone **(that issue)**.

Bergen K. J, & Beroza, G. C. (2019). Earthquake fingerprints: extracting waveform features for similarity-based earthquake detection **(that issue)**.

Gu, N., Wang, K., Gao, J., Ding, N., Yao, H., & Zhang, H. (2019). Shallow crustal structure of the Tanlu fault zone near Chao Lake in eastern China by direct surface wave tomography from local dense array ambient noise analysis **(that issue)**.

Hauksson, E, & Meier, M. (2019). Applying Depth Distribution of Seismicity to Determine Thermo-mechanical Properties of the Seismogenic Crust in Southern California: Comparing Litho-tectonic Blocks **(that issue)**.

Huang T, & Yao, H. (2019). Coseismic Radiation of the 2008 Mw 7.9 Wenchuan Earthquake and its Relationship with Fault Complexities **(that issue)**.

Ide, S, & Yabe, S. (2019). Two-dimensional probabilistic cell automaton model for broadband slow earthquakes **(that issue)**.

Jiang, J., Li, J., & Fu, H. (2019). Seismicity analysis of the 2016 Ms5.0 Yunlong earthquake, Yunnan, China and its tectonic implications **(that issue)**.

Kaduri, M., Gratier, J.-P., Lasserre, C., Çakir, Z., & Renard, F. (2019). Quantifying the partition between seismic and aseismic deformation along creeping and locked sections of the North Anatolian Fault, Turkey **(that issue)**.

Kurzon, I., Lyakhovsky, V., & Ben-Zion, Y. (2019). Dynamic rupture and seismic radiation in a damage-breakage rheology model **(that issue)**.

Li, T.-J. (2019). Long-term evolution of the deviatoric background stress in the focal region of the 2011 Tohoku-Oki earthquake **(that issue)**.

Madariaga, R., Ruiz, S., Rivera, E., Leyton, F., & Baez, J. C. (2019). Near-field Spectra of Large Earthquakes **(that issue)**.

Obermann, A., Planes, T., Larose, E., & Campillo, M. (2019). 4-D imaging of subsurface changes with coda waves: numerical studies of 3-D combined sensitivity kernels and applications to the Mw7.9, 2008 Wenchuan earthquake **(that issue)**.

Renard, F., McBeck, J., Cordonnier, B., Zheng, X., Kandula, N., Sanchez, J. R., Kobchenko, M., Noiriel, C., Zhu, W., Meakin, P., Fusseis, F., & Dysthe, D. K. (2019). Dynamic in situ three-dimensional imaging and digital volume correlation analysis quantify strain localization and fracture coalescence in sandstone **(that issue)**.

Share, P.-E., Allam, A, A., Ben-Zion, Y., Lin, F.-C., Vernon, F. L. (2019a). Structural properties of the San Jacinto fault zone at Blackburn Saddle from seismic data of a dense linear array **(that issue)**.

Share, P.-E, Guo, H., Thurber, C. H., Zhang, H., & Ben-Zion, Y. (2019b). Seismic imaging of the Southern California plate boundary around the South-Central Transverse Ranges using double-difference tomography **(that issue)**.

Wollin, C., Bohnhoff, M., Vavryčuk, V., & Martinez-Garzon, P. (2019). Stress inversion of regional seismicity in the Sea of Marmara Region, Turkey **(that issue)**.

Yang, C., Li, G., Niu, F., & Ben-Zion. Y. (2019). Significant effects of shallow seismic and stress properties on phase velocities of Rayleigh waves up to 20 s **(that issue)**.

Yoshida, K., Hasegawa, A., Yoshida, T., & Matsuzawa, T. (2019) Heterogeneities in stress and strength in Tohoku and its relationship with earthquake sequences triggered by the 2011 M9 Tohoku-Oki earthquake **(that issue)**.

(Published online February 26, 2019)

Pure Appl. Geophys. 176 (2019), 983–1001

https://doi.org/10.1007/s00024-018-1983-x

Pure and Applied Geophysics

Near-Field Spectra of Large Earthquakes

Raul Madariaga,[1] Sergio Ruiz,[2] Efrain Rivera,[2,3] Felipe Leyton,[4] and Juan Carlos Baez[4]

Abstract—We have studied the spectra of large subduction earthquakes in Chile at short epicentral distances. In this work, we concentrate on the M_w 8.2, Iquique earthquake of 1 April 2014. For such a large event, in most stations we cannot separate P and S waves; which arrive very close together and interfere to produce the static displacement field. We find that the displacement spectra observed in all the accelerograms are significantly different from the usual far-field Brune spectrum. Displacement spectra have clear omega-1 decay at low frequencies that we prove to be associated with the dominant role of near and intermediate field waves in the ground motion. The origin of the omega-1 decay is that at short distances, displacement contains a finite static displacement. We confirm this spectral behavior by comparing the spectra computed from accelerograms integrated to velocity with co-located GNSS recordings. Both spectra are flat at low frequencies and the low-frequency asymptote of the velocity spectrum is proportional to the static displacement determined from GNSS. We explore the transition from these near-field records to far-field ones and find that the usual omega-2 spectrum appears only at large distances compared to the size of the dynamic rupture zone. For large earthquakes, the displacement spectra have a term proportional to the moment spectrum and another proportional to moment rate spectrum. The term proportional to moment dominates the low-frequency behavior of accelerograms in the near and intermediate field, so that they do not follow the omega-square decay of the moment rate spectrum.

Key words: Earthquakes, Chile, seismic spectra.

1. Introduction

Large earthquakes in subduction zones are particularly dangerous because they produce extensive damage, but their spectral characteristics in the near field have been only recently studied because of lack of local unclipped data. In Chile, we can use a combination of new geodetic and acceleration data to study the waves radiated by large earthquakes at low frequencies in addition to the best known high frequencies.

Northern Chile is an active seismic zone, where large-magnitude earthquake occur frequently (Ruiz and Madariaga 2018). The intraplate intermediate-depth event of 2005 Tarapacá M_w 7.8 (Delouis and Legrand 2007; Peyrat et al. 2006) prompted the deployment of seismological stations within the Integrated Plate Boundary Observatory Chile—IPOC, a multi-component network created by German, French and Chilean researchers (IPOC 2006). Since then, northern Chile earthquakes have been well recorded by GNSS, broadband and strong motion stations located on hard rock sites (Leyton et al. 2018a, b; Báez et al. 2018). The first large earthquake after the installation of IPOC was the M_w 7.7 Tocopilla earthquake of 2007 (Delouis et al. 2009; Peyrat et al. 2010; Schurr et al. 2012; Fuenzalida et al. 2013). Lancieri et al. (2012) studied the spectral characteristics of this earthquake and its aftershocks. They found that all aftershocks had a typical omega-squared spectrum (Aki 1967; Brune 1970; Madariaga 1976; Madariaga and Ruiz 2016). The main shock, on the other hand, had a different kind of spectrum that diverged at low frequencies, increasing like omega-1. At that time few accelerometers were available and these authors proposed that this behavior was probably due to near-field waves, but that it could also be due to the complexity of this double event. These observations suggested that the ground motion spectrum of large Chilean earthquakes was different in the near and intermediate field from the usual omega-squared

[1] Laboratoire de Géologie, PSL Université, Ecole Normale Supérieure and CNRS, 24 rue Lhomond, 75230 Paris Cedex 05, France. E-mail: madariag@geologie.ens.fr
[2] Department of Geophysics, Universidad de Chile, Blanco Encalada 2002, Santiago, Chile.
[3] Department of Geology, Universidad de Chile, Plaza Ercilla, Santiago, Chile.
[4] Centro Sismológico Nacional, Universidad de Chile, Blanco Encalada 2002, Santiago, Chile.

model proposed by Aki (1967) and Brune (1970) for events recorded in the far field.

The Aki–Brune spectrum has been successfully used to invert the dynamic characteristics of earthquake ruptures using far-field P or S waves (Abercrombie 1995; Ide and Beroza 2001; Prieto et al. 2004; Allmann and Shearer 2009; Oth et al. 2010; Lancieri et al. 2012; Archuleta and Ji 2016; Denolle and Shearer 2016; and many others). Here, we focus on a different aspect of spectral records, characterizing the behavior of large-magnitude earthquakes recorded with both GNSS and strong motion instruments in the near field.

The first very large Chilean earthquake recorded by modern seismological data was the 2010 M_w 8.8, Maule megathrust earthquake, which produced excellent continuous GNSS records that were used by Vigny et al. (2011) as seismograms to model this event. The displacement records of this earthquake had large static co-seismic displacements. Unfortunately, no digital good-quality accelerometers were located close to the epicenter of the Maule 2010 earthquake (Ruiz et al. 2012). After this earthquake, the Centro Sismologico National (CSN) of the University of Chile was created and deployed a large network of broadband, accelerometers and GNSS stations (Barrientos et al. 2018; Leyton et al. 2018a, b; Báez et al. 2018). These stations that include now those installed by IPOC have recorded several large earthquakes, such as the M_w 8.2 Iquique event of 1 April 2014, the M_w 8.3 Illapel earthquake of 15 September 2015 and the M_w 7.6 Chiloé earthquake of 25 December 2016. These events provide excellent recordings that have been largely used to model the events and to study their principal characteristics (e.g., Ruiz et al. 2014, 2016, 2017a, b; Schurr et al. 2014; Hayes et al. 2014; Lay et al. 2014; Melgar et al. 2016, 2017; Tilmann et al. 2016; Ye et al. 2016; Lange et al. 2017). These studies have been centered on the slip distribution, tsunami effects, nucleation process and their relation with slow slip events, but overlook the spectral properties of strong motion records in the near field. Here, we examine the general properties of these accelerograms. We use the records written by the Iquique earthquake of 1 April 2014, because this is the Chilean earthquake with the largest number of co-located GNNS and strong motion records on hard rock sites (Leyton et al. 2018a; Báez et al. 2018). Our goal is to understand the basic features of displacement spectra, the relative role of low and high frequencies of the spectrum and their relation with the seismic moment and seismic moment rate histories.

2. *The Iquique Earthquake and Its Near Field*

On 1 April 2014, a magnitude M_w 8.2 hit the northern Chile region of Tarapacá near the City of Iquique, and 2 days later the largest aftershock with magnitude M_w 7.7 occurred (see Fig. 1) (Ruiz et al. 2014; Schurr et al. 2014; Hayes et al. 2014; Lay et al. 2014; Duputel et al. 2015). This event has been studied carefully because it was preceded by a long series of precursory shocks that began several years before 2014 and culminated in an intense, but intermittent series of foreshocks that started in July 2013 (León-Ríos et al. 2016; Cesca et al. 2016). A slow slip event was observed before the main shock that has been carefully documented (Ruiz et al. 2014; Kato and Nakagawa 2014; Socquet et al. 2017). The main event was studied by a number of researchers using a combination of far- and near-field data that were reviewed by Duputel et al. (2015) and, recently, its post-seismic deformation has been analyzed by Hoffmann et al. (2018).

The Iquique seismic sequence was recorded by a large set of multi-parameter stations of the IPOC networks. In addition to these data, several other acceleration and GNSS stations deployed by the CSN (Leyton et al. 2018a; Báez et al. 2018) are available. The full set of stations considered in this work are listed in Table 1 and plotted in Fig. 1.

The Iquique main shock started with an immediate foreshock located some 30 km ocean-ward of the main rupture. In Fig. 2, we plot the EW accelerograms at two stations, PSGCX and TA01, and their locations can be found in Fig. 1. The two records show that the Iquique event started with an immediate foreshock that occurred at 23:46:46 (T_1) and was followed by the main event at time T_2 = 23:47:03 (Ruiz et al. 2014). The difference between T_1 and T_2 (17 s) is larger than the arrival time of the S waves from the first event at many stations, so that after a

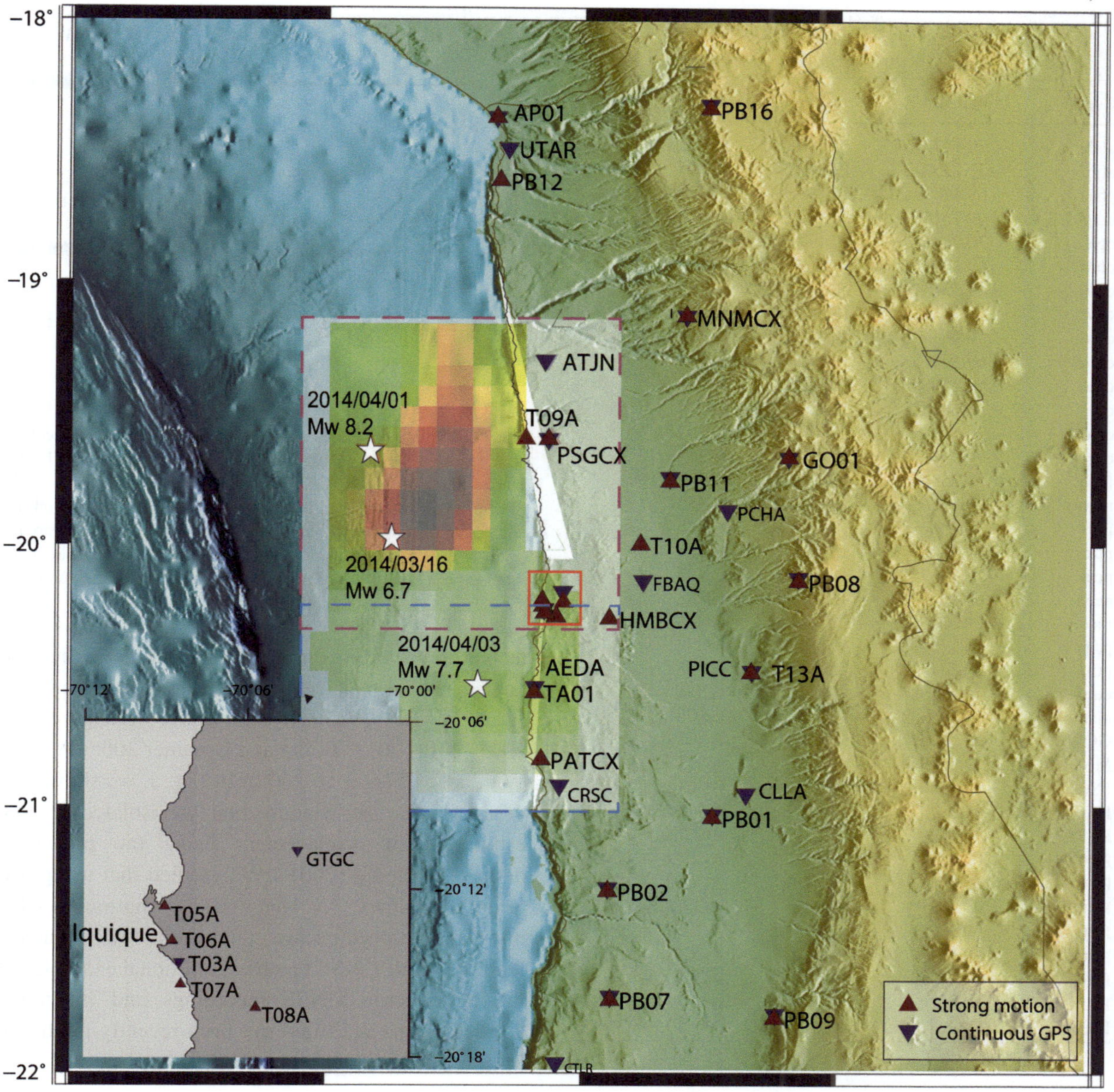

Figure 1
Map of northern Chile near the Iquique earthquake of 1 April 2014. The slip distribution of the main event and its main aftershock of 3 April 2014 are shown with colors (from Ruiz et al. 2014). White stars show the epicenters of both events in addition to the 16 March 2014 foreshock. The events are situated between the trench and the Chilean coastline along the plate interface. The main stations used in the present study are shown on the continent. The inset shows the instruments available around the City of Iquique

few seconds the P and S waves were mixed in the accelerograms. This prevents us from separating the P and S far-field waves associated with the rupture of the main slip patch (main asperity) of this large-magnitude event.

First, we focus on the recordings of the accelerometer at station PSGCX and the collocated PSGA GNSS site, both situated near the City of Pisagua in northern Chile. These stations are the closest to the epicenter of the nucleation foreshock as

Table 1

Northern Chile instruments

Code	Longitude	Latitude	Type
aeda	− 70.1777	− 20.54614	GNSS
atjn	− 70.136738	− 19.3008 0	GNSS
cbaa	− 68.448198	− 22.746394	GNSS
cdlc	− 69.761555	− 22.189982	GNSS
cgtc	− 70.068623	− 20.177139	GNSS
chm2	− 69.194134	− 19.668575	GNSS
clla	− 69.35664	− 20.95448	GNSS
chyt	− 70.342112	− 18.3708285	GNSS
cgtc	− 70.068623	− 20.177139	GNSS
chm2	− 69.194134	− 19.668575	GNSS
clla	− 69.35664	− 20.95448	GNSS
colc	− 68.638617	− 19.27621	GNSS
crsc	− 70.07976	− 20.91768	GNSS
ctlr	− 70.096824	− 21.964205	GNSS
fbaq	− 69.755396	− 20.134643	GNSS
mnmi	− 69.595661	− 19.131416	GNSS
pccl	− 70.106751	− 18.45766	GNSS
ptre	− 69.57451	− 18.19452	GNSS
pccl	− 70.106751	− 18.45766	GNSS
pcha	− 69.431944	− 19.869444	GNSS
picc	− 69.334617	− 20.489858	GNSS
psga	− 70.12301	− 19.59738	GNSS
ptre	− 69.57451	− 18.19452	GNSS
uapf	− 70.141388	− 20.243055	GNSS
utar	− 70.296388	− 18.490556	GNSS
GO01	− 69.1942	− 19.6686	ACCEL
MNMCX	− 69.5955	− 19.1311	ACCEL
PSGCX	− 70.1230	− 19.5972	ACCEL
HMBCX	− 69.8879	− 20.2782	ACCEL
PATCX	− 70.1529	− 20.8207	ACCEL
PB01	− 69.4874	− 21.0432	ACCEL + GNSS
PB02	− 69.8960	− 21.3197	ACCEL + GNSS
PB03	− 69.7531	− 22.0485	ACCEL + GNSS
PB07	− 69.8862	− 21.7267	ACCEL + GNSS
PB08	− 69.1534	− 20.1411	ACCEL + GNSS
PB09	− 69.2419	− 21.7964	ACCEL + GNSS
PB11	− 69.6558	− 19.7610	ACCEL + GNSS
PB12	− 70.3281	− 18.6141	ACCEL + GNSS

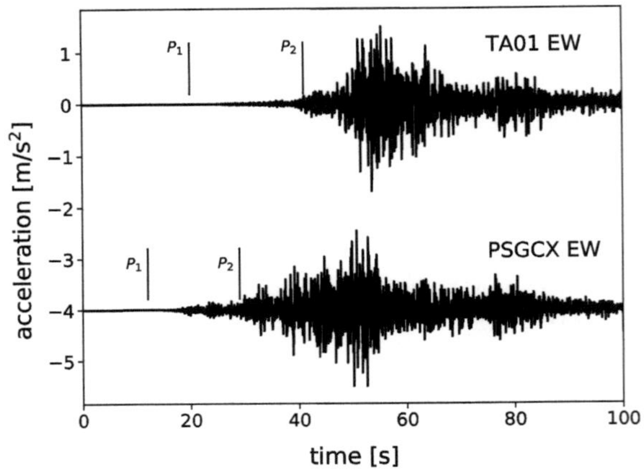

Figure 2
The 2014 Iquique main shock recorded at two near-field accelerograms located in Pisagua (PSGCX) and the Iquique airport (TA01). The zero line of the PSGCX record has been displaced by 4 s. The P1 and P2 arrows denote the times of first P-wave arrival from the immediate foreshock and the P-wave arrival from the main rupture, respectively. At these stations the P2 wave arrived at about the same time as the S wave generated by the immediate foreshock (P1)

shown in Fig. 1. Figure 3 shows the displacement record obtained at the EW component of the GNSS geodetic station on top of which we have superposed the displacement field doubly integrated from the EW accelerogram recorded at the PSGCX accelerometer. The accelerogram was integrated using the method proposed by Boore (2001) which consists of isolating a central time window in the signal where the strongest motion is concentrated, and then removing the trend outside the window. Variations of this technique have been proposed some time earlier and later extended by other authors (Graizer 1979, 2010; Iwan et al. 1985; Boore 2001; Wang et al. 2003, 2011, 2013; Boore and Bommer 2005; Wu and Wu 2007; Chao et al. 2009). In Fig. 3, we show with vertical lines the window that we isolated for the estimation of displacement using the procedure defined by Boore (2001). We verified that using different windows for the strong motion double integration procedure, we could fit the low-frequency features of the GNSS records. Additional examples of the fit between GNSS recordings and integrated accelerograms in Chile have been recently published by Báez et al. (2018).

3. *Computing Ground Motion Spectra from Accelerograms*

We will now compute the spectra of the records shown in Fig. 3. For large earthquakes in the near field, it is not possible to separate P and S waves because these waves are delayed by less time than the total seismic source duration of the earthquake. For the Iquique earthquake, the total duration is about 50 s (Figs. 2, 3). Figure 3 shows that the near-field

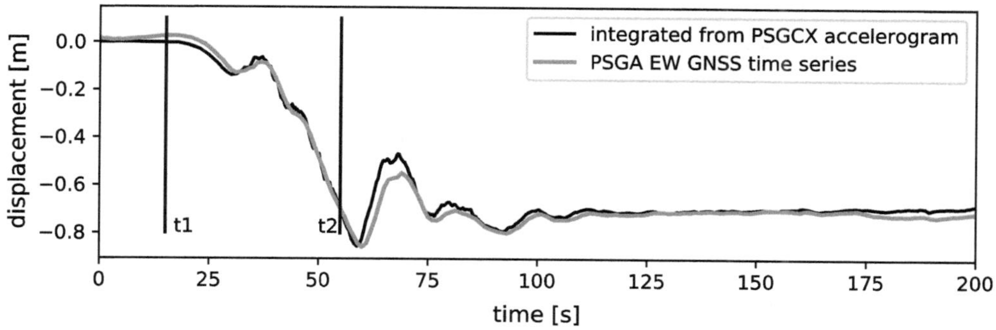

Figure 3
Comparison of the GNSS EW displacement at station PSGA (Pisagua) and that obtained from the integration of the collocated PSGCX accelerogram using Boore's (2001) method. The vertical lines define the beginning t_1 and end t_2 of the strong motion window used to integrate the accelerograms. The GNNS data has been corrected for leap seconds

term starts immediately after the arrival of P waves and decreases significantly before the S-wave arrival. After the end of the S waves (and possibly of surface waves), displacement becomes constant and approximately equal to the static displacement observed in the GNSS station PSGA.

Unfortunately, accelerograms processed by the Boore (2001) procedure cannot be used to compute spectra, because the removal of the base line is nonlinear. The spectra of the displacement signals computed in this way will contain the effect of the linear trends and the transition of the zero line in the strong-motion window. It is possible to improve the integration of accelerograms using displacement records obtained at collocated GPS instruments as proposed by several authors (e.g., Bock et al. 2011; Wang et al. 2013, among others). Unfortunately, even if the integration of accelerograms is improved, the displacement records still cannot be used to compute displacement spectra because of the finite displacement jump at the end of the record. The reason is that the numerical Fourier transform used to compute the spectra from signals like those of Fig. 3 assumes that the time series is periodic with a period equal to the duration of the record. Thus, the computed Fourier transform sees a jump in displacement at the end of the record and is contaminated at all frequencies. The Fourier transform of such a jump is simply $u_s/i\omega$, where u_s is the static displacement. All the other spectral information contained in the accelerogram is hidden by this jump. Many techniques have been proposed in the literature to remove this effect of the finite time window. Some of them consist in using a window to multiply the time signal, but these windows contaminate the low-frequency contents. In Fig. 4, we show the displacement spectrum computed directly from the displacement signal integrated by the Boore (2001) procedure shown in Fig. 3. In this figure, the numerical Fourier transform of the displacement trace is shown in black. As expected, it is a straight line of slope − 1 with amplitude proportional to the static displacement. For this reason we have chosen a method that exploits the finiteness of the velocity time series. In the same figure, we plot in gray the displacement spectrum computed from the velocity traces using the method described in the next section

We propose a simple method to compute the displacement spectrum that uses a property of the velocity time series. The accelerograms integrated once to determine the ground velocity have been shown to be very well fitted at low frequencies by the ground velocity derived from GNSS signals (e.g., Báez et al. 2018). Figure 5 shows the three components of the ground velocity integrated from accelerograms at the PSGCX station, and the velocities derived from the PSGA GNSS recordings. The instrument response of the accelerograms was removed, and the records have been low-pass filtered with a causal Butterworth filter of order 2 and corner frequency 0.5 Hz. The coincidence between the records is excellent. This representation is better than comparisons between displacement records to show that the frequency content of the two records is the

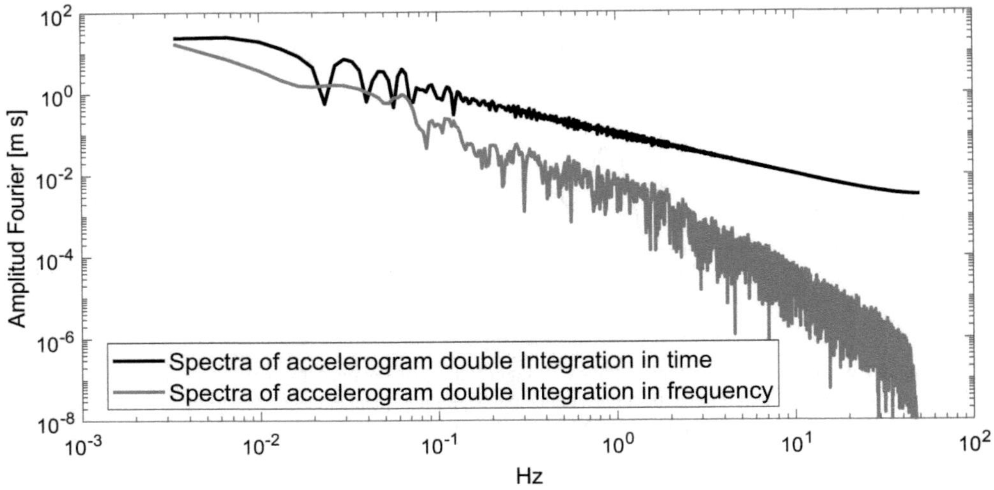

Figure 4
Finite Fourier transform of the displacement trace at station PSGCX shown in Fig. 3. The black spectrum was computed directly from displacement obtained by doubly integrating the accelerogram shown in Fig. 3, without correction for the jump in displacement at the end of the time series. In gray, we show the spectrum computed for the displacement obtained by double integration in the spectral domain (i.e., dividing the acceleration spectrum by ω^{-2})

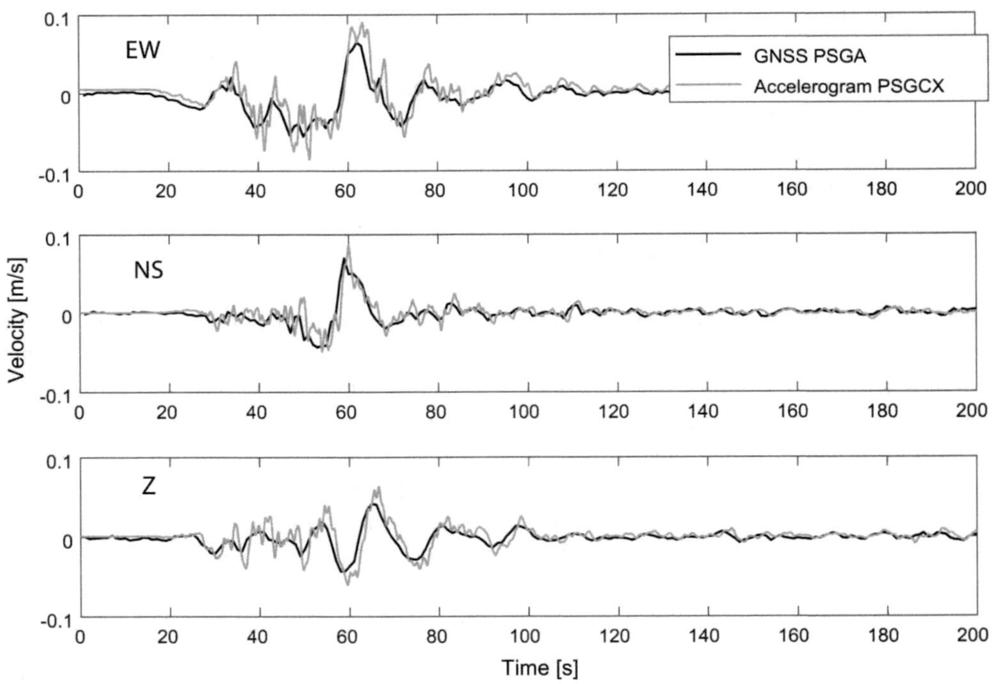

Figure 5
Comparison of ground velocity determined at the Pisagua stations from GNSS records (PSGA) and co-located integrated accelerogram (PSGCX). The velocity records obtained from accelerograms have been low-pass filtered at 0.5 Hz to enhance the similarity between these records in their common band pass

same in the frequency range where they both coincide (conservatively estimated as above 0.15 Hz).

In Fig. 6, we show the Fourier spectra of the velocity traces at station PSGCX plotted in Fig. 5. Since velocity returns to zero at the end of traces, except for seismic noise, we can compute the velocity spectrum without the problems of the finite displacement at the end of the time window discussed above for Figs. 3 and 4. The spectra computed using regular FFT routines display a common property of ground velocity for the Iquique and other large subduction earthquakes in Chile. The spectral trend is flat at low frequencies in contrast to predictions of far-field velocity radiation for finite sources (Brune 1970; Madariaga 1976) that should linearly increase with ω at low frequencies. A more detailed explanation will be provided in the next section, but it is relatively easy to understand. At close distances from the source, the near-field terms of the Green function dominate the radiated spectrum. The near field is dominated by the moment time function, not the moment rate, so that at close distances the velocity spectrum at low frequencies resembles that of the integral of the far-field source time function. We can now compute displacement spectra from the velocity spectra by simply dividing them by $i\omega$, that is, by the infinite integration operator in the frequency domain. The spectra thus calculated remove the spurious high-frequency ω^{-1} spectral decay of the Fourier transforms computed from a finite displacement time series in the near field. The displacement spectrum of the ground motion is shown in gray in Fig. 4.

A particular property of the velocity spectra in Fig. 6 is that the Fourier transform of the ground velocity in the near field must approach at low frequencies the static displacements observed in Fig. 3. From classical Fourier transform theory, the spectrum of ground velocity near zero frequency approaches

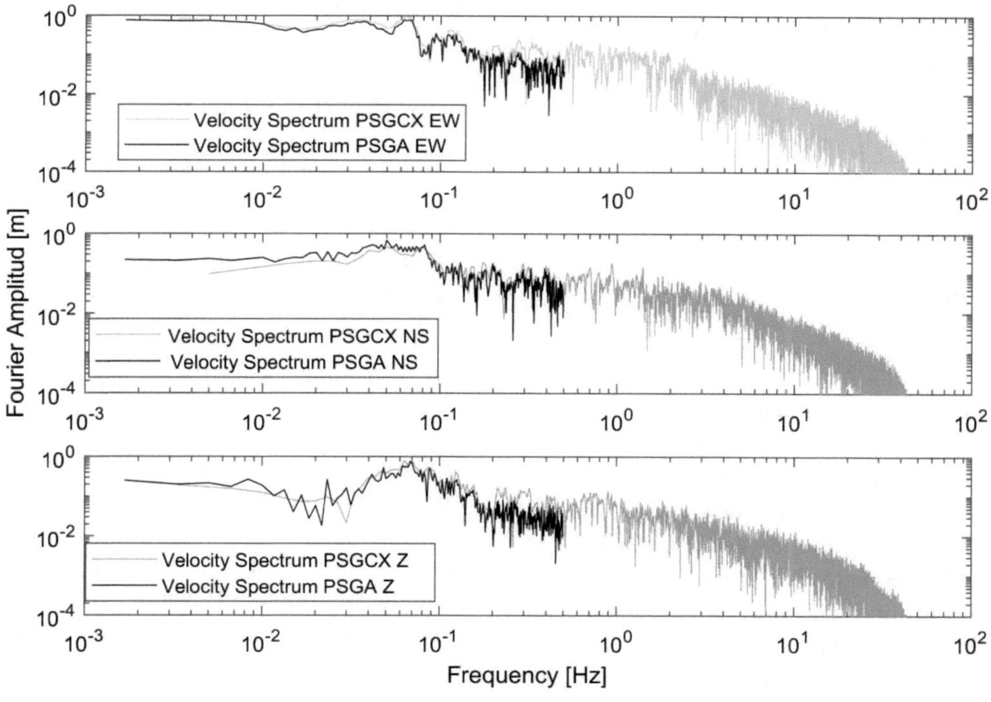

Figure 6
Spectra of the velocity records from the PSGA GNSS station and the PSGCX accelerogram. The time domain velocity signals are shown in Fig. 5. These spectra were directly computed using a fast Fourier transform algorithm. We observe that at low frequencies, the accelerogram and GNSS instrument share an almost identical spectrum in spite of the different noise levels in these two types of instrument. All three components have almost flat spectra at low frequencies. These low-frequency velocity spectra converge to co-seismic displacement values observed in that site

the integral of velocity, which is the static displacement:

$$\lim_{\omega\to 0}\tilde{u}(\omega)=\int_{-\infty}^{\infty}\dot{u}(t)\mathrm{d}t=u(\infty). \quad (1)$$

Although this property is verified in the PSGCX records it may be difficult to use in practice because of noise and limited frequency band. Anyway, the high quality of northern Chile accelerographs and the high signal/noise ratio allow us to obtain in some cases the static displacement directly from velocity spectra using Eq. (1); see Fig. 6.

4. Why Omega-1?

Radiation from seismic sources can be approximated by the radiation from a point double couple inside a homogeneous medium (see Aki and Richards 2002 and the Appendix). Modifying slightly the first term in Equation (7), the radiation from a point seismic moment source is

$$\begin{aligned}u_i(r,t)&=\frac{1}{4\pi\rho\beta^2}A^N\frac{1}{r^2}\int_{\frac{\beta}{\alpha}}^{1}\tau M_0\left(\frac{\beta t}{r}-\tau\right)\mathrm{d}\tau\\&+\frac{1}{4\pi\rho\alpha^2}A^{\mathrm{IP}}\frac{1}{r^2}M_0\left(t-\frac{r}{\alpha}\right)+\frac{1}{4\pi\rho\beta^2}A^{\mathrm{IS}}\frac{1}{r^2}M_0\left(t-\frac{r}{\beta}\right)\\&+\frac{1}{4\pi\rho\alpha^3}A^{\mathrm{FP}}\frac{1}{r}\dot{M}_0\left(t-\frac{r}{\alpha}\right)+\frac{1}{4\pi\rho\beta^2}A^{\mathrm{FS}}\frac{1}{r}\dot{M}_0\left(t-\frac{r}{\beta}\right),\end{aligned} \quad (2)$$

where r is distance between the source and observer, and t is the time. The terms in the equation are usually called near, intermediate and far field. The A^i terms are the radiation patterns defined by Aki and Richards (2002). ρ, α, β are the density, P-wave and S-wave speeds, respectively. $M_0(t)$ is the seismic moment time history. The dot above the moment denotes moment rate. We wrote the first term of Eq. (2) in a slightly different form from Aki and Richards (2002) because we want to stress that the near and intermediate terms both decay as r^{-2} and cannot be separated in the study of radiation.

For the study of the spectrum of radiated waves, we rewrite the Fourier transform of (2) in the self-similar form:

$$\begin{aligned}\tilde{u}_i(r,\omega)&=\frac{M_0(\omega)}{r^2}\left[\frac{1}{4\pi\rho\alpha^2}F^{\mathrm{INP}}(\omega r/\alpha)e^{-\frac{i\omega r}{\alpha}}\right.\\&\left.+\frac{1}{4\pi\rho\beta^2}F^{\mathrm{INS}}(\omega r/\beta)e^{-\frac{i\omega r}{\beta}}\right]\\&+\frac{\dot{M}_0(\omega)}{r}\frac{1}{4\pi\rho\alpha^3}A^{\mathrm{FP}}e^{-\frac{i\omega r}{\alpha}}+\frac{1}{4\pi\rho\beta^3}A^{\mathrm{FS}}e^{-\frac{i\omega r}{\beta}},\end{aligned} \quad (3)$$

where $F^{\mathrm{INP}}(\omega r/\alpha)$ and $F^{\mathrm{INS}}(\omega r/\beta)$ are self-similar functions of frequency scaled by propagation time. They are defined in the appendix.

Expression (3) for the Green function in the frequency domain clearly shows that the near and intermediate field that decay as r^{-2} are proportional to moment, while the far field that decays like r^{-1} is proportional to moment rate. Since moment rate is known to be flat at low frequencies (i.e., proportional to ω^0), the near field is then inversely proportional to frequency, i.e., it behaves at low frequencies like ω^{-1}. We can show this property very simply by adopting Brune's (1970) source time function. Brune's moment rate is

$$\dot{M}(t)=M_0\omega_0^2te^{-\omega_0 t}H(t), \quad (4)$$

whose Fourier transform is:

$$\dot{M}_0(\omega)=M_0\frac{1}{\left(1+\frac{i\omega}{\omega_0}\right)^2}. \quad (5)$$

Here, ω_0 is the corner frequency. The Fourier transform of $M_0(\omega)$ is that of (5) divided by $i\omega$.

4.1. A Simplified Model for S Waves

We can illustrate the properties of strong motion in the near field with a simplified model of S waves that we derived from an equivalent model for P waves proposed by Vidale et al. (1995). From the previous theoretical considerations, we write a very simplified version of (2) that retains its main features. We consider only S waves that have Brune's (1970) waveform (4) in the time domain:

$$u(r,t)=\frac{C^{\mathrm{NF}}}{r^2}M_0\left(t-\frac{r}{\beta}\right)+\frac{C^{\mathrm{FF}}}{r}\dot{M}_0\left(t-\frac{r}{\beta}\right). \quad (6)$$

C^{NF} contains the near-field radiation pattern and normalization. C^{FF} contains the far-field radiation pattern and normalization. Expression (6) is a simple way of writing the main features of the radiated waves from a point dislocation including a Brune finite source time function. Of course any other source time function can be used in this approximation.

In Fig. 7, we show a wave form in the near field computed using (6) that simulates quite well those observed in the EW component for large subduction earthquakes in Chile. We assumed a corner frequency $\omega_0 = 4(f_0 = 0.637\text{Hz})$ and a ratio $C = \frac{C_{\rm NF}}{rC_{\rm FF}} = 0.75$ between the near- and far-field coefficients in (6). In the same figure, we show the spectrum computed using the exact transforms of the moment and moment rate. We observe that for these particular parameters, the peak displacement—dominated by the moment rate—is about twice the static displacement—dominated by the moment—in the time domain. Any other ratio of peak to static amplitude can be obtained by changing the ratio of the coefficients or the corner frequency. On the other hand, for this particular set of parameters (ω_0, C), the static field dominates the spectrum in most of the frequency range displayed in the figure. Moment rate becomes dominant only at frequencies larger than 2 Hz, well beyond the corner frequency. The interplay between the near-field and far-field terms is complex and depends on several parameters: the radiation pattern and distance that determine the coefficient C; and the corner frequency ω_0, that is inversely proportional to the duration. Thus, as we will elaborate in the discussion, the relative weight of near- and far-field terms depends on the relative values of the duration and the P–S time.

5. *Ground Motion Spectrum in Northern Chile After the Iquique Earthquake*

In the previous section, we studied the spectrum at station PSGCX (Pisagua) in northern Chile. This is

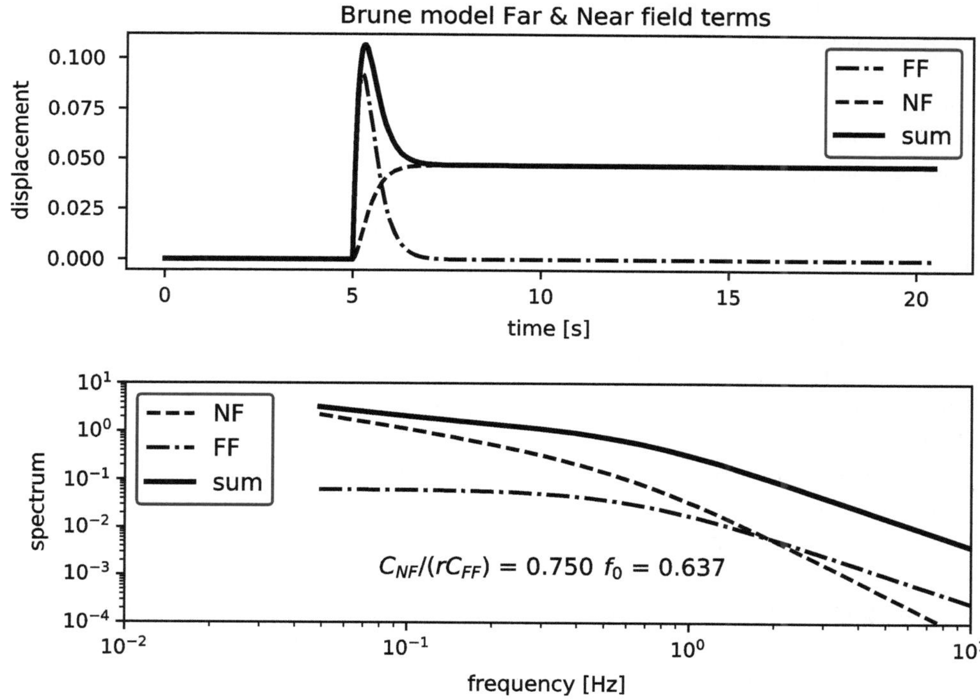

Figure 7
Spectra of a simplified near-field displacement spectrum computed for an S wave that contains both near- and far-field components. The near field is plotted with a dashed line and the far-field (Brune) signal is plotted with a dash-dotted line. The thick black line is the full near-field displacement signal. In this case, the ratio between the peak displacement and the static displacement is 2. The displacement spectrum is shown at the bottom with the same convention for the far- and near-field components. We observe that the near field (in black) dominates the spectrum for a very broad range of frequencies up to the corner frequency, which here is 0.637 Hz ($\omega_0 = 4\text{rad/s}$). The ratio between the coefficients in Eq. (6) is 0.75

the closest station to the earthquake epicenter and, as shown in the previous section, its displacement spectrum is dominated by the ω^{-1} low-frequency decay (Fig. 4) due to the finite jump in displacement observed in Fig. 3. At this distance, the seismic moment-dependent terms in the Green function, first term in (3), dominate the spectrum. As shown in Fig. 1 and Table 1, the Iquique earthquake was recorded at more than 20 accelerometric stations in northern Chile.

Let us first consider the PB11 station, which, as observed from Fig. 1, is situated some 60 km inland from Pisagua and located 119 km from the immediate foreshock of the earthquake. Figure 8 shows the EW component of displacement at this station. Displacement was computed by the Boore (2001) procedure discussed earlier. Superimposed on the seismic displacement trace, we plot the ground displacement determined at the same station from the GNSS antenna of the PB11 IPOC multi-parametric station. The traces for both the integrated accelerogram and the geodetic data are very similar and have a large static component. It is obvious that in this station the field is dominated by the moment time function. In the two bottom panels of Fig. 8, we show the ground displacement spectrum determined from the accelerogram spectrum by double Fourier integration (division by omega-2) and the velocity spectrum. The displacement spectrum shows the characteristic omega-1 behavior at low frequency, as expected from the static displacement observed in the displacement. The velocity spectrum at the bottom has the typical flat spectrum observed in northern Chile for the Iquique earthquake. The level of the flat part of the velocity spectrum is very close to the static displacement observed in the GNSS instrument trace (GPS) at the top of Fig. 8. This confirms relation (1) derived earlier.

As we move away from the source, the level of ground displacement decreases, yet the spectra are still dominated by the static displacement, and thus by the moment time function, rather than moment rate. We illustrate this with the record obtained at station PB01 of the IPOC network (see Fig. 1). This station is located near 21°S, about 219 km from the initial foreshock of the Iquique earthquake. Figure 9 shows the displacement records obtained from the accelerogram and the GNSS instruments of the PB01 IPOC station. Both coincide very well in the large amplitude part in the 60–100 s window and they diverge near the end of the record where the displacement is less than 2.5 cm. The spectra shown in the two bottom panels are very similar to those of the PB11 station. The flat part of the ground velocity spectrum is about 0.05 m, as shown with horizontal line in Fig. 9. Clearly, the velocity spectrum integrated from the accelerograms is determined by the static displacement produced by the earthquake at this station.

The similarity of the ground velocity spectra recorded at different distances from the Iquique earthquake is striking. In Fig. 10, we show a stack of the ground velocity spectra computed for several stations in northern Chile that recorded the Iquique earthquake.

The stations chosen for the plot of velocity spectra in Fig. 10 range in distance from 70 to 250 km, yet the spectra are very similar in the intermediate frequency range from 0.1 to 1 Hz. At lower frequencies, the spectra are dominated by the static field and thus by the moment time function. These spectra scale with the static displacement as shown in Figs. 9 and 10 for stations PB01 and PB11. In the high-frequency range, from 1 to 10 Hz, on the other hand, these spectra decay as omega-1 as predicted by Brune's model indicating that the high-frequency part is dominated by the moment rate–time function. In most of northern Chile, the spectra from the Iquique earthquake resemble those shown in the previous pictures, with strong contribution from the static near field. It is surprising that even at 219 km distance from the epicenter, at station PB01, shown in Fig. 9, the velocity spectra are still flat at the lowest resolvable frequencies.

Why do we observe the near field at such long distances? This is because the permanent co-seismic displacement induced by the moment tensor is still large at distances of the order of six times the size of the event. Although the near field decreases as r^{-2}, the ratio between far- and near-field terms decrease only as r^{-1} as shown in previous sections. Thus, the effect of the moment time function is still important at those distances.

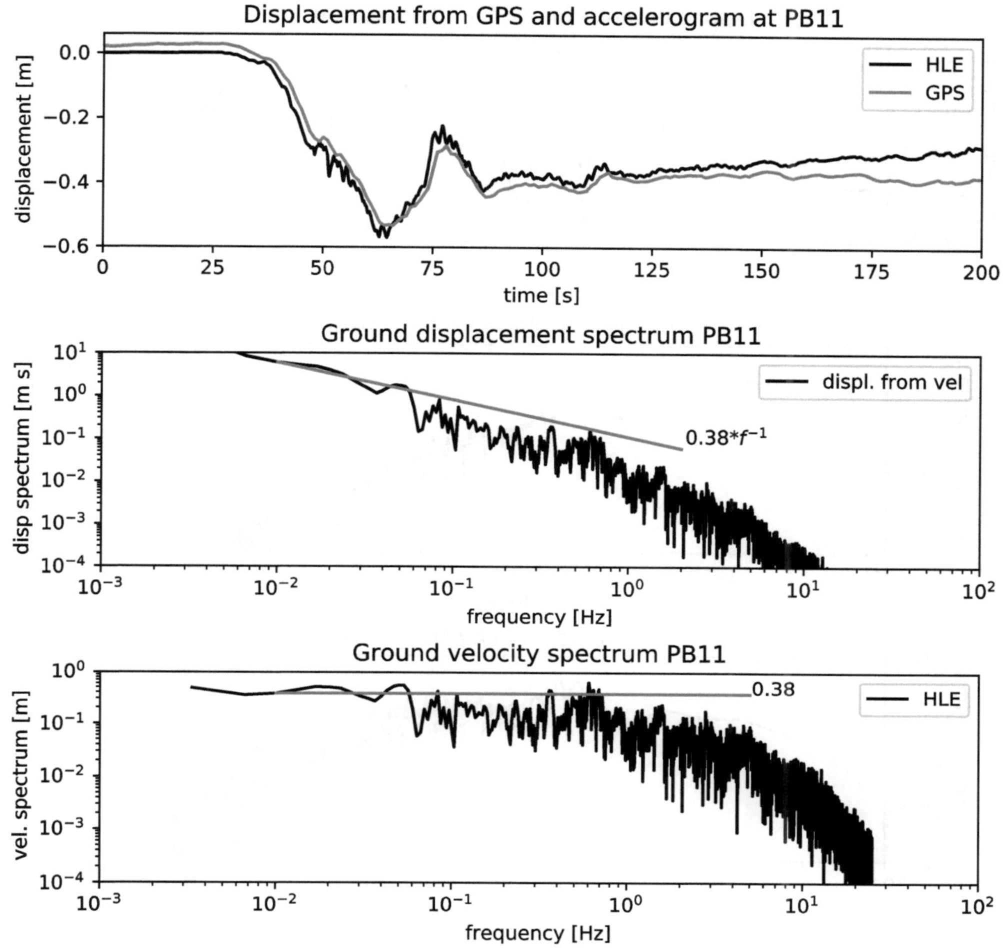

Figure 8
EW component of displacement integrated from the accelerogram of the PB11 station located 119 km from the epicenter of the 1 April 2014 Iquique earthquake of M_w 8.1. At the top, the displacement waveform computed by integrating the accelerogram is compared with the computed displacement from the co-located GNSS station. At the bottom, the displacement and velocity spectra computed from the accelerogram. The flat portion of the velocity spectrum fits very well the displacement jump of 0.38 m observed in the GNSS record. The ω^{-1} trend in the displacement spectrum fits nicely the trend derived from the velocity spectrum

6. Transition to Aki–Brune Spectrum at Far Distances from the Source

The previous section shows that in a very broad region around Iquique, the ground velocity spectra of the 1 April 2014 earthquake have similar low-frequency spectra dominated by the moment time function. We expected that as we move away from the source we should observe the emergence of a Brune like spectrum peaked at the corner frequency, i.e., proportional to ω at low frequencies and decreasing as ω^{-1} at higher frequencies. This is not obvious in the velocity spectra of Fig. 10 except that of station PB07, 257 km from the epicenter, which decreases at high frequencies as predicted by Aki (1967) and Brune (1970), but still has a strong low-frequency component.

To observe Brune's spectrum more clearly, we take a smaller event. We chose the foreshock of Iquique earthquake that occurred on 16 March 2014 near the south-west edge of the main rupture zone of the Iquique earthquake (see Fig. 1 for location). This is an M_w 6.7 event that had a somewhat different source mechanism than the main event (Ruiz et al.

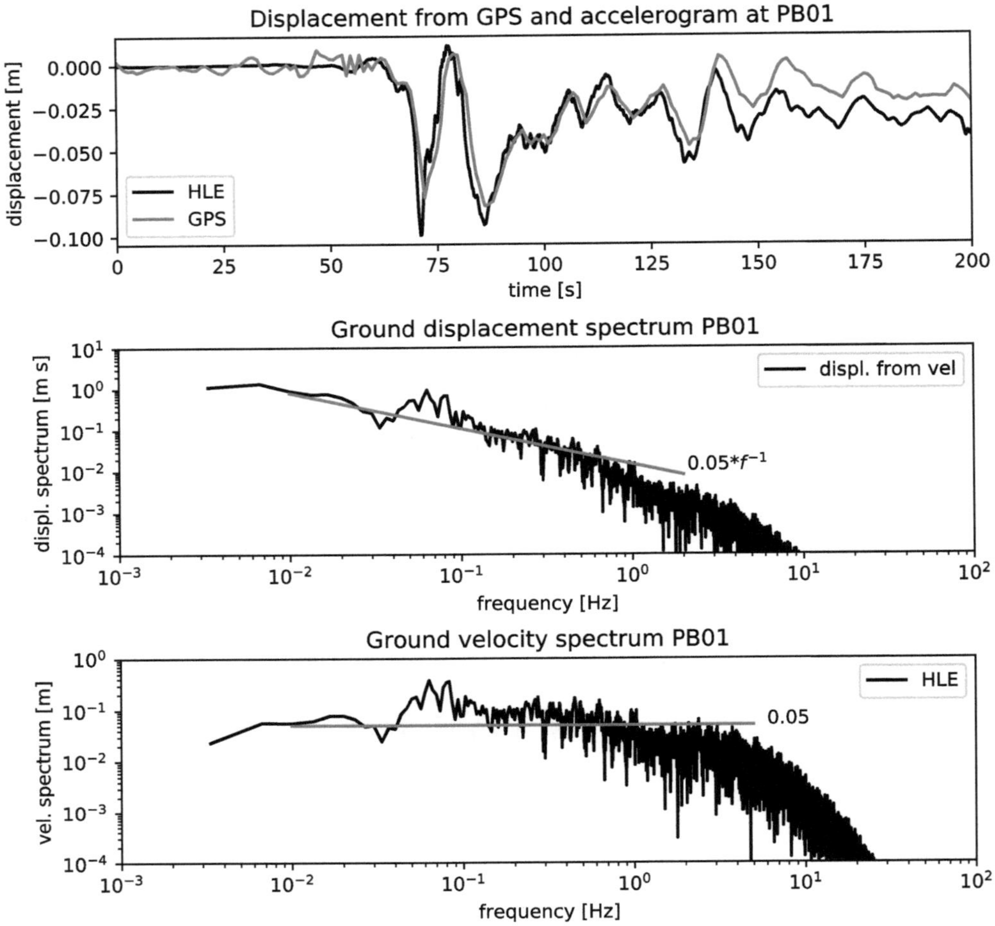

Figure 9
EW component of displacement integrated from the accelerogram of PB01 station located 219 km from the epicenter of the 1 April 2014 Iquique earthquake, M_w 8.1. At the top, the displacement integrated from the accelerograms is compared with the computed displacement from the collocated GNSS station. At the bottom, the displacement and velocity spectra computed from the accelerogram. The flat portion of the velocity spectrum fits very well the velocity jump observed in displacement. The ω^{-1} trend in the displacement spectrum fits nicely the trend derived from the velocity spectrum

2014; Leon-Ríos et al. 2016; Cesca et al. 2016). In Fig. 11, we plot two accelerograms generated by this earthquake. Stations PB11 and PB01 provide excellent recordings. The first is at 121 km, and the other at 150 km distance from an event that probably had a source size of about 10 km. For station PB11, it is still difficult to clearly separate the P- and S-wave trains in the integrated displacement trains. On the PB01 records, the P and S waves are very well separated so that we could identify an S window. This is indicated in red in the PB01 accelerogram and was isolated using multitaper spectral techniques (Park et al. 1987; Prieto et al. 2009). At central northern Chile stations, this event has the typical ω^{-2} high-frequency decay.

In Fig. 12, we plot the displacement spectra for the two stations of Fig. 11. At the PB01 station, we computed spectra both for the entire record and for the S-wave window shown here in red. The two spectra resemble many other spectra published in the literature for the largest local event recordings. The high-frequency decay is ω^{-2} beyond 1 Hz. They also have a flat spectrum plateau at low frequencies, for frequencies less than 0.1 Hz for the PB01

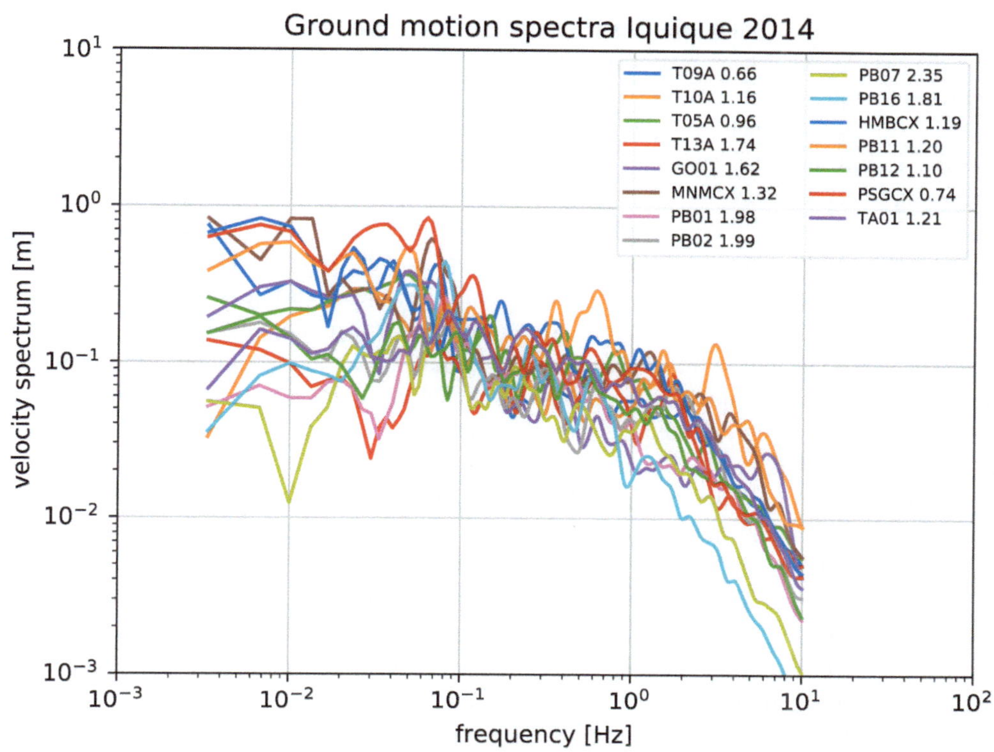

Figure 10

Stack of the EW component of velocity spectra observed at a set of selected strong motion stations in northern Chile. The spectra were smoothed using a Konno–Ohmachi filter. All other records resemble this group of spectra. The numbers next to the station codes are the epicentral distances of each station in degrees

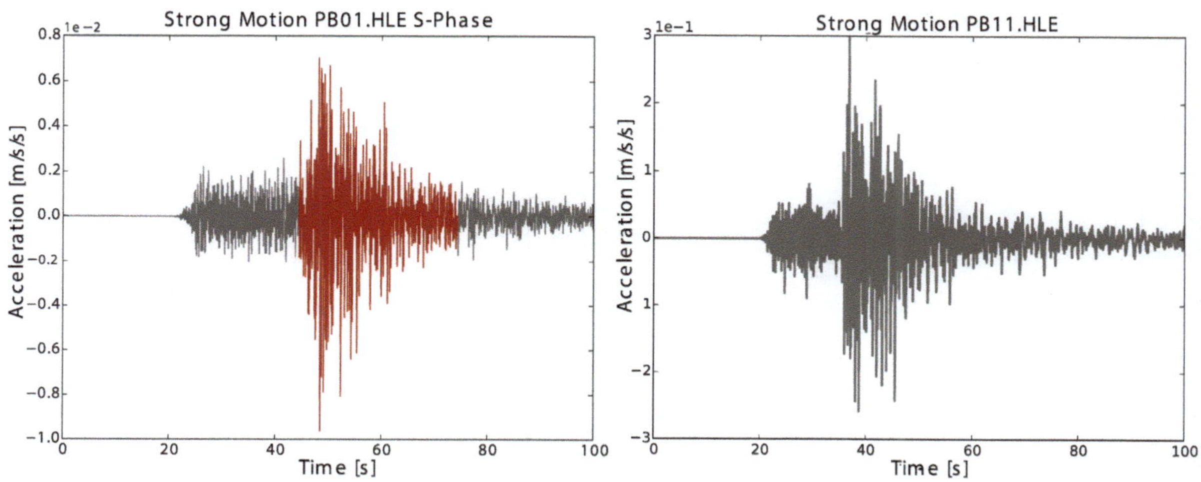

Figure 11

Accelerograms recorded at two IPOC stations PB01 and PB11 for the 16 March 2014 precursor of the Iquique earthquake of 1 April 2014. PB11 is at 121 km and PB01 at 256 km from the source. This event of M_w 6.7 is much smaller than the main shock accelerogram. In Fig. 12 we show the spectrum computed for the whole record in black and the spectrum computed for the S-wave window in red. As expected, since S waves are much stronger than P waves, the whole record spectrum and the S-wave window spectrum are very similar.

The most important feature of these records is the presence of a broad spectral range where the spectra behave like ω^{-1}, as indicated by the thin grid superimposed on the spectra. The spectra do not exactly decay as ω^{-1}, but just have a broad transition region from 0.1 to 10 Hz. Intermediate frequency range with ω^{-1} decay have been reported by many authors [e.g., Denolle and Shearer (2016), Archuleta and Ji (2016), among others]. Archuleta and Ji (2016) observed those intermediate frequency ranges in the far field, while for the Iquique foreshock we cannot assume that the near field can be neglected. They explained the presence of the omega-1 trend with a modification of Brune's source time function (4). To properly model wave propagation in northern Chile and determine the relative contributions of the moment rate and moment time functions, we have to do a more detailed modeling of wave propagation. We are in the process of studying acceleration, GNSS and broadband records for the Valparaiso earthquake of 24 April 2017 that had excellent coverage in Central Chile (Ruiz et al. 2017a, 2018).

7. Discussion

It is surprising that the effect of the near field on the seismic spectra computed from the accelerograms of the 2014 Iquique earthquake has not been reported earlier. Many authors used these records for modeling the main event filtering out the low frequencies. One reason may be that the problems of integration of accelerograms (Boore 2001) discouraged the study of the spectral properties of seismic sources in the near field. An additional possible problem is that many standard routines used for processing accelerograms remove the trend assuming that the average velocity field is zero. In the near field, the average velocity is non-zero as proven by expression (1).

Our main observation is that the spectra of ground motion in the vicinity of an earthquake source contain two types of contributions: (1) the near-field terms that decay as r^{-2} with distance and are proportional to the seismic moment time function; (2) the other contribution is the far field that is proportional to moment rate and decays like r^{-1}.

While the P and S waves can be studied separately in the far field, in the near field they cannot be separated because they interfere to produce the static field. The terms that contribute to P or S waves decay individually like r^{-4} (Aki and Richards 2002), but interfere at low frequencies to produce a single r^{-2}

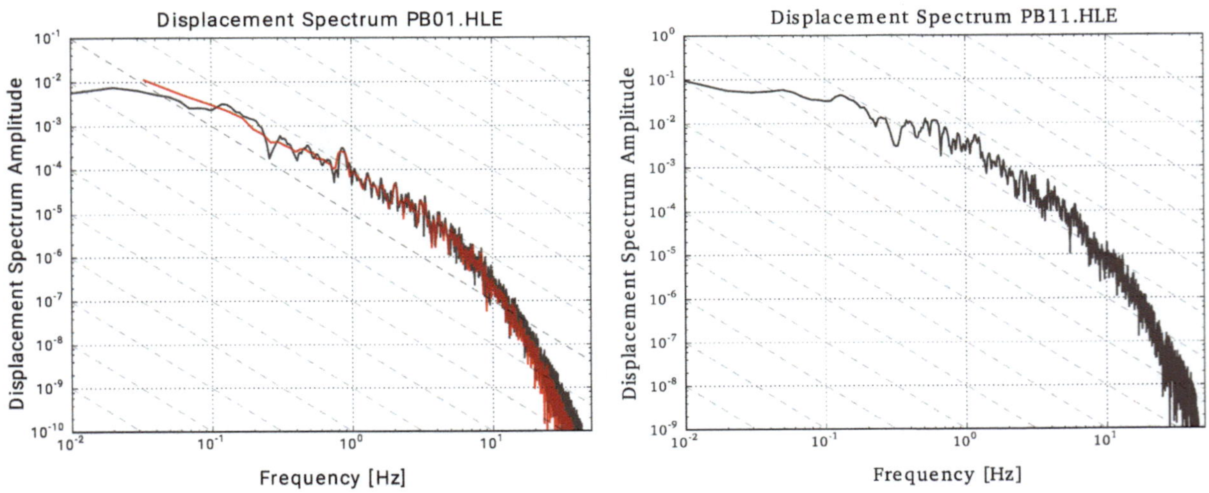

Figure 12
Displacement spectra of the two records shown in Fig. 11. Black spectra consider all records and red spectra the S-wave window

near-field contribution to strong motion (Eq. 3). This is the term that dominates the spectrum of all the accelerograms recorded after the 2014 Iquique earthquake in a range of less than 200 km from the epicenter of the earthquake. Further study of the properties of the near field requires more accurate numerical simulations including the free-surface effects and structural properties of wave propagation. The simple point source model used here provides a clear proof that near- and far-field terms in the accelerograms are equally important, but the effects of wave propagation must be taken carefully into account. Observations of the relative weight of near and far field have been reported earlier (Vidale et al. 1995; Jiao et al. 1995), but a systematic study of the spectral properties has not been done to our knowledge.

At what distance does the far-field spectrum become dominant? For the M_w 8.4, 2014 Iquique earthquake studied here, the near field dominates all the records out to 219 km. Only for smaller events studied in Sect. 5, a far-field spectrum resembling the Brune-Aki model appears at distant stations. Thus, the relative value of the near- and far-field terms depends not just on distance, but also on the duration of the source. Although we cannot provide a simple rule for the transition distance, we believe that the transition occurs beyond the distance at which the seismic source duration of the event is of the same order as the separation between P and S waves, for the Iquique earthquake with duration of 50 s, this distance is close to 400 km assuming an average P-wave speed of 6 km/s and an P/S velocity ratio of 1.73. A more detailed study including careful modeling of the interaction of surface reflections and guided waves is necessary to provide an accurate answer. A more accurate estimate will be the object of future work.

We proved from observations that, as expected from the properties of Fourier transform, the low-frequency velocity spectral amplitude converges to the co-seismic displacement observed in collocated GNSS instruments. Thus, it may be possible, in principle, to use velocity spectra to determine the static displacements as complement to GNSS data. Unfortunately, this requires acceleration traces with a good signal/noise ratio, located in hard rock sites to avoid non-linear site effects.

The stack of the displacement spectra observed in northern Chile after the Iquique earthquake (Fig. 10) shows that the velocity spectra in the frequency range between 0.1 and 1 Hz are very similar in all the stations. In the low-frequency range less than 0.1 Hz the spectra have variable amplitudes that scale with the static displacement at the site. This confirms the theoretical prediction that the level of the flat part of the velocity spectrum is proportional to the static ground displacement. These static levels can be computed by the usual techniques to predict static ground motion using Okada's (1985) formulas. As shown in Figs. 8 and 9 for stations PB01 and PB11, the low-frequency level of the velocity spectrum fits quite well with the static displacement observed by collocated GNSS instruments.

One of the main consequences of this study is that the classical scaling of the earthquake spectra (Aki 1967) does not hold for large earthquakes, for which the displacement spectra decay as omega-1 at low frequencies. This was first observed by Lancieri et al. (2012) in their study of the Tocopilla earthquake of 2007 and its aftershocks. Although few stations were available at the time, the aftershocks less than M_w 6 had spectra that matched the scaling law of Aki (1967) and Prieto et al. (2004).

The results reported here for the Iquique earthquake hold for most large events in Chile for which we have very good broadband accelerograms and GNSS recordings. We have observed omega-1 low-frequency displacement spectra for the 2010 Maule earthquake in two good-quality accelerograms and for other large events including the 2007, M_w 7.8 Tocopilla earthquake; the 2015, M_w 8.3 Illapel earthquake and the 2016 Chiloe earthquake of M_w 7.6. For events less than 7.0, we have explored just a few events, typically aftershocks of the larger events. The best data is for the Valparaíso earthquake of M_w 6.9 that occurred on 23 April 2017 (Ruiz et al. 2017a, 2018). The ground velocity spectrum for this event resembles that of a scaled down version of the Iquique earthquake studied here at close distances and then becomes similar to that of Brune's spectrum at longer distances from the source. The origin of the intermediate ω^{-1} slope in displacement spectra needs further study and comparison with other events as reported by Archuleta and Ji (2016) and Denolle and Shearer (2016), among others, for far-field observations.

What are the consequences for ground motion prediction? A usual assumption about the spectrum of large earthquakes is that they have an omega-2 displacement spectrum and the spectra for large earthquakes is generated by summing—often random—combinations of small events. This assumption must be revised for large-magnitude synthetic strong motion simulations by adding a near-fault term. This problem was analyzed, among others, by Tumarkin et al. (1994) who showed that it was very difficult to produce an omega-2 spectrum by adding smaller point-like sources.

8. Conclusions

We have shown that the spectrum of ground motion in the vicinity of very large subduction earthquakes in northern Chile is quite different from that proposed by Aki and Brune, the so-called omega-2 spectrum, which is flat at low frequencies and decays as omega-2 at high frequencies. The spectrum in the near field is dominated by the seismic moment time function that contains static terms. This produces an omega-1 displacement spectrum at low frequencies. This kind of spectrum dominates radiation surprisingly far from the source, up to 250 km in the present case. We also find that separation of P and S waves is not possible because the P- and S-like near-field terms interfere to produce the static field of the earthquake.

9. Data

The data used for this study were downloaded mainly from Centro Sismologico Nacional of the University of Chile (CSN). For large Chilean earthquakes, the significant earthquake database of the CSN provides corrected records at the website (http://edb.csn_uchile.cl). Geodetic data are available also from the CSN at the GPS site (http://gps.csn.uchile.cl/). Additional information and data for the IPOC stations of northern Chile are available from the Geofon pages of the GFZ data center (https://geofon.gfz-potsdam.de/) and from the Wilber3 pages of the IRIS data center (http://ds.iris.edu/wilber3/).

Acknowledgements

This study was supported by FONDECYT contract No. 1170430 and by PRS (Programa Riesgo Sísmico of Universidad de Chile).

Appendix

From Aki and Richards (2002), the Green function for a point moment tensor source in a homogeneous elastic space is:

$$
\begin{aligned}
u_i(r,t) &= \frac{1}{4\pi\rho}\frac{1}{r^4}A^N \int_{\frac{r}{\alpha}}^{\frac{r}{\beta}} \tau M_0(t-\tau)\mathrm{d}\tau \\
&+ \frac{1}{4\pi\rho\alpha^2}A^{\mathrm{IP}}\frac{1}{r^2}M_0\left(t-\frac{r}{\alpha}\right) + \frac{1}{4\pi\rho\beta^2}A^{\mathrm{IS}}\frac{1}{r^2}M_0\left(t-\frac{r}{\beta}\right) \\
&+ \frac{1}{4\pi\rho\alpha^3}A^{\mathrm{FP}}\frac{1}{r}\dot{M}_0\left(t-\frac{r}{\alpha}\right) + \frac{1}{4\pi\rho\beta^3}A^{\mathrm{FS}}\frac{1}{r}\dot{M}_0\left(t-\frac{r}{\beta}\right).
\end{aligned} \tag{7}
$$

It is usually assumed that the near-field term (the first one) decays as r^{-4}, but this is actually not correct because the time integral on the left grows like r^2 with distance r from the source. It is possible to rewrite the integral changing variables to $\tau = \vartheta\frac{r}{\beta}$, where ϑ is non-dimensional. Equation (7) becomes (3) of the main text that clearly shows that the first term behaves as r^{-2}, like those of the intermediate field.

We can now determine the spectrum of (7) using the continuous Fourier transform:

$$
\tilde{u}(\omega) = \int_{-\infty}^{\infty} u(t)e^{-i\omega t}\mathrm{d}t. \tag{8}
$$

We observe that the integral of the first line in (7) is a convolution, so that its Fourier transform is

$$\tilde{M}_0(\omega)\int_{\frac{r}{\alpha}}^{\frac{r}{\beta}}\tau e^{-i\omega\tau}\mathrm{d}\tau=\tilde{M}_0(\omega)\frac{1}{\omega^2}\left[\left(\frac{i\omega r}{\alpha}+1\right)e^{-i\omega r/\alpha}-\left(\frac{i\omega r}{\beta}+1\right)e^{-i\omega r/\beta}\right]. \quad (9)$$

That can be compared to similar expressions from Aki and Richards (2002), taking into account that they used a different definition of the Fourier transform (8) than us. Taking the Fourier transform of (7), using (9) and collecting the terms that depend on $\tilde{M}_0(\omega)$ we get:

$$\begin{aligned}\tilde{u}_i(r,\omega)=&\frac{1}{4\pi\rho\beta^2}\frac{1}{r^2}\tilde{M}_0(\omega)\frac{A^N\beta^2}{\omega^2r^2}\left[\left(\frac{i\omega r}{\alpha}+1\right)e^{-i\omega r/\alpha}-\left(\frac{i\omega r}{\beta}+1\right)e^{-i\omega r/\beta}\right]\\&+\frac{1}{4\pi\rho\alpha^2}\frac{1}{r^2}\tilde{M}_0(\omega)A^{\mathrm{IP}}e^{-\frac{i\omega r}{\alpha}}+\frac{1}{4\pi\rho\beta^2}\frac{1}{r^2}\tilde{M}_0(\omega)A^{\mathrm{IS}}e^{-\frac{i\omega r}{\beta}}\\&+\frac{1}{4\pi\rho\alpha^3}\frac{1}{r}\tilde{\dot{M}}_0(\omega)A^{\mathrm{FP}}e^{-\frac{i\omega r}{\alpha}}+\frac{1}{4\pi\rho\beta^3}\frac{1}{r}\tilde{\dot{M}}_0(\omega)A^{\mathrm{FS}}e^{-\frac{i\omega r}{\beta}}.\end{aligned} \quad (10)$$

Expression (10) shows that there are two types of terms, those that depend on moment which are also inversely proportional to r^2 and those that depend on moment rate and are inversely proportional to radius r. Gathering all the moment-dependent terms we get the expression of Eq. (4) in the text. The non-dimensional coefficients F^{INP} and F^{INS} are given by:

$$F^{\mathrm{NIP}}\left(\frac{\omega r}{\alpha}\right)=A^N\frac{\alpha^2}{\omega^2r^2}\left(\frac{i\omega r}{\alpha}+1\right)+A^{\mathrm{IP}},$$

$$F^{\mathrm{NIS}}\left(\frac{\omega r}{\beta}\right)=-A^N\frac{\beta^2}{\omega^2r^2}\left(\frac{i\omega r}{\beta}+1\right)+A^{\mathrm{IS}}.$$

These are non-dimensional functions that depend only on non-dimensional frequency $\omega r/a$ or $\omega r/\beta$ and the radiation patterns.

References

Abercrombie, R. E. (1995). Earthquake source scaling relationships from − 1 to 5 ML using seismograms recorded at 2.5-km depth. *Journal of Geophysical Research, 100,* 24015–24036.

Aki, K. (1967). Scaling law of seismic spectrum. *Journal of Geophysical Research, 73,* 5359–5376.

Aki, K., & Richards, P. G. (2002). *Quantitative seismology*. USA: University Science Books. ISBN 978-1-891389-63-4.

Allmann, B. P., & Shearer, P. M. (2009). Global variations of stress drop for moderate to large earthquakes. *Journal of Geophysical Research, 114,* B01310. https://doi.org/10.1029/2008jb005821.

Archuleta, R. J., & Ji, C. (2016). Moment rate scaling for earthquakes 3.3 ≤ *M* ≤ 5.3 with implications for stress drop. *Geophysical Research Letters, 43*(23), 12004–12011.

Báez, J. C., Leyton, F., Troncoso, C., del Campo, F., Bevis, M., Vigny, C., et al. (2018). The Chilean GNSS network: Current status and progress toward early warning applications. *Seismological Research Letters, 89,* 1546–1554.

Barrientos, S., & National Seismological Center (CSN). (2018). The seismic network of Chile. *Seismological Research Letters, 89,* 467–474.

Bock, Y., Melgar, D., & Crowell, B. W. (2011). Real-time strong-motion broadband displacements from collocated GPS and accelerometers. *Bulletin of the Seismological Society of America, 101,* 2904–2925.

Boore, D. M. (2001). Effect of baseline corrections on displacements and response spectra for several recordings of the 1999 Chi-Chi, Taiwan, earthquake. *Bulletin of the Seismological Society of America, 91,* 1199–1211.

Boore, D. M., & Bommer, J. J. (2005). Processing of strong-motion accelerograms: Needs, options and consequences. *Soil Dynamics and Earthquake Engineering, 25,* 93–115.

Brune, J. (1970). Tectonic stress and the spectra of seismic shear waves from earthquakes. *Journal of Geophysical Research, 75,* 4997–5009.

Cesca, S., Grigoli, F., Heimann, S., Dahm, T., Kriegerowski, M., Sobiesiak, M., et al. (2016). The M_w 8.1 2014 Iquique, Chile, seismic sequence: A tale of foreshocks and aftershocks. *Geophysical Journal International, 204,* 1766–1780.

Chao, W., Wu, Y., & Zhao, L. (2009). An automatic scheme for baseline correction of strong-motion records in coseismic deformation determination. *Journal of Seismology, 14,* 495–504.

Delouis, B., & Legrand, D. (2007). M_w 7.8 Tarapaca intermediate depth earthquake of 13 June 2005 (northern Chile): Fault plane identification and slip distribution by waveform inversion. *Geophysical Research Letters*. https://doi.org/10.1029/2006gl028193.

Delouis, B., Pardo, M., Legrand, D., & Monfret, T. (2009). The M_w 7.7 Tocopilla earthquake of 14 November 2007 at the southern edge of the northern Chile seismic gap: Rupture in the deep part of the coupled plate interface. *Bulletin of the Seismological Society of America, 99,* 87–94.

Denolle, M. A., & Shearer, P. M. (2016). New perspectives on self-similarity for shallow thrust earthquakes. *Journal of Geophysical Research, 121,* 6533–6565.

Duputel, Z., Jiang, J., Jolivet, R., Simons, M., Rivera, L., Ampuero, J. P., et al. (2015). The Iquique earthquake sequence of April 2014: Bayesian modeling accounting for prediction uncertainty. *Geophysical Research Letters, 42,* 7949–7957.

Fuenzalida, A., Schurr, B., Lancieri, M., & Madariaga, R. (2013). High resolution relocation and mechanism of aftershocks of the 2007 Sobiesiak M. and Tocopilla (Chile) earthquake. *Geophysical Journal International, 194,* 1216–1238.

Graizer, V. M. (1979). Determination of the true displacement of the ground from strong-motion recordings. *Izv. USSR Academy of Sciences, Physics Solid Earth, 15,* 875–885.

Graizer, V. (2010). Strong motion recordings and residual displacements: What are we actually recording in strong motion seismology? *Seismological Research Letters, 81*, 635–639.

Hayes, G. P., Herman, M. W., Barnhart, W. D., Furlong, K. P., Riquelme, S., Benz, H. M., et al. (2014). Continuing megathrust earthquake potential in Chile after the 2014 Iquique earthquake. *Nature, 512*(7514), 295–298.

Hoffmann, F., Metzger, S., Moreno, M., Deng, Z., Sippl, C., Ortega-Culaciati, F., & Oncken, O. (2018). Characterizing afterslip and ground displacement rate increase following the 2014 Iquique-Pisagua M_w 8.1 earthquake, Northern Chile. *Journal of Geophysical Research, 123*, 4171–4192. https://doi.org/10.1002/2017JB014970.

Ide, S., & Beroza, G. C. (2001). Does apparent stress vary with earthquake size? *Geophysical Research Letters, 28*, 3349–3352.

IPOC. (2006). GFZ German Research Centre for Geosciences; Institut des Sciences de l'Univers-Centre National de la Recherche CNRS-INSU: IPOC Seismic Network. Integrated Plate boundary Observatory Chile—IPOC. Other/Seismic Network. https://doi.org/10.14470/pk615318.

Iwan, W., Moser, M., & Peng, C. (1985). Some observations on strong motion earthquake measurement using a digital acceleration. *Bulletin of the Seismological Society of America, 75*, 1225–1246.

Jiao, W., Wallace, T. C., Beck, S., Silver, P. G., & Zandt, G. (1995). Evidence fir static displacements from the June 9, 1994 deep Bolvian earthquake. *Geophysical Research Letters, 22*, 2285–2288.

Kato, A., & Nakagawa, S. (2014). Multiple slow-slip events during a foreshock sequence of the 2014 Iquique, Chile M_w 8.1 earthquake. *Geophysical Research Letters, 41*, 5420–5427.

Lancieri, M., Madariaga, R., & Bonilla, F. (2012). Spectral scaling of the aftershocks of the Tocopilla 2007 earthquake in northern Chile. *Geophysical Journal International, 188*, 469–480.

Lange, D., Ruiz, J., Carrasco, S., & Manríquez, P. (2017). The Chiloé M_w 7.6 earthquake of 25 December 2016 in Southern Chile and its relation to the M_w 9.5 1960 Valdivia earthquake. *Geophysical Journal International, 213*, 210–277.

Lay, T., Yue, H., Brodsky, E. E., & An, C. (2014). The 1 April 2014 Iquique, Chile, M_w 8.1 earthquake rupture sequence. *Geophysical Research Letters, 41*, 3818–3825.

Leon-Ríos, S., Ruiz, S., Maksymowicz, A., Leyton, F., Fuenzalizada, A., & Madariaga, R. (2016). Diversity of the Iquique's foreshocks and aftershocks: A clue about complex rupture process of a M_w 8.1 earthquake. *Journal of Seismology, 20*, 1059–1073.

Leyton, F., Leopold, A., Hurtado, G., Pasten, C., Ruiz, S., Montalva, G., et al. (2018a). Geophysical characterization of the Chilean seismological stations: First results. *Seismological Research Letters*. https://doi.org/10.1785/0220170156.

Leyton, F., Pasten, C., Ruiz, S., Idini, B., & Rojas, F. (2018b). Empirical site classification of CSN network using strong-motion records. *Seismological Research Letters*. https://doi.org/10.1785/0220170167.

Madariaga, R. (1976). Dynamics of an expanding circular fault. *Bulletin of the Seismological Society of America, 65*, 163–182.

Madariaga, R., & Ruiz, S. (2016). Earthquake dynamics on circular faults: a review 1970–2015. *Journal of Seismology, 20*, 1059–1073.

Melgar, D., Fan, W., Riquelme, S., Geng, J., Liang, C., Fuentes, M., et al. (2016). Slip segmentation and slow rupture to the trench during the 2015, M_w 8. 3 Illapel, Chile earthquake. *Geophysical Research Letters, 43*, 961–966.

Melgar, D., Riquelme, S., Xu, X., Baez, J. C., Geng, J., & Moreno, M. (2017). The first since 1960: A large event in the Valdivia segment of the Chilean Subduction Zone, the 2016 *M*7.6 Melinka earthquake. *Earth and Planetary Science Letters, 474*, 68–75.

Okada, Y. (1985). Surface deformation due to shear and tensile faults in a half-space. *Bulletin of the Seismological Society of America, 75*, 1135–1154.

Oth, A., Bindi, D., Parolai, S., & Di Giacomo, D. (2010). Earthquake scaling characteristics and the scale-(in) dependence of seismic energy-to-moment ratio: Insights from KiK-net data in Japan. *Geophysical Research Letters, 37*, L19304. https://doi.org/10.1029/2010gl044572.

Park, J., Lindberg, C. R., & Vernon, F. L. (1987). Multitaper spectral analysis of high-frequency seismograms. *Journal of Geophysical Research, 92*(B12), 12675–12674.

Peyrat, S., Campos, J., De Chabalier, J. B., Perez, A., Bonvalot, S., Bouin, M. P., et al. (2006). Tarapacá intermediate-depth earthquake (M_w 7.7, 2005, northern Chile): A slab-pull event with horizontal fault plane constrained from seismologic and geodetic observations. *Geophysical Research Letters*. https://doi.org/10.1029/2006gl027710.

Peyrat, S., Madariaga, R., Buforn, E., Campos, J., Asch, G., & Vilotte, J. P. (2010). Kinematic rupture process of the 2007 Tocopilla earthquake and its main aftershocks from teleseismic and strong-motion data. *Geophysical Journal International, 182*, 1411–1430.

Prieto, G. A., Parker, R. L., & Vernon, F. L, I. I. I. (2009). A Fortran 90 library for multitaper spectrum analysis. *Computers & Geosciences, 35*(8), 1701–1710.

Prieto, G. A., Shearer, P. M., Vernon, F. L., & Kilb, D. (2004). Earthquake source scaling and self-similarity estimation from stacking P and S spectra. *Journal of Geophysical Research*. https://doi.org/10.1029/2004jb003084.

Ruiz, S., Aden-Antoniow, F., Baez, J. C., Otarola, C., Potin, B., del Campo, F., et al. (2017a). Nucleation phase and dynamic inversion of the M_w 69 Valparaíso 2017 earthquake in Central Chile. *Geophysical Research Letters*. https://doi.org/10.1002/2017gl075675.

Ruiz, J. A., Contreras-Reyes, E., Ortega-Culaciati, F., & Manríquez, P. (2018). Rupture process of the April 24, 2017, M_w 6.9 Valparaíso earthquake from the joint inversion of teleseismic body waves and near-field data. *Geophysical Research Letters, 279*, 1–14.

Ruiz, S., Klein, E., del Campo, F., Rivera, E., Poli, P., Metois, M., et al. (2016). The seismic sequence of the 16 September 2015, Illapel M_w 8.3 earthquake. *Seismological Research Letters, 87*, 789–799.

Ruiz, S., & Madariaga, R. (2018). Historical and recent large Megathrust earthquakes in Chile. *Tectonophysics, 733*, 37–56.

Ruiz, S., Madariaga, R., Astroza, M., Saragoni, G. R., Lancieri, M., Vigny, C., et al. (2012). Short period rupture process of the 2010 M_w 8.8 Maule earthquake in Chile. *Earthquake Spectra, 28*(S1), 1–18.

Ruiz, S., Metois, M., Fuenzalida, A., Ruiz, J., Leyton, F., Grandin, R., et al. (2014). Intense foreshocks and a slow slip event preceded the 2014 Iquique M_w 8.1 earthquake. *Science, 345*, 1165–1169.

Ruiz, S., Moreno, M., Melnick, D., Campo, F., Poli, P., Baez, J. C., et al. (2017b). Reawakening of large earthquakes in South-

Central Chile: The 2016 M_w 76 Chiloe event. *Geophysical Research Letters*. https://doi.org/10.1002/2017gl074133.

Schurr, B., Asch, G., Hainzl, S., Bedford, J., Hoechner, A., Palo, M., et al. (2014). Gradual unlocking of plate boundary controlled initiation of the 2014 Iquique earthquake. *Nature, 512,* 299–302.

Schurr, B., Asch, G., Rosenau, M., Wang, R., Oncken, O. & Barrientos, S. (2012). The 2007 M7. 7 Tocopilla northern Chile earthquake sequence: Implications for along-strike and downdip rupture segmentation and megathrust frictional behavior. Journal of Geophysical Research: Solid Earth, 117 (B5).

Socquet, A., Valdes, J. P., Jara, J., Cotton, F., Walpersdorf, A., Cotte, N., et al. (2017). An 8 month slow slip event triggers progressive nucleation of the 2014 Chile megathrust. *Geophysical Research Letters, 44,* 4046–4053.

Tilmann, F., Zhang, Y., Moreno, M., Saul, J., Eckelmann, F., Palo, M., et al. (2016). The 2015 Illapel earthquake, central Chile: A type case for a characteristic earthquake? *Geophysical Research Letters, 43,* 574–583.

Tumarkin, A. G., Archuleta, R., & Madariaga, R. (1994). Basic scaling rules for composite earthquake models. *Bulletin of the Seismological Society of America, 54,* 1279–1883.

Vidale, J., Goes, S., & Richards, P. G. (1995). Near-field deformation seen on distant broad band seismograms. *Geophysical Research Letters, 22,* 1–4.

Vigny, Ch., Socquet, A., Peyrat, S., Ruegg, J.-C., Metois, M., Madariaga, R., et al. (2011). The 2010 M_w 8.8 Maule mega-thrust earthquake of central Chile, monitored by GPS. *Science, 331,* 1417–1421.

Wang, G., Boore, D. M., Igel, H., & Zhou, X. (2003). Some observations on colocated and closely spaced strong ground-motion records of the 1999 Chi-Chi, Taiwan, earthquake. *Bulletin of the Seismological Society of America, 93,* 674–693.

Wang, R., Parolai, S., Ge, M., Jin, M., Walter, T. R., & Zschau, J. (2013). The 2011 M_w 9.0 Tohoku earthquake: Comparison of GPS and strong-motion data. *Bulletin of the Seismological Society of America, 103,* 1336–1347.

Wang, R., Schurr, B., Milkereit, C., Shao, Zh, & Jin, M. (2011). An improved automatic scheme for empirical baseline correction of digital strong-motion records. *Bulletin of the Seismological Society of America, 101,* 2029–2044.

Wu, Y., & Wu, C. (2007). Approximate recovery of coseismic deformation from Taiwan strong-motion records. *Journal of Seismology, 11,* 159–170.

Ye, L., Lay, T., Kanamori, H., & Koper, K. D. (2016). Rapidly estimated seismic source parameters for the 16 September 2015 Illapel, Chile M_w 8.3 earthquake. *Pure and Applied Geophysics, 173,* 321–332.

(Received June 1, 2018, revised August 17, 2018, accepted August 22, 2018, Published online September 13, 2018)

Pure Appl. Geophys. 176 (2019), 1003–1020

https://doi.org/10.1007/s00024-018-2060-1

Pure and Applied Geophysics

Dynamic Rupture and Seismic Radiation in a Damage–Breakage Rheology Model

Ittai Kurzon,[1] Vladimir Lyakhovsky,[1] and Yehuda Ben-Zion[2]

Abstract—We present simulations of dynamic ruptures in a continuum damage–breakage rheological model and waves radiated by the ruptures and observed in the far field. The propagating rupture produces rock damage and granulation in the process zone ahead of the rupture front. An expansion–compaction process in the process zone leads to an isotropic source term, while shear motion that accumulates behind the propagating front produces a deviatoric source term and shear heating behind the rupture front. The process zone dissipation due to the damage–breakage mechanism, and the isotropic source component, significantly affect the *S*/*P* energy partitioning and the radiation patterns of the waves. The calculated *S*/*P* seismic energy ratio can be significantly lower than results of standard models with no damage–breakage mechanism and no source volume components. The *P* radiation pattern becomes more isotropic compared with the classical deviatoric solution, with increased lobes at 45° to the direction of rupture propagation. The *S* radiation pattern is affected more strongly by the damage–breakage process in the source volume, mainly within the process zone, and is significantly different from classical deviatoric results. The *S* waves propagate from the rupture front through the process zone (unlike the *P* waves), experiencing stronger dissipation, so the *S* radiation pattern is more affected than the *P* radiation pattern. Hence, analysis of *P* waves can provide more reliable results on rupture directivity than *S* waves.

Key words: Fault zone rheology, dynamic rupture, deviatoric and isotropic source terms, radiation patterns, *P*/*S* energy ratio.

1. Introduction

Seismological analyses of earthquake source properties are based on fitting observed seismograms to theoretical solutions of seismic radiation from assumed sources. Basic source models have far-field displacement spectra with a flat part at low frequencies proportional to the seismic potency or moment, and a decaying portion beyond a corner frequency, which is inversely proportional to the source dimension for uniform rupture propagation over a continuous fault (e.g., Aki 1967; Brune 1970). Details of the corner frequency and spectral decay depend strongly on various aspects of the source process, including geometrical and material heterogeneities, type of faulting (shear, tensile, mixed), wave type (*P* or *S*), and source–receiver direction (e.g., Haskell 1964; Savage 1972; Molnar et al. 1973; Sato and Hirasawa 1973; Madariaga 1976). Despite these complexities, corner frequencies have been widely used to estimate the dimensions and stress (or strain) drops of earthquake sources (e.g., Prieto et al. 2004; Yang et al. 2009; Oth 2013; Trugman and Shearer 2017), along with the directivity of earthquake ruptures from azimuthal variations of f_c (e.g., Lengliné and Got 2011; Kane et al. 2013; Calderoni et al. 2015; Ross and Ben-Zion 2016).

Various studies have tried to clarify properties of seismic radiation in models with increased realism; For example, Peyrat et al. (2001) analyzed the source process of the Landers earthquake (1992) using a dynamic fault model. Shi and Ben-Zion (2009) calculated wave radiation from shear and tensile dislocations between similar and dissimilar solids. Causse et al. (2014) examined source properties of inferred kinematic models of large earthquakes. Kaneko and Shearer (2015) analyzed source properties in cohesive circular and elliptical models of ruptures. However, these and many other studies were limited to rupture processes on a preexisting fault that do not include rock fracturing during rupture propagation. The high stresses generated during failure around the main fault are expected to produce intense cracking at the rupture zone that may have significant effects on the local physics and radiated seismic waves (e.g., Sammis et al. 2009; Ben-Zion

[1] Geological Survey of Israel, Jerusalem, Israel. E-mail: vladi@gsi.gov.il

[2] Department of Earth Sciences, University of Southern California, Los Angeles, USA.

and Ampuero 2009; Bhat et al. 2012; Xu et al. 2015; Wang et al. 2018).

In the present paper, we use simulations of dynamic rupture in a continuum damage–breakage model (e.g., Lyakhovsky et al. 2016), and related calculations of elastic waves generated by the ruptures, to analyze properties of seismic waves in cases that include coseismic generation of brittle rock damage in the source volumes. In the next section, we review key aspects of the model and the numerical approach used for the presented calculations. Section 3 describes results associated with several cases of dynamic ruptures in the damage–breakage model and corresponding calculations of radiated P and S waves. The results show that the generated rock damage significantly affects the energy partitioning between S and P and the radiation pattern of the S waves. This is largely an outcome of mechanical dissipation and the isotropic source term generated by the damage process. The implications of the results for earthquake source physics are discussed in the final Sect. 4.

2. *Theoretical Model and Numerical Approach*

There is a long history of using various damage rheology models in material sciences, engineering, and physics to describe the deformation of solids beyond the elastic limit (e.g., Kachanov 1958, 1986; Rabotnov 1959, 1988; Valanis 1990; Lyakhovsky et al. 1993, 1997, 2015; Hansen and Schreyer 1994; Krajcinovic 1996; Allix and Hild 2002; Hamiel et al. 2004; Alava et al. 2006, and others). Applications of damage rheology models to problems of earthquakes and faults in the brittle crust include studies by Ashby and Sammis (1990), Ben-Zion et al. (1999), Turcotte et al. (2003), Suzuki (2013), Veveakis and Regenauer-Lieb (2014), and related works. Studies using damage rheology models for problems associated with dynamic rupture and seismic radiation are rather limited. Bhat et al. (2012) used the Ashby–Sammis damage model with embedded microcracks that can interact and grow to analyze various quantities associated with dynamic propagation of ruptures in relation to experimental results. Xu et al. (2015) used the nonlinear viscoelastic continuum damage model of Lyakhovsky et al. (1997, 2001) to study dynamic feedback mechanisms between rupture and evolving elastic properties that modify rupture properties, seismic radiation, and fault zone structure (Fig. 1a). Lyakhovsky et al. (2016) used an updated damage–breakage continuum damage model including a solid–granular transition at the onset of dynamic instabilities and rupture front to study further properties of ruptures, evolving damage, and generated motion. Thomas and Bhat (2018) used the Ashby–Sammis damage model to study similar topics. Here, we follow Lyakhovsky et al. (2016) with a focused study of radiated seismic waves. In the next subsections, we outline key aspects of the model relevant for the results presented in the paper.

2.1. *Damage–Breakage Rheology Model*

Lyakhovsky and Ben-Zion (2014a, b) and Lyakhovsky et al. (2016) developed a thermodynamically based model that combines a continuum damage framework for brittle rock deformation (Lyakhovsky et al. 1997, 2011, and references therein) with a continuum breakage framework for granular flow (Einav 2007a, b). The formulation is done in the overall framework of the Landau phenomenological mean-field theory of phase transitions (e.g., Chaikin and Lubensky 2000). The model assumes that materials with brittle damage may contain a mixture of cracked solids and granular media. The density of distributed cracking in the solid phase is represented with a scalar damage parameter α ($0 \leq \alpha \leq 1$), while the grain size distribution of the granular phase is addressed with a breakage parameter B ($0 \leq B \leq 1$). The damage parameter provides a measure of the microcrack density, and the breakage parameter reflects the grain size distribution. The damage–breakage model accounts for two coupled time-dependent processes controlling the source dynamics: damage accumulation in a process zone in front of the propagating rupture front, and Couette-like viscous flow of granular material within a gouge zone generated behind the rupture front (Fig. 1b). General relations controlling the rate of the fracture process and granular flow are briefly discussed below.

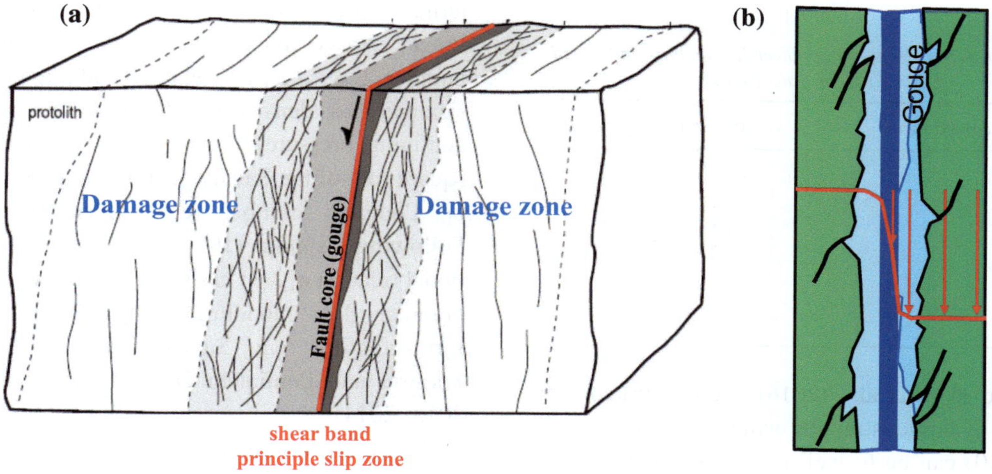

Figure 1
a Schematic fault architecture and **b** conceptual cross-section of a three-body configuration of a fault based on laboratory observations (modified after Boneh et al. 2014)

2.2. *Kinetics of the Fracture Process*

Following Griffith's (1920) energy balance concept, the growth rate of a fracture with length l is assumed to be proportional to the square of the stress intensity factor and thus proportional to the fracture length ($\frac{\mathrm{dl}}{\mathrm{d}t} \sim K^2 \sim l$). This relation corresponds to "complete similarity", in terms of Barenblatt (1996), with exponential growth in time and lack of any length scale. However, complete similarity is hardly ever achieved, except in some aluminum alloys. The rates of subcritical crack growth are most commonly represented by a power-law equation (Charles 1958; Paris and Erdogan 1963):

$$\frac{\mathrm{d}l}{\mathrm{d}t} \sim l^n. \tag{1}$$

Experimental observations of fracture propagation (e.g., Meredith and Atkinson 1985; Collins 1993) and indirect calculations of n values from microseismic data (Main et al. 1992; 1993) indicate that propagation starts with an n value between 1 and 2 in the very slow crack corrosion regime, increasing up to about 5 or even 10. The rate of accelerating fracture is certainly limited by the speed of elastic waves. Therefore, Eq. (1) should be corrected for the dynamic regime.

Based on energy and entropy balance considerations, and adopting Onsager reciprocal relations between thermodynamic forces and fluxes, the equation for the damage evolution has the form (Lyakhovsky et al. 1997):

$$\frac{\mathrm{d}\alpha}{\mathrm{d}t} = -C\frac{\partial F}{\partial \alpha}, \tag{2}$$

where C is a positive function of the state variables describing the temporal rate of the damage process, and F is the free energy of the damaged material, being a second-order function of strain, thus $\frac{\partial F}{\partial \alpha} \sim \varepsilon^2$. The average damage of the rock volume scales as the size of the highly damaged region ($\alpha \sim l$), and the strain concentrated around this region scales as $\varepsilon \sim \sqrt{l}$, as in linear elasticity. Substitution of these scaling relations into the damage evolution equation (2) leads to the same relation ($\frac{\mathrm{dl}}{\mathrm{d}t} \sim l$) as in the Griffith energy balance concept. This linear scaling holds until the coefficient C in Eq. (2) is constant. However, the rock strength calculated using a constant C value significantly overestimates strengths measured at high strain rates for different rocks (see Fig. 2 in Lyakhovsky et al. 2016). This may be solved following Grady and Kipp (1980), who preserved the same form of the kinetic equations for damage accumulation under high strain rates ($e > 10^{-4}$ 1/s), and fit the experimental data by increasing the kinetic coefficient proportionally to a certain power of the strain rate, $C \sim \left(\frac{\mathrm{d}\varepsilon}{\mathrm{d}t}\right)^m$.

Table 1

Relation between Grady–Kipp power-law damage index and Paris–Erdogan fracture index

Power-law damage index	Power-law fracture index
0.0	1.0
0.5	1.5
0.6	1.8
0.7	2.2
0.8	3.0
0.9	5.5
0.95	10.5

Lyakhovsky et al. (2016) demonstrated that observed dynamic strength values reported by Lu et al. (2010) can be fit well to the model strain rates with m values between 0.75 and 0.9. For $m = 0.8$, the model provides a reasonable general fit to the observed strength values for different rocks. With this strain-rate correction, the rate of damage accumulation is proportional to $\frac{d\alpha}{d_t} \sim \left(\frac{d\varepsilon}{d_t}\right)^m \varepsilon^2$. Using the same scaling between elastic strain and fracture length leads to the following power-law relation:

$$\frac{dl}{dt} \sim l^{\frac{2-m}{2(1-m)}}. \tag{3}$$

Similarly to Eq. (1), relation (3) should also be corrected after the fracture propagation rate for the dynamic regime. The power-law scaling relations (1) and (3) enable connection of the fracture index n (1) with the strain-rate index m (3) of the damage mechanics. Table 1 shows that typical n values for accelerated fractures between $n \approx 2$ to $n \approx 10$ correspond to strain-rate damage index values $m \approx 0.65$ to $m \approx 0.95$, which fall well within the range of index values estimated based on rock strength experiments. This scaling analysis connects fracture mechanics observations and the empirical Paris–Erdogan crack growth law with the Grady–Kipp strain-rate correction in damage mechanics. The numerical simulations described below show that the strain-rate index is a crucial factor controlling the front speed of the dynamic rupture.

2.3. Granular Flow

Various approaches based on experiments, dimensional arguments, and numerical simulations lead to a continuum description of dense granular flows in terms of a functional relation between the Drucker–Prager stress ratio, $\sigma_{DP} = \tau/P$, and the inertial number, $I = e\,d\sqrt{\rho/P}$ (where d and ρ are the particle size and mass density, e is the strain rate, and P is the confining pressure). This is also referred to as the square root of the Savage number (e.g., Savage 1984) or Coulomb number (e.g., Ancey et al. 1999). Following Bagnold (1954), the dimensionless functional relation $\sigma_{DP}(I)$ has become widely accepted for granular flow modeling with different functional forms for σ_{DP} (e.g., Savage 1998; Jop et al. 2006; Forterre and Pouliquen 2008; Henann and Kamrin 2013, 2014).

Using typical conditions for seismic events leads to surprisingly low values $I \approx 10^{-3}$. This estimate is obtained by adopting the following parameters: Based on laboratory results (e.g., Henann and Kamrin 2013, 2014) and field observations (e.g., Chester et al. 1993, 2005), the width of the principal slip zone (red line in Fig. 1a and zone of localized strain in Fig. 1b) falls between 5 and 10 particle sizes ($N \sim 5$–$10d$). For such width, the strain rate is $e \sim v_{slip}/10d$, where the typical slip velocity $v_{slip} \sim 1$ m/s. Material density $\rho \sim 3 \times 10^3$ kg/m^3 and confining pressure $P \sim 10^8$ Pa were used for the inertial number calculation. Henann and Kamrin (2013, 2014) demonstrated that the linear relation $\sigma_{DP}(I) = bI$ (where b is a nondimensional coefficient) provides a good fit for dense granular flow under low I values typical for the slip distribution within the gouge zone under seismogenic conditions. Lyakhovsky et al. (2016) adopted this linear relation for their modeling of the dynamic rupture process and defined the effective viscosity of the gouge zone with thickness h as

$$\eta_{gouge} = \frac{h}{N} b\sqrt{\rho P}, \tag{4a}$$

or as the fluidity, which is the inverse of viscosity:

$$\phi_{gouge} = \frac{N}{h\,b\sqrt{\rho P}}. \tag{4b}$$

Relations (4) show that the gouge fluidity scales with the thickness of the gouge zone and the square root of the confining pressure. In the numerical simulations, we predefine the thickness of the gouge zone, introduce the reference fluidity value, ϕ_0, for the initial pressure value P_0, and use the following relation for the gouge fluidity:

$$\phi_{\text{gouge}} = \phi_0 \cdot \sqrt{\frac{P_0}{P}}. \quad (5)$$

2.4. Numerical Modeling of Dynamic Ruptures and Radiated Waves

The implementation of the damage–breakage rheology model into a DBRM-3D code (Lyakhovsky et al. 2016) uses the explicit finite difference Lagrangian method, based on the Fast Lagrangian Analysis of Continua (FLAC) algorithm originally developed by Cundall (1989) for elastoplastic rheology and implemented in the ITASCA software. The FLAC algorithm was modified for viscoelastic media (Poliakov et al. 1993). A modified version of this code incorporating heat transport is known as PAROVOZ and is widely used by many researchers. Lyakhovsky et al. (2001) developed their own 3D code for quasistatic viscoelastic damage rheology modeling, which was used in many geodynamic applications and simulations of the coupled evolution of earthquakes and faults. Recently, the code was modified for dynamic processes, by reducing the force damping to realistic values corresponding to wave attenuation and implementing breakage evolution in the damage kinetic module. The general procedure involves solving the force balance equation (Newton's second law) for every grid point with mass M. During integration, the previously calculated force vector, F_i, is damped with factor d:

$$M\frac{\partial v_i}{\partial t} = F_i - d|F_i| \cdot sign(v_i). \quad (6)$$

This algorithm utilizes the model of Knopoff and MacDonald (1960) for frequency-independent attenuation (see also Knopoff 1964). In terms of lumped parameters, Knopoff and MacDonald (1960) suggested representing the linear model by a mass–spring system which is attenuated by some sort of "sandpaper" in which the attenuation depends not only on the sign and value of the force vector, F_i, but also on the sign of the velocity vector v_i. They also connected the numerical damping factor with the quality factor Q:

$$Q = \pi\frac{1 + \text{d}}{2d} \sim \frac{\pi}{2d}. \quad (7)$$

The DBRM-3D code may run in two different modes. In the first mode, the numerical grid is refined around the source zone having details of damage–breakage processes. The size of the tetrahedral elements within the core zone is half of the zone thickness and gradually increases away from the fault zone. Initial results and technical details of these simulations were presented by Lyakhovsky et al. (2016). Velocity–time functions are collected at the numerical nodes surrounding the source zone during the first run. In the second mode (second calculation step), these velocities serve as source functions for upscaled simulations which include distinct elements with independently prescribed velocity functions on both sides. The rectangular model domain of 8 km × 8 km × 4 km is divided into (approximately) 1-m-sized regular tetrahedral elements. This two-step multigrid approach allows resolution of the detailed damage–breakage evolution in the source volume as well as generation of far-field synthetic seismograms using a significantly larger domain size. The model can naturally incorporate fault zone structures with complex geometries and evolving properties during failure, as well as basic ingredients such as gradient zones and the free surface.

3. Results

3.1. Seismic Source Modeling

The initial configuration consists of a narrow elliptic homogeneous damage zone with several predefined width values (h = 20, 40, or 60 cm) and α = 0.7 (tabular fault zone) inside a damage-free volume (see Fig. 2 for a map view). The damage zone is surrounded by an edge region with high gradient of damage, which serves as a low-damage barrier terminating the rupture propagation. A small perturbation with elevated damage (α = 0.8) close to the left edge of the damage zone serves as a nucleation site for the dynamic rupture. The simulation starts with 50 MPa confining pressure. The lateral edges of the model volume are sheared slowly to increase the shear stress component, σ_{xz}, while keeping all other components constant away from the fault zone. The loading rate is several orders of magnitude below typical rates of dynamic processes, so fixed boundary conditions are practically realized during the dynamic failure. The remote loading produces stress concentration around the region with high initial damage ($\alpha = 0.8$), leading to additional damage, breakage increase, and transition to a granular phase in a "nucleation zone." The failure process in the nucleation zone redistributes stresses and produces a radially propagating crack-like rupture with a granular phase behind the rupture front (Fig. 3a). When the edges of the rupture encounter the low-damage barrier that prevents additional propagation in the vertical direction, a pulse-like rupture with a few-meter process zone in front is formed and continues to propagate to the right along the damage zone (Fig. 3b, c).

During the rupture process, there is shear heating behind the propagating front generated by the failure process. This is illustrated in the right panels of Fig. 3 with evolving temperature values in a 1-mm core layer inside the damage zone. The shear heat production is integrated over the simulation time and is calculated as deviatoric stress multiplied by strain rate, $\tau_{ij}e_{ij}$, divided by ρC_p, and multiplied by the zone thickness. Cooling associated with heat transport away from the fault zone is not considered, so the presented temperature maps are upper-bound values. The results demonstrate that heating starts at the rupture front with the onset of slip, but the zone

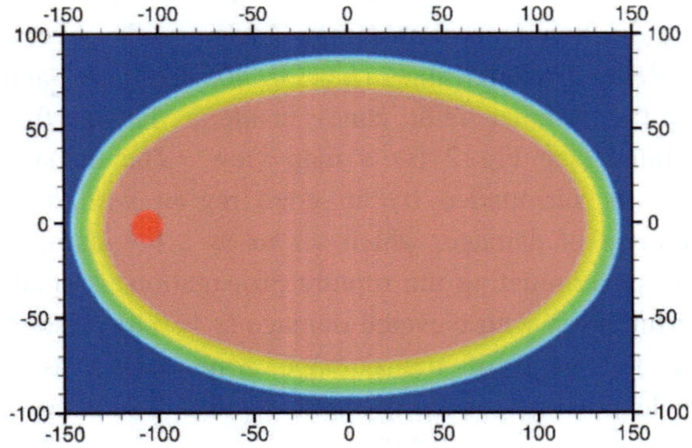

Figure 2
Cross-section of the source geometry with uniform initial damage distribution and small round zone with slightly elevated damage

with high temperature is located well behind the rupture front where slip is accumulated. This implies that the physical processes at the rupture front controlling rupture propagation involve rock fracturing with little temperature-assisted weakening. Elevated temperature and possible generation of melt products may modify processes behind the propagating rupture front. However, such effects are not considered at the current stage of modeling.

The numerical simulations provide information on characteristics of the failure process, such as the inelastic strain tensor at the source volume, speed of the rupture front, and slip rate, which control the properties of each event and radiation from the source. For the source geometry of Fig. 2, thickness $h = 0.4$ m, and fluidity $\phi_0 = 0.2$ (MPa s)$^{-1}$, the speed of the rupture front changes significantly for different values of the Grady–Kipp strain-rate correction index m. Figure 4 shows the rupture front velocity normalized by the S-wave speed (V_{rp}/V_S) as a function of distance from the center of the nucleation zone for m values from 0.7 to 0.9. At m values below 0.7, the rupture does not propagate dynamically; the rupture front advances quasistatically at rates well below the seismic wave velocity and without any observable radiation. At m values above 0.9, the damage growth is very fast and the front propagates with supershear velocity. The present study focuses on the standard, subshear propagation regime.

The speed of the rupture front depends on the thickness of the source volume and the product of the thickness multiplied by the fluidity of the granular material behind the front. Figure 5 shows modeling results with the same strain-rate correction index $m = 0.8$. In the employed damage–breakage model, the rate of damage accumulation does not directly depend on the source volume thickness, but is strongly affected by the stress concentration in front of the propagating rupture. The stress is weakly localized around the relatively thick source volume, and the stress concentration in the process zone is significantly enhanced for thinner source volume. The stress concentration in the process zone increases the damage growth rate and promotes faster rupture dynamics, even for the same slip rate of the granular material behind the rupture front that is controlled by

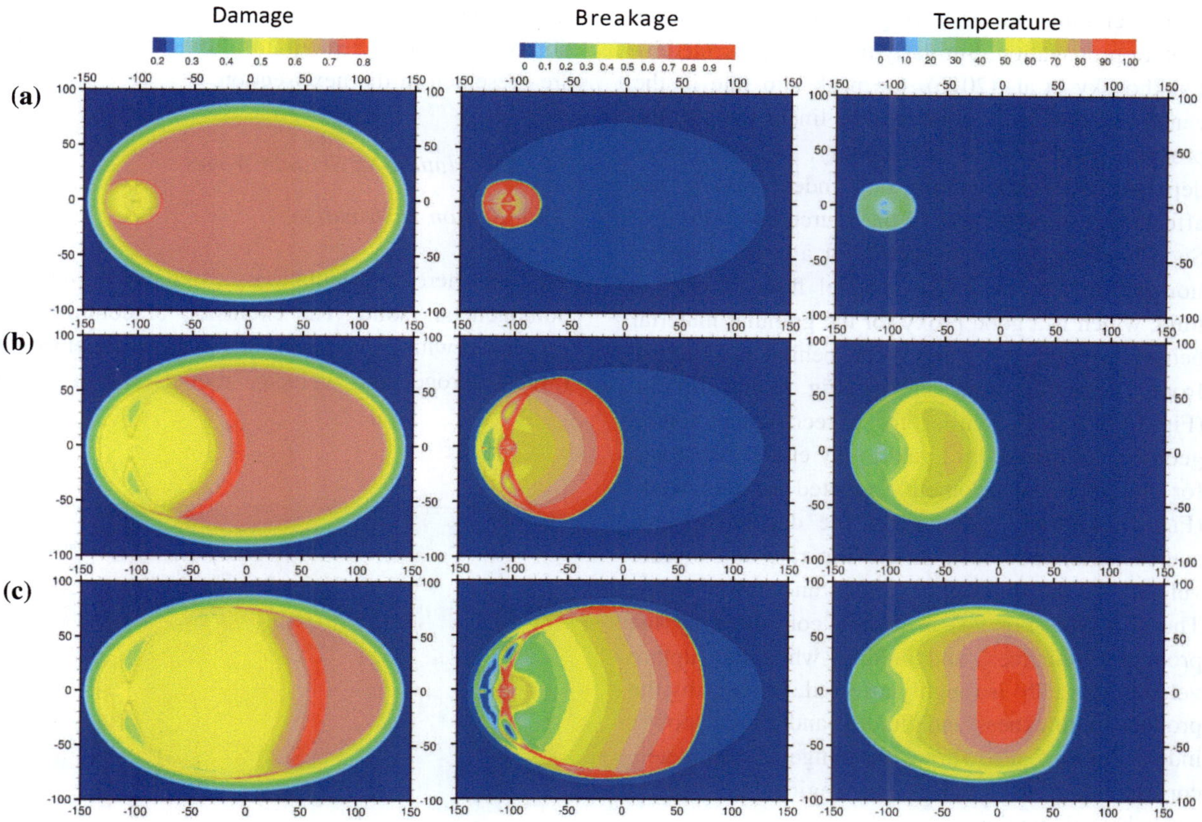

Figure 3
Three time snapshots with cross-sections for damage, breakage, and temperature for **a** 0.03 s, **b** 0.06 s, and **c** 0.09 s after the onset of rupture propagation

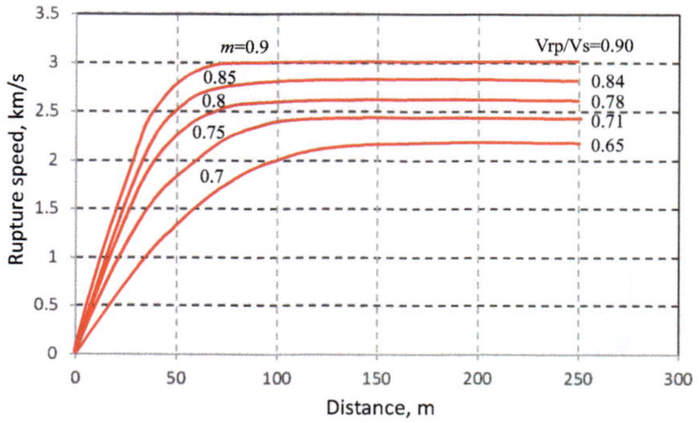

Figure 4
Rupture speed as a function of distance from the center of the nucleation zone for different values of the Grady–Kipp index m for thickness h = 40 cm and fluidity 0.2 (MPa s)$^{-1}$

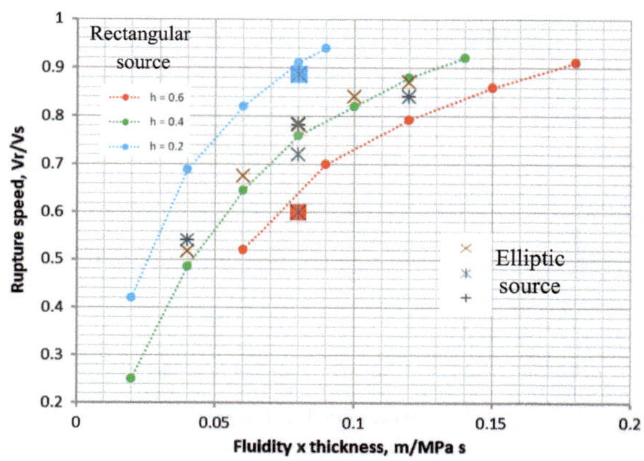

Figure 5
Rupture velocity versus fluidity multiplied by thickness for different predefined gouge zone thicknesses and shapes. The same Grady–Kipp index m = 0.8 is used. Results for the elliptic source are obtained in this study, and results for the rectangular source are taken from Lyakhovsky et al. (2016)

the product of thickness and fluidity (Fig. 6). Elevated values of fluidity multiplied by thickness allow more efficient stress relaxation and higher strain rates of the granular material (Fig. 6), leading to faster rupture propagation. Similarly to results presented by Lyakhovsky et al. (2016), the peak slip rate in the center of the fault zone scales linearly with the product of thickness and fluidity, exhibits weak dependence on the strain rate index, and is not affected by the thickness of the source zone (Fig. 6). Such linear scaling is expected for a stress distribution within a Couette-like channel flow of viscous fluid, which is a good proxy for the granular material behavior within the gouge zone behind the rupture front. The rupture front velocity (Fig. 5) and slip rate (Fig. 6) are both similar for a rectangular source geometry discussed by Lyakhovsky et al. (2016) and for the elliptical geometry adopted in this study (Fig. 2). Figure 7 summarizes the dependences of source parameters such as the slip rate and rupture velocity on the fluidity, thickness, and power index. The slip rate (Fig. 7a) is mainly controlled by the product of thickness and fluidity, while the rupture velocity (Fig. 7b) is roughly affected similarly by the product of thickness and fluidity and by the power index. Figure 7 also shows the range of parameters corresponding to a quasistatic regime as well as supershear dynamics.

During the simulations of dynamic ruptures, velocity–time recordings are collected at the numerical nodes surrounding the source zone, for upscaled simulations of radiated seismic waves and calculations of synthetic seismograms. Some of these results are presented in the next section.

3.2. Calculations of Seismic Waves

3.2.1 Station Distribution

We focus here on calculations of radiated waves in a homogeneous solid. The exclusion of complexities associated with curved ray paths in three-dimensional (3D) heterogeneous velocity models allows more

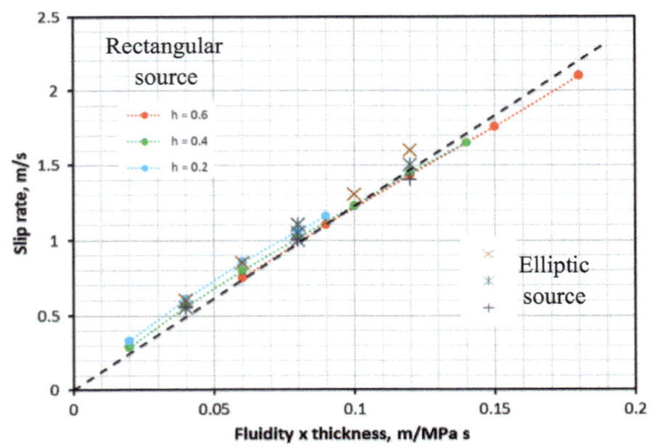

Figure 6
Slip rate versus fluidity multiplied by thickness for Grady–Kipp index m = 0.8 and different thicknesses and shapes. The slip rate shows strong linear dependence on fluidity times thickness. As in Fig. 5, results for the elliptic source are obtained in this study, and results for the rectangular source are taken from Lyakhovsky et al. (2016)

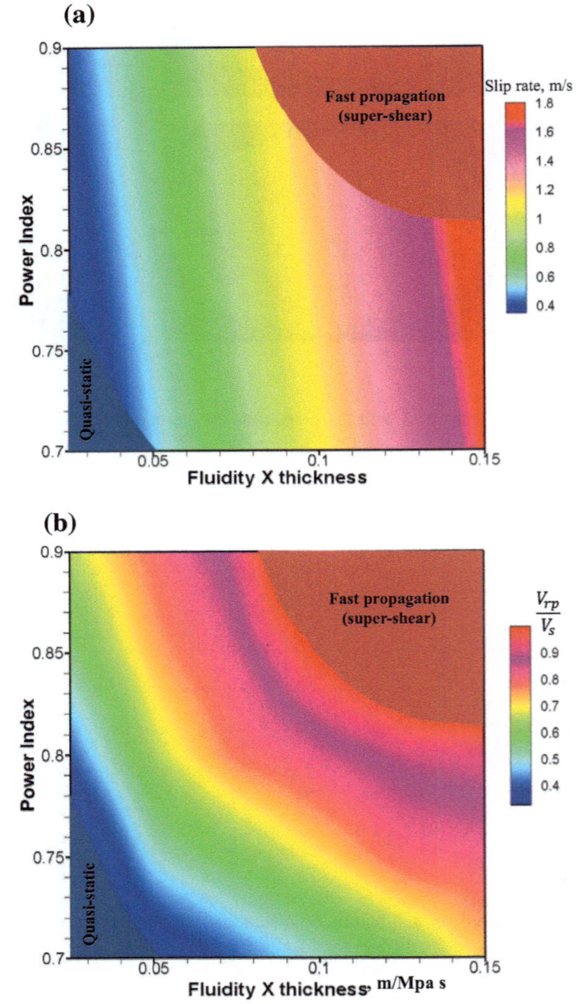

Figure 7
Maps summarizing combined effects of fluidity times thickness and Grady–Kipp index (m value) on the slip rate (**a**) and rupture speed (**b**) for predefined thickness h = 40 cm

reliable comparisons with theoretical results. For analysis of radiated waves, we use two circles of receivers distributed at 5° station interval at the hypocenter depth at two radial distances, 1500 m and 3000 m, from the hypocenter. For a 200 m rupture length, these stations are far enough to be treated as far-field stations without near-field terms. Altogether we have 144 stations, 72 stations per circle (Fig. 8). At each station, we record several channels: three velocity, $v_i(t)$, and three displacement, $u_i(t)$, components with i denoting the x, y, and z (vertical) mesh directions. In addition, four additional channels are recorded: one for the P-wave calculated by taking the divergence of the displacement vector, $v_P(t) = V_P\, div(u_i)$, and three for the S-wave vector calculated by taking the curl of the displacement vector, $v_S(t) = V_S\, curl(u_i)$.

3.2.2 Analysis of Seismic Motion

The main measured quantities in analysis of seismic data are ground velocities and accelerations, depending on the type of sensor. Ground displacement is obtained by integrating the velocity recorded or geodetic data and is analyzed less often, although it is now used more frequently in coupled analyses of seismic and geodetic observations. In this study we focus on analyzing the velocity channels recorded by the model seismic stations. The velocity channels, $v_i(t)$, originally oriented along the mesh axes (Fig. 9), are rotated to a radial–transverse coordinate system (Fig. 10). The rotated components should correspond to the different body waves radiated from the source; the radial component, $v_r(t)$, reflects the P wave, and the transversal component, $v_t(t)$, reflects the S wave. The additional channels, $v_P(t)$ and $v_S(t)$, calculated by taking the divergence and curl of the displacement vector, separate directly the contribution of the velocity field to the P and S waves. In analysis of seismic data, the separation of the wavefield into P and S waves is typically done by rotating records to radial–transverse coordinates (e.g., Rosenberger 2010; Kurzon et al. 2014). Comparison of $v_r(t)$ and $v_P(t)$ shows mainly the P wave packet arriving at the site at two radial distances, 1500 m and 3000 m, from the hypocenter; comparison of $v_t(t)$ and $v_S(t)$ shows similar results for the S wave packet (Fig. 10). In the analysis below we use the $v_P(t)$ and $v_S(t)$ records.

3.2.3 Radiation Pattern and Rupture Directivity

Far-field point-source solutions for vertical strike-slip ruptures are associated with P and S wave radiation patterns. For pure deviatoric sources, the radiation patterns have four symmetrical amplitude lobes, at 90° to each other (e.g., Aki and Richards 1980). Tensile components of faulting significantly modify

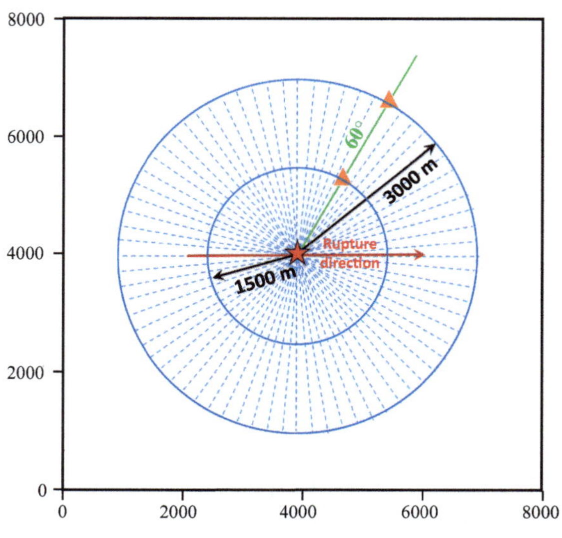

Figure 8
Map view showing the horizontal positioning of the receivers at the depth level of the source. There are two circles of receivers, at 5° interval, one at 1500 m from the source, and the other at 3000 m, resulting in 72 receivers per circle

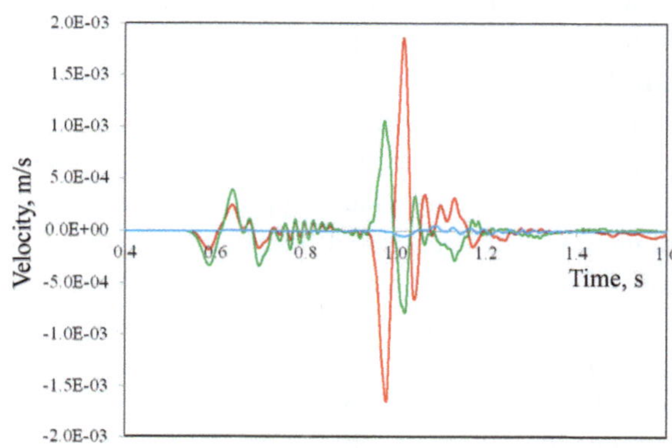

Figure 9
Calculated velocity waveforms, V_x (red), V_y (green), and V_z (blue), at 3000 m from the hypocenter at 60° to the direction of rupture propagation

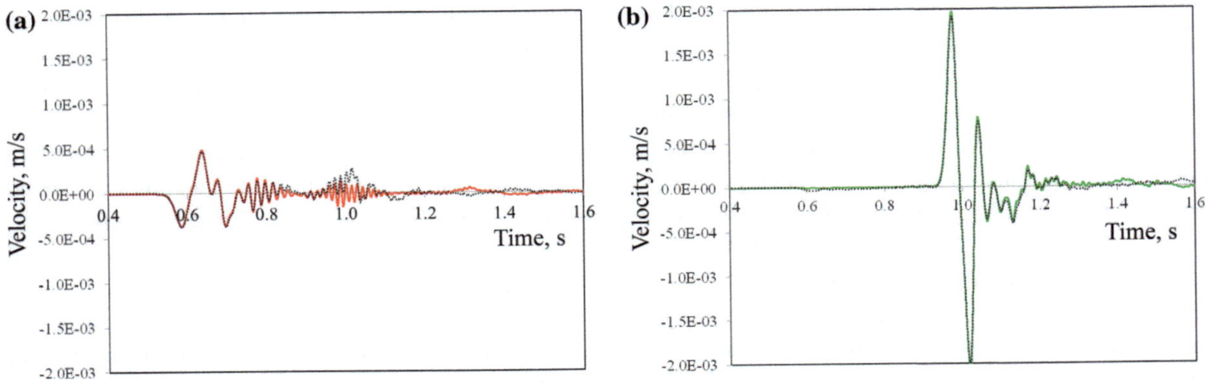

Figure 10
Radial (**a**) and tangential (**b**) components (dotted lines) of the seismograms from Fig. 9 versus $v_P(t)$ (red) and $v_S(t)$ (green) components based on application of the divergence and curl operators to the displacement field

the radiation patterns (e.g., Vavryčuk 2001; Kwiatek and Ben-Zion 2013). However, since earthquakes are dominated by shear faulting and most studies on earthquake source parameters assume pure deviatoric processes, we discuss explicitly radiation patterns for shear faulting. In such cases, the lobes of *S* radiation are oriented parallel and perpendicular to the rupture surface, while for the *P* radiation pattern they are rotated 45° to the rupture surface. In spherical coordinates, the far-field *P* and *S* radiation patterns are given by $A_p = \cos\varphi \cdot \sin 2\theta$ and $A_S = \cos 2\theta \cos \varphi\theta - \cos\theta \sin \varphi\theta$, respectively. At the surface level of the hypocentral depth, $\varphi = 0°$, reducing the equations to simpler forms: $A_P = \sin 2\theta$ and $A_S = \cos 2\theta$.

More elaborate source models include space–time variations of slip, where rupture is nucleated at a given location and then propagates along the fault until being arrested. Rupture propagation can be unilateral (e.g., Haskell 1964) or approximately symmetric in the different directions (e.g., Sato and Hirasawa 1973). Large faults typically have bimaterial interfaces that separate different elastic solids and promote preferred rupture direction (e.g., Weertman 1980; Andrews and Ben-Zion 1997; Shlomai and Fineberg 2016), hence we focus on unilateral ruptures. In such cases, rupture directivity produces amplification of wave amplitudes and shifting of source spectra to higher frequencies in the direction of rupture propagation, and opposite effects in the other direction. This is associated with superposition of Doppler effects operating on all frequencies of the finite source spectra (Douglas et al. 1988) and modification of the radiation patterns. For *P* waves, the amplitude lobes at ±45° to the rupture propagation ($\theta = \pm 45°$) are increased by directivity, while the opposite lobes ($\theta = \pm 135°$) are reduced. For *S* waves, the amplitude lobe in the direction of rupture propagation ($\theta = 0°$) is increased, while the opposite lobe ($\theta = 180°$) is reduced.

A basic description of the directivity effect in the far field is associated with a shift of the corner frequency f_c of the *P* source spectra (Ben-Menahem and Singh 1981)

$$f_c(\theta) = \frac{f_c^0}{\left(1 - \frac{V_{rp}}{V_P} \cdot \cos(\theta)\right)}. \tag{8}$$

The modified radiation pattern of velocities or displacement amplitude for *P* and *S* waves are

$$A_P = \frac{\sin(2\theta)}{\left(1 - \frac{V_{rp}}{V_P} \cdot \cos(\theta)\right)} \tag{9a}$$

$$A_S = \frac{\cos(2\theta)}{\left(1 - \frac{V_{rp}}{V_S} \cos(\theta)\right)}. \tag{9b}$$

The dependence on the polar angle φ is omitted, since all the receivers (Fig. 8) are located in a plane that is in the middle of and perpendicular to the source zone. Fitting $f_c(\theta)$ in Eq. 8 provides an estimate for f_c and the asymptotic rupture velocity V_{rp} after initial rupture acceleration (Fig. 4).

Following the classical model of Brune (1970) for a frequency falloff at f^{-2} in the displacement spectra, the velocity spectrum is written in the form

$$\Omega(f) = \frac{\Omega_0}{1 + (f/f_c)^2} 2\pi f \qquad (10)$$

and has a maximum at $f = f_c$ (see also Abercrombie 1995). Thus, the corner frequency $f_c(\theta)$ at each orientation is taken from the maximum of the corresponding velocity spectra.

Figure 11a illustrates the directivity effect of variable azimuthal shifting of the corner frequency of radiated P waves generated by a unilateral rupture with a source model having index $m = 0.8$, thickness $h = 0.4$ m, and fluidity $\phi_0 = 0.2$ (MPa s)$^{-1}$. Similar corner frequency shifts are obtained for radiation generated by source models with different thicknesses ($h = 0.2$ and 0.6 m) but the same product of fluidity times thickness (Fig. 11b). In both cases, the shifts of the corner frequency fit well with values obtained by Eq. 8 with rupture velocity $V_{rp} \sim 2.5$ km/s and $f_c^0 \sim 5.3$ Hz. The rupture velocity estimated from the frequency shift is in good agreement with the asymptotic rate of rupture front obtained directly from the source model (Figs. 4, 7). Most of the points fit well the theoretical curve (dashed line), but in some directions the corner frequency value is not well resolved because of low wave amplitude in these directions.

The amplitudes obtained from $V_P(t)$ and $V_S(t)$ are shown in Fig. 12 together with the theoretically predicted radiation patterns (Eqs. 9a and 9b). The P wave radiation pattern of the simulation results (red lobes, Fig. 12a) is similar to the theoretical solution (blue lobes, Fig. 12a, Eq. 9a). The S wave radiation pattern of the synthetic seismograms (red lobes) is affected more than the P radiation and deviates significantly from the theoretical result (blue lobes, Fig. 12b, Eq. 9b). The simulated source model (Fig. 13) includes, in addition to the common fault-parallel slip velocity component (red line), a fault-normal expansion–compaction component (blue line), and a process zone, in front of the rupture front, experiencing fast damage–breakage evolution (Lyakhovsky and Ben-Zion 2014a). Significant rigidity reduction inside the process zone leads to reduced S-wave speed and energy dissipation of waves passing through this zone. The effect of reduced wave speed can hardly be recognized in the far field, since the zone size is a small fraction of the propagation path that is dominated by the surrounding unchanged rock. On the other hand, wave attenuation is very small along most of the propagating path and mainly occurs within the process zone. The zone in front of the rupture front has two significant effects: (1) dissipation of waves propagating from the rupture front through the process zone, which consist mainly of S waves propagating in the direction of rupture, and (2) dynamic expansion–compaction that produces isotropic radiation within and somewhat behind the process zone. The former dissipation mechanism leads to significant absorption of S energy (Fig. 12b), and together with the latter effect reduces the overall S/P energy ratio.

To illustrate the separate effects on the radiated energy of (a) expansion–compaction components of motion within the process zone, and (b) dissipation due to damage–breakage processes, we artificially modify the source-time functions generated by the damage–breakage model. The simulated source produces velocity components in all three directions, including a fault-normal velocity component associated with expansion–compaction of the source. To eliminate these volumetric effects, instead of allowing discontinuous normal velocity components, we prescribe a continuous normal component while keeping the fault-parallel velocity components unchanged. The conditions in the normal direction include continuity of both the normal stress component, $[\sigma_n]_\pm = 0$, and the normal velocity component, $[V_n]_\pm = 0$, and the momentum equation is solved for the direction normal to the outer boundaries of the source volume. This situation is common for models with a 2D fault surface and mostly eliminates volumetric source effects, while preserving the same rapture velocity and slip behind the rupture front without volume change. Figure 14 shows P- and S-wave directivity patterns for three models with different thicknesses ($h = 0.2$, 0.4, and 0.6 m) used previously for Fig. 11. The red lobes stand for models with source-time functions generated by the damage model including the fault-normal velocity component, while green lobes are for models eliminating the volumetric changes. The shape of the theoretical

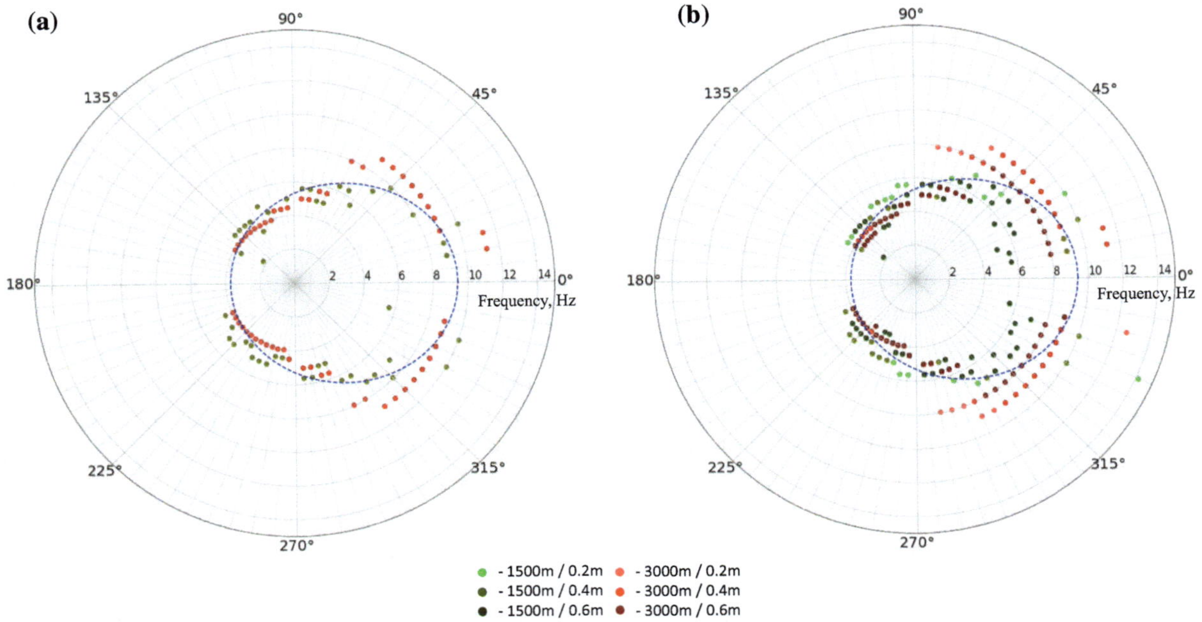

Figure 11
P-wave corner frequency at 3000 m (red dots) and 1500 m (green dots) for different directions from the rupture propagation. Blue dashed line marks the best fit using Eq. 8 with $f_c^0 \sim 5.3$ Hz and $V_{rp} = 2.5$ km/s. **a** Model results for h = 0.4 m, and **b** results of three models with h = 0.2, 0.4, and 0.6 m; wider faults are correlated with darker color shade as indicated in the legend

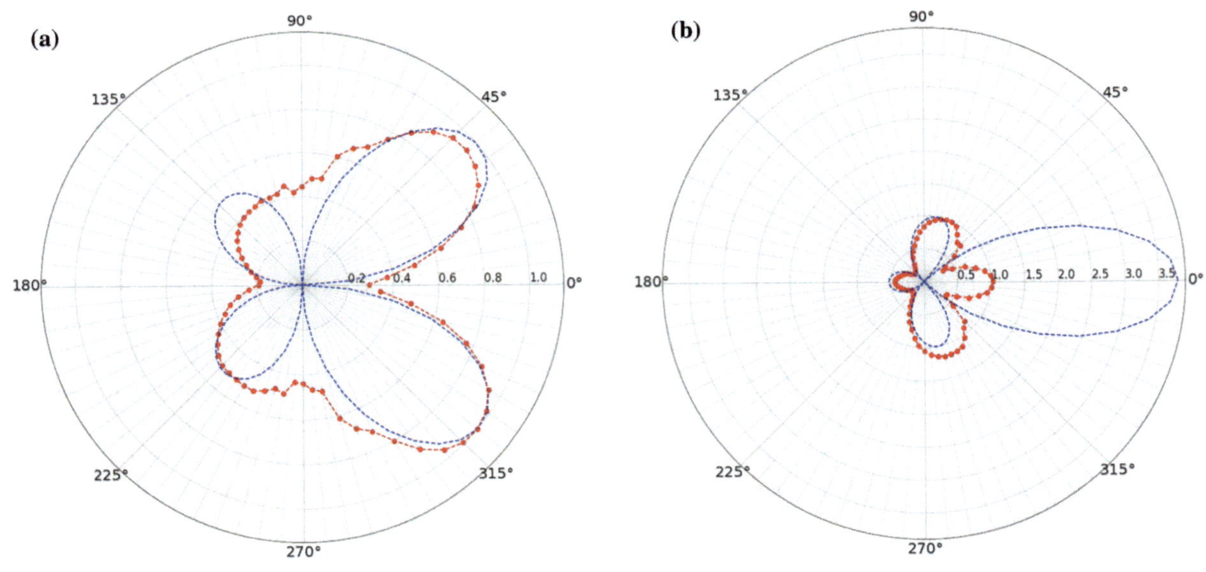

Figure 12
Radiation patterns for **a** *P*-waves and **b** *S*-waves for simulation with h = 0.4 m. The amplitude of each wave type is normalized to the maximum value. Blue dashed lines represent the radiation pattern predicted by Eq. 9

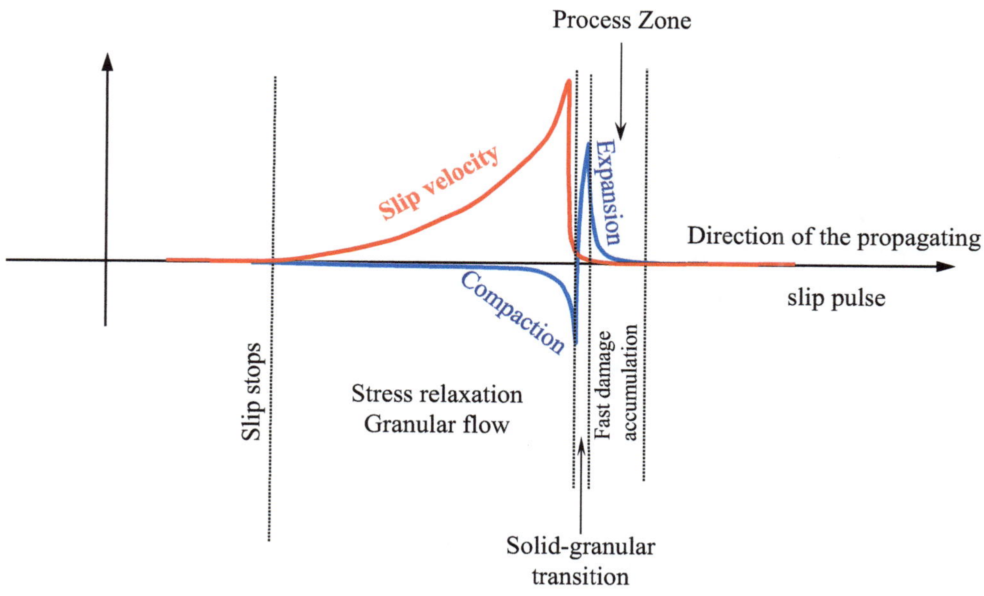

Figure 13
Schematic slip pulse characterizing the simulated dynamic ruptures

lobes (blue lobes) for standard shear sources fit reasonably well with the model results associated with the green and red lobes for *P* waves (Fig. 14a), with some deviation from the blue lines mainly due the contribution of the volumetric components. A comparison between the green and blue lobes for the *S* radiation (Fig. 14b) shows the significant effect of dissipation on the *S* radiation in the direction of rupture propagation. The additional effect of compaction–expansion is illustrated by comparing the green and red lobes in Fig. 14b.

As mentioned, the reduction of elastic moduli during the damage process and expansion–compaction in the process zone change the ratio of *S*-wave energy to *P*-wave energy, $q = E_S/E_P$. For a shear point-source model and a Poisson solid ($\upsilon = 0.25$), the energy ratio is predicted to be $q = 1.5 \cdot (V_P/V_S)^5 \sim 23.4$. Observations generally suggest average *q* values between 9 and 25, with a large amount of scatter for individual earthquakes (Shearer 2009). The simulated $q \sim 16.6$ (Fig. 15a), calculated by integrating the squared velocities of the *P*- and *S*-waves in the time domain (e.g., Shearer 2009), fits well with the observed range between 9 and 25 for the three simulations that eliminated volumetric effects and the process zone, by applying the normal stress condition, $[\sigma_n]_{\pm} = 0$. However, the model with the same slip function and normal velocity components (complete time-source function) produces significantly lower E_S/E_P ratios between 4 and 9 with average *q* value ~ 7.1 (Fig. 15a). The estimated *q* values are similar to the estimates based on the synthetic seismograms recorded at 1500 or 3000 m from the hypocenter. Reduced *q* values with even lower average ~ 6.5 are obtained for all models with different source properties (Fig. 15b). These values are similar to those estimated for earthquakes in a South Africa mine associated with significant rock fracturing (Kwiatek and Ben-Zion 2013).

4. Discussion

We study dynamic ruptures in a continuum damage–breakage rheological model that associates the brittle instability with a phase transition between a damaged solid with distributed cracks and a granular medium within the generated rupture zone. The waves radiated from the source volumes are

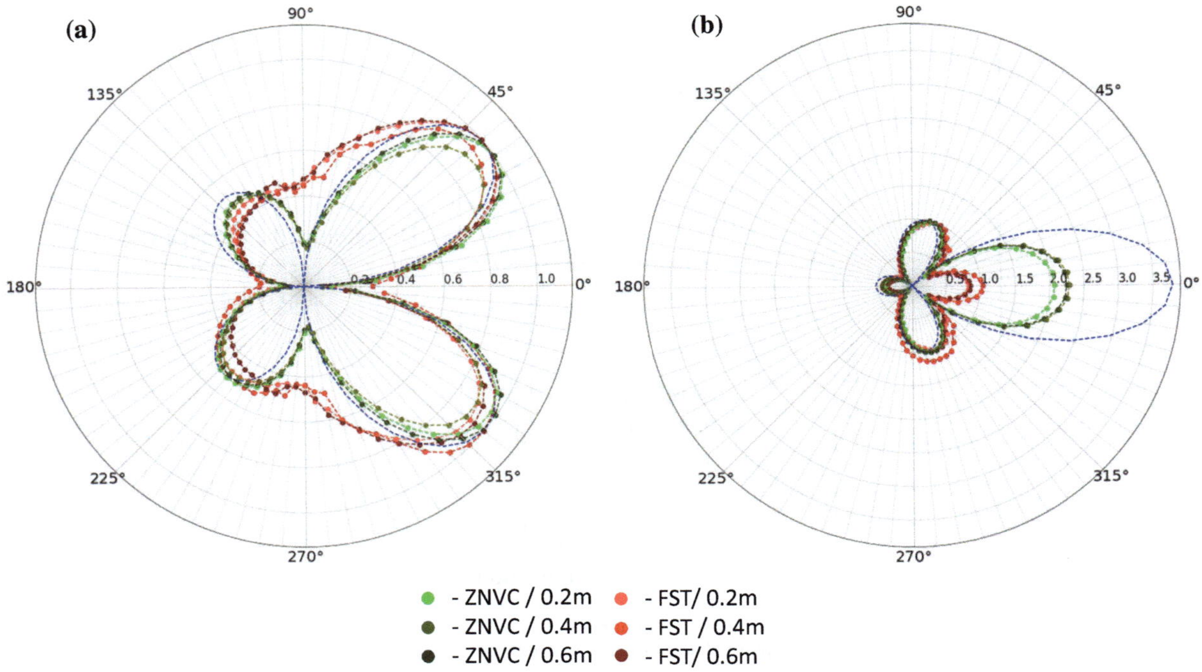

Figure 14
Radiation pattern for three models with h = 0.2, 0.4, and 0.6 m (FST—full source time—red lines) and model results eliminating the normal velocity component (ZNVC—zero normal velocity components—green lines) for **a** P waves, and **b** S waves. Blue dashed lines (as in Fig. 12) represent the radiation pattern predicted by Eq. 9. Wider fault correlated with darker color shade (see color legend in the figure)

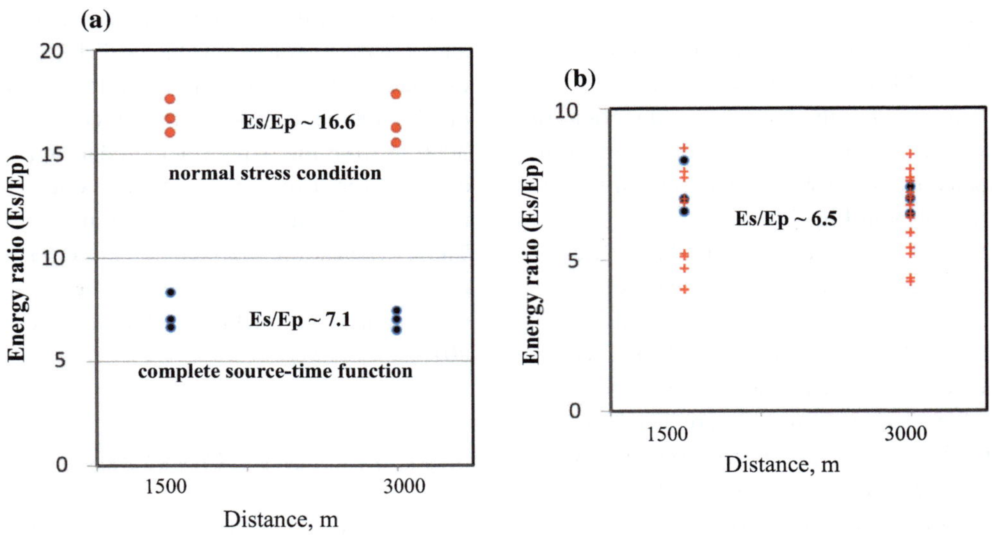

Figure 15
Energy ratio (E_S/E_P) for **a** three models with h = 0.2, 0.4, and 0.6 m (blue dots) and model results eliminating the normal velocity component (red dots). **b** Energy ratio values for all the models of this study

calculated at receiver positions in the far field. In contrast to many 2D source models, the rupture process in this work is associated with a 3D source volume and it includes dissipation to damage processes and generation of both deviatoric and isotropic source terms.

The damage–breakage framework significantly extends the ability to model brittle processes in fault zones of finite thickness in relation to various key dynamic variables. A local dependence of the strain energy function on damage and breakage parameters leads to singular localization; this is avoided by adopting a more general, nonlocal model formulation (e.g., Lyakhovsky et al. 2011, and references therein). The used gradient-type nonlocal model formulation produces a finite length scale for damage diffusion that eliminates the unrealistic singular localization. A damage diffusion process (which is slow relative to the dynamic rupture) controls the width of the zone, with damage approaching a critical value associated with the fault core thickness (Lyakhovsky et al. 2011). The rupture simulation results in this paper demonstrate effects of a predefined fault core thickness on the dynamic and final properties of simulated seismic events. An explicit study of effects associated with realistic (nonzero) thickness of the fault core is an important advantage of the used formulation compared with other dynamic earthquake rupture models assuming a frictional interface.

Two main properties of the simulated dynamic ruptures, viz. the propagation rate of the rupture front and the slip rate behind it, are controlled by two coupled time-dependent processes of the damage–breakage model. Figure 7a shows that the slip rate is mostly controlled by the fluidity of granular material within a gouge zone times its thickness. Such scaling is expected for Couette-like viscous flow within a straight channel of constant width. Deviations from the linear scaling shown in Fig. 6 are expected for more complex settings that include varying gouge zone thickness and fluidity of a granular material depending on strain rate, temperature, and other properties. The rupture front rate (Fig. 7b) is roughly equally controlled by the rate of stress relaxation due to shear motion in the gouge zone and the strain-rate correction of the damage accumulation rate. The strain-rate correction in a certain sense is similar to dynamic weakening of frictional models. The physical origin of the weakening is that the material becomes weaker as it moves (deforms) faster. The weakening effect of the damage rheology is defined through the Grady–Kipp strain-rate correction index m. m values estimated from rock strengths measured in laboratory experiments vary between 0.7 and 0.9. Adopting scaling relations (Eq. 3 and Table 1), these m values correspond to Paris–Erdogan fracture index values between 2.2 and 5.5; this range is widely accepted in fracture mechanics. At lower Grady–Kipp index and low gouge fluidity, the dynamic rupture fails to propagate, while at high values of these parameters the rupture propagates in supershear regime.

A third key property of the simulated ruptures is expansion–compaction within and behind the process zone. This is a generic expectation for volumetric effects associated with the reduction of elastic moduli (e.g., Ben-Zion and Ampuero 2009), although some aspects of the simulated results are specific to the used damage–breakage rheology model (Lyakhovsky et al. 2016). The energy dissipation and entropy production in the damage–breakage model result from all the dissipative irreversible processes in the medium, including internal friction and damage–breakage evolution. Dunn and Serrin (1985) noted that, for complex materials, the balance equation for the internal energy should include an extension of the classical balance of energy through an additional rate of mechanical energy supply, the interstitial working. For granular materials, the interstitial work flux is associated with the surface energy of the grain boundaries and its evolution during collisions and crushing of grains (e.g., Giovine 1999). Following these ideas, the heat flux in the continuum damage–breakage model is assumed to be associated with internal friction or shear heating, while energy dissipation due to damage–breakage evolution (interstitial flux) is associated with the surface energy change of nucleated and growing microcracks and grains (Lyakhovsky and Ben-Zion 2014b). The modeling results show (Fig. 3) that most of the heat generated by shear motion in the core damage zone is produced behind the rupture front where the solid–granular transition occurs (Fig. 13). Significant shear heating and elevated temperatures occur only at locations that are relatively far behind the rupture front (Fig. 3), and therefore do not affect the strain localization, damage–breakage processes, and other dynamic effects in the process zone. Subsequent temperature changes can affect the damaged rocks when they continue to accumulate shear motion behind the rupture front. This can change the fluidity and other properties of the core damage zone and indirectly affect the rupture dynamics.

The radiation pattern and rupture directivity reflect the mechanical processes occurring within the process zone, and specifically the generated isotropic source terms and energy dissipation primarily caused by interstitial work. The generated *P* energy is increased relative to the *S* energy (Fig. 15), and the *P* lobes are smeared due to the additional isotropic radiation (Figs. 12, 14). The *S* lobe in the direction of rupture propagation is reduced significantly, while the other *S* lobes remain similar. The strong modification of the *S* waves in different directions compared with classical theoretical calculations (Figs. 12b, 14b) can significantly influence analyses of rupture directivity based on *S* waves. The shape of the *P* radiation pattern is less affected by the damage process, so analysis of *P* waves can provide more reliable directivity results than *S* waves (although they have generally smaller amplitude).

The calculated seismic energy ratio, q, shows lower values ($q \sim$ 5–10) for cases in which the process zone is significant, and can explain observations of low q values, such as those reported for earthquakes in mines that are likely associated with significant rock fracturing (Kwiatek and Ben-Zion 2013). The dense and homogeneous station coverage in the simulated results allows analysis of directivity effects with the classical Eq. (8). Taking the corner frequency at each orientation from the maximum of the corresponding velocity spectra, Eq. 8 is fit with rupture velocity of $V_{\mathrm{rp}} \sim$ 2.5 km/s and $f_{\mathrm{c}}^{0} \sim 5.3$ Hz. Since a source length scale may be estimated by the ratio $L \sim 0.5 V_{\mathrm{rp}}/f_{\mathrm{c}}$, this provides a good proxy for the source length of L = 250 m used in the simulations. However, performing corresponding analysis with observed results would be far more challenging due to poor station coverage, more complex wave propagation effects, noise, etc. Additional analyses of waves radiated by dynamic ruptures in the damage–breakage model in relation to various source properties will be the subject of a follow-up study.

Acknowledgements

We thank A. Ilchev and O. Salomon for help with the simulations and acknowledge support by the US–Israel Bi-national Science Foundation (BSF Grant 2016043). The paper benefited from useful comments of two anonymous referees.

References

Abercrombie, R. E. (1995). Earthquake source scaling relationships from − 1 to 5 ML using seismograms recorded at 2.5-km depth. *Journal of Geophysical Research, 100,* 24015–24036.

Aki, K. (1967). Scaling law of seismic spectrum. *Journal of Geophysical Research, 30,* 1217–1231.

Aki, K., & Richards, P. G. (1980). *Quantitative seismology: theory and methods.* New York: W. H. Freeman & Co.

Alava, M. J., Nukala, P., & Zapperi, S. (2006). Statistical models of fracture. *Advances in Physics, 55,* 349–476. (**10.1080**).

Allix, O., & Hild, F. (2002). *Continuum damage mechanics of materials and structures* (p. 396). Amsterdam: Elsevier.

Ancey, C., Coussot, P., & Evesque, P. A. (1999). A theoretical framework for very concentrated granular suspensions in steady simple shear flow. *Journal of Rheology, 43,* 1673–1699.

Andrews, D. J., & Ben-Zion, Y. (1997). Wrinkle-like slip pulse on a fault between different materials. *Journal of Geophysical Research, 102,* 553–571.

Ashby, M. F., & Sammis, C. G. (1990). The damage mechanics of brittle solids in compression. *Pure and Applied Geophysics, 133*(3), 489–521.

Bagnold, R. A. (1954). Experiments on a gravity-free dispersion of large solid spheres in a Newtonian fluid under shear. *Proceedings of the Royal Society London, 225,* 49–63.

Barenblatt, G. I. (1996). *Scaling, self-similarity and intermediate asymptotics.* Cambridge: Cambridge University Press.

Ben-Menahem, A., & Singh, S. J. (1981). *Seismic waves and sources.* New York: Springer.

Ben-Zion, Y., & Ampuero, J.-P. (2009). Seismic radiation from regions sustaining material damage. *Geophysical Journal International, 178,* 1351–1356. https://doi.org/10.1111/j.1365-246X.2009.04285.x.

Ben-Zion, Y., Dahmen, K., Lyakhovsky, V., Ertas, D., & Agnon, A. (1999). Self-driven mode switching of earthquake activity on a fault system. *Earth and Planetary Science Letters, 172,* 11–21.

Bhat, H. S., Rosakis, A. J., & Sammis, C. G. (2012). A micromechanics based constitutive model for brittle failure at high strain rates. *Journal of Applied Mechanics, 79*(3), 031016. https://doi.org/10.1115/1.4005897.

Boneh, Y., Chang, J.C., Lockner, D. A., Reches, Z. (2014). Evolution of wear and friction along experimental faults. *Pure and Applied Geophysics, 171*(11):3125–3141.

Brune, J. N. (1970). Tectonic stress and the spectra of seismic shear waves from earthquakes. *Journal of Geophysical Research, 75,* 4997–5009.

Calderoni, G., Rovelli, A., Ben-Zion, Y., & DiGiovambattista, R. (2015). Along-strike rupture directivity of earthquakes of the 2009 L'Aquila, central Italy, seismic sequence. *Geophysical Journal International, 203,* 399–415. https://doi.org/10.1093/gji/ggv275.

Causse, M., Dalguer, L. A., & Mai, P. M. (2014). Variability of dynamic source parameters inferred from kinematic models of past earthquakes. *Geophysical Journal International, 196,* 1754–1769. https://doi.org/10.1093/gji/ggt478.

Chaikin, P. M., & Lubensky, T. C. (2000). *Principles of condensed matter physics* (p. 720). Cambridge: Cambridge University Press.

Charles, R. J. (1958). Static fatigue of glass. *Journal of Applied Physics, 29,* 1549–1560.

Chester, J. S., Chester, F. M., & Kronenberg, A. K. (2005). Fracture surface energy of the Punchbowl Fault, San Andreas system. *Nature, 437,* 133–135.

Chester, F. M., Evans, J. P., & Biegel, R. L. (1993). Internal structure and weakening mechanisms of the San Andreas fault. *Journal of Geophysical Research, 98,* 771–786.

Collins, J. A. (1993). *Failure of materials in mechanical design.* New York: Wiley.

Cundall, P. A. (1989). Numerical experiments on localization in frictional materials. *Ingenieur-Archiv, 59,* 148–159.

Douglas, A., Hudson, J. A., & Pearce, R. G. (1988). Directivity and the Doppler effect. *Bulletin of the Seismological Society of America, 78*(3), 1367–1372.

Dunn, J. E., & Serrin, J. (1985). On the thermodynamics of interstitial working. *Archive for Rational Mechanics and Analysis, 88,* 95–133.

Einav, I. (2007a). Breakage mechanics—Part I: Theory. *Journal of the Mechanics and Physics of Solids, 55,* 1274–1297.

Einav, I. (2007b). Breakage mechanics—Part II: Modeling granular materials. *Journal of the Mechanics and Physics of Solids, 55,* 1298–1320.

Forterre, Y., & Pouliquen, O. (2008). Flows of dense granular media. *Annual Review of Fluid Mechanics, 40,* 1–24.

Giovine, P. (1999). Nonclassical thermomechanics of granular materials. *Mathematical Physics, Analysis and Geometry, 2,* 179–196.

Grady, D. E., & Kipp, M. E. (1980). Continuum modeling of explosive fracture in oil shale. *International Journal of Rock Mechanics and Mining Sciences & Geomechanics Abstracts, 17,* 147–157.

Griffith, A. A. (1920). The phenomena of rupture and flow in solids. *Philosophical Transactions of the Royal Society London, 221,* 582–593.

Hamiel, Y., Liu, Y., Lyakhovsky, V., Ben-Zion, Y., & Lockner, D. (2004). A visco-elastic damage model with applications to stable and unstable fracturing. *Geophysical Journal International, 159,* 1–11.

Hansen, N. R., & Schreyer, H. L. (1994). A thermodynamically consistent framework for theories of elasticity coupled with damage. *International Journal of Solids and Structures, 31,* 359–389.

Haskell, N. A. (1964). Total energy and energy density of elastic wave radiation from propagating faults. *Bulletin of the Seismological Society of America, 54,* 1811–1841.

Henann, D. L., & Kamrin, K. (2013). A predictive, size-dependent continuum model for dense granular flows. *Proceedings of the National Academy of Sciences of the United States of America, 110,* 6730–6735.

Henann, D. L., & Kamrin, K. (2014). Continuum thermomechanics of the nonlocal granular rheology. *International Journal of Plasticity, 60,* 145–162.

Jop, P., Forterre, Y., & Pouliquen, O. (2006). A constitutive law for dense granular flows. *Nature, 441,* 727–730. https://doi.org/10.1038/nature04801.

Kachanov, L. M. (1958). On the time to rupture under creep conditions. *Izvestiya Akademii Nauk SSSR, OTN 8,* 26–31. **(in Russian)**.

Kachanov, L. M. (1986). *Introduction to continuum damage mechanics* (p. 135). Leiden: Martinus Nijhoff.

Kane, D. L., Shearer, P. M., Goertz-Allmann, B. P., & Vernon, F. L. (2013). Rupture directivity of small earthquakes at Parkfield. *Journal of Geophysical Research, 118,* 212–221. https://doi.org/10.1029/2012JB009675.

Kaneko, Y., & Shearer, P. M. (2015). Variability of seismic source spectra, estimated stress drop and radiated energy, derived from cohesive-zone models of symmetrical and asymmetrical circular and elliptical ruptures. *Journal of Geophysical Research.* https://doi.org/10.1002/2014jb011642.

Knopoff, L. (1964). Q. *Reviews of Geophysics, 2*(4), 625–660.

Knopoff, L., & MacDonald, G. J. F. (1960). Models for acoustic loss in solids. *Journal of Geophysical Research, 65,* 2191–2197.

Krajcinovic, D. (1996). *Damage mechanics.* Amsterdam: Elsevier.

Kurzon, I., Vernon, F. L., Rosenberger, A., & Ben-Zion, Y. (2014). Real-time automatic detectors of P and S waves using singular value decomposition. *Bulletin of the Seismological Society of America, 104,* 1696–1708.

Kwiatek, G., & Ben-Zion, Y. (2013). Assessment of P and S wave energy radiated from very small shear-tensile seismic events in a deep South Africa mine. *Journal of Geophysical Research, 118,* 3630–3641. https://doi.org/10.1002/jgrb.50274.

Lengliné, O., & Got, J.-L. (2011). Rupture directivity of microearthquake sequences near Parkfield, California. *Geophysical Research Letters, 38,* L08310. https://doi.org/10.1029/2011GL047303.

Lu, Y. B., Li, Q. M., & Ma, G. W. (2010). Numerical investigation of the dynamic compressive strength of rocks based on split Hopkinson pressure bar tests. *International Journal of Rock Mechanics and Mining Sciences, 47,* 829–838.

Lyakhovsky, V., & Ben-Zion, Y. (2014a). A continuum damage–breakage faulting model and solid-granular transitions. *Pure and Applied Geophysics.* https://doi.org/10.1007/s00024-014-0845-4.

Lyakhovsky, V., & Ben-Zion, Y. (2014b). Damage–breakage rheology model and solid-granular transition near brittle instability. *Journal of the Mechanics and Physics of Solids, 64,* 184–197.

Lyakhovsky, V., Ben-Zion, Y., & Agnon, A. (1997). Distributed damage, faulting, and friction. *Journal of Geophysical Research, 102,* 27635–27649.

Lyakhovsky, V., Ben-Zion, Y., Ilchev, A., & Mendecki, A. (2016). Dynamic rupture in a damage–breakage rheology model. *Geophysical Journal International, 206,* 1126–1143. https://doi.org/10.1093/gji/ggw183.

Lyakhovsky, V., Hamiel, Y., & Ben-Zion, Y. (2011). A non-local visco-elastic damage model and dynamic fracturing. *Journal of the Mechanics and Physics of Solids, 59,* 1752–1776. https://doi.org/10.1016/j.jmps.2011.05.016.

Lyakhovsky, V., Ilchev, A., & Agnon, A. (2001). Modeling of damage and instabilities of rock mass by means of a non-linear rheological model. In G. van Aswegen, R. J. Durrheim, & W. D. Ortlepp (Eds.), *Rockbursts and seismicity in mines: dynamic rock mass response to mining* (pp. 413–420). Johannesburg: RaSiM-5, South African Institute of Mining and Metallurgy.

Lyakhovsky, V., Podladchikov, Y., & Poliakov, A. (1993). Rheological model of a fractured solid. *Tectonophysics, 226,* 187–198.

Lyakhovsky, V., Zhu, W., & Shalev, E. (2015). Visco-poroelastic damage model for brittle-ductile failure of porous rocks. *Journal of Geophysical Research: Solid Earth.* https://doi.org/10.1002/2014jb011805.

Madariaga, R. (1976). Dynamics of an expanding circular fault. *Bulletin of the Seismological Society of America, 66,* 639–666.

Main, I. G., Meredith, P. G., & Sammonds, P. R. (1992). Temporal variations in seismic event rate and b-values from stress corrosion constitutive laws. *Tectonophysics, 211,* 233–246.

Main, I. G., Meredith, P. G., & Sammonds, P. R. (1993). Application of a modified Griffith criterion to the evolution of fractal damage during compressional rock failure. *Geophysical Journal International, 115,* 367–380.

Meredith, P. G., & Atkinson, B. K. (1985). Fracture toughness and subcritical crack growth during high-temperature tensile deformation of Westerly granite and Black gabbro. *Tectonophysics, 39,* 33–51.

Molnar, P., Tucker, B. E., & Brune, J. N. (1973). Corner frequencies of P and S waves and models of earthquake sources. *Bulletin of the Seismological Society of America, 63,* 2091–2104.

Oth, A. (2013). On the characteristics of earthquake stress release variations in Japan. *Earth and Planetary Science Letters, 377,* 132–141.

Paris, P. C., & Erdogan, F. (1963). A critical analysis of crack propagation laws. *Journal of Basic Engineering, 85,* 528–534.

Peyrat, S., Olsen, K. B., & Madariaga, R. (2001). Dynamic modelling of the 1992 Landers earthquake. *Journal of Geophysical Research, 106*(B11), 26467–26482. https://doi.org/10.1029/2001JB000205.

Poliakov, A., Cundall, P., Podladchikov, Y., & Lyakhovsky, V. (1993). An explicit inertial method for the simulation of viscoelastic flow: an evaluation of elastic effects on diapiric flow in two- and three-layers model. In K. E. Runcorn & D. Stone (Eds.), *Dynamic modeling and flow in the earth and planets, proceedings of the NATO advanced study institute* (p. 175). Dordrecht: Kluwer.

Prieto, G., Shearer, P. M., Vernon, F. L., & Kilb, D. (2004). Earthquake source scaling and self–similarity estimation from stacking P and S spectra. *Journal of Geophysical Research, 109,* B08310. https://doi.org/10.1029/2004JB003084.

Rabotnov, Y.N. (1959). A mechanism of a long time failure. In Y.N. Rabotnov (Ed.), *Creep problems in structural members.* USSR Acad. of Sci. Publ. (pp. 5–7), Amsterdam: North-Holland.

Rabotnov, Y. N. (1988). *Mechanics of deformable solids* (p. 712). Moscow: Science.

Rosenberger, A. (2010). Real-time ground motion analysis: Distinguishing P and S arrivals in a noisy environment. *Bulletin of the Seismological Society of America, 100,* 1252–1262.

Ross, Z. E., & Ben-Zion, Y. (2016). Towards reliable automated estimates of earthquake source properties from body wave spectra. *Journal of Geophysical Research, 121,* 4390–4407. https://doi.org/10.1002/2016JB013003.

Sammis, C. G., Rosakis, A. J., & Bhat, H. S. (2009). Effects of off-fault damage on earthquake rupture propagation: experimental studies. *Pure and Applied Geophysics, 166,* 1629–1648. https://doi.org/10.1007/s00024-009-0512-3.

Sato, T., & Hirasawa, T. (1973). Body wave spectra from propagating shear cracks. *Journal of Physics of the Earth, 21,* 415–432.

Savage, J. C. (1972). Relation of corner frequency to fault dimensions. *Journal of Geophysical Research, 77,* 3788–3795.

Savage, S. B. (1984). The mechanics of rapid granular flows. *Advances in Applied Mechanics, 24,* 289–366.

Savage, S. B. (1998). Analyses of slow high-concentration flows of granular materials. *Journal of Fluid Mechanics, 377,* 1–26.

Shearer, P. M. (2009). *Introduction to seismology* (p. 396). Cambridge: Cambridge University Press.

Shi, Z., & Ben-Zion, Y. (2009). Seismic radiation from tensile and shear point dislocations between similar and dissimilar solids. *Geophysical Journal International, 179,* 444–458. https://doi.org/10.1111/j.1365-246X.2009.04299.x.

Shlomai, H., & Fineberg, J. (2016). The structure of slip-pulses and supershear ruptures driving slip in bi-material friction. *Nature Communications, 7,* 11787. https://doi.org/10.1038/ncomms11787.

Suzuki, T. (2013). Damage-tensor-based nondimensional parameters governing secondary faulting behavior. *Tectonophysics, 600,* 205–216.

Thomas, M. Y., & Bhat, H. S. (2018). Dynamic evolution of off-fault medium during an earthquake: A micromechanics based model. *Geophysical Journal International, 214,* 1267–1280.

Trugman, D. T., & Shearer, P. M. (2017). Application of an improved spectral decomposition method to examine earthquake source scaling in southern California. *Journal of Geophysical Research.* https://doi.org/10.1002/2017jb013971.

Turcotte, D. L., Newman, W. I., & Shcherbakov, R. (2003). Micro and macroscopic models of rock fracture. *Geophysical Journal International, 152,* 718–728.

Valanis, K. C. (1990). A theory of damage in brittle materials. *Engineering Fracture Mechanics, 36,* 403–416.

Vavryčuk, V. (2001). Inversion for parameters of tensile earthquakes. *Journal of Geophysical Research, 106*(B8), 16339–16355. https://doi.org/10.1029/2001JB000372.

Veveakis, E., & Regenauer-Lieb, K. (2014). The fluid dynamics of solid mechanical shear zones. *Pure and Applied Geophysics, 171,* 3159–3174.

Wang, Z., Ning, J., & Ren, H. (2018). Frequency characteristics of the released stress wave by propagating cracks in brittle materials. *Theoretical and Applied Fracture Mechanics.* https://doi.org/10.1016/j.tafmec.2018.04.004.

Weertman, J. (1980). Unstable slippage across a fault that separates elastic media of different elastic constants. *Journal of Geophysical Research, 85,* 1455–1461.

Xu, S., Ben-Zion, Y., Ampuero, J.-P., & Lyakhovsky, V. (2015). Dynamic ruptures on a frictional interface with off-fault brittle damage: Feedback mechanisms and effects on slip and near-fault motion. *Pure and Applied Geophysics.* https://doi.org/10.1007/s00024-014-0923-7.

Yang, W., Peng, Z., & Ben-Zion, Y. (2009). Variations of strain-drops of aftershocks of the 1999 İzmit and Düzce earthquakes around the Karadere-Düzce branch of the North Anatolian Fault. *Geophysical Journal International, 177,* 235–246. https://doi.org/10.1111/j.1365-246X.2009.04108.x.

(Received May 30, 2018, revised October 14, 2018, accepted November 26, 2018, Published online December 7, 2018)

Pure Appl. Geophys. 176 (2019), 1021–1036

https://doi.org/10.1007/s00024-018-1976-9

Pure and Applied Geophysics

Two-Dimensional Probabilistic Cell Automaton Model for Broadband Slow Earthquakes

SATOSHI IDE[1] and SUGURU YABE[2]

Abstract—Slow tectonic events, such as deep tectonic tremor, low-frequency earthquakes, very low frequency earthquakes, and slow slip events, are regarded as "broadband slow earthquakes". Various observational properties of these events have previously been explained using a one-dimensional (1D) stochastic model that mimics Brownian motion in the time domain. Here we present a simple two-dimensional (2D) slow earthquake model using a probabilistic cell automaton (PCA), where each cell possesses one of two states (slip or hold) at a given time step. The status of each cell is updated as a Markov process dependent on its own status and that of neighboring cells. The seismic moment rate of the event in the 2D PCA model is assumed to be proportional to the number of slipping cells, which is approximated using a stochastic differential equation equivalent to the previous 1D Brownian model approach. This 2D PCA model is, therefore, able to explain the various observational properties of slow earthquakes. Moreover, it includes an additional degree of freedom to separate the effects of loading and dissipation, which enables us to explain the variations in episodic slow slip activity. The size of the slow earthquake is limited by either the system size or the characteristic size, which is controlled by the degree of dissipation. Such a simple representation of slow earthquakes is useful for characterizing the broad spectrum of slow earthquake behavior in different environments.

Key words: Slow earthquake, slow slip events, tectonic tremor, Brownian motion, probabilistic cellular automaton, stochastic differential equation.

1. Introduction

Slow slip events (SSEs) have been discovered in various subduction zones and along major faults over the past two decades, often producing seismologically observable signals (e.g., Schwartz and Rokosky 2007; Beroza and Ide 2011; Obara and Kato 2016). Tectonic or non-volcanic tremor is observable at frequencies above 1 Hz (Obara 2002) and is characterized by seismic waves radiated from successive small shear slip events, each of which can be regarded as a low-frequency earthquake (LFE, e.g., Shelly et al. 2006). Very low frequency (0.02–0.05 Hz range) seismic signals radiated from shear slip motion are also often observed, which are interpreted as radiated from very low frequency (VLF) events (Ito et al. 2007). These tremor and VLF signals are separated by the 0.1–1 Hz frequency range, where large microseism noise is usually dominant. However, in special environments with very small microseism noise, seismic signals are continuously observed across the entire frequency range between tremor and VLF ranges, spanning 0.01–10 Hz, with the sources of these signals located at approximately the same place as the tremors and VLF events (Kaneko et al. 2018). It would, therefore, be natural to assume a broadband shear slip phenomenon, at least in the seismological frequency range (> 0.01 Hz), rather than an independent phenomenon for each discrete frequency range. We propose the term "broadband slow earthquakes" for such phenomena.

Over much wider frequency range that spans the 3×10^{-8} to 10 Hz range, or periods of 0.1 s to 1 year, the uniformity of geodetic SSEs and seismic slow earthquakes has not been confirmed. Tremors are distributed across only a part of some SSE regions where long-term SSEs occur, such as the Bungo Strait in Japan and Guerrero in Mexico (e.g., Yoshioka et al. 2015; Maury et al. 2016). Nevertheless, we may consider a statistically uniform very broadband slow slip phenomenon occurring in the space–time domain as an extreme case. As a representation of such broadband slow earthquake in the time domain, Ide (2008, 2010a) proposed a stochastic model that explains the various characteristics of tremors, VLF

[1] Department of Earth and Planetary Science, The University of Tokyo, 7-3-1 Hongo, Bunkyo, Tokyo 113-0033, Japan. E-mail: ide@eps.s.u-tokyo.ac.jp

[2] Department of Solid Earth Geochemistry, Japan Agency for Marine-Earth Science and Technology, 2-15, Natsushima-cho, Yokosuka-City, Kanagawa 237-0061, Japan.

events, and SSEs, which we will review in the next section. This model assumes random expansion and shrinkage of the slow slip area, which can be considered as the spatially distributed LFE activity, using a stochastic process similar to a Brownian walk. This Brownian slow earthquake (BSE) model has been applied to explain the regional differences in slow earthquakes observed in Japan, Cascadia, and Mexico (Ide and Maury 2018). Using a similar but slightly different statistical approach that superimposed many slip events with varying duration, Hawthorne and Bartlow (2018) reproduced the broadband characteristics of slow earthquakes. Furthermore, as a more specific application, Gomberg et al. (2016) presented a similar model to explain the co-existence of tremor and VLF events.

Previous researchers have proposed physical models to explain the two-dimensional (2D) aspects of such broadband slow earthquakes. One idea is a combination of binary materials, such as brittle patches that radiate seismic waves and the surrounding slowly slipping viscous region (Ando et al. 2010, 2012; Nakata et al. 2011). This model can explain various behaviors, ranging from slow to fast (ordinary) earthquakes, by altering the spatial distribution of the patches and the viscosity. While the idea of binary materials has been used in many studies to discuss the diverse behavior of shear slip (Skarbek et al. 2012; Dublanchet et al. 2013; Luo and Ampuero 2017; Yabe and Ide 2017), the combination of binary materials is not a necessary condition to model slow earthquakes. Ben-Zion (2012) demonstrated that a discrete fault system with almost zero weakening produces a random sequence of slow slip that occurs near the bottom of the shallow locked part of a large strike-slip fault. Zigone et al. (2015) applied essentially the same model to reproduce SSEs in a subduction environment. The occurrence of random slow slip sequences with negligible stress drop has also been demonstrated using earthquake cycle simulations with the computationally efficient version of the rate and state-dependent friction law (Colella et al. 2013).

The numerical model of Ben-Zion (2012) employed a critical depinning transition (Fisher et al. 1997) that considers a discrete elastic system where the external driving force competes with the resistive pinning force, with cell rupture occurring when the driving force exceeds the pinning force. While the elastic material is continuum, a fault system behaves in a discrete manner when the heterogeneity of the system, i.e., pinning force in this case, dominates the long-range interaction (Rice and Ben-Zion 1996; Shaw and Rice 2000). The deformation of the ruptured cell interacts with other cells through elastic forces, which may then trigger further ruptures. Randomness in the system, such as the heterogeneous spatial distribution of the pinning force, may generate complex avalanches of ruptures. The critical depinning transition is important in various phenomena from a statistical physics perspective. For example, Barkhausen noise in a magnet, which is excited by changes to the external magnetic field, is apparently similar to tremor signals and is interpreted as a phenomenon that occurs due to the critical behavior of interactive discrete units (spin) as the external magnetic field changes (e.g., Sethna et al. 2001). The idea of interactive discretized units has been essential in reproducing tremor-like signal in previous 2D models of broadband slow earthquakes (Ben-Zion 2012; Colella et al. 2013; Gershenzon and Bambakidis 2015).

Here we attempt to bridge the gap between these 2D statistical models and the one-dimensional (1D) representation of the BSE model, which has previously been applied to explain various characteristics of broadband slow earthquakes. The primary controlling physics of slow movement must be elastic deformation and brittle failure, similar to ordinary earthquakes. Furthermore, the slow earthquake process is supposed to be driven by a microscopic mechanism, such as velocity strengthening friction (Shibazaki and Shimamoto 2007), porosity change and fluid diffusion (Suzuki and Yamashita 2009; Segall et al. 2010), or viscoelastic deformation (Yamaguchi et al. 2011; Namiki et al. 2014). However, the dominant microscopic mechanism remains debated. To avoid the complexities due to the various microscopic mechanisms, here we focus on the probabilistic interactions among discrete units of a fault. A convenient method for this purpose is the probabilistic (stochastic) cellular automaton (PCA), which is commonly applied to research topics in the natural and social sciences (e.g., Schiff 2011; Fernández et al. 2018).

We begin with a short review of the 1D BSE model and the slow earthquake characteristics explained by this model. We then introduce a PCA model as a 2D expansion of the 1D BSE model. We demonstrate that the PCA model is identical to the 1D BSE model after simplification, through the representation of the stochastic differential equation (SDE). We also discuss the relationship between the 2D PCA model and the Burridge and Knopoff (BK) (1967) model, as well as the applicability of the model in explaining the diversity of slow earthquakes. We note that there are some known discrepancies between this very simple model and observed slow earthquake phenomena, some of which can be overcome by introducing slight complexities to the presented model.

2. *Brief Review of the 1D BSE Model*

We may safely state that a SSE is a phenomenon in which some part of the fault plane is slipping at a slow speed. While we do not know the shape of the slipping region, the total slip area can be measured as a time-dependent function $S(t)$. The linear dimension corresponding to S is defined as $r = \sqrt{S/C_0}$, where C_0 is a constant that represents the geometry (e.g., $C_0 = \pi$ for a circle). The 1D BSE model assumes that this size r changes randomly, following a simple SDE (Ide 2008), such that

$$\mathrm{d}r = -\alpha r \mathrm{d}t + \sigma \mathrm{d}B, \quad (1)$$

where α and σ are the characteristic frequency (α^{-1} is the characteristic time) and magnitude of the fluctuation, respectively, and dB is Gaussian white noise with zero mean and a variance of time increment Δt. The SDE in Eq. (1) is converted into another SDE for S, using Ito's rule (e.g., Øksendal 1998, Appendix 1), where

$$\mathrm{d}S = (C_0\sigma^2 - 2\alpha S)\mathrm{d}t + 2\sigma\sqrt{C_0 S}\mathrm{d}B. \quad (2)$$

This formulation is more convenient here, as Eq. (2) is directly comparable to the equations for the 2D model in the following sections. If we assume that the slip speed V_{slip} is constant, the seismic moment of the slow earthquake is then proportional to the area as follows:

$$\dot{M}_0 = \mu V_{\mathrm{slip}} S, \quad (3)$$

where μ is t rigidity of the medium. The long-term average of $\dot{M}_0$ is calculated as $E[\dot{M}_0] = \mu V_{\mathrm{slip}} C_0 \sigma^2 / 2\alpha$ for periods longer than α^{-1}. Figure 1a, b shows examples of the calculated moment rate functions with different characteristic times $\alpha^{-1} = 100$ and 1000s, with a time step $\Delta t = 0.1$s.

This model can explain various features of slow earthquakes (Ide 2008, 2010a; Ide and Maury 2018). The seismic moment is proportional to the duration T of a given event when it is longer than α^{-1}, with a proportionality constant $E[\dot{M}_0]$. For shorter durations, the seismic moment is proportional to the square of the duration (Fig. 1c). The seismic moment rate function of this model is proportional to the seismic energy rate function, as demonstrated by the observation of tectonic tremors (Ide et al. 2008). The moment rate spectrum of this model is analytically written as

$$\widetilde{\dot{M}_0}(f) = \frac{M_0'}{\sqrt{1 + (f/f_L)^2}\sqrt{1 + (f/f_H)^2}}, \quad (4)$$

where $f_L = \alpha/\pi$ and $f_H = (\pi\Delta t)^{-1}$ are the lower and higher corner frequencies, respectively, assuming that the spectral fall-off at higher frequencies is similar to that of the omega-square model for ordinary earthquakes (e.g., Aki and Richards 2002). Note that the flat level of this spectrum $M_0' = 2E[\dot{M}_0]\sqrt{(T/2\alpha)}$ is not equal to the total seismic moment $M_0 = E[\dot{M}_0]T$ at zero frequency (Fig. 1d, e). The observed velocity spectrum is approximately proportional to the seismic moment acceleration spectrum, which has a constant amplitude for a wide frequency range between f_L and f_H.

The 1D BSE model also explains the size–frequency statistics of slow earthquakes, which have been debated in previous studies using various definitions of the events and event sizes. If we assume each single tremor pulse is an event, then the cumulative density function (cdf) for the amplitude of the pulse can be utilized to measure the statistics. The cdf of the amplitude of the tremor envelope generally follows an exponential distribution (e.g., Watanabe

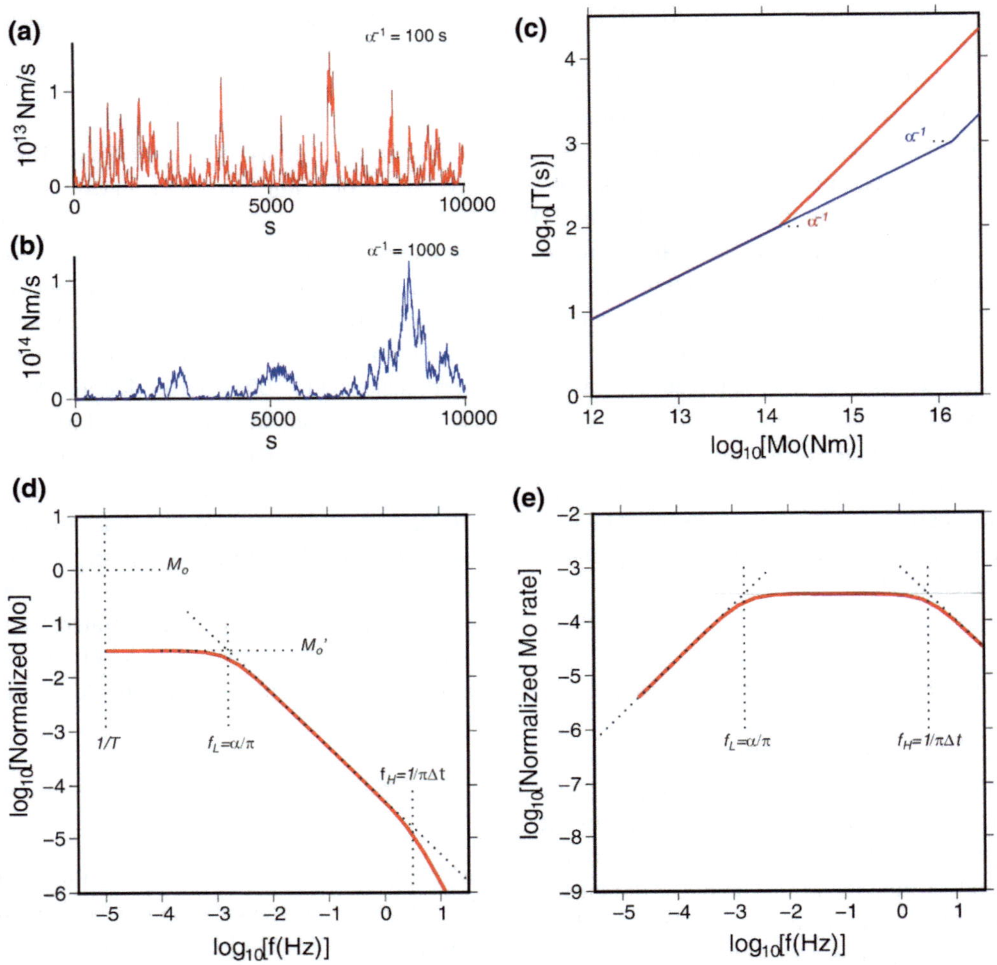

Figure 1
Characteristics of the 1D BSE model. **a** Moment rate functions for $\alpha^{-1} = 100$ s. **b** Moment rate functions for $\alpha^{-1} = 1000$ s. **c** Scaling relation between the seismic moment and duration for the two cases shown in **a** and **b**. **d** Moment rate spectrum for **a**. **e** Moment acceleration spectrum for **a**

et al. 2007). This exponential distribution is also reported for LFEs along the San Andreas Fault (Shelly and Hardebeck 2010) and Alpine Fault (Chamberlain et al. 2014). However, if we regard a sequence of successive tremors as a single event and the event size is measured for the length of the sequence, then the frequency–size statistics obey a power law distribution, similar to the Gutenberg–Richter relationship (Wech et al. 2010). The cdf of the moment rate function of the 1D BSE model was approximated to an exponential law (Ide 2010a), while the duration–frequency statistics of the events in the 2D PCA model followed a power law, which we discuss in the following section.

3. 2D Slow Earthquake Model

3.1. Basic Rules for Slip Propagation in the 2D PCA Model

We assume that a slow earthquake occurs on a planar fault in an elastic medium. The fault is divided into $N_a = N_1 \times N_2$, $(N_1 \geq N_2)$ square cells with length Δx (Fig. 2). We model slip on this fault using

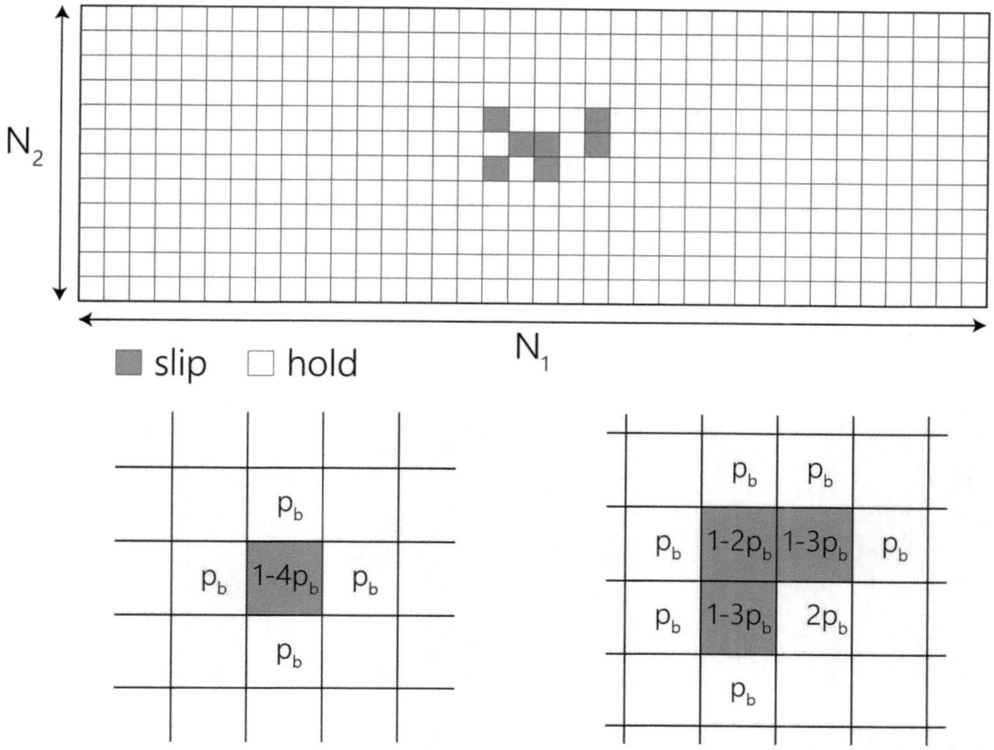

Figure 2
Illustration of the PCA. The model space is divided into $N_1 \times N_2$ cells, with each cell set to 1 (slip, gray) or 0 (hold, white). The bottom panels show the slip probability distributions for two simple cases

the PCA model and assume that the status of the slip at each cell v_i is binary, where $v_i = 0$ (hold) or 1 (slip). The slip speed V_{slip} is assumed to be constant in the same manner as in the 1D model. The seismic moment rate is then given as

$$\dot{M}_0 = \mu(\Delta x)^2 V_{\text{slip}} N_s, \quad (5)$$

where $N_s = \sum v_i$ is the number of slipping cells at a given time step. $\mu, \Delta x$ and V_{slip} are required only to convert the result of PCA into a physical quantity like seismic moment, and their value does not affect the system behavior.

When the status of the slip is known for every cell at a given time step, the status at the next time step is probabilistically determined for each cell based on its own status and those of the four surrounding cells. This is, therefore, a Markov process determined solely by the current status and all the possible states of v_i comprise a Markov chain with a transition probability among the different states. We assume that the expectation of the moment rate (the number of slipping cells) at a given time step is equal to that at the previous time step, based on the fact that SSE continues for a long time, without the large acceleration observed in ordinary earthquakes or the gradual decrease observed in afterslip. The size of time step is another parameter required to convert the result of PCA into physical quantity.

This assumption introduces the following probabilistic rules to update the slip status at the next time step (Fig. 2).

Rule 1: Each "slip" cell stops and becomes a "hold" cell at probability $(4 - N_b)p_b$, or it continues to "slip".

Rule 2: Each "hold" cell becomes a "slip" cell at probability $N_b p_b$, or it continues to be a "hold" cell.

N_b is the number of surrounding cells in the "slip" state, which is an integer between 0 and 4 defined for each cell at each time step. The number of slipping cells at the next step then changes from N_s to

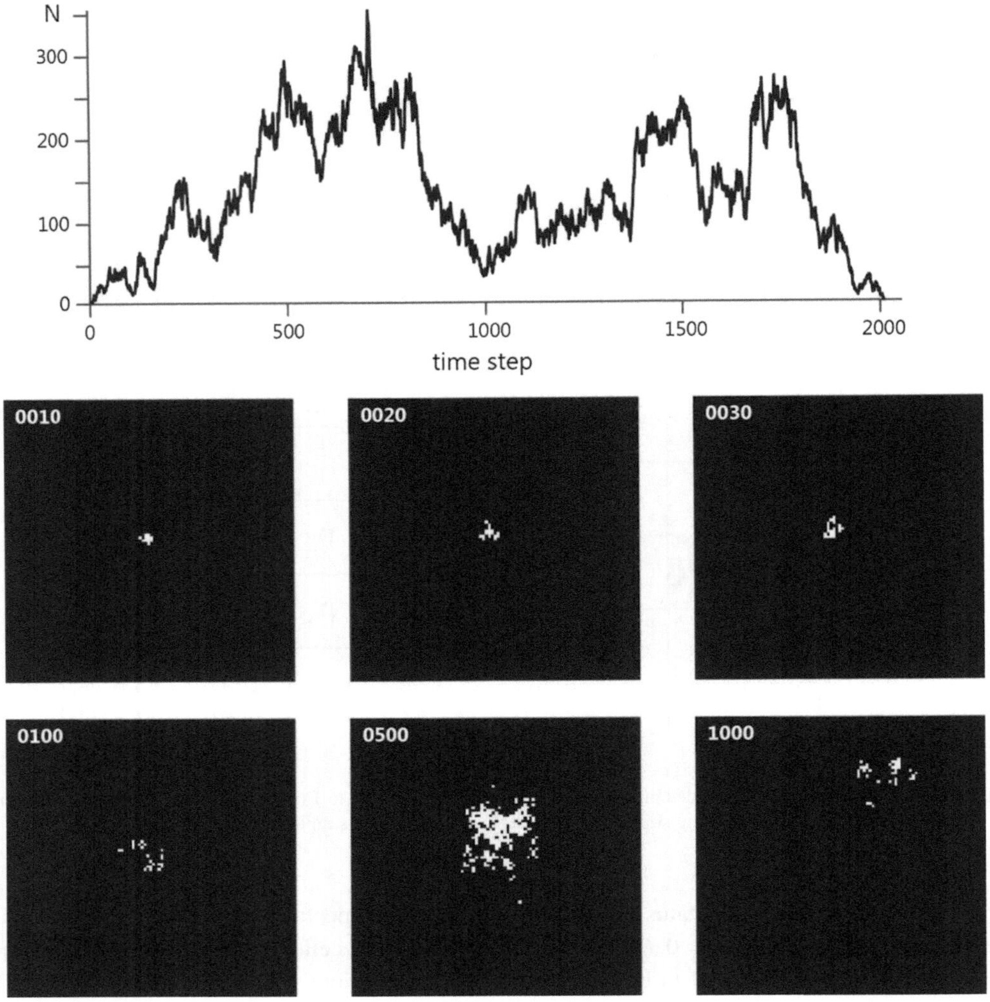

Figure 3
Example PCA computations. The top panel shows the moment rate function during an event. The bottom panels show the distributions of the slip (white) and hold (black) cells at six selected time steps

$-4N_s p_b + \sum N_b p_b$. The expected change is zero since the total sum of N_b in the entire system is $4N_s$, with a balance between the expected numbers of newly slipping and newly stopping cells. Each of above rules represents a Bernoulli process at each cell, which allows these rules to be written using a random number ξ with a uniform distribution between 0 and 1. The status of each cell at the $k+1$th time step, v_i^{k+1}, is determined by the status of cells at the kth step, v_j^k, such that

$$v_i^{k+1} = H\left(v_i^k + p_b \sum_{NN}(v_j^k - v_i^k) - \xi\right), \tag{6}$$

where *NN* represents the four nearest neighbor cells and *H* is the Heaviside function.

Consider a 2D model where all the boundary cells are always stopping. If we start from the state where only one cell is slipping in the center of the system, then the slip region randomly expands and shrinks, and all the cells eventually stop. Figure 3 shows an example of this process for $p_b = 0.2$ and $N_1 = N_2 = 101$. The SSE sequence in this example

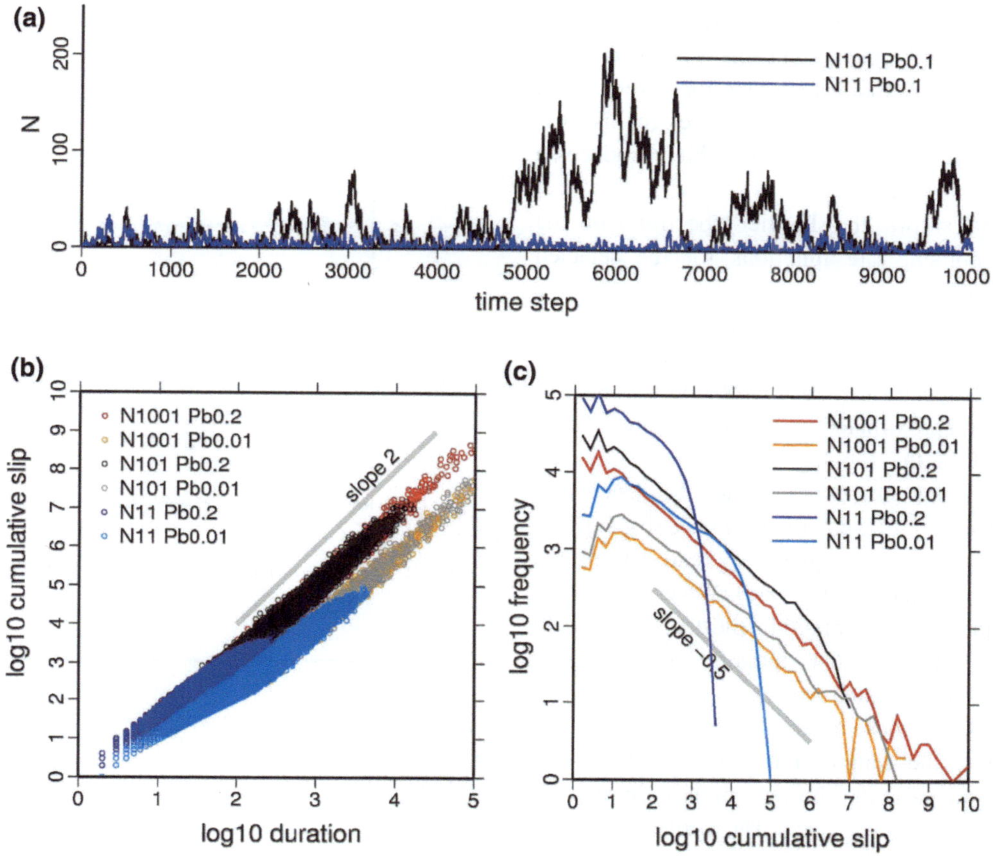

Figure 4

Numerical simulation results for the PCA model, following Eq. (6). **a** The number of slipping cells, which is equivalent to the moment rate function. The black and blue curves show the results for small ($N_1 = N_2 = 11$) and medium ($N_1 = N_2 = 101$) systems, respectively. **b** Duration and cumulative number of slipped cells, measured between two succeeding time steps of zero slip. The different colors represent different system sizes $N = N_1 = N_2$ and propagation probabilities p_b, with the thick gray line possessing a slope of 2. **c** Size–frequency statistics of the event size, measured by the cumulative number of slipped cells. The different colors represent different system sizes $N = N_1 = N_2$ and propagation probabilities p_b. The thick gray line has a slope of -0.5

lasted for $\sim$ 2000 time steps, with the slip area undergoing almost circular expansion during the first 500 steps, shifting to the upper-right sector and with a decreasing slip area by 1000 steps, and then expanding again and shrinking before termination. The shape of slip area is always complex and possesses a fractal geometry.

Figure 4 summarizes the results from a series of simulations with $p_b = 0.2, 0.1, \text{and } 0.01$ on fault surfaces, where $N_1 = N_2 = 11, 101, \text{and } 1001$. We start from a single slipping cell at the center of the model space and measure both the time steps and the cumulative number of slipped cells before the termination to determine the duration and size of each event. The calculation is then repeated after each termination to produce a continuous history of the number of slipping cells, which is considered as the moment rate function in the 1D BSE model. Figure 4a shows that large events, such as that in Fig. 3, are rare, with most events visualized as small fluctuations slightly above zero. The cumulative number of slipped cells is approximately proportional to the square of the duration (Fig. 4b). The least-squares fit of the events with durations longer than 1000 time steps yields exponents of 1.94 and 1.82 for $p_b = 0.2$and 0.01, respectively. Figure 4c shows the size–frequency statistics of the events. The distribution resembles a power law distribution with an exponent of approximately − 0.5 and a large event cutoff determined by the system size. This behavior is strongly limited by the system size for a very small system; however, the system size effect is less significant in the case of small p_b, which produces a slow expansion speed.

Figure 5a shows the Fourier spectrum of the simulated moment rate functions for the selected parameter sets, normalized by the cumulative moment. Since this system has no characteristic time, such as α^{-1} in the 1D BSE model, for a sufficiently large model space (i.e., $N_a = 1001^2$), the Fourier spectrum is approximated by a straight line with a slope of −1 in the log–log domain. However, the finite system size introduces corner frequencies f_c, and the spectrum is well approximated by Eq. (4), with a lower corner frequency f_L and an infinite higher corner frequency ($f_H = \infty$). We estimated the corner frequencies by obtaining the least-squares fit to Eq. (4) in the log–log domain. The estimated values are well correlated with the line $f_c = 4.5p_b/N_a$ (Fig. 5b). While these examples are for a square model space, the length that controls the corner frequency is actually the shorter length N_2 in the case of a rectangular model space. The corner frequency decreases as N_1 increases when N_2 is fixed as the shorter length. However, a fivefold increase in N_1 only yields a twofold decrease in the corner frequency, with further increases in N_1 resulting in no significant changes in the corner frequency. Therefore, the corner frequency depends mainly on the shorter length of the model space, and is approximately proportional to $C_f p_b/N_2^2$ for $C_f = 2.0–4.5$, depending on the aspect ratio of the system.

The behavior of this system is also expressed by a SDE. While it is expected that the number of slipping cells is always equal to that in the previous time step, the actual value fluctuates stochastically. These difference variations are analytically written for some simple cases. For example, when only one cell is slipping in the system, the variance is given as a summation of the variances for the five Bernoulli processes at the slipping cell and its four neighbor cells:

$$4p_b(1 - 4p_b) + 4 \times p_b(1 - p_b). \quad (7)$$

This variance is approximately $8p_b$ for small p_b, which increases to approximately $12p_b$ for two

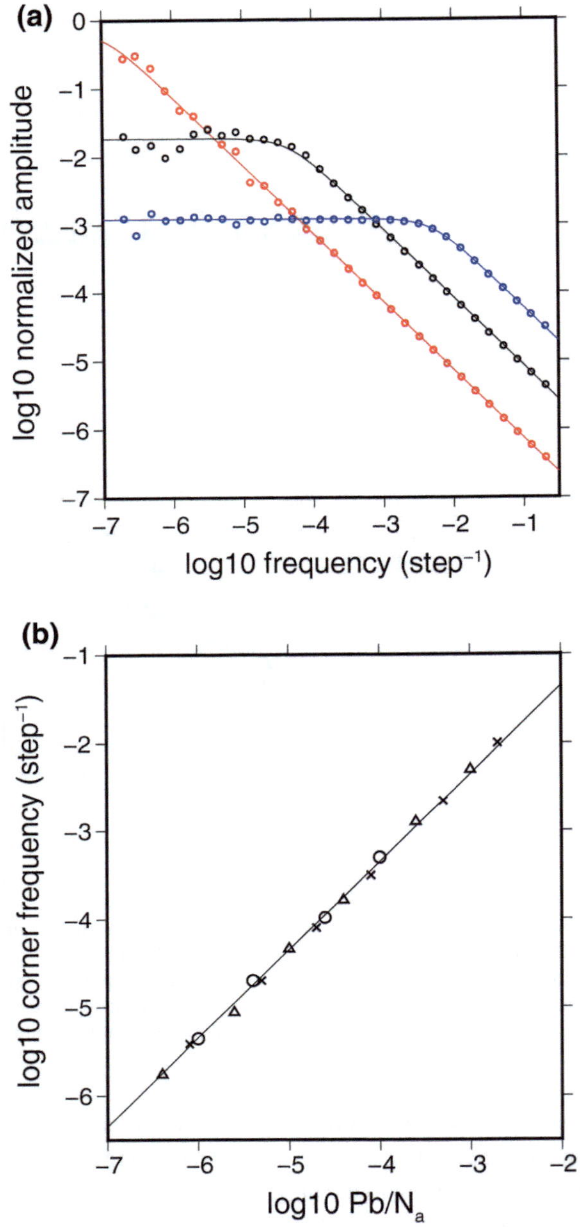

Figure 5
Fourier spectra for various PCA models. **a** Circles indicate the Fourier spectra, averaged every 0.2 in the log–log domain. The red, black, and blue data sets correspond to the spectra for $N_a = 1001^2, 101^2$, and 11^2, respectively. The thin curves are the least-squares fits to the calculated spectra in the log–log domain, which are used to determine the corner frequency f_c of each PCA model. **b** Comparison between p_b/N_a and f_c for various values of p_b and N_a. Circles, triangles, and crosses correspond to $p_b = 0.01, 0.1$, and 0.2, respectively. The line represents the relation $f_c = 4.5p_b/N_a$

connected slipping cells. The variance depends on the geometry of the slipping region for many slipping cells, and becomes difficult to generalize for large regions of slip. Although we cannot provide an analytical form, we can empirically show that the variance increases in an approximately linear fashion as $p_b N_s^{\gamma}$, where the exponent γ is approximately one. Here we estimated γ as 0.89 and 0.88 for $p_b = 0.1$and 0.01, respectively, via the least-squares method (Fig. 6). The change in the number of slipping cells during a time step $\mathrm{d}N_s$ is, therefore, approximated as a normal distribution $N(0, C_b p_b N_s)$, which is written as a SDE, where

$$\mathrm{d}N_s = \sqrt{C_b p_b N_s}\mathrm{d}B, \tag{8}$$

and C_b is a slowly decreasing function of N_s that may be regarded as a constant with a value in the range 3–4 in the following applications.

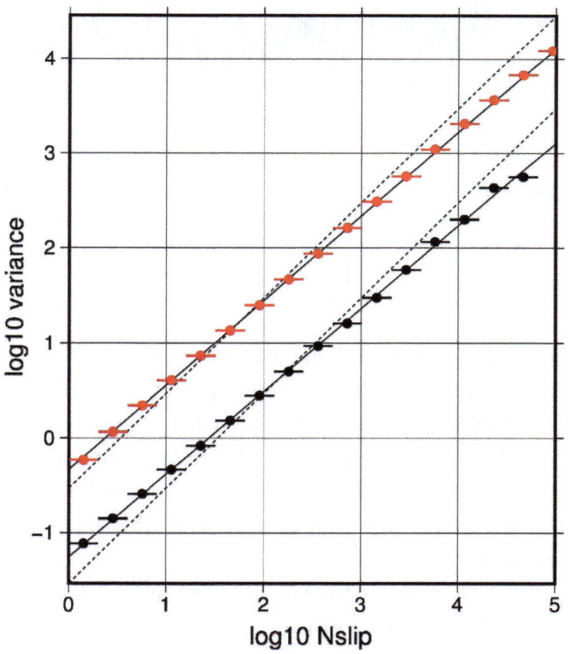

Figure 6
Variance of the fluctuation in the number of slipping cells. The variance is calculated for all time steps when the number of slipping cells is in the range indicated by each horizontal bar, with the circles representing the average values. The black and red data sets correspond to the simulations with $p_b = 0.1$ and 0.01, respectively. The solid lines are the least-squares fits to the data sets, where $\gamma = 0.89$ and 0.88 for $p_b = 0.1$ and 0.01, respectively, while the dashed lines each have a slope of 1

3.2. *PCA with Random Ignition*

The above system was artificially started many times, but a natural system requires an ignition process. For example, the ignition of a fault system must be provided by the probabilistic failure of a small region in response to the slow loading from the continuous deformation of the surrounding region. We, therefore, assume that there is a uniform ignition probability at each cell during each time increment p_l, and modify Eq. (8) as

$$\mathrm{d}N_s = p_l N_a \mathrm{d}t + \sqrt{C_b p_b N_s}\mathrm{d}B. \tag{9}$$

This equation is mathematically equivalent to the governing equation of the 1D BSE model in Eq. (2), with either $\alpha = 0$ or an infinite time constant α^{-1}, and $p_l = C_b p_b / 4N_a$. It is also obvious that the first term of Eq. (9) continuously increases the number of slipping cells, regardless of the fluctuation term, which means that the slip area will increase to infinity in the long run. In any PCA simulation with a finite number of cells, almost all cells slip after a sufficient time, which is not a realistic model.

4. *PCA with Ignition and Dissipation*

In a realistic system, the energy dissipates during slip and the slip area subsequently decreases, which introduces an additional characteristic size to the system. Therefore, in addition to the ignition process, we include a dissipation process by assuming that the stopping probability of each slipping cell is slightly larger than the value given in rule (1). We modify the update rules, including the effect of loading, for the excess of the stopping probability p_v as follows.

Rule 1′: Each "slip" cell stops and becomes a "hold" cell at probability $(4 - N_b)p_b + p_v - p_l$, or it continues to "slip".

Rule 2′: Each "hold" cell becomes a "slip" cell at probability $N_b p_b + p_l$, or it continues to be a "hold" cell.

The rule is again written using the Heaviside function as

$$v_i^{k+1} = H\left(v_i^k + p_b \sum_{NN}(v_j^k - v_i^k) + p_l - p_v v_i^k - \xi\right), \quad (10)$$

and the corresponding SDE is

$$dN_s = (p_l N_a - p_v N_s)dt + \sqrt{C_b p_b N_s}dB. \quad (11)$$

This is formally equivalent to the governing equation of the 1D BSE model in Eq. (2). The comparison of these two equations yields a relationship between the newly introduced excess stopping probability p_v and the characteristic time constant of the 1D BSE model, where $\alpha^{-1} = 2p_v^{-1}$. This mathematical equality ensures that the PCA models following this rule are also able to explain all the characteristics of slow earthquakes that are explained by the 1D BSE model. In the 1D BSE model in Eq. (2), the $C_s\sigma^2 dt$ term corresponds to the loading term $p_l N_a dt$ in Eq. (11), which is dependent on the fluctuation term, $2\sigma\sqrt{C_s S}dB$, whereas the two terms are independent in Eq. (11). This difference introduces a variety of behaviors in the model, which are captured as different average levels in the time function of N_s.

Figure 7 shows three examples of temporal functions of N_s with different loading terms. The loading term equivalent to that in the 1D BSE model is given as $p_l = C_b p_b / 4N_a$. The case in Fig. 7b is close to this situation, with the time function being similar to the moment rate function in Fig. 1. The system is not always ignited and has long quiescent periods for a very small p_l (Fig. 7a), with such a time function being appropriate to explain the episodic activity of SSEs and tremors. On the other hand, when the loading is too large, the slip does not stop and stabilizes around the long-term average of the moment rate (Fig. 7c), which is calculated from Eq. (11) as

$$E[N_s] = \frac{p_l N_a}{p_v}. \quad (12)$$

When p_b is small, the fluctuation becomes negligible relative to this stable value. This condition can mimic slow slip without seismologically detected signals (i.e., long-term SSEs). The differences in the loading conditions are, therefore, independent of the fluctuation term, which potentially explains the broad

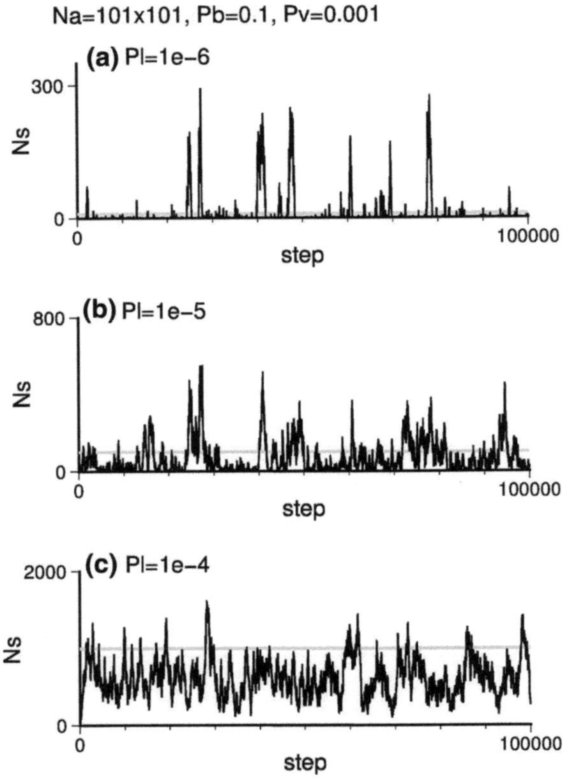

Figure 7
Number of slipping cells (moment rate function) for different p_l values, which determine the probability of starting a new process for the PCA model via Eq. (10). **a–c** $p_l = 10^{-6}, 10^{-5}$, and 10^{-4}, respectively. $N_1 = N_2 = 101$, $p_b = 0.1$, and $p_v = 10^{-3}$. The horizontal gray line shows the average for each case

spectrum of slow earthquake activity. Note that Fig. 7 shows the number of slipping cells, while the observable quantities are related to moment rate in (5), which include v_{slip}, too. We have assumed constant v_{slip}, but it may correlate or anti-correlate with p_l. Therefore, it is not easy to tell which is more detectable.

Each sequence also has a common characteristic time that is defined by the excess stopping probability or energy dissipation $2p_v^{-1}$, with a pulse-like increase and decrease of the slip area typically occurring with this time constant, which is 2000 time steps in Fig. 7. Figure 8 shows a space–time diagram of the slipping cells for the simulations presented in Fig. 7. The spatial characteristic sizes of the clusters of slipping cells are almost independent of the p_l values, with the characteristic size being smaller than the system size in these models. However, if the system size is

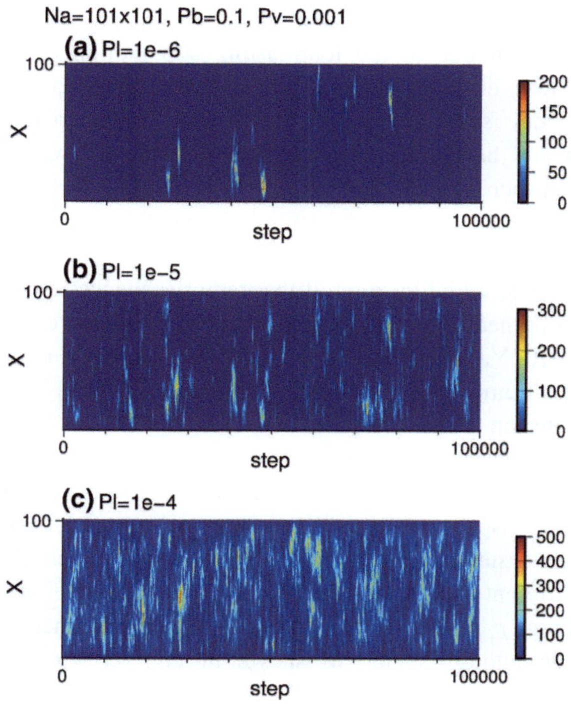

Figure 8

Spatio-temporal distribution of slipping cells. The number of slipping cells is summed in one direction of the model space and averaged every 10 time steps. **a-c** $p_l = 10^{-6}, 10^{-5}, \text{and} 10^{-4}$, respectively. $N_1 = N_2 = 101$, $p_b = 0.1$, and $p_v = 10^{-3}$. These settings are the same as those in Fig. 7a–c

smaller than the characteristic size, then the typical duration is controlled by the characteristic frequency determined by the shortest system size $C_f p_b / N_2^2$, as discussed in Sect. 3.1 (Fig. 5). It is difficult to distinguish which of these two factors is dominant in slow earthquake observations. For the SSEs in Japan, Cascadia, and Mexico, the width of the slow earthquake zone appears to correlate with the characteristic time in controlling the long-term moment rate (Ide and Maury 2018).

5. Discussion

5.1. Relation to the Spring-Slider Model

The two-dimensional extension of 1D BSE model, which we may call 2D BSE, is provided in both 2D PCA and SDE forms, following Eqs. (10) and (11), respectively, although its physical meaning might not be obvious. We, therefore, consider the relation between the 2D PCA model and a spring-slider model (Burridge and Knopoff (BK) 1967; Carlson et al. 1991), which is well known and possesses a clear physical meaning.

Consider a 2D system of blocks, where the blocks are connected to each other with springs, as well as to the loading surface. The blocks slide on the plate where there is viscous friction. When the inertia term is neglected, the equilibrium equation is written using the displacement of ith block U_i as follows:

$$\eta V_i = k_l(U_l - U_i) + k_b \sum_{NN} (U_j - U_i), \tag{13}$$

where k_b and k_l are the spring constants for the inter-slider and loading springs, respectively; η is the viscosity; U_l is the displacement of the loading surface; and V_i is the velocity. Fisher et al. (1997) studied the continuum version of Eq. (13) with a pinning force without dynamic weakening to demonstrate that the slip region extends as a fractal. The SSEs modeled by Ben-Zion (2012) and Zigone et al. (2015) possessed similar characteristics.

The relation between the equilibrium equation and the PCA model is demonstrated by taking the derivative of Eq. (13), such that

$$\eta \frac{dV_i}{dt} = k_l(V_l - V_i) + k_b \sum_{NN} (V_j - V_i) - \in, \tag{14}$$

where $\in$ represents any fluctuation due to the differentiation of Eq. (13). This can be considered as a random fluctuation of the viscous forces, such as a pinning force, which is neglected in Eq. (13). When the viscosity is sufficiently large, the velocity and its time derivative are small enough to be assumed using the PCA binary states. Equation (14) may be regarded as an equation that provides a rule for updating the velocity from the kth to $k+1$th steps in this instance, as

$$V_i^{k+1} = H\left(V_i^k + p'_b \sum_{NN} \left(V_j^k - V_i \right) + p'_l - p'_v V_i^k - \xi' \right), \tag{15a}$$

$$p'_b = \frac{k_b \mathrm{d}t}{\eta}, p'_l = \frac{k_l \mathrm{d}t}{\eta} V_l, p'_v = \frac{k_l \mathrm{d}t}{\eta}, \xi' = \frac{\mathrm{d}t}{\eta} \in. \tag{15b}$$

This equation is formally identical to Eq. (10). However, this transformation is not possible for all types of BK models since the random fluctuation term in Eq. (15a) is not necessarily the uniform distribution that is assumed in Eq. (10). Nevertheless, this identity suggests that the entire seismic nature of slow earthquakes can be explained using the BK model and a certain set of conditions.

5.2. *Spatial Heterogeneity*

We have thus far considered the temporal change in the slip distribution on homogeneous cells. However, slow earthquakes generally occur on quite complex and heterogeneous fault systems, which indicates that some degree of heterogeneity needs be introduced in the models to explain realistic spatial heterogeneities, such as the segmentation of SSEs and tremors (e.g., Brudzinski, and Allen 2007; Obara 2010), and the finer heterogeneities recognized from the observed seismic energy and duration of tectonic tremors (e.g., Ide 2010b; Yabe and Ide 2014). This heterogeneity can be introduced through the probabilities p_b, p_l, and p_v, or in the slip velocity V_{slip} when employing the 2D PCA model.

When a single cell with large V_{slip} is surrounded by cells with much smaller V_{slip}, the slip may be recognized as an impulsive event, which resembles an LFE family occurring with an approximately constant duration (Bostock et al. 2015; Thomas et al. 2016). The time-independent distribution of high-V_{slip} cells can be viewed as stationary tremor sources according to a high-resolution imaging technique (Rubin and Armbruster 2013). Nevertheless, as long as we consider that the macroscopic properties of slow earthquakes span many cells, such small-scale heterogeneities would be smoothed out due to the central limit theorem, as demonstrated by Ide and Maury (2018) for the 1D BSE model.

However, some systematic heterogeneity must be macroscopically visible as anisotropy in the model, even after smoothing. Tremor sources have striation structures in the direction parallel to the fault slip (e.g., Ide 2010b), with short-term migration tending to occur in this direction (Ide 2012) at a velocity that is orders of magnitude higher than the average migration velocity in the slip-perpendicular direction (e.g., Shelly et al. 2007; Ghosh et al. 2010). This is not surprising, as the irregularity of natural faults is smoothed out in the slip direction by continuous abrasion due to the long-lasting slip motion (e.g., Power et al. 1987; Sagy et al. 2007). One way to model such anisotropic migration characteristics would be to introduce anisotropy in p_b, where we can increase and decrease p_b in the slip-parallel and slip-perpendicular directions, respectively, while maintaining the same average value.

The initial location of the tremor sequences tends to be near the edge of the tremor zone (Obara et al. 2010; Yabe and Ide 2014), and the episodic nature of the tremor activity changes with depth (Shelly and Johnson 2011; Wech and Creager 2011). Such differences in tremor activity are usually explained by the differences in elastic loading from outside the model space, which is dependent on the geometry of the system. However, a PCA model with spatially different p_l values can also mimic such features. Large p_l values ignite the system in an approximately continuous manner, as shown in Fig. 7c, whereas small p_l values yield episodic activity. The large heterogeneity in p_l indicates that the migrating tremor activity tends to start from a large p_l cell due to the increased probability of starting a slip process, whereas smaller p_l cells slip rather passively. While we may also change p_v spatially, the effect of this parameter is difficult to distinguish from that of the finiteness of the model region, as both control the characteristic time of the slow earthquake process, which is only evaluated macroscopically.

It is important to reiterate that the 2D PCA model is a Markov process, which is independent of the previous state of the system. Therefore, while it is an appropriate model for explaining the characteristics of radiated seismic waves, it is not fully consistent in explaining some of the characteristics of slow earthquake activity. For example, tremor activity sometimes migrates over distances of > 100 km (e.g., Dragert et al. 2004; Ito et al. 2007), which suggests that the migration direction is uniformly maintained by the stress gradient produced by the shear slip of the slow earthquakes. The recurrence interval of SSEs is quasi-regular, suggesting that some characteristic times are required to accumulate the strain energy. These features might be mimicked by introducing temporal changes to the probabilities in the PCA model, but they are better explained by more deterministic approaches.

6. Conclusion

Slow earthquakes are observed as very broadband tectonic phenomena, ranging from sub-second tectonic tremors to multi-year SSEs. The temporal and spectral characteristics of these broadband signals have been modeled using a SDE as the 1D BSE model (Ide, 2008, 2010a) or by superimposing many slip events with various durations (Hawthorne and Bartlow 2018). The numerical representation of slow earthquakes in 2D was also proposed with many interacting discrete units slipping in an approximately random fashion (e.g., Ben-Zion 2012; Colella et al. 2013). Here the 1D BSE model is expanded to the 2D domain using PCA simulation, with mathematical equality observed between the two representations. The proposed 2D PCA model is also similar to the spring-slider model of Burridge and Knopoff (BK) (1967) in terms of its mathematical expression.

In the PCA model, a slow earthquake is represented by the spatio-temporal distribution of slipping cells on a finite 2D area with the seismic moment being proportional to the number of slipping cells. The state of slip at each cell is determined probabilistically, depending on the state of slip on the cell and its four surrounding cells. The PCA model assumes that the expectation of the number of slipping cells at the next step is equal to the current number for a given spatial slip distribution at a given time step. The number of slipping cells, therefore, fluctuates around a constant value, and the magnitude of the fluctuation increases in an approximately linear fashion with the number of cells. When this increase is approximated as linear, the SDE representation of the PCA model is mathematically equivalent to the SDE in the 1D BSE model. Therefore, the PCA model can explain various characteristics of slow earthquakes, such as their constant seismic moment, constant seismic energy, near-flat velocity spectrum, and diffusive migration. Different frequency–size statistics based on the different definitions of the event size are also explained by this model.

While the 1D BSE model has just one characteristic size, which is defined as α^{-1} in SDE, the 2D PCA model has two sizes. One size is defined as the characteristic time of energy dissipation, and the other is determined by the shortest size of the model space. The shorter timescale determines the typical duration of the pulse-like increase of the moment rate, as well as the corner frequencies of the spectrum. The loading condition of the model is arbitrarily defined to capture the broad spectrum of slow earthquake behavior, ranging from episodic bursts to continuous movement. These complex parameter settings may explain the additional features of slow earthquakes, such as the existence of LFE families, anisotropic migration speed, and spatially heterogeneous activity.

However, it may not be worth complicating the models since the simple form of the 2D PCA model is suitable to characterize the diversity of slow earthquakes. While we observe many similarities among slow earthquakes in various tectonic environments, non-negligible differences exist in the amplitude of signals of tremor, VLF, and SSE, the characteristic duration, the velocity and extent of migration, etc. The relation between slow and fast (ordinary) earthquakes is diverse as well. They often occur independently with spatial separation, but in some regions we observe mixed activity including tremor, VLF, SSE, and earthquakes of various sizes including swarm. The 2D PCA model, or even the original 1D BSE model, in the present paper can extract basic features of broadband phenomena with handful parameters and work as a standard model to characterize diversity of slow earthquakes and discover hidden relations with tectonic conditions.

Acknowledgements

Comments from two anonymous reviewers helped to improve the manuscript. This research was supported by the Japanese Society for the Promotion of Science (JSPS) Kakenhi (16H02219), the Ministry of Education, Culture, Sports, Science and Technology (MEXT) Kakenhi (16H06477), and the Earthquake and Volcano Hazards Observation and Research Program of MEXT.

Appendix 1

Derivation of (2) from (1).

For an Ito process X defined by

$$dX = udt + vdB,$$

with random processes u and v, $Y = g(t, X)$ satisfies the Ito's lemma (e.g., Øksendal 1998),

$$dY = \frac{\partial g}{\partial t}(t, X)dt + \frac{\partial g}{\partial x}(t, X)dX + \frac{1}{2}\frac{\partial^2 g}{\partial x^2}(t, X)(dX)^2. \tag{A1}$$

In case

$$dr = -\alpha r dt + \sigma dB, \tag{1}$$

and $S = C_0 r^2$, we can write

$$\frac{\partial S}{\partial t} = 0, \frac{\partial S}{\partial r} = 2C_0 r, \frac{\partial^2 S}{\partial r^2} = 2C_0. \tag{A2}$$

Since $(dr)^2 = \sigma^2 dt$, substituting (A2) into (A1), we obtain

$$dS = \left(C_0\sigma^2 - 2\alpha S\right)dt + 2\sigma\sqrt{C_0 S}dB. \tag{2}$$

References

Aki, K., & Richards, P. G. (2002). *Quantitative seismology* (2nd ed., p. 700). Sausalito: University Science Books.

Ando, R., Nakata, R., & Hori, T. (2010). A slip pulse model with fault heterogeneity for low-frequency earthquakes and tremor along plate interfaces. *Geophysical Research Letters, 37*(10), L10310. https://doi.org/10.1029/2010GL043056.

Ando, R., Takeda, N., & Yamashita, T. (2012). Propagation dynamics of seismic and aseismic slip governed by fault heterogeneity and Newtonian rheology. *Journal of Geophysical Research: Solid Earth, 117*(11), B11308. https://doi.org/10.1029/2012JB009532.

Ben-Zion, Y. (2012). Episodic tremor and slip on a frictional interface with critical zero weakening in elastic solid. *Geophysical Journal International, 189*(2), 1159–1168. https://doi.org/10.1111/j.1365-246X.2012.05422.x.

Beroza, G. C., & Ide, S. (2011). Slow earthquakes and nonvolcanic tremor. *Annual Review of Earth and Planetary Sciences, 39*(1), 271–296. https://doi.org/10.1146/annurev-earth-040809-152531.

Bostock, M. G., Thomas, A. M., Savard, G., Chuang, L., & Rubin, A. M. (2015). Magnitudes and moment-duration scaling of low-frequency earthquakes beneath southern Vancouver Island. *Journal of Geophysical Research: Solid Earth, 121*(9), 6329–6350. https://doi.org/10.1002/2015JB012195.

Brudzinski, M. R., & Allen, R. M. (2007). Segmentation in episodic tremor and slip all along Cascadia. *Geology, 35*(10), 907–910. https://doi.org/10.1130/G23740A.1.

Burridge, R., & Knopoff, L. (1967). Model and theoretical seismicity. *Bulletin of the Seismological Society of America, 57*(3), 341–371.

Carlson, J. M., Langer, J. S., Shaw, B. E., Tang C. (1991). Intrinsic properties of a Burridge-Knopoff model of an earthquake fault. *Physical Review A, 44*, 884. https://doi.org/10.1103/PhysRevA.44.884

Chamberlain, C. J., Shelly, D. R., Townend, J., & Stern, T. A. (2014). Low-frequency earthquakes reveal punctuated slow slip on the deep extent of the Alpine fault, New Zealand. *Geochemistry, Geophysics, Geosystems, 15*(7), 2984–2999. https://doi.org/10.1002/2014GC005436.

Colella, H. V., Dieterich, J. H., & Richards-Dinger, K. (2013). Spatial and temporal patterns of simulated slow slip events on the Cascadia megathrust. *Geophysical Research Letters, 40*(19), 5101–5107. https://doi.org/10.1002/grl.50984.

Dragert, H., Wang, K., & Rogers, G. (2004). Geodetic and seismic signatures of episodic tremor and slip in the northern Cascadia subduction zone. *Earth, Planets and Space, 56*(12), 1143–1150. https://doi.org/10.1186/BF03353333.

Dublanchet, P., Bernard, P., & Favreau, P. (2013). Interactions and triggering in a 3-D rate-and-state asperity model. *Journal of Geophysical Research: Solid Earth, 118*(5), 2225–2245. **10.1002/jgrb.50187**.

Fernández R., Louis PY., Nardi F.R. (2018) Overview: PCA Models and Issues. In: Louis PY., Nardi F. (eds) Probabilistic Cellular Automata. Emergence, Complexity and Computation, vol 27 (pp. 1–30). Springer, Cham

Fisher, D. S., Dahmen, K., Ramanathan, S., & Ben-Zion, Y. (1997). Statistics of earthquakes in simple models of heterogeneous faults. *Physical Review Letters, 78*(25), 4885. https://doi.org/10.1103/PhysRevLett.78.4885.

Gershenzon, N. I., & Bambakidis, G. (2015). Model of deep nonvolcanic tremor part II: episodic tremor and slip. *Bulletin of the Seismological Society of America, 105*(2A), 816–830. https://doi.org/10.1785/0120140225.

Ghosh, A., Vidale, J. E., Sweet, J. R., Creager, K. C., Wech, A. G., Houston, H., et al. (2010). Rapid, continuous streaking of tremor in Cascadia. *Geochemistry, Geophysics, Geosystems, 11*(12), Q12010. https://doi.org/10.1029/2010GC003305.

Gomberg, J., Agnew, D. C., & Schwartz, S. Y. (2016). Alternative source models of very low frequency events. *Journal of Geophysical Research: Solid Earth, 121*(9), 6722–6740. https://doi.org/10.1002/2016JB013001.

Hawthorne J.C., & Bartlow, N.M. (2018). Observing and modeling the spectrum of a slow slip event. *Journal of Geophysical Research: Solid Earth, 123*, 4243–4265. https://doi.org/10.1029/2017jb015124.

Ide, S. (2008). A Brownian walk model for slow earthquakes. *Geophysical Research Letters, 35*(17), 3–7. https://doi.org/10.1029/2008GL034821.

Ide, S. (2010a). Quantifying the time function of nonvolcanic tremor based on a stochastic model. *Journal of Geophysical Research: Solid Earth, 115*(8), 1–13. https://doi.org/10.1029/2009JB000829.

Ide, S. (2010b). Striations, duration, migration and tidal response in deep tremor. *Nature, 466*, 356–359. https://doi.org/10.1038/nature09251.

Ide, S. (2012). Variety and spatial heterogeneity of tectonic tremor worldwide. *Journal of Geophysical Research: Solid Earth, 117*(3), B03302. https://doi.org/10.1029/2011JB008840.

Ide, S., Imanishi, K., Yoshida, Y., Beroza, G. C., & Shelly, D. R. (2008). Bridging the gap between seismically and geodetically detected slow earthquakes. *Geophysical Research Letters, 35*(10), 2–7. https://doi.org/10.1029/2008GL034014.

Ide, S., & Maury, J. (2018). Seismic moment, seismic energy, and source duration of slow earthquakes: application of Brownian slow earthquake model to three major subduction zones. *Geophysical Research Letters, in press.*. https://doi.org/10.1002/2018GL077461.

Ito, Y., Obara, K., Shiomi, K., & Sekine, S. (2007). Slow earthquakes coincident with episodic tremors and slow slip events. *Science, 315*(5811), 503–506. https://doi.org/10.1126/science.1134454.

Kaneko, L., Ide, S., & Nakano, M. (2018). Slow earthquakes in the microseism frequency band (0.1–1.0 Hz) off Kii Peninsula, Japan. *Geophysical Research Letters, 45,* 2618–2624. https://doi.org/10.1002/2017GL076773.

Luo, Y., & Ampuero, J. P. (2018). Stability of faults with heterogeneous friction properties and effective normal stress. *Tectonophysics, 733,* 257–272. https://doi.org/10.1016/j.tecto.2017.11.006.

Maury, J., Ide, S., Cruz-Atienza, V. M., Kostoglodov, V., Gonzáles-Molina, G., & Péres-Campos, X. (2016). Comparative study of tectonic tremor locations: characterization of slow earthquakes in Guerrero, Mexico. *Journal of Geophysical Research: Solid Earth, 121*(7), 5136–5151. https://doi.org/10.1002/2016JB013027.

Nakata, R., Ando, R., Hori, T., & Ide, S. (2011). Generation mechanism of slow earthquakes: numerical analysis based on a dynamic model with brittle-ductile mixed fault heterogeneity. *Journal of Geophysical Research: Solid Earth, 116*(8), B08308. https://doi.org/10.1029/2010JB008188.

Namiki, A., Yamaguchi, T., Sumita, I., Suzuki, T., & Ide, S. (2014). Earthquake model experiments in a viscoelastic fluid: a scaling of decreasing magnitudes of earthquakes with depth. *Journal of Geophysical Research: Solid Earth, 119*(4), 3169–3181. https://doi.org/10.1002/2014JB011135.

Obara, K. (2002). Nonvolcanic deep tremor associated with subduction in Southwest Japan. *Science, 296*(5573), 1679–1681. https://doi.org/10.1126/science.1070378.

Obara, K. (2010). Phenomenology of deep slow earthquake family in southwest Japan: spatiotemporal characteristics and segmentation. *Journal of Geophysical Research: Solid Earth, 115*(8), B00A25. https://doi.org/10.1029/2008JB006048.

Obara, K., & Kato, A. (2016). Connecting slow earthquakes to huge earthquakes. *Science, 353*(6296), 253–257. https://doi.org/10.1126/science.aaf1512.

Obara, K., Tanaka, S., Maeda, T., & Matsuzawa, T. (2010). Depth-dependent activity of non-volcanic tremor in south-west Japan. *Geophysical Research Letters, 37*(13), L13306. https://doi.org/10.1029/2010GL043679.

Øksendal, B. K. (1998). *Stochastic Differential Equations* (5th ed.). Berlin: Springer.

Power, W. L., Tullis, T. E., Brown, S. R., Boitnott, G. N., & Scholz, C. H. (1987). Roughness of natural fault surfaces. *Geophysical Research Letters, 14*(1), 29–32. https://doi.org/10.1029/GL014i001p00029.

Rice, J. R., & Ben-Zion, Y. (1996). Slip complexity in earthquake fault models. *Proceedings of the National academy of Sciences of the United States of America, 93,* 3811–3818.

Rubin, A. M., & Armbruster, J. G. (2013). Imaging slow slip fronts in Cascadia with high precision cross-station tremor locations. *Geochemistry, Geophysics, Geosystems, 14*(12), 5371–5392. https://doi.org/10.1002/2013GC005031.

Sagy, A., Brodsky, E. E., & Axen, G. J. (2007). Evolution of fault-surface roughness with slip. *Geology, 35*(3), 283–286. https://doi.org/10.1130/G23235A.1.

Schiff, J. L. (2011). *Cellular automata: a discrete view of the world* (Vol. 45). Hoboken, NJ: Wiley.

Schwartz, S. Y., & Rokosky, J. M. (2007). Slow slip events and seismic tremor at circum-Pacific subduction zones. *Review of Geophysics, 45*(3), RG3004. https://doi.org/10.1029/2006RG000208.

Segall, P., Rubin, A. M., Bradley, A. M., & Rice, J. R. (2010). Dilatant strengthening as a mechanism for slow slip events. *Journal of Geophysical Research: Solid Earth, 115*(12), B12305. https://doi.org/10.1029/2010JB007449.

Sethna, J. P., Dahmen, K. A., & Myers, C. R. (2001). Crackling noise. *Nature, 410*(6825), 242–250. https://doi.org/10.1038/35065675.

Shaw, B. E., & Rice, J. R. (2000). Existence of continuum complexity in the elastodynamics of repeated fault ruptures. *Journal of Geophysical Research: Solid Earth, 105*(10), 23791–23810. https://doi.org/10.1029/2000JB900203.

Shelly, D. R., Beroza, G. C., & Ide, S. (2007). Non-volcanic tremor and low-frequency earthquake swarms. *Nature, 446*(7133), 305–307. https://doi.org/10.1038/nature05666.

Shelly, D. R., Beroza, G. C., Ide, S., & Nakamula, S. (2006). Low-frequency earthquakes in Shikoku, Japan, and their relationship to episodic tremor and slip. *Nature, 442*(7099), 188–191. https://doi.org/10.1038/nature04931.

Shelly DR, Hardebeck JL (2010) Precise tremor source locations and amplitude variations along the lower-crustal central San Andreas Fault. *Geophysical Research Letters, 37*(14), L14301. https://doi.org/10.1029/2010GL043672.

Shelly, D. R., & Johnson, K. M. (2011). Tremor reveals stress shadowing, deep postseismic creep, and depth-dependent slip recurrence on the lower-crustal San Andreas fault near Parkfield. *Geophysical Research Letters, 38*(13), L13312. https://doi.org/10.1029/2011GL047863.

Shibazaki, B., & Shimamoto, T. (2007). Modelling of short-interval silent slip events in deeper subduction interfaces considering the frictional properties at the unstable—stable transition regime. *Geophysical Journal International, 171*(1), 191–205. https://doi.org/10.1111/j.1365-246X.2007.03434.x.

Skarbek, R. M., Rempel, A. W., & Schmidt, D. A. (2012). Geologic heterogeneity can produce aseismic slip transients. *Geophysical Research Letters, 39*(21), L21306. https://doi.org/10.1029/2012GL053762.

Suzuki, T., & Yamashita, T. (2009). Dynamic modeling of slow earthquakes based on thermoporoelastic effects and inelastic generation of pores. *Journal of Geophysical Research: Solid Earth, 114*(B6), B00A04. https://doi.org/10.1029/2008JB006042.

Thomas, A. M., Beroza, G. C., & Shelly, D. R. (2016). Constraints on the source parameters of low-frequency earthquakes on the San Andreas Fault. *Geophysical Research Letters, 43,* 1464–1471. https://doi.org/10.1002/2015GL067173.

Watanabe, T., Hiramatsu, Y., & Obara, K. (2007). Scaling relationship between the duration and the amplitude of non-volcanic deep low-frequency tremors. *Geophysical Research Letters, 34*(7), L07305. https://doi.org/10.1029/2007GL029391.

Wech, A. G., & Creager, K. C. (2011). A continuum of stress, strength and slip in the Cascadia subduction zone. *Nature Geoscience, 4,* 624–628. https://doi.org/10.1038/ngeo1215.

Wech, A. G., Creager, K. C., Houston, H., & Vidale, J. E. (2010). An earthquake-like magnitude-frequency distribution of slow slip in northern Cascadia. *Geophysical Research Letters, 37*(22), L22310. https://doi.org/10.1029/2010GL044881.

Yabe, S., & Ide, S. (2014). Spatial distribution of seismic energy rate of tectonic tremors in subduction zones. *Journal of Geophysical Research: Solid Earth, 119*(11), 8171–8185. https://doi.org/10.1002/2014JB011383.

Yabe, S., & Ide, S. (2017). Slip-behavior transitions of a heterogeneous linear fault. *Journal of Geophysical Research: Solid Earth, 122*(1), 387–410. https://doi.org/10.1002/2016JB013132.

Yamaguchi, T., Morishita, M., Doi, M., Hori, T., Sakaguchi, H., & Ampuero, J.-P. (2011). Gutenberg-Richter's law in sliding friction of gels. *Journal of Geophysical Research: Solid Earth, 116*(12), B12306. https://doi.org/10.1029/2011JB008415.

Yoshioka, S., Matsuoka, Y., & Ide, S. (2015). Spatiotemporal slip distributions of three long-term slow slip events beneath the Bungo channel, southwest Japan, inferred from inversion analyses of GPS data. *Geophysical Journal International, 201*(3), 1437–1455. https://doi.org/10.1093/gji/ggv022.

Zigone, S., Ben-Zion, Y., & Campillo, M. (2015). Modelling non-volcanic tremor, slow slip events and large earthquakes in the Guerrero subduction zone (Mexico) with space-variable frictional weakening and creep. *Geophysical Journal International, 202*(1), 653–669. https://doi.org/10.1093/gji/ggv174.

(Received May 7, 2018, revised July 22, 2018, accepted August 9, 2018, Published online August 17, 2018)

Pure Appl. Geophys. 176 (2019), 1037–1059

https://doi.org/10.1007/s00024-018-1995-6

Pure and Applied Geophysics

Earthquake Fingerprints: Extracting Waveform Features for Similarity-Based Earthquake Detection

Karianne J. Bergen[1,2] and Gregory C. Beroza[3]

Abstract—Seismologists are increasingly adopting data mining and machine learning techniques to detect weak earthquake signals in large seismic data sets. The detection performance of these new methods, especially their sensitivity and false detection rate, depends on the choice of feature representation for waveform data. We have previously introduced Fingerprint and Similarity Thresholding (FAST), a new method for waveform-similarity-based earthquake detection that uses a pattern mining approach to detect earthquake signals without template waveforms. FAST has two key steps: fingerprint extraction and efficient indexing for similarity search. In this work, we focus on FAST fingerprint extraction: the method used to map short-duration waveforms to a set of features, called waveform fingerprints, used for detection. We describe the FAST fingerprint extraction method, a data-adaptive variation on the Waveprint audio fingerprinting method tailored for use in continuous seismic data. We compare the performance of the FAST fingerprint extraction method with existing fingerprinting techniques designed for audio identification. To overcome the challenges associated with using limited or incomplete event catalogs to evaluate detection algorithms, we propose a framework for quantifying the performance of different fingerprint extraction methods in the context of blind similarity-based detection. Our framework uses computational experiments on benchmark data sets, constructed with known event waveforms, to compute a measure of fingerprint effectiveness. We use this framework to show that, among the audio fingerprinting schemes considered in this work, our proposed FAST fingerprint extraction method achieves the most consistent performance in distinguishing similar, low signal-to-noise earthquake waveforms from noise in waveform data sets from eight stations in the Northern California Seismic Network.

Key words: Earthquake detection, seismology, audio fingerprinting, feature extraction, time-series analysis, similarity search.

1. Introduction

Earthquake detection, the task of identifying earthquake signals in continuously recorded ground motion data from one or more stations in a seismic network, is a challenging and fundamental task in seismology. Waveform cross-correlation, also referred to as template matching, is a widely used and highly sensitive method for detecting weak earthquake signals (Gibbons and Ringdal 2006). Waveform cross-correlation is limited to detecting events with waveforms similar to those of known events, and these template waveforms are derived from earthquake catalogs that are often incomplete. To overcome this limitation, Yoon et al. (2015) introduced Fingerprint and Similarity Thresholding (FAST), a general detector based on waveform similarity. FAST[1] is the first detector based on waveform similarity that is capable of identifying events with unknown sources in long-duration (large-T) data sets of up to 10 years (Rong et al. 2018).

FAST is inspired by algorithms used for content-based audio search and retrieval tasks, such as identifying songs directly from noisy audio waveform data recorded on a mobile phone (Wang 2003; Baluja and Covell 2008). FAST is an uninformed (unsupervised) detector and does not require template waveforms. The FAST detection model, like waveform cross-correlation, is based on the observation that earthquakes with similar sources produce similar waveforms. In the absence of template waveforms, FAST labels any pair of waveforms with high similarity as candidate earthquakes, relying on the assumption that noise waveforms are mutually

[1] Institute for Computational and Mathematical Engineering, Stanford University, Stanford, CA 94305, USA. E-mail: karianne_bergen@fas.harvard.edu

[2] Department of Earth and Planetary Sciences, Harvard University, Cambridge, MA 02138, USA.

[3] Department of Geophysics, Stanford University, Stanford, CA 94305, USA. E-mail: beroza@stanford.edu

[1] FAST code available at https://github.com/stanford-futuredata/FAST.

dissimilar (in contrast, waveform cross-correlation requires high similarity to a *known* template waveform). FAST identifies candidate earthquakes by using locality-sensitive hashing (LSH) (Andoni and Indyk 2006), a randomized algorithm for computationally efficient similarity search, to find pairs of similar waveforms in single-channel continuous data.

The FAST detection pipeline is made up of algorithmic modules, each of which can be independently tuned or optimized. Single-station FAST has two key steps: fingerprint extraction and efficient similarity search (see Fig. 1); an overview of the single-station FAST detection method is presented in Yoon et al. (2015). In the fingerprint extraction step, FAST computes a set of features, called a *waveform fingerprint*, for each short-duration sliding window in the single-channel continuous waveform data, and performs similarity search over this collection of waveform fingerprints. Waveform fingerprints are used as proxies for the raw waveform signals in the FAST similarity search step. Therefore, the performance of FAST will depend on our ability to extract appropriate features as inputs to similarity search for unsupervised earthquake detection. This work focuses on the fingerprint extraction step and extends the preliminary results of Bergen et al. (2016).

The efficient similarity search step, which is optimized for large-T detection in Rong et al. (2018), depends only on the choice of similarity metric for the waveform fingerprints, and not on the feature extraction algorithm itself. Bergen and Beroza (2018) present a multi-station extension to FAST, which combines the single-station FAST detection results to reduce false detections among the candidate events. Extending FAST over a network requires running FAST separately on data from each station, so the FAST feature extraction method is designed for use on single-station data [multi-station feature extraction is addressed in the supplement of Bergen and Beroza (2018)].

Earthquake detection is an active area of research, and in recent years there has been growing interest in machine learning and data mining approaches for event detection and seismic signal analysis. The performance of these methods can be critically dependent on the selection of appropriate waveform features. In the related field of audio signal processing, there is an extensive body of literature dedicated to methods for extracting features, called audio fingerprints, from raw audio waveform data for audio classification and identification (Cano et al. 2005). However, the topic of feature extraction for seismic

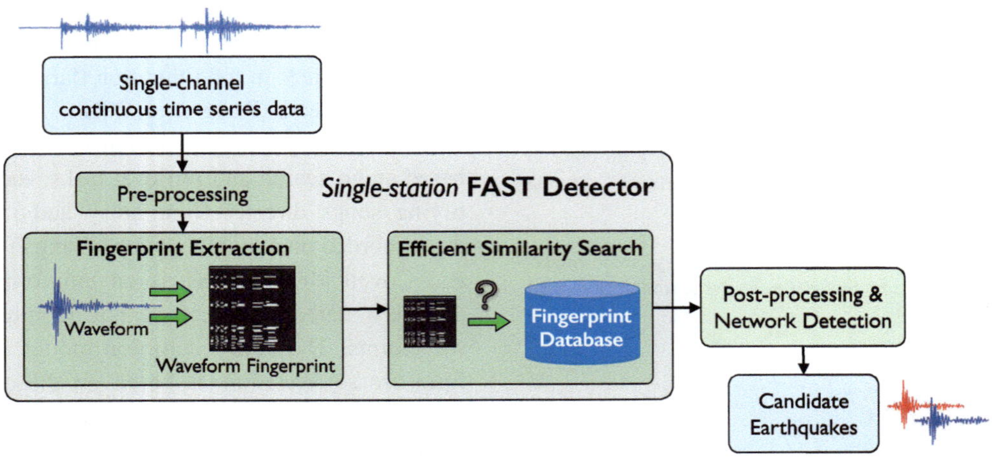

Figure 1
Main processing steps in the FAST earthquake detection method (Yoon et al. 2015). The input to FAST is continuous ground motion data from a single channel, usually with some basic pre-processing such as bandpass filtering. The FAST detector has two key steps: Fingerprint Extraction, which is the subject of this work, and Efficient Similarity Search. The Efficient Similarity Search step uses locality-sensitive hashing (Andoni and Indyk 2006) to build an index or "database" of waveform fingerprints and query the database to identify similar waveforms (Yoon et al. 2015; Bergen et al. 2016; Bergen and Beroza 2018; Rong et al. 2018)

signals has received less attention. Valentine and Trampert (2012) analyze the appropriateness of autoencoder neural networks for extracting compact waveform features from which the earthquake waveforms can be reconstructed. Holtzman et al. (2018) use an unsupervised learning approach to extract fingerprints for a set of earthquake waveforms that reveal subtle differences and relationships between events. Both studies apply their analysis only to event waveforms rather than to continuous data, and neither focuses on the task of discriminating earthquake signals from noise.

The focus of this work is feature extraction for the specific task of earthquake detection via blind similarity search in long-duration continuous seismic waveform data sets. We discuss the desirable properties of waveform fingerprints in the context of similarity search, and present the fingerprint extraction method used in FAST and alternate methods from the audio fingerprinting literature. One of the inherent challenges in developing new earthquake detection algorithms is the lack of ground truth data, an objective reference data set for measuring algorithm performance. Existing event catalogs, the closest thing to such a reference data set that is available, are known to be incomplete and thus do not accurately label all segments in the continuous waveform data as either earthquake signals or noise/non-earthquake signals. Without an objective performance metric, it is difficult to assess and compare the performance of different detection algorithms. Thus, in order to evaluate the relative performance of different fingerprinting schemes, we introduce a framework for quantifying the relative effectiveness of different feature extraction methods for similarity-based detection using computational experiments. We use our proposed framework to demonstrate that the procedure used to extract FAST waveform fingerprints produces a lower false alarm rate and enables the detection of lower signal-to-noise events than the alternative methods considered. Our analysis framework provides a template for other researchers who wish to optimize and validate feature representations of seismic waveform data for use in machine learning or data mining techniques.

2. *Background*

2.1. *Fingerprinting*

Feature extraction refers to the process of mapping an input data object to a set of features. A *fingerprint* is a compact signature that represents an object, such as a data file or high-dimensional data vector (Broder 1993). *Fingerprinting* or *fingerprint extraction* is a form of feature extraction that follows two general properties: (1) if two fingerprints, $f(x_1)$ and $f(x_2)$, corresponding to data objects x_1 and x_2, are different from each other (i.e. not identical), then the original data objects x_1 and x_2 are also different, and (2) if x_1 and x_2 are different, then there should be a low probability that they map to the same fingerprint:

$$f(x_1) \neq f(x_2) \Rightarrow x_1 \neq x_2 \tag{1}$$

$$\mathbb{P}[f(x_1) = f(x_2)] \ll 1 \quad \text{if } x_1 \neq x_2. \tag{2}$$

These properties allow fingerprints to act as nearly unique identifiers, similar to the most commonly understood definition of (human) fingerprints, the ridge patterns on human fingers (Pankanti et al. 2002). Fingerprints are often used in conjunction with hashing in information search and retrieval tasks; applications include document fingerprints for detecting plagiarism and duplicate webpages (Manber 1994; Broder 1997), *acoustic* or *audio fingerprints* for identifying audio recordings directly from the waveforms rather than from metadata (Cano et al. 2005), and molecular fingerprints for finding similar molecular structures in chemical databases (Willett et al. 1998).

2.2. *Waveform Fingerprints for Similarity Search*

In this work, we are interested specifically in *waveform fingerprints*, features that represent short-duration seismic waveform data, and are used as inputs to the efficient similarity search step in FAST for waveform-similarity-based earthquake detection. Waveform fingerprints should be tailored for use in similarity search, and to the particular characteristics of seismic data and the earthquake detection problem.

- Waveform fingerprints should distinguish earthquake signals from noise. In earthquake detection via similarity search, earthquake signals are distinguished from noise by high similarity to one or more other earthquake waveforms. The fingerprinting scheme should map similar earthquake waveforms to similar fingerprints and map noise waveforms to fingerprints with low similarity to other noise fingerprints; we describe fingerprinting schemes that satisfy this property as *discriminative* for similarity search. Note that features that work well for supervised waveform classification, where features should map earthquakes and noise waveforms into different regions in the feature space, may not be suitable features for use in similarity-based detection.

- The fingerprinting scheme must be paired with an appropriate similarity measure. The similarity measure should be appropriate based on the characteristics of the fingerprints (e.g. real-valued vs. binary, sparse vs. dense) and suitable for use in an efficient search method [e.g. LSH or *k*-d tree (Bentley 1975)]; examples include the Jaccard, Hamming, and cosine similarities (Leskovec et al. 2014), and l_p norms (Datar et al. 2004).
- A good fingerprinting scheme needs to be robust to the types of distortions that are expected in the data set of interest. For earthquake detection, fingerprints should be robust to high levels of noise and differences in signal amplitude. In this work we focus on methods that extract features from the time-frequency representation (spectrogram) of the signal, as these are more robust to small phase shifts than time-domain features.
- The fingerprinting scheme should produce fingerprints that maintain similarity over long time periods, as activity along faults persists over long timescales, and ideally should be fixed for the full time duration. Care should be taken with adaptive or dynamic fingerprinting schemes to ensure that similar waveforms separated by long time periods will produce fingerprints with high similarity.
- The fingerprinting scheme should not require prior waveform information, since the FAST detection method is designed to be effective even when no template waveforms are available. The fingerprinting scheme may be either data-agnostic (i.e. not dependent on any properties or characteristics of the data) or data-driven, so long as it does not incorporate any label information.

3. Methods

3.1. Audio Fingerprinting

Audio fingerprints are a compact representation of audio waveform data, including music, speech, or other sound recordings (Alías et al. 2016). Among previous work on fingerprinting for information retrieval and similarity search tasks, audio fingerprinting in particular provides a good starting point for the development of earthquake waveform fingerprints—there are structural similarities between the data (oscillatory signals with frequent zero-crossings and temporal variations in amplitude and frequency content), and both applications require fingerprints that are robust to small variations and additive noise.

In this work we focus on three well-known, effective audio fingerprinting methods: the landmark-based (Wang 2003), Philips (Haitsma and Kalker 2002), and Waveprint (Baluja and Covell 2008) methods. These methods and the FAST fingerprint extraction method are summarized in Fig. 2. In Sect. 5 we compare the performance of these fingerprinting schemes for seismic waveform data in the context of earthquake detection by blind similarity search.

3.1.1 Landmark Method

The landmark-based algorithm (Wang 2003, 2006), developed to identify audio clips recorded by a cell phone microphone for the Shazam mobile app, relies on the assumption that energy peaks in the time-frequency representation of the audio signal are key features that can be used to identify an audio clip and are invariant to additive noise. The algorithm identifies key points, defined as local maxima in the spectrogram, and builds hashes from the constellations based on the relative positions of these key points. The search and identification step scans a database to find sequences of multiple matching hashes.

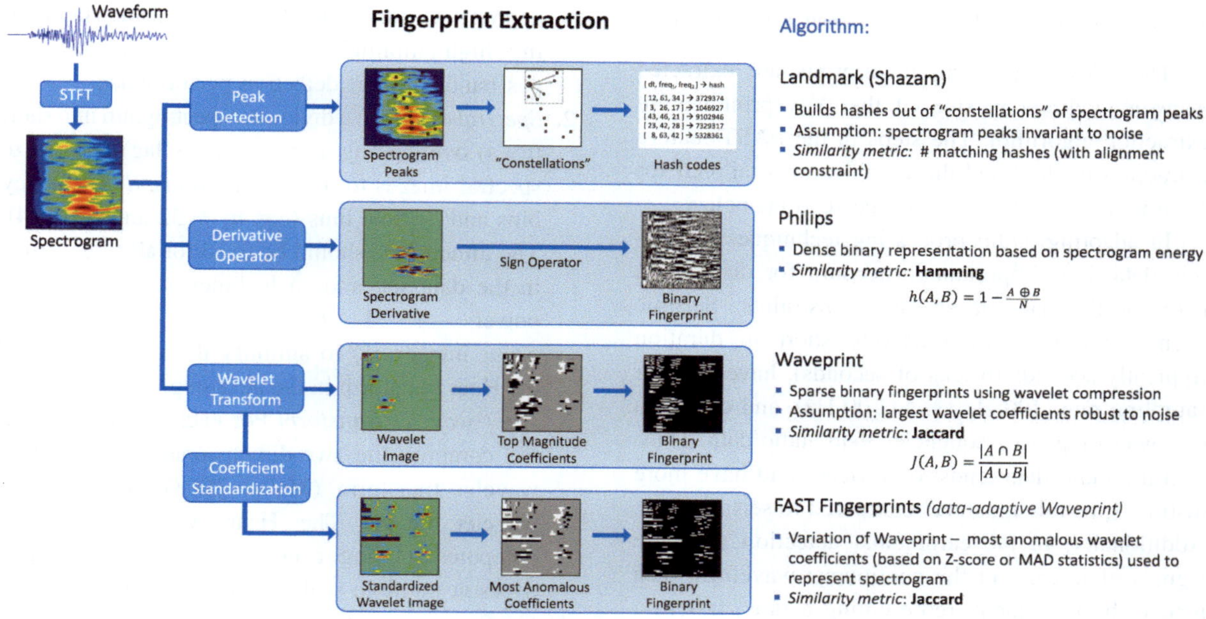

Figure 2
A comparison of three audio fingerprinting methods and the FAST feature extraction method. The landmark algorithm is similar to that used in the Shazam audio identification system (Wang 2003) and uses hash codes derived from peaks in the spectrogram. The Philips algorithm (Haitsma and Kalker 2002) is a dense binary fingerprint that uses the sign of the time and frequency derivatives to compute the fingerprint. The FAST feature extraction uses a modified version of the Waveprint (Baluja and Covell 2008) method

3.1.2 Philips Method

The Philips audio fingerprinting method (Haitsma and Kalker 2002) converts short segments of audio waveforms to a dense binary fingerprint. The Philips scheme computes the spectrogram, E, and then obtains a binary representation, F, from the sign of the derivatives of the signal energy; from Eq. (1) in Haitsma and Kalker (2002):

$$F[n,m] = \begin{cases} 1 & \text{if } (E[n,m] + E[n-1,m+1]) \\ & > (E[n,m+1] + E[n-1,m]) \\ 0 & \text{otherwise} \end{cases} \tag{3}$$

This binary representation is then divided into fingerprints with a sliding window along the time axis. The similarity between these dense binary fingerprints (those with roughly equal numbers of `0`s and `1`s) is measured by the *Hamming similarity*. The Hamming similarity of bit strings (fingerprints) A and B, each of dimension N, is the fraction of the N bits that are identical in both strings:

$$\begin{aligned} \texttt{Hamming}(A,B) &= \frac{\#\{i:\ A[i]=B[i],\ i=1,\ldots,N\}}{N} \\ &= 1-\frac{A\oplus B}{N}. \end{aligned} \tag{4}$$

3.1.3 Waveprint Method

The Waveprint method (Baluja and Covell 2008) for audio identification takes an image processing approach to extracting audio fingerprints. Waveprint compresses sliding windows along the time dimension of the spectrogram (*spectral images*) using a discrete wavelet transform, a technique for image compression and denoising (Donoho and Johnstone 1994). The sign values of the largest wavelet coefficients are encoded in a sparse, binary fingerprint. This approach is robust against additive noise because the wavelet transform typically concentrates the signal energy in a relatively small number of wavelet coefficients, with the noise captured by *detail coefficients* that have smaller values and are not retained when the largest coefficients are selected.

3.2. Our Approach: FAST Waveform Fingerprints

The FAST waveform fingerprints are extracted using a modified version of the Waveprint feature extraction algorithm. The steps in the FAST feature extraction method and the modifications tailored for seismic data are discussed in the sections below.

In adapting audio processing techniques to seismic data, it is important to account for differences between the data, tasks, and constraints. Seismic event waveforms are relatively short in duration (typically seconds to tens of seconds), have a lower sampling rate (100 Hz vs. 44,100 Hz), and contain a narrower range of frequencies than audio data. Thus, overall, audio data tends to be richer and have more distinct spectral signatures compared to seismic data. Additionally, in the earthquake detection task, the signals of interest in the continuous waveform data tend to be infrequent, representing a rare class in a data set dominated by noise, while in audio identification tasks we do not expect a large number of fingerprints (either in the database or queries) to contain only noise.

3.2.1 FAST Fingerprint Extraction

The input to FAST is single-channel continuous waveform data. No template waveforms or label information is available to the algorithm. Each short-duration waveform is converted into a sparse binary fingerprint using the procedure described below (see Bergen 2018 for additional details).

0. *Data pre-processing* We apply a fixed bandpass filter to the continuous waveform data to remove frequencies that contain high levels of noise, especially narrow-band noise, or frequencies that do not contain useful information for the detection task.
1. *Spectrogram* We transform the time series data to the spectrogram, a time-frequency representation computed with the short-time Fourier transform. The spectrogram has a window length of w_s s and a lag between windows of length ℓ_s s. Although the original Waveprint algorithm uses logarithmically spaced frequency bins, for seismic data we use linear spacing in the frequency domain. We truncate the spectrogram frequency range when appropriate; inclusion of frequency bands containing high-amplitude noise or those outside the passband can hurt detection performance.
2. *Spectral images* We divide the spectrogram into short (w_f s) overlapping segments (ℓ_f s lag) and resize spectral images to fixed dimensions: d_1 frequency bins and d_2 time bins (e.g. $d_1 = 32$ and $d_2 = 64$). The dimensions should be fixed for all fingerprints in the data set, and each dimension should be a power of 2 (i.e. $d_1, d_2 \in \{x \mid x = 2^n$ for some integer $n\}$) to simplify the computation of the Haar transform in the next step.
3. *Haar wavelet transform* For each spectral image, we compute the two-dimensional discrete Haar wavelet transform (Mallat 2008) to produce a wavelet image. The Haar wavelet transform computes moving averages and moving differences at multiple scales in a signal or image.
4. *Coefficient standardization* We standardize the value of each wavelet coefficient, based on the distribution of values the coefficient takes over the full data set. For the ith fingerprint, the value of the jth wavelet coefficient, $X_j^{(i)}$, is replaced with a standardized value:

$$X_j^{(i)} \leftarrow \frac{X_j^{(i)} - a_j}{b_j}, \tag{5}$$

where a_j and b_j are measures of the center and spread, respectively, of the distribution of the jth wavelet coefficient. We consider two standardization schemes, one using the mean and standard deviation (Z score), and the other using the median and median absolute deviation (MAD) (Hampel 1974):

$$\begin{aligned} Z\text{ score:}\quad a_j &= \mu_j = \frac{1}{N_{fp}} \sum_{i=1}^{N_{fp}} X_j^{(i)}, \\ b_j &= \sigma_j = \sqrt{\frac{1}{N_{fp}} \sum_{i=1}^{N_{fp}} \left(X_j^{(i)} - \mu_j\right)^2}, \end{aligned} \tag{6}$$

$$\begin{aligned} \text{MAD:}\quad a_j &= \underset{i}{\texttt{median}}\left(X_j^{(i)}\right), \\ b_j &= \underset{i}{\texttt{median}}\left(\left|X_j^{(i)} - \underset{i}{\texttt{median}}\left(X_j^{(i)}\right)\right|\right), \end{aligned} \tag{7}$$

where N_{fp} is the number of fingerprints in the data set. Before the statistics are computed, each of the

wavelet images should be scaled to Frobenius norm 1. Because the properties of the noise-dominated continuous waveform data vary by station, it is necessary to perform the coefficient standardization using Z score/MAD statistics computed separately for each channel.
Coefficient standardization is a new step introduced for FAST feature extraction and is not included in the original Waveprint method. Based on the results presented in Sect. 5, the MAD statistics are used by default for coefficient standardization in FAST fingerprint extraction.

5. *Coefficient selection* We select the K *standardized* wavelet coefficients that are largest in magnitude; these correspond to the *most anomalous* Haar coefficients for each spectral image. K is typically selected in the range of 10–40% (which will produce a fingerprint sparsity of 5–20%).
6. *Conversion to binary* For the selected coefficients, we retain only the sign value and set all other coefficients to zero. We convert the sign values to binary using two bits per coefficient ($+ \rightarrow$ `01`, $- \rightarrow$ `10`, and $0 \rightarrow$ `00`), resulting in sparse binary fingerprints of dimension $D = 2 \cdot d_1 \cdot d_2$ with K non-zeros.

Because this fingerprinting scheme generates sparse, binary fingerprints, we measure the similarity between the resulting waveform fingerprints using a metric called the *Jaccard similarity*:

$$\texttt{Jaccard}(A, B) = \frac{|A \cap B|}{|A \cup B|} \tag{8}$$

Jaccard similarity is a similarity measure for comparing sets, here the sets of active (non-zero) coefficients in the fingerprints.

3.2.2 Haar Coefficient Standardization

The FAST fingerprint extraction method largely follows the Waveprint method, with the addition of the coefficient standardization step that is applied to the Haar wavelet coefficients prior to coefficient selection. We explored the use of other wavelet representations, but none of them exceeded the performance we obtained using the Haar basis. The necessity of the coefficient standardization step for the FAST fingerprints comes down to a key difference between seismic and audio data: in continuous seismic data, noise-only signals make up the majority of the data set (and, by extension, the majority of waveform fingerprints), while the signals of interest, earthquakes, are infrequent. FAST identifies candidate earthquakes by searching for all pairs of similar waveform fingerprints. Since most of the fingerprints extracted from continuous data correspond to noise signals, it is critical that noise fingerprints have low mutual similarity for FAST to be effective.

The original Waveprint method represents a waveform using the largest-magnitude wavelet coefficients, but when applied to noise-dominated seismic data sets this representation is inefficient; a small subset of wavelet coefficients regularly take large values and are frequently selected, while a larger number of coefficients are almost never selected (see Fig. 3 for examples of coefficient distributions). In a representative data set with 10% sparsity, 16% of the coefficients are active in at least 1 in 4 fingerprints, while half of all coefficients are active in fewer than 1 in 100 fingerprints (Bergen et al. 2016). This inefficiency results in higher similarity between pairs of noise fingerprints (Fig. 4), and negatively impacts detection sensitivity by making it more difficult to identify pairs of similar low signal-to-noise (SNR) earthquake waveforms.

Coefficient standardization can be viewed as a balancing of the coefficient distributions, transforming the distribution of each coefficient to be centered around zero and have roughly the same spread. This increases the probability of selection for coefficients that originally were centered at zero with a small variance (or MAD), but decreases the probability of selection for coefficients that were originally centered away from zero or had large variance. The updated FAST fingerprint representation is more efficient: after MAD standardization is applied to the example data set described above, each coefficient is active in 5–15% of all waveform fingerprints, and there are no longer coefficients that are frequently or rarely active. The underlying assumption when we perform this balancing of the coefficient distributions is that it will not hurt detection performance by substantially reducing the similarity between the fingerprints of similar earthquake waveforms; we show in Sect. 5

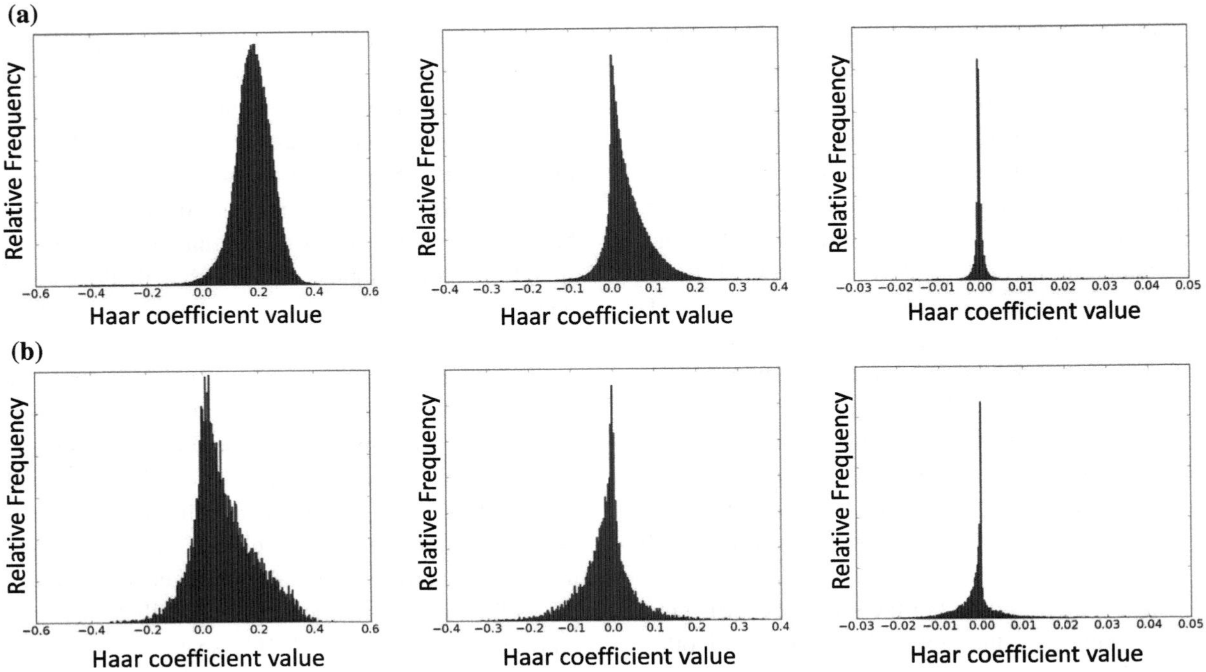

Figure 3
Distribution of wavelet coefficient values across all **a** noise and **b** earthquake fingerprints extracted for station NC.CCOB for 3 of the 2048 wavelet coefficients. Note that the same 3 wavelet coefficients (left to right: coefficient nos. 64, 322, and 273) are shown for both noise (top) and earthquake (bottom) waveforms; noise and earthquake waveform data were separated for comparison. For some coefficients there is not a significant difference between the coefficient distributions for noise and earthquake waveforms, but often the distribution for earthquakes has a different spread (right panel) or a different center or skew (center and left panels). For the purposes of coefficient standardization, each of these distributions is modeled by two parameters: the mean and standard deviations (*Z* score) or median and median absolute deviation (MAD). **a** Examples of distributions of wavelet coefficient values for noise waveforms. Most of the distributions are strongly peaked (right), but some are less strongly peaked (left) or asymmetrical (center). **b** Examples of distributions of wavelet coefficient values for earthquake waveforms. As in the case of noise waveforms, most of the wavelet coefficient distributions for earthquake waveforms are strongly peaked. The wavelet coefficient distributions for earthquakes often differ from the corresponding distribution for noise

that, empirically, this approach does preserve similarity for earthquakes (Fig. 4).

4. Experiments

In order to select the best feature extraction method and optimize any necessary parameter values (e.g. the sparsity parameter, *K*, for Waveprint and FAST fingerprints), we require a means of quantifying the performance of each method in the context of similarity search. In this section we describe a test for comparing the performance of different feature extraction methods. In the next section we use this benchmark test to compare three audio fingerprinting methods (Sect. 3.1) and the FAST feature extraction method for use in waveform similarity search.

4.1. Benchmark Data Set

To compare the effectiveness of different fingerprinting schemes for waveform similarity search, we need examples of known earthquake and noise waveforms for performance evaluation. The best fingerprint extraction method and optimal parameter settings can vary by station depending on the local noise properties, so we use multiple data sets to reduce bias and determine the method with the best overall performance. We compile a benchmark data set containing earthquake waveforms and segments of background noise from each of eight different

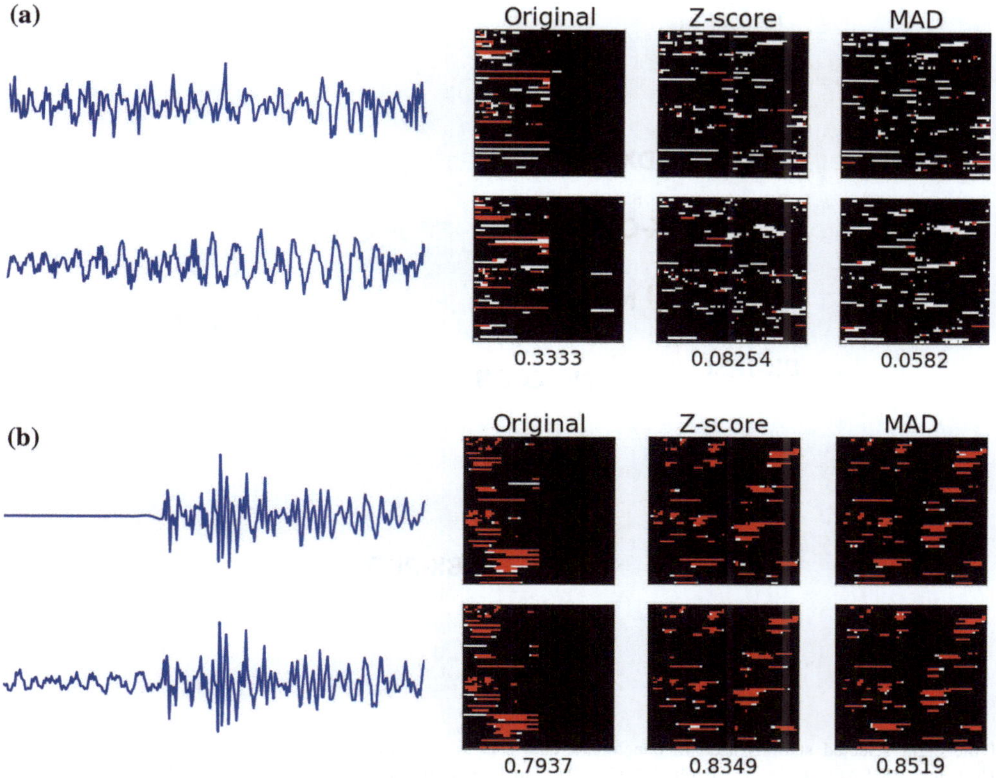

Figure 4
Comparison of Waveprint and FAST fingerprinting schemes applied to **a** background noise and **b** similar earthquake waveforms. Each row displays the fingerprints extracted from corresponding waveform data, and each column represents a different fingerprinting scheme: Original refers to the Waveprint scheme with no coefficient standardization step; *Z* score and MAD refer to standardization with the mean and standard deviation or median and median absolute deviation, respectively. Pixels in white and red represent the active/selected coefficients, with those in red indicating the intersection of fingerprints for each of the two distinct waveforms under the same fingerprinting scheme. All waveform data below are from station BK.SAO and represent a duration of 16 s. **a** Fingerprints corresponding to two noise signals. The Jaccard similarities between fingerprints are: 0.33 (original), 0.08 (*Z* score), and 0.06 (MAD). Fingerprints corresponding to noise waveforms have lower similarity after coefficient standardization compared to the original fingerprints. **b** Fingerprints corresponding to two similar earthquake signals: the first waveform corresponds to a real event; the second waveform is a copy of the event waveform embedded in noise at SNR 5.0 (as illustrated in Fig. 6). The Jaccard similarities between fingerprints are: 0.79 (original), 0.83 (*Z* score), and 0.85 (MAD). Fingerprints corresponding to similar waveforms maintain high similarity after coefficient standardization

stations. The stations, shown in Fig. 5, are selected for geographic diversity within the northern California region and to ensure availability of a relatively high number of event waveforms.

For each of the eight stations, we create an "earthquake" and a "noise" waveform data set. All data are taken from the 8.5-year period from 1 January 2008 to 31 May 2016. For both earthquake and noise waveform data, we use the vertical channel data, apply a 1–10 Hz bandpass filter, and decimate to 20 Hz. We apply the same bandpass filter for each station because we want to ensure good performance even if the optimal bandpass filter is not selected.

The earthquake waveform data is taken from catalog events in the Northern California Seismic Network (NCSN) phase-pick catalog (NCEDC 2014). We include only events magnitude 2.0 and larger. For each event in the NCSN catalog, we download a 2-min segment of waveform data (when available): the 60 s leading up to the P-wave arrival and the 60 s after the P arrival. To ensure that the earthquake data set contains events with clear waveforms, we include events only if (1) the wave arrival appears close to the

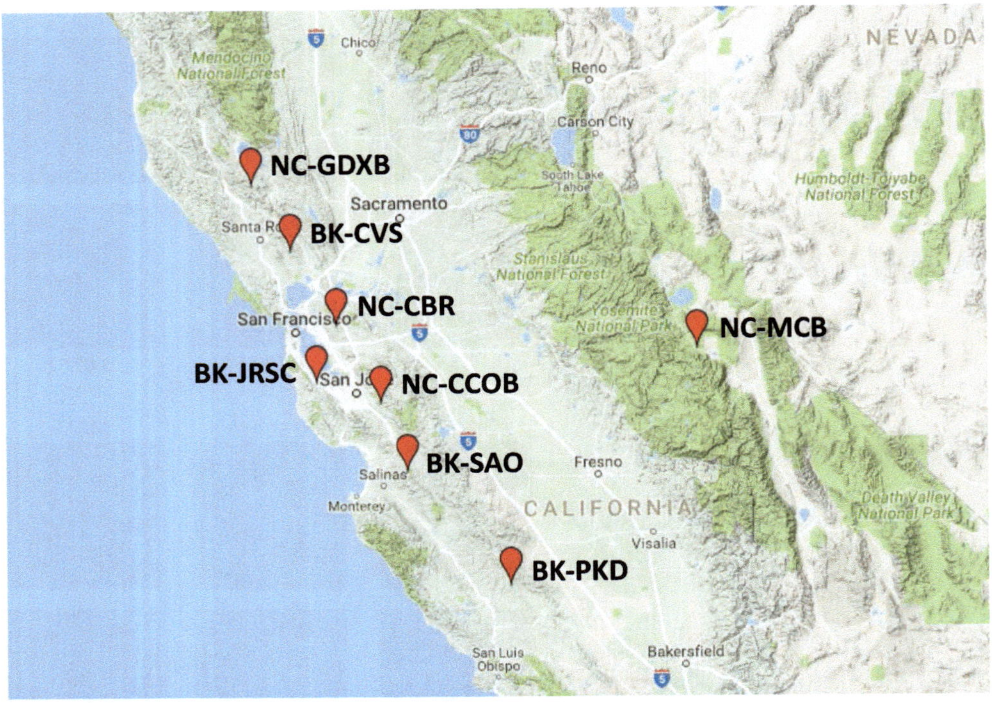

Figure 5
Locations of the eight selected stations used in benchmark data sets to compare performance of feature extraction methods: Parkfield (BK.PKD), Hollister (BK.SAO), Calaveras Fault (NC.CCOB), San Francisco Peninsula (BK.JRSC), San Ramon (NC.CBR), Napa Valley (BK.CVS), Mammoth (NC.MCB), and Geysers geothermal field (NC.GDXB). Created with Google Maps

Table 1

Benchmark waveform data sets

Station	Number of waveforms in data set	
	Earthquakes	Noise ($\mathcal{N}_{\mathcal{C}}$ and $\mathcal{N}_{\mathcal{L}}$)
BK.PKD	1972	3083
BK.SAO	2271	3094
NC.CCOB	1691	2851
BK.JRSC	1824	3092
NC.CBR	1069	2469
BK.CVS	3012	2974
NC.MCB	1276	3051
NC.GDXB	1991	3080
Total	15,106	23,694

"Earthquakes" refers to the number of medium-to-high SNR catalog events in the data set for each station. "Noise" refers to the number of "clean" or "lively" background noise segments (with likely earthquakes excluded) in the data set for each station

expected P arrival time and (2) the STA/LTA ratio (Allen 1982) (with window lengths 1 and 30 s) exceeds a threshold value of 5.0 at some time up to 15 s after the P arrival. The number of earthquake waveforms included in the benchmark from each of the selected stations is given in Table 1, and the event epicenter locations are shown in Figs. 12 and 13 in the Appendix.

The noise waveform data are sampled from the continuous data. For each day in the 8.5-year period, we select a 2-min time interval at random and download the continuous waveform data (when available). These "noise" segments may contain regional or teleseismic earthquakes or other transient signals, so we apply an additional data cleaning step. We label each segment as either "clean noise" ($\mathcal{N}_{\mathcal{C}}$), "lively noise" ($\mathcal{N}_{\mathcal{L}}$), or "non-noise" ($\mathcal{N}_{\mathcal{S}}$) (see Appendix 7.2). Non-noise are segments of data that likely contain signals from regional earthquake arrivals, earthquake codas, or teleseismic waves; these segments are removed from the noise data set. Clean noise segments are those with relatively uniform signal energy across the interval, while lively noise segments have more variation in signal energy.

4.2. *Extracting Waveform Fingerprints*

All of the fingerprint extraction methods initially convert the continuous waveform data into the time-frequency domain. The computation and processing of the spectrogram is the same across all methods, with the exception of the choice of window function. All spectrograms use a window length $w_s = 10.0$ s and a window lag $\ell_s = 0.1$ s, and have 32 linearly spaced frequency bins (33 for Philips to produce 32 bins in the derivative).

In the Philips, Waveprint, and FAST fingerprints we use a fingerprint window $w_f = 10.0$ s. The Waveprint and FAST fingerprints have dimension $D = 4096$ and the Philips fingerprints have dimension $D = 3200$. For the Waveprint and FAST fingerprints we vary the value of sparsity parameter, K, from 50 to 800 for parameter testing, but we fix the value $K = 400$ when comparing different fingerprint extraction methods. In the benchmark data sets, earthquakes are oversampled, so we use only the noise waveforms to compute the Z score and MAD statistics for the FAST fingerprints.

The landmark-based method computes hashes rather than fingerprints. Key points are selected by identifying local maxima in the spectrogram using a maximum filter (with a minimum distance of 1.5 Hz along the frequency axis and 1 s along the time axis). The hash constellations are formed between each key point and other local maxima within a target region that includes any local maxima that both occurs 1–10 s after the key point and has a frequency coordinate within a 4 Hz range.

4.3. *Performance Metrics*

To evaluate the performance of different fingerprint extraction methods for earthquake detection, we require a metric to quantify the degree to which a fingerprinting scheme is discriminative for similarity search. A fingerprinting scheme is discriminative if (1) under the fingerprinting scheme, two similar earthquake waveforms are mapped to fingerprints that have high similarity, and (2) an arbitrary pair of noise waveforms are mapped to fingerprints that have low similarity. We quantify these criteria using two metrics: fingerprint accuracy and baseline similarity, respectively.

4.3.1 *Quantifying Accuracy and Baseline Similarity*

In the discussion that follows, let $x^{(i)} \in \mathbb{R}^M$ be the ith earthquake waveform, represented in the time domain with M samples, and let $n^{(j)} \in \mathbb{R}^M$ be the jth noise waveform. Let $\mathcal{F} : \mathbb{R}^M \rightarrow \{0,1\}^D$ represent the fingerprint extraction operation, and let α be a scaling factor that controls the signal-to-noise ratio (α is not a fixed value but varies to produce the desired signal-to-noise ratio; see Appendix 7.1).

Fingerprint accuracy is a measure of the quality of the fingerprints of earthquake waveforms for similarity-based detection under additive noise. In our benchmark test we consider the challenging detection task of identifying two similar event waveforms, both at low SNR, because we would like our fingerprinting scheme to be discriminative for weak earthquake signals. We compare the fingerprints of two versions of the same earthquake waveform, $x^{(i)}$, (high signal-to-noise, from "earthquake" waveform data set) embedded in two different noise segments, $n^{(j)}$ and $n^{(k)}$, both at low SNR (see Fig. 6):

$$accuracy(i,j,k) = \mathtt{sim}\Big(\mathcal{F}(\alpha_j x^{(i)} + n^{(j)}),\ \mathcal{F}(\alpha_k x^{(i)} + n^{(k)})\Big), \tag{9}$$

where noise segments $n^{(j)}, n^{(k)} \in \mathcal{N}_C$ are only drawn from the "clean noise" data set from the same station as the earthquake waveform $x^{(i)}$. Here, $\mathtt{sim}(\cdot,\cdot)$ represents the relevant similarity measure: number of matching hashes for landmark method, Hamming similarity for the Philips method, and Jaccard similarity for the Waveprint and FAST fingerprints.

Waveform fingerprints should be effective even when the earthquake waveforms are not perfectly aligned; this is necessary for good performance in continuous data when the lag between adjacent fingerprints may be large compared to the sampling rate. To make our tests more representative of the detection challenges in continuous data, we compare fingerprints for similar waveforms with a time offset

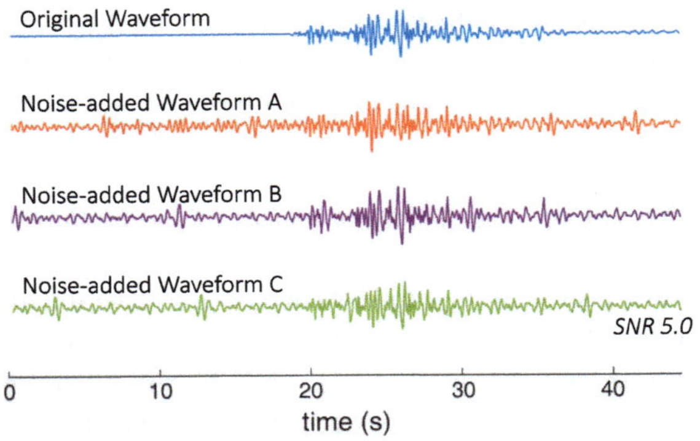

Figure 6
Original earthquake waveform is shown in the top row (blue). This original waveform is scaled down and embedded in three different intervals containing only sensor-recorded background noise, forming the three "noise-added" copies of the waveform in the bottom three rows, at a lower signal-to-noise ratio than the original waveform (SNR 5.0)

(applied before extracting fingerprints). We use offsets ranging from 0 to 10 samples, corresponding to the maximum offset for continuous data sampled at 20 Hz with a fingerprint lag of 1.0 s. These offsets only apply to the Philips, Waveprint, and FAST fingerprints, as the notion of a fingerprint lag is not applicable to the landmark-based hash method.

To measure the similarity between fingerprints extracted from noise waveforms, we define the *baseline similarity* between the kth and ℓth noise waveforms as:

$$\text{baseline }(k,\ell) = \mathtt{sim}\Big(\mathcal{F}(n^{(k)}),\ \mathcal{F}(n^{(\ell)})\Big), \qquad (10)$$

where noise segments $n^{(k)}, n^{(\ell)} \in (\mathcal{N}_C \cup \mathcal{N}_L)$ are drawn from both "clean " and "lively" noise. The baseline similarity is an important statistic because the threshold for what will be considered high accuracy for earthquake waveform fingerprints is determined relative to typical similarities between fingerprints for noise waveforms.

To obtain the baseline similarity distribution, we compare 1,000,000 pairs of noise fingerprints (some noise fingerprints may belong to multiple pairs) per station. For measuring the accuracy distribution, the number of pairs of fingerprints is proportional to the number of earthquake waveforms available from a given station (see Table 1). A separate distribution of values for fingerprint accuracy and baseline similarity are computed using waveform data from each of the eight stations.

Figure 7 ▶
Distributions used to characterize the trade-off between true positives and false positives. Green distributions show values of baseline similarity for pairs of noise fingerprints; blue distributions show accuracy measure for pairs of earthquakes embedded in noise at SNR 2.0. Data shown are from station BK.SAO. **a** Original Waveprint fingerprints (no coefficient standardization) with sparsity parameter $K = 400$. **b** FAST fingerprints (modified Waveprint with Z score coefficient standardization) with $K = 400$. **c** FAST fingerprints (modified Waveprint with MAD coefficient standardization) with $K = 400$. **d** Philips fingerprints. Because these fingerprints are dense binary fingerprints, the similarity of pairs of noise fingerprints is centered around 0.5, which is the expected Hamming similarity for two dense binary fingerprints with bits selected at random. **e** Landmark-based hashes. Similarity is quantified by the number of matching hash values, and is represented as counts rather than a quantity between 0 and 1 (as in the Jaccard similarity for Waveprint or Hamming similarity for Philips)

4.3.2 ROC Curve and Truncated-AUC

For a fingerprint extraction scheme to be effective for similarity-based detection in long-duration data, we require both high accuracy for fingerprints and low baseline similarity to limit false detections. We characterize the trade-off between false detections and missed detections using a receiver operating characteristic (ROC) curve. For a given Jaccard similarity threshold, τ, we define the *true positive rate* (TPR_τ) as the fraction of earthquake waveform pairs for which the accuracy exceeds the threshold, and we define the *false positive rate* (FPR_τ) as the fraction of noise waveform pairs for which the baseline similarity exceeds the same threshold (see Fig. 7). The true and false positive rates should be interpreted with caution with respect to detection in continuous data, as TPR and FPR do not account for the non-linear relationship between Jaccard similarity and the probability of detection in LSH-based similarity search. The ROC curve traces the values (FPR_τ, TPR_τ) for all choices of threshold τ between 0 and 1 for Philips and Waveprint variants (for landmark-based features, the threshold values are counts).

One commonly used measure to summarize the overall detection performance across the full range of threshold values is the area under the ROC curve (AUC) (Bradley 1997). Because high values for the FPR are not appropriate for the earthquake detection

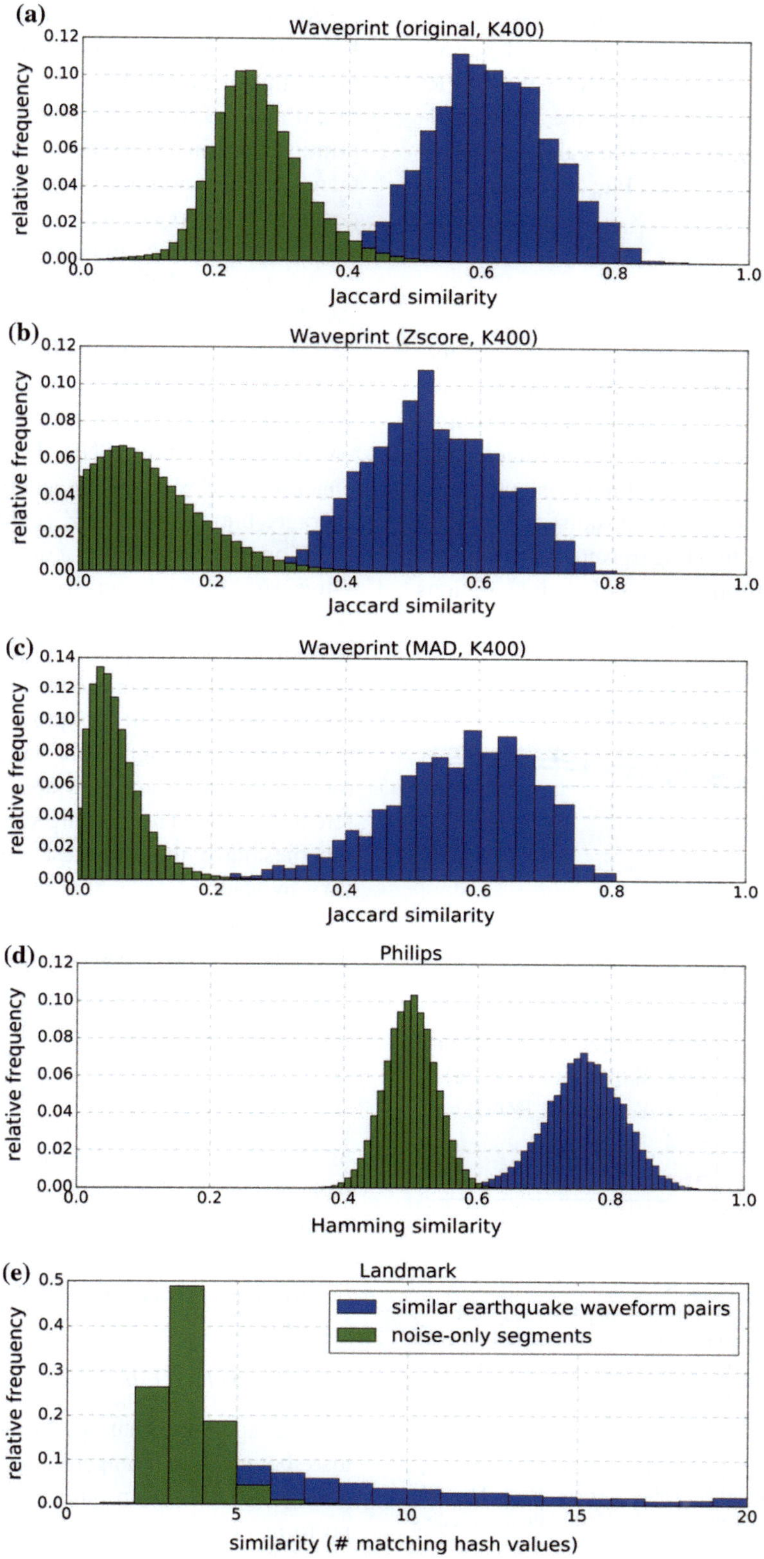
(a)
Waveprint (original, K400)
relative frequency
Jaccard similarity
(b)
Waveprint (Zscore, K400)
relative frequency
Jaccard similarity
(c)
Waveprint (MAD, K400)
relative frequency
Jaccard similarity
(d)
Philips
relative frequency
Hamming similarity
(e)
Landmark
relative frequency
similar earthquake waveform pairs
noise-only segments
similarity (# matching hash values)

task, we use a truncated version of the AUC that considers only the area under the section of the curve corresponding to small values of the FPR: $\mathrm{FPR} \in [0, \mathrm{FPR}_{\max}]$; we normalize the truncated-AUC so that its maximum possible value is always 1 for any choice of $\mathrm{FPR}_{\max}$ (see Fig. 8).

5. Results

We compare the performance of audio fingerprinting and FAST fingerprinting schemes for earthquake detection: (1) landmark-based features, (2) Philips fingerprints, (3) Waveprint fingerprints, and FAST fingerprints with coefficient standardization using (4) *Z* score or (5) MAD statistics. Figure 9 shows a comparison of five fingerprinting schemes on the benchmark waveform data sets for eight stations using the truncated-AUC metric.

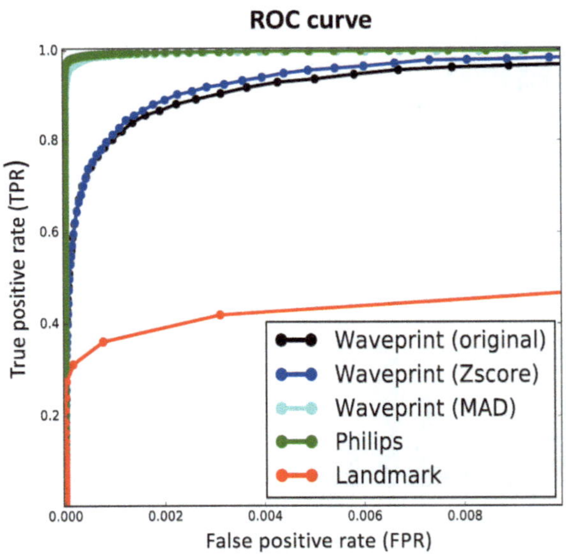

Figure 8
Comparison of ROC curve for different fingerprinting schemes. The area under each curve (with appropriate normalization) gives the truncated-AUC (FPR ≤ 0.01) statistic used in Fig. 9 for each method. Data shown are for the data set from station BK.SAO with earthquake waveforms at SNR 2.0. A separate ROC curve is computed for each station and method. ROC curves are generated from distributions in Fig. 7 by plotting the fraction of baseline similarity (green) distribution exceeding a given threshold to the fraction of the fingerprint accuracy (blue) distribution exceeding the same threshold, for all thresholds from 0 to 1 (or 0–20 for landmark method) for each method

At a higher SNR of 5.0, the Philips method and all Waveprint variants perform consistently well across the eight benchmark data sets. As the SNR is lowered to 2.0 and 1.0, the performance of all of the fingerprinting schemes degrades, and for some schemes there is more variation in performance, with a large performance gap between the best- and worst-case performance on the eight benchmark data sets.

At lower SNR (1.0–2.0), FAST fingerprints (MAD) achieve the best and most consistent performance; this scheme has the best worst-, average-, and best-case performance compared to the other methods. The Philips fingerprints also perform well in the low SNR case. At low SNR, the Waveprint and FAST fingerprints (*Z* score) achieve similar performance on average, but the FAST (*Z* score) scheme has more variation in performance compared to the original Waveprint method; on some data sets the FAST (*Z* score) scheme performs well, but on others its performance is very poor. In all cases, the landmark-based method performs poorly, even for signals in the higher SNR case. The box plots in Fig. 9 demonstrate that there can be significant variation in performance for fingerprinting schemes applied to data from different stations. The ranking of which fingerprinting schemes have the strongest performance may also vary by station.

Note that caution should be taken in interpreting the false and true positive rates for continuous data. The TPR and FPR are computed on the true Jaccard similarity values, and do not account for the fact that the probability of detection varies with the Jaccard similarity in FAST's LSH-based similarity search. Using the Jaccard similarity values gives a measure of the fingerprinting scheme effectiveness that is independent of the choice of the similarity search method or parameters, making it possible to draw general conclusions about the relative effectiveness of different approaches.

5.1. Parameter Testing

The approach that we have outlined above can also be used to compare and optimize parameter values for a given fingerprinting scheme. The performance of the FAST (MAD) fingerprinting scheme is not particularly sensitive to changes in

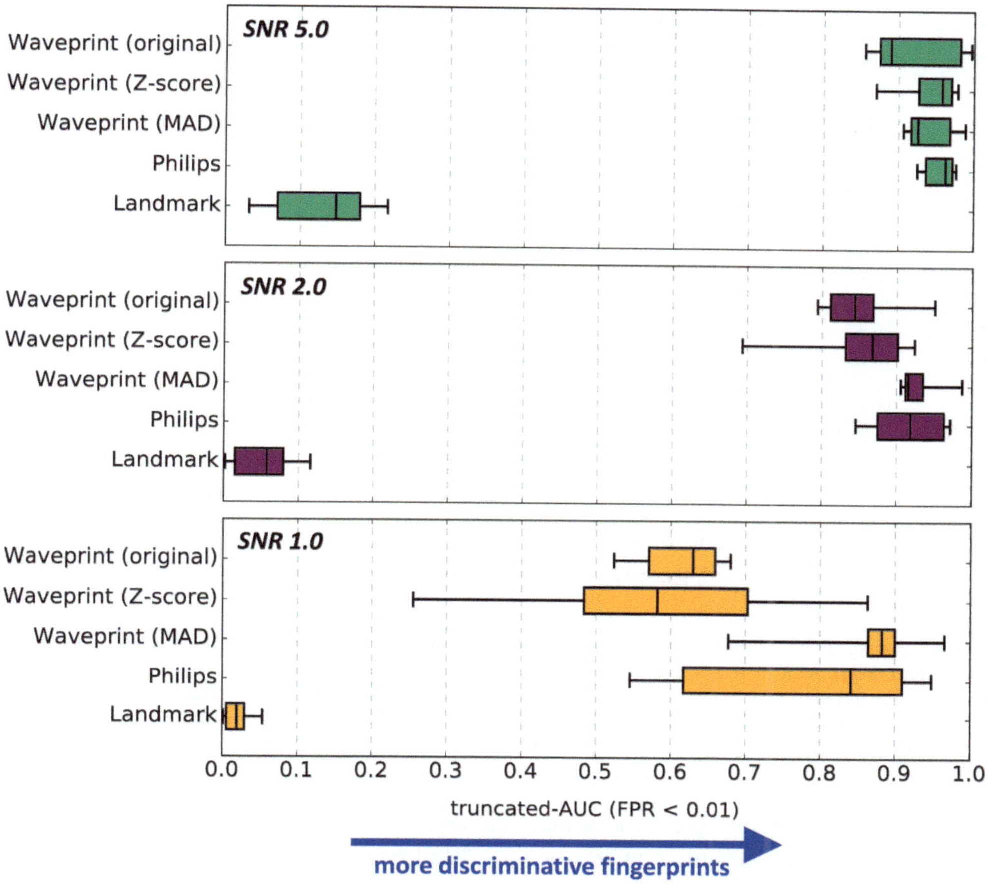

Figure 9
Relative performance of five fingerprinting schemes on the benchmark data sets. Box plots show variation in truncated-AUC measure over the data sets from eight stations. Each panel shows results for different signal-to-noise ratio (SNR) values for the earthquake signals: 5.0, 2.0, and 1.0 (top to bottom)

most of the parameter values (e.g. the choice of wavelet basis), with the exception of one key parameter: the sparsity parameter K. Ideally, fingerprints should be as sparse as possible (i.e. have few non-zero values, low K) to enable efficient search, but must remain sensitive to weak earthquake signals. A comparison of the performance with different values of K (Fig. 10) shows that values in the range of $K = 300$–500 (out of 4096 coefficients in binary fingerprint), corresponding to 10% sparsity, are optimal for the FAST fingerprints with MAD-adjusted coefficients.

6. Discussion and Conclusions

6.1. Audio Fingerprints for Earthquake Detection

Among the three audio fingerprinting methods, the landmark-based features perform poorly on seismic data even at high SNR, while both the Philips and Waveprint fingerprints perform well when applied to seismic data. Each of these three approaches captures spectrogram features at different scales; the landmark-based features only capture a few prominent points, the Philips fingerprints contain only local information (derivatives) in the spectrogram, and Waveprint captures information about different-sized patches (from local to global) within the spectrogram.

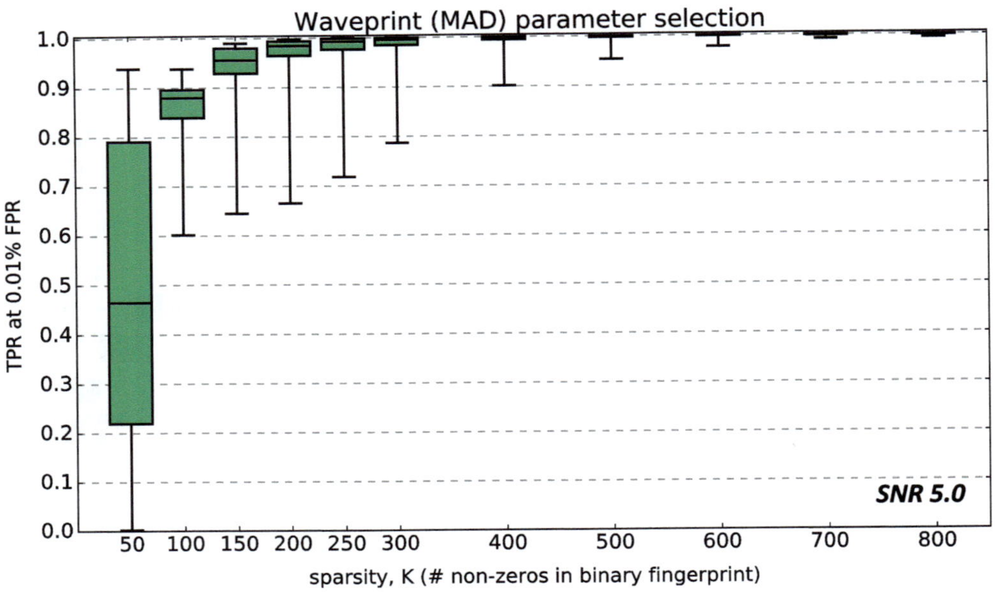

Figure 10

Parameter testing: the horizontal axis is the value of the sparsity parameter, K, which corresponds to the number of non-zero values in the binary fingerprint (out of 4096). As K increases, the fingerprints are less sparse, which makes them less compact to store, and the runtime of the efficient similarity search step of FAST tends to increase. The optimal range for K is 300–500, which corresponds to 7–12% sparsity. The vertical axis is a measure of how discriminative the resulting fingerprints are, as measured by the true positive rate (TPR) for a fixed false positive rate of 0.01%. Box plots show variation in performance over the waveform data sets from eight different stations

The landmark-based features perform poorly because local maxima in the spectrogram are not appropriate features for characterizing earthquake waveforms. Even for two earthquake waveforms that are nearly identical in the time domain, there may be limited correspondence between the spectrogram peaks for the two events (Fig. 11). This approach misses important information for detection, as P-wave arrivals correspond to intervals with increasing signal energy rather than energy maxima, and thus generate very few key points.

The Philips fingerprints are also essentially local features based on the derivatives of the spectrogram. An advantage of the Philips method is that each feature contains only information from two neighboring frequency bins, so relatively few features will be affected by the presence of persistent narrow-band noise, making this approach robust to a poor choice of bandpass filter. In contrast, because the Waveprint fingerprints and FAST variant effectively treat the spectrogram as an image, and search for similar images using features at multiple scales, these approaches may be more prone to false detections or missed detections in the presence of strong narrow-band noise. The dense binary Philips fingerprints are associated with Hamming similarity measure, so it requires a different similarity search implementation from that shared by the Waveprint and FAST fingerprints.

The fingerprint extraction method that demonstrates the most consistent performance at low SNR is the FAST fingerprints, a modified version of the Waveprint method, with wavelet coefficients standardized using the MAD statistics. The standardization step improves detection performance of the original Waveprint method because coefficient standardization induces a more uniform distribution of fingerprints in the continuous data set, with the effect of lowering the similarity between fingerprints corresponding to noise while maintaining high similarity between earthquake fingerprints. Although the standardization was originally implemented using the mean and standard deviation (Yoon et al. 2015), our recommendation is to use the median and MAD statistics for coefficient standardization in FAST fingerprints.

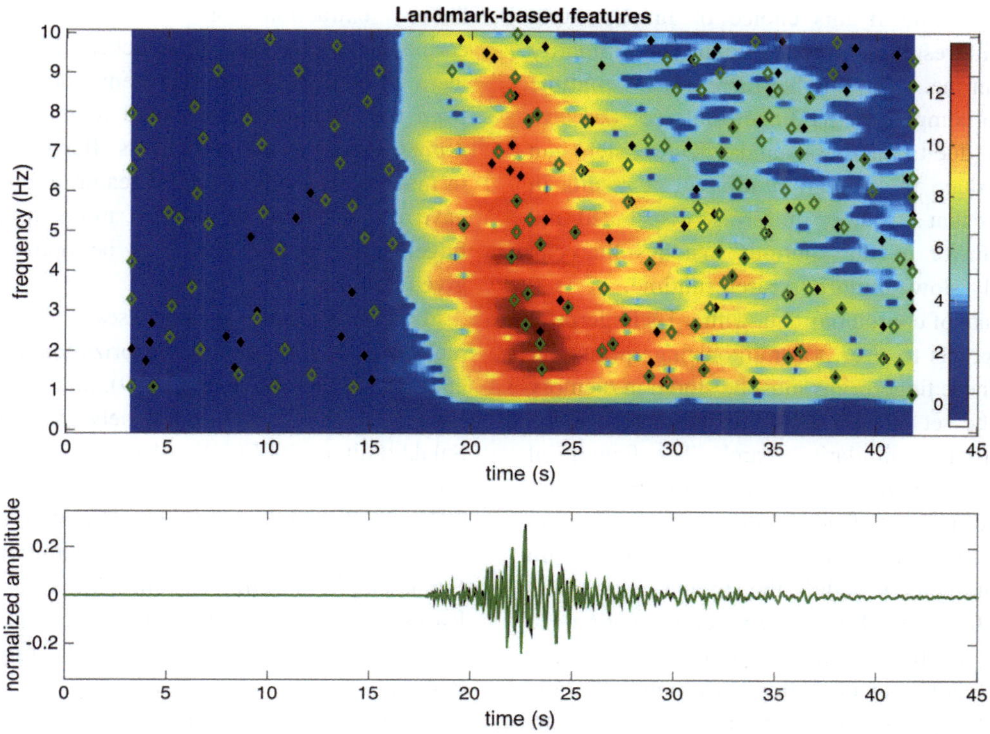

Figure 11
Illustration of the appropriateness of local maxima in the spectrogram as features for waveform similarity search for the earthquake detection problem. The lower panel shows two events with nearly identical waveforms, plotted in green (foreground) and black (background), and normalized to equal amplitude. The upper panel shows the spectrogram corresponding to the green event waveform, with the local maxima, or key points, identified by the landmark-based algorithm marked with green and black diamonds (corresponding to the green and black event waveforms, respectively). Although the green and black waveforms are nearly identical in the time domain, the key points for these two events have limited correspondence with each other. Local maxima in the spectrogram work well for music audio signals, which are typically localized in frequency and time, but these features are not as useful for identifying similar earthquake waveforms. Colormap is plotted on a log scale

6.1.1 Computational Concerns

The metric used to compare different feature extraction methods (Sect. 4.3) is a measure of how discriminative the resulting fingerprints are for waveform similarity search, but this measure does not capture other advantages or disadvantages of these methods with respect to our task. Our ultimate aim is to select a feature representation that will enable earthquake detection via blind search for similar earthquake waveforms in long-duration data. An important factor in selecting a feature extraction method is whether the chosen approach can be scaled to large data sets; the computational cost of feature extraction, the compactness of the feature representation, the effect on similarity search memory usage, and runtime must also be considered in selecting a fingerprinting scheme for use in waveform similarity search. The runtime for fingerprint extraction should not be prohibitive for data sets with durations on the order of months to years, though a slower feature extraction method that can be computed in a parallel or distributed manner would also be acceptable.

The need to estimate the median and MAD statistics for the distribution of each wavelet coefficient is a disadvantage of the FAST fingerprint extraction scheme from a computational standpoint. For large data sets, this can be partially addressed by computing the statistics on a sample of fingerprints rather than the entire data set. However, since the computational bottleneck in the FAST detection pipeline is often the similarity search step, it may be desirable to select a slightly slower feature extraction method if this choice of fingerprinting scheme will result in a speed-up in the similarity search runtime. The slight increase in runtime required to compute FAST fingerprints compared to Waveprint (due to the extra standardization step) can actually reduce the overall runtime if it results in fingerprints that are more uniformly distributed and produces fewer of the spurious matches that can significantly slow similarity search runtime.

Ten years of continuous data with a 1-s fingerprint lag corresponds to over 300 million fingerprints per channel. These fingerprints must be stored in memory to create the set of hash tables for the search index used in efficient similarity search. The fingerprint representation should be compact, requiring a relatively small number of bits to store each fingerprint. For FAST fingerprints of dimension $D = 4096$ with sparsity parameter $K = 400$, the storage requirement is $2 \times 4096 = 8192$ bits (binary representation) or $16 \times 400 = 6400$ bits (integer representation) per second of continuous data; this corresponds to 250–325 GB (gigabytes) to store fingerprints for 10 years of data. The Philips fingerprints have a computational advantage over the Waveprint and FAST fingerprints in this respect. While each Philips fingerprint requires $2 \times 32 \times 100 = 6400$ bits to store, there is significant overlap between the representation of adjacent fingerprints, so that in practice the storage requirement is only $2 \times 32 \times 10 = 640$ bits per second of continuous data, or 25GB for 10 years of continuous data, a factor of 10 reduction compared to the FAST fingerprints.

6.2. *Feature Learning for Waveform Similarity Search*

All of the fingerprinting methods discussed in this work are examples of "hand-engineered" features; they apply a set of fixed transformations that were selected based on expert knowledge of the properties of audio data and that have been demonstrated to perform well in practice. The three audio fingerprinting methods are data-independent, while the FAST fingerprints are modified to include an additional data-adaptive component. As an extension of the work presented in the previous sections, we investigated the use of fully data-driven approaches to feature extraction for waveform-similarity-based earthquake detection. Approaches that attempt to learn an optimal feature representation for a data set from the data itself are called *feature learning* or *representation learning* methods (Bengio et al. 2013) in machine learning. Learned feature representations have the potential to produce more discriminative fingerprints because they can be optimized for a given data set.

Feature learning encompasses a range of techniques, including matrix factorization (Deerwester et al. 1990; Lee and Seung 1999), dictionary learning (Lee et al. 2007), and neural networks (Hinton and Salakhutdinov 2006). We also considered a related set of methods called *learning to hash* (Wang et al. 2014, 2018), a data-dependent alternative to locality-sensitive hashing for approximate similarity search. For example, spectral hashing (Weiss et al. 2009) learns a set of data-dependent projections instead of using random projections as in locality-sensitive hashing.

There are a number of challenges that arise when applying feature learning or learning-to-hash methods to uninformed waveform-similarity-based earthquake detection (Bergen 2018).

It may be possible to learn high-quality features for similarity search using a labeled training set (collection of known earthquake waveforms). However, FAST is intended to be an uninformed method for earthquake detection. One of the properties that makes FAST particularly useful is that it can be applied in cases when there is a limited record of past seismic activity (i.e. the available catalog of template waveforms is limited or incomplete). Thus the feature learning approach should not require or assume the availability of known template waveforms; in the context of learning algorithms, this means limiting the approach to unsupervised methods, i.e. those that do not require labeled training data.

There are many unsupervised methods for feature learning and learning to hash that do not require template waveforms, but seismic data sets pose an additional challenge. Seismic data sets are imbalanced, meaning that the events of interest represent a minority of the signals in continuous data sets dominated by long periods of noise. Machine learning tasks with imbalanced data often result in poor performance on the minority class (in this case earthquake signals) because the learning algorithm has less information about the minority class, and the training samples from the minority class carry less weight overall in the learned model (He and Garcia 2009). As a result, it is difficult to learn a feature representation that produces low similarity between noise waveforms and high similarity between earthquake waveforms with unsupervised feature learning.

The challenges posed by the lack of labeled training data and the imbalanced, noise-dominated data make it difficult to provide FAST users with a single approach or framework for learning features that can be applied to diverse data sets. Therefore, our proposed FAST fingerprints, the data-adaptive variant of the Waveprint with MAD-standardization of the wavelet coefficients, are designed to be applied to any data set regardless of the availability of template waveforms, and take into account the data set imbalance problem. The coefficient standardization step in the FAST fingerprint extraction method is similar to the approach taken by some learning-to-hash methods: creating balanced hash codes that distribute the data more evenly in the hash tables (Weiss et al. 2009; Wang et al. 2010).

6.3. *Benchmarking for Performance Evaluation*

The approach outlined in Sect. 4 of this work can be used to evaluate alternative feature extraction schemes for waveform data beyond the audio fingerprinting methods presented in this work. Our framework for evaluating performance includes two key elements: benchmark data sets and an appropriate performance measure. The use of benchmark waveform data sets, containing known examples of noise and earthquake waveforms, provides a means of determining the detection performance for weak events. This allows us to overcome one challenge in evaluating the performance of algorithms for detecting weak events often missed in existing catalogs: the lack of ground truth, an objective standard for measuring detection performance. This framework provides a way to quantify the effectiveness of each feature extraction method for separating weak earthquake signals from noise in the context of similarity search. The experiments presented in this work only test for the distortions in the waveforms due to additive noise, but a similar framework can be used to compare fingerprinting schemes with respect to other criteria.

While our approach allows us to validate the feature extraction algorithm used in FAST, there remains the ongoing challenge of assessing detection performance in continuous data without ground truth. The goal of developing new earthquake detection algorithms is to outperform existing methods, but the nature of our task, discovering new events and sources, makes it inherently difficult to measure performance. Therefore, an open challenge in earthquake detection research is how to validate new candidate events, specifically how to distinguish true earthquake signals from transient signals due to local noise sources such as vehicles, transportation systems, and air traffic (Díaz et al. 2017; Meng and Ben-Zion 2018). The development of appropriate benchmark data sets and performance measures for comparing and validating earthquake detection algorithms will be critical for leveraging new developments in machine learning and data mining for seismology research.

Acknowledgements

This research is supported by National Science Foundation Grants EAR-1551462 and EAR-1818579. Computing resources were provided by the Center for Computational Earth and Environmental Science, Stanford University. Waveform data, metadata, or data products for this study were accessed through the Northern California Earthquake Data Center (NCEDC 2014). The ObsPy Python toolbox (Beyreuther et al. 2010) was used to retrieve and process waveform data.

Appendix

Signal-to-Noise Ratio

In this work, the signal-to-noise ratio (SNR) is computed using a 15-s interval of waveform data following the P-wave arrival. For signal x and noise

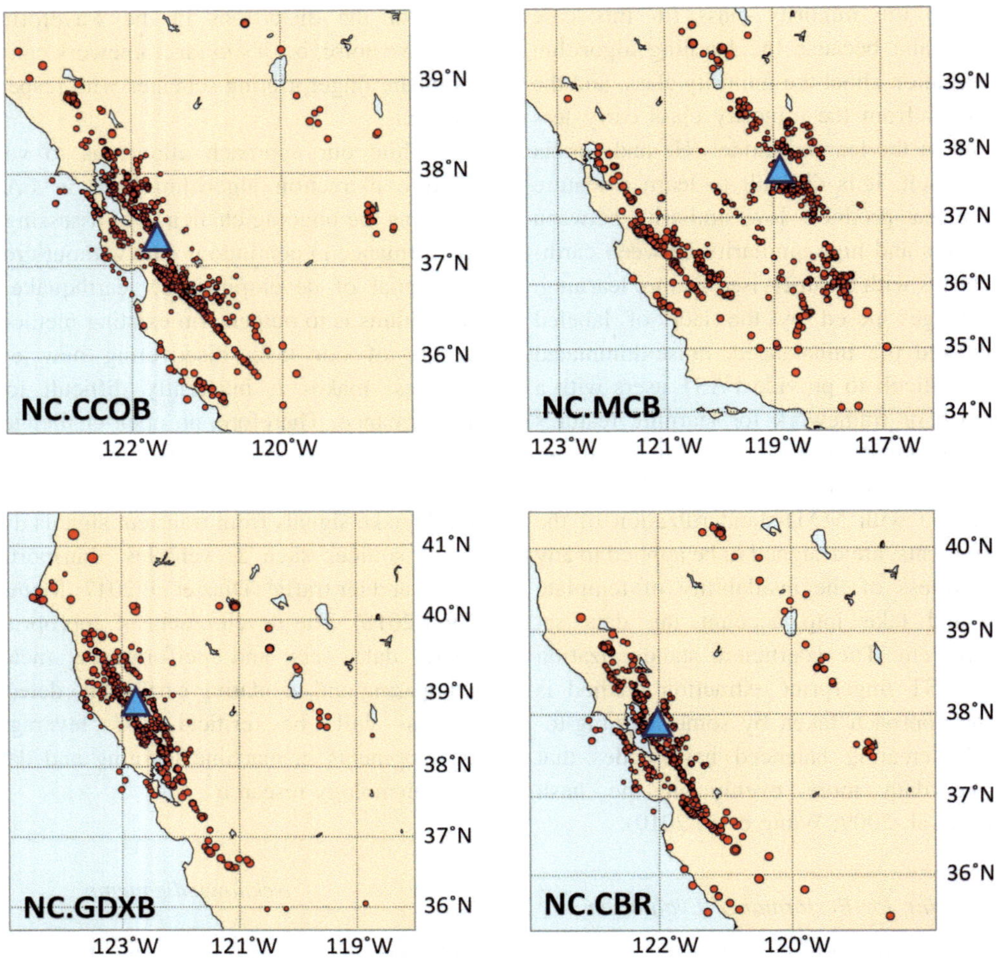

Figure 12
Map plots show the epicenter locations for the events in the earthquake waveform benchmark data set for stations NC.CCOB, NC.MCB, NC.GDXB, and NC.CBR. The location of each station is marked with a blue triangle, and the event epicenters are shown in red. The set of earthquakes is selected independently for each station depending on whether a P-phase arrival was recorded in the NCSN phase-pick catalog for a given event

n, each of duration $M = 300$ samples (15 s of data sampled at 20 samples per second), the SNR is given by:

$$SNR = \frac{P_{\text{signal}}}{P_{\text{noise}}}, \quad \text{where} \quad P_{\text{signal}} = \frac{1}{M}\sum_{i=0}^{M} |x[i]|^2, \text{ and } P_{\text{noise}} = \frac{1}{M}\sum_{i=0}^{M} |n[i]|^2. \tag{11}$$

Noise Segment Classification

Noise segments are assigned one of these three labels based on two criteria: (1) an STA/LTA threshold, and (2) a uniform energy criterion that quantifies how uniform the energy is across the 2-min interval. A noise segment is labeled as "clean" noise if the maximum STA/LTA ratio in the interval is below 3.0 and the uniform energy score is below 0.1. A noise segment is labeled as "non-noise" if the maximum STA/LTA ratio in the interval exceeds 6.0 or the uniform energy score exceeds 0.2. All other noise segments are labeled "lively" noise. The parameters for short and long windows used in the STA/LTA ratio are 3 and 45 s, respectively. The uniform energy score, $u^{(j)}$, associated with a noise segment $n^{(j)}$ of length M samples is defined as:

$$u^{(j)} = \frac{1}{M}\sum_{k=1}^{M}\left| s^{(j)}[k] - \frac{k}{M}\right|, \quad \text{where} \quad s^{(j)}[k] = \sum_{i=1}^{k}\left(n^{(j)}[i]\right)^2, \tag{12}$$

and $n^{(j)}$ is normalized such that $||n^{(j)}||_2^2 = 1$. This value represents the difference between the cumulative signal energy over the interval and the cumulative energy for a signal with uniform energy, and u takes values between 0 and 0.5, with larger values associated with larger deviations from uniform signal energy.

Events in Benchmark Data Set, by Station

The list of events in the benchmark data set and the corresponding earthquake waveform and noise data matrices, by station, are available upon request (see Figs. 12, 13).

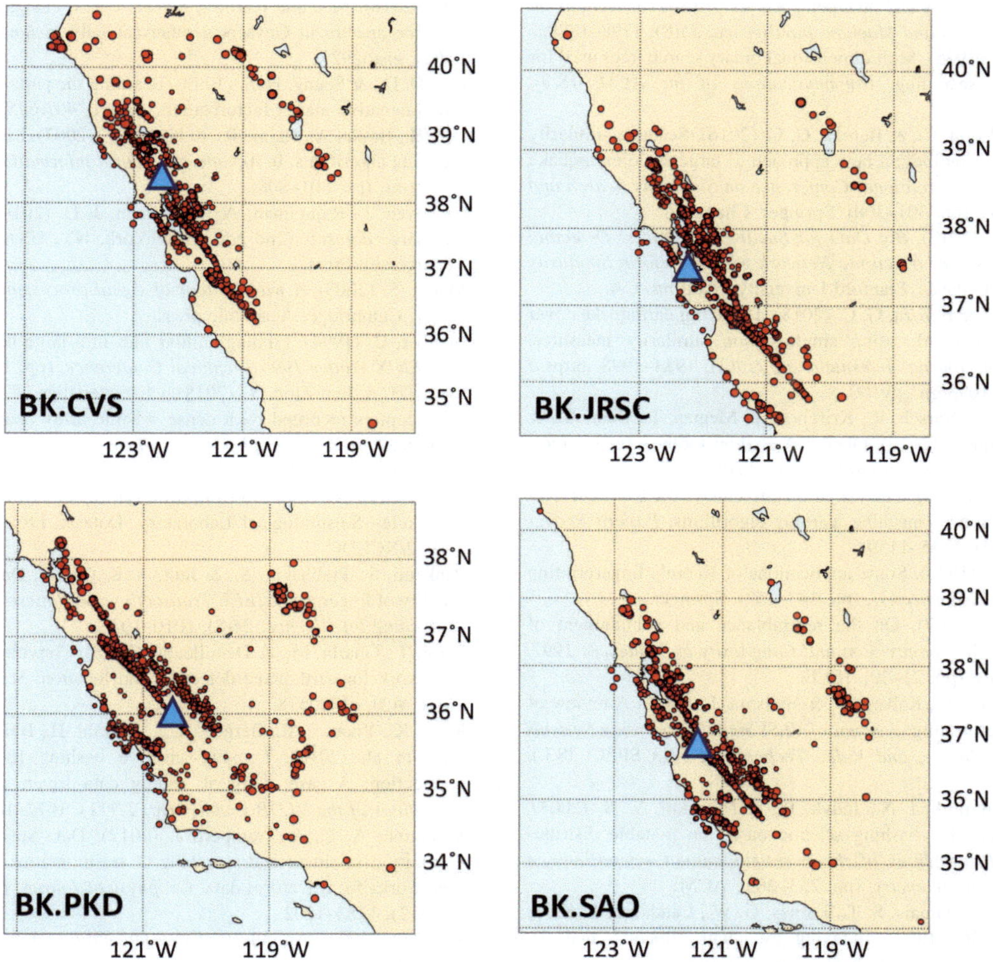

Figure 13

Map plots show the epicenter locations for the events in the earthquake waveform benchmark data set for stations BK.CVS, BK.JRSC, BK.PKD, and BK.SAO. The location of each station is marked with a blue triangle, and the event epicenters are shown in red. The set of earthquakes is selected independently for each station depending on whether a P-phase arrival was recorded in the NCSN phase-pick catalog for a given event

References

Alías, F., Socoró, J. C., & Sevillano, X. (2016). A review of physical and perceptual feature extraction techniques for speech, music and environmental sounds. *Applied Sciences, 6*(5), 143.

Allen, R. (1982). Automatic phase pickers: Their present use and future prospects. *Bulletin of the Seismological Society of America, 72*(6B), S225–S242.

Andoni, A., & Indyk, P. (2006). Near-optimal hashing algorithms for approximate nearest neighbor in high dimensions. In *Foundations of Computer Science, 2006. FOCS'06. 47th Annual IEEE Symposium on*, (pp. 459–468). IEEE.

Baluja, S., & Covell, M. (2008). Waveprint: Efficient wavelet-based audio fingerprinting. *Pattern Recognition, 41*(11), 3467–3480.

Bengio, Y., Courville, A., & Vincent, P. (2013). Representation learning: A review and new perspectives. *IEEE Transactions on Pattern Analysis and Machine Intelligence, 35*(8), 1798–1828.

Bentley, J. L. (1975). Multidimensional binary search trees used for associative searching. *Communications of the ACM, 18*(9), 509–517.

Bergen, K., Yoon, C., & Beroza, G. C. (2016). Scalable similarity search in seismology: a new approach to large-scale earthquake detection. In *International Conference on Similarity Search and Applications* (pp. 301–308). Springer, Cham.

Bergen, K. J. (2018). *Big Data for Small Earthquakes: Detecting Earthquakes over a Seismic Network with Waveform Similarity Search.* PhD thesis, Stanford University, Stanford, CA.

Bergen, K. J., & Beroza, G. C. (2018). Detecting earthquakes over a seismic network using single-station similarity measures. *Geophysical Journal International, 213*(3), 1984–1998. https://doi.org/10.1093/gji/ggy100.

Beyreuther, M., Barsch, R., Krischer, L., Megies, T., Behr, Y., & Wassermann, J. (2010). ObsPy: A Python toolbox for seismology. *Seismological Research Letters, 81*(3), 530–533.

Bradley, A. P. (1997). The use of the area under the ROC curve in the evaluation of machine learning algorithms. *Pattern Recognition, 30*(7), 1145–1159.

Broder, A. Z. (1993). Some applications of Rabin's fingerprinting method. In *Sequences II*, (pp. 143–152). Springer.

Broder, A.Z. (1997). On the resemblance and containment of documents. In *Compression and Complexity of Sequences 1997. Proceedings*, (pp. 21–29). IEEE.

Cano, P., Batlle, E., Kalker, T., & Haitsma, J. (2005). A review of audio fingerprinting. *Journal of VLSI Signal Processing Systems for Signal, Image, and Video Technology, 41*(3 SPEC. ISS.), 271–284.

Datar, M., Immorlica, N., Indyk, P., & Mirrokni, V. S. (2004). Locality-sensitive hashing scheme based on p-stable distributions. In *Proceedings of the twentieth annual symposium on computational geometry*, (pp. 253–262). ACM.

Deerwester, S., Dumais, S. T., Furnas, G. W., Landauer, T. K., & Harshman, R. (1990). Indexing by latent semantic analysis. *Journal of the American Society for Information Science, 41*(6), 391.

Díaz, J., Ruiz, M., Sánchez-Pastor, P. S., & Romero, P. (2017). Urban seismology: On the origin of earth vibrations within a city. *Scientific Reports, 7*(1), 15296.

Donoho, D. L., & Johnstone, J. M. (1994). Ideal spatial adaptation by wavelet shrinkage. *Biometrika, 81*(3), 425–455.

Gibbons, S. J., & Ringdal, F. (2006). The detection of low magnitude seismic events using array-based waveform correlation. *Geophysical Journal International, 165*(1), 149–166.

Haitsma, J., & Kalker, T. (2002). A highly robust audio fingerprinting system. *Proceedings of the 3rd international society for music information retrieval conference (ISMIR02)*, (pp. 107–115).

Hampel, F. R. (1974). The influence curve and its role in robust estimation. *Journal of the American Statistical Association, 69*(346), 383–393.

He, H., & Garcia, E. A. (2009). Learning from imbalanced data. *IEEE Transactions on Knowledge and Data Engineering, 21*(9), 1263–1284.

Hinton, G. E., & Salakhutdinov, R. R. (2006). Reducing the dimensionality of data with neural networks. *Science, 313*(5786), 504–507.

Holtzman, B. K., Paté, A., Paisley, J., Waldhauser, F., & Repetto, D. (2018). Machine learning reveals cyclic changes in seismic source spectra in Geysers geothermal field. *Science Advances, 4*(5), eaao2929.

Lee, D. D., & Seung, H. S. (1999). Learning the parts of objects by non-negative matrix factorization. *Nature, 401*(6755), 788.

Lee, H., Battle, A., Raina, R., & Ng, A. Y. (2007). Efficient sparse coding algorithms. In *Advances in neural information processing systems*, (pp. 801–808).

Leskovec, J., Rajaraman, A., & Ullman, J. D. (2014). *Mining of massive datasets* (2nd ed.). New York, NY, USA: Cambridge University Press.

Mallat, S. (2008). *A wavelet tour of signal processing: the sparse way.* Cambridge: Academic press.

Manber, U. (1994). Finding similar files in a large file system. In *USENIX Winter 1994 Technical Conference*, (pp. 1–10).

Meng, H. & Ben-Zion, Y. (2018). Characteristics of airplanes and helicopters recorded by a dense seismic array near Anza California. *Journal of Geophysical Research: Solid Earth.* https://doi.org/10.1029/2017JB015240.

NCEDC. (2014). *Northern California Earthquake Data Center.* UC Berkeley Seismological Laboratory. Dataset. https://doi.org/10.7932/NCEDC.

Pankanti, S., Prabhakar, S., & Jain, A. K. (2002). On the individuality of fingerprints. *IEEE Transactions on Pattern Analysis and Machine Intelligence, 24*(8), 1010–1025.

Perol, T., Gharbi, M., & Denolle, M. (2018). Convolutional neural network for earthquake detection and location. *Science Advances, 4*(2), e1700578.

Rong, K., Yoon, C. E., Bergen, K. J., Elezabi, H., Bailis, P., Levis, P., et al. (2018). Locality-sensitive hashing for earthquake detection: A case study of scaling data-driven science. *Proceedings of the VLDB Endowment, 11*(11), 1674–1687.

Valentine, A. P., & Trampert, J. (2012). Data space reduction, quality assessment and searching of seismograms: Autoencoder networks for waveform data. *Geophysical Journal International, 189*(2), 1183–1202.

Wang, A. (2003). An industrial strength audio search algorithm. In *ISMIR*, Vol. 2003, (pp. 7–13). Washington, DC.

Wang, A. (2006). The Shazam music recognition service. *Communications of the ACM, 49*(8), 44–48.

Wang, J., Kumar, S., & Chang, S. -F. (2010). Semi-supervised hashing for scalable image retrieval. In *Computer vision and pattern recognition (CVPR), 2010 IEEE conference on*, (pp. 3424–3431). IEEE.

Wang, J., Shen, H.T., Song, J., & Ji, J. (2014). Hashing for similarity search: A survey. arXiv preprint arXiv:1408.2927.
Wang, J., Zhang, T., Song, J., Sebe, N., & Shen, H. T. (2018). A survey on learning to hash. *IEEE Transactions on Pattern Analysis and Machine Intelligence*, PP(99):1–1.
Weiss, Y., Torralba, A., & Fergus, R. (2009). Spectral hashing. In *Advances in neural information processing systems*, (pp. 1753–1760).
Willett, P., Barnard, J. M., & Downs, G. M. (1998). Chemical similarity searching. *Journal of Chemical Information and Computer Sciences*, *38*(6), 983–996.
Yoon, C. E., O'Reilly, O., Bergen, K. J., & Beroza, G. C. (2015). Earthquake detection through computationally efficient similarity search. *Science Advances*, *1*(11), e1501057.

(Received April 13, 2018, revised August 1, 2018, accepted September 10, 2018, Published online October 8, 2018)

Pure Appl. Geophys. 176 (2019), 1061–1081

https://doi.org/10.1007/s00024-018-1981-z

Pure and Applied Geophysics

Applying Depth Distribution of Seismicity to Determine Thermo-Mechanical Properties of the Seismogenic Crust in Southern California: Comparing Lithotectonic Blocks

Egill Hauksson[1] and Men-Andrin Meier[1]

Abstract—We analyze waveform-relocated seismicity (1981–2016) and other geophysical and geological datasets from 16 lithotectonic crustal blocks in southern California. We explore how earthquake depth histograms (EDH) are related to crustal strength, lithology, and temperature of the crust. First, we calculate relative EDHs to quantify the depth distribution of seismicity for each lithotectonic block. Second, we calculate depth profiles of maximum differential stress ("yield strength envelopes", YSEs) using Byerlee's law and a non-linear dislocation creep law. We use observed average heat flow values, strain rates, and states of stress to parameterize YSEs for five different crustal candidate lithologies in each lithotectonic block. We assume that seismicity ceases where the mechanical rock strength falls below a critical threshold level, and identify the YSE that best predicts the depth extent of seismicity in each block. The lithologies of the best matching YSEs are found to agree well with expectations from past tectonics: they are mostly quartz-dominated except for the feldspar-rich diorite lithologies in the Great Valley, the southernmost western Sierra Nevada, Inner Continental Borderland, and Rifted crust in the Salton Trough. Similarly, the inferred thermo-mechanical properties, including differential stress, lithology, and geotherms reflect the previously mapped tectonic variability between the 16 lithotectonic blocks. On average, the differential yield stress is smaller and peaks at a shallower depth in hotter and more quartz rich crust but is larger and peaks at greater depths for colder and predominantly diorite crust. The good agreement between the modeled YSEs, the EDHs and tectonic considerations suggests that EDHs indeed reflect long-term geophysical properties of the crust and can be used to infer thermo-mechanical properties at depth. In contrast, shallow seismicity may be more likely to reflect short-term strain transients from fluid flow or recent anthropogenic disturbances.

Key words: Earthquake depth histograms, yield strength envelopes, lithology, crustal temperature, crustal thermo-mechanics.

[1] Seismological Laboratory, Division of Geological and Planetary Sciences, California Institute of Technology, Pasadena, CA 91125, USA. E-mail: hauksson@caltech.edu; mmeier@caltech.edu

1. Introduction

We use available seismicity and other geophysical datasets to develop a new understanding of the thermo-mechanics, including differential yield stress and the geotherm, of the seismogenic part of the southern California crust. We focus on the thermo-mechanics of blocks rather than individual faults to put our results into the context of the lithotectonics of the southern California crust (Fig. 1). The current lithotectonic fabric of the crust was formed during large-scale crustal extension dating back as far as $\sim$ 30 or 40 m.y. (e.g., Crouch and Suppe 1993; Chapman 2017). During this rifting, many blocks rotated and moved 10 s of kilometers relative to stable eastern North America plate, including the western Transverse Ranges that rotated clockwise more than 90° (Luyendyk et al. 1980).

We use a representation of this tectonic fabric to group the seismicity into the depth distributions within each lithotectonic belt or block as modified by Oskin et al. (2016) based on the work of previous investigators such as (Crouch and Suppe 1993) (Fig. 1). This spatially heterogeneous seismicity is mostly associated with long-term loading of major fault systems that constitute the Pacific-North America plate boundary, as well as significant release of tectonic strain along the boundary, which occurs in major mainshock–aftershock sequences (Hauksson et al. 2012). Secondary processes such as gravitational collapse of the high Sierra Nevada, mid-crustal delamination, or anthropogenic activity related to geothermal energy also cause a small fraction of the seismicity.

The depth distribution of southern California seismicity has been the subject of several previous

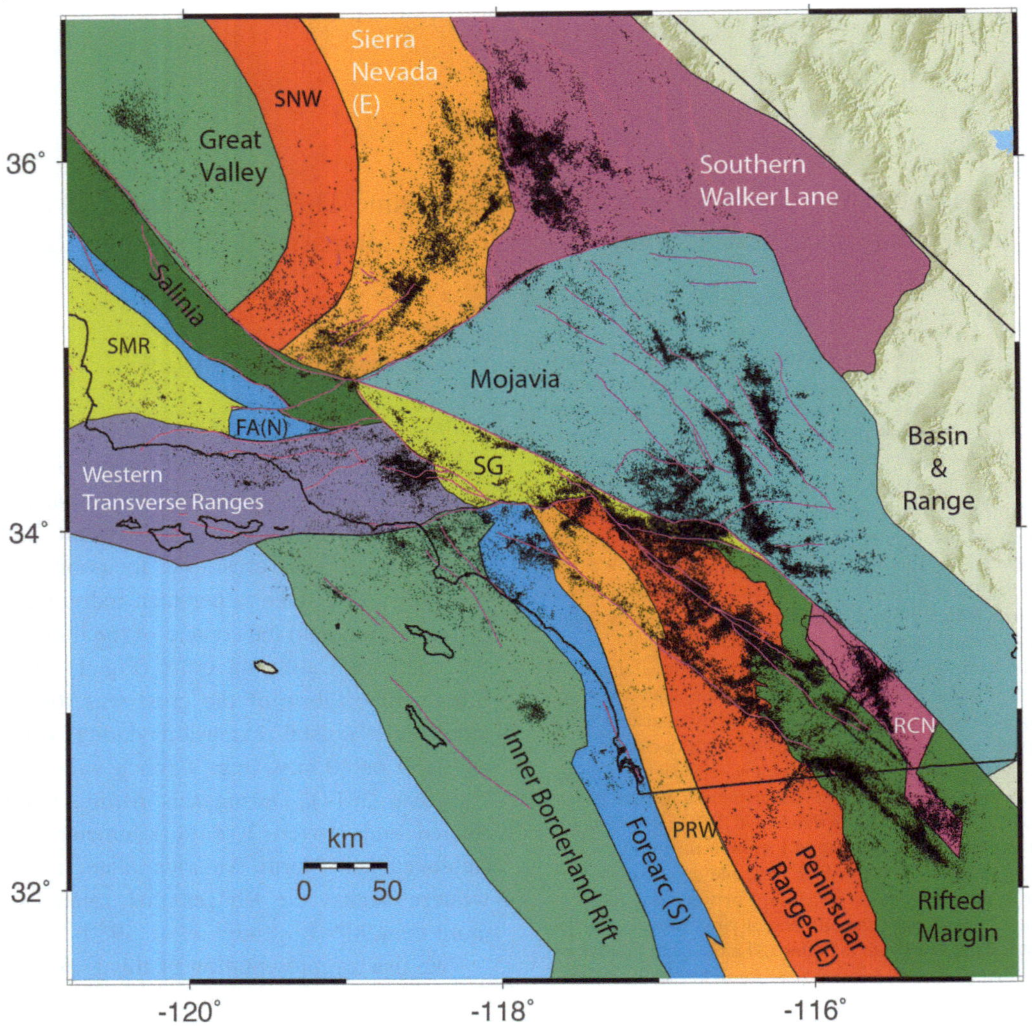

Figure 1
Map showing lithotectonic belts for southern California. Modified from Crouch and Suppe (1993) by Oskin et al. (2016). *PRW* Peninsular Ranges west, *SNW* Sierra Nevada west, *SMR* Santa Maria Rift, *FA(N)* Forearc north, *RCN* Rifted Crust north, *SG* San Gabriel. Waveform relocated hypocenter from Hauksson et al. (2012) and major faults from Jennings and Bryant (2010) are also shown. The block called "Rifted Crust south", which is located outside southern California, has insufficient seismicity and thus is not included in this analysis. The basement of the Inner Continental Borderland rift consists of high-grade blueschist metamorphic rocks, which differ significantly from continental granite (Crouch and Suppe 1993). Both the Sierra Nevada and Peninsular batholiths have similar tectonic origin with quartz diorite, gabbro, and tonalite to the west (Chapman 2017). On the east side plutons with tonalite, granodiorite, and granite dominate. However, the arrangement of the lithotectonic belts in the southern Sierra Nevada no longer exists and batholith affinity assemblages dominate the region (Chapman 2017)

studies (Doser and Kanamori 1986; Magistrale and Zhou 1996; Williams 1996; Magistrale 2002; Bonner et al. 2003). They argued that varying combinations of lithology and temperature controlled the depth distribution of the seismicity. Nazareth and Hauksson (2004) used relocated seismicity to show that the seismogenic thickness ranged from 10 km in the Salton Trough to 25 km beneath the southern Great Valley. They also inferred that the maximum moment release in moderate–large earthquakes extended to similar depths as the 99% moment release of the background seismicity, suggesting that the depth distribution of background seismicity is not magnitude dependent for ($M \leq 7.3$) southern California mainshocks. In contrast, Rolandone et al. (2004) have used seismicity depth observations to suggest that the transient strain rate increase caused by the 1992 Mw7.3 Landers, California earthquake may have temporally increased the yield strength at depth, and led to seismic activity at depths that were aseismic without such perturbations.

Thermo-mechanical or rheology properties relate the response of crustal rocks to external forces under different pressure, temperature, and fluid content conditions. In the upper most 10–15 km, the mechanical strength of the crust is limited by the brittle rock strength, and can be described by Byerlee's law (Byerlee 1967). At greater depths, where the weakest minerals start to deform plastically, non-linear viscous creep laws can be used to describe the maximum strength (Sibson 1982; Albaric et al. 2009). Yield strength envelopes (YSEs) combine the two formulations to describe how the maximum mechanical strength varies with depth, where the overall yield strength is limited by the weaker of the two deformation mechanisms.

Besides strain rate and temperature, yield strength estimates are also primarily affected by the bulk rock composition and the style of deformation (Sibson 1984; Afonso and Ranalli 2004). In southern California, all four of these properties, including their lateral variations, are relatively well constrained from decades of high-quality geophysical (e.g., Hauksson 2011) and geological (e.g., Crouch and Suppe 1993; Chapman 2017) observations. This allows us to make well-informed predictions of how strength should change with depth in different places across southern California, and to compare them against observed seismicity.

Observed seismicity provides a direct observational constraint for lithospheric strength in that sufficient strength is a necessary, albeit not a sufficient, precondition for the occurrence of earthquakes. It provides information down to what depth level sufficient strength is present to store elastic strain energy. Previously, Déverchère et al. (2001) compared YSEs to EDHs from the Baikal rift system and estimated the thermo-mechanics of the seismogenic crust of predominantly mafic composition. Similarly, Albaric et al. (2009) analyzed YSEs and EDHs from the east African rift system. Both studies suggested that by selecting the appropriate lithology and matching the shape of the EDHs and the YSEs they were able to determine local absolute differential stress depth profiles. There is, however, no clear physical motivation for why increased strength would result in higher seismicity, and hence we would not necessarily expect the YSEs and the EDHs to have the same shape. We, therefore, only make the much milder assumption that seismicity should be limited to depths where the YSE predicts substantial mechanical strength.

In each lithotectonic block, we make yield strength profile predictions for different candidate lithologies. We measure at what depths the strength decreases towards zero, and compare these estimates to the observed lower depth limit of seismicity. We explore to what extent the observed lateral changes in earthquake depths can be explained with changes in yield strength profiles and, in turn, to what extent seismicity observations can be used to constrain mechanical strength distributions at depth.

2. *Lithotectonic Provinces*

We use a lithotectonic model of southern California that divides the region into 16 lithotectonic provinces to group the seismicity (Fig. 1); (Oskin et al. 2016). This model identifies blocks by their lithotectonic assemblage of rocks, which have similar tectonic history. It draws heavily from the regional tectonic synthesis by Crouch and Suppe (1993) to delineate schist terranes, accretionary prisms, forearcs, rigid blocks, rifts, and batholiths. In particular, the forearc blocks were created by past tectonic processes such as arc-derived graywacke sedimentation on top, and scraped-off accretion from below. The western batholith belts (collision accretion), forearc belts (subsidence), and Santa Maria and Inner Borderland (Neogene rifting) have their own tectonic history and are characterized by their distinct lithology, ranging from quartz rich to feldspar rich, and geophysical properties such as heat flow (Crouch and Suppe 1993).

Some of the major late-Quaternary faults such as the SAF and the Garlock Fault form the major lithotectonic boundaries but a few of the model block boundaries are preliminary such as the eastern boundary of Mojavia, which is a region of complex crustal deformation. It consists of attenuated batholith-derived crust underlain by Rand–Orocopia schist (Oskin et al. 2016). The northeastern boundary of Mojavia is based on Miller et al. (2000) but the eastern boundary and interior structure are highly uncertain. Its southeastern boundary is drawn to encompass most of schist outcrops described by Jacobson et al. (2011). Active from 12 Myr or later, the Southern Walker Lane in eastern California is a region of older lithospheric removal, and mantle upwelling characterized by several young volcanic rock fields, like Coso, and transtensional rifting (Oskin et al. 2016; Putirka et al. 2012). It also includes the Stateline Fault and crustal extension of Cretaceous granites and older rocks, which extends to the California–Nevada boundary in the east (Mahan et al. 2009).

Past subduction or rifting processes have added complexity to some of the lithotectonic blocks. Pelona schist underplates the shortened and deformed San Gabriel terraine that is bounded by San Andreas Fault, Sierra Madre Fault, and the Banning Fault (Matti and Morton 1993; and Oskin et al. 2016). The Rifted Crust north that floors the Salton Sea graben extends along the rift axis to the south across the Salton Sea, to about 30 km south of the US Mexico boarder. The Rifted Margin that extends across the western flank of the Salton Trough forms the edge of significant post-8 Ma transtensional rifting and highly attenuated continental crust on the margins of the Gulf of California (Oskin et al. 2016). The Pelona–Orocopia–Rand schist may be derived from shallow subduction complex (Chapman 2017).

3. *Seismicity and Analytics*

As part of our data analysis, we have reduced the southern and northern extents of the lithotectonic blocks to match the available seismicity distribution. Below we describe the datasets and the modeling approach used to determine the thermo-mechanical model of the crust.

3.1. Datasets

We analyze the relocated earthquake catalog from Hauksson et al. (2012), covering the time period from 1981 to 2016. We only include about 65% of the events that qualified for waveform cross-correlation and have high-quality hypocenters. Implicitly, we have excluded most of the $M > 4$ events because their waveforms do not cross-correlate with smaller events.

We determine a depth histogram for the whole dataset and for the seismicity of each individual block. We use 2-km-depth intervals to match the average uncertainties in the focal depths (Fig. 2). Implicitly, we assume that the number of earthquakes in each depth range is sufficient to determine a representative EDH.

We included aftershock sequences caused by the occurrence of a major mainshock, such as the 1992 Mw7.3 Landers, 1994 Mw6.7 Northridge, and Mw7.1 Hector Mine earthquakes, although these may have increased the cutoff depth by $\sim$ 2 km, which is within the range of the depth uncertainty (Rolandone et al. 2004). Depending on the modeling approach, Beeler et al. (2018) showed that the coseismic deepening of slip is small or on the order of 10%, which would be 1–2 km in depth in this study. Thus, such changes in the depth of the 95%D would be within the depth uncertainties and are not easily detectable. If the frictional coefficient in the brittle regime is low, the coseismic deepening of the 95%D may be somewhat larger (Beeler et al. 2018). Ben-Zion and Lyakhovsky (2006) performed 3D

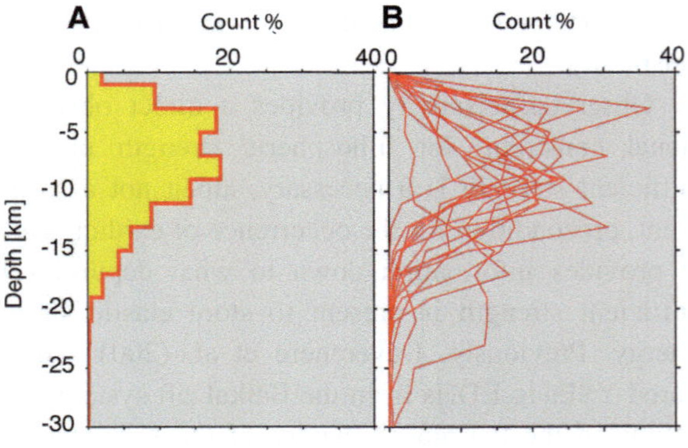

Figure 2
a Earthquake depth histogram (EDH) for the waveform relocated southern California (1981–2016) catalog of seismicity. **b** Average EDHs for each of the 16 lithotectonic blocks

numerical simulations in a model consisting of a seismogenic crust governed by viscoelastic damage rheology over a viscoelastic upper mantle. Their results show that mainshock ruptures can cause deepening of a small fraction of early aftershocks by $\sim$ 5 km and up to 20 km in the cases considered. However, such cases of deep events are not observed in the aftershocks sequences of the analyzed 1981–2016 SCSN relocated catalog.

Using the method of Zaliapin and Ben-Zion (2013), we declustered the complete catalog (Hauksson et al. 2012) to isolate the effects of the long-term loading from short-term transient effects during aftershock sequences. However, the declustered catalog had almost identical EDHs for each block as the complete waveform relocated catalog, and so we proceed by the EDHs from the complete catalog.

We also determine the value of the geophysical parameters for each block by averaging values measured at each epicenter for: (1) USGS heat flow (*C. Williams, written communication*, 2016), (2) geodetic strain rate (Holt et al. 2010), and (3) style of faulting (Yang and Hauksson 2013), as shown in Fig. 3. To analyze the seismicity and the geophysical parameters, we determine the average value of each of these parameters for each block.

3.2. Determining Yield Strength Envelopes

A yield strength or yield stress is the material property defined as the stress at which a material begins to deform plastically. We use the empirical deformation laws for the brittle and ductile behavior of the crust to determine the peak differential stress as a function of depth (Sibson 1974, 1983).

3.2.1 Brittle Regime

Using modified Anderson theory, Sibson (1974) described the brittle behavior as Coulomb friction law that applies to favorably oriented existing faults when assuming negligible cohesion (less than 1 or 2 MPa):

$$(\sigma_1 - \sigma_3)(z) = \alpha\rho(z)gz(1 - \lambda), \qquad (1)$$

where $(\sigma_1 - \sigma_3)$ is the difference between maximum and minimum principal stresses, ρ is average rock density at depths shallower than z, α depends on faulting type and frictional coefficient, $\lambda \cong 0.4$ is the ratio of pore fluid pressure to overburden pressure, and g is the gravitational constant. The choice of the value of λ being hydrostatic is appropriate for strike-slip tectonic settings (Sibson 1984). The value of α depends on the style of faulting and has values of 3.0, 1.2, and 0.75 for thrust, strike-slip, and normal faulting (Afonso and Ranalli 2004). The average style of faulting within each lithotectonic block is determined using data from Yang and Hauksson (2013).

By Byerlee's law, we assume that the brittle crust is populated by randomly oriented faults within the stress field of each block. Byerlee's law is only experimentally verified to mid-crustal pressures but at higher pressures, the differential stress is no longer dependent on lithostatic load (Ranalli 1997).

3.2.2 Ductile Regime

At depth, we assume that the deformation of the ductile part of the crust consists of power law creep (dislocation recovery), which is highly temperature and composition dependent (Ranalli 1997). We use a power-law dislocation creep to model the ductile regime (Sibson 1983; Afonso and Ranalli 2004):

$$\sigma_1 - \sigma_3 = \left(\frac{\dot{\varepsilon}}{A}\right)^{\frac{1}{n}} \exp\left(\frac{H}{nRT}\right), \qquad (2)$$

$\dot{\varepsilon}$ is the surface strain rate estimated from Holt et al. (2010), T is temperature in degrees Kelvin, R is universal gas constant. The creep parameters are n, A, and H, which is the creep activation enthalpy. The material parameters for the selected rock types are listed in Table 1. These parameters can easily vary by $\pm$ 25% (Ranalli 1997).

The viscous creep parameters have been assigned mineral composition names, such as wet quartzite, quartz, granite, and quartz diorite. As discussed by Harden (1998) and in many other textbooks, these rocks can be described as follows. Quartzite was originally pure quartz sandstone and through heat and pressure was converted into metamorphic rock, which is characterized by low Vp and low Poisson's ratio. Granite is a coarse grained plutonic igneous rock that consists of 20–60% of quartz and at least 35% feldspar by volume. Diorite is an intrusive

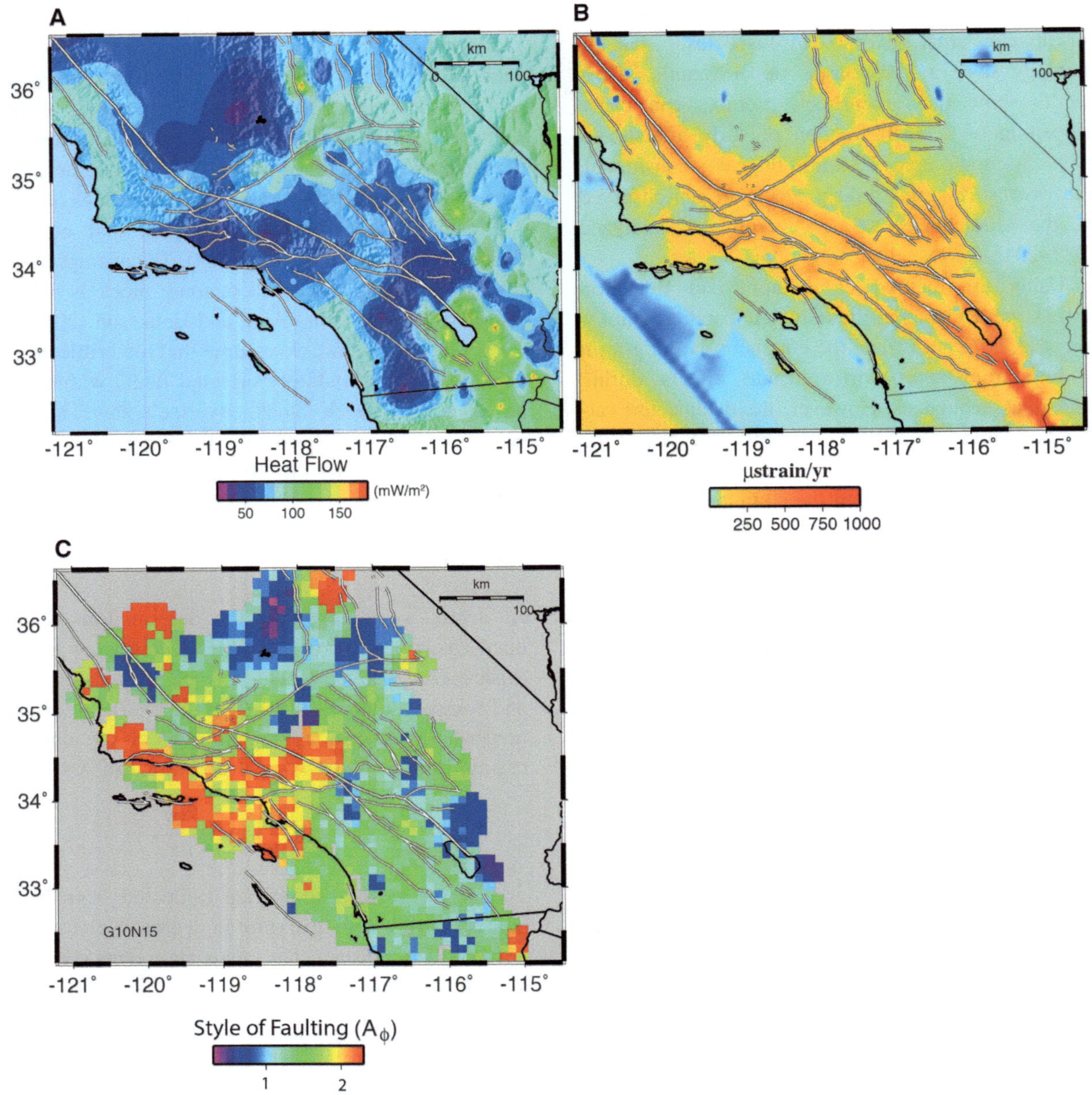

Figure 3
Southern California geophysical crustal parameters for **a** heat flow (C. Williams, written communication, 2016), **b** geodetic strain rate (Holt et al. 2010), and **c** style of faulting (Yang and Hauksson 2013)

batholithic igneous rock with a chemical composition that intermediates between gabbro and felsic granite. If diorite contains more than 5 wt% of quartz, it is generally referred to as quartz diorite but in this study, we are unable to resolve this difference. For completeness, we include parameters for a more felsic rock, mafic granulite, in our analysis but our results indicate that felsic rocks are not present in significant quantities to affect our results. These choices agree with previously inferred crustal lithology profiles for the main physiographic provinces in California (Mooney and Weaver 1989).

Table 1

Viscous creep parameters and average crustal density as compiled by Ranalli (1997), Turcotte and Schubert (2002), Afonso and Ranalli (2004), and Albaric et al. (2009)

[a]Mineral composition	A (MPa^{-n} s^{-1})	n	E (kJ mol^{-1})	Rho (kg m^{-3})
Wet quartzite	3.2×10^{-4}	2.3	154	2640
Quartz	1.0×10^{-3}	2.0	167	2640
Granite	1.8×10^{-9}	3.2	123	2650
Quartz diorite	1.3×10^{-3}	2.4	219	2800
Mafic granulite	1.4×10^{4}	4.2	445	2880

[a]See text for explanation of A, n, E, and Rho

These designations of mineral composition should not be taken literally because the weakest of the most abundant minerals controls the mechanical properties of a rock (Burov 2011). In granite or quartzite, this is quartz, and poly-phase aggregate rocks are weaker than their constitutive minerals. For instance, adding micas to andesitic rocks lowers viscosity more than adding water (Shinevar et al. 2015). They also showed that the lower crustal viscosity of andesitic rocks (64 wt% quartz) was most consistent with rapid tectonic deformation processes such as post-seismic relaxation.

3.3. Determining Temperature Depth Profiles

Yield strengths generally vary strongly with temperature. We use the 1-D steady state heat conduction equation to determine the geotherm, *T(z)* (Sass et al. 1997).

$$T(z) = T_0 + \left(\left(q - bA_0e^{-z/b}\right)z + b^2A_0\left(1 - e^{-z/b}\right)\right)/\lambda, \tag{3}$$

where z is depth in (km), T_0 is the surface temperature at 15 °C, A_0 is the surface value of heat production of 1.5 μWm^{-3}, b is assumed to be 15 km, λ is the thermal conductivity of 2.5 $Wm^{-1}K^{-1}$, q is the surface heat flow (mWm^{-2}) which varies across southern California. The average heat flow values for each region are from the USGS database provided by C. F. Williams (*Written communication*, 2016). The assumed depth dependence of radiogenic heat production becomes less certain with depth, and the crustal temperature may vary by several tens of degrees Celcius near the brittle ductile transition (Sass et al. 1997; Bonner et al. 2003).

3.4. Yield Stress Envelope (YSE)

The estimation of yield stress requires several assumptions because we know neither the true rheology nor the details of the applied tectonic strain field. To determine the yield stress profiles, we calculate the brittle and ductile yield stress as a function of depth using Eqs. (1), (2) and (3). The YSE at each depth is given by weaker of the two strength measures. This differential stress envelope depends on crustal density, stress state, heat flow, strain rate, and lithology, and describes the onset of brittle and ductile deformation in the material as a function of depth. In geodynamic applications, the differential stress envelopes are used to determine the strength of the whole crust by integrating the differential stress over the thickness of the crust (e.g., Afonso and Ranalli 2004).

To obtain relative strength envelopes of individual crustal blocks, we normalize the absolute YSEs by following Albaric et al. (2009) and integrate the differential stress in 2-km-depth increments, which corresponds to the average depth uncertainty of the seismicity, down to 30-km depth (Fig. 4). Because the area under the relative YSE curve is normalized, crust with a weak layer would exhibit a relative peak in the YSE at that depth, while a uniformly strong crust would exhibit a low amplitude YSE for the whole crustal thickness.

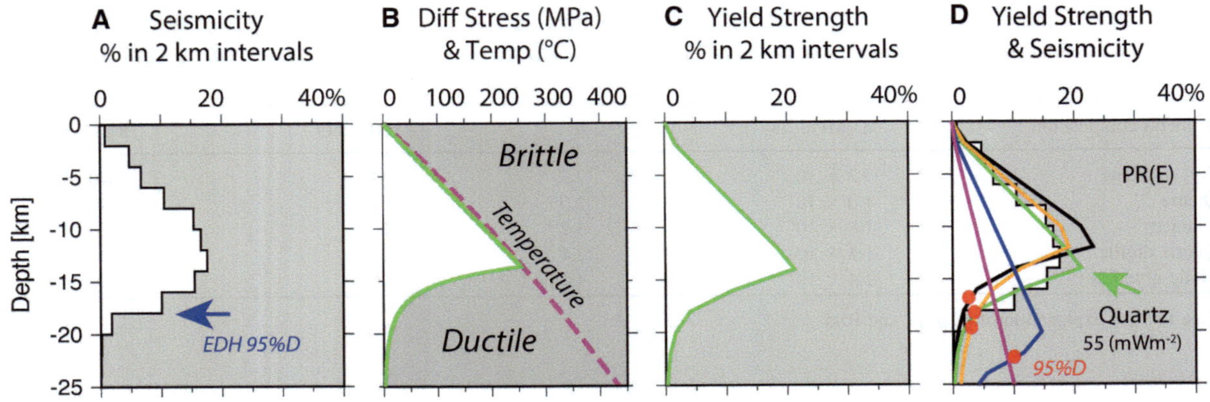

Figure 4

The four panels in this figure explain the data analysis method applied in this paper. **a** First we determine an earthquake depth histogram (EDH) and the 95% depth (95%D) shown as a blue arrow; **b** we calculate maximum differential stress and temperature depth profiles; **c** a sample integrated yield strength envelope (YSE) for quartz composition; and **d** comparison of the EDH and five YSEs for quartzite, granite, quartz, diorite, and mafic granulite. The 95% depth of each YSE is labeled with a red solid circle to facilitate comparison of the minimum distance of the EDH 95%D and the respective YSE 95%D. The 95%D of the quartz YSE has the best minimum distance fit to the EDH 95%D with heat flow of 55 mWm^{-2}

3.5. *Effects of Crustal Properties on Relative YSEs*

We selected a sample EDH for a region near the south end of the San Jacinto Fault to compare the effect of different crustal properties on the shape and depth offset of the corresponding relative YSEs (Fig. 5). We vary one of the four crustal parameters at a time within the expected range, while the other three are held fixed. Because the area under each yield curve is normalized, their peak value is inversely proportional to their total depth range. The heat flow and lithology have the strongest influence on the depth distribution of the YSEs, while the effects of the tectonic strain rate and style of faulting are less prominent. The three quartz-rich felsic compositions have similar YSEs (all other parameters being equal), which implies that we likely will not be able to distinguish between them with the resolution of our method.

4. *Results*

In this study we assume that the relative earthquake depth histogram (EDH) and the corresponding YSE for each lithotectonic block remain fairly stationary in space and time over centuries or longer. Thus, we can analyze them to infer the long-term geophysics of the crustal blocks located adjacent to the plate boundary. These assumptions are in part supported by the Pacific and North America constant rate of relative strike-slip movement over the last several million years (Harden 1998; Chapman 2017).

4.1. *Earthquake Depth Histograms (EDH)*

In very general terms, the EDHs increase with depth, reaching a peak at different depths, and then decaying rapidly (Fig. 6). The rapid decay with depth is related to increasing temperature and associated onset of ductility. In this study, we use the 95% cutoff depth of the EDHs (95%D) as markers, because these are located in the depth range, where the crustal strength, stressing rate, and availability of nucleations converge to a minimum. This is justified for most of southern California where earthquakes are caused by plate boundary loading stress. However, it may not apply in limited areas that accommodate recent shallow-induced seismicity, located mostly in high heat flow geothermal areas (Hauksson et al. 2012). By calculating block averages, we implicitly assume that the lower crustal deformation under each crustal block is accommodated by distributed flow of viscous lower crust as opposed to localized ductile shear with associated localized changes in ductility and thermal

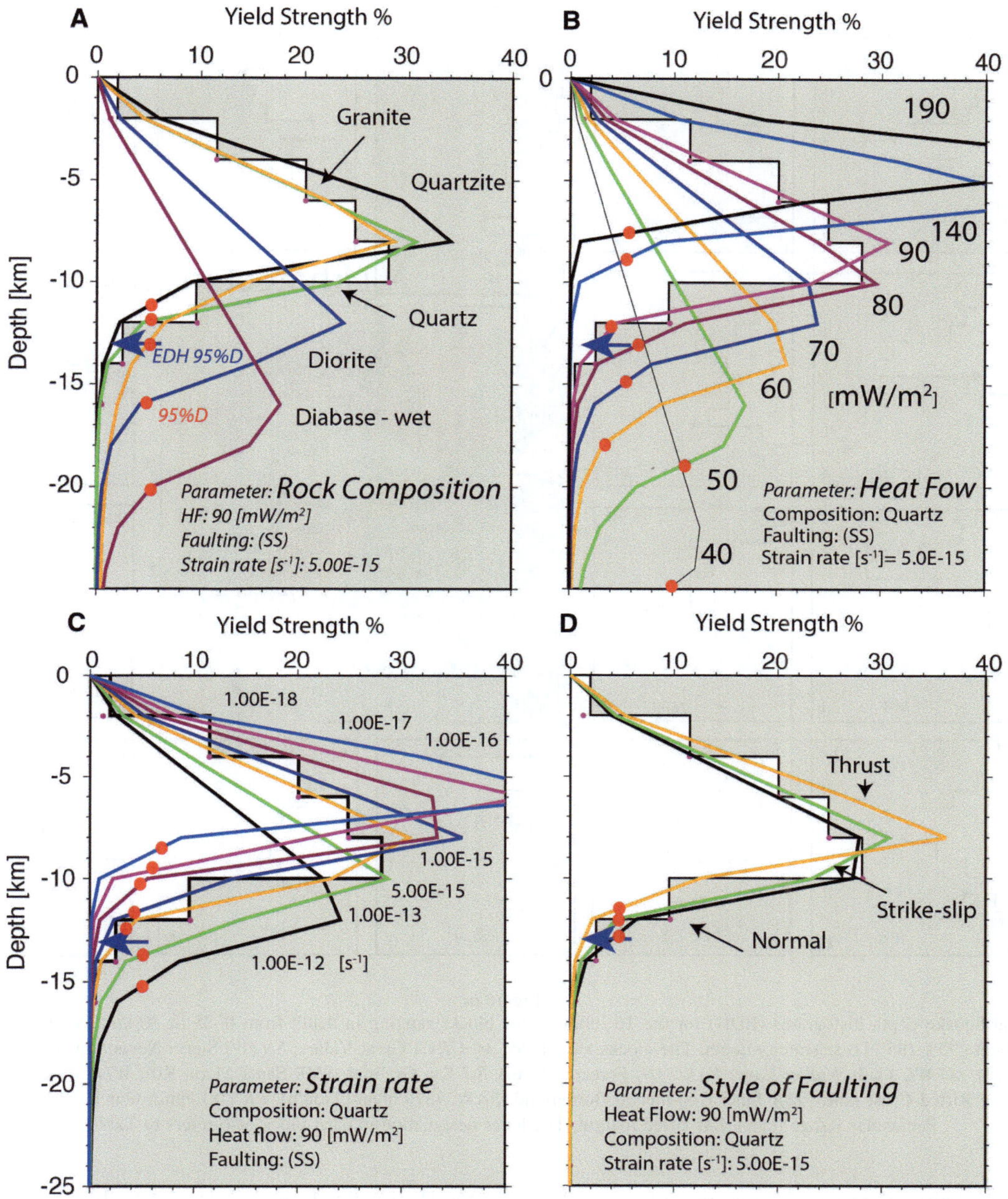

Figure 5

For a random earthquake depth histogram (EDH) with the 95% depth (95%D) shown as a blue arrow; we calculate the effects of the four geophysical parameters; **a** average lithology as labeled for different YSEs; SS—strike-slip faulting; **b** surface heat flow as labeled for different heat flow values; **c** surface GPS strain rate values and **d** styles of faulting as derived from the stress state. The 95% depth of each YSE is labeled with a red solid circle to facilitate comparison of the minimum distance of the EDH 95%D and the respective YSE 95%D

anomalies (Burgmann and Dresen 2008). In other words, we do not account for possible spatial variations in the ductile shear in the lower crust.

In four cases the EDHs appear to have double peaks (Fig. 6). These double peaks are likely caused by superimposing distinct seismicity clusters in these blocks. The SG block, for instance, combines the 1994 Northridge earthquake sequence with the deep seismicity beneath the San Gorgonio region, where crustal delamination may play a role. These double

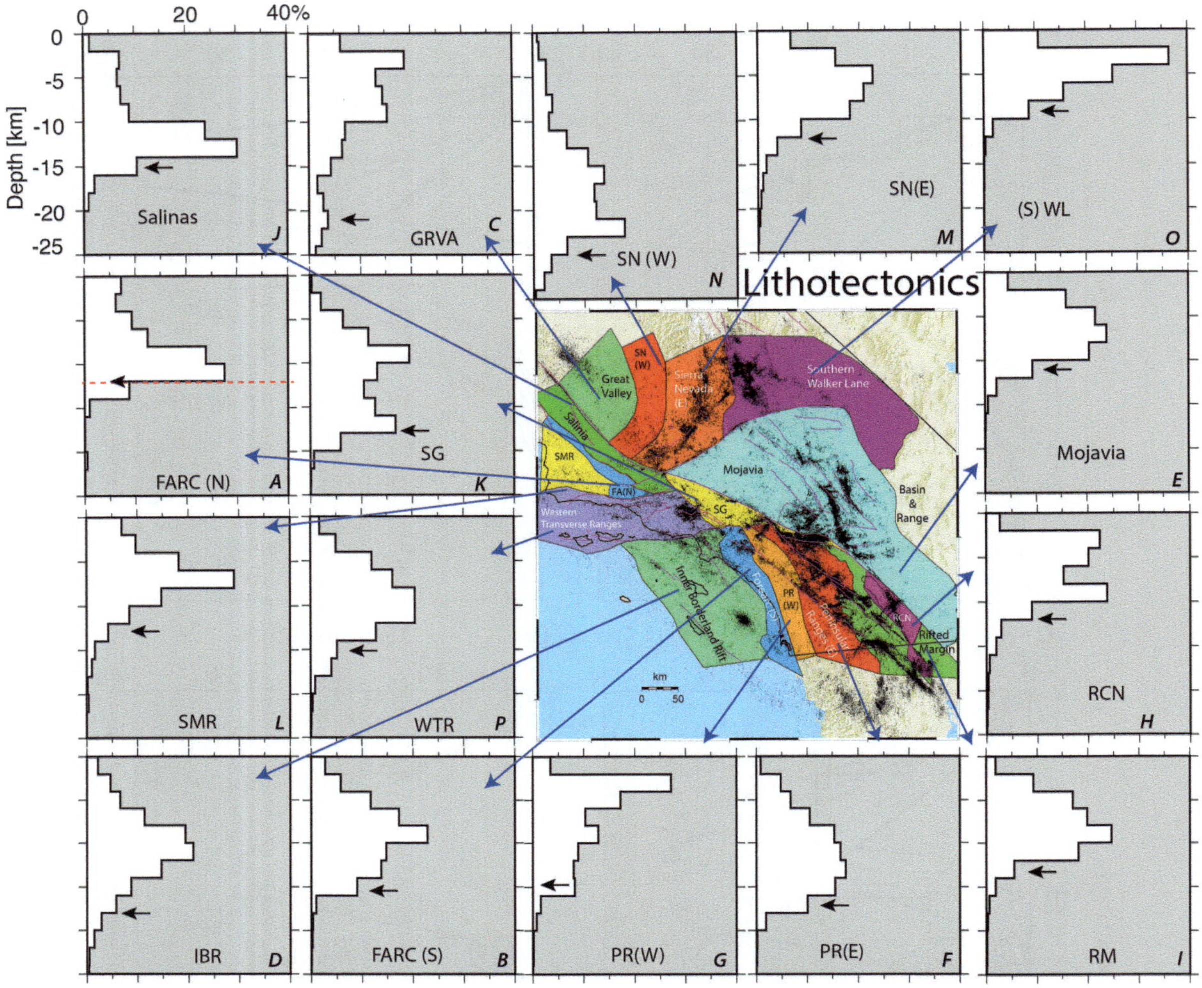

Figure 6
Relative earthquake depth histograms (EDH) for the 16 lithotectonic blocks ranging in depth from 0–25 or 30 km. The horizontal black arrows indicate 95% (95%D) seismicity depth. The blocks are labeled as: *GRVA* Great Valley, *SN (W)* Sierra Nevada (west), *SN (E)* Sierra Nevada (east), *(S) WL* south Walker Lane, *FARC (N)* Forearc (north), *SG* San Grabriel, *SMR* Santa Maria Rift, *WTR* Western Transverse Ranges, *RCN* Rifted Crust north, *IBR* Inner Continental Borderland, *FARC (S)* Forearc (south), *PR (W)* Peninsular Range (west), *PR (E)* Peninsular Range (east), *RM* rifted margin. The letter designation in each block also refers to Table 2

peaks are likely related to lateral variations within blocks but do not require depth-layered lithology or strength variations. In some cases, they may be related to different seismogenic processes, such as induced seismicity or complex plate boundary deformation such as crustal thickening. As an example, the deeper seismicity in the Rifted Crust north (RCN) was previously identified by Shearer (2002), as being more representative of the regional tectonic loading while most of the shallow seismicity is connected to geothermal processes. Most of the seismicity within the Peninsular Range west (PRW) block is located in the northern part of the block. The deeper seismicity is located along the Elsinore Fault while the shallower part is to the north forming the Fontana seismicity trend (Hauksson et al. 2012).

In all 16 blocks we observe a dearth of seismicity at very shallow depths. Unless this is caused by systematic vertical location errors for very shallow events, this may indicate a velocity-strengthening friction regime, or the presence of aseismic deformation processes, such as solution-transfer.

4.2. *Matching the Depth Extent of EDHs and YSEs (D95%)*

The relative YSEs of the five possible lithologies calculated using the strain rate and heat flow for each lithotectonic block are shown in Fig. 7, along with the EDHs. In general, there are shape similarities between the EDHs and the YSEs with increasing values at shallow depth and deceasing values at greater depths. This similarity has prompted Déverchère et al. (2001) and Albaric et al. (2009) to assume that EDHs are directly proportional to the YSE, i.e., that the number of earthquakes grows with yield strength. There is, however, no physical model that would predict such a relation.

Here, we make the much weaker assumption that seismicity should cease when the yield strength approaches zero. To this end, we determine the 95th percentile depths ("D95%") for both EDHs and YSEs. For the EDHs the D95% is the depth at which 95% of events locate above. For the YSEs, D95% marks the depth at which 95% of the YSE function mass is above that depth. From the five candidate

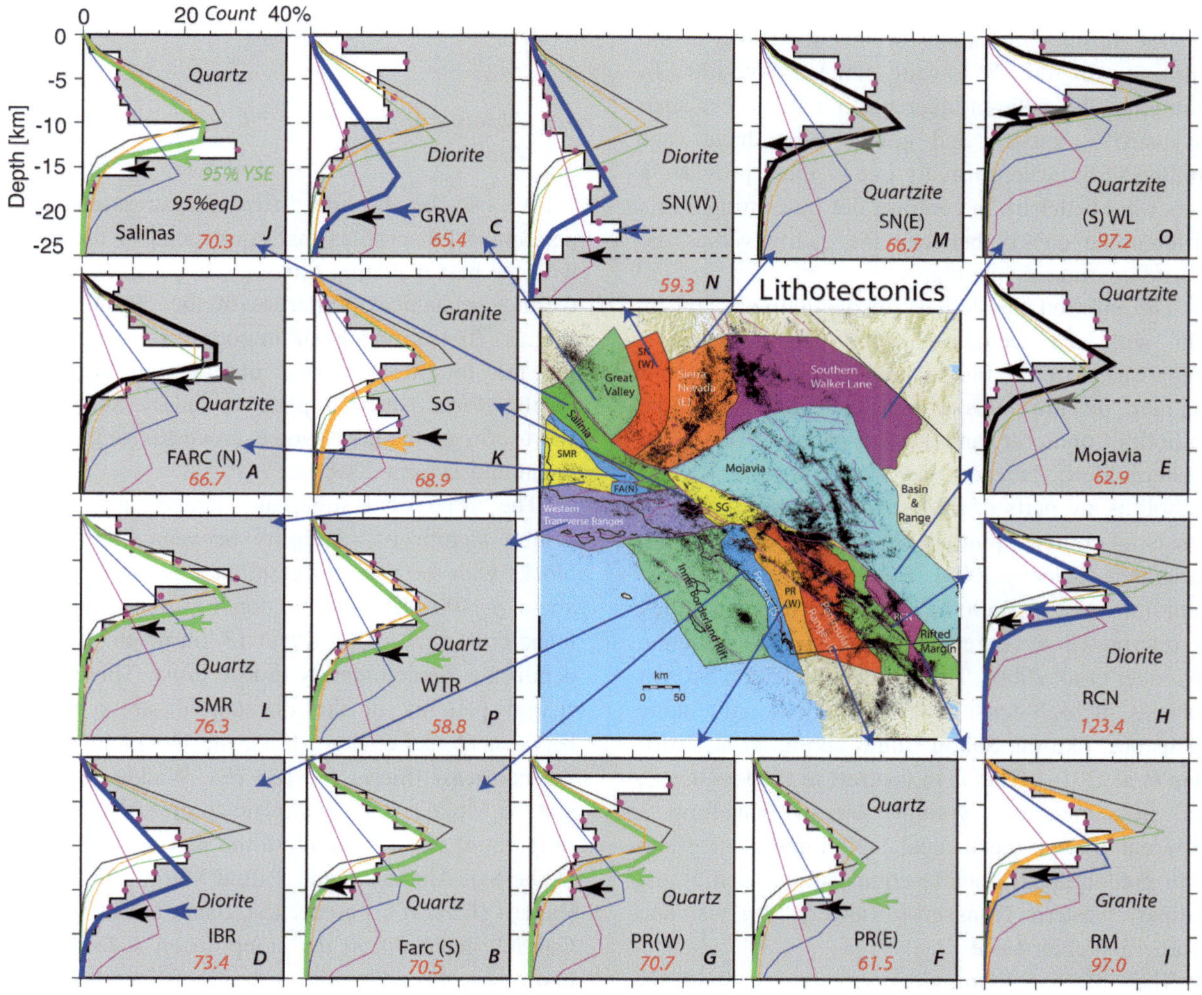

Figure 7

Relative yield strength envelopes (YSE) and earthquake depth histograms (EDH) for the 16 lithotectonic blocks in southern California. The 95%D for each EDH and YSE for each block are shown as arrows; with the 95%D for YSEs colored. The two cases with the 95Ds about ~ 3 km apart, are indicated with dashed lines. The YSEs are color coded with: Quartzite—black; Quartz—green; Granite—orange; Diorite—blue; and Mafic granulite—magenta. The average heat flow value within each block is shown as a red number in mWm^{-2}. See Table 2 and Fig. 6 for block names and labels. The letters *A*, *B*, *C* etc. enable cross reference to Table 2 lithologies we identify the best matching lithology as the one that produces a YSE with a 95%D depth most similar to the respective EDH 95%D.

The YSEs for the different lithologies in each block are shown in Fig. 7. In 12 out of the 16 blocks, the best matching lithologies are felsic (Granite, Quartzite or Quartz). The depth extent of seismicity in the Great Valley, the Western Sierra Nevada, the Inner Borderland Rift and the Rifted Crust North, on the other hand, is much more consistent with a dioritic lithology. There are no examples where envelopes of more dense rocks such as mafic granulite are required to fit the EDHs D95%.

One of the limitations of this modeling is the similarity of the YSEs for some lithologies. In particular, the different quartz-dominated rocks such as quartzite, quartz, and granite often have very similar YSEs, where usually one of these provides a somewhat better fit. In contrast, our best fitting YSEs differentiate clearly between the quartz versus the diorite-dominated lithologies.

The inferred lithologies are qualitatively consistent with the tectonic history and petrological composition of the blocks. Our study separates out the more dioritic (feldspar-rich) blocks, including the batholiths, ophiolite, and nascent gabbroic crust. The western Sierra Nevada and the Peninsular Ranges batholiths are parts of a Mesozoic magmatic arc and consist of complex plutons rich in mafic and tonalitic rocks (e.g., Saleeby 2003). The high-density gabbroic ophiolite crust beneath the southeastern Great Valley is interpreted as a remnant from past subduction (Fliedner et al. 2000). The crust in the Salton Trough consists of high-density gabbroic rocks with mafic intrusions, like the Salton Buttes (Barak et al. 2015; Han et al. 2016). This is in contrast to the less dense, and in some cases younger blocks that form a different thermo-mechanical and lithological group with predominantly quartz composition, such as the Salinas, western Transverse Ranges, Mojavia, and southern Walker Lane.

The inferred lithologies are also qualitatively consistent with the 3D seismic velocity models of Lee et al. (2014). At mid-crustal depths beneath the parts of the Great Valley, batholiths [Sierra Nevada (W)], the Inner continental Borderland, and Peninsular Ranges, the Lee et al. (2014) model has elevated *Vp*, *Vs*, and *Vp*/*Vs* values. This is in agreement with the elevated values found for dioritic rocks in empirical studies (Christensen 1996; Brocher 2005). Similarly, their model has *Vp*, *Vs*, and *Vp*/*Vs* values that are below average beneath the eastern Mojave, Walker Lane, and the western Transverse Ranges. These values agree with the quartz-rich compositions we infer, which often have average or slightly lower *Vp*/*Vs* ratios because their *Vp* is lower than average. Thus, although the southern California geology is complex, our inferred block lithologies are consistent with independently determined regional geological considerations, and with state-of-the-art 3D velocity models.

4.3. *Absolute Differential Yield Stress and Geotherm Profiles*

We use the absolute differential stress and temperature depth profiles that correspond to the YSEs with the best matching lithology, to parameterize the thermo-mechanical properties of the crustal blocks (Fig. 8). These models are important for understanding the long-term crustal deformation along the Pacific-North America plate boundary as well as short-term earthquake source processes (e.g., Burgmann and Dresen 2008).

The differential stress and temperature profiles exhibit significant variability between lithotectonic blocks with the peak differential stress varying from about $\sim$ 100 MPa at 8-km depth in hotter and more felsic crust to $\sim$ 500 MPa at 18-km depth at greater depths for predominantly diorite crust. Three major spatial patterns in the differential stress profiles between blocks emerge. First, the blocks located to the northeast [Sierra Nevada (E), Walker Lane (s), and the Mojavia] all have felsic composition and low differential stress. Second, to the west and southwest of the San Andreas Fault [Rifted Margin, Peninsular Ranges (E & W), Forearc (S), Santa Maria rift, San Gabriel, and Salinas] the composition is felsic with two cases of granite with average differential stress. Third, parts of the Great Valley, batholiths [Sierra Nevada (W)] and the Inner continental Borderland all exhibit diorite composition and high differential stress. These variations in spatial pattern of lithology and stress are consistent with the lithotectonic model

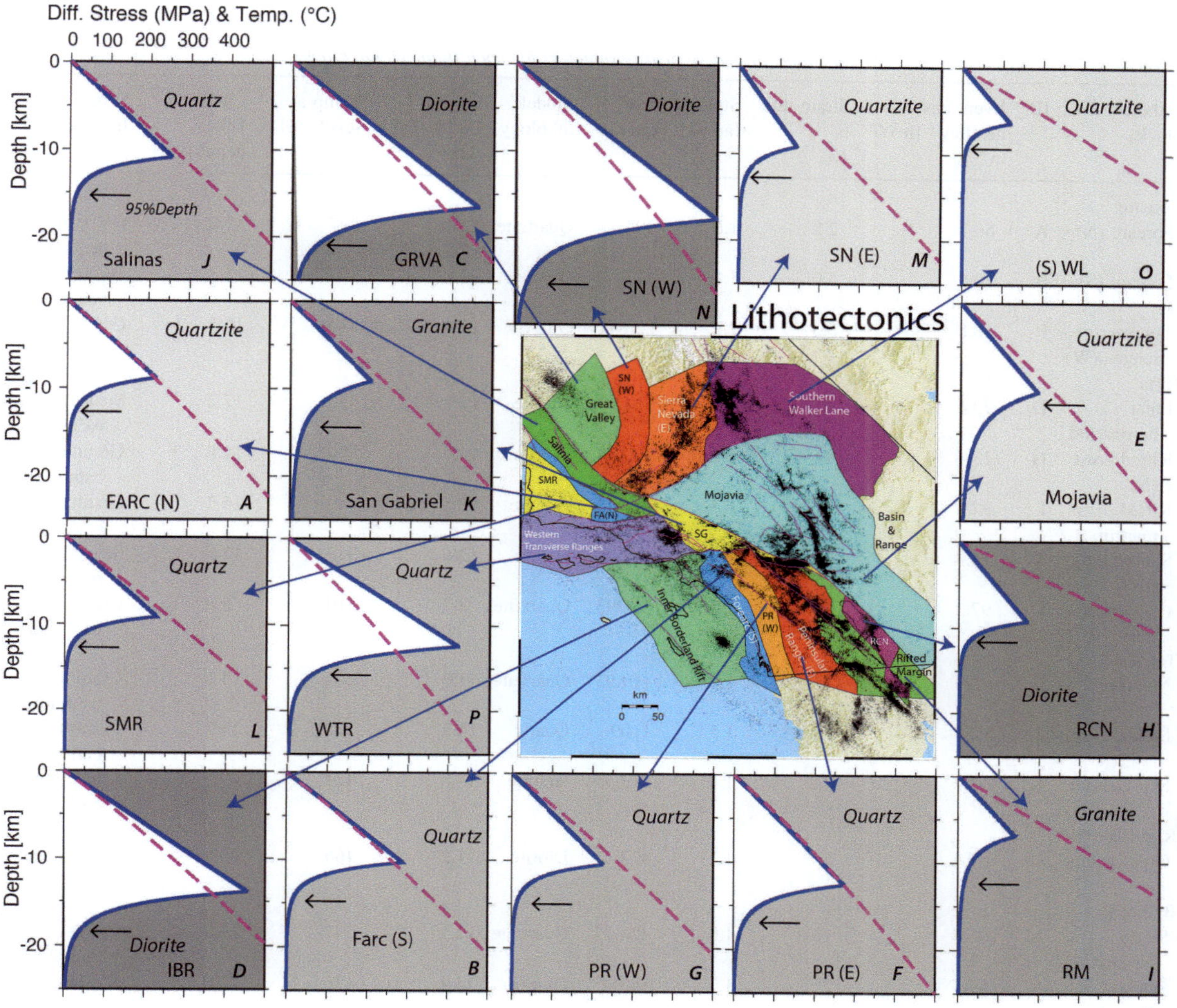

Figure 8
The differential stress and temperature plotted versus depth for the 16 lithotectonic blocks. Lithologies are indicated by background shades of gray and are also labeled. The two-layered models are also shown in Fig. 8. The dashed magenta lines represent the temperature gradient with depth. The dashed horizontal black lines represent the 95% depth of seismicity. See Fig. 6 for block names and labels

presented by Crouch and Suppe (1993), and probably not caused by uncertainties in the geothermal gradient or other geophysical parameters.

In some cases, current tectonics explain the observed differences in differential stress. In particular, there is a clear difference in differential stress between the Great Valley block and San Gabriel blocks as expected because they are separated by the San Andreas Fault, which has a large cumulative offset of 100 s of kilometers (Fig. 8). Previously, the change in stress or strength across the San Andreas Fault has been identified in stress inversion studies (e.g., Yang and Hauksson 2013). Using geodetic data Huang and Johnson (2012) showed that the Salinas block, located on the west side of the San Andreas Fault, has a thinner and stiffer elastic thickness as compared to the Great Valley block on the east side. Their result differs from ours because our version of the Great Valley block includes the deep ophiolite to the east. Similarly, Bird (2017) pointed out the stress guide effects of the Peninsular Ranges and Sierra Nevada batholiths as they accommodate high stress and low strain as opposed to low stress and high strain along the tectonically active plate boundary.

Table 2

Mean values of geophysical parameters for each lithotectonic block

Lithotectonic block	ID[c]	Event based heatflow[a] [mW/ m^{-2}]	Strain rate [s^{-1}] 10^{-15}	Style of faulting	# of quakes	Model lithology	Seism. D95% depth [km]	Temp at seis 95%D °C	Litho. D95% depth [km]	C & S 1993[b] lithology
Forearc										
Forearc (N)	A	66.7	2.8	1.6	193	Quartzite	12.5	297	12.0	GV franciscan sequence
Forearc (S)	B	70.5	5.0	1.8	4771	Quartz	14.5	362	13.1	GV franciscan sequence
Transverse ranges (W)	P	58.8	5.5	2.2	22,641	Quartz	15.4	310	15.7	GV franciscan sequence
Rifts										
Inner borderland	D	73.4	3.3	2.0	5642	Diorite	17.7	453	17.2	Schist/igneous rocks
Rifted crust (N)	H	123.4	19.5	1.1	30,133	Diorite	12.0	559	11.6	Granitic batholith
Rifted margin	I	97.0	9.0	1.3	58,184	Granite	12.5	449	15.2	Granitic batholith
Santa maria rift	L	76.3	2.1	1.6	462	Quartz	12.6	348	11.6	Schist/igneous rocks
(S) walker lane	O	97.2	2.9	1.2	88,293	Quartzite	8.7	319	8.8	Granitic batholith
Terrane										
Mojave	E	62.9	7.0	1.2	119,937	Quartzite	11.0	248	14.1	Granitic batholith
Salina	J	70.3	9.2	1.5	1109	Quartz	15.2	377	13.7	Granitic batholith
San Gabriel	K	68.9	7.2	1.4	27,655	Granite	17.8	424	19.1	Schist/granitic batholith
Rigid Block										
Great valley	C	65.4	6.7	1.7	872	Diorite	21.2	466	20.5	GV sequence/ ophiolite
Batholith										
Sierra Nevada (E)	M	66.7	1.9	1.0	42,377	Quartzite	12.5	297	12.7	Granitic batholith
Sierra Nevada (W)	N	59.3	2.7	1.7	580	Diorite	24.9	476	22.0	Metamorphic belt/ ophiolite
Peninsular ranges (E)	F	61.5	8.0	1.4	104,702	Quartz	16.7	351	15.8	Granitic batholith
Peninsular ranges (W)	G	70.7	4.0	1.4	4124	Quartz	14.8	370	13.2	Metamorphic schist/ volcanics

[a]Average heat flow derived from heat flow values at each epicenter; from USGS data

[b]C&S 1993 refers to Crouch and Suppe (1993)

[c]Reference block identifier (ID) letter also used in Figs. 6, 7, 8, and 9

Overall, our results agree with previous studies that showed the transition from brittle to ductile regime depends on lithology with 300° ± 50 °C for quartz-rich rocks, 450° ± 100 °C for feldspar-rich diorite rocks, and 650° ± 100 °C for olivine rich rocks (e.g. Sibson 1983; Magistrale and Zhou 1996; Ranalli 1997; Bonner et al. 2003).

4.4. Crustal Thermo-mechanics

To synthesize the overall thermo-mechanical properties, we analyze the geophysical crustal parameters at the D95% depth of seismicity: (1) the D95% depth of the YSE, (2) predicted temperature, and (3) peak differential stress (Fig. 9). All of these

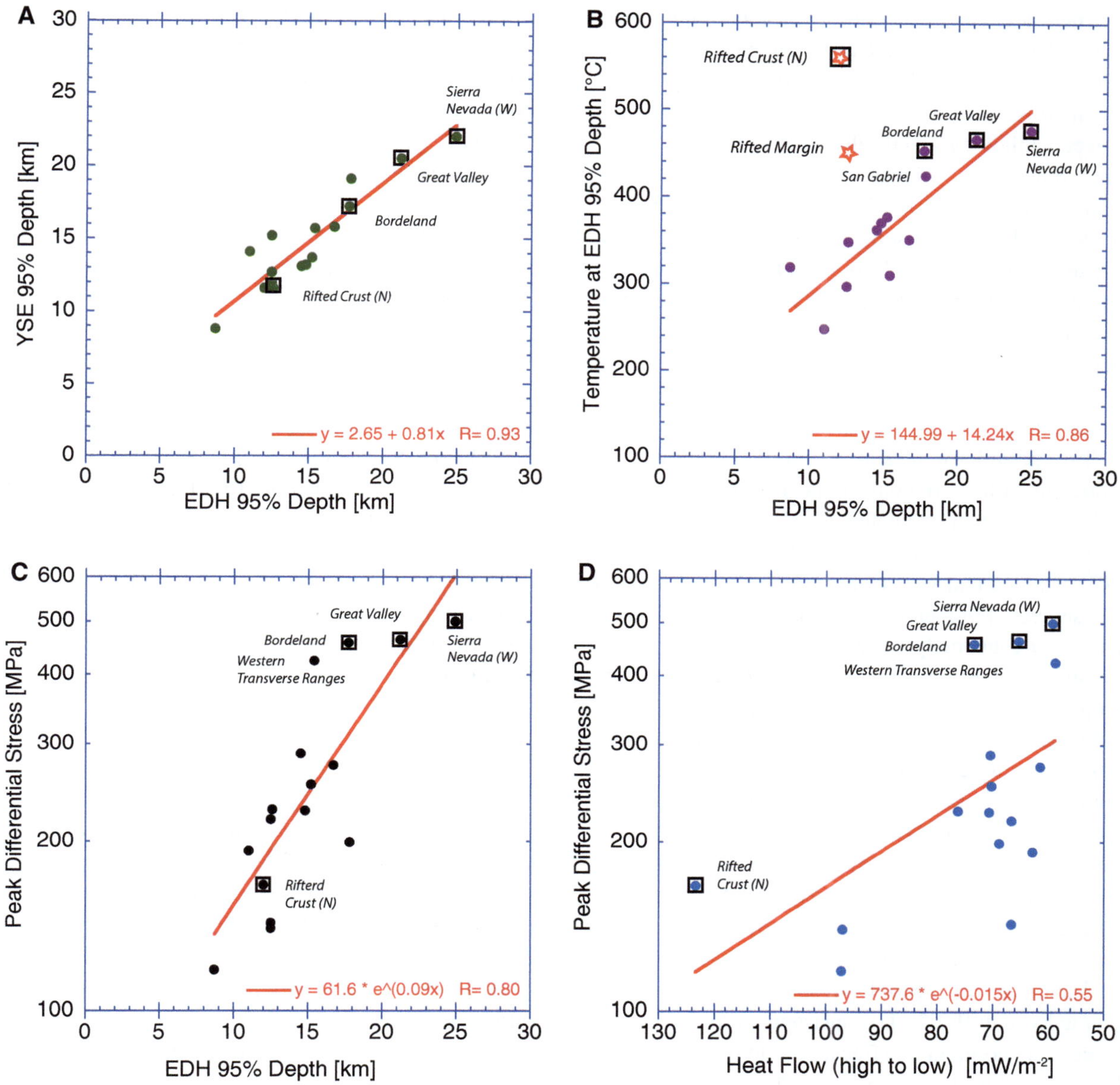

Figure 9
Comparison of geophysical crustal parameters. The data points for the four diorite blocks are labeled and marked as with black squares. A few other data points of special interest are also labeled. The red line shows a linear or exponential fit as calculated with the included formula; the correlation factor, R is also shown. **a** Preferred YSE 95% depth versus EDH 95% depth. **b** The estimated temperature at the EDH 95%D versus the corresponding EDH 95%D. Two anomalous data points that did not fit the regression curve are shown as red stars, Rifted Crust North, and Rifted Margin. Both cover parts of the Salton Trough suggesting that heat flow is overestimated where the seismicity is occurring. In **c** and **d**, logarithm of peak differential stress versus **c** EDH 95%D and **d** heat flow plotted from high to low values

parameters exhibit significant variations with values of the EDH 95%D for different blocks. We also compare the heat flow to the corresponding peak differential stress. We used solid circle symbols except for the blocks of diorite composition, which we overlay by a square symbol.

As expected, the lithology 95%D and seismicity 95%D are well correlated with a regression coefficient of 0.93 as shown in Fig. 9a. The seismicity 95%D increases in depth about 19% faster than the lithology 95%D, which is in part expected from the denser and colder diorite lithology.

There are some uncertainties in each temperature profile depending on local variations in the heat flow and the radiogenic heat production at depth, but nonetheless we observe significant differences between lithologies (Fig. 9b). The quartz composition dominated blocks range in temperature from ~ 100 to ~ 300 °C at depths ranging from 9 to 18 km. In comparison, the peak temperatures in diorite rich blocks are close to 500 °C except for the Rifted Crust north, which peaks at ~ 550 °C. The temperature profiles reach 500° ± 50 °C at depths ranging from 10 to 26 km, which is considered to be the maximum temperature where crustal earthquakes can occur.

The seismicity 95%D could be shallower or deeper in limited regions if the tectonic strain rate varies more than predicted by the geodetic model used in this study. For instance, Takeuchi and Fialko (2012) argued that viscous heating in shear zones that are several kilometers wide and may sustain a temperature anomaly of 200–400 °C. The data that are shown in Fig. 9b show such anomalously high temperatures only within the Rifted Margin and Rifted Crust (*N*) blocks. Alternatively, this anomalous temperature could be caused by the heat flow being overestimated where the seismicity occurs within these blocks. By smoothing the heat flow in the Salton Trough, we smear the very high heat flow anomalies in the geothermal areas into the surrounding regions, which are like to have much lower heat flow and a steeper temperature gradient.

The peak differential stress increases with the seismicity cutoff 95%D consistent with its dependence on combined effects of temperature and lithology with depth (Fig. 9c). Three of the four diorite blocks and the Transverse Ranges block accommodate the highest stresses. When ignoring the depth dependence of the differential stress, and only investigating the heat flow dependence, the same four blocks have consistently high stress and low heat flow. The anomalous diorite Rifted Crust North that exhibits the highest heat flow, probably has its heat flow overestimated by at least 30% as discussed above (Fig. 9d).

Apart from pore pressures, the parameters affecting the decay of the YSEs and EDHs at depth are known from field or laboratory observations. The differential stress profiles may be upper bound estimates if high fluid pressures exist (higher than the hydrostatic assumed here) because these may lower the effective strength (e.g., Sibson, 1974). However, any strength reduction caused by fluids may be more important at shallow depths where effective contact areas on faults are small, than at large depths, and where creep or slow plastic deformation increases the effective contact area (Hirth and Beeler, 2015).

In summary, the peak differential stress is small for shallow seismicity in high heat flow regions such as Rifted Crust north, and large for deep seismicity in low heat flow regions, such as Sierra Nevada west. The change in block lithology from predominantly quartz- to feldspar-rich diorite increase the depth range of the seismicity by 7 km or down to maximum block averaged depth of 26 km. This is consistent with the 95%D of the pairs of EDHs and YSEs representing the transition zone where there is rapid decrease in strength, stressing rate, and availability of nucleations.

5. Discussion

Our thermo-mechanical model consists of depth profiles of absolute differential stress, temperature, and lithology of each block. This model is a step towards improved understanding of the constitutive relations that control the long-term tectonic deformation of southern California. It shows peak differential stresses that the crustal material in each block can possibly withstand, ranging from low ~ 150 MPa to very high ~ 500 MPa. Locally the earthquakes are occurring under these stress conditions, which also correspond to the local yield strength of the rocks. The limitations of the model include various averaging of geophysical variables as well as scarce knowledge of the presence of fluids and deformation processes in the lower crust. Below, we discuss the model limitations and compare the model to independent models of absolute crustal stress in southern California.

5.1. *Effects of Heat Flow and, Strain Rate*

Regional variations in both heat flow and tectonic strain rate affect the details of the thermo-mechanical model but style of faulting has a much smaller effect. The heat flow that is a proxy for the temperature gradient with depth exhibits gradual spatial variations across the plate boundary region. In contrast, the tectonic strain rates are usually concentrated along major faults but facilitate some scattered seismicity across the region.

In general, our model that is based on minimizing the difference between the EDH and YSE 95%D values is consistent with the general results from previous studies that reported onset of plasticity in schist rocks of quartz and micas at $\sim$ 260 $\pm$ 40 °C, while feldspar-rich rocks become ductile at $\sim$ 450 $\pm$ 50 °C (e.g., Sibson 1982; Bonner et al. 2003). Our model also agrees with Magistrale and Zhou (1996) who argued that this variability in the composition of rocks caused 4–10 km changes in depth to the base of the seismicity. They showed that areas inferred to be underlain by schist rocks had shallower seismicity than areas possibly floored by basement rocks with higher feldspar content. In a later study, Magistrale (2002) concluded that heat flow and lithology affected similarly the depth distribution of earthquakes in southern California, which is also confirmed by our study.

The largest mismatch between the 95%D values is observed in the Mojave Desert where the seismicity is shallower than expected from the measured heat flow and strain rates (Fig. 7), for all the possible lithologies. Previously Williams (1996) showed that the heat flow varied from $\sim$ 50 mWm^{-2} in the south to 70 mWm^{-2} in the north along the 1992 Joshua Tree and Landers ruptures. We used an average heat flow of 62.9 mWm^{-2} for the whole Mojave block but a value of 70 mWm-2 would raise the YSEs by 2–3 km, and thus make the EDH and the YSE more consistent. Thus we infer that the heat flow in the Mojave may be underestimated. For instance, regional aquifers beneath the Mojave Desert may perturb the surficial heat flow measurements. The heat flow in the Sierra Nevada west block where the 95%D values also differ may be overestimated by a similar amount because of the difficulty in finding suitable measurement sites.

The geodetically measured tectonic strain field is explicitly accounted for in the determination of the thermo-mechanical profiles or the YSEs. In a complementary approach, Smith-Konter et al. (2011) determined the depth of the seismogenic zone by treating the locking depth estimates from both geodetic and seismological data independently. They found consistent geodetic locking depths of 6–22 km and 95% seismicity cutoff depths of 11–20 km, which agrees with the 95%D estimates in this study. In one case, they found inconsistent locking depths in the Salton Trough region with much shallower geodetically inferred locking depths on fault segments as compared to the local EDHs. In contrast, we find a good match between the EDH and YSE for the Salton Trough when assuming diorite lithology, which is consistent with geological observations (Mooney and Weaver 1989).

5.2. *Absolute Crustal Stress Field*

In general, the crustal stress field consists of three main components: lithostatic pressure load, topographic stress deviations related to lithological density variations, and tectonic stress variations (Bird 2017). Our study provides a model of the differential stress field as a function of depth for each block by averaging lateral variations within blocks. The variations with depth in each block and between blocks reflect the regional tectonic stress variations, but smooth out variations near major fault zones (Takeuchi and Fialko 2012). The average crustal stress field determined with our block approach is consistent with the results of Townend and Zoback (2000) who argued that in general crustal strength is maintained by critically stressed faults with hydrostatic pore pressure.

Our model is consistent with the absolute stress field in the southern California crust as determined by Bird (2017) who showed that low heat flow regions are strong and act as stress guides with a very low long-term strain rate. In contrast, areas of elevated or high heat flow are characterized by low stresses and high long-term strain rate, often expressed by faults with high slip rate. We find similar high differential stresses within the low heat flow and low strain rate Sierra Nevada and Peninsular Ranges batholiths and low stresses in areas of elevated heat flow and strain rate, such as the Salton Trough. The eastern Sierra Nevada seismicity exhibits similar behavior as the off-plate boundary seismicity found in northern California, to the east of the San Andreas Fault, away from the plate boundary (Miller and Furlong 1988). In particular, the off-plate boundary region beneath the Sierra Nevada that has a large crustal thickness of 37–43 km is floored by seismicity with a cutoff depth at ~ 250 °C (Hauksson 2011).

In a recent study, Luttrell and Smith-Konter (2017) determined absolute differential stresses in the southern California crust by balancing topographic stresses and stress orientations from focal mechanisms. They argued that at seismogenic depths, the stresses were larger than 62 MPa at ~ 5-km depth and that pore pressures were probably close to hydrostatic with average coefficient of friction. We find a much wider range in average differential stress levels, extending up to ~ 150 MPa at 5-km depth. In general, the peak differential stress varies significantly both in value and depth between blocks (Fig. 9). We do not find an obvious lower differential stress limit for driving seismicity, but our results are consistent with those of Luttrell and Smith-Konter (2017) who identified 20 MPa as the minimum differential stress required to drive the stressing rate causing seismicity.

5.3. *Block Models and Their Limitations*

Our approach of analyzing the thermo-mechanics of lithotectonic blocks provides regional parameters for geodynamic models. The thermo-mechanics of the blocks describes how they respond to long-term tectonic deformation as they participate in the strain loading and release over the numerous earthquake cycles. Our results are compatible with the results of Chéry (2008) who analyzed regional variations of the elastic thickness of the crust, which included several strike-slip faults. He showed that the regional elastic thickness played a major role in determining the strength of faults within blocks. In contrast most previous studies have focused on modeling the rheology of thin fault zones as 1D or 2D structures. These studies usually provide only information about the specific fault and assume the same rheology of the blocks on either side (e.g., Thatcher and England 1998). In this study we implicitly assume that the major faults are locked and determine the thermo-mechanics of the blocks that are in many cases bounded by major faults such as the San Andreas Fault.

The shapes of the EDHs can be affected by localized variations in heat flow or the stressing rate that may not be captured in our analysis. In regions with very high heat flow geothermal areas, the smoothed regional heat flow may be overestimated, and earthquakes are actually occurring at lower temperatures than suggested by our model. Similarly, small-scale variations in the stressing rate in the ductile zone can modulate seismicity rates within the upper brittle layer (Sammis et al. 2016) but our block averaging may minimize these effects. Because > 30 Ma have passed since these lithotectonic blocks were formed, their stress and temperature profiles are likely to have been overprinted several times. Thus, we do not expect a simple extrapolation of stress and temperature between blocks. These, as well as other model limitations, may affect the differential stress, temperature, and lithology profiles but calculating formal uncertainties is beyond the scope of this paper.

6. *Conclusions*

The long-term thermo-mechanical properties of the crust vary across the southern California plate boundary. To map these variations, we minimize the depth offset of EDHs and YSEs near their respective cutoff 95%D depths, and determine a best matching average lithology for each of the 16 lithotectonic blocks. In general the inferred lithology is mostly of quartz-rich composition within only four blocks out of 16 of predominantly feldspar-rich diorite composition, and consistent with the tectonic history of the blocks.

In agreement with previous studies, the absolute differential stress profiles derived from the forward modeling of the EDHs show a general pattern of high differential stresses within the low heat flow Sierra Nevada and Peninsular Ranges batholiths and low stresses in areas of elevated heat flow, such as the Salton Trough. We find highly variable average differential stress levels between blocks, extending up to $\sim$ 500 MPa, and the depth of the stress peak varies significantly between blocks. Within the diorite-dominated crustal blocks the lithology is colder and denser, and the differential peak stress is deeper and larger.

Acknowledgements

We thank J.-P. Avouac, N. Beeler, L. M. Jones, J. P. Ampuero, N. van der Elst, and J. Stock for discussions; J. Albaric for sharing her matlab code for calculating the YSEs, and Z. Ross for the declustered catalog. This research was supported by USGS/NEHRP grants G16AP00147 & G18AP00028; NSF award: EAR-1550704; and by the Southern California Earthquake Center, which is funded by NSF Cooperative Agreement EAR- 1033462 and USGS Cooperative Agreement G12AC20038. This paper is SCEC contribution # 8085. We used GMT from Wessel et al. (2013) to make the Figures. We have used waveforms and parametric data from the Caltech/USGS Southern California Seismic Network (SCSN); https://doi.org/10.7914/sn/ci; stored at the Southern California Earthquake Data Center. https://doi.org/10.7909/c3wd3xh1.

References

Afonso, J. C., & Ranalli, G. (2004). Crustal and mantle strengths in continental lithosphere: is the jelly sandwich obsolete? *Tectonophysics, 394,* 221–232.

Albaric, J., Déverchère, J., Petit, C., Perrot, J., & Le Gall, B. (2009). Crustal rheology and depth distribution of earthquakes: insights from the central and southern East African rift system. *Tectonophysics, 468,* 28–41. https://doi.org/10.1016/j.tecto.2008.05.021.

Barak, S., Klemperer, S. L., & Lawrence, J. F. (2015). San andreas fault dip, peninsular ranges mafic lower crust and partial melt in the Salton trough, Southern California, from ambient-noise tomography. *Geochemistry, Geophysics, Geosystems, 16,* 3946–3972. https://doi.org/10.1002/2015GC005970.

Beeler, N. M., Hirth, G., Tullis, T. E. & Webb, C. H. (2018). On the depth extent of coseismic rupture. *Bulletin of the Seismological Society of America, 108*(2), 761–780.

Ben-Zion, Y., & Lyakhovsky, V. (2006). Analysis of aftershocks in a lithospheric model with seismogenic zone governed by damage rheology. *Geophysical Journal International, 165*(1), 197–210. https://doi.org/10.1111/j.1365-246X.2006.02878.x.

Bird, P. (2017). Stress field models from Maxwell stress functions: southern California. *Geophysical Journal International, 210,* 951–963. https://doi.org/10.1093/gji/ggx207.

Bonner, J. L., Blackwell, D. D., & Herrin, E. T. (2003). Thermal constraints on earthquake depths in California. *Bulletin of the Seismological Society of America, 93,* 2333–2354. https://doi.org/10.1785/0120030041.

Brocher, T. M. (2005). Empirical relations between elastic wavespeeds and density in the earth's crust. *Bulletin of the Seismological Society of America, 95*(6), 2081–2092. https://doi.org/10.1785/0120050077.

Burgmann, R., & Dresen, G. (2008). Rheology of the lower crust and upper mantle: evidence from rock mechanics, geodesy, and field observations. *Annual Review of Earth and Planetary Sciences, 36*(1), 531–567. https://doi.org/10.1146/annurev.earth.36.031207.124326.

Burov, E. B. (2011). Rheology and strength of the lithosphere. *Marine and Petroleum Geology.* https://doi.org/10.1016/j.marpetgeo.2011.05.008.

Byerlee, J. D. (1967). Frictional characteristics of granite under high confining pressure. *Journal of Geophysical Research, 72,* 3639–3648.

Chapman, A. D. (2017). The Pelona–Orocopia–Rand and related schists of southern California: a review of the best-known archive of shallow subduction on the planet. *International Geology Review, 59*(5–6), 664–701. https://doi.org/10.1080/00206814.2016.1230836.

Chéry, J. (2008). Geodetic strain across the San Andreas fault reflects elastic plate thickness variations (rather than fault slip rate). *Earth and Planetary Science Letters, 269*(3–4), 352–365.

Christensen, N. I. (1996). Poisson's ratio and crustal seismology. *Journal of Geophysical Research, 101*(B2), 3139–3156.

Crouch, J. K., & Suppe, J. (1993). Late cenozoic tectonic evolution of the Los Angeles basin and inner California borderland: a model for core complex-like crustal extension. *GSA Bulletin, 105*(11), 1415–1434. https://doi.org/10.1130/0016-7606(1993)105<1415:LCTEOT>2.3.CO;2.

Déverchère, J., Petit, C., Gileva, N., Radziminovitch, N., Melnikova, V., & San'kov, V. (2001). Depth distribution of earthquakes in the Baikal rift system and its implications for the rheology of the lithosphere. *Geophysical Journal International, 146,* 714–730.

Doser, D. I., & Kanamori, H. (1986). Depth of seismicity in the imperial valley region (1997–1983) and its relationship to heat flow, crustal structure, and the October 15, 1979, earthquake. *Journal of Geophysical Research, 91,* 675–688. https://doi.org/10.1029/JB091iB01p00675.

Fliedner, M. M., Klemperer, S. L., & Christensen, N. I. (2000). Three-dimensional seismic model of the Sierra Nevada arc, California, and its implications for crustal and upper mantle composition. *Journal of Geophysical Research, 105*(B5), 10899–10921. https://doi.org/10.1029/2000JB900029.

Han, L., Hole, J. A., Stock, J. M., Fuis, G. S., Kell, A., Driscoll, N. W., et al. (2016). Continental rupture and the creation of new crust in the Salton trough rift, Southern California and northern Mexico: results from the salton seismic imaging project. *Journal of Geophysical Research, 121*(10), 7469–7489.

Harden, D. R. (1998). *California Geology*, Upper Saddle River, NJ: Prentice-Hall Inc., pp. 479. ISBN 0-02-350042-5.

Hauksson, E. (2011). Crustal geophysics and seismicity in southern California. *Geophysical Journal International, 186,* 82–98. https://doi.org/10.1111/j.1365-246X.2011.05042.x.

Hauksson, E., Yang, W., & Shearer, P. M. (2012). Waveform relocated earthquake catalog for Southern California (1981 to June 2011). *Bulletin of the Seismological Society of America, 102*(5), 2239–2244. https://doi.org/10.1785/0120120010.

Hirth, G., & Beeler, N. M. (2015). The role of fluid pressure on frictional behavior at the base of the seismogenic zone. *Geology, 43,* 223–226.

Holt, W. E., Klein, E. & Flesch, L. M. (2010). GPS strain rates, optimal fault slip rates, and predicted moment rates in Western U.S. Plate Boundary Zone, In *Proceedings of the Workshop on Incorporating Geodetic Surface Deformation Data into UCERF3,* Palm Spring, California. https://files.scec.org/s3fs-public/AgendaAbstracts.pdf. Accessed 21 Aug 2018.

Huang, W.-J., & Johnson, K. M. (2012). Strain accumulation across strike-slip faults: investigation of the influence of laterally varying lithospheric properties. *Journal of Geophysical Research, 117,* B09407. https://doi.org/10.1029/2012JB009424.

Jacobson, C. E., Grove, M., Pedrick, J. N., Barth, A. P., Marsaglia, K. M., Gehrels, G. E., et al. (2011). Late cretaceousearly cenozoic tectonic evolution of the southern California margin inferred from provenance of trench and forearc sediments. *Geological Society of America Bulletin, 123*(3–4), 485–506. https://doi.org/10.1130/B30238.1.

Jennings, C. W., & W. A. Bryant (2010). Fault activity map of California. Geologic Data Map No. 6, California Geological Survey, Sacramento, California, US. http://maps.conservation.ca.gov/cgs/fam/. Accessed 21 Aug 2018.

Lee, E.-J., Chen, P., Jordan, T. H., Maechling, P. B., Denolle, M. A. M., & Beroza, G. C. (2014). Full-3-D tomography for crustal structure in Southern California based on the scattering-integral and the adjoint-wavefield methods. *Journal of Geophysical Research: Solid Earth, 119,* 6421–6451. https://doi.org/10.1002/2014JB011346.

Luttrell, K., & Smith-Konter, B. (2017). Limits on crustal differential stress in southern California from topography and earthquake focal mechanisms. *Geophysical Journal International, 211*(1), 472–482. https://doi.org/10.1093/gji/ggx301.

Luyendyk, B. P., Kamerling, M. J., & Terres, R. R. (1980). Geometric model for neogene crustal rotations in southern California. *Geological Society of America Bulletin, 91,* 211–217.

Magistrale, H. (2002). Relative contributions of crustal temperature and composition to controlling the depth of earthquakes in Southern California. *Geophysical Research Letters, 29*(10), 1447. https://doi.org/10.1029/2001GL014375.

Magistrale, H., & Zhou, H. (1996). Lithologic control of the depth of earthquakes in Southern California. *Science, 273,* 639–642. https://doi.org/10.1126/science.273.5275.639.

Mahan, K. H., Guest, B., Wernicke, B., & Niemi, N. A. (2009). Low-temperature thermochronologic constraints on the kinematic history and spatial extent of the Eastern California shear zone. *Geosphere, 5*(6), 483–495. https://doi.org/10.1130/GES00226.1.

Matti J. C., & Morton D. M. (1993). Paleogeographic evolution of the San Andreas Fault in southern California: a reconstruction based on a new cross-fault correlation: In R. E. Powell, R. J. Weldon, II & J. C. Matti (Eds.), *The San Andreas Fault system: displacement, palinspastic reconstruction, and geologic evolution.* Geological Society of America, Memoir 178 (Chapter 2, special volume, pp. 107–160). Geological Society of America.

Miller, C. K., & Furlong, K. P. (1988). Thermal-mechanical controls on seismicity depth distributions in the San Andreas fault zone. *Geophysical Research Letters, 15,* 1429–1432. https://doi.org/10.1029/GL015i012p01429.

Miller, J. S., Glazner, A. F., Farmer, G. L., Suayah, I. B., & Keith, L. A. (2000). A Sr, Nd, and Pb isotopic study of mantle domains and crustal structure from miocene volcanic rocks in the Mojave Desert, California. *Geological Society of America Bulletin, 112,* 1264–1279. https://doi.org/10.1130/00167606(2000)112<1264: ASNAPI>2.0.CO;2.

Mooney, W., & Weaver, C. (1989). Regional crustal structure and tectonics of the Pacific coastal states; California, Oregon, and Washington. In L. Pakiser & W. Mooney (Eds.), *Geophysical framework of the continental United States GSA memoir 172* (pp. 129–162). Boulder: Geological Society of America.

Nazareth, J. J., & Hauksson, E. (2004). The seismogenic thickness of the Southern California crust. *Bulletin of the Seismological Society of America, 94,* 940–960. https://doi.org/10.1785/0120020129.

Oskin, M. E., Behr, W. M., Hearn, L., & Thatcher, W. (2016). Community Rheology model workshop: establishing a geologic framework. Southern California Earthquake Center (SCEC), workshop report, Los Angeles, CA. https://www.scec.org/proposal/report/16183. Accessed 21 Aug 2018.

Putirka, K., Jean, M., Cousens, B., Sharma, R., Torrez, G., & Carlson, C. (2012). Cenozoic volcanism in the sierra Nevada and walker lane, California, and a new model for lithosphere degradation. *Geosphere, 8*(2), 265–291. https://doi.org/10.1130/GES00728.1.

Ranalli, G. (1997). Rheology and deep tectonics. *Annals of Geophysics, 40*(3), 671–680.

Rolandone, F., Bürgmann, R., & Nadeau, R. M. (2004). The evolution of the seismic- aseismic transition during the earthquake cycle: constraints from the time- dependent depth distribution of aftershocks. *Geophysical Research Letters, 31,* L23610. https://doi.org/10.1029/2004GL021379.

Saleeby, J. (2003). Production and loss of high-density batholithic root, southern Sierra Nevada, California. *Tectonics, 22*(6), 1064. https://doi.org/10.1029/2002TC001374.

Sammis, C. G., Smith, S. W., Nadeau, R. M., & Lippoldt, R. (2016). Relating transient seismicity to episodes of deep creep at Parkfield, California. *BSSA, 106*(4), 1887–1899. **(225)**.

Sass, J. H., Williams, C. F., Lachenbruch, Arthur H., Galanis, S. P., Jr., & Grubb, F. V. (1997). Thermal regime of the San Andreas fault near Parkfield, California. *Journal of Geophysical Research, 102*(B12), 27575–27585.

Shearer, P. M. (2002). Parallel fault strands at 9-km depth resolved on the imperial fault, Southern California, *Geophysical Research Letters,* 29(14), 19-1–19-4. https://doi.org/10.1029/2002gl015302.

Shinevar, W. J., Behn, M. D., & Hirth, G. (2015). Compositional dependence of lower crustal viscosity. *Geophysical Research Letters, 42,* 8333–8340. https://doi.org/10.1002/2015GL065459.

Sibson, R. H. (1974). Frictional constraints on thrust, wrench and normal faults. *Nature, 249*(5457), 542–544. https://doi.org/10.1038/249542a0. **(1035)**.

Sibson, R. H. (1982). Fault zone models, heat flow, and the depth distribution of earthquakes in the continental crust of the United States. *Bulletin of the Seismological Society of America, 72*(1), 151–163.

Sibson, R. H. (1983). Continental fault structure and the shallow earthquake source. *Journal of the Geological Society, 140*(5), 741–767. https://doi.org/10.1144/gsjgs.140.5.0741.

Sibson, R. H. (1984). Roughness at the base of the seismogenic zone: contributing factors. *Journal of Geophysical Research, 89,* 5791. https://doi.org/10.1029/JB089iB07p05791.

Smith-Konter, B. R., Sandwell, D. T., & Shearer, P. (2011). Locking depths estimated from geodesy and seismology along the San Andreas fault system: implications for seismic moment release. *Journal of Geophysical Research, 116,* B06401. https://doi.org/10.1029/2010JB008117.

Takeuchi, C. S., & Fialko, Y. (2012). Dynamic models of interseismic deformation and stress transfer from plate motion to continental transform faults. *Journal of Geophysical Research, 117,* B05403. https://doi.org/10.1029/2011JB009056.

Thatcher, W., & England, P. C. (1998). Ductile shear zones beneath strikeslip faults: implications for the thermomechanics of the San Andreas fault zone. *Journal of Geophysical Research, 103,* 891–905. https://doi.org/10.1029/97JB02274.

Townend, J., & Zoback, M. D. (2000). How faulting keeps the crust strong. *Geology, 28*(5), 399–402.

Turcotte, D. L., & Schubert, G. (2002). *Geodynamics* (2nd ed., p. 456). Cambridge, New York, Melbourne: Cambridge University Press. ISBN:0-521-66186-2.

Wessel, P., Smith, W. H. F., Scharroo, R., Luis, J. F., & Wobbe, F. (2013). Generic mapping tools: improved version released. *Eos, Transactions American Geophysical Union, 94*(45), 409–410.

Williams, C. F. (1996). Temperature and the seismic/aseismic transition: observations from the 1992 landers earthquake. *Geophysical Research Letters, 23,* 2029–2032. https://doi.org/10.1029/96GL02066.

Yang, W., & Hauksson, E. (2013). The tectonic crustal stress field and style of faulting along the Pacific North America plate boundary in Southern California. *Geophysical Journal International, 194*(1), 100–117. https://doi.org/10.1093/gji/ggt113.

Zaliapin, I., & Ben-Zion, Y. (2013). Earthquake clusters in southern California I: identification and stability. *Journal of Geophysical Research: Solid Earth, 118*(6), 2847–2864.

(Received March 26, 2018, revised August 16, 2018, accepted August 20, 2018, Published online September 3, 2018)

Pure Appl. Geophys. 176 (2019), 1083–1115

https://doi.org/10.1007/s00024-018-2003-x

Pure and Applied Geophysics

Dynamic In Situ Three-Dimensional Imaging and Digital Volume Correlation Analysis to Quantify Strain Localization and Fracture Coalescence in Sandstone

François Renard,[1,2] Jessica McBeck,[1] Benoît Cordonnier,[1,3] Xiaojiao Zheng,[1] Neelima Kandula,[1] Jesus R. Sanchez,[1] Maya Kobchenko,[1] Catherine Noiriel,[4] Wenlu Zhu,[5] Paul Meakin,[6] Florian Fusseis,[7] and Dag K. Dysthe[1]

Abstract—Advances in triaxial compression deformation apparatus design, dynamic X-ray microtomography imaging, data analysis techniques, and digital volume correlation analysis provide unparalleled access to the in situ four-dimensional distribution of developing strain within rocks. To demonstrate the power of these new techniques and acquire detailed information about the micromechanics of damage evolution, deformation, and failure of porous rocks, we deformed 3-cm-scale cylindrical specimens of low-porosity Fontainebleau sandstone in an X-ray-transparent triaxial compression apparatus, and repeatedly recorded three-dimensional tomograms of the specimens as the differential stress was increased until macroscopic failure occurred. Experiments were performed at room temperature with confining pressure in the range of 10–20 MPa. Distinct grayscale subsets, indicative of density, enabled segmentation of the three-dimensional tomograms into intact rock matrix, pore space, and fractures. Digital volume correlation analysis of pairs of tomograms provided time series of three-dimensional incremental strain tensor fields throughout the experiments. After the yield stress was reached, the samples deformed first by dilatant opening and propagation of microfractures, and then by shear sliding via grain rotation and strain localization along faults. For two samples, damage and dilatancy occurred by grain boundary opening and then a sudden collapse of the granular rock framework at failure. For the third sample, a fault nucleated near the yield point and propagated in the sample through the development of transgranular microfractures. The results confirm findings of previous experimental studies on the same rock and provide new detailed quantifications of: (1) the proportion of shear versus dilatant strain in the sample, (2) the amount of dilatancy due to microfracture opening versus pore opening when a fault develops, and (3) the role of grain boundaries and pore walls in pinning microfracture propagation and slowing down the rate of damage accumulation as failure is approached. Our study demonstrates how the combination of high-resolution in situ dynamic X-ray microtomography imaging and digital volume image correlation analysis can be used to provide additional information to unravel brittle failure processes in rocks under stress conditions relevant to the upper crust.

Key words: Faulting, rupture nucleation, damage, dynamic X-ray microtomography, Fontainebleau sandstone, digital volume correlation.

1. Introduction

Acquiring detailed observations about the processes that control the propagation and coalescence of microfractures that lead to system-size failure is critical for robust understanding of borehole and tunnel stability, the geometry of fractures and faults in crustal reservoirs, and earthquake physics. Microfractures may coalesce and lead to macroscopic failure (Scholz 1968; Wawersik and Fairhurst 1970; Mogi 1971; Peng and Johnson 1972; Tapponnier and Brace 1976; Lockner et al. 1991; Dresen and Guéguen 2004; Paterson and Wong 2005), and evolving microfractures within fault damage zones alter the stress field surrounding faults (Otsuki and Dilov 2005; Faulkner et al. 2006). Fracture networks may influence fluid flow near major faults, which can

Electronic supplementary material The online version of this article (https://doi.org/10.1007/s00024-018-2003-x) contains supplementary material, which is available to authorized users.

[1] Departments of Geosciences and Physics, The Njord Centre, PGP, University of Oslo, Oslo, Norway. E-mail: francois.renard@geo.uio.no

[2] Université Grenoble Alpes, Université Savoie Mont Blanc, CNRS, IRD, IFSTTAR, ISTerre, 38000 Grenoble, France.

[3] ESRF, The European Synchrotron Radiation Facility, Grenoble, France.

[4] Géosciences Environnement Toulouse, Observatoire Midi-Pyrénées, Université de Toulouse, Université Paul Sabatier, CNRS, IRD, CNES, 14 avenue Edouard Belin, 31400 Toulouse, France.

[5] Department of Geology, University of Maryland, College Park, MD 20742, USA.

[6] Department of Physics, Temple University, Philadelphia, PA, USA.

[7] School of Geosciences, University of Edinburgh, Edinburgh EH9 3JW, UK.

decrease the effective stress on the fault plane and lower the shear stress required to trigger an earthquake (Miller et al. 2004). In addition, growing microfracture networks that evolve as rocks approach macroscopic failure may change the mechanical properties of the rock (Heap and Faulkner 2008), upon which accurate seismic imaging of fault damage zones depends.

Many laboratory studies have relied on acoustic emission monitoring to provide insights into deformation preceding failure. This technique was the first method used to probe inside rocks during in situ deformation, and it has very successfully provided information about failure modes and the approximate spatial and temporal distribution of damage events in crystalline rocks and porous sedimentary rocks (e.g. Scholz 1968; Lockner et al. 1991, 1992; Cox and Meredith 1993; Wu et al. 2000; Stanchits et al. 2006; Fortin et al. 2009; Ghaffari et al. 2014). A limitation of using acoustic emission recording is that it only captures events that emit acoustic waves with frequencies and intensities that can be detected in the presence of background noise. Furthermore, the velocities, attenuation, scattering, and diffraction of acoustic waves depend on spatially varying heterogeneities that evolve with the developing fracture networks. Consequently, the error in the acoustic source location is on the order of several millimeters (Stanchits et al. 2006), which is typically larger than the grain size of the rock. Additional information about the acoustic source size and orientation can be obtained from moment tensor analysis (Kwiatek et al. 2014). This analysis requires high-quality data, a large number of well-calibrated and high-dynamic-range acoustic sensors, and knowledge of the elastodynamic tensor Green's function. Like other mechanical properties, the Green's function changes during fracturing, and can be accurately determined for heterogeneous materials only if its local variations can be measured. Consequently, accurate information about the fracture sizes, shapes, and orientations can be obtained from acoustic emission experiments only when high-quality and low-noise data are available and heavy data processing is performed (e.g. Kwiatek et al. 2014).

In situ three-dimensional dynamic X-ray microtomography imaging combined with digital volume correlation (DVC) analysis is complementary to acoustic emission tomography because it provides detailed information about the evolution of the microscopic and macroscopic strain fields at micrometer-scale spatial resolution and precision. Data may include aseismic deformation that cannot be measured by acoustic emissions. Here, we describe this recent experimental technique, discuss its potential limits, and show how it can be applied to study the initiation of faulting in rocks. We focus on the application of this technique to an experimental investigation of the microscopic deformation of three specimens of Fontainebleau sandstone under triaxial compression. Using in situ three-dimensional X-ray tomography, digital volume correlation analysis, and scaling statistics, we quantified the evolution of: (1) porosity, (2) nucleation, growth, and coalescence of microfractures, and (3) interactions between local microscopic dilation, contraction, and shear strain that leads to system-size shear failure with increasing stress. Segmentation of the three-dimensional tomograms into rock matrix, pore space, and fractures provided four-dimensional spatiotemporal information about the evolving pore space and fracture network. The results provide insights into the deformation mechanisms at the grain scale, and how small-scale strain concentrations evolved preceding macroscopic failure, confirming the results of previous studies of sandstones (Handin et al. 1963; Menéndez et al. 1996; Wu et al. 2000; El Bied et al. 2002; Schubnel et al. 2007; Nasseri et al. 2014; Goodfellow et al. 2015). In addition, digital volume correlation analysis enabled calculation of the three-dimensional incremental strain fields between successive tomograms at a strain resolution one order of magnitude higher than the imaging resolution. Information about the magnitude and distribution of local volumetric and shear strains preceding failure was used to determine the probability density distribution of the incremental strain magnitudes. Combining analysis of segmented three-dimensional tomograms with DVC analysis improved our understanding of deformation mechanisms preceding macroscopic failure by enabling quantification of the evolving four-dimensional pore network and strain field. Postfailure scanning electron microscopy (SEM) provided supplementary information concerning deformation and failure mechanisms.

In the present study, we illustrate the potential of this emergent experimental technique that can provide data with unprecedented spatial resolution, that is complementary to other experimental techniques such as acoustic emission analysis, and that can contribute to better understanding of deformation processes in rocks. Because this technique enables imaging of both aseismic and seismic deformation and damage in the sample with unprecedented spatial resolution, questions that existing experimental techniques can hardly address may be answered. These include: What is the proportion and spatial distribution of shear relative to volumetric strain events inside the rock prior to failure? What is the proportion of microfracture opening relative to pore opening during dilation? How does the rate of aseismic and seismic damage accumulation evolve as failure is approached?

2. *Background*

2.1. *Failure of Intact Rocks*

In rock deformation experiments, the evolution of a specimen is usually characterized while either a constant strain rate or increasing stress is imposed at the boundaries. In true triaxial compression tests, all three principal global strains or stresses, $\sigma_1 > \sigma_2 > \sigma_3$, are controlled. In the experiments reported here, $\sigma_1 > \sigma_2 = \sigma_3 = P_c$, where P_c is the confining pressure, and the maximum principal stress, σ_1, was increased from an initial value of $\sigma_1 = P_c$ at constant P_c, until the sample failed. The sign convention that compressive stress is positive and compressive strain (shortening or a decrease in volume) is positive, which is most commonly used in rock physics, is adopted in this article.

Previous experiments indicate that, for confining pressures on the order of 10 MPa, tensile microcracks nucleate and then undergo dilation, propagation, and coalescence as the differential stress increases, and thereby promote macroscopic failure (e.g., Peng and Johnson 1972; Tapponnier and Brace 1976; Paterson and Wong 2005). Early experiments suggested that dilating microfractures produce macroscopic dilation of rock samples preceding failure (Brace et al. 1966), and that a critical density of microfractures develops preceding macroscopic shear failure (Scholz 1968; Lockner et al. 1991). The spatial distribution of acoustic emissions indicates that microfractures initially nucleate and grow at apparently random locations in Westerly granite (Lockner et al. 1991). Reches and Lockner (1994) proposed that, as a granite approaches failure, microfractures form elongated arrays inclined at an angle of about 30° relative to the direction of the maximum principal stress until they coalesce into a macroscopic fault. Using high-resolution two-dimensional image correlation analysis, Tal et al. (2016) observed that both local compaction and local dilation occur before failure in a Carrara marble. Experiments on crystalline rocks such as granite, experiments on other brittle materials with preexisting fractures [e.g., poly(methyl methacrylate)], and numerical models suggest that macroscopic dilation occurs through the development of microfractures that are dominated by tensile wing cracks (Horii and Nemat-Nasser 1986; Ashby and Sammis 1990; Kemeny and Cook 1991). Laboratory experiments on analogous rock material that contains frictional flaws suggest that preexisting microfractures coalesce through the propagation and linkage of such tensile wing cracks and secondary shear fractures (e.g., Wong et al. 2001; Dresen and Guéguen 2004). Experiments have also documented how the geometries of preexisting fractures control the coalescence pattern of the resulting wing cracks and shear fractures (e.g., Wong and Einstein 2009). Discrete element method models (Hazzard et al. 2000) and analytical and numerical damage models (Ashby and Sammis 1990; Lyakhovsky et al. 1997; Girard et al. 2010) have produced macroscopic fracture propagation via tensile crack development and long-range elastic interactions.

Analysis of time-lapse three-dimensional X-ray tomograms acquired during deformation of a quartz monzonite rock specimen indicate that the total volume of microfractures, the rate of damage accumulation, and the size of the largest microfracture all increase as power laws and diverge with increasing differential stress as failure is approached (Renard et al. 2018). This behavior suggests that fracture growth within low-porosity crystalline rocks such as monzonite evolves as a critical phenomenon in which an acceleration of damage accumulation precedes system-size failure, confirming models developed in statistical physics (Dahmen et al. 2009; Girard et al. 2010). In these experiments, most of the fractures formed within grains (i.e., by transgranular fracturing), likely because the rock had low initial porosity (< 1%).

2.2. *Micromechanical Models of Sandstone Deformation*

Several experimental and analytical studies have characterized the micromechanisms of deformation in sandstones. With increasing effective stresses, the deformation of sandstones transitions from brittle faulting to cataclastic processes (i.e., grain comminution) (Handin et al. 1963; Wong et al. 1997). Depending on the initial rock porosity, in the brittle regime, dilation or shear-enhanced compaction may initiate near a yield point, defined as the point at which a significant deviation from linear elastic behavior occurs (point C′ defined in Brace et al. 1966). At higher effective confining pressures (between ~ 100 and 200 MPa), failure occurs through grain crushing and pore collapse facilitated by microscopic fractures (Wong et al. 1997). At high temperature (900 °C), the failure envelope may transition from a dilatant Mohr–Coulomb relationship to a failure envelope with an elliptical shape as the confining stress increases (e.g., Kanaya and Hirth, 2018).

For sandstones with porosities above ~ 13%, deformation may be dominated by three different mechanisms: shear localization, compaction localization, or cataclastic deformation. Compaction may dominate strain localization, depending on the loading path (e.g., Fortin et al. 2009). Localized compaction bands may be formed in sandstones under high confining pressure conditions (Fortin et al. 2009); For example, at an effective confining pressure of 10 MPa, strain softening and brittle failure occurred in Bleurswiller sandstone specimens with porosities of 23.5–25.3% as the differential stress increased (Baud et al. 2015). At larger effective confining pressures (30–50 MPa), shear-enhanced compaction and shear bands oriented at ~30° to the maximum principal stress promoted macroscopic failure, and at even higher effective confining pressures (70–90 MPa), compaction bands developed (Baud et al. 2015). At the grain scale, fractures within grains and along grain contacts, as well as the collapse of high-porosity volumes, produced the deformation bands.

Menéndez et al. (1996) conducted a series of experiments on Berea sandstone with 21% porosity at constant pore pressure of 10 MPa and confining pressures of 20, 50, and 260 MPa, and characterized postfailure damage in thin sections prepared after failure. At low confining stresses (< 50 MPa), the breakage of grain contacts produced the majority of the damage, and few microfractures propagated through grains. As failure approached, pore collapse and transgranular fractures provided a greater contribution to the overall damage. Tensile fracture along grain contacts dominated the fracture of quartz grains, and shear localization occurred through the coalescence of groups of microfractures. The few observed transgranular fractures led to the conclusion that the development of tensile wing cracks did not play a significant role during the initiation of failure (Menéndez et al. 1996).

During the deformation of a Gosford sandstone with grain size in the range of 0.1–1 mm and 13% porosity, an acceleration of acoustic emissions as failure was approached was measured, as well as a power-law distribution of acoustic emission sizes (Cox and Meredith 1993). A relationship between these acoustic emission parameters and crack dimensions was developed to reconstruct the strain–stress curve and the weakening of the rock as failure was approached.

Zhang et al. (1990) developed a model for the micromechanics of grain crushing under hydrostatic conditions based on the Hertzian contact concept. This model describes the failure of porous rocks using the maximum tangential tensile stress at the edge of the contact area between spherical grains, and linear elastic fracture mechanics assuming a high density of microscopic crack-like flaws with a characteristic length at the grain surfaces (Johnson 1982). This model successfully described the failure envelopes of several sandstones (Baud et al. 2000).

2.3. *Deformation of Fontainebleau Sandstone*

The relatively homogeneous mineralogy and microstructure of Fontainebleau sandstone at scales greater than the grain size make it an ideal target for deformation experiments. These properties also make this sandstone an ideal candidate for three-dimensional X-ray microtomography imaging studies, and so it was one of the first rocks imaged using this technique (e.g., Auzerais et al. 1996; Coker et al. 1996). Separation of the rock matrix from the pore structure and fracture network by segmentation enabled the first quantitative analysis of the topological and geometrical properties of the pore structure of a sandstone (Lindquist et al. 2000), and led to important insights into the distribution of multiphase fluids in the pore space, and modeling of flow and permeability (Auzerais et al. 1996). These analyses initiated the new field of digital rock physics (e.g., Andrä et al. 2013 and references therein).

Previous studies have analyzed how elastic properties and damage of this rock evolve during deformation (El Bied et al. 2002; Schubnel et al. 2007; Nasseri et al. 2014; Ghaffari et al. 2014; Goodfellow et al. 2015). During triaxial deformation of a 14% porosity Fontainebleau sandstone, Schubnel et al. (2007) recorded series of acoustic emission events preceding macroscopic failure by attaching sensors directly to the rock sample. These recorded precursory events reflect the acoustic energy released by nucleating and propagating microfractures. The acoustic emissions highlighted distinct failure stages: (1) clustered acoustic emissions and strain localization along an incipient fault plane, (2) lack of acoustic emissions within an aseismic nucleation zone, (3) unstable rupture propagation within the previously aseismic, and perhaps locked, zone, and (4) triggering of multiple sets of aftershocks, with the second set of aftershocks emitted from the rupture plane as the pore pressure rapidly dropped (Schubnel et al. 2007). The increasing number of acoustic emissions prior to failure followed an inverse Omori law with a power-law time dependence with an exponent close to 1 (Schubnel et al. 2007). In another series of experiments using a true triaxial stress apparatus with independent control of the three normal stress magnitudes, the failure of a 4.52% porosity Fontainebleau sandstone included: (1) initial compaction of the rock via crack closure, (2) macroscopic dilation of the sample due to microfractures that opened perpendicular to the minimum stress direction, and (3) accumulation of microfractures until macroscopic failure occurred (Goodfellow et al. 2015).

In the present study, we used X-ray microtomography and three-dimensional DVC analysis to obtain quantitative information about the evolving microfracture networks and pore structure throughout a series of triaxial compression experiments on three air-saturated Fontainebleau sandstone cores. The X-ray attenuation contrast between air and quartz allowed the evolving pore volumes and open microfractures to be distinguished from the rock matrix, thus enabling investigation of deformation via image segmentation (Videos S1 and S2). Three-dimensional digital volume correlation analysis of tomograms can identify diffuse deformation zones, as well as fractures that do not dilate sufficiently to locally decrease the X-ray attenuation (Videos S3, S4, and S5). The main goal of this work is to analyze evolving porosity, and transgranular and intergranular fractures, using segmentation of dynamic X-ray tomograms and digital volume correlation. The results provide new experimental observations of the micromechanisms of deformation and strain localization in sandstone. The results confirm those obtained in previous studies on Fontainebleau sandstone and other porous sandstones (Handin et al. 1963; Menéndez et al. 1996; Wu et al. 2000; El Bied et al. 2002; Schubnel et al. 2007; Nasseri et al. 2014; Goodfellow et al. 2015), and new information was also obtained.

3. *Methods and Material*

3.1. *X-Ray-Transparent Triaxial Deformation Apparatus*

The samples were deformed in the X-ray-transparent HADES triaxial apparatus (Renard et al. 2016, 2017), installed on the X-ray microtomography beamline ID19 at the European Synchrotron Radiation Facility. This apparatus enables time-lapse imaging of the sample during compressive deformation. With the full white beam of the beamline with X-ray energies up to 200 keV, the average energy of X-rays that cross the sample is close to 120 keV after the attenuation of X-rays by the wall of the triaxial rig. Acquisition of a three-dimensional dataset of two-dimensional radiographs required about 1.5 min. The experiments were performed at room temperature (24 °C) on dry cylindrical specimens of 10 mm in length and 5 mm in diameter, corresponding to ~50 grains in the axial direction and ~20 grains in the radial direction. The voxel size was 6.5 μm. The table in Fig. 1 describes the imposed loading conditions used in these experiments.

The specimens were installed in the rig between two stainless-steel pistons. The lower piston was immobile, and the axial load was imposed on the specimen by displacement of the upper piston. The interfaces between the rock sample and the pistons were not lubricated. Two independent pumps controlled the axial load and confining pressure. In most

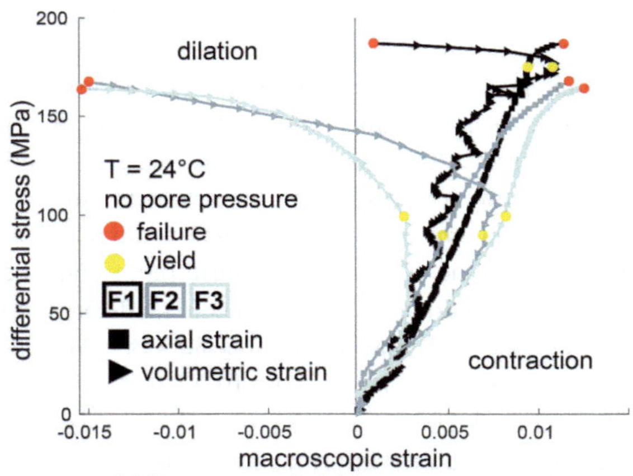

	σ_D at failure	σ_D at yield	σ_2	# scans
F1	185 MPa	165 MPa	20(14) MPa	184
F2	167 MPa	90 MPa	10 MPa	49
F3	159.5 MPa	100 MPa	10 MPa	55

Figure 1
Experimental axial and volumetric strains as functions of the differential stress for the three Fontainebleau sandstone experiments. Each point corresponds to the acquisition of a three-dimensional X-ray tomography dataset. Yellow and red circles show the yield and failure points. The table lists the experimental conditions. In experiment F1, the confining pressure was reduced from 20 MPa after the axial stress reached 199 MPa, and failure occurred when the confining pressure was reduced below 15 MPa at a constant axial stress of 199 MPa (i.e., at a differential stress of 185 MPa). In experiments F2 and F3, the confining pressure was held constant. In experiment F1, the oscillations in the volumetric curve arose from instabilities in the confining pressure pump

laboratory experiments on the deformation and failure of rocks, a constant strain rate is imposed as principal stresses and other quantities of interest, such as acoustic emissions, are measured. This approach cannot be used with X-ray tomography because the resulting tomogram would capture average density contrast data as the specimen deforms over the data acquisition time interval, or 1.5 min in these experiments. Instead, to capture snapshots of deformation, the stress is increased in small steps and X-ray attenuation data are acquired while the strain or stress is held constant. In principle, the structure of the specimen might change during the acquisition time under constant loading conditions if processes such as creep or subcritical fracture propagation occur. However, in these experiments, we did not find evidence for significant changes in structure during data acquisition, such as blurring of the distinct edges between void space and quartz in the tomograms. For other materials or under different conditions, the mechanical behavior may be different, and unacceptable image blurring could occur.

Three experiments were conducted by increasing the differential stress (the difference between the axial stress, σ_1, and the confining pressure) in steps of 2 or 5 MPa far from failure, and steps of 0.5 or 1 MPa close to failure. A jacket made of Viton® fluoropolymer elastomer encased each rock sample, and silicone oil applied the confining pressure to this jacket. In experiment F1, instability of the confining pressure pump produced undulations in the volumetric strain curve (Fig. 1). In this experiment, the confining pressure was held at 20 MPa until the axial stress reached 199 MPa, which closely approached the maximum permissible axial stress of the HADES rig (200 MPa). Under these loading conditions, the sandstone core had not macroscopically failed, so the differential stress was then increased by reducing the confining pressure in steps of 1 MPa from 20 MPa to 14 MPa, at which failure occurred. In this experiment, failure occurred at a differential stress between 185 MPa and 186 MPa. Experiments F2 and F3 were conducted with a constant confining pressure of 10 MPa, and the samples failed before the axial loading reached the limit of the HADES rig.

After each differential stress increase, three-dimensional X-ray microtomography imaging of the specimen was performed. After failure, both the axial stress and confining pressure were decreased, allowing voids to relax and open, and a final imaging step was performed. Each X-ray imaging acquisition required 1.5 min, and each stress increase between scans required 1 min.

Between 49 and 184 datasets, each consisting of 1600 radiographs, were acquired during each deformation experiment, with a total of 288 for the three experiments. From the two-dimensional radiographs, three-dimensional volumes of the specimens (1600 × 1600 × 1600 voxels) were reconstructed in 16-bit grayscale using a phase-contrast retrieval algorithm (Mirone et al. 2014). During the reconstruction, filters were applied to remove ring artifacts and other spurious patterns, and to correct for the fluctuation of the X-ray source intensity. The grayscale value of each voxel in the three-dimensional volume is proportional to the X-ray attenuation coefficient, itself proportional to local density. Low gray levels (dark shade of gray) correspond to low-attenuating materials (i.e., air), and high gray levels (light shades of gray) correspond to highly attenuating materials (i.e., quartz), while intermediate gray levels correspond to voxels that contain both air and solid (i.e., voxels that are intersected by void boundaries).

In experiments with strain rate loading conditions, the differential stress typically reaches a maximum and then decreases if the confining pressure is not too large. Under stress loading conditions, similar to those used in our experiments, macroscopic failure typically occurs within one stress increment step in brittle rocks. Failure is associated with a macroscopic stress drop. In our experiments, macroscopic failure pulverized the sandstones, producing sand-like material, as well as discrete core-spanning fractures within one stress increment step, and so strain softening following macroscopic failure could not be investigated. The stress–strain relationship and the micromechanical processes that occur prior to failure do not strongly depend on whether the imposed loading are strains or stresses on the time scales of our experiments. Furthermore, our stress loading conditions provided a rich dataset of coalescing fractures prior to macroscopic failure.

3.2. *Macroscopic Stress and Strain*

The three-dimensional tomograms were used to calculate the axial strain, radial strain, and volumetric strain curves prior to failure (Figs. 1, S1). The macroscopic axial strain was calculated from the distance between the two pistons visible in the tomograms at two locations in an axial plane perpendicular to the piston faces. The average value of these two distances was taken as the height of the sample. The macroscopic radial strain (Fig. S1, inset) was determined by measuring the lengths of two mutually perpendicular transects within horizontal cross sections that intersected the vertical axis at heights of 1/3, 1/2, and 2/3 of the sample height, providing a total of six measurements of the sample diameter, from which the mean value of radial strain was calculated (Fig. S1b). The volumetric strain was calculated from the average height and average radius of the sample during deformation, assuming a cylindrical shape (Fig. 1).

Because of the friction between the ends of the sandstone specimens, we expected that the sandstone specimens would have a slight barrel-like shape under imposed differential stresses, and this was observed (Fig. 4a). So, the volumetric dilatational strain will be somewhat smaller than the volumetric strain calculated with our method that uses core diameters at least 3 mm from the piston–sandstone interfaces. The resolution (and the error) of this macroscopic strain measurement was $10^{-4} l_0$ for the axial strain and $\sim 3 \times 10^{-4} r_0$ for the axial and radial strains, where l_0 and r_0 are the initial length and radius. Consequently, the resolution for the volumetric strain was $\sim 5 \times 10^{-4} V_0$, where V_0 is the initial volume. The resolution was higher for the axial strain because there was a higher X-ray attenuation contrast between the pistons and the sample, which controlled the axial strain measurement, than between the jacket and the sample, which controlled the radial strain measurement.

3.3. *Segmentation Procedure and Microscopy Imaging*

To determine the pore and fracture sizes and shapes from the tomograms (Fig. 2b), the following

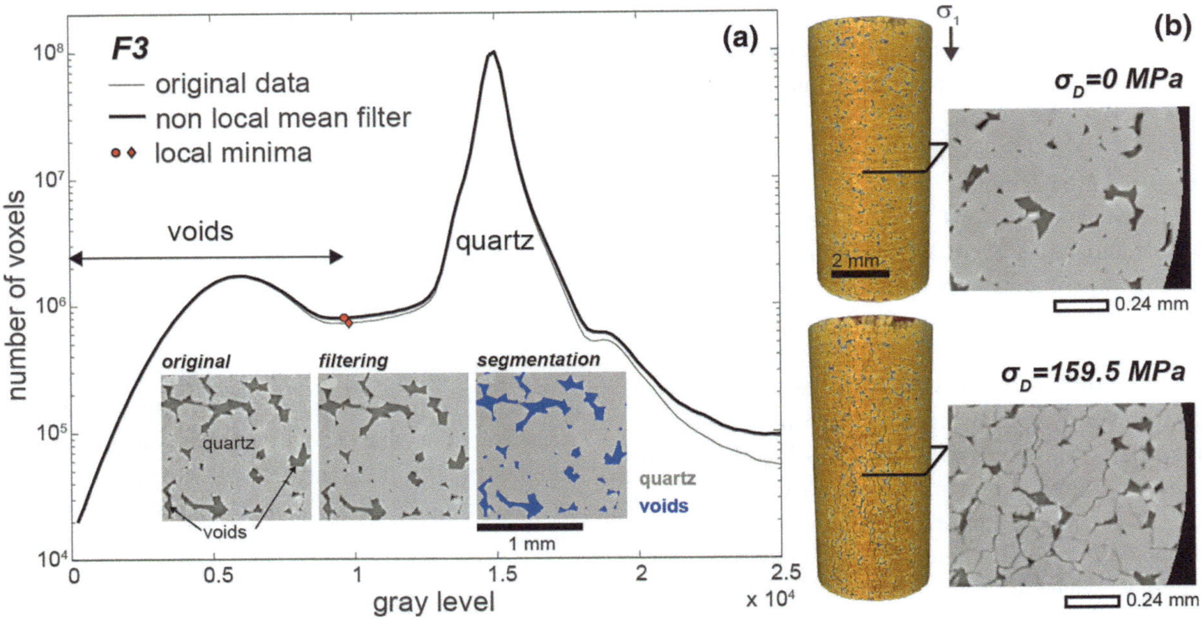

Figure 2

Thresholding procedure used to segment the three-dimensional datasets into quartz matrix and air-filled voids, including pores and fractures. **a** Distribution of the gray levels of experiment F3 for the original three-dimensional dataset (thin line) at a differential stress of 10 MPa, before and after applying the nonlocal mean filter (thick line). Insets show two-dimensional slices of the sample with the unfiltered data (left), filtered data (middle), and segmented data (right). We selected the local minima of the histogram between the peaks arising from voids (left peak) and from quartz grains (right peak) to segment the tomograms into quartz grains and voids. The red symbols show the minima of the original (diamond) and filtered (circle) data, which do not differ significantly. **b** Three-dimensional rendered view of the specimen before and after experiment, and two-dimensional horizontal cross-sections. Video S2 shows the time-lapse evolution of the specimen in this experiment

procedure was applied using the software AvizoFire®: (1) denoising of the three-dimensional volumes with a nonlocal mean filter (Buades et al. 2005), (2) application of a mask to remove the jacket, pistons, and confining oil around the rock sample, (3) application of a nonlocal median filter to sharpen the boundaries between intact rock material and voids (pore space or fractures), and (4) thresholding the data to partition the voxels occupied primarily by air from those occupied primarily by quartz (Fig. 2a).

Representative histograms from a three-dimensional dataset (experiment F3) show how thresholding can differentiate between voids and the rock matrix, which is composed primarily of quartz (Fig. 2). The grayscale values of the tomograms correspond to the X-ray attenuation produced by materials, which is lower for air than for quartz grains. Both the original and filtered histograms show two peaks corresponding to the voids (low X-ray attenuation) and the quartz grains (high X-ray attenuation). The local minimum in the histogram of the filtered data was used to separate the voids from the grains (red markers in Fig. 2). We also tested segmentation of the data using the midpoint between the two peaks, and the results were similar to those using the minima. We conclude that at least two criteria could be used to select a threshold gray value to separate voids and grains. Here, we chose the minimum value between the two peaks of the histogram to segment the data.

Filtering the tomograms reduced the noise and enhanced the contrast between pores and grains, but it produced only small differences in the histogram and minimum. Thresholding with the histogram minimum enabled extraction of pores and major microfractures from the three-dimensional volumes. For the complete series of tomograms acquired during each experiment, the same threshold value was used to segment the data. However, this thresholding technique did not unambiguously capture microcracks with apertures less than or approximately equal to the imaging resolution, although they could be identified by inspection of the tomograms by eye.

Some parts of the specimens were recovered in their jackets at the end of experiments F1 and F2 after unmounting them from the HADES rig. For experiment F2, almost the entire specimen could be recovered, but specimen F3 was mostly reduced to quartz particles. The specimens from experiments F1 and F2 were impregnated with liquid epoxy resin and then cut along the axial plane after the liquid epoxy had formed a cross-linked polymeric solid. The exposed surface was polished, coated with 10 nm gold, and imaged using a Hitachi SU5000 scanning electron microscope at the University of Oslo with a voltage of 15 kV.

3.4. Digital Volume Correlation Analysis

Three-dimensional digital volume correlation (DVC) analysis was initially developed for applications in mechanics and engineering (e.g., Bay et al. 1999 and references therein). DVC analysis finds the translations and rotations that best map subvolumes within a three-dimensional dataset onto subvolumes in another three-dimensional dataset by identifying similar patterns within those subvolumes. This technique has been used to study the development of shear bands in soils (Viggiani et al. 2004), and the compaction of a Rothbach sandstone specimen with a strain resolution of 10^{-3} (Louis et al. 2007). We used the open-source DVC analysis software TomoWarp2 for our DVC analyses (Hall et al. 2010; Tudisco et al. 2015, 2017).

DVC analysis is based on finding the displacement field that maximizes the correlation between voxel intensity subsets (subvolumes) within pairs of sequential tomograms. Interpolation methods enable subvoxel-scale displacement resolution to be obtained. However, if there is little variation in X-ray attenuation coefficients within the subvolumes, DVC analysis may lead to unreliable incremental displacement fields; For example, there is little or no variation in the X-ray attenuation coefficients within individual quartz grains, except perhaps lower-density fluid inclusions. Consequently, the parameters of DVC analyses must be tuned to capture displacement fields that include subvolumes that include several grains or grain boundaries, or other contrasts in the X-ray attenuation coefficient fields.

By identifying similar patterns across successive volumes, digital volume correlation produces three-dimensional displacement fields from which the six independent components of the second-rank three-dimensional strain tensor may be calculated. These incremental strain fields reveal strain localization that occurred within the time interval between the acquisitions of the pair of microtomograms. The cumulative strain from the onset of loading could also be determined. Following the approach of McBeck et al. (2018), we calculated three-dimensional incremental displacement fields between pairs of three-dimensional tomograms corresponding to approximately constant increments in the macroscopic axial strain, ε_{zz}^{M}, throughout each experiment. We calculated ε_{zz}^{M} from the change in axial length of the sandstone specimen relative to the initial length l_0, as described in Sect. 3.2. In addition, to investigate in more minute detail the dominant failure modes near the onset of yielding in experiment F2, we used DVC analysis to calculate the incremental strain fields surrounding the yield point, with the highest possible temporal (i.e., differential stress) resolution.

We calculate changes in the local volumetric and shear strain fields using the first invariant of the incremental strain tensor, $I_1(\Delta\varepsilon) = \varepsilon_{xx} + \varepsilon_{yy} + \varepsilon_{zz}$, and the second invariant of the incremental strain tensor, $I_2(\Delta\varepsilon) = (\varepsilon_{xy})^2 + (\varepsilon_{xz})^2 + (\varepsilon_{yz})^2 - (\varepsilon_{xx}\varepsilon_{yy} + \varepsilon_{xx}\varepsilon_{zz} + \varepsilon_{yy}\varepsilon_{zz})$, where ε_{ij} is a component of the local incremental strain tensor, i.e., the local strain calculated by comparing two successive tomograms in the DVC analysis. In this coordinate system, the z axis is vertical (parallel to σ_1) and the x–y plane is horizontal (perpendicular to σ_1). $I_1(\Delta\varepsilon)$ provides information about the volumetric strain in coordinate systems independent of the principal axis system. To characterize the incremental shear strain, we calculated the second invariant of the incremental deviatoric strain, $J_2(\Delta\varepsilon)$. Since $(3J_2(\varepsilon))^{\frac{1}{2}}$ is the von Mises yield criterion equivalent strain, we refer to $(3J_2(\Delta\varepsilon))^{\frac{1}{2}}$ as the von Mises incremental strain. The second invariant of the incremental deviatoric strain, $J_2(\Delta\varepsilon)$, is related to the first invariant $I_2(\Delta\varepsilon)$, and the second

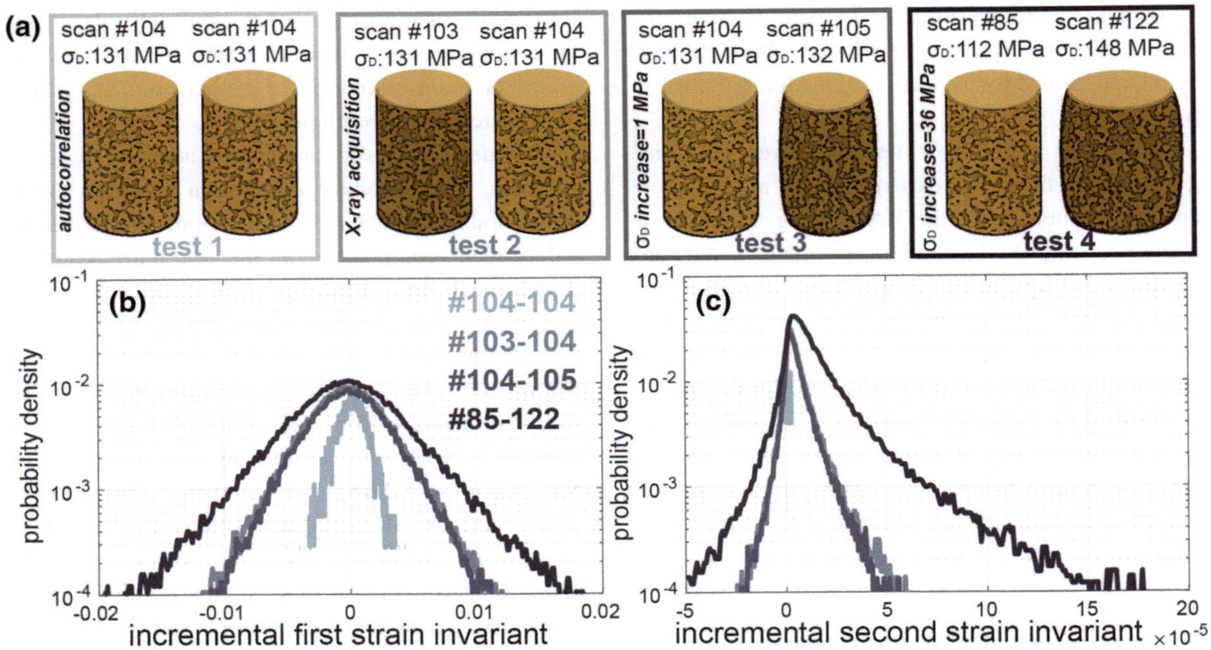

Figure 3
Characterization of the resolution of the digital volume correlation technique using four tests. **a** Pairs of three-dimensional datasets used in the four tests of digital volume correlation analysis. **b**, **c** Results of the tests. The gray level of the boxes surrounding each pair of tomograms in (**a**) matches the gray levels of the probability densities shown in (**b**) and (**c**). The strain resolution is taken as twice the standard deviation found in test 2 between scans #103–104, a more conservative estimate of the error than test 1

invariant, $I_2(\Delta\varepsilon)$, of the incremental strain tensor by $J_2(\Delta\varepsilon) = \frac{1}{3}(I_1(\Delta\varepsilon))^2 - I_2(\Delta\varepsilon)$. To be able to compare the values of $I_1(\Delta\varepsilon)$ (first order in the strain) and $J_2(\Delta\varepsilon)$ (second order in the strain), we report in the following the values of $(3J_2(\Delta\varepsilon))^{1/2}$. In the adopted sign convention, negative values of $I_1(\Delta\varepsilon)$ indicate volumetric dilation, and positive values of $I_1(\Delta\varepsilon)$ indicate volumetric contraction.

The windows used to perform the DVC correlations had a cubic shape with four faces parallel to the main compressive stress σ_1 and two faces perpendicular to it. Following tests on the influence of the correlation window size and node spacing size on the resolution, computation time, and robustness of the resulting displacement fields, we selected a correlation window size of 10 voxels (65 μm) and node spacing size of 20 voxels (130 μm). This choice ensured that each window contained at least a subvolume of one grain and a subvolume of one pore, maximizing the contrast for volume correlation. To characterize the resolution of strain values obtained from DVC analysis in these experiments, and thus the lower limit of robust strain values, we performed DVC analysis on: (1) the same tomogram (i.e., autocorrelation), (2) two tomograms that were acquired at the same differential stress, (3) two tomograms that were separated by 1 MPa of differential stress (131 MPa and 132 MPa), and (4) two tomograms that were separated by 36 MPa of differential stress (112 MPa and 148 MPa) (Fig. 3a).

The resulting strain populations show that the autocorrelation (test 1) produced smaller strain magnitudes than the other tests, as expected (Fig. 3b, c). The nonzero displacement field produced in this autocorrelation calculation arises from interpolation between integer (pixel) displacements in the subpixel resolution method used in the TomoWarp2 software. The strain populations of test 2 and test 3 were similar to each other, indicating that a differential stress increase of 1 MPa did not produce significant deformation of the sandstone in this increment of the experiment. The larger strain magnitudes produced in test 4 indicate that, as expected, more strain accumulated during the 36 MPa of differential stress increase than during the 1 MPa of differential stress increase of test 3.

The incremental strain populations produced in the autocorrelation test (test 1) suggest a characteristic error in the calculated strains (Fig. 3b, c). Twice the standard deviation of the $I_1(\Delta\varepsilon)$ and $I_2(\Delta\varepsilon)$ populations produced in test 2 were used as the thresholds for the populations of these two invariants. These thresholds are more conservative than those produced in the autocorrelation of test 1, and so produce strain populations with lower ratios of signal to noise. Invariants with magnitudes below these thresholds, which are on the order of 10^{-3} for $I_1(\Delta\varepsilon)$ and $I_2(\Delta\varepsilon)$, were removed from the incremental strain fields.

3.5. *Fontainebleau Sandstone Samples*

Fontainebleau sandstone is a quartz arenite of Oligocene age outcropping around Fontainebleau City near Paris, France (Bourbie and Zinszner 1985). This sandstone is considered an ideal reservoir rock because it has a relatively homogeneous mineralogical composition (> 99% quartz), well-sorted grain size with an average diameter of 0.25 mm, and a wide range of porosities (3–30%) depending on the degree of quartz cementation. As expected for a clean sandstone with a characteristic grain diameter of 0.25 mm, the high-porosity specimens have a high permeability (Bourbie and Zinszner 1985). In specimens with porosities higher than 5%, most of the pores are connected in three dimensions (Fredrich et al. 1993). For samples with porosities smaller than 4%, the pore space is less connected, so the permeability is at least one order of magnitude lower than that of samples with higher porosities (Fredrich et al. 1993). In addition, the P-wave velocity in this sandstone decreases almost linearly from 5500 to 3000 m s^{-1} as the porosity increases from 3 to 30% (Bourbie and Zinszner 1985).

In the present study, three specimens of diameter 5 mm and length 10 mm were cored perpendicular to bedding from a single $10 \times 10 \times 10$ cm^3 Fontainebleau sandstone block with a matrix made of quartz and a minor amount (< 1%) of iron oxides. The mean grain size was 0.25 mm. The porosity derived from the initial three-dimensional datasets following segmentation was in the range of 5.5–7%. The porosity was also measured by weighing ten dry specimens before and after imbibition with water, and a value of 6 ± 1% (standard deviation) was obtained, consistent with the porosity measured using X-ray tomography. Three-dimensional imaging indicates that the initial pore structure was almost entirely connected in three dimensions. There was almost no microporosity in these specimens (i.e., porosity within grains), except for a few micrometer-size fluid inclusions in the quartz grains that were not connected to the bulk porosity.

4. *Results*

4.1. *Macroscopic and Microscopic Deformation*

In each experiment, the macroscopic axial strain increased faster than a linear trend (i.e., superlinearly) relative to the applied stress until the differential stress reached about 20 MPa. This initial nonlinear phase arose from the closure of the weakest preexisting microfractures and pores, grain reorientation, and the deformation of weak intergrain contacts. In experiments on Mount Scott granite, Katz and Reches (2004) observed inconsistencies in the initial, low differential stress behavior (an increase in the deformation modulus, $\partial\sigma_D/\partial\varepsilon_M|_{P_c}$) with increasing strain for some specimens and a decrease in the deformation modulus for others. Consequently, we cannot conclude that the initial softening ($\partial^2\sigma_D/\partial\varepsilon_M^2|_{P_c} < 0$) is generic for low-porosity Fontainebleau sandstone under a confining stress of 10 MPa. In addition, the closure of microfractures and pores may cause hardening rather than softening. However, the initial softening was more pronounced in our experiments than those of Katz and Reches (2004). We did not investigate these initial mechanisms in detail because the focus of this work was on the damage preceding macroscopic failure.

After the initial nonlinear phase, the macroscopic mechanical behavior transitioned gradually from a quasilinear phase at intermediate differential stresses (20–80 MPa), to significant deviation from

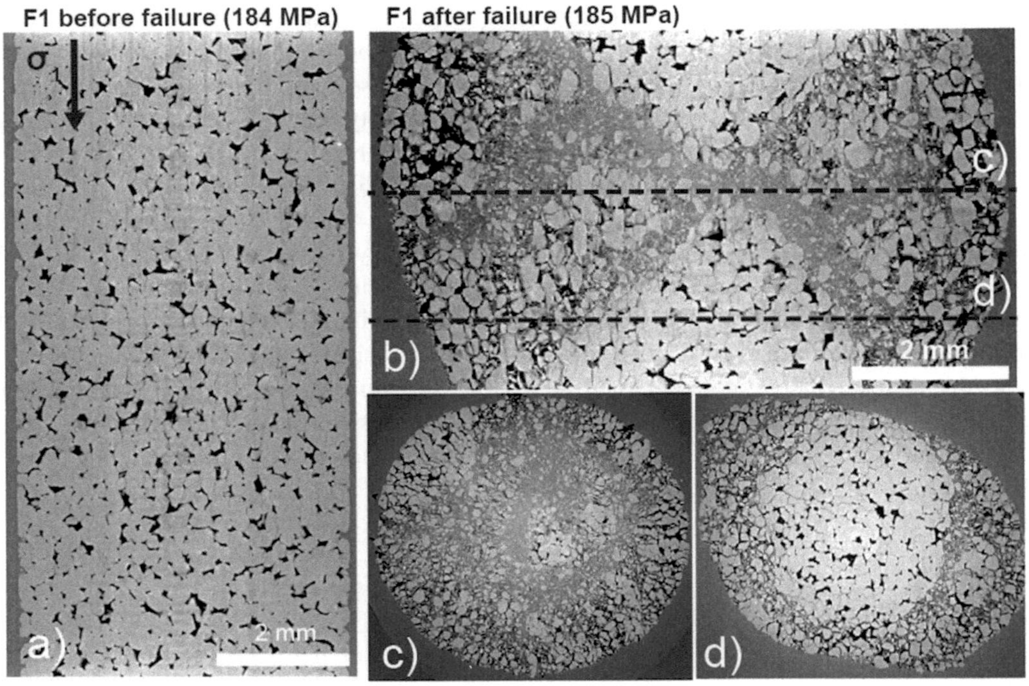

Figure 4
Two-dimensional views of experiment F1 immediately preceding and following macroscopic failure at differential stresses of 184 and 185 MPa. **a** Vertical axial transect at the onset of failure showing quartz grains, pores, and microfractures oriented subparallel to the main compressive stress direction (vertical). **b** Vertical transect of the sample after macroscopic failure. Compared with **a**, the sample has shortened and widened significantly because of failure. Dashed lines indicate the location of the two perpendicular (horizontal) cross sections shown in **c** and **d**. The sample failed through the formation of conical fracture zones in which grains were comminuted

quasilinear behavior associated with significant dilation, and finally to macroscopic failure (Fig. 1). We chose to define the yield point as the point at which the measured volumetric strain differed by 3% from the strain predicted by linear regression of the mean stress–strain data over the range $20 \leq \sigma_D \leq 80$ MPa (also named point C′ in Brace et al. 1966). After the yield point was reached, the relationship between the differential stress and axial strain became increasingly nonlinear until macroscopic failure occurred. The yield point corresponded to an axial strain close to 0.08 for experiments F1 and F3 and to 0.04 for experiment F2 where a fault localized near the yield point. The volumetric strain curves revealed that the sample compacted (volumetric strain increased) preceding the yield point, and then began to dilate (volumetric strain decreased) near the yield point in each experiment (Fig. 1). Following yielding and preceding macroscopic failure, microfractures propagated through grains as well as along grain boundaries (Figs. 2b, 4a, 5a–c, 6c, 7, Videos S1 and S2).

Consistent with the similar stress–strain curves of experiments F1 and F3, the evolving microfracture distributions were also similar in these experiments. In particular, damage remained diffuse and distributed throughout much of each experiment (Fig. 4a), and only concentrated into a narrower zone close to failure ($\sim$ 95% of the failure stress) (Video S2). In contrast, in experiment F2, damage localized early ($\sim$ 75% of the failure stress) along a narrow planar zone that evolved into a fault (Fig. 5, Video S1). These differing behaviors are also revealed in the spatial distribution of high strains from the DVC analyses. The incremental strains above the 95th percentile of each strain population were more spatially diffuse in experiments F1 (Video S3) and F3 (Video S5) than in experiment F2 (Video S4). Despite these differences in strain localization preceding failure, the development of conical faults, in

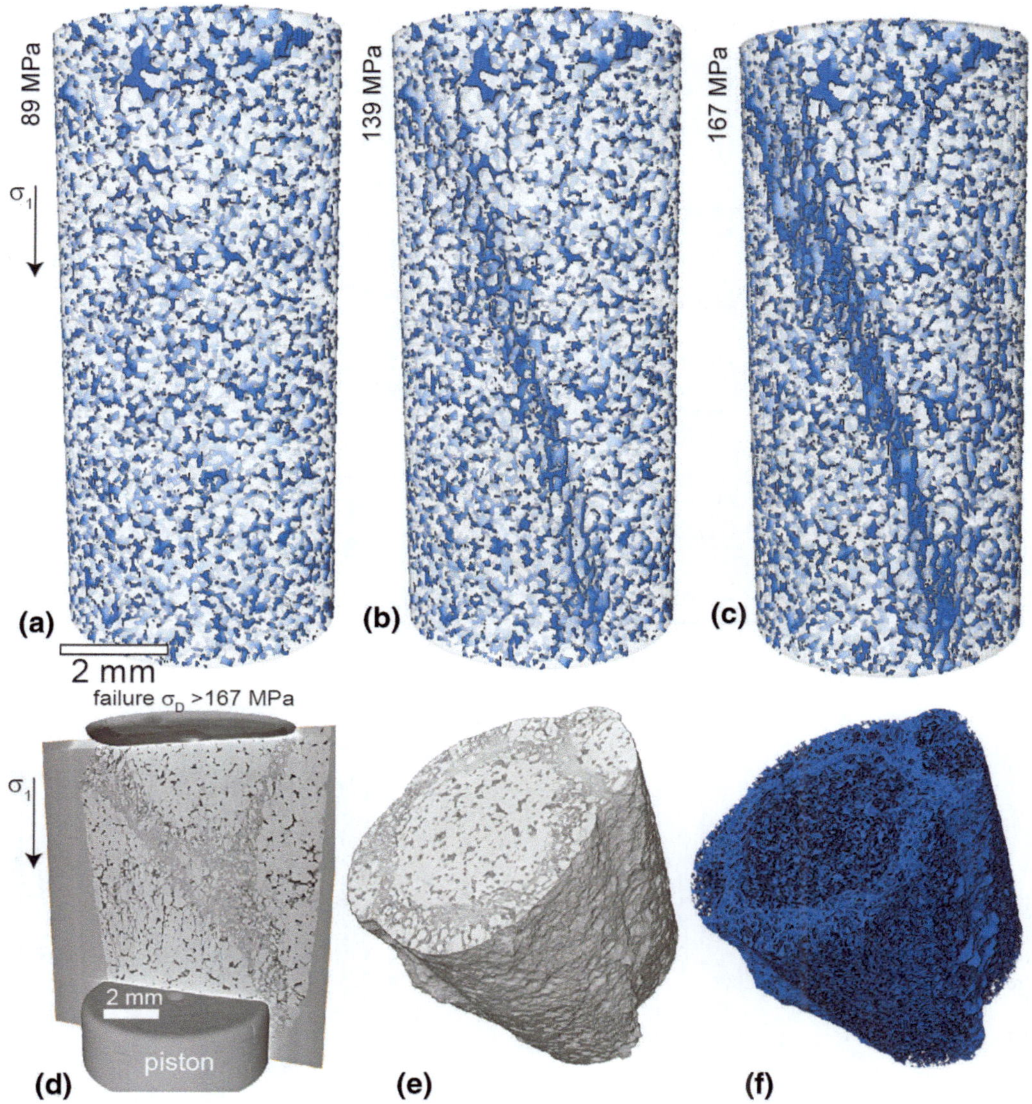

Figure 5
Evolution of pore space and fracture networks in experiment F2. **a–c** Three-dimensional rendered views of the specimen at increasing differential stresses. Porosity and fractures are shown in blue, and the quartz matrix is shown in white. **c** The onset of failure (167 MPa). Video S1 shows a complete time-lapse three-dimensional rendered view of the specimen throughout the experiment. **d** Sample after failure ($\sigma_D > 167$ MPa) with the two pistons of the HADES rig displayed. **e** Three-dimensional rendered view of the sample after failure with comminuted grains. **f** Structure of the porous network, which was 99% connected in three dimensions both preceding and following failure. Following failure, the porosity in the fracture zones that host comminuted grains was smaller than the porosity in the other parts of the rock, where the pore space is similar to that in the rock before loading

which grain comminution and porosity reduction occurred, ultimately caused the macroscopic failure of each sandstone sample (Figs. 4b–d, 5d–f). Dilatational shear failure was the main mechanism of faulting, and significant compaction, grain comminution, and porosity reduction were observed in the final fault zone. The fractures developed into conical shapes because the pistons of the rig were not lubricated. This lack of lubrication produced frictional resistance to lateral movement of the sandstone in contact with the pistons, which localized shear strain and so promoted fracture nucleation near the

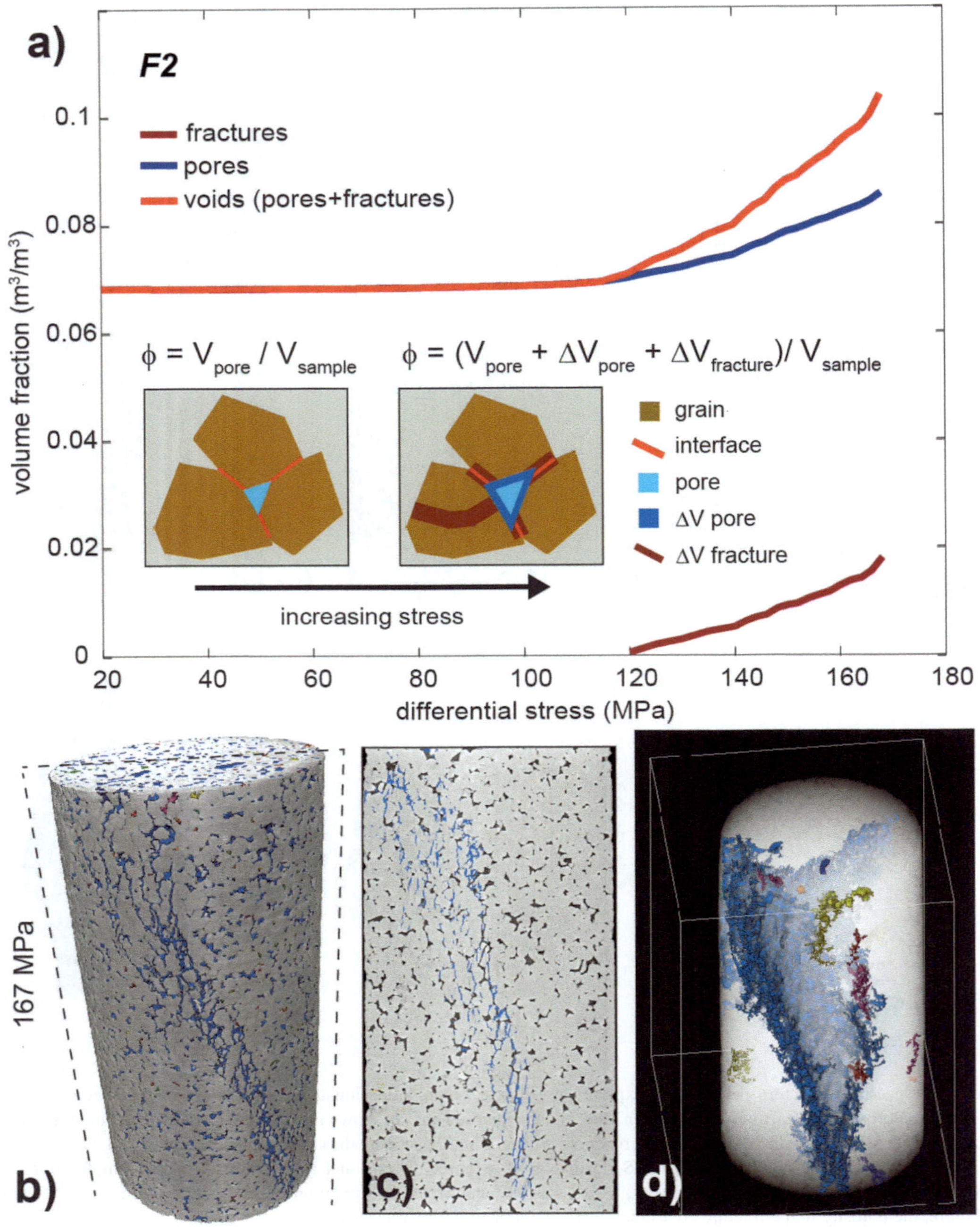

Figure 6

Pores and fractures identified after segmentation of the three-dimensional datasets of experiment F2. **a** Evolution of the pore, fracture, and total void (i.e., pore + fracture) volume fractions as a function of the differential stress. The total void volume fraction (red curve) can be separated into propagating fractures (blue curve) and expanding pores (dark-red curve) in this experiment. The inset sketch shows how the pore and fracture volumes increased. **b–d** Results of segmentation for experiment F2 at a differential stress of 167 MPa. **b** Three-dimensional rendered view of the specimen at the onset of failure with quartz shown in light gray, and pores and fractures shown in colors. **c** Vertical transect of the previous image, with fractures shown in blue, pore space shown in black, and quartz shown in light gray. **d** Three-dimensional view of the microfracture network shown in blue and other colors. Each color shows a unique, unconnected, fracture network edges of the top and bottom sandstone–piston interfaces (e.g., Peng and Johnson 1972). Consequently, the boundary condition at the pistons may have controlled the final angle of the faults at failure.

In all the experiments, more than 99% of the voids, including the pores and fractures, were connected in three dimensions (Fig. 6b), both preceding and following faulting (Fig. 5). In experiment F2, the gray levels (X-ray attenuation coefficients) of the fractures were slightly higher than those of the pores, enabling segmentation of the fractures from the pore space (Fig. 6). This segmentation revealed that a core-spanning fracture developed through the linkage of subvertical, dilating microcracks (Fig. 6b, c). The core-spanning fracture did not immediately result in macroscopic failure denoted by a reduction in the axial stress. This segmentation also enabled separation of the contributions to the total porosity increase from the growth of preexisting pores, and the propagation and opening of new fractures (Fig. 6a).

Observations of microstructures at the grain scale using scanning electron microscopy (SEM) and visual inspection of the tomograms at the micrometer scale enabled various deformation mechanisms to be differentiated, including intergranular fracturing (i.e., grain boundary opening), transgranular fracturing, and grain comminution (Fig. 7). Each of the sandstone cores failed macroscopically through the development of conjugate faults (Figs. 4b, 5d, 7a). Cracks propagated both within the quartz grains (i.e., transgranular fractures) (Fig. 7b, d) and between grains (i.e., intergranular fractures) (Fig. 7c). In some cases, these cracks stopped at a pore interface. In experiment F2, the cracks self-organized into an incipient fault in which shear displacements could be observed (Fig. 7g–i).

In summary, the three samples have similar macroscopic stress–strain relationships characterized by: (1) initial macroscopic compaction arising from the closing of voids, (2) macroscopic dilation arising from microfracture development and pore opening, (3) transgranular and intergranular propagation of fractures that drives the strain–stress relationship from linear to nonlinear, and (4) macroscopic shear failure due to the coalescence of microscopic fractures. These macroscopic behaviors arose from microscopic deformation processes that the four-dimensional strain tensors revealed. In experiments F1 and F3, damage localized into a narrow deformation zone near macroscopic failure. In contrast, in experiment F2, damage began to localize into a narrow zone near the macroscopic yield point, and the core-spanning fault grew slowly enough to be captured in several tomograms preceding macroscopic failure.

4.2. Global Evolution of Damage toward Failure

To characterize the evolution of cumulative damage toward failure, we extracted the voids from the quartz grains in each tomogram. Segmentation into rock matrix and voids (pores and fractures) provided the volume fraction of voids as a function of increasing differential stress (Figs. 6, 8, S2). To quantify the accumulation of damage, including opening pores and propagating fractures, Renard et al. (2018) employed a damage index, $D_\phi = \frac{\phi - \phi_i}{1-\phi_i}$, where ϕ_i is the initial void fraction of the sample under the initial confining pressure, preceding axial loading, and ϕ is the void fraction measured at a given differential stress. In this approach, the normalized distance to failure in stress space is defined as $\Delta = \left(\frac{\sigma_f - \sigma}{\sigma_f}\right)$, where σ_f is the differential stress at which failure occurred, and σ is the differential stress when the tomogram was acquired. When $\Delta = 0$, the rock is at failure, and when $\Delta = 1$, no differential stress is applied to the sample. Renard et al. (2018) applied this normalization to show that damage accelerates as a power law in quartz monzonite ($\partial D_\phi / \partial \sigma \sim \Delta^{-\beta}$), in agreement with mechanical models that consider failure as a critical phenomenon (Dahmen et al. 2009; Girard et al. 2010). This behavior suggests that failure is a critical phenomenon in crystalline rocks such as monzonite.

However, in our sandstone experiments, D_ϕ did not increase as a negative power of Δ as failure was approached. In these experiments, the total porosity, ϕ, and the damage index, D_ϕ, remained near their initial values until the differential stress was several MPa smaller than the differential stress at yield point, and then changed rapidly until failure occurred (Fig. 8a). This acceleration of D_ϕ could not be fitted by a power law or an exponential relationship

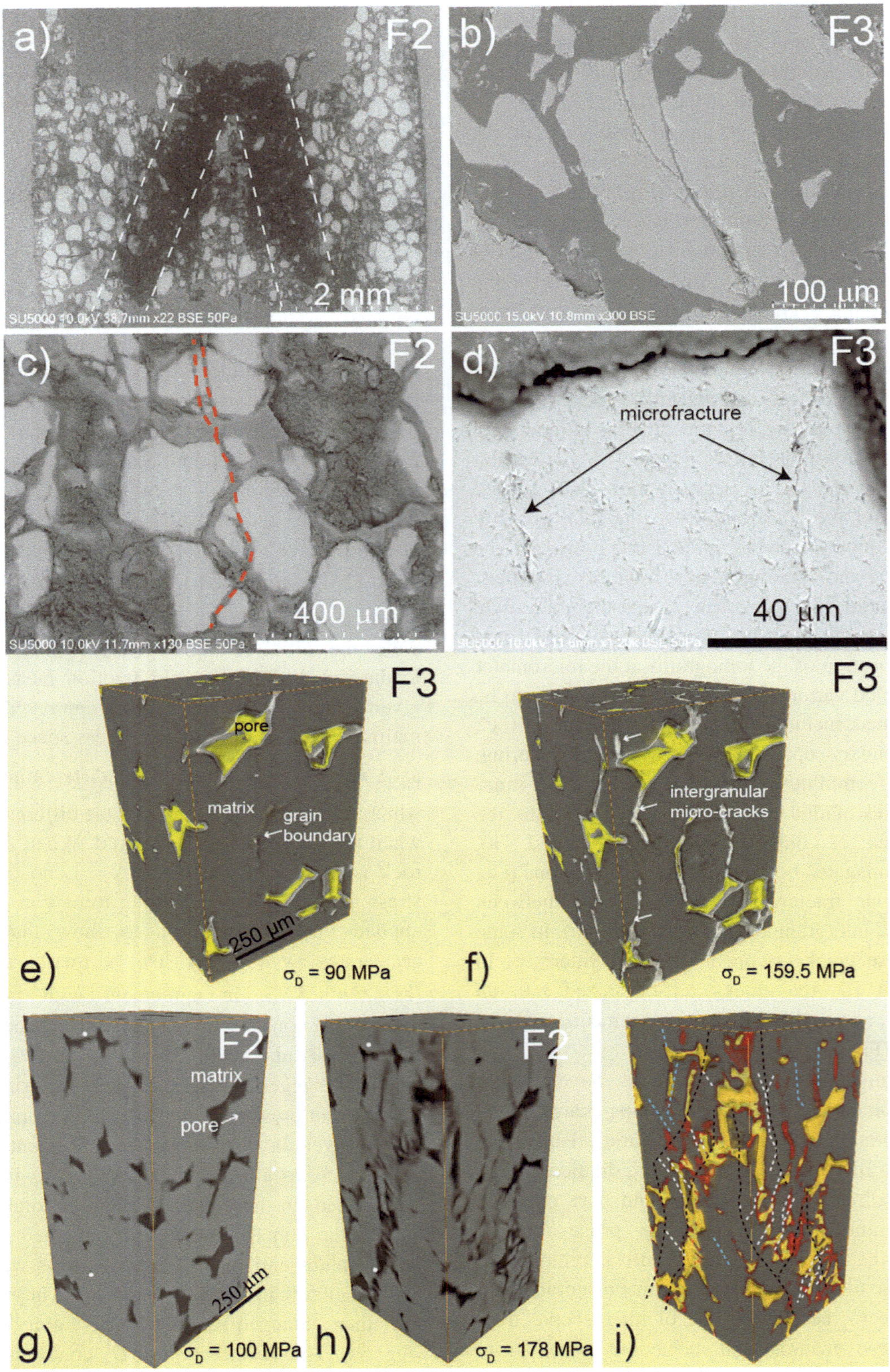
a)
F2
2 mm
b)
F3
100 μm
c)
F2
400 μm
d)
F3
microfracture
40 μm
F3
pore
matrix
grain boundary
250 μm
e)
σD = 90 MPa
F3
intergranular micro-cracks
f)
σD = 159.5 MPa
F2
matrix
pore
250 μm
g)
σD = 100 MPa
F2
h)
σD = 178 MPa
i)

◀Figure 7
Microstructures of deformation. **a–d** Scanning electron microscopy images of samples F2 and F3 after failure. **a** Sample F2 with fault zones highlighted by white dashed lines. **b** Transgranular fracture. **c** Transgranular fractures that may have been produced by a directed force network, or chain (dashed red lines). **d** Onset of microfractures in a quartz grain. **e, f** Three-dimensional rendering of X-ray microtomograms from experiment F3 **e** before and **f** after failure showing mostly intergranular fractures. Yellow shows porosity, and white shows intergranular fractures. Three-dimensional rendered views of experiment F2 **g** before and **h, i** after failure showing both intergranular and transgranular fractures. **i** Red highlights damaged areas. **g, h** White dots show strain markers in both images. **i** White dotted lines highlight shear bands. Blue dotted lines show intergranular and transgranular fractures. Dark dotted lines highlight the fault zones

(Fig. 8b), so this relationship differs from the power-law acceleration observed for the crystalline quartz monzonite (Renard et al. 2018). Due to the instability of the confining pressure pump in experiment F1, we focus on the evolving damage index in experiments F2 and F3 (Fig. 8). Figure S2 shows the results from all three experiments. We interpret the absence of power-law behavior as the pinning of microfractures when they reach pore walls: the presence of pores screens the stress concentration at the fracture tips, which may reduce the long-range elastic interactions necessary to develop a power-law increase of damage as failure is approached (Dahmen et al. 2009; Girard et al. 2010).

4.3. Evolving Statistics of Incremental Strains

To further characterize the evolution of microscopic damage, we performed DVC analysis of sequential pairs of tomograms in each experiment. The resulting series of three-dimensional incremental strain tensor fields, normalized by division by the incremental macroscopic axial strain, $\Delta\varepsilon_{zz}^{M}$,, enabled quantitative assessment of the evolving volumetric and shear strain localization (Fig. 9). The normalization by $\Delta\varepsilon_{zz}^{M}$ was performed because the macroscopic strain increase between pairs of tomograms was not constant (Fig. 10a, b, g). The top 5% of the negative $\frac{\Delta I_1(\Delta\varepsilon)}{\Delta\varepsilon_{zz}^{M}}$ (dilatational), positive $\frac{\Delta I_1(\Delta\varepsilon)}{\Delta\varepsilon_{zz}^{M}}$ (contractive), and $\frac{(3\Delta J_2(\Delta\varepsilon))^{1/2}}{\Delta\varepsilon_{zz}^{M}}$ (shear) populations in five differential stress increments of each experiment (Fig. 9, Videos S3–5) revealed the localization of the core-spanning fracture observed in experiment F2 (Figs. 5a–c, 6b–d,

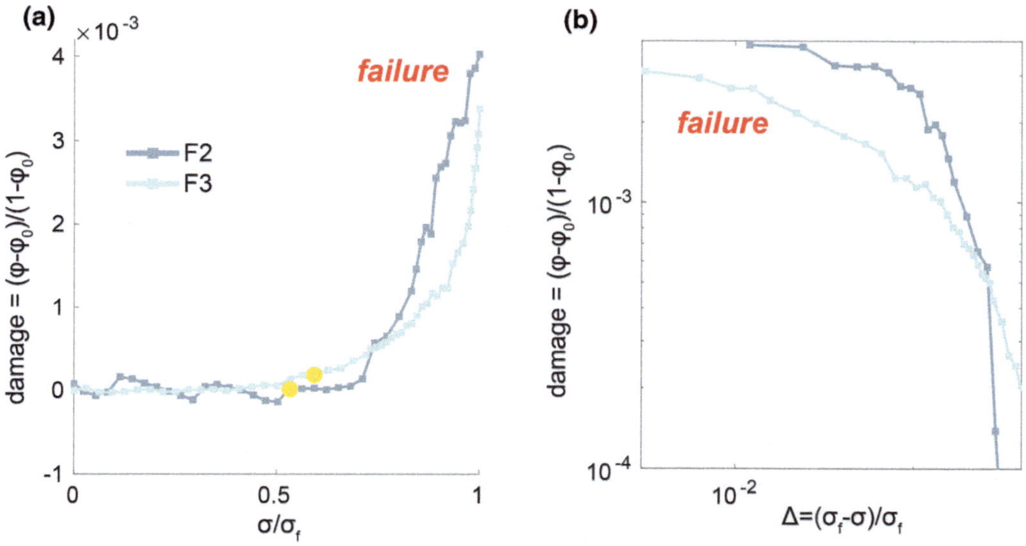

Figure 8
Evolution of the damage index as the sandstone specimens approached failure in experiments F2 and F3. Figure S2 shows data from experiment F1. **a** Evolution of the damage index as a function of the differential stress normalized by the differential stress at failure, σ/σ_f. The porosity and damage index increased as the differential stress approached the differential stress at failure. **b** Evolution of the damage index as a function of the normalized distance to failure, $\Delta = (\sigma_f - \sigma)/\sigma_f$, in log–log space. Failure occurred when $\sigma/\sigma_f = 1$ in **a** and when $\Delta = 0$ in **b**. Yellow circles show the yield points

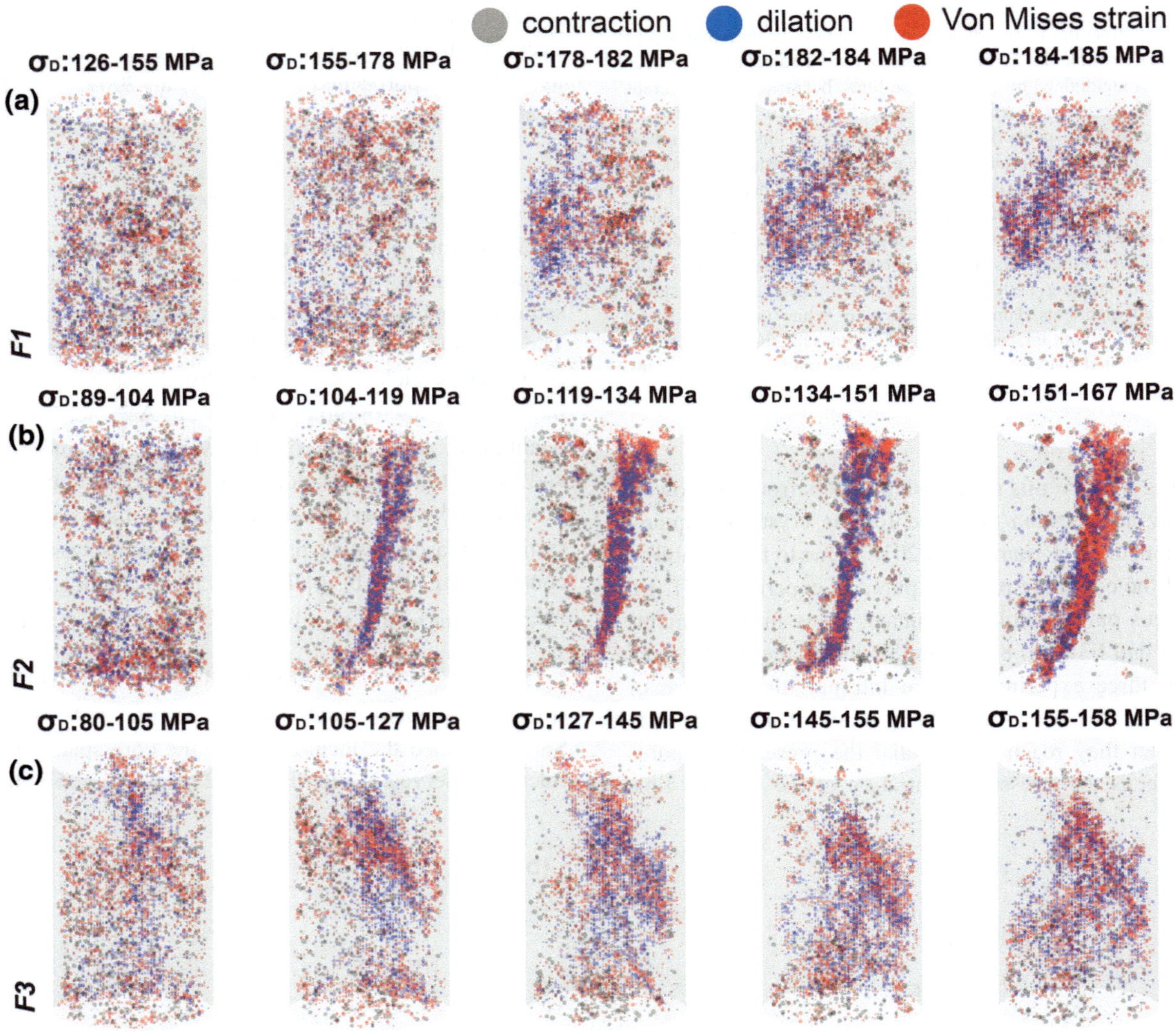

Figure 9

Spatial distribution of incremental strain invariant magnitudes above the 95th percentile of each incremental strain population for experiments F1 (**a**), F2 (**b**), and F3 (**c**). The sizes of the points are proportional to the magnitude of the incremental strain invariants. Gray dots show the contractive $\frac{I_1(\Delta\varepsilon)}{\Delta\varepsilon_{zz}^{M}}$, blue dots show dilatational $\frac{I_1(\Delta\varepsilon)}{\Delta\varepsilon_{zz}^{M}}$, and red dots show the Von Mises incremental strain, $\frac{(3J_2(\Delta\varepsilon))^{1/2}}{\Delta\varepsilon_{zz}^{M}}$, used here to characterize shear strain. The differential stresses above each figure indicate the stresses at which the pairs of tomograms used in the DVC analysis were acquired. Videos S3–S5 show the spatial distribution of high strains throughout each experiment

Video S2). While segmentation of the tomograms provided information about the evolving dilatational, contractional, and shear strain as well as pore volumes, fractures, and their connectivity, DVC analysis provided displacement fields from which strain fields were calculated with a strain resolution close to 10^{-3} for these scans.

We calculated the incremental strain fields between pairs of three-dimensional datasets that were separated by an approximately constant change in the macroscopic axial strain, $\Delta\varepsilon_{zz}^{M}$. For each experiment, ten incremental strain fields that encompassed each complete experiment were calculated (Fig. 10). To track the interplay between nondeviatoric and deviatoric strains, we report histograms of the first

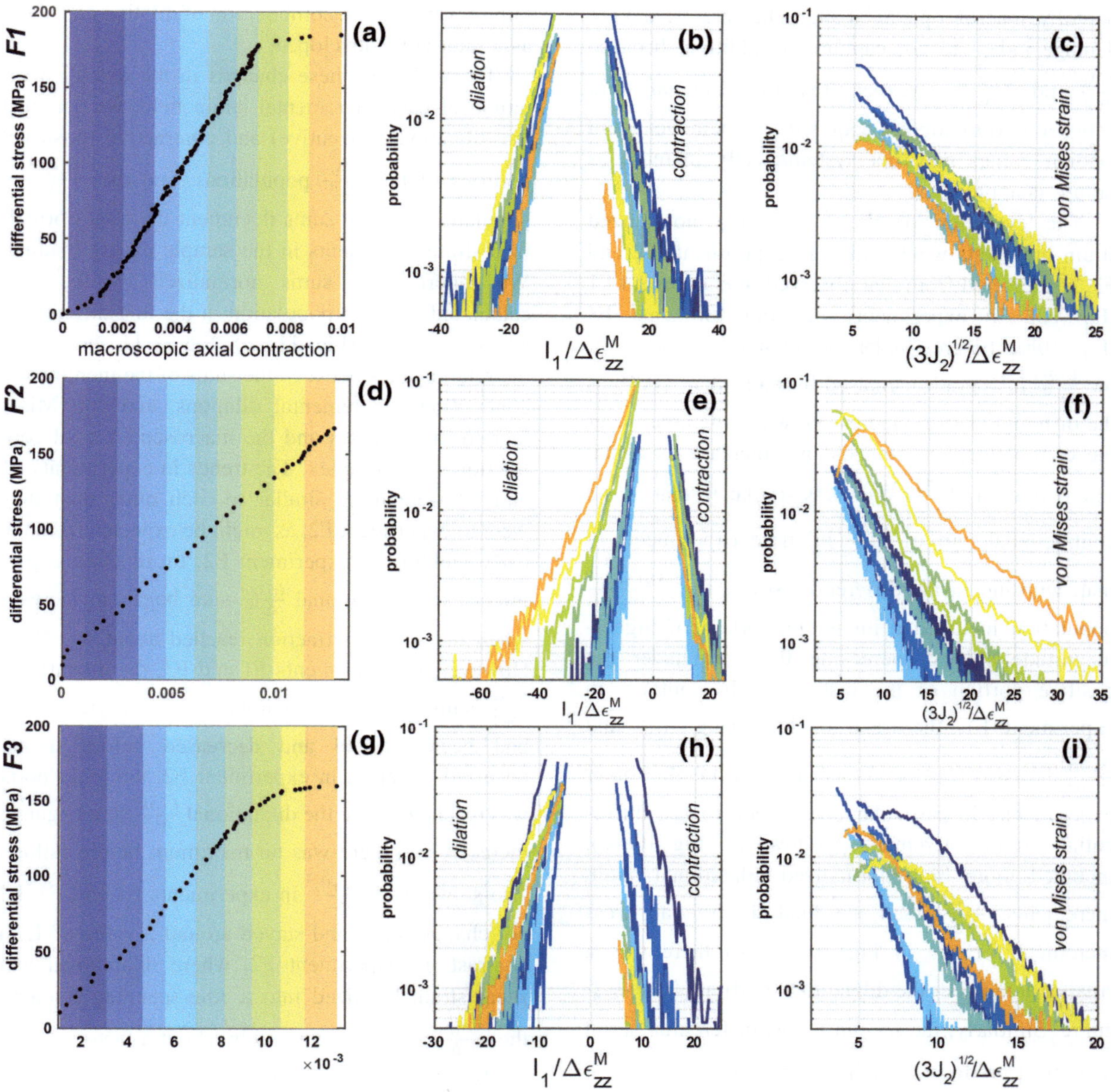

Figure 10
Histograms of incremental normalized strain invariants determined by DVC analysis of pairs of three-dimensional datasets for experiments F1 (**a–c**), F2 (**d–f**), and F3 (**g–i**). **a**, **d**, **g** Differential stress–axial strain curves. The rectangles show the strain intervals between which we performed digital image correlation. **b**, **e**, **h** Histograms of $\frac{I_1(\Delta\varepsilon)}{\Delta\varepsilon_{zz}^M}$. **c**, **f**, **i** Histograms of $\frac{(3J_2(\Delta\varepsilon))^{1/2}}{\Delta\varepsilon_{zz}^M}$. The color of each histogram corresponds to the strain interval displayed in the first column. The gap in the histogram curves at small magnitudes arises because strain values below the identified threshold (Fig. 3) were removed

invariant of the incremental strain, $I_1(\Delta\varepsilon)$, divided by $\Delta\varepsilon_{zz}^M$, and the square root of the von Mises incremental strain, $(3J_2(\Delta\varepsilon))^{1/2}$, divided by $\Delta\varepsilon_{zz}^M$ (Fig. 10). Because the incremental normalized strain invariants $\frac{I_1(\Delta\varepsilon)}{\Delta\varepsilon_{zz}^M}$ and $\frac{(3J_2(\Delta\varepsilon))^{1/2}}{\Delta\varepsilon_{zz}^M}$ were calculated at many locations (> 50,000 points per tomogram pair), they provided additional information about deformation within the sandstone specimens that global measures such as porosity, macroscopic axial and radial strains, and the damage index do not capture. In addition, because $\frac{I_1(\Delta\varepsilon)}{\Delta\varepsilon_{zz}^{M}}$ and $\frac{(3J_2(\Delta\varepsilon))^{1/2}}{\Delta\varepsilon_{zz}^{M}}$ are measures of incremental strain between two tomogram acquisitions, their mean and median values may not systematically increase as failure is approached.

The evolution of the incremental normalized strain invariants were similar in experiments F2 and F3, but differed from that observed in experiment F1 (Fig. 10). In experiments F2 (Fig. 10b) and F3 (Fig. 10h), the area under the dilatational portion of the $\frac{I_1(\Delta\varepsilon)}{\Delta\varepsilon_{zz}^{M}}$ histogram increased by more than 200% as the differential stress increased, while the area under the contractional portion did not increase as much. The evolution of $\frac{(3J_2(\Delta\varepsilon))^{1/2}}{\Delta\varepsilon_{zz}^{M}}$ was similar to that of the contractional part of $\frac{I_1(\Delta\varepsilon)}{\Delta\varepsilon_{zz}^{M}}$ in these experiments in that, with increasing differential stress, the $\frac{(3J_2(\Delta\varepsilon))^{1/2}}{\Delta\varepsilon_{zz}^{M}}$ population included a higher proportion of higher-magnitude values, increasing the area under the positive portion of the histogram. In contrast, in experiment F1, the trend of increasing $\frac{I_1(\Delta\varepsilon)}{\Delta\varepsilon_{zz}^{M}}$ and $\frac{(3J_2(\Delta\varepsilon))^{1/2}}{\Delta\varepsilon_{zz}^{M}}$ was not as consistent as in the other experiments. In particular, the second to last differential strain increment (yellow in Fig. 10a–c), included more high normalized dilatational strain increments, $\frac{I_1(\Delta\varepsilon)}{\Delta\varepsilon_{zz}^{M}}$, than the final differential stress increment (orange in Fig. 10a–c), so there was a smaller area under the dilatational portion of the $\frac{I_1(\Delta\varepsilon)}{\Delta\varepsilon_{zz}^{M}}$ curve. Similarly, the second to last differential strain increment produced a higher area under the $\frac{(3J_2(\Delta\varepsilon))^{1/2}}{\Delta\varepsilon_{zz}^{M}}$ curve than the final differential strain increment, indicating a higher magnitude of overall shear strain (Fig. 10c). The difference in behavior could be related to the difference of loading path, since the confinement was decreased as failure was approached in experiment F1 but remained constant in experiments F2 and F3 (Fig. 1). To summarize the results of Fig. 10: for samples F1 and F3, most of the strain in the sample was accommodated by local dilation events that we relate to the opening of grain boundaries, whereas for sample F2, the road to failure was due to a combination of dilation and shear as a fault zone developed.

To synthesize these changes in the overall magnitude of each incremental strain field, we summed the dilatational (negative) and contractive (positive) portions of the $\frac{I_1(\Delta\varepsilon)}{\Delta\varepsilon_{zz}^{M}}$ populations, and the $\frac{(3J_2(\Delta\varepsilon))^{1/2}}{\Delta\varepsilon_{zz}^{M}}$ populations. These sums document the total normalized strain increments in the sample during loading. We report these sums normalized by the sum calculated for each population in the first differential stress increment (Fig. 11a–c). Figure 11 shows that the relationships between the sums of the incremental contractions, incremental dilations, and von Mises incremental stresses and the macroscopic axial contraction were complex. The trends in experiments F1 and F3 were more similar to each other than they were to experiment F2, as might be expected from the early yielding in experiment F2. In all three experiments, the dilatational $\frac{I_1(\Delta\varepsilon)}{\Delta\varepsilon_{zz}^{M}}$ sum began to increase when the axial contraction reached about 6×10^{-3}. However, in experiments F1 and F3, the dilatational $\frac{I_1(\Delta\varepsilon)}{\Delta\varepsilon_{zz}^{M}}$ sum reached a maximum after the yield point had been reached and decreased before failure occurred, whereas in experiment F2, the yield point was reached before the dilatational $\frac{I_1(\Delta\varepsilon)}{\Delta\varepsilon_{zz}^{M}}$ sum began to increase, and there was no maximum before failure. The sums of $\frac{(3J_2(\Delta\varepsilon))^{1/2}}{\Delta\varepsilon_{zz}^{M}}$ in experiments F1 and F3 did not change much and stayed around a value of 1. In contrast, in experiment F2, where dilatational and shear strain localized into a core-spanning fracture, the $\frac{(3J_2(\Delta\varepsilon))^{1/2}}{\Delta\varepsilon_{zz}^{M}}$ sum increased by about a factor of two.

To track the overall strain increments, we extracted the number of local incremental strain values above a threshold (Fig. 11d–f), and then normalized by the sum calculated for each population in the first differential stress increment. The evolution of this quantity is similar to that of the sum of the increments shown in Fig. 11a–c.

The evolution of the spatial distributions of the normalized incremental strain invariants (Fig. 9) illuminates the morphology of the coalescing microfractures that produced the trends observed in the histograms and sums. Consistent with the

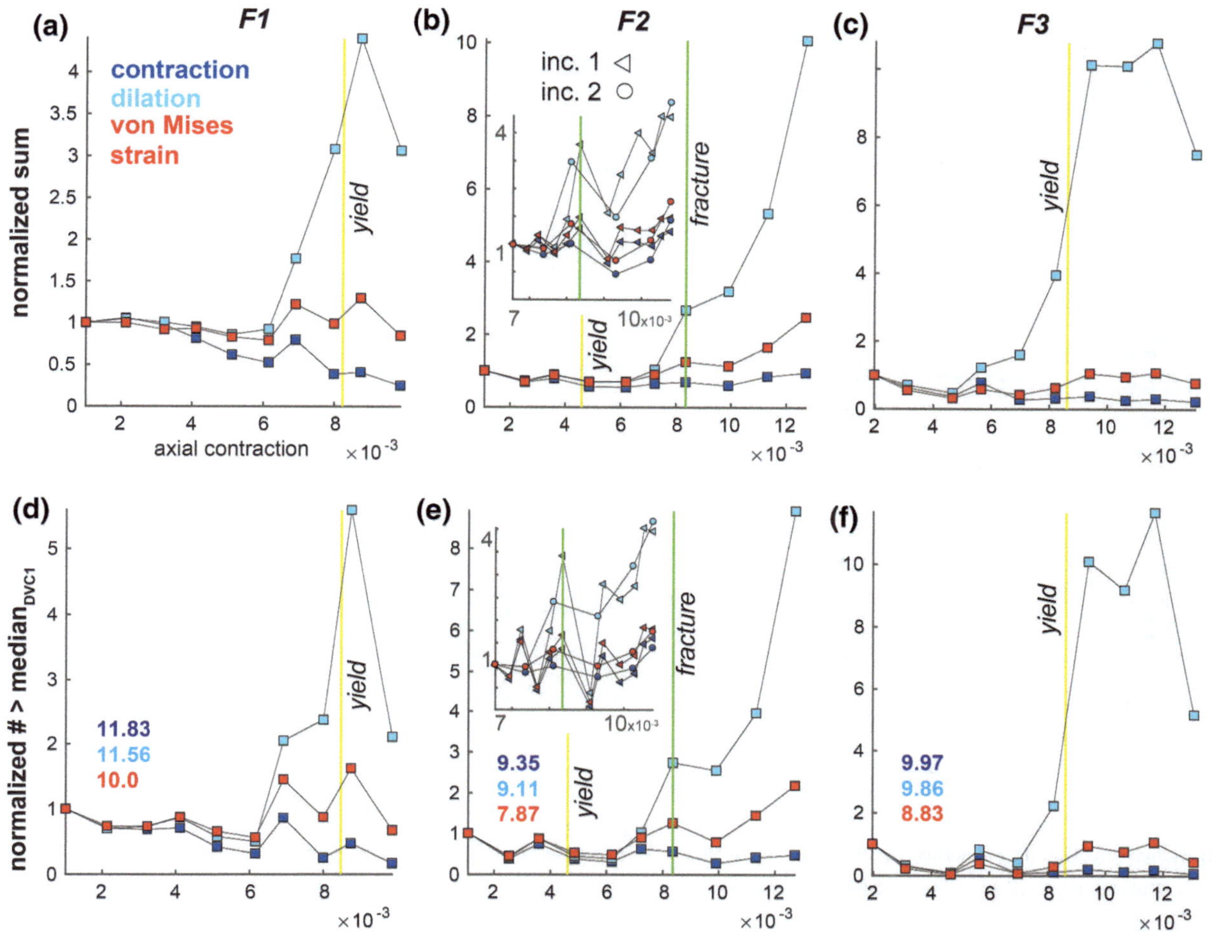

Figure 11

Evolution of sums of incremental strain invariants (**a–c**) and number of values above the median identified in the first increment, $\mathrm{median_{DVC1}}$, (**d–f**) in experiments F1 (**a**, **d**), F2 (**b**, **e**), and F3 (**c**, **f**). Dark-blue, light-blue, and red squares show values of contractive $\frac{I_1(\Delta\varepsilon)}{\Delta\varepsilon^{M}_{zz}}$, dilatational $\frac{I_1(\Delta\varepsilon)}{\Delta\varepsilon^{M}_{zz}}$, and $\frac{(3J_2(\Delta\varepsilon))^{1/2}}{\Delta\varepsilon^{M}_{zz}}$ (shear strain). Each sum is normalized by the sum calculated in the first increment. Each number of strain values above $\mathrm{median_{DVC1}}$ is normalized by that number in the first increment. Inset of (**b**) shows sums near the yield point of experiment F2 where DVC analyses were performed on 12 pairs of tomograms with successive stress step increase. Yellow lines show the yield point for each experiment. The green line shows the onset of fracture development in sample F2. The value of $\mathrm{median_{DVC1}}$ for each strain invariant population is shown in each of the plots, corresponding to the normalized incremental strain invariant. The dark-blue, light-blue, and red numbers indicate the median values of contraction, dilation, and von Mises incremental strains, respectively

evolution of the contractive vales of $\frac{I_1(\Delta\varepsilon)}{\Delta\varepsilon^{M}_{zz}}$ observed in the histograms and sums, high values of contractive $\frac{I_1(\Delta\varepsilon)}{\Delta\varepsilon^{M}_{zz}}$ did not localize within or around the fracture. This lack of localization of the contractive $\frac{I_1(\Delta\varepsilon)}{\Delta\varepsilon^{M}_{zz}}$ values produced only small changes in the sums and histogram shape, whereas the opening of and slip along fractures produced fracture-localized dilation and shear strain, thus increasing the sums and areas under the dilatational portion of the $\frac{I_1(\Delta\varepsilon)}{\Delta\varepsilon^{M}_{zz}}$ histogram and the $\frac{(3J_2(\Delta\varepsilon))^{1/2}}{\Delta\varepsilon^{M}_{zz}}$ histogram.

To track the interplay between dilatation, contraction, and shear strain, we show the dilative part of $I_1(\Delta\varepsilon)$, the contractive part of $I_1(\Delta\varepsilon)$, and $(3J_2(\Delta\varepsilon))^{1/2}$, without normalization by $\Delta\varepsilon^{M}_{zz}$, as a function of each value (Fig. 12). The curves outline pairs of strain values with a bivariate kernel density >

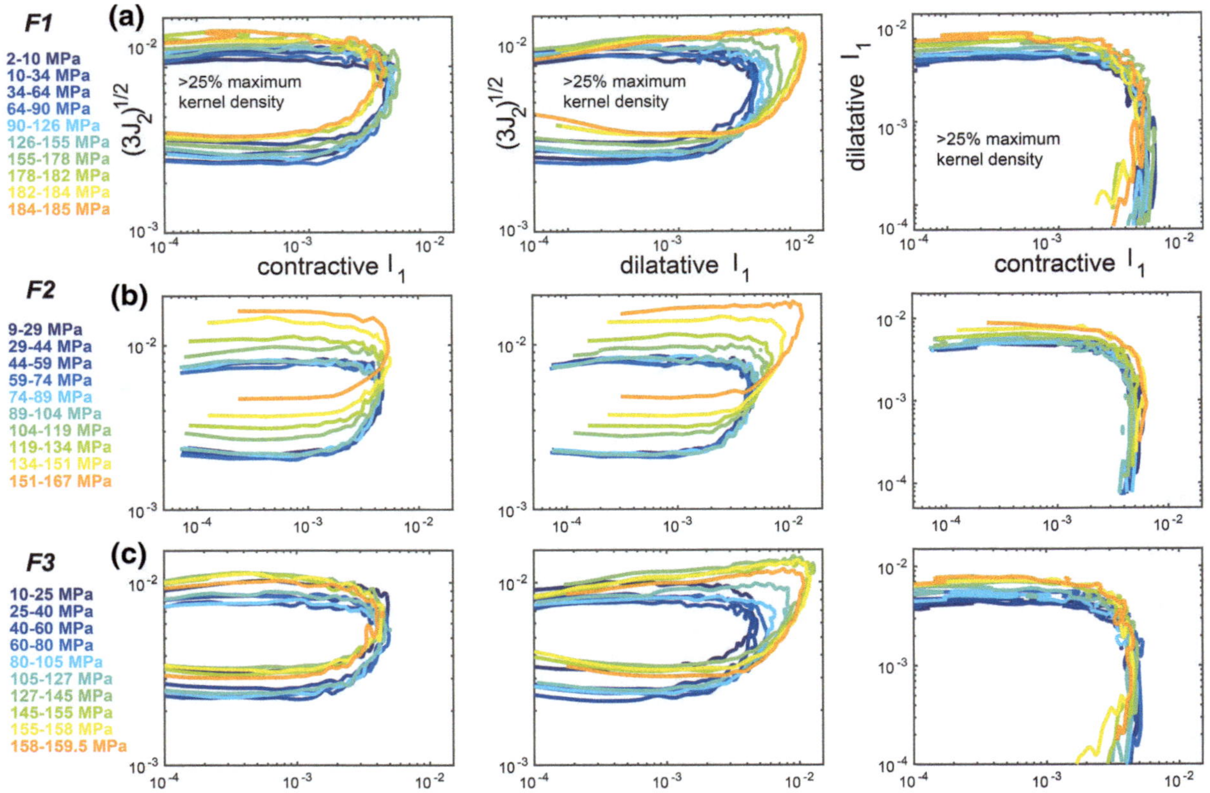

Figure 12

Evolution of deviatoric and normal incremental strain invariants for experiments F1 (**a**), F2 (**b**), and F3 (**c**). For each incremental strain field, we calculated the dilative $\frac{I_1(\Delta\varepsilon)}{\Delta\varepsilon_{zz}^{M}}$, contractive $\frac{I_1(\Delta\varepsilon)}{\Delta\varepsilon_{zz}^{M}}$, and $\frac{(3J_2(\Delta\varepsilon))^{1/2}}{\Delta\varepsilon_{zz}^{M}}$ value at each point. The bivariate kernel density of each pair of strain values was estimated assuming a Gaussian distribution: contractive $\frac{I_1(\Delta\varepsilon)}{\Delta\varepsilon_{zz}^{M}}$ versus $\frac{(3J_2(\Delta\varepsilon))^{1/2}}{\Delta\varepsilon_{zz}^{M}}$ (left), dilative $\frac{I_1(\Delta\varepsilon)}{\Delta\varepsilon_{zz}^{M}}$ versus $\frac{(3J_2(\Delta\varepsilon))^{1/2}}{\Delta\varepsilon_{zz}^{M}}$ (middle), and contractive $\frac{I_1(\Delta\varepsilon)}{\Delta\varepsilon_{zz}^{M}}$ versus dilative $\frac{I_1(\Delta\varepsilon)}{\Delta\varepsilon_{zz}^{M}}$ (right). Lines outline the strain values with a kernel density > 25% of the maximum kernel density

25% of the maximum kernel density. For the three specimens, these data show that: (1) incremental dilation increased much more than incremental compaction as failure was approached, (2) shear strain increased as failure was approached, and (3) the increase in shear strain acted in concert with dilation and was not strongly correlated with contraction (Fig. 12).

5. Discussion

5.1. Micromechanical Models of Brittle Failure

Micromechanical models including the wing-crack model, the pore-emanated crack model, the Hertzian fracture concept, and the pore collapse model have linked grain-scale microscopic processes to macroscopic strain in rocks (see Sect. 2.2). Pore collapse was not observed in the tomograms of these experiments, so the pore collapse model may not adequately describe the failure of these Fontainebleau sandstone specimens. Development of mode I cracks and their dilation preceded the onset of strain localization (samples F1 and F3, Fig. 11a, c) or coincided with this localization (sample F2, Fig. 11b), indicating that relatively high magnitudes of dilation began to localize before high magnitudes of shear strain. We found no evidence for wing crack formation. Wing cracks develop through localized shear displacement that then produces regions of high tensile stresses, so we would expect to observe localized high shear strain regions before localized dilatational regions in the DVC data, rather than the observed opposite trend. In addition, visual inspection of the tomograms, before and after segmentation, provided no evidence for wing cracks, though it is possible that wing cracks with apertures that were too small to detect were formed.

The pore-emanated crack model, in which cracks open along grain boundaries, could be consistent with the experimental observations of microscale strain due to opening of grain boundaries (Fig. 7e, f). However, the Hertzian fracture concept also matches the experimental observations of the development of transgranular cracks that develop within grains, which the pore-emanated crack model does not predict. The Hertzian fracture concept uses the maximum tangential tensile stress within the contact area between spherical grains to predict grain-scale failure that leads to macroscopic failure (Johnson 1982). Consistent with this concept, we observed that, at small differential stresses, a higher number of intergranular fractures than transgranular fractures developed. Then, as the differential stress increased, and particularly after the yield stress was reached, transgranular fracturing began to dominate deformation until extensive grain comminution occurred (Figs. 4, 5). During early-stage deformation, intergranular fractures may preferentially develop instead of transgranular fractures because the cement that binds sandstone grains may be weaker under tension than the intact grains. Consequently, intergranular fractures may be able to propagate under lower differential stresses than transgranular fractures. In addition, as the differential stress increases, the effect of the difference in strength between the cement and the grains may diminish as preexisting pores close and intergranular contacts are clamped shut under higher normal stresses.

The geometry and volume of the pore network play an important role in controlling the deformation and failure of sandstones because the concentration of tensile stress near pore surfaces initiates the growth of fractures. For a spherical cavity under a compressive uniaxial far-field stress embedded in homogeneous isotropic linear elastic material (e.g., Goodier 1933; Sadowsky and Sternberg 1949; Eshelby 1957; Mura 1982), the stress at the surface of the cavity depends only on the far-field stress and the polar angle, θ, relative to the direction of σ_1. As this angle varies from $\pi/2$ at the equator of the pore to 0 or π at the poles, the resulting compressive stress at the pore surface decreases and becomes tensile near the poles. Because mineral grains and cement are weaker under tensile stress than under compressive stress, fractures are expected to nucleate near the poles of a spherical cavity and propagate preferentially along planes that are oriented subparallel to σ_1. Shear fractures may nucleate at cavity walls at orientations parallel to σ_1, and then propagate at angles that are oblique to σ_1 (Davis et al. 2017). These analytical and modeling approaches have been extended to plastic materials (Monchiet et al. 2008), multiple voids (Tandon and Weng 1986), and polygonal voids. However, exact results cannot usually be obtained, and the effects of the necessary approximations are challenging to assess. Our experiments provide additional insights because these three-dimensional data show that pores have shapes that differ from the spherical geometry often assumed in numerical models (Fig. 7).

Numerical methods such as finite element models (e.g., Eggers et al. 2006; Avazmohammadi and Naghdabadi 2013) and boundary element models (Davis et al. 2017), as well as micromechanical modeling coupled with analogue experiments (Sammis and Ashby 1986), provide additional insights that complement analytical approximations; For example, Nadimi et al. (2015) conducted two-dimensional finite element model simulations of the compression of Fontainebleau sandstone. In these simulations, load-bearing columns developed subparallel to the direction of the compressive stress, consistent with finite element simulations performed by Nadimi et al. (2015), and a model of the failure of brittle porous solids proposed by Sammis and Ashby (1986). In our experimental work, we also observed the formation of subvertical (subaxial) columns separated by subvertical fractures (Figs. 5, 6, 7g–i). However, the columns did not become well developed until vertical strains were substantially larger than those used in the simulations of Nadimi et al. (2015). The geometric difference between the two-dimensional model and the three-dimensional experiment, the strength of the simulated cement bonds, and the nondetection of fractures with small aperture widths in our X-ray microtomography experiments may have produced this apparent discrepancy between the experiments and the simulations. Features with characteristic scales below the resolution of X-ray microtomography, such as flaws within pore surfaces, mineral grains, and cement, play an important role in the nucleation of microfractures.

5.2. Dilation and Acceleration of Damage When Approaching Failure

In our Fontainebleau sandstone specimens, the total damage, D_ϕ, accelerated toward failure (Fig. 8a) and occurred concomitantly with dilation. In a study of crystalline rock, the damage accelerated as a power law of the normalized stress, $\Delta = \frac{\sigma - \sigma_f}{\sigma_f}$, where σ_f is the differential stress at failure, such that $D_\phi \sim \frac{1}{\beta-1}[\Delta^{-(\beta-1)} - 1]$ with an exponent β between 1.4 and 1.8 (Fig. 5 in Renard et al. 2018). In these experiments, the cumulated damage was measured, so that the power-law exponent is $-(\beta - 1)$, where β is the exponent that characterized the power-law divergence of the rate at which the damage increases with increasing differential stress as failure is approached. Based on two-dimensional simulations of uniaxial compression with no confining pressure, Girard et al. (2010) found evidence for power-law divergence of the damage correlation length with decreasing Δ as Δ approached failure, thus suggesting that failure is a critical phenomenon. This model of brittle failure indicates that elastic interactions between developing fractures over long-range, system-size, distances can explain the power-law divergence of incremental damage as Δ decreases toward failure in low-porosity crystalline rock (Girard et al. 2010). A power-law acceleration of acoustic emission prior to failure was also observed in the deformation of heterogeneous materials (Vasseur et al. 2015).

In contrast, in our experiments on Fontainebleau sandstone, the acceleration of damage occurred at a slower than power-law rate (Fig. 8b). This slower acceleration of damage may occur in more porous rocks compared with crystalline rocks because, when the propagating tip or edge of a microfracture reaches a pore, the local stress concentrations at the fracture tip or edge diminish. This may prevent further propagation of the fracture until the local stress field increases. Although the rate of damage accumulation differed in our recent monzonite experiments (Renard et al. 2018) and these sandstone experiments, in all five experiments (three sandstone and two monzonite experiments) we observed an acceleration of damage as failure was approached. This acceleration of damage was a precursor to shear failure in our experiments, as rock damage models (Lyakhovsky et al. 1997; Dahmen et al. 2009; Girard et al. 2010) and acoustic emission experiments on sandstones (Cox and Meredith 1993; Wu et al. 2000; Schubnel et al. 2007; Fortin et al. 2009; Nasseri et al. 2014; Ghaffari et al. 2014; Goodfellow et al. 2015) indicate. Our analysis reveals that this damage increase is slower than the power-law divergence predicted by some damage models (Fig. 8). Furthermore, this analysis is the first to separate this damage increase into new microfracture development and pore dilation (Fig. 6a). Segmentation of X-ray tomograms enabled this distinction because it captures aseismic and seismic strain, whereas acoustic emissions record only seismic strain.

Proposed micromechanical models of failure of porous sedimentary rock depend on the processes that produce opening-mode failure and dilation at the grain scale. In particular, the global evolution of D_ϕ likely differs between porous and crystalline rocks because of the pervasive pore network in more porous rocks. Macroscopic dilation has been observed during the deformation of low-porosity crystalline rocks (e.g., Brace 1978) and porous sedimentary rocks (e.g., Baud et al. 2000). However, the degree of dilation in porous sedimentary rocks differs from that in crystalline rocks because, in porous rocks, microfractures may nucleate at grain boundaries, break them, and cause pore collapse (e.g. Handin et al. 1963; Wong et al. 1997). In our Fontainebleau sandstone experiments, the majority of cracks nucleated at grain contacts and opened along grain boundaries (Fig. 7), similar to the findings of some other studies on sandstones (Menéndez et al. 1996; Wu et al. 2000). Some cracks also propagated through grains, as observed in other studies (Handin et al. 1963; Zhang et al. 1990; Wu et al. 2000; El Bied et al. 2002). The tomograms do not reveal clear compaction or pore collapse at the grain scale prior to failure. Although there was a global axial shortening of the sample, the overall porosity increased from the onset of yield to failure. The low initial porosity of this rock (5–7%) may have inhibited pore collapse because the potential maximum tensile stress that may develop at pore surfaces decreases with decreasing pore volume (e.g., Wong and Baud 2012).

5.3. Implications of DVC Strain Analysis

Videos S3–S5 show that some high values of strain were detected by DVC analysis inside the sample from the beginning of loading. This could be related to the presence of weak zones inside the samples that deformed even under low differential stress. Such behavior has also been observed when acoustic emissions were used to monitor deformation in porous sandstones (e.g., Fortin et al. 2009).

The more diffuse distribution of the high incremental strains in experiments F1 and F3 (Fig. 9a, c, Videos S3 and S5) compared with experiment F2 (Fig. 9b, Video S4) suggests why the sums of the incremental von Mises strains did not change as significantly in experiments F1 and F3 compared with experiment F2 (Fig. 11). A core-spanning fracture did not develop prior to macroscopic failure in these experiments as it did in experiment F2. Instead, preceding macroscopic failure, the high values of the incremental strain remained relatively diffuse, although there was some localization of the high strain values into tighter clusters (Fig. 9a) and narrower bands (Fig. 9c) preceding macroscopic failure. Consequently, much of the high incremental strain population remained outside of these strain localization zones preceding macroscopic failure. However, propagation, opening, and sliding along faults produced the final stage of macroscopic failure in all three experiments. This suite of observations is similar to what was observed in experiments that monitored acoustic emissions (e.g. Wu et al. 2000; Fortin et al. 2009) and in which a surge of acoustic emissions occurred from the yield point to failure in sandstones during brittle deformation. The DVC analyses of experiments F1 and F3 show that strain events occurred homogeneously in the volume (Videos S3 and S5), similar to the cataclastic compaction of a porous sandstone (top panel of Fig. 5 in Fortin et al. 2009) and was attributed to the opening of microfractures along grain boundaries (Menéndez et al. 1996). The DVC analysis of experiment F2, in which a fault developed, is also similar to observations of shear localization in a porous sandstone (middle panel of Fig. 5 in Fortin et al. 2009; Wu et al. 2000). In contrast to the experiments of Fortin et al. (2009), we did not observe pore collapse in the tomography data and our DVC results do not show significant volumetric compaction, likely due to the low porosity of the Fontainebleau sandstone that we investigated.

The evolving distribution of the higher incremental strain magnitudes revealed by the DVC analyses of experiment F2 indicates that both dilation and shear strain localized along the core-spanning fracture (Fig. 9b). The incremental strain fields suggest that, for differential stresses of 104 MPa to 151 MPa, incremental dilatational strains dominated the strain field surrounding and within the developing shear fault zone, and above 151 MPa, incremental shear strain became more localized than dilatational strains around the developing fault zone (Figs. 9b, S3). These observations are consistent with post mortem scanning electron microscopy images of deformed sandstones that showed shear cracks dominating within the fault zone (e.g., Wu et al. 2000). Initially, dilation, and perhaps tensile failure, localized along the incipient fault zone, and then localized shear strain occurred within the developing fault only after some dilation occurred.

To investigate in more detail the dominant failure modes that facilitated the initiation of the fracture observed in F2, we calculated the incremental strain fields following the yield point with the highest possible temporal (differential stress) resolution. Comparing the distribution of the high differential strain magnitudes above the 95th percentile, or the top 5% of values, indicated that the high magnitudes of dilatational $\frac{I_1(\Delta\varepsilon)}{\Delta\varepsilon_{zz}^{\mathrm{M}}}$ began to concentrate around the protofracture before the high magnitudes of $\frac{(3J_2(\Delta\varepsilon))^{1/2}}{\Delta\varepsilon_{zz}^{\mathrm{M}}}$ (Fig. S3). This observation is consistent with the more localized distribution of dilatational $\frac{I_1(\Delta\varepsilon)}{\Delta\varepsilon_{zz}^{\mathrm{M}}}$

relative to the spatial distribution of $\frac{(3J_2(\Delta\varepsilon))^{1/2}}{\Delta\varepsilon_{zz}^{\mathrm{M}}}$ from 104–151 MPa, and the more localized distribution of $\frac{(3J_2(\Delta\varepsilon))^{1/2}}{\Delta\varepsilon_{zz}^{\mathrm{M}}}$ relative to dilatational $\frac{I_1(\Delta\varepsilon)}{\Delta\varepsilon_{zz}^{\mathrm{M}}}$ from 151–167 MPa observed in the lower temporal resolution analysis (Fig. 9). The evolution of the sums above the yield point (insets in Fig. 11b, e) indicate that the normalized sum of $\frac{(3J_2(\Delta\varepsilon))^{1/2}}{\Delta\varepsilon_{zz}^{\mathrm{M}}}$ is generally slightly higher than the normalized sum of the dilatational part of the $\frac{I_1(\Delta\varepsilon)}{\Delta\varepsilon_{zz}^{\mathrm{M}}}$ field in the differential stress increments immediately preceding the yield point. These properties of the dilatational $\frac{I_1(\Delta\varepsilon)}{\Delta\varepsilon_{zz}^{\mathrm{M}}}$ began to exceed those of $\frac{(3J_2(\Delta\varepsilon))^{1/2}}{\Delta\varepsilon_{zz}^{\mathrm{M}}}$ when dilatational strains began to localize along the core-spanning fracture (109–114 MPa differential stress in Fig. 9; green lines in Fig. 11).

In the two experiments (F2 and F3) in which the incremental strain invariant sums generally increase (Fig. 11), and the areas under the histograms generally increase (Fig. 10), the cumulative strain magnitude distributions generally progressed toward higher strain magnitudes with increasing differential stress (Fig. S4). In experiment F1, this trend was less consistent than in the other experiments. In particular, the magnitude–frequency distributions shifted toward lower magnitudes of strain from the second to last differential stress increment to the final differential stress increment, consistent with the observed decrease in the sums (Fig. 11).

To characterize the distribution of strain magnitudes close to the macroscopic shear failure, we searched for scaling relationships within the populations of normalized incremental dilatational $\frac{I_1(\Delta\varepsilon)}{\Delta\varepsilon_{zz}^{\mathrm{M}}}$, contractional $\frac{I_1(\Delta\varepsilon)}{\Delta\varepsilon_{zz}^{\mathrm{M}}}$, and $\frac{(3J_2(\Delta\varepsilon))^{1/2}}{\Delta\varepsilon_{zz}^{\mathrm{M}}}$ strain increments shown in Fig. 10. Such scaling relationships may reflect the long-range elastic interactions of the stress field produced by individual local microfractures and the presence of initial mechanical heterogeneities (e.g., Renard et al. 2018). The cumulative frequency distributions of the incremental strain invariants show a linear trend on a log–log scale covering 1–2 orders of magnitude (Fig. S4). The incremental strain invariant curves flatten toward lower magnitudes near the resolution limit of the incremental strain measurements, below which the values were removed (Fig. S4). The linear trend is indicative of a power-law scaling behavior with numerous small incremental strain values, and fewer large incremental strain values. In log–log space, the slope ($\beta + 1$) of the cumulative histogram gives the exponent of the power-law scaling relationship, β, of the strain increments (Fig. S5). We calculated the power-law exponents using the maximum-likelihood approach of Clauset et al. (2009). These exponents varied during the loading in the three experiments (Fig. S5). However, we observed no significant trend in the exponent value throughout deformation. This relative stability of the exponents could be due to the small range of spatial scales over which effective power-law behavior was observed (Fig. S4).

5.4. *Transition from Microfracture Coalescence to Localized Shear Failure*

The SEM images revealed that, after a distinct shear band formed in experiment F2, extensive grain comminution produced a layer of gouge within the shear zone (Fig. 7). The time resolution of the experiments ($\sim$1.5 min between scans) prevented determination of whether the core-spanning fracture in experiment F2 developed through (1) a localized nucleation and propagation mechanism or (2) the formation of a dense array of microfractures that coalesced and ultimately formed a volume of unconsolidated, comminuted granular material. Tomograms acquired beyond the yield point of experiment F2 showed progressive nucleation, growth, and coalescence of an array of opening cracks oriented subparallel to σ_1 (Figs. 5, 7g–i).

Mechanical breakdown within rupture zones may initially produce a sand-like granular material at these confining stresses. As comminution broadens the grain size distribution within a shear zone, the local shear zone porosity may decrease as small grains fill spaces between larger grains. The rounding and reorientation of large survivor grains (Cladouhos 1999) may also contribute to porosity reduction in shear zones following larger shear displacements. Such shear zone gouges may also enable local dilation. However, the development of unconsolidated sand grains in the shear zone alone cannot produce the observed macroscopic dilation (negative macroscopic volumetric strain) of about − 0.15 following failure. This macroscopic dilation of the specimen suggests that microfractures in the material surrounding the shear zone contributed to the postfailure porosity, in addition to dilation within the shear zone.

In experiments in which unconsolidated granular material was subjected to shear deformation with confining stress perpendicular to the shear plane, shear bands developed with enhanced comminution (e.g., Marone and Scholz 1989). When a broad grain size distribution developed within the fault gouge, the local porosity within the gouge decreased (Marone and Scholz 1989). In the case of the low-porosity Fontainebleau sandstone used in the experiments reported here, dilation occurred within the incipient shear rupture zone because of the formation of a dense array of subvertical microfractures (Fig. 6). The opening of these microfractures may lead initially to the formation of a high-porosity band of unconsolidated quartz particles that evolves into a low-porosity zone of highly comminuted quartz particles with a broad size distribution. However, we did not observe a low-porosity zone in these experiments, perhaps because the shear strain within the rupture zone and/or the confining stress perpendicular to the shear plane were too small to reorient, round, or crush the grains within the incipient fault gouge, and thereby decrease porosity within the zone. Perhaps if we had continued the experiment beyond macroscopic failure and allowed more evolution of the fault gouge within the shear zone, the local porosity within the gouge would have decreased. In additional, the short (multiple hour) time span and dry conditions in our experiments prevented chemical processes that would further reduce fault zone porosity in the crust.

Differences in the rate of fracture coalescence and fault propagation may have caused a core-spanning fracture to localize earlier in experiment F2 (closer to the yield point) than experiments F1 and F3 (immediately before macroscopic failure). For the same rate of deformation, fault propagation tends to progress at a slower rate in more macroscopically brittle rocks than in less brittle rocks that have a higher component of ductile deformation (Ougier-Simonin and Zhu 2013, 2015). The imposed confining stress of experiment F1 (20 MPa) was larger than that of experiment F2 (10 MPa), and the nominal strain rate of experiment F1 (184 scans, each of 1.5 min duration) was lower than that of experiment F2 (49 scans, each of 1.5 min duration). Higher confining stresses and lower strain rates tend to allow more ductile deformation to accommodate the applied strain, and rocks behave in a less brittle manner with lower fault propagation rates. Consequently, the experimental conditions produced more brittle behavior in experiment F2 than in experiment F1, which promoted slower fault propagation in experiment F2 compared with experiment F1. Similarly, fault propagation may have occurred at slower rates in experiment F2 than experiment F3 because the Young's modulus and failure stress of experiment F3 were lower than those of experiment F2, signaling more brittle behavior during experiment F2 than F3.

In each of our experiments, the angle of inclination, θ, between the maximum compression direction and the core-spanning fracture was smaller than that predicted by the Mohr–Coulomb failure criterion. We cannot rule out the possibility that this was caused by the absence of lubrication of the piston–sandstone interface. However, this observation was also reported in previous experiments on Vosges sandstone (22% porosity) in which failure occurred via propagation of subaxial fractures (e.g., Bésuelle et al. 2000). In these experiments, increasing confining stress increased the orientation of the through-going fractures with respect to σ_1. At the highest confining pressures, compaction bands developed subperpendicular to σ_1, thereby following the observed trend. The subaxial microfractures appeared to control how strain localized into core-spanning fractures. As the density of subvertical microfractures increased, the mechanical properties of the rock became increasingly anisotropic, which may have modified the orientation of new fractures and how fractures coalesced. These fractures had orientations between mode I subaxial fractures when $P_c \ll \sigma^f$, and shear failure consistent with the orientation predicted by the Mohr–Coulomb failure criterion when $P_c < \sigma^f$ (i.e., when P_c is a few tens of MPa), where σ^f is the differential stress at failure. As the differential stress increased and the subaxial fractures increased in size, the unfractured columns between them became more unstable (e.g., Fig. 41 in Peng and Johnson 1972). If one column collapses, the forces acting on it will be transferred to other columns, and a cascade of column collapses may be responsible for strain localization that forms fractures at higher angles than those predicted by the Mohr–Coulomb criterion. Such conceptual description of the failure process is quantified in statistical physics models. As failure is approached, the sizes of the cascades (or avalanches) will increase in size as observed in numerical simulations (e.g., Dahmen et al. 2009; Girard et al. 2010). The preferential orientation of subaxial microfractures suggests that the long-range function that describes how the rate of damage evolution at one point influences the damage evolution at other points should depend on both the direction and length of the vector between the two points because damage develops preferentially in certain directions, and is thus anisotropic, as proposed for crystalline rocks in the model of Lockner and Madden (1991).

The final core-spanning fault in our experiments was not planar, but curved (Figs. 5d, 6d). Such curved fault surfaces have also been observed in crystalline rocks (Peng and Johnson 1972). Curved strain concentration zones are not uncommon; For example, Sulem and Ouffroukh (2006) mentioned measuring the orientation of the shear band on the central part of the specimen because the shear band often warps close to the extremities.

The development of a new micromechanical model for sandstone is out of the scope of the present study. A sound understanding of the micromechanics of complex materials should help justify the forms of macroscopic constitutive equations that predict the mechanical and rheological behaviors of rocks. Such macroscopic laws are useful because of their small number of parameters (e.g., Rudnicki and Rice 1975; see also the discussion in Dresen and Guéguen 2004) and provide a complementary approach to micromechanical modeling. A micromechanical constitutive model should incorporate processes that cause the local increase of damage (e.g., opening of grain boundaries and formation of transgranular microfractures) and the effect of damage on the local stress distribution (Cox and Meredith 1993). It should predict macroscopic properties such as strength, amount of dilation, and evolution of elastic parameters and also whether a fault will develop early, near the yield point (like the core-spanning fracture in experiment F2) or later, near failure (experiments F1 and F3). However, the materials and experimental conditions were essentially the same in experiments F2 and F3, and this suggests that experiments F2 and F3 were conducted very near the boundary between two quite different deformation and failure regimes or that prediction of deformation and failure will be very challenging if not impossible for rocks like low-porosity Fontainebleau sandstone.

5.5. Limitations of Digital Volume Correlation Analysis Near Failure

Our DVC analyses quantified local strain components (dilation, compaction, shear strain) that may be compared with the evolving pore and fracture network revealed through segmentation of the tomograms, and the macroscopic strain. Histograms of incremental strain components (Fig. 10) and the sum of the increments of each strain component (Fig. 11) at increasing differential stress steps showed that components of local strain concentrations can be up to one order of magnitude larger than the macroscopic axial strain. The power-law scaling analysis of the cumulated distribution of the incremental strain components showed that the exponent did not vary significantly as failure is approached (Fig. S5), whereas the total damage increased (Fig. 8).

During macroscopic failure and the postfailure phase when comminution occurred, the structure of the incipient fault zone changed rapidly. Consequently, DVC analysis of the evolution of the incipient fault zone may not produce meaningful results because changes in the sizes, shapes, and orientations of quartz grains, and grain fragmentation, during the acquisition time may blur the tomograms. In addition, large differences in microstructures between tomograms may inhibit adequate correlation of subvolumes across sequential tomograms. Time-lapse X-ray microtomography of deformation with small increments of shear strain could reveal the comminution process within the incipient fault zone in more detail than these experiments. However, strain fields obtained from DVC analysis of tomograms separated by these small strain increments will have lower signal-to-noise ratios than the strain fields presented in this study because the signal (incremental strain) decreases when the the macroscopic strain between scans decreases (e.g., McBeck et al. 2018).

5.6. Neutron Tomography

Neutron tomography has been used with DVC analysis to investigate strain localization and compressive failure of Bentheim sandstone (Tudisco et al. 2015). However, the technology of neutron tomography is not as far advanced as the technology of X-ray tomography. Neutron microtomography requires several hours for three-dimensional acquisitions at 30 μm spatial resolution, whereas the HADES deformation apparatus installed at beamline ID19 at ESRF enables 1.5 min acquisition times at 6.5 μm spatial resolution for cores 1 cm tall and 0.5 cm wide. Consequently, time-lapse neutron tomography has been only rarely used to investigate the micromechanics of rocks (e.g., Stavropoulou et al. 2018). The attenuation coefficients of neutrons are not as high as those of X-rays, which is an advantage for triaxial rigs with thick walls that are required for applying higher stresses than the 200 MPa differential stress limit of the HADES rig. In addition, neutron and X-ray contrasts are quite different, and in particular, the high neutron cross section of hydrogen would be advantageous when studying fluids such as water and hydrocarbons within rocks.

5.7. Temperatures and Stress Limits

The HADES rig was fabricated from titanium, and experiments can be conducted in the rig at stresses up to 200 MPa and temperatures up to 250 °C. This enables processes that occur at depths up to about 7 km to be investigated. While a very broad range of geological processes and most geotechnical applications occur under these conditions, there is a clear need to extend the range of temperatures and stresses. It should be possible to achieve this by using advanced materials that are strong at high temperatures and have low X-ray attenuation coefficients. Candidate materials include alloys composed primarily of lithium, beryllium, magnesium, and aluminum, metal-matrix composites formed from these light alloys and carbon or ceramic fibers, and carbon-fiber-reinforced carbon. However, most, if not all, of these and similar materials are difficult and expensive to fabricate, and few companies have experience with the design and manufacture of apparatus constructed from these materials.

6. Conclusions

Segmentation and digital volume correlation analysis of micrometer-resolution tomograms acquired through in situ dynamic synchrotron X-ray microtomography during triaxial compression loading revealed the damage and strain preceding brittle macroscopic failure in three experiments on Fontainebleau sandstone. In one experiment, a system-size fracture nucleated at 78% of the differential stress at failure and accommodated an increasing proportion of incremental shear strain relative to incremental dilatational strain. In the other two experiments, opening of grain boundaries was the dominant dilation mechanism and localization occurred later, at 95% of the differential stress at failure. Increases in dilatancy and shear strain at the micro- and macroscales were observed during the loading stage preceding failure for all three experiments.

Our experiments are consistent with experiments performed on Fontainebleau sandstone (El Bied et al. 2002; Schubnel et al. 2007; Nasseri et al. 2014; Goodfellow et al. 2015) and on other sandstones (Handin et al. 1963; Menéndez et al. 1996; Wu et al. 2000; Fortin et al. 2009). Failure of Fontainebleau sandstone did not produce discontinuous shear displacement along an idealized plane separated by two essentially undamaged sandstone bodies. Instead, progressive damage accumulated with increasing differential stress as microfractures propagated along grain boundaries as well as through grains (Videos S1 and S2). After the yield stress was reached, damage accumulated at increasing rates until macroscopic shear failure occurred. The microscopic damage developed as subaxial fractures that propagated, coalesced, and became increasingly concentrated into a high-aspect-ratio system-spanning volume.

Our experimental approach provided new information on the nucleation of failure and on the transitions between microfracture nucleation, microfracture coalescence, and macroscopic failure. Incremental strain components revealed by DVC analysis indicated that the high-density subaxial fracture zone did not form a plane but a curved surface. This progressive damage enhanced heterogeneity in the stress and strain fields, as well as the local distribution of mechanical properties. Time-lapse X-ray microtomography and DVC analysis provided detailed information about the evolving microstructure of the Fontainebleau sandstone cylinders as they deformed and approached macroscopic failure. The dilation of the samples during loading was due to both crack opening and pore opening, with an almost equal volumetric contribution of these two processes to the increase of void volume (Fig. 6a). The difference in micromechanical behavior between experiments F1 and F3 on the one hand, and F2 on the other hand did not appear clearly in the stress–strain curves (Fig. 1), which showed that the ability to "look inside" rock specimens provided by X-ray tomography is critical to identify the route to failure. Another important result was obtained by segmenting the voids during loading. We observed a slowdown of the rate at which damage, characterized by the void volume fraction based damage index, D_{ϕ}, increased with increasing differential stress as failure was approached (Fig. 8), which is not observed in crystalline rocks (Renard et al. 2018). We attribute this to the pinning of microfractures by pores that may reduce the stress concentration at crack tips. These new results demonstrate that very few, if any, other experimental methods enable such precise correlations between the macroscopic mechanical behavior of rock specimens and the micromechanical processes that occur within them.

Acknowledgements

The deformation apparatus was built by Sanchez Technology. Elodie Boller, Paul Tafforeau, and Alexander Rack provided advice on the design of the tomography setup. This study received funding from the Norwegian Research Council (project ARGUS, grant 272217) and the European Union's Horizon 2020 Research and Innovation Program under ERC advanced grant agreement no. 669972, "Disequilibrium Metamorphism" ("DIME"). W.Z. acknowledges partial support from the US National Science Foundation EAR-1761912. Beam time was allocated at the European Synchrotron Radiation Facility (long-term proposal ES-295). Data storage was provided by UNINETT Sigma2-the National Infrastructure for High Performance Computing and Data Storage in Norway (project NS9073 K). Médard Thiry (Mines ParisTech) is acknowledged for providing the Fontainebleau sandstone. We thank the Editor, Yehuda Ben-Zion and two anonymous referees for constructive suggestions during the review process.

References

Andrä, H., Combaret, N., Dvorkin, J., Glatt, E., Han, J., Kabel, M., et al. (2013). Digital rock physics benchmarks—Part I: Imaging and segmentation. *Computers and Geosciences, 50,* 25–32.

Ashby, M. F., & Sammis, C. G. (1990). The damage mechanics of brittle solids in compression. *Pure and Applied Geophysics, 133*(3), 489–521.

Auzerais, F. M., Dunsmuir, J., Ferreol, B. B., Martys, N., Olson, J., Ramakrishnan, T. S., et al. (1996). Transport in sandstone: A study based on three dimensional microtomography. *Geophysical Research Letters, 23*(7), 705–708.

Avazmohammadi, R., & Naghdabadi, R. (2013). Effective behavior of porous elastomers containing aligned spheroidal voids. *Acta Metallurgica, 224,* 1901–1915.

Baud, P., Reuschlé, T., Ji, Y., Cheung, C. S., & Wong, T. F. (2015). Mechanical compaction and strain localization in Bleurswiller sandstone. *Journal of Geophysical Research: Solid Earth, 120,* 6501–6522.

Baud, P., Zhu, W., & Wong, T. F. (2000). Failure mode and weakening effect of water on sandstone. *Journal of Geophysical Research: Solid Earth, 105,* 16371–16389.

Bay, B. K., Smith, T. S., Fyhrie, D. P., & Saad, M. (1999). Digital volume correlation: Three-dimensional strain mapping using X-ray tomography. *Experimental Mechanics, 39*(3), 217–226.

Bésuelle, P., Desrues, J., & Raynaud, S. (2000). Experimental characterisation of the localisation phenomenon inside a Vosges sandstone in a triaxial cell. *International Journal of Rock Mechanics and Mining Science, 37,* 1223–1237.

Bourbie, T., & Zinszner, B. (1985). Hydraulic and acoustic properties as a function of porosity in Fontainebleau sandstone. *Journal of Geophysical Research: Solid Earth, 90*(B13), 11524–11532.

Brace, W. F. (1978). Volume changes during fracture and frictional sliding: A review. *Pure and Applied Geophysics, 116,* 603–614.

Brace, W. F., Paulding, B. W., & Scholz, C. H. (1966). Dilatancy in the fracture of crystalline rocks. *Journal of Geophysical Research, 71*(16), 3939–3953.

Buades, A., Coll, B., & Morel, J. M. (2005). A non-local algorithm for image denoising, in computer vision and pattern recognition. *IEEE Computer Society Conference, 2,* 60–65.

Cladouhos, T. T. (1999). Shape preferred orientations of survivor grains in fault gouge. *Journal of Structural Geology, 21*(4), 419–436.

Clauset, A., Shalizi, C. R., & Newman, M. (2009). Power-law distributions in empirical data. *Society for Industrial and Applied Mathematics Review, 51*(4), 661–703.

Coker, D. A., Torquato, S., & Dunsmuir, J. H. (1996). Morphology and physical properties of Fontainebleau sandstone via a tomographic analysis. *Journal of Geophysical Research: Solid Earth, 101*(B8), 17497–17506.

Cox, S. J. D., & Meredith, P. G. (1993). Microcrack formation and material softening in rock measured by monitoring acoustic emissions. *International Journal of Rock Mechanics and Mining Sciences and Geomechanics Abstracts, 30*(1), 11–24.

Dahmen, K. A., Ben-Zion, Y., & Uhl, J. T. (2009). A micromechanical model for deformation in disordered solids with universal predictions for stress-strain curves and related avalanches. *Physical Review Letters, 102,* 175501.

Davis, T., Healy, D., & Bubeck, A. (2017). Stress concentrations around voids in three dimensions: The roots of failure. *Journal of Structural Geology, 102,* 193–207.

Dresen, G., & Guéguen, Y. (2004). Damage and rock physical properties. In Y. Guéguen & M. Bouteca (Eds.), *Mechanics of fluid-saturated rocks* (pp. 169–217). Amsterdam: Elsevier Academic.

Eggers, C. G., Berli, M., Accorsi, M. L., & Or, D. (2006). Deformation and permeability of aggregated soft earth materials. *Journal of Geophysical Research, 111,* B10204. https://doi.org/10.1029/2005JB004123.

El Bied, A., Sulem, J., & Martineau, F. (2002). Microstructure of shear zones in Fontainebleau sandstone. *International Journal of Rock Mechanics and Mining Sciences, 39,* 917–932.

Eshelby, J. D. (1957). The determination of the elastic field of an ellipsoidal inclusion and related problems. *Proceedings of the Royal Society of London Series A—Mathematical and Physical Sciences, 241,* 376–396.

Faulkner, D. R., Mitchell, T. M., Healy, D., & Heap, M. J. (2006). Slip on 'weak' faults by the rotation of regional stress in the fracture damage zone. *Nature, 444*(7121), 922.

Fortin, J., Stanchits, S., Dresen, G., & Guéguen, Y. (2009). Acoustic emissions monitoring during inelastic deformation of porous sandstone: Comparison of three modes of deformation. *Pure and Applied Geophysics*. https://doi.org/10.1007/s00024-009-0479-0.

Fredrich, J. T., Greaves, K. H., & Martin, J. W. (1993). Pore geometry and transport properties of Fontainebleau sandstone. *International journal of rock mechanics and mining sciences, 30*(7):691–697.

Ghaffari, H. O., Nasseri, M. H. B., & Young, R. P. (2014). Faulting of rocks in a three-dimensional stress field by micro-anticracks. *Scientific Report, 4,* 5011.

Girard, L., Amitrano, D., & Weiss, J. (2010). Failure as a critical phenomenon in a progressive damage model. *Journal of Statistical Mechanics: Theory and Experiment*. https://doi.org/10.1088/1742-5468/2010/01/P01013.

Goodfellow, S. D., Tisato, N., Ghofranitabari, M., Nasseri, M. H. B., Maxwell, S. C., & Young, R. P. (2015). Attenuation properties of Fontainebleau sandstone during true triaxial deformation using active and passive ultrasonics. *Rock Mechanics and Rock Engineering, 48,* 2551–2566.

Goodier, J. N. (1933). Concentration of stress around spherical and cylindrical inclusions and pores. *Journal of Applied Mechanics, 55,* 39–44.

Hall, S. A., Bornert, M., Desrues, J., Pannier, Y., Lenoir, N., Viggiani, G., et al. (2010). Discrete and continuum analysis of localised deformation in sand using X-ray μCT and volumetric digital image correlation. *Géotechnique, 60*(5), 315–322.

Handin, J., Hager, R. V., Friedman, M., & Feather, J. N. (1963). Experimental deformation of sedimentary rocks under confining pressure: Pore pressure tests. *AAPG Bulletin, 47,* 717–755.

Hazzard, J. F., Young, R. P., & Maxwell, S. C. (2000). Micromechanical modeling of cracking and failure in brittle rocks. *Journal of Geophysical Research: Solid Earth, 105,* 16683–16697.

Heap, M. J., & Faulkner, D. R. (2008). Quantifying the evolution of static elastic properties as crystalline rock approaches failure. *International Journal of Rock Mechanics and Mining Sciences, 45*(4), 564–573.

Horii, H., & Nemat-Nasser, S. (1986). Brittle failure in compression: splitting faulting and brittle-ductile transition. *Philosophical Transactions of the Royal Society of London A, 319*(1549), 337–374.

Johnson, K. L. (1982). One hundred years of Hertz contact. *Proceedings of the Institution of Mechanical Engineers, 196*(1), 363–378.

Kanaya, T., & Hirth, G. (2018). Brittle to semibrittle transition in quartz sandstone: Energetics. *Journal of Geophysical Research: Solid Earth, 123,* 84–106.

Katz, O., & Reches, Z. (2004). Microfracturing, damage, and failure of brittle granites. *Journal of Geophysical Research: Solid Earth, 109,* B01206.

Kemeny, J. M., & Cook, N. G. (1991). Micromechanics of deformation in rocks. In *Toughening mechanisms in quasi-brittle materials* (pp. 155–188). Dordrecht: Springer.

Kwiatek, G., Charalampidou, E. M., Dresen, G., & Stanchits, S. (2014). An improved method for seismic moment tensor inversion of acoustic emissions through assessment of sensor coupling and sensitivity to incidence angle. *International Journal of Rock Mechanics and Mining Sciences, 65,* 153–161.

Lindquist, W. B., Venkatarangan, A., Dunsmuir, J., & Wong, T. F. (2000). Pore and throat size distributions measured from synchrotron X-ray tomographic images of Fontainebleau sandstones. *Journal of Geophysical Research: Solid Earth, 105*(B9), 21509–21527.

Lockner, D., Byerlee, J. D., Kuksenko, V., Ponomarev, A., & Sidorin, A. (1991). Quasi-static fault growth and shear fracture energy in granite. *Nature, 350,* 39–42.

Lockner, D. A., Byerlee, J. D., Kuksenko, V., Ponomarev, A., & Sidorin, A. (1992). Observations of quasistatic fault growth from acoustic emissions. In B. Evans & W. F. Brace (Eds.), *Fault mechanics and transport properties of rocks* (Vol. 1, pp. 3–31). Cambridge: Academic Press.

Lockner, D., & Madden, T. (1991). A multiple-crack model of brittle fracture. 1. Non-time-dependent simulations. *Journal of Geophysical Research, 96,* 19623–19642.

Louis, L., Wong, T. F., & Baud, P. (2007). Imaging strain localization by X-ray radiography and digital image correlation: Deformation bands in Rothbach sandstone. *Journal of Structural Geology, 29*(1), 129–140.

Lyakhovsky, V., Ben-Zion, Y., & Agnon, A. (1997). Distributed damage, faulting, and friction. *Journal of Geophysical Research: Solid Earth, 102,* 27635–27649.

Marone, C., & Scholz, C. H. (1989). Particle-size distribution and microstructures within simulated fault gouge. *Journal of Structural Geology, 11,* 799–814.

McBeck, J., Kobchenko, M., Hall, S. A., Tudisco, E., Cordonnier, B., Meakin, P., & Renard, F. (2018). Investigating the onset of strain localization within anisotropic shale using digital volume correlation of time-resolved X-ray microtomography images. *Journal of Geophysical Research: Solid Earth.* https://doi.org/10.1029/2018JB015676.

Menéndez, B., Zhu, W., & Wong, T. F. (1996). Micromechanics of brittle faulting and cataclastic flow in Berea sandstone. *Journal of Structural Geology, 18,* 1–16.

Miller, S. A., Collettini, C., Chiaraluce, L., Cocco, M., Barchi, M., & Kaus, B. J. (2004). Aftershocks driven by a high-pressure CO_2 source at depth. *Nature, 427*(6976), 724–727.

Mirone, A., Brun, E., Gouillart, E., Tafforeau, P., & Kieffer, J. (2014). The PyHST2 hybrid distributed code for high speed tomographic reconstruction with iterative reconstruction and a priori knowledge capabilities. *Nuclear Instruments and Methods in Physics Research, Section B: Beam Interactions with Materials and Atoms, 324,* 41–48.

Mogi, K. (1971). Fracture and flow of rocks under high triaxial compression. *Journal of Geophysical Research, 76*(5), 1255–1269.

Monchiet, V., Kondo, D., Charkaluk, E., & Oana, C. (2008). Macroscopic yield criteria for plastic anisotropic materials containing spheroidal voids. *International Journal of Plasticity, 24,* 1158–1189.

Mura, T. (1982). *Micromechanics of defect in solids* (p. 588). The Hague: Martinus Nijhoff.

Nadimi, S., Fonseca, J., Bésuelle, P. & Viggiani, G. (2015). A microstructural finite element analysis of cement damage on Fontainebleau sandstone. Paper presented at the ICTMS 2015, 2nd International Conference on Tomography of Materials and Structures, 29 June–3 July 2015, Québec, Canada.

Nasseri, M. H. B., Goodfellow, S. D., Lombos, L., & Young, R. P. (2014). 3-D transport and acoustic properties of Fontainebleau sandstone during true-triaxial deformation experiments. *International Journal of Rock Mechanics and Mining Sciences, 69,* 1–18.

Otsuki, K., & Dilov, T. (2005). Evolution of hierarchical self-similar geometry of experimental fault zones: implications for seismic nucleation and earthquake size. *Journal of Geophysical Research: Solid Earth, 110,* B03303. https://doi.org/10.1029/2004JB003359.

Ougier-Simonin, A., & Zhu, W. (2013). Effects of pore fluid pressure on slip behaviors: An experimental study. *Geophysical Research Letters, 40*(11), 2619–2624.

Ougier-Simonin, A., & Zhu, W. (2015). Effect of pore pressure buildup on slowness of rupture propagation. *Journal of Geophysical Research: Solid Earth, 120*(12), 7966–7985.

Paterson, M. S., & Wong, T. F. (2005). *Experimental rock deformation-the brittle field.* Berlin: Springer.

Peng, S., & Johnson, A. M. (1972). Crack growth and faulting in cylindrical specimens of Chelmsford granite. *International Journal of Rock Mechanics and Mining Science, 9,* 37–86.

Reches, Z., & Lockner, D. A. (1994). Nucleation and growth of faults in brittle rocks. *Journal of Geophysical Research, 99,* 18159–18173.

Renard, F., Cordonnier, B., Dysthe, D. K., Boller, E., Tafforeau, P., & Rack, A. (2016). A deformation rig for synchrotron microtomography studies of geomaterials under conditions down to 10 km depth in the Earth. *Journal of Synchrotron Radiation, 23,* 1030–1034.

Renard, F., Cordonnier, B., Kobchenko, M., Kandula, N., Weiss, J., & Zhu, W. (2017). Microscale characterization of rupture nucleation unravels precursors to faulting in rocks. *Earth and Planetary Science Letters, 476,* 69–78. https://doi.org/10.1016/j.epsl.2017.08.002.

Renard, F., Weiss, J., Mathiesen, J., Ben-Zion, Y., Kandula, N., & Cordonnier, B. (2018). Critical evolution of damage toward system-size failure in crystalline rock. *Journal of Geophysical Research: Solid Earth.* https://doi.org/10.1002/2017JB014964.

Rudnicki, J. W., & Rice, J. R. (1975). Conditions for the localization of deformation in pressure-sensitive dilatant materials. *Journal of the Mechanics and Physics of Solids, 23,* 371–394.

Sadowsky, M. A., & Sternberg, E. (1949). Stress concentration around a triaxial ellipsoidal cavity. *Journal of Applied Mechanics–Transactions of the ASME, 16,* 149–157.

Sammis, C. G., & Ashby, M. F. (1986). The failure of brittle porous solids under compressive stress states. *Acta Metallurgica, 34*(3), 511–526.

Scholz, C. H. (1968). Microfracturing and the inelastic deformation of rock in compression. *Journal of Geophysical Research, 73,* 1417–1432.

Schubnel, A., Thompson, B. D., Fortin, J., Guéguen, Y., & Young, R. P. (2007). Fluid-induced rupture experiment on Fontainebleau sandstone: Premonitory activity, rupture propagation, and aftershocks. *Geophysical Research Letters, 34,* L19307.

Stanchits, S., Vinciguerra, S., & Dresen, G. (2006). Ultrasonic velocities, acoustic emission characteristics and crack damage of basalt and granite. *Pure and Applied Geophysics, 163,* 975–994.

Stavropoulou, E., Andò, E., Tengattini, A., Briffaut, M., Dufour, F., Atkins, D., et al. (2018). Liquid water uptake in unconfined Callovo Oxfordian clay-rock studied with neutron and X-ray imaging. *Acta Geotechnica.* https://doi.org/10.1007/s11440-018-0639-4.

Sulem, J., & Ouffroukh, H. (2006). Shear-banding in drained and undrained triaxial tests on a saturated sandstone; porosity and permeability evolution. *International Journal of Rock Mechanics and Mining Sciences, 43,* 292–310.

Tal, Y., Evans, B., & Mok, U. (2016). Direct observations of damage during unconfined brittle failure of Carrara marble. *Journal of Geophysical Research: Solid Earth.* https://doi.org/10.1002/2015JB012749.

Tandon, G. P., & Weng, G. J. (1986). Stress distribution in and around spheroidal inclusions and voids at finite concentration. *Journal of Applied Mechanics, 53,* 511–518.

Tapponnier, P., & Brace, W. F. (1976). Development of stress-induced microcracks in Westerly granite. *International Journal of Rock Mechanics and Mining Science and Geomechanical Abstracts, 13,* 103–112.

Tudisco, E., Andò, E., Cailletaud, R., & Hall, S. A. (2017). TomoWarp2: A local digital volume correlation code. *SoftwareX, 6,* 267–270.

Tudisco, E., Hall, S. A., Charalampidou, E. M., Kardjilov, N., Hilger, A., & Sone, H. (2015). Full-field measurements of strain localisation in sandstone by neutron tomography and 3D-volumetric digital image correlation. *Physics Procedia, 69,* 509–515.

Vasseur, J., Wadsworth, F. B., Lavallée, Y., Bell, A. F., Main, I. G., & Dingwell, D. B. (2015). Heterogeneity: The key to failure forecasting. *Scientific Reports, 5,* 13259.

Viggiani, G., Lenoir, N., Bésuelle, P., Di Michiel, M., Marello, S., Desrues, J., et al. (2004). X-ray microtomography for studying localized deformation in fine-grained geomaterials under triaxial compression. *Comptes Rendus Mécanique, 332*(10), 819–826.

Wawersik, W. R., & Fairhurst, C. (1970). A study of brittle rock fracture in laboratory compression experiments. *International Journal of Rock Mechanics and Mining Sciences and Geomechanics Abstracts, 7*(5), 561–575.

Wong, T. F., & Baud, P. (2012). The brittle-ductile transition in porous rock: A review. *Journal of Structural Geology, 44,* 25–53.

Wong, R., Chau, K., Tang, C., & Lin, P. (2001). Analysis of crack coalescence in rock-like materials containing three flaws—Part I: Experimental approach. *International Journal of Rock Mechanics and Mining Sciences, 38*(7), 909–924.

Wong, T. F., David, C., & Zhu, W. (1997). The transition from brittle faulting to cataclastic flow in porous sandstones: Mechanical deformation. *Journal of Geophysical Research: Solid Earth, 102*(B2), 3009–3025.

Wong, E., & Einstein, H. (2009). Crack coalescence in modeled gypsum and Carrara marble: Part 1. Macroscopic observations and interpretation. *Rock Mechanics and Rock Engineering, 42*(3), 475–511.

Wu, X. Y., Baud, P., & Wong, T. F. (2000). Micromechanics of compressive failure and spatial evolution of anisotropic damage in Darley Dale sandstone. *International Journal of Rock Mechanics and Mining Science, 37,* 143–160.

Zhang, J., Wong, T.-F., & Davis, D. M. (1990). Micromechanics of pressure-induced grain crushing in porous rocks. *Journal of Geophysical Research: Solid Earth, 95,* 341–352.

(Received April 16, 2018, revised August 5, 2018, accepted September 21, 2018, Published online October 2, 2018)

Pure Appl. Geophys. 176 (2019), 1117–1143

https://doi.org/10.1007/s00024-018-2042-3

Pure and Applied Geophysics

Seismic Imaging of the Southern California Plate Boundary around the South-Central Transverse Ranges Using Double-Difference Tomography

Pieter-Ewald Share,[1] Hao Guo,[2] Clifford H. Thurber,[3] Haijiang Zhang,[2] and Yehuda Ben-Zion[1]

Abstract—We derive high-resolution *P* and *S* seismic velocities (V_P and V_S) within the South-Central Transverse Ranges section of the San Andreas Fault (SAF), using a new double-difference tomography algorithm incorporating both event-pair and station-pair differential times. The addition of station-pair data allows for better absolute event locations and higher model resolution at shallow depths. Velocities within a 222 km × 164 km region are inverted using > 1,000,000 *P* and *S* arrival (picked with an automatic detection algorithm) and differential times from > 10,000 local events recorded by > 250 stations. Similarly large *P* and *S* datasets lead to high-quality V_P/V_S estimates of the region. The resulting models include low velocities along major fault segments and across-fault velocity contrasts. They also show very high V_P/V_S anomalies near shallow damaged rock, whereas fault zones exhibit either low (< 1.73) or high (> 1.73) V_P/V_S characteristics at greater depth. The variations in amplitude of these anomalies along the SAF through San Gorgonio Pass (SGP) suggest abrupt west-to-east changes in elastic properties. Moreover, their geometries imply near-vertical SAF segments northwest of SGP and northeast dipping faults southeast of that area. The SAF near Coachella Valley is estimated to dip by 57°. Regional-scale low and high V_P/V_S values are related to relative abundances of crystalline or metamorphic rocks. Near-fault V_P/V_S anomalies at depth are likely associated with changes in wet crack geometries. The obtained results can improve future calculations of seismic motion from large earthquakes in the area and related seismic hazard estimates.

Key words: Double-difference tomography, seismic velocities, fault zone structure, South-Central Transverse Ranges, San Gorgonio Pass.

Electronic supplementary material The online version of this article (https://doi.org/10.1007/s00024-018-2042-3) contains supplementary material, which is available to authorized users.

[1] Department of Earth Sciences, University of Southern California, Los Angeles, CA 90089, USA. E-mail: pshare@usc.edu
[2] School of Earth and Space Sciences, University of Science and Technology of China, Hefei 230026, Anhui, China.
[3] Department of Geoscience, University of Wisconsin-Madison, Madison, WI 53706, USA.

1. Introduction

The South-Central Transverse Ranges (SCTR) section of the plate boundary in Southern California is highly complex and includes intersections between the San Andreas Fault (SAF), San Jacinto Fault Zone (SJFZ), and Eastern California Shear Zone (ECSZ) (Fig. 1). Understanding the geometry and continuity of fault structure through the region is key to assessing the potential for large earthquake rupture along the southern SAF (Yule 2009; Porter et al. 2011) and SJFZ (Fialko 2006; Lindsey and Fialko 2013; Lozos 2016). Geological mapping (e.g., Allen 1957; Sharp 1967; Matti et al. 1992; Yule and Sieh 2003) and paleoseismic surveys (e.g., McGill et al. 2013; Rockwell et al. 2015 and references therein) show a variety of fault and earthquake properties. The SAF, for example, has a significant discontinuity through San Gorgonio Pass (SGP), where it steps 15 km to the left over a distance of 20 km (e.g., Allen 1957; Matti et al. 1992) and transfers strain to a network of dextral-normal and dextral-reverse faults (SGP fault zone—SGPFZ and Banning/Garnet Hill faults—BF/GHF) south of the step-over (Fig. 1, bottom left inset). Fault complexity at depth, such as the proposed northward dip of the SGPFZ and BF/GHF (e.g., Magistrale and Sanders 1996; Yule and Sieh 2003), is determined through analyses of earthquake hypocenters (e.g., Jones et al. 1986; Magistrale and Sanders 1996; Nicholson et al. 2018), focal mechanisms (Yang et al. 2012), and potential field data (e.g., Fuis et al. 2012). Fault dip and continuity are also assessed by analysis of fault zone head waves that refract along major fault bimaterial interfaces (Ben-Zion 1989). Continuous segments of the SAF via the inactive Mission Creek fault (MCF, Share and

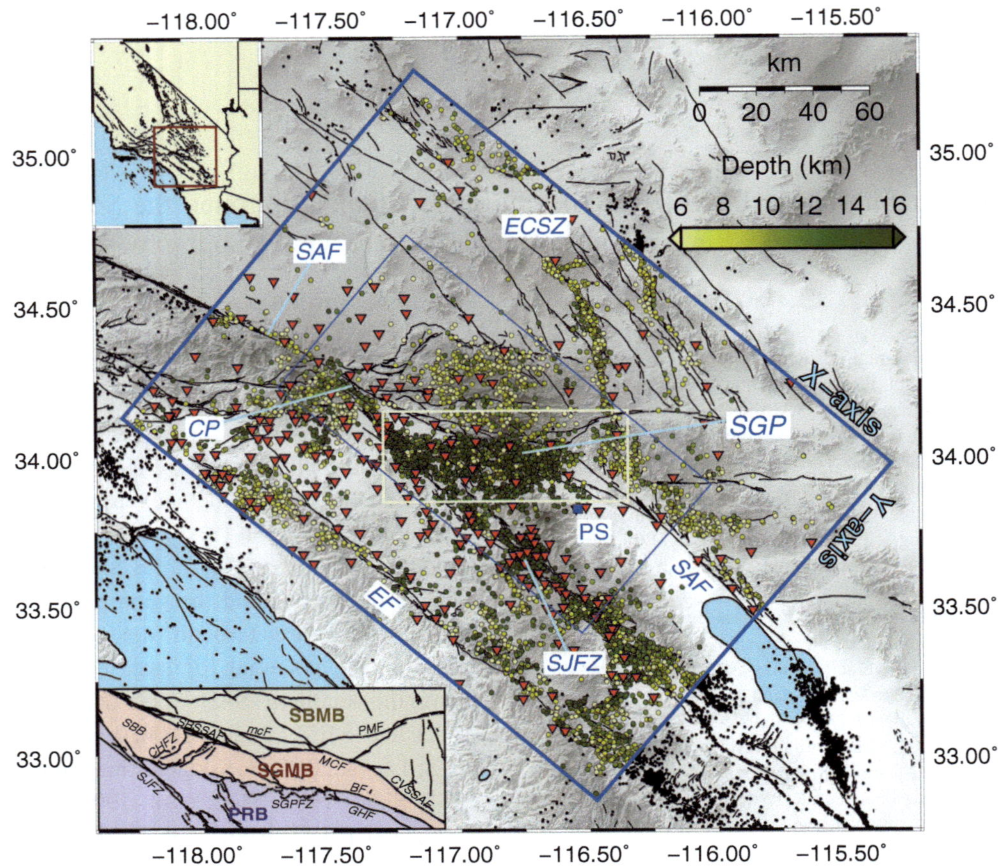

Figure 1
Location of the South-Central Transverse Ranges (SCTR) where the San Andreas Fault (SAF), San Jacinto fault zone (SJFZ), and Eastern California Shear Zone (ECSZ) intersect. Shown for reference is Cajon Pass (CP), Elsinore fault (EF), and the town of Palm Springs (PS). Depicted by circles are local earthquakes ($M_W > 1$) from 01/01/10 to 06/30/2015 (Hauksson et al. 2012). Small black dots are all events in the area, and colored larger circles and red triangles are the 12,674 events and 259 stations used during inversion, respectively. The inversion grid consists of nodes spaced 1 km apart in the center (small blue box) and down to 15 km depth and node spacing as large as 2 km outside that region (large blue box). Beige rectangle highlights San Gorgonio Pass (SGP) at the center of the SCTR, and the bottom left inset shows a zoom-in of this area with its complex network of faults. *SBB* San Bernardino Basin, *CHFZ* Crafton Hills fault zone, *SBSSAF* San Bernardino strand of the San Andreas Fault, *SGPFZ* San Gorgonio Pass fault zone, *mcF* Mill Creek fault, *MCF* Mission Creek fault, *BF* Banning fault, *GHF* Garnet Hill fault, *PMF* Pinto Mountain fault, *CVSSAF* Coachella Valley strand of the San Andreas Fault. The geological basement blocks that make up the region from south to north are the Peninsular Ranges batholith (PRB—purple colors), the San Gabriel Mountains block (SGMB—red colors), and the San Bernardino Mountains block (SBMB—brown colors)

Ben-Zion 2016) and the SJFZ (Share et al. 2017) through the Anza seismic gap (Sanders and Kanamori 1984) have been established through such analysis.

Tomographic images of *P* and *S* wave velocities (V_P and V_S) in volumetric elements within a study area give regional context and insights on large-scale fault structures. Tomography is effective at highlighting anomalous properties of the derived velocity models that can be associated with major faults such as boundaries between different lithological units and damage zones. Fault damage zones, which also act as fluid conduits (e.g., Ben-Zion and Sammis 2003; Faulkner et al. 2010), usually have anomalously low seismic velocities and high V_P/V_S ratios. Seismic tomography and studies of fault zone head and trapped waves have been used to image bimaterial interfaces and localized damage zones along several faults in California (e.g., Ben-Zion and Malin 1991;

Li et al. 1994; McGuire and Ben-Zion 2005; Thurber et al. 2006; Lewis and Ben-Zion 2010; Bennington et al. 2013) and elsewhere (e.g., Mizuno and Nishigami 2006; Calderoni et al. 2012; Najdahmadi et al. 2016). Similar tools have been used to image fault structures in the Southern California plate boundary including parts of the SCTR (e.g., Allam et al. 2014; Zigone et al. 2015; Fang et al. 2016; Share and Ben-Zion 2016; Qiu et al. 2017; Share et al. 2017; Qin et al. 2018). Larger-scale velocity models of the entire Southern California (Tape et al. 2010; Lee et al. 2014; Barak et al. 2015; Shaw et al. 2015) have too low resolution to capture such fault zone features.

In this work, we image V_P and V_S velocity models around the SCTR, with a focus on SGP at high-resolution ($\geq$ 1 km), using arrival time data of local earthquakes and a newly developed seismic double-difference tomography (DDT) method (Guo et al. 2018). The use of an advanced automatic detector (Ross and Ben-Zion 2014; Ross et al. 2016) to pick the onset of *P* and *S* waves allows for rapid extraction of large arrival time datasets from only a few years of data recorded in recent years by a high number of stations of the regional seismic networks. The detector produces similar amounts of *P* and *S* picks (Ross et al. 2016), which leads to improved V_S models in particular and high-quality estimates of V_P/V_S in the area. The new DDT method can simultaneously determine high-precision absolute and relative earthquake locations, and high-resolution velocity structures from shallow near-surface zones to deep seismogenic crustal sections (Guo et al. 2018).

The resulting velocity models provide improved high-resolution images of fault structures and seismic properties in the Southern California plate boundary around the SCTR. Incorporating these velocity models into future calculations of seismic motion from large earthquakes in the area will provide improved estimates of the seismic shaking hazard (Porter et al. 2011 and references therein). In the next section, known large-scale geology and tectonics of the SCTR are discussed in some detail. Sections 3 and 4 describe the new DDT methodology and the data selection process, respectively. Section 5 begins with a discussion on model resolution and continues with V_P and V_S inversion results. At the end of Sect. 5, prominent features in the resulting models are highlighted. These features are interpreted and discussed in the final Sect. 6.

2. *Geology and Seismotectonic Setting*

Figure 1 (bottom left inset) shows three geologic basement types (blocks) that exist in distinctly different regions within the SCTR but complexly interact around SGP (Yule 2009). The Peninsular Ranges batholith (PRB), located south of the northern SJFZ, SGPFZ, BF/GHF, and Coachella Valley strand of the SAF (CVSSAF), consists of Jurassic and Cretaceous granitoid and prebatholithic metamorphic rocks (Matti et al. 1992; Gutierrez et al. 2010). These rocks, especially in the western PRB, have some of the highest densities (Langenheim et al. 2004) and seismic velocities (Barak et al. 2015) in the SCTR. Located north of the San Bernardino strand of the SAF (SBSSAF), MCF, and CVSSAF is the San Bernardino Mountains block (SBMB), which contains Cretaceous plutonic and Paleozoic to Proterozoic metamorphic rocks. The metamorphic rocks are more prominent in the eastern SBMB (Gutierrez et al. 2010; Fuis et al. 2017). Also, SBMB rocks adjacent to the SAF are characterized by high magnetic susceptibilities (Bankey et al. 2002; Langenheim et al. 2005). The San Gabriel Mountains block (SGMB) is located between the PRB and SBMB and is divided into lower and upper (outcrops mostly in the east) plates by the Vincent Thrust (Matti et al. 1992). The lower plate consists of Paleogene to Cretaceous Pelona schist, whereas the upper plate is made up of plutonic (Late Cretaceous to Triassic) and Proterozoic crystalline rocks (Matti et al. 1992; Langenheim et al. 2005; Gutierrez et al. 2010). The lower plate rocks have below-average densities (Anderson et al. 2004; Langenheim et al. 2005) and low seismic velocities (Allam and Ben-Zion 2012), which are amplified within the sediment-filled and faulted San Bernardino Basin (SBB) near the surface (Anderson et al. 2004). These across-fault geological contrasts suggest that major faults in the area separate regions of different velocities and are therefore important targets for seismic imaging.

The oblique orientation of several active SAF strands (e.g., SBSSAF, SGPFZ, and BF/GHF) with respect to northwest relative plate motion (McGill et al. 2015) and the presence of inactive ancestral strands of the SAF (e.g., western BF and MCF) drive structural complexity around SGP and partitioning of strain to surrounding faults. The SBSSAF and CVSSAF only accommodate limited portions of plate motion (< 25 mm/year, McGill et al. 2013; Gold et al. 2015) and are proposed to dip northward (Dair and Cooke 2009; Lindsey and Fialko 2013; Barak et al. 2015) by as much as 37° and 65°, respectively (Fuis et al. 2012). Some of the remaining strain is released by frequent low to moderate-sized events ($M_W < 5.0$, Magistrale and Sanders 1996; Hauksson et al. 2012) and infrequent moderately large earthquakes ($M_W > 5.0$, Jones et al. 1986; Nicholson 1996) along the BF/GHF as well as uplift of the surrounding mountains (Cooke and Dair 2011). Structural complexity within SGP is highlighted by these diverse earthquake focal mechanisms (Bailey et al. 2009; Yang et al. 2012) and variable stress drops (Goebel et al. 2015).

The ECSZ (Spinler et al. 2010) and SJFZ (e.g., Meade and Hager 2005; Dair and Cooke 2009) also accommodate significant amounts of plate boundary-related strain. The ECSZ has relatively low average slip rates (< 5 mm/year, Oskin et al. 2008) but high rates of seismicity (Hauksson et al. 2012), which is mostly concentrated along fault segments that ruptured during the 1992 M_W 6.1 Joshua Tree, 1992 M_W 7.3 Landers, and 1999 M_W 7.1 Hector Mine earthquakes. The SJFZ is the most seismically active component of the plate boundary in Southern California in recent times (Hauksson et al. 2012; Ross et al. 2017) and has experienced several $M_W > 7.0$ earthquakes in the past 4000 years (Rockwell et al. 2015 and references therein). Compared with the nearby southern SAF, the SJFZ has lower than average slip rate (< 20 mm/year versus > 20 mm/year) despite being more optimally aligned with present-day plate motion (e.g., Harden and Matti 1989; Meade and Hager 2005; Blisniuk et al. 2013; Onderdonk et al. 2015).

The intricate fault structures and seismic activity distributed throughout the SCTR suggest that extensive deformation and damage is present. High-resolution imaging and analysis of velocity anomalies associated with these structural complexities will help map them and give insight into fault zone properties in the region.

3. Double-Difference Tomography

The DDT method is a widely used seismic tomography technique for high-resolution imaging of crustal structures. It makes use of absolute arrival times to constrain the large-scale velocities, and event-pair differential times to better constrain the relative location between events and near-source velocity structures (Zhang and Thurber 2003, 2006). In this study, we use a newly developed DDT method that further includes in the inversion station-pair differential times to better constrain absolute event locations and details of the velocity model in the shallow crust (Guo and Zhang 2017; Zhang et al. 2017; Guo et al. 2018).

The DDT inverse problem is solved iteratively using a nonlinear least-squares approach, where, per iteration, the approximate solution to a system of linearized equations is used to update model parameters. This process is then repeated for a user-defined number of iterations. An example linear equation corresponding to an absolute arrival time at station k from event i is (Zhang and Thurber 2003)

$$r_k^i = \sum_{m=1}^{3} \frac{\partial T_k^i}{\partial x_m^i} \Delta x_m^i + \int_i^k \delta u \mathrm{d}s + \Delta \tau^i + S_k, \quad (1)$$

where r_k^i is the residual between observed and calculated arrival times, T is travel time, x and Δx are hypocenter coordinates and their perturbations, τ is origin time, δu is perturbation to slowness, ds is an element of path length, and S is station correction. Station corrections are added to help compensate for unmodeled slowness structure near the surface. Equations for residual r_k^j corresponding to the same station k (as in Eq. 1) but different event j and residual r_l^i for the same event i (as in Eq. 1) but different station l are

$$r_k^j = \sum_{m=1}^{3} \frac{\partial T_k^j}{\partial x_m^j} \Delta x_m^j + \int_j^k \delta u ds + \Delta\tau^j + S_k, \quad (2)$$

$$r_l^i = \sum_{m=1}^{3} \frac{\partial T_l^i}{\partial x_m^i} \Delta x_m^i + \int_i^l \delta u ds + \Delta\tau^i + S_l. \quad (3)$$

By subtracting Eqs. (2) and (3) from Eq. (1) we obtain, respectively,

$$\begin{aligned} d_k^{i,j} &= r_k^i - r_k^j \\ &= \sum_{m=1}^{3} \left(\frac{\partial T_k^i}{\partial x_m^i} \Delta x_m^i - \frac{\partial T_k^j}{\partial x_m^j} \Delta x_m^j \right) + \int_i^k \delta u ds \\ &\quad - \int_j^k \delta u ds + \Delta\tau^i - \Delta\tau^j, \end{aligned} \quad (4)$$

$$\begin{aligned} d_{k,l}^i &= r_k^i - r_l^i \\ &= \sum_{m=1}^{3} \left(\frac{\partial T_k^i}{\partial x_m^i} - \frac{\partial T_l^i}{\partial x_m^i} \right) \Delta x_m^i + \int_i^k \delta u ds - \int_i^l \delta u ds \\ &\quad + S_k - S_l, \end{aligned} \quad (5)$$

where $d_k^{i,j}$ and $d_{k,l}^i$ are called event-pair double differences (DD) (Waldhauser and Ellsworth 2000; Zhang and Thurber 2003) and station-pair DD (Guo and Zhang 2017; Zhang et al. 2017), respectively. Note that the station correction terms and the event origin time terms can be canceled in event-pair DD and station-pair DD, respectively.

Equations 1, 4, and 5 corresponding to all available data are weighted from 0 to 1 (based on data quality and misfit per iteration) and constitute the rows of

$$\boldsymbol{Wd} = \boldsymbol{WGm}, \quad (6)$$

where $\boldsymbol{W}$ is an $N \times N$ diagonal matrix containing the weights, $\boldsymbol{d}$ is the $N \times 1$ residual vector, $\boldsymbol{m}$ is the $M \times 1$ model perturbation vector, and $\boldsymbol{G}$ is the $N \times M$ matrix of Fréchet derivatives. The $\boldsymbol{WG}$ matrix in Eq. (6) is augmented with Tikhonov regularization operators with smoothing (η) and damping (ε) factors. Sharp spatial changes in slowness are penalized more for larger η, whereas increasing ε leads to smaller changes in model parameters per iteration. Finally, the model perturbation vector $\boldsymbol{m}$ is solved using the LSQR algorithm that includes damping (Paige and Saunders 1982), and results after final iterations are reported in Sect. 5.

The event-pair DD are sensitive to the relative locations between pairs of events (Waldhauser and Ellsworth 2000) and the velocity structure near the source region (Zhang and Thurber 2003), whereas the station-pair DD can better resolve the absolute event locations (Guo and Zhang 2017) and the structure outside the source region (Guo et al. 2018). Guo et al. (2018) further combine the event-pair DD and the station-pair DD into one inversion system to simultaneously determine high-precision absolute event locations and relative locations between events, as well as high-resolution velocity structure both near and outside the source region.

4. Data Selection

We chose a model space with dimensions of $X = 222$ km, $Y = 164$ km, and $Z = 34$ km (3 km above to 31 km below sea level), centered on SGP and rotated clockwise by 40° (Fig. 1). Waveform data corresponding to 19,332 $M_W > 1$ events occurring from 01/01/2010 to 06/30/2015 and recorded by all available stations located within this space were downloaded from the Southern California Earthquake Data Center (SCEDC 2013) and ANZA-YN seismic network database (Vernon 1982; Vernon and Ben-Zion 2010). Accelerometer and geophone data recorded by available short-period and broadband channels were extracted in 70-s windows. The windows start 10 s before and end 60 s after the origin times reported in the relocated Southern California catalog (Hauksson et al. 2012).

Next, an automatic picking algorithm (Ross and Ben-Zion 2014; Ross et al. 2016) was used to detect the onset of P and S arrivals in the extracted waveforms. The algorithm uses short-term average to long-term average detectors together with kurtosis- and skewness-based detectors to identify and pick the onset times. The picker produces a signal-to-noise ratio (SNR) and a kurtosis-based metric (Ross et al. 2016) for P and S picks, respectively, which were used to select the highest-quality pick for each event–

station combination and to assign a weight to the selected data point. Outlier picks and redundant data were systematically removed by application of the following steps:

1. We discarded picks more than 1 s off from predicted arrival times, using the CVM-H15.1.0 velocity model (Shaw et al. 2015, hereinafter referred to as the Harvard model).
2. Only one *P* and one *S* arrival time is allowed for each event–station combination. To this end, the *P* and *S* picks with the largest SNR and kurtosis-based metric, respectively, were selected.
3. The model space was divided into 8-km^3 cubes, azimuths to observing stations were calculated for events within each volume, and event–station combinations were randomly selected such that there was a maximum of seven *P* and seven *S* arrival times per 10° azimuth.
4. In cases where one or two stations were located within a 10° bin, only four *P* and four *S* data were allowed per station. This step helps remove redundant data when, for example, an isolated station records multiple events from a dense cluster.

Steps 1–2 produced 586,367 *P* and 590,421 *S* picks, corresponding to 18,875 events and 262 stations. The choice of the three-dimensional (3D) model used to remove outliers in step 1 is not crucial, as using the CVM-S4.26 (Lee et al. 2014) and Fang et al. (2016) velocity models (hereinafter referred to as SCEC and Fang2016 models), for example, produced 88 % and 90 % of the same picks. The maximum allowed data in step 3 was derived from the average number of stations located within an arbitrary 10° bin measured from the center of the model space; That is, 10° × (262/360°) ~ 7 stations. This step limits the number of rays with similar azimuths and therefore homogenizes azimuthal ray coverage through the model space, which is key to producing undistorted anomalies in travel time inversions. It also ensures that the V_P and V_S models have similar ray coverage, which facilitates later comparison between the two and improves V_P/V_S model quality. Application of steps 3–4 reduces the amount of data by ~ 55 %. The remaining data ($P = 242{,}122$ and $S = 243{,}445$) associated with 12,674 events and 259 stations (Fig. 1) constitute the absolute arrival times incorporated during inversion (Eq. 1).

The criteria used to select event-pair differential times from the absolute arrival times are summarized in Table 1 (top) and are similar to those used by Waldhauser and Ellsworth (2000). These criteria produce 469,278 *P* and 534,741 *S* event-pair data. Station-pair differential times were selected from absolute arrival times in a similar fashion, with some control parameters including the minimum and maximum interstation offset, minimum interstation epicentral distance, minimum and maximum number of station-pair observations per event, and maximum number of neighboring stations attached to a single station (Guo and Zhang 2017). The values used during the station-pair data selection process appear in Table 1 (bottom). The total selected *P* and *S* station-pair data were 816,976 and 837,337.

5. Results

We discretized the velocity model such that nodes were 1 km apart in a 142 km by 72 km area centered on SGP and down to 15 km depth, and as much as 2 km apart outside that region (Figs. 1 and S1). The inversion used 20 iterations to ensure satisfactory convergence of the final result. During the first 10 iterations, emphasis was placed on fitting the absolute data, whereas during the last 10 iterations emphasis was placed on DD data. Both sets of 10 iterations consisted of 7 joint inversions of velocity and location, and 3 inversions of relocation only. In Sect. 5.1, we test the resolution of structure in different parts of the model space by inverting synthetic data using the described settings. In Sect. 5.2, we invert the observed seismic data.

5.1. Model Resolution

Model resolution was first assessed using the well-known checkerboard test, where synthetic data generated from a regular grid of positive (above average) and negative (below average) anomalies are inverted and the resultant model (recovered model) is compared with the checkerboard model. Resolution is

Table 1

Selection criteria for event-pair (top) and station-pair (bottom) DD from arrival time data

Event-pair selection criteria	
Maximum distance between event-pair and station	100 km
Maximum hypocentral separation between events	5 km
Maximum number of neighbors per event	10
Minimum number of links required to define a neighbor	8
Minimum number of event-pair observations per station	8
Maximum number of event-pair observations per station	30
Station-pair selection criteria	
Minimum distance between stations per event	2 km
Maximum distance between stations per event	100 km
Minimum epicentral distance difference between stations	2 km
Maximum number of neighbors per station per event	40
Minimum number of station-pair observations per event	4
Maximum number of station-pair observations per event	200

said to be good in areas where the two models agree and poor where they are not sufficiently similar. This test is useful, but amongst several limitations (Lévêque et al. 1993) there is no one-to-one mapping between resolution in the recovered model and the model obtained by inverting actual data (inverted model). This is because the two models have different velocities, thus rays associated with the same event–station combinations do not necessarily sample the same parts of the model space. Given this, we additionally analyzed model quality using derivate weighted sums (DWS, Thurber and Eberhart-Phillips 1999) and the diagonal of the resolution matrix (e.g., Rawlinson et al. 2014), which are both more direct indicators of resolution in the inverted model.

To test the various length scales resolvable with the available data, we used checkerboard models with anomalies that extended 2–10 grid nodes in X, Y, and Z directions. These anomalies alternate between 4.75 km/s and 5.25 km/s (5 km/s $\pm$ 5 %) in the V_P models, and those values divided by 1.73 (average regional V_P/V_S ratio) in the V_S models. As examples, checkerboard models with anomalies spanning 2 and 6 grid nodes are shown at 10 km depth in Figs. 2a, b for V_P and 3a, b for V_S, and in cross-section at $X = 8$ km in Figs. 4a, b for V_P and 5a, b for V_S. Synthetic travel times corresponding to the data discussed in Sect. 4 were extracted from the checkerboard models. DD were then calculated from the travel times using the parameters in Table 1. Next, we gave all synthetic data a weight of 1 and inverted them using homogeneous half-space starting models of 5 km/s for V_P and 2.89 km/s for V_S (5 km/s divided by 1.73). To recover as best possible small-scale anomalies and in general resolve sharp velocity contrasts, we set a small smoothing parameter $\eta = 1$. We then tested different damping parameters (ε) and chose one that ensured that the ratio between the largest and smallest eigenvalues (condition number) of the regularized ***WG*** matrix (Eq. 6) lay between 100 and 200. It is important to keep this range low but not too narrow, because the true condition number is different, and likely lower, than the one calculated here with LSQR (Zhang and Thurber 2007). We found that using $\varepsilon = 1400$ and 1450 during the first and second sets of seven joint inversions, respectively, and $\varepsilon = 200$ during relocation-only inversion produced condition numbers in the specified range. The results show that, with synthetic data, we could resolve checkerboard anomalies spanning as few as two grid nodes over a broad region (Figs. 2, 3, 4, 5c).

To facilitate comparison with the inverted models, we repeated this process using the same synthetic data, but the regularization parameters used in the inversion of actual data (see Sect. 5.2). The larger smoothing applied ($\eta = 45$) results in poorer resolution of small-scale velocity structures (not shown)

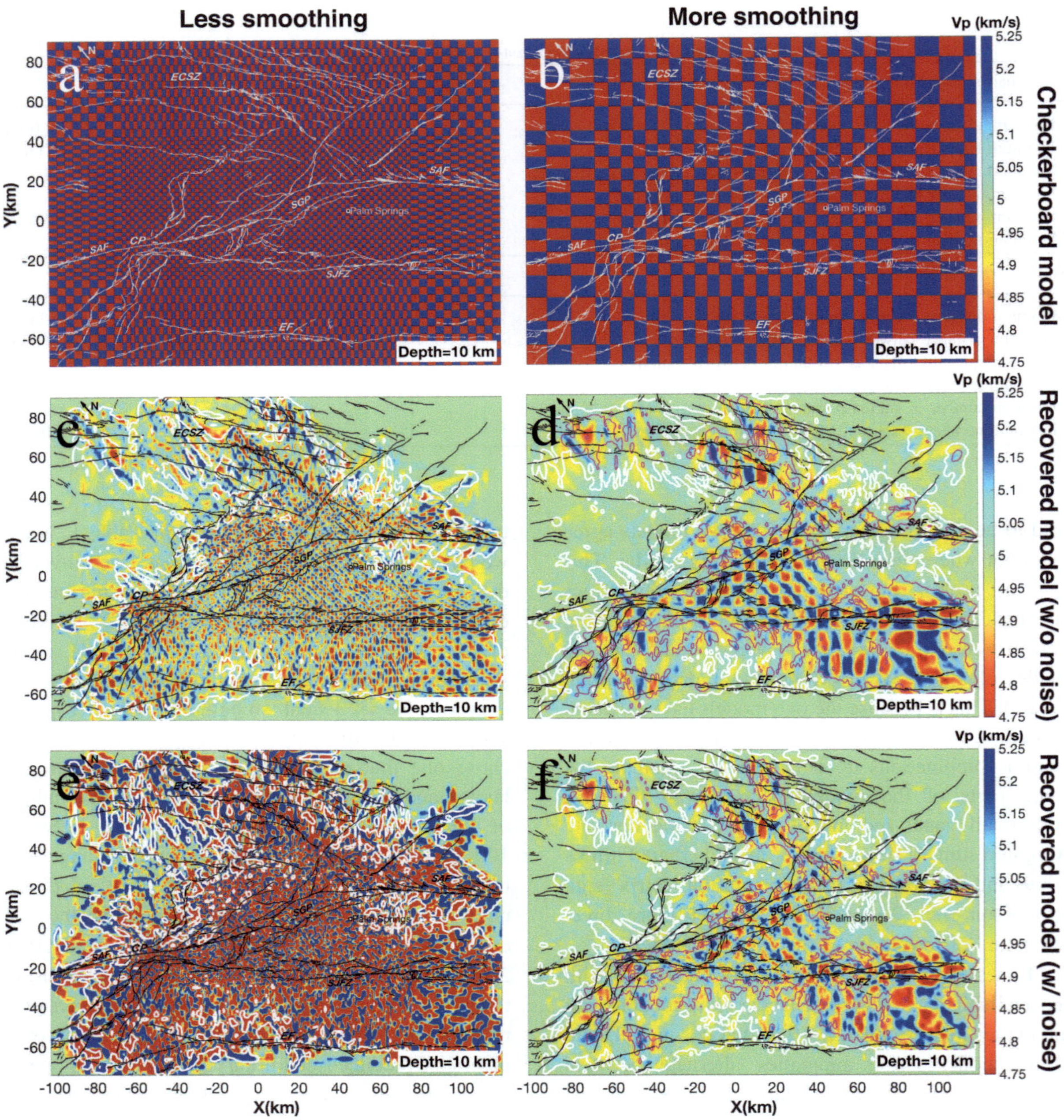

Figure 2
Checkerboard recovery results at 10 km depth (V_P). Checkerboard model used to generate synthetic data consisting of $\pm$ 5 % anomalies spanning 2 (**a**) and 6 (**b**) grid nodes. **c–d** Recovered models corresponding to checkerboard models in **a** and **b** and noise-free synthetic data. **e–f** Recovered models corresponding to checkerboard models in **a** and **b** and noisy data. More smoothing is applied during inversion of **d** and **f**. White contours depict DWS = 20, and magenta contours show r_s = 0.008

and increased misfit. Specifically, we found that only checkerboard anomalies spanning > 4 grid nodes were sufficiently recovered in regions with adequate ray coverage. As examples, results corresponding to 6 grid node checkerboard models are shown in Figs. 2, 3, 4, and 5d. The addition of the station-pair data to

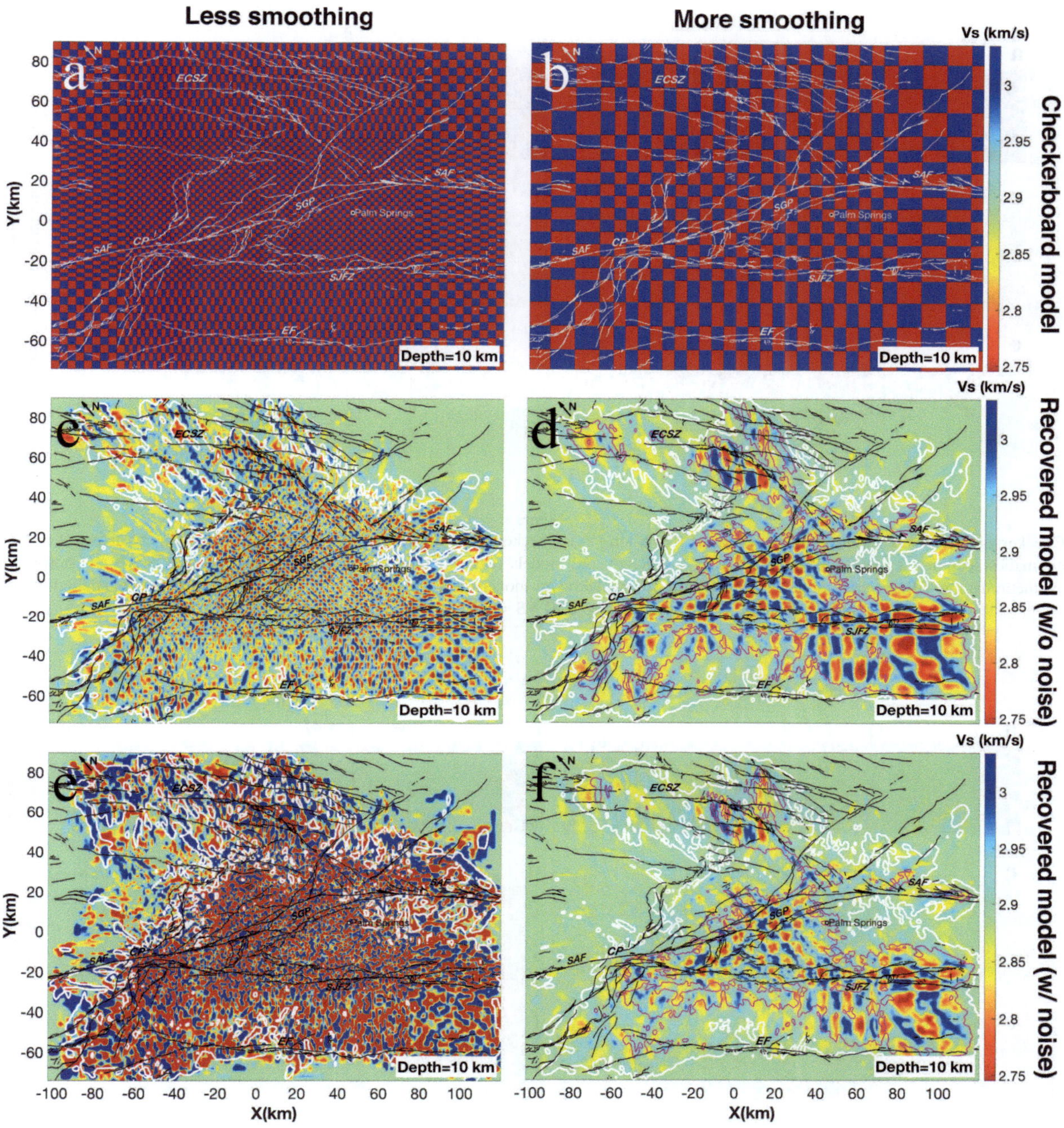

Figure 3
Similar to Fig. 2 but for V_S checkerboard tests at 10 km depth

the inversion dramatically improved the ability to recover (i.e., resolve) these checkerboard anomalies at shallow depths (Fig. S2).

Arrival time uncertainty is greater in the present study compared with similar earlier studies (e.g., Allam and Ben-Zion 2012), because an automated detector instead of a trained analyst was used to pick arrivals. To test the resolvable length scales when data contain errors, we inverted the same synthetic data again but now with noise added. The amount of

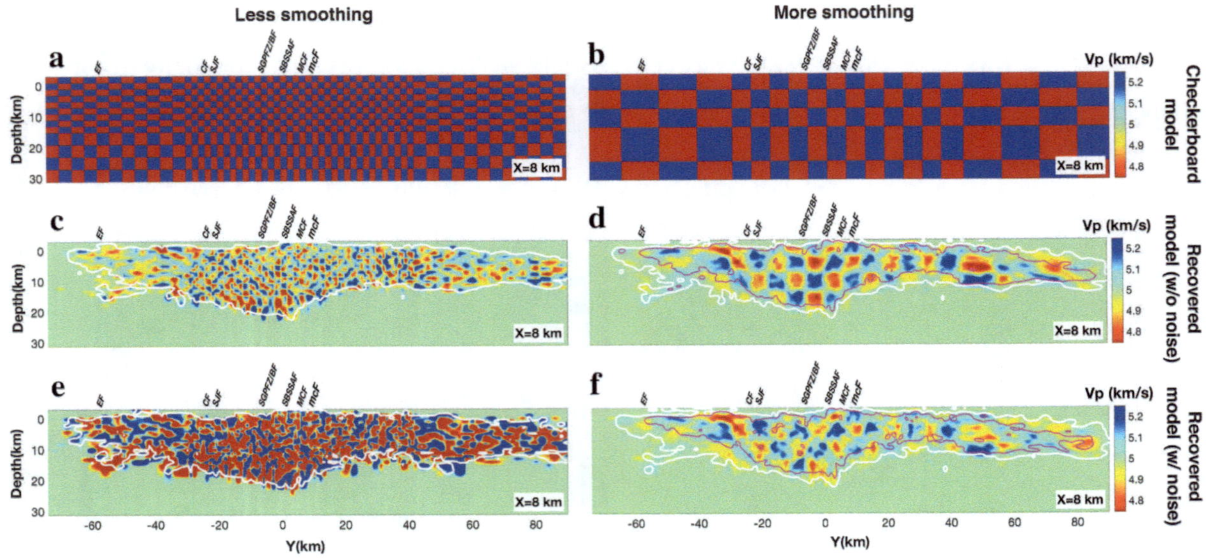

Figure 4
Checkerboard recovery results in cross-section at X = 8 km (V_P). Checkerboard model used to generate synthetic data consisting of ± 5 % anomalies spanning 2 (**a**) and 6 (**b**) grid nodes. **c–d** Recovered models corresponding to checkerboard models in **a** and **b** and noise-free synthetic data. **e–f** Recovered models corresponding to checkerboard models in **a** and **b** and noisy data. More smoothing is applied during inversion of **d** and **f**. White contours depict DWS = 20, and magenta contours show r_s = 0.008

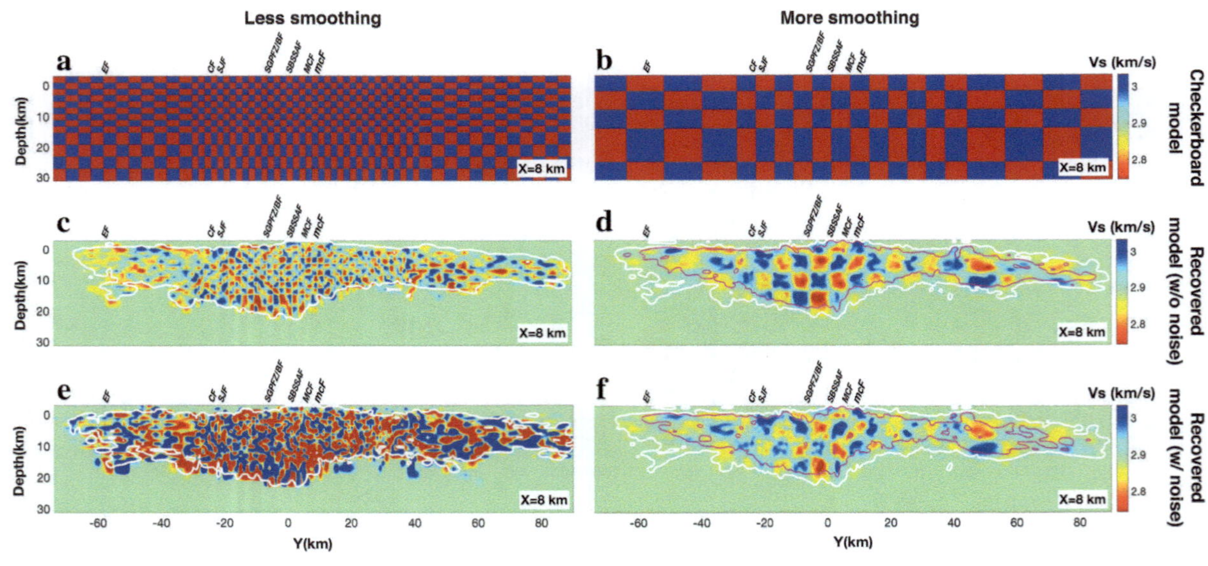

Figure 5
Similar to Fig. 4 but for V_S checkerboard tests at X = 8 km

noise added to synthetic travel time data was based on the picking errors normally produced by the automatic detector. The short-term average (STA)/-long-term average (LTA) detector used to pick P arrivals (Ross and Ben-Zion 2014) performs similarly to other STA/LTA detection algorithms (e.g., Gentili and Michelini 2006; Nippress et al. 2010) where the distribution of the difference between "true" and automatic picks has a standard deviation of $\sim$ 0.1 s. In terms of S arrivals, the employed detector produces automatic picks that are within 0.16 s of manual picks 75 % (1.15 standard deviations for normal distribution) of the time (Ross et al. 2016). Subsequently, zero-mean Gaussian noise with standard deviation equal to 0.1 s and 0.14 s (0.16 s divided by 1.15) was added to the synthetic P and S travel time data. In addition, we assigned lower weights to data with more noise added. To simultaneously test how uncertainty in event location affected structural resolution, we displaced hypocenters by amounts equal to the absolute relocation uncertainties in the Hauksson et al. (2012) catalog. That is, for 90 % (1.64 standard deviations for normal distribution) of the events in the catalog, the horizontal and vertical uncertainties are less than 0.8 km and 1.5 km. Therefore, horizontal and vertical displacements sampled from zero-mean Gaussian distributions with standard deviation of 0.5 km and 0.9 km (0.8 km and 1.5 km divided by 1.64) were added to event hypocenters.

Noisy synthetic data were inverted using the same regularization parameters as before, and the resulting models compared with those obtained from inverting noise-free synthetic data. First, we inverted using $\eta = 1$ and $\varepsilon = 1400$ and 1450 during the first and second set of 7 joint inversions, respectively, and $\varepsilon = 200$ during relocation-only inversion. The results show that these regularization parameters were inappropriate to use when data contained noise, because they led to distorted and incoherent recovered models (compare Figs. 2, 3, 4, and 5c with 2, 3, 4, and 5e). We subsequently inverted the noisy data using settings that included more smoothing ($\eta = 45$, see Sect. 5.2). In addition, to suppress the unwanted effects of noise, outliers were removed per iteration by only allowing data residuals within a certain number of median absolute deviations (MAD) from the median to be inverted. The number of allowed MAD was gradually decreased from 10 in the first iteration to 3 in the last iteration. Medians and MAD were calculated independently for P and S data per iteration, as these data have characteristically different errors given the employed detectors. This stricter regularization and outlier removal process leads to more coherent anomalies being recovered. As with synthetic data, acceptable recovery is achieved for anomalies spanning > 4 grid nodes (Figs. 2, 3, 4, 5f). However, the anomalies are slightly distorted due to the influence of noise, and the recovered area is smaller than before (compare Figs. 2, 3, 4, and 5d with 2, 3, 4, and 5f). The latter mostly relates to lower data weights and the removal of outliers per iteration during noisy data inversion.

To further quantify model resolution, we compared the checkerboard results with DWS values computed per grid node for each model. These values were calculated by summing the $\boldsymbol{WG}$ matrix elements (prior to regularization, Eq. 6) column-wise. They therefore represent the sum of all fractional ray lengths (times data weight) in the vicinity of a given grid node. Comparison of the DWS and checkerboard results shows that the two are complementary. Specifically, DWS = 20 values (white contours in Figs. 2, 3, 4, 5) enclose regions where checkerboard anomalies are well recovered and appear in subsequent figures. Generally, near the edges of these regions, checkerboard anomalies are only partially recovered (< 50 % of original amplitudes), whereas at the center recovered amplitudes are > 80 % of the original checkerboard amplitudes. For nodes where both V_P and V_S models have DWS ≥ 20, the mean $\pm$ standard deviation of recovered V_P/V_S values was 1.73 $\pm$ 0.01 (noise-free synthetic data) and 1.73 $\pm$ 0.02 (noisy synthetic data) (dark-blue lines in Figs. S3b-c and S4b-c). These means are equal to the true checkerboard model V_P/V_S (Figs. S3a and S4a). Model deviations from 1.73 relate to small-scale ($\leq$ 3 km) V_P/V_S anomaly artifacts caused by the different resolutions of the P and S data. These artifacts become especially pronounced when noisy data are inverted (Figs. S3c and S4c). To suppress the small-scale artifacts in the V_P/V_S model inverted using real data (next section), we apply a 3 by 3 median filter to the depth slices and cross-sections presented. This inevitably decreases the resolution in V_P/V_S models but helps avoid incorrect interpretations due to model artifacts. Figures S3d and S4d demonstrate the effects of applying a median filter to a model containing small-scale artifacts (Figs. S3c and S4c).

In general, DWS is an effective tool for indicating regions of acceptable resolution. However, structures with similar DWS (and ≥ 20) are not necessarily equally well resolved; For example, at 10 km depth, the model space around Palm Springs ($X = 50$ to 70 km, $Y = -20$ to 0 km) was not as well resolved as around the SJFZ trifurcation area ($X = 70$ to 110 km, $Y = -30$ to -10 km), while both regions are at the edge of acceptable ray coverage (bordered by DWS = 20 contours in Figs. 2d, f and 3d, f). One reason for this is that DWS relates to ray density but is insensitive to the azimuthal distribution of those rays. As a consequence, if a region has large DWS but all through-going rays have similar azimuths, then imaged anomalies in the region will be distorted and thus poorly resolved.

We used the resolution matrix to quantify this variability in resolution. Using the notations in Sect. 3, the $N \times N$ resolution matrix $\boldsymbol{R}$ is defined as

$$\boldsymbol{R} = \left(\boldsymbol{G}^{\mathrm{T}}\boldsymbol{W}^{-1}\boldsymbol{G} + \eta^2\boldsymbol{D}^{\mathrm{T}}\boldsymbol{D} + \varepsilon^2\boldsymbol{I}\right)^{-1}\boldsymbol{G}^{\mathrm{T}}\boldsymbol{W}^{-1}\boldsymbol{G}, \quad (7)$$

where $\boldsymbol{D}$ is the finite difference approximation of the first spatial derivative of slowness and $\boldsymbol{I}$ is the identity matrix. Present-day state-of-the-art algorithms and computers are sufficient to calculate the full $\boldsymbol{R}$ (Bogiatzis et al. 2016), however, to save on time and computational cost, we used a stochastic method of estimating the resolution matrix diagonal only (Bekas et al. 2007; MacCarthy et al. 2011). This technique is one of several (e.g., An 2012; Trampert et al. 2013; Fichtner and van Leeuwen 2015) that utilize random vectors to quantify model resolution. Consider a sequence of s random vectors, $\boldsymbol{v}_1, \ldots, \boldsymbol{v}_s$, each $N \times 1$ in size, sampled from a standard normal distribution. An estimate of the diagonal $\boldsymbol{r}_s$ of the resolution matrix $\boldsymbol{R}$ is (MacCarthy et al. 2011)

$$\boldsymbol{r}_s = \left[\sum_{k=1}^{s} \boldsymbol{v}_k . \times \boldsymbol{R}\boldsymbol{v}_k\right] ./ \left[\sum_{k=1}^{s} \boldsymbol{v}_k . \times \boldsymbol{v}_k\right], \quad (8)$$

where .× and ./ denote element-wise vector multiplication and division, respectively. To compute $\boldsymbol{r}_s$, we therefore need to be able to generate random vectors, which is trivial, and the product $\boldsymbol{R}\boldsymbol{v}_k$ of the resolution matrix and the random vectors. The latter is obtained by multiplying Eq. (7) with $\boldsymbol{v}_k$, and is given by

$$\boldsymbol{R}\boldsymbol{v}_k = \left(\boldsymbol{G}^{\mathrm{T}}\boldsymbol{W}^{-1}\boldsymbol{G} + \eta^2\boldsymbol{D}^{\mathrm{T}}\boldsymbol{D} + \varepsilon^2\boldsymbol{I}\right)^{-1}\boldsymbol{G}^{\mathrm{T}}\boldsymbol{W}^{-1}\boldsymbol{G}\boldsymbol{v}_k, \quad (9)$$

which is simply the normal equations solution for

$$\min \left\| \begin{bmatrix} \boldsymbol{W}\boldsymbol{G} \\ \eta\boldsymbol{D} \\ \varepsilon\boldsymbol{I} \end{bmatrix} \boldsymbol{R}\boldsymbol{v}_k - \begin{bmatrix} \boldsymbol{W}\boldsymbol{G}\boldsymbol{v}_k \\ 0 \end{bmatrix} \right\|_2, \quad (10)$$

and can be solved using LSQR. As s becomes large, $\boldsymbol{r}_s$ approaches the true diagonal of the resolution matrix $\boldsymbol{R}$. To ensure satisfactory convergence of $\boldsymbol{r}_s$, we used the upper bound of $s = 512$ suggested by MacCarthy et al. (2011) and ran 20 realizations to compute 20 independent estimates of $\boldsymbol{r}_s$, taking in the end the median of those estimates to produce a single robust value for $\boldsymbol{r}_s$.

We computed the resolution matrix diagonal in this manner for the recovered models with checkerboard anomalies spanning > 4 grid nodes. This was done for the regularized $\boldsymbol{W}\boldsymbol{G}$ matrix generated during the final iteration. Comparisons of recovered models and associated resolution values (r_s) show that, during final iteration, regions with $r_s \geq 0.008$ have the best-resolved checkerboard anomalies. Figures 2, 3, 4, 5d, f and S3b, c and S4b, c show $r_s = 0.008$ contours for the recovered models with 6 grid node checkerboard anomalies (magenta lines). r_s values calculated here are relatively low compared with earlier studies using DDT (e.g., Zhang and Thurber 2007; Zhang et al. 2017), and this stems from the large η and ε employed to suppress the effects of data noise. However, the checkerboard tests show that, after 20 iterations, anomalies were sufficiently recovered despite relatively low r_s per iteration. Moreover, changes in r_s highlight well the variability in resolution here, which was the original purpose of calculating the resolution matrix diagonals. Recovered anomalies within regions with $r_s \geq 0.008$ have amplitudes > 50 % of the original checkerboard anomalies with a maximum recovery of ~ 90 %. Outside these regions, there is a rapid decrease in these percentages.

To summarize, the checkerboard tests indicate that, when using the same regularization as in the inversion of actual data (Sect. 5.2), we could resolve structures > 4 km. Moreover, analyses of DWS and resolution matrix diagonals revealed that > 4 km anomalies were only well constrained if they also had associated DWS ≥ 20 and were best resolved if they had $r_s \geq 0.008$. Lastly, regularization inevitably produced recovered V_P and V_S anomalies lower in amplitude when compared with the checkerboard models from which the data were generated. However, original V_P/V_S values were well recovered, because the P and S datasets were similar in coverage and size and the same regularization was applied to both datasets. In Sects. 5.2 and 6, these findings are used as guidelines when interpreting the inverted model.

5.2. Inversion of Observed Data

Starting with the Harvard model (Shaw et al. 2015) shown in Fig. S5a at 13 km depth, we systematically tested different combinations of η and ε to find values that produced inverted models containing coherent and geologically feasible features. Simultaneously, we required the condition number to be around 100 for all iterations. This is near the lower bound of the range used during inversion of synthetic data and was done to ensure that data uncertainty did not significantly diminish model quality. Similar to the checkerboard tests using noisy data, we removed outliers per iteration by discarding data more than 10 MAD from the model median during the first iteration, gradually decreasing this number to 3 during the last iteration. After extensive testing, we obtained satisfactory results for regularization parameters around $\eta = 45$ and $\varepsilon = 800$ and 900 during the first and second set of 7 joint inversions, respectively, and $\varepsilon = 300$ during relocation-only inversion. The initial data RMS for P and S arrival times (0.29 s and 0.42 s) was reduced to $P = 0.10$ s and $S = 0.14$ s after 20 iterations (Fig. 6a). The latter are equal to the expected uncertainties in P and S arrival time picks (Sect. 5.1), providing confidence that the chosen regularization parameters did not lead to significant overfitting of the data during inversion. Figure 6b shows the inversion results for V_P and V_S structures at 13 km depth.

Next, we illustrate that the obtained results are independent of the starting model by repeating the same data selection and inversion processes but starting instead with the SCEC (Lee et al. 2014) and Fang2016 (Fang et al. 2016) models. Figures S5b and c show these models at 13 km depth. After applying the selection criteria in Sect. 4, the number of arrival time data to be used during inversion was 93.2 % (SCEC) and 93.1 % (Fang2016) of those described in Sect. 4. Therefore, according to our criteria, more data were considered outliers relative to the SCEC and Fang2016 models compared with the Harvard model. Related to this is the relatively large initial data RMS for P and S data associated with the SCEC and Fang2016 models, namely, $P = 0.44$ s and $S = 0.52$ s (Fig. 6c) and $P = 0.43$ s and $S = 0.56$ s (Fig. 6e), respectively. From these data, we extracted DD using the criteria in Table 1. Then, following the same inversion procedure as before (Fig. 6b), we inverted the respective data, and after 20 iterations their arrival time misfits decreased to $P = 0.11$ s and $S = 0.15$ s (SCEC, Fig. 6c) and $P = 0.09$ s and $S = 0.13$ s (Fang2016, Fig. 6e). Figure 6d and f contain, respectively, updates to the SCEC and Fang2016 models at 13 km depth following inversion. Generally, in regions with adequate ray coverage (DWS ≥ 20), there are several correlated structures between the different inversion results (compare Fig. 6b, d, and f) that were not present in the starting models (Fig. S5). For example, at 13 km depth, all the inverted models showed very low V_P anomalies near faults in the southeastern SJFZ, SGP, and ECSZ (and not so much in V_S) and a clear transition from the eastern (slow) to western PRB (fast), the latter being the most prominent high-velocity anomaly in these models.

We compared the depth-dependent variations between the different models by exploring their respective average velocities at each depth [one-dimensional (1D) models] pre- and post-inversion. Figure 7 shows these profiles (P = blue; S = red) before (a) and after (b) inversion from 5 to 15 km depth. This depth range was chosen because it is where we have greatest sensitivity given the data and methodology employed. The Harvard and Fang2016 profiles prior to inversion are similar, but both differ from the SCEC profiles, which generally have higher velocities in the study region and at the depths presented (Figs. 7a and S5). Following inversions, the 1D profiles (calculated only for nodes with

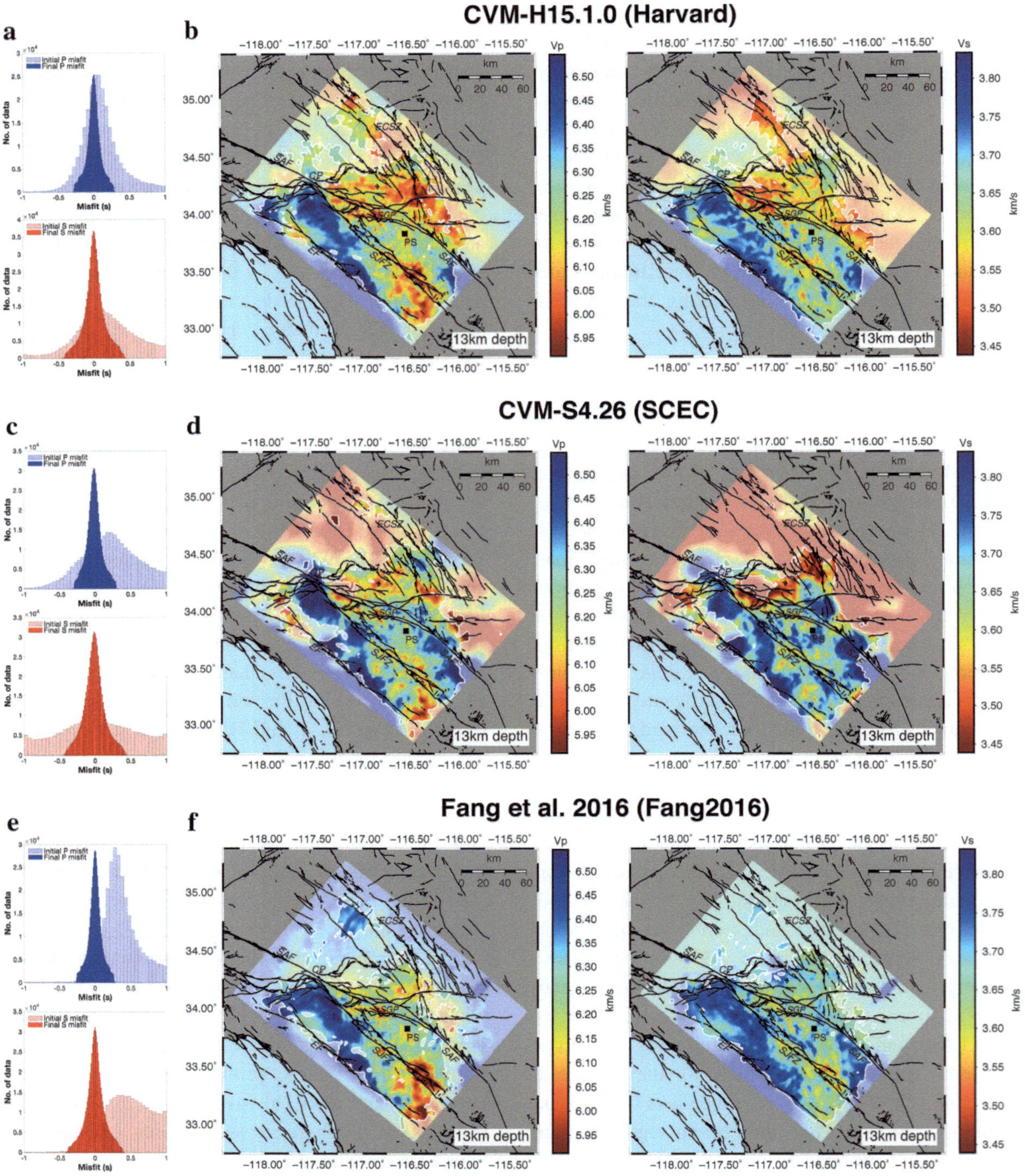

Figure 6
Inversion results at 13 km depth for different starting models. **a** Initial (transparent) and final (not transparent) *P* (blue) and *S* (red) model misfits for the Harvard model. **b** Inverted V_P (left) and V_S (right) structures using the Harvard model as starting model. White contours depict DWS = 20. **c–d** Same layout but for the SCEC model. **e–f** Same layout but for the Fang2016 model

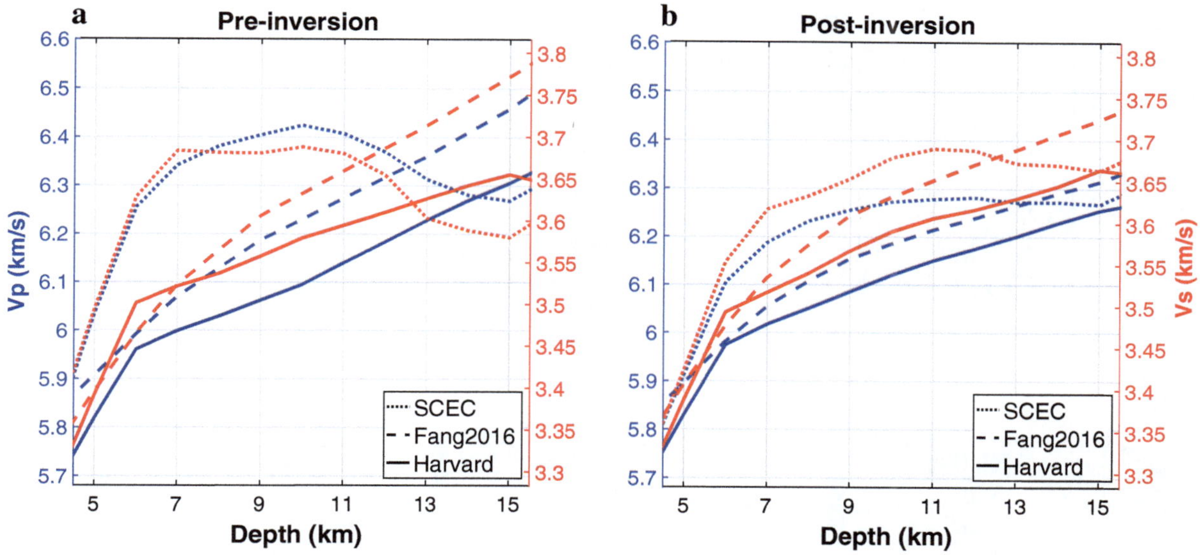

Figure 7
Average V_P (blue) and V_S (red) versus depth before (**a**) and after (**b**) inversion using the Harvard (solid lines), SCEC (dotted lines), and Fang2016 (dashed lines) as starting model

DWS $\geq$ 20) agreed more. For example, there is a significant decrease in the velocities of the SCEC profiles, especially in V_P, as they start to converge to the Harvard and Fang2016 profiles. The Harvard and Fang2016 profiles, in turn, are closer to each other still, and show a parallel increasing trend in velocity with depth. The small differences that still exist between the profiles are consequences of a nonlinear inverse problem linearized about different starting models at iteration 1. These comparisons demonstrate that, where we have acceptable ray coverage and resolution, the results converged to similar solutions irrespective of the starting model and the inherent nonlinearity of the problem. Given this, and that it is best practice in linearized inversions to use a starting model as close as possible to the final model, we chose to interpret features in results generated with the Harvard model. This model had the lowest initial data RMS, and prominent features observed in the inverted model also appeared in results obtained using the SCEC or Fang2016 model (compare Fig. 6b, d, and f).

Figure 8 shows final inversion results (using the Harvard model) at various depths for the entire model space. Relocated events are compared with locations from the Hauksson et al. (2012) catalog in Fig. 9. The mean epicentral difference between the two sets of events is 0.08 $\pm$ 1.4 km, and the mean difference in depth is 0.28 $\pm$ 2.3 km (with relocated events being on average deeper). The majority of relocated and catalog epicenters differ by small amounts (< 0.5 km). Moreover, apart from a few outlier events, large discrepancies (> 1.0 km) in epicenters are only observed for events near the edges of the model space, where data constraints are weaker. This is true for the differences in depth between the two sets of events also. The largest differences in depth are associated with events near the surface or at the base of the seismogenic zone. If we exclude these poorly relocated events, then the differences between the relocated and catalog events are of the same order as the absolute uncertainty in the latter (see Sect. 5.1). This means the absolute locations in the Hauksson et al. (2012) catalog are of high quality and did not need to be displaced by large amounts to achieve a sufficiently low misfit during inversion. Various cross-sections through the 142 km by 72 km box centered on SGP (Fig. 1) appear in Figs. 10, 11, and 12. The resulting models show various large-scale features also observed in other smooth

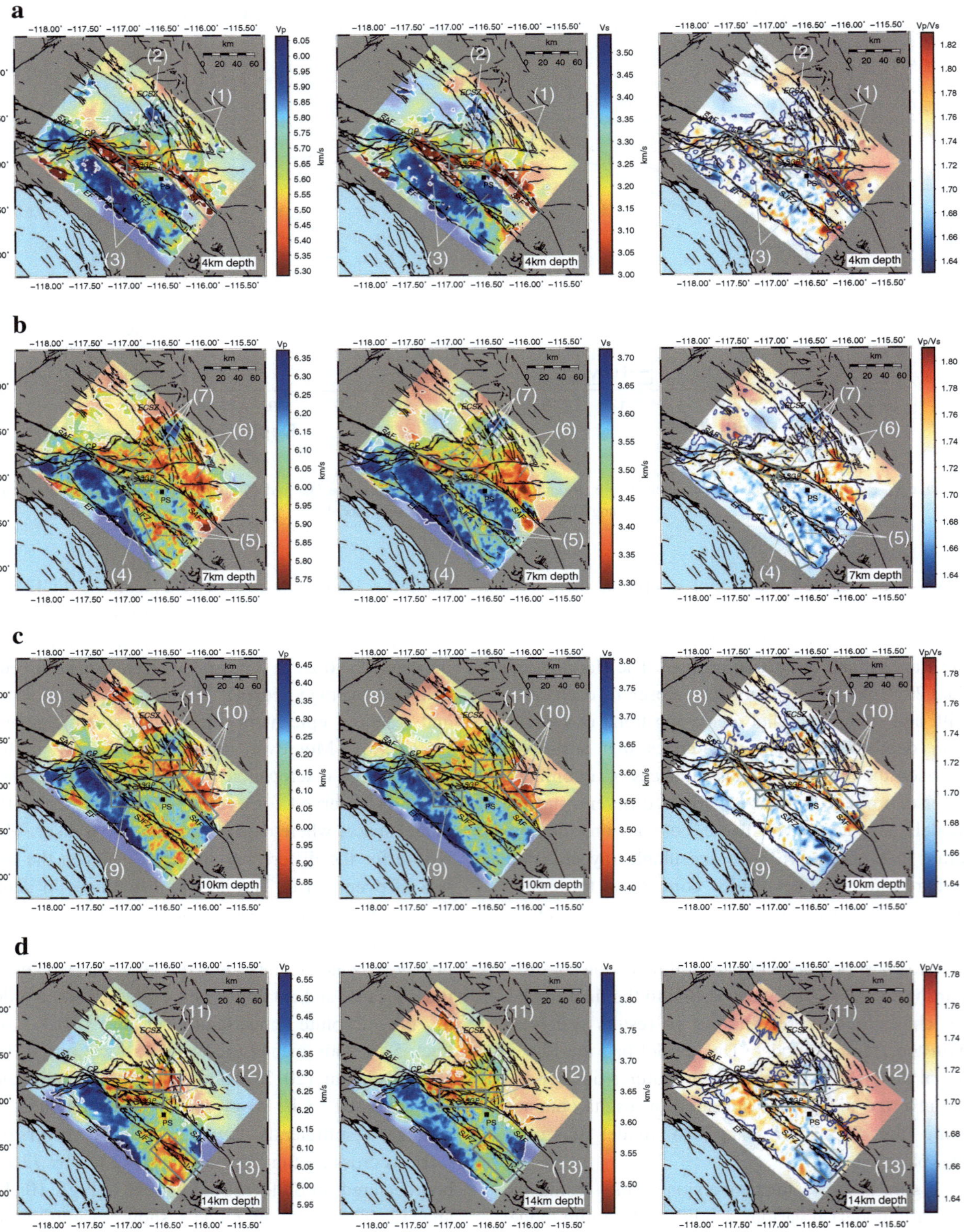
a
b
c
d
Vp
Vs
Vp/Vs
km/s
km
0 20 40 60
SAF
ECSZ
CP
SGP
PS
EF
SJFZ
4km depth
7km depth
10km depth
14km depth
(1)
(2)
(3)
(4)
(5)
(6)
(7)
(8)
(9)
(10)
(11)
(12)
(13)
−118.00° −117.50° −117.00° −116.50° −116.00° −115.50°
35.00°
34.50°
34.00°
33.50°
33.00°

◀Figure 8
Final V_P (left), V_S (middle), and V_P/V_S (right) inversion results at depth of 4 km (**a**), 7 km (**b**), 10 km (**c**), and 14 km (**d**). Prominent features are highlighted and numbered (see text for descriptions). White contours depict DWS = 20, and regions with DWS < 20 are shaded. Note the color bars change with depth

tomography models of the area (e.g., Fig. S5). We list below several other large- and small-scale features that were either not observed in previous works or not observed in as much detail and number them accordingly in Figs. 8, 10, 11, and 12. From shallow to greater depth, the highlighted features are:

1. At depths ≤ 5 km, the lowest velocities and highest V_P/V_S values (≥ 1.8) are associated with the San Bernardino Basin (SBB), San Jacinto Basin (SJB), and Coachella Valley.
2. Faults such as the San Bernardino strand of the SAF (SBSSAF) and Pinto Mountain fault (PMF) have similarly low velocities and high V_P/V_S values, especially where they intersect with other faults. In contrast, the Mission Creek fault (MCF) and SGP fault zone (SGPFZ) locate within regions of average to high velocities and average V_P/V_S values.
3. Rocks of the PRB have the highest velocities, and at shallow depth, abut the southwest sides of

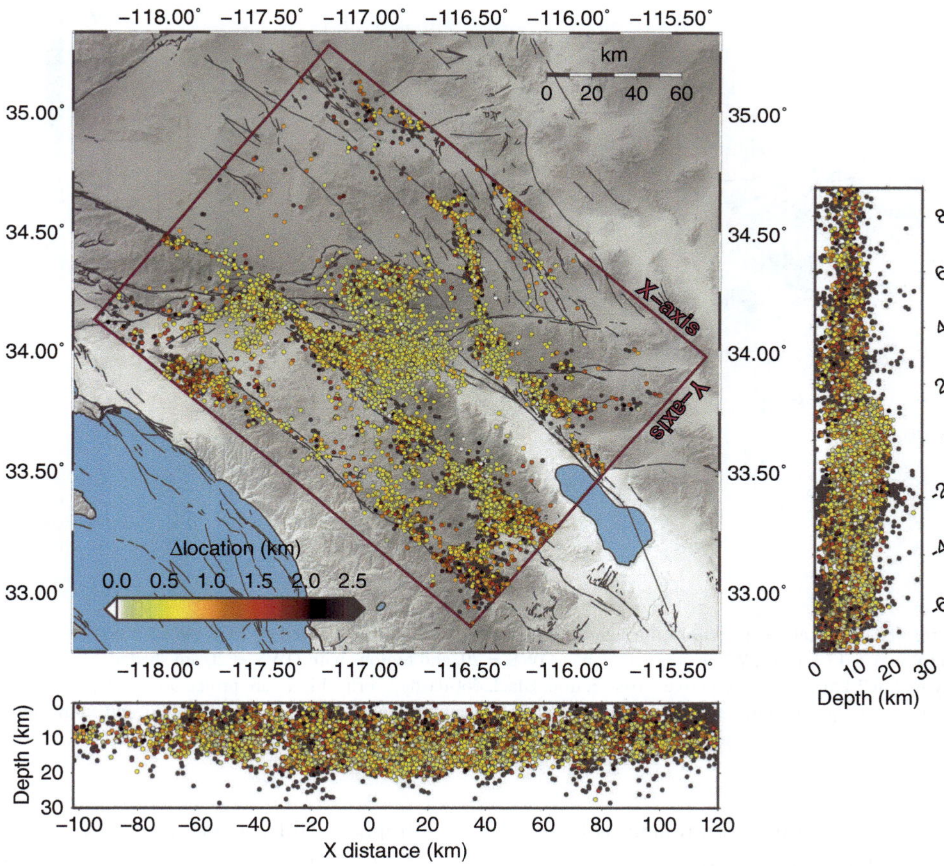

Figure 9
Comparison between the Hauksson et al. (2012) catalog and relocated events. The main plot shows differences in event epicenters, with warmer colors representing larger differences. The cross-sections depict differences in depth between the two sets of events projected along the x-axis (bottom) and y-axis (right). The same color scale as in the main plot applies

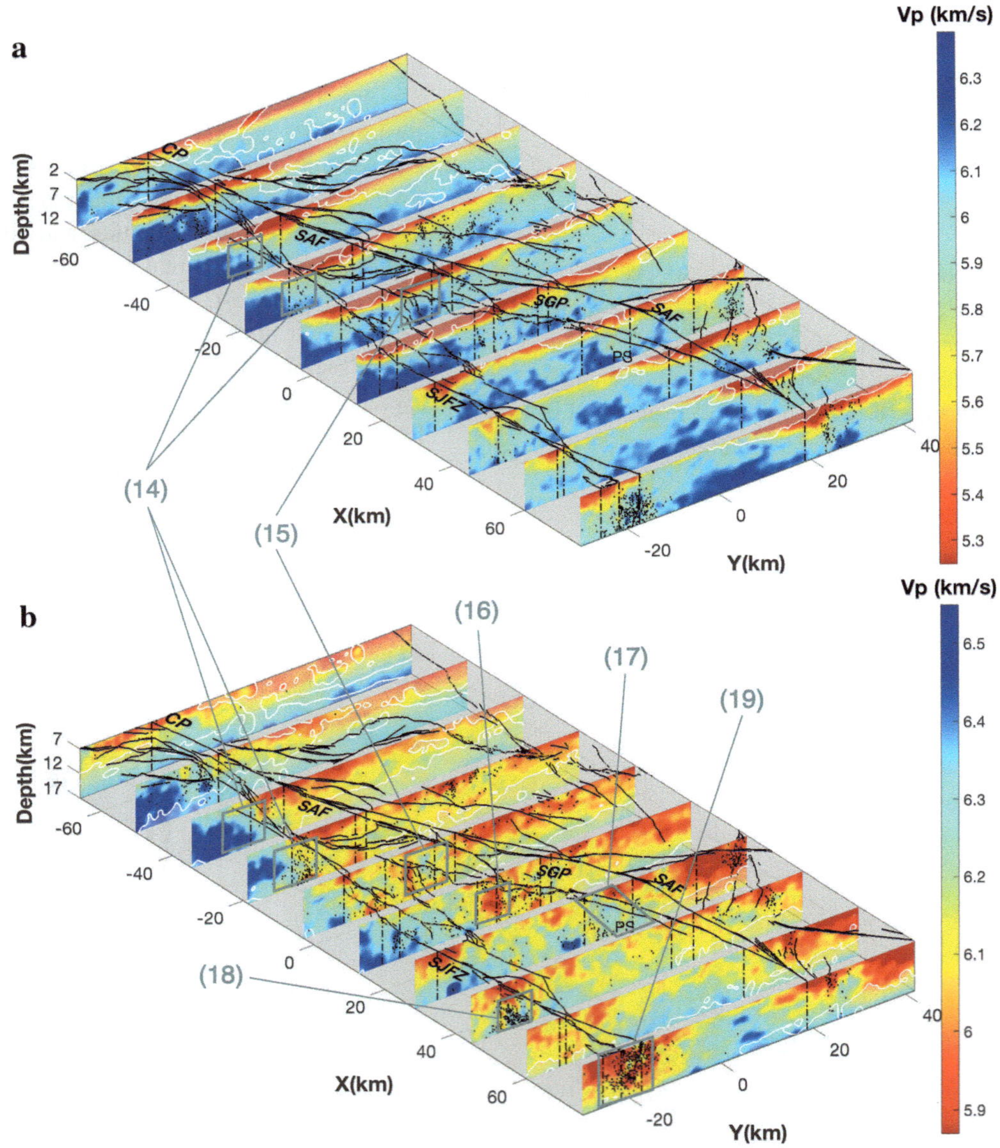

Figure 10

Final V_P inversion results at several cross-sections through the 142 km by 72 km area centered on SGP (small blue box in Fig. 1) for depth ranges of 2–12 km (**a**) and 7–17 km (**b**). White contours depict DWS = 20, and regions with DWS < 20 are shaded. Small black dots denote relocated seismicity within 2 km of each respective cross-section. Black dot-dashed vertical lines are projections in depth of major fault traces at the surface and help highlight dipping structures in the model. Some prominent features are highlighted and numbered (see text for descriptions). Note the color bar changes with depth

the SBB and SJB, forming prominent velocity contrasts across the northern SJFZ.

4. At depths ≥ 6 km, the transition from the western to eastern PRB manifests as a sharp decrease in velocity from west to east. This contrast is more prominent in V_P than in V_S, emphasized by a decrease in average V_P/V_S from 1.72 (west) to 1.69 (east). Also, the contrast does not migrate with depth, showing that it is a vertical feature at least down to depths where we have adequate resolution.

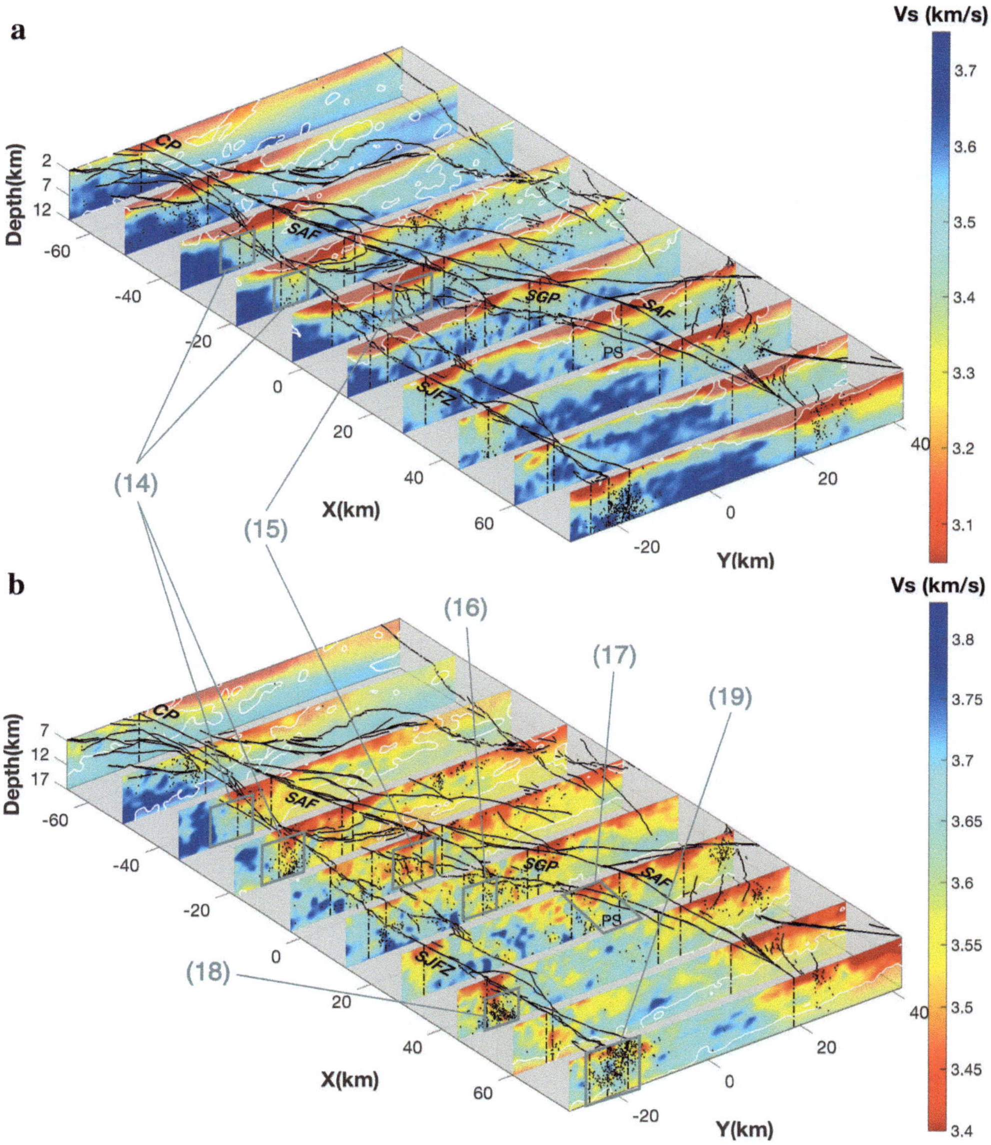

Figure 11
Same as Fig. 10 but for V_S inversion results

5. The lowest velocities in the eastern PRB are associated with the SJFZ and elongated fault-normal features that extend to the Elsinore fault (EF). These elongated features are clearest in the V_P model at depths $\geq$ 6 km.
6. Rocks of the SGMB lower plate (outcrops mostly in west) and eastern SBMB have low velocities and high V_P/V_S values (> 1.75).
7. The SGMB upper plate and western SBMB have variable velocities but average to low V_P/V_S values.
8. At greater depth, the velocity contrast across the northern SJFZ becomes confined to where rocks of the western PRB and SGMB lower plate meet.
9. Rocks of the western PRB, which border the SJB at shallow depth (3), intersect the SJFZ at greater

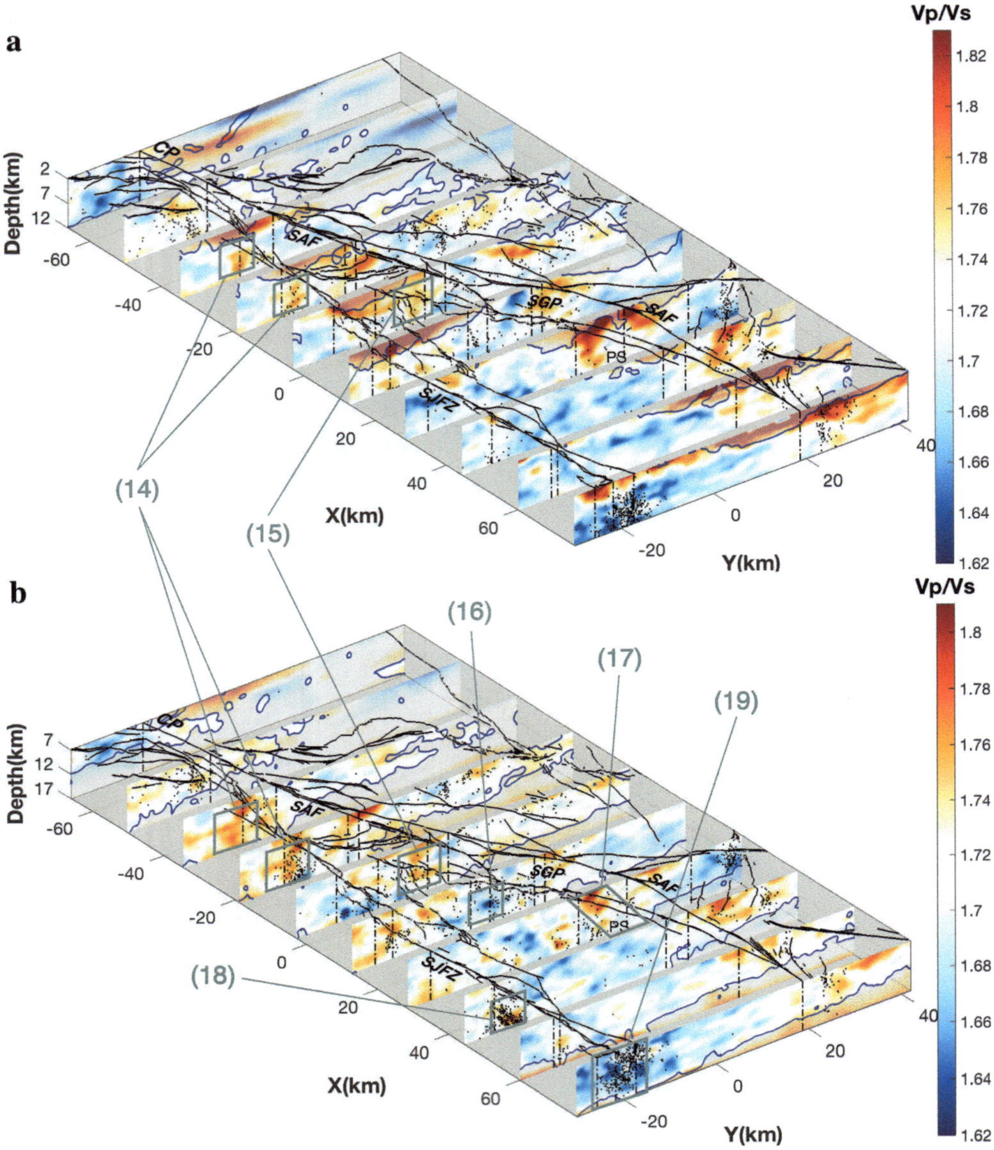

Figure 12
Same as Fig. 10 but for V_P/V_S structure

depth (> 7 km) and occupy the region known as the Hemet step-over (Marliyani et al. 2013).

10. The lowest velocities around 10 km depth are associated with the eastern SBMB. East of Palm Springs, these low velocities undergo a sharp transition to fast velocities in the southwest along a bimaterial contact. This contact extends for > 50 km and is parallel to the Coachella Valley strand of the SAF (CVSSAF) but offset to the northeast by ~ 7 km.
11. At these same depths, low V_P and low V_P/V_S anomalies relate to fault strands of the ECSZ north of the PMF.
12. From 13 km downwards, low V_P and V_P/V_S anomalies develop beneath the SGPFZ.

13. Low V_P and the lowest V_P/V_S anomalies ($\sim$ 1.6) are associated with the SJFZ trifurcation area at depths $\geq$ 10 km.

Focusing next on the region around SGP, we observe the following additional features:

14. At various depths, the northern SJFZ and its intersection with the Crafton Hills fault zone (CHFZ) are characterized by anomalously low V_S and high V_P/V_S ($\geq$ 1.78). These anomalies extend vertically downwards and have no apparent dip.
15. The mapped surface trace of the SBSSAF near where it intersects with the MCF has similarly low V_S and high V_P/V_S at various depths. These anomalies are also near vertical.
16. Furthermore, the earlier mentioned deep low V_P and V_P/V_S anomaly beneath the SGPFZ (12) concentrates near its intersection with the SBSSAF.
17. In contrast, a high V_P/V_S anomaly (> 1.76) with apparent northward dip is associated with the Banning fault/Garnet Hill fault (BF/GHF) system east of the SGPFZ.
18. At depths $\geq$ 13 km, the Hot Springs area of the SJFZ is also characterized by high V_P/V_S values.
19. Finally, the earlier mentioned deep low V_P and V_P/V_S anomalies beneath the SJFZ trifurcation area (13) concentrate beneath the Clark fault (CF) and Buckridge fault to the northeast and southwest of the Coyote Creek fault.

All listed features have length scales > 4 km, acceptable resolution (DWS $\geq$ 20), and for the most part $r_s \geq 0.008$ (Figs. S6–S9). In the next section, we discuss these features and explore their implications.

6. Discussion

We obtained detailed tomographic images of seismic velocities in the Southern California plate boundary around the SCTR using a newly developed DDT algorithm (Guo and Zhang 2017; Zhang et al. 2017; Guo et al. 2018) that incorporates station-pair DD with the absolute arrival times and event-pair DD traditionally used. The station-pair data allowed for structural resolution near the surface ($\leq$ 2 km) and better absolute locations of relocated events. We used an automatic detection algorithm to extract P and S arrival times from event waveforms. This allows for rapid extraction of large P and S datasets from only a few years of recordings (01/01/2010–06/30/2015) associated with a large number of seismic stations. Using these data, we inverted for high-resolution V_P and V_S models and derived from them high-quality V_P/V_S models. These models contained various interesting features, which are discussed below in light of what they reveal about fault zone structures. We note in parenthesis and in italics the feature number as listed in Sect. 5.2 (Figs. 8, 10, 11, 12) that is the topic of discussion.

Our results show evidence of contrasting elastic properties across major faults in the area as well as anomalous velocities along these faults. At shallow depths ($\leq$ 5 km), the low-density basins (Anderson et al. 2004; Langenheim et al. 2004) have very low velocities (*1*). For the SBB in particular, these low velocities are juxtaposed against high velocities of the dense mafic western PRB (*3*) (Silver and Chappell 1987; Langenheim et al. 2004), and crystalline rocks of the western SBMB (Gutierrez et al. 2010), forming prominent velocity contrasts across the northern SJFZ and SBSSAF, respectively. At similarly shallow depths, low velocities associated with the SBSSAF and CVSSAF probably relate to damage and deformation along these relatively young and active fault zones (Matti et al. 1992). Similar low velocities are not observed for the MCF (especially between the SBSSAF and PMF) or at the center of the SGPFZ/BF/GHF system (*2*). The relatively high velocities around these faults are also imaged at lower resolution in other tomography results (e.g., Zigone et al. 2015; Fang et al. 2016). We relate such lack of low velocities to the recent inactivity along these faults (Kendrick et al. 2015) where damage is no longer consistently produced. We also attribute the higher velocities to crystalline rocks that persist from the western SBMB in the north through the upper plate of the SGMB to the PRB in the south (Gutierrez et al. 2010).

Fault-related anomalies are observed near SGP at greater depth. Contrasts across the northern SJFZ and SBSSAF persist at > 7 km (albeit weaker) as velocities beneath the SBB remain low given the presence of Pelona schist within the SGMB lower plate (*8*). Velocities within the SGMB upper plate remain high (*7*), whereas a clear contrast develops across the CVSSAF where crystalline rocks of the PRB to the south (fast) meet metamorphosed rocks of the eastern SBMB in the north (slow) (*10*). The observed sharp change in velocity at ~ 10 km depth may correspond to this lithological change across the CVSSAF, and the fault is dipping to the northeast if this inference is correct. A northeast dipping CVSSAF has been proposed by several other authors (e.g., Fialko 2006; Lindsey and Fialko 2013; Fattaruso et al. 2014; Share and Ben-Zion 2016; Fuis et al. 2017). Using our V_P model, a plane dipping by 57° is formed by connecting the main surface trace of the CVSSAF with the region of maximum change in velocity within the relocated seismicity cluster at 10 km depth (Fig. 9). This value is similar to the 65° proposed by Fuis et al. (2017). At even greater depth ($\geq$ 13 km), high velocities around the MCF persist. In contrast, the SGPFZ is characterized by a low V_P anomaly that disappears at the transition to the BF/GHF in the west (*12*). This anomaly is amplified near the intersection between the SGPFZ and SBSSAF and is less prominent in the V_S model (*12*).

Several imaged features south of SGP are worth discussing. The decrease in velocity from the western to eastern PRB (*4*) relates to the lower density and more felsic composition of the latter (Silver and Chappell 1987; Langenheim et al. 2004). This transition has been observed in potential field results (e.g., Langenheim et al. 2004) and other tomography studies (e.g., Hearn and Clayton 1986; Allam and Ben-Zion 2012; Barak et al. 2015). However, here it is a vertical feature (at depths $\geq$ 7 km) and does not dip to the east as has been suggested (Langenheim et al. 2004; Barak et al. 2015). East of this transition, the imaged elongated low velocity features (*5*) correlate with abnormally high seismic activity between the SJFZ and EF (Fig. 9). This suggests high strain localization in an intrafault region that may be related to basin structures (Axen and Fletcher 1998) or counterclockwise block rotation from bookshelf faulting (e.g., Ron et al. 1984; Tapponnier et al. 1990) that can produce complex geodetic signals along and in between the SJFZ and EF.

The obtained V_P/V_S models provide additional constraints on lithology and fault structure. V_P/V_S values generally reflect bulk chemical composition (e.g., Christensen 1996; Brocher 2005), rock damage, and fluid content (e.g., Lockner et al. 1992; Mavko et al. 1998; Hamiel et al. 2004). Most crustal rocks have V_P/V_S values in the range 1.7–1.85 (Brocher 2005). There are, however, two notable exceptions (Christensen 1996): First, pure quartz has a V_P/V_S of 1.5, and therefore rocks abundant in quartz (> 55% of total volume) also have anomalously low V_P/V_S values. Second, anorthosite has very high V_P/V_S (> 1.9), and metamorphic rocks, which contain anorthosite-rich plagioclase feldspar, usually have increased V_P/V_S values. These values are temperature and pressure dependent (Christensen 1996), but their effects on V_P/V_S can be neglected here given the depths under investigation. With regard to less intact rock, very low V_S and high V_P/V_S (~ 2 to 5, Brocher 2005) have been reported for highly porous or fractured fluid-filled rocks, such as those found in sedimentary basins. Conversely, highly fractured dry rock near the surface is expected to have low V_P and low V_P/V_S (O'Connell and Budiansky 1974). At greater depth, where fracturing is expected to be less intense, variations in V_P/V_S are more influenced by the aspect ratio (height divided by radius) of fluid-filled cracks. Anomalously low V_P and low V_P/V_S values near deep fault damage zones have been linked to cracks with relatively large aspect ratio (thick cracks) (e.g., O'Connell and Budiansky 1974; Lin and Shearer 2009). V_P/V_S is therefore expected to vary greatly across faults separating predominantly crystalline and metamorphic units, and even greater along fault zones as the degree and geometry of cracking and the amount of fluids occupying those cracks changes laterally and in depth.

Some general trends in V_P/V_S can be related to lithology in the SCTR. The PRB (*3*), SGMB upper plate, and western SBMB (*7*), with their abundant quartz-rich crystalline rocks, are regions with below-average V_P/V_S. The eastern PRB in particular contains granitic rocks with 60–70 % quartz (Silver and Chappell 1987) and has the lowest V_P/V_S values in the study region (*5* and *13*). The low V_P/V_S anomaly beneath the SGPFZ at depths $\geq$ 13 km (*12*) suggests that this fault system intersects similarly quartz-rich crystalline rocks that may be derived from the western SBMB and/or the SGMB upper plate (Langenheim et al. 2005). In contrast, metamorphic rocks present in the SGMB lower plate and eastern SBMB have, as expected, above-average V_P/V_S (*6*).

Overprinted on large-scale variations in V_P/V_S are anomalies concentrated around major faults that give insight into the extent and type of damage present. Very high V_P/V_S values near the surface are caused by intensely fractured and porous rocks within the SBB, SJB, and Coachella Valley basin (*1*) and damaged wet rocks near major faults (e.g., O'Connell and Budiansky 1974; Mavko et al. 1998). Near-vertical low V_S and high V_P/V_S anomalies extend downwards from the surface traces of the northern SJFZ, CHFZ, and SBSSAF (*14* and *15*). This suggests that these faults are near vertical also and represent regions of relatively intense damage that may contain fluid-filled cracks. The pronounced low V_P and low V_P/V_S anomaly present beneath the mapped intersection of the eastern end of the SBSSAF and the SGPFZ (*16*) similarly suggest near-vertical fault structure. Moreover, the low V_P/V_S values point to a damage regime consisting of thick fluid-filled cracks (e.g., Lin and Shearer 2009). Near-vertical faults around this intersection contrast with inferences of dipping faults from analysis of potential field data (Fuis et al. 2012) and numerical modeling to reproduce geologic uplift and slip rates (Cooke and Dair 2011). However, this is supported by suggested fault orientations from studies on seismicity distributions (Carena et al. 2004). The northeast-dipping low V_S and high V_P/V_S anomalies beneath the BF/GHF (*17*) corroborate earlier suggestions (e.g., Jones et al. 1986; Nicholson 1996) that this fault system is dipping in that direction.

It is of some interest to note the correlation between near-fault V_P/V_S anomalies at depth and the basement rocks within which they locate. Very low V_P/V_S structures are only observed near faults located within quartz-rich crystalline rocks [e.g., SGPFZ, SJFZ trifurcation area (*19*) and ECSZ north of PMF (*11*)], but not near faults bounding or intersecting metamorphic rock units (e.g., northern SJFZ, GHFZ, SBSSAF, BF/GHF, and ECSZ south of PMF). This may indicate a dependence of the geometry of fluid-filled cracks on lithology. Thick cracks might preferentially form in the isotropic mineral texture of crystalline rocks such as granite. Such low V_P/V_S values have been measured in cracked silica-rich rocks under high pressure from the KTB-drill holes (Popp and Kern 1994). Alternatively, cracks forming in metamorphic rocks may predominantly be flat and long (low aspect ratio–high V_P/V_S, O'Connell and Budiansky 1974), as they are parallel to the planar foliation in these rocks. Dry cracks can help explain some of these variations, but several studies suggest fluid-rich fault zones in the region (e.g., Hardebeck and Hauksson 1999). To further investigate V_P/V_S variations near faults within different lithologies, it would be useful to employ other geophysical tools, such as magnetotellurics (Becken and Ritter 2012), that are sensitive to the same properties but have inherently different resolutions.

In conclusion, the highlighted features (*1–19*) suggest that structural complexity of the SGP at the surface continues at depth. No continuous low-velocity anomaly (potential damage zone) is observed along the SAF system through SGP. Likewise, a continuous bimaterial structure with a single velocity polarity is also not detected. Instead, a fast rock body cuts across the SAF system near the surface, whereas at depth, abrupt west-to-east changes in elastic properties characterize the transition from the SBSSAF and SGPFZ to the BF/GHF. This significant structural heterogeneity through SGP is corroborated by recent results from other seismological studies (e.g., Share and Ben-Zion 2016). Our results also show strong evidence that faults west of SGP have near-vertical dips, whereas faults to the east of that area dip to the northeast. These changes in elastic properties and fault geometry can help explain the high rates of small to moderate-sized events in the area (Hauksson et al. 2012), large variations of earthquake focal mechanisms and stress drops (Bailey et al. 2009; Yang et al. 2012; Goebel et al. 2015), and strain partitioning in general. It also provides a potential structural barrier for through-going earthquake ruptures, as suggested by the 1948 M_W 6.0 Desert Hot Springs (Nicholson 1996) and 1986 M_W 5.8 North Palm Springs (Jones et al. 1986) earthquakes, which both ruptured large sections of the BF/GHF but did not extend into the SGPFZ.

The results presented also give insight into the role of the SJFZ in large earthquakes in Southern California. The sharp change in velocity across the northern SJFZ (*8*) points to a potential bimaterial structure at the core of this fault section. Based on theoretical models and laboratory experiments (e.g., Weertman 1980; Andrews and Ben-Zion 1997; Shlomai and Fineberg 2016), ruptures on bimaterial faults tend to have preferred propagation direction. An earthquake with strong directivity can produce an order of magnitude larger ground motion in the direction of propagation than in the opposite direction. Directivity is therefore a key component in estimations of expected ground motion from large earthquakes. On the other hand, the extent of potential earthquake rupture along the SAF and northern SJFZ is affected by the continuity within and between the two fault systems. Our results point to along-fault structural heterogeneity in the seismic zone around the Hemet step-over (*9*). To inform further both properties of potential bimaterial structures and fault continuity through CP, analysis of fault zone head waves should be applied to local earthquake data acquired in the region. The results of this study can also be improved by performing joint inversions of earthquake arrival times and surface wave dispersion data (e.g., Fang et al. 2016), as well as body and head wave data (Bennington et al. 2013). Work on some of these topics will be the subject of follow-up studies.

Acknowledgements

The seismic data used in this work were obtained from the Southern California Earthquake Data Center (SCEDC 2013) and the ANZA-YN seismic network database (Vernon 1982; Vernon and Ben-Zion 2010). The study was supported by the Earthquake Hazards Program of the USGS (grant G16AP00105). Several maps were made using the Generic Mapping Tools (Wessel et al. 2013). The manuscript benefited from useful comments by two anonymous reviewers.

References

Allam, A., & Ben-Zion, Y. (2012). Seismic velocity structures in the southern California plate boundary environment from double difference tomography. *Geophysical Journal International, 190,* 1181–1196.

Allam, A. A., Ben-Zion, Y., Kurzon, I., & Vernon, F. L. (2014). Seismic velocity structure in the Hot Springs and Trifurcation Areas of the San Jacinto Fault Zone, California, from double-difference tomography. *Geophysical Journal International, 198,* 978–999. https://doi.org/10.1093/gji/ggu176.

Allen, C. (1957). San Andreas fault zone in San Gorgonio Pass, southern California. *Geological Society of America Bulletin, 68,* 315–350.

An, M. (2012). A simple method for determining the spatial resolution of a general inverse problem. *Geophysical Journal International, 191,* 849–864.

Anderson, M., Matti, J., & Jachens, R. (2004). Structural model of the San Bernardino basin, California, from analysis of gravity, aeromagnetic and seismicity data. *Journal of Geophysical Research, 109,* B04404.

Andrews, D. J., & Ben-Zion, Y. (1997). Wrinkle-like slip pulse on a fault between different materials. *Journal of Geophysical Research, 102,* 553–571.

Axen, G. J., & Fletcher, J. M. (1998). Late Miocene-Pleistocene extensional faulting, northern Gulf of California, Mexico and Salton Trough, California. *International Geology Review, 40*(3), 217–244.

Bailey, I. W., Becker, T. W., & Ben-Zion, Y. (2009). Patterns of co-seismic strain computed from Southern California focal mechanisms. *Geophysical Journal International, 177,* 1015–1036. https://doi.org/10.1111/j.1365-246X.2009.04090.x.

Bankey, V., Cuevas, A., Daniels, D., Finn, C. A., Hernandez, I., Hill, P., Kucks, R., Miles, W., Pilkington, M., Roberts, C., Roest, W., Rystrom, V., Shearer, S., Snyder, S., Sweeney, R., Velez, L., Phillips, J. D., & Ravat, D. (2002). Digital data grids for the magnetic anomaly map of North America, U.S. Geological Survey Open-File Report 02-414.

Barak, S., Klemperer, S. L., & Lawrence, J. F. (2015). San Andreas Fault dip, Peninsular Ranges mafic lower crust and partial melt in the Salton Trough, Southern California, from ambient-noise tomography. *Geochemistry, Geophysics, Geosystems, 16,* 3946–3972.

Becken, M., & Ritter, O. (2012). Magnetotelluric studies at the San Andreas Fault Zone: Implications for the role of fluids. *Surveys in Geophysics, 33*(1), 65–105.

Bekas, C., Kokiopoulou, E., & Saad, Y. (2007). An estimator for the diagonal of a matrix. *Applied Numerical Mathematics, 57*(11–12), 1214–1229.

Bennington, N. L., Thurber, C., Peng, Z., Zhang, H., & Peng, Z. (2013). Incorporating fault zone head wave and direct wave secondary arrival times into seismic tomography: Application at Parkfield, California. *Journal of Geophysical Research, 118,* 1008–1014.

Ben-Zion, Y. (1989). The response of two joined quarter spaces to SH line sources located at the material discontinuity interface. *Geophysical Journal International, 98,* 213–222.

Ben-Zion, Y., & Malin, P. (1991). San Andreas fault zone head waves near Parkfield, California. *Science, 251,* 1592–1594.

Ben-Zion, Y., & Sammis, C. G. (2003). Characterization of fault zones. *Pure and Applied Geophysics, 160,* 677–715.

Blisniuk, K., Oskin, M., Me Riaux, A.-S., Rockwell, T., Finkel, R. C., & Ryerson, F. J. (2013). Stable, rapid rate of slip since inception of the San Jacinto fault, California. *Geophysical Research Letters, 40,* 4209–4213.

Bogiatzis, P., Ishii, M., & Davis, T. A. (2016). Toward using direct methods in seismic tomography: Computation of the full resolution matrix using high performance computing and sparse QR factorization. *Geophysical Journal International, 205*(2), 830–836.

Brocher, T. M. (2005). Empirical relations between elastic wavespeeds and density in the Earth's crust. *Bulletin of the Seismological Society of America, 95*(6), 2081–2092.

Calderoni, G., Di Giovambattista, R., Vannoli, P., Pucillo, S., & Rovelli, A. (2012). Fault-trapped waves depict continuity of the fault system responsible for the 6 April 2009 M_W 6.3 L'Aquila earthquake, central Italy. *Earth and Planetary Science Letters, 323–324,* 1–8.

Carena, S., Suppe, J., & Kao, H. (2004). Lack of continuity of the San Andreas Fault in southern California: Three-dimensional fault models and earthquake scenarios. *Journal of Geophysical Research, 109,* B04313.

Christensen, N. I. (1996). Poisson's ratio and crustal seismology. *Journal of Geophysical Research, 101*(B2), 3139–3156.

Cooke, M. L., & Dair, L. (2011). Simulating the recent evolution of the southern big bend of the San Andreas fault, Southern California. *Journal of Geophysical Research, 116,* B04405.

Dair, L., & Cooke, M. L. (2009). San Andreas fault geometry through the San Gorgonio Pass, California. *Geology, 37*(2), 119–122.

Fang, H., Zhang, H., Yao, H., Allam, A., Zigone, D., Ben-Zion, Y., et al. (2016). A new three-dimensional joint inversion algorithm of body-wave and surface-wave data and its application to the Southern California Plate Boundary Region. *Journal of Geophysical Research, 121*(5), 3557–3569.

Fattaruso, L. A., Cooke, M. L., & Dorsey, R. J. (2014). Sensitivity of uplift patterns to dip of the San Andreas fault in the Coachella Valley, California. *Geosphere, 10*(6), 1557–1575.

Faulkner, D. R., Jackson, C. A. L., Lunn, R. J., Schlische, R. W., Shipton, Z. K., Wibberley, C. A. J., et al. (2010). A review of recent developments concerning the structure, mechanics and fluid flow properties of fault zones. *Journal of Structural Geology, 32,* 1557–1575.

Fialko, Y. (2006). Interseismic strain accumulation and the earthquake potential on the southern San Andreas fault system. *Nature, 441,* 968–971.

Fichtner, A., & van Leeuwen, T. (2015). Resolution analysis by random probing. *Journal of Geophysical Research, 120,* 5549–5573.

Fuis, G. S., Bauer, K., Goldman, M. R., Ryberg, T., Langenheim, V. E., Scheirer, D. S., et al. (2017). Subsurface geometry of the San Andreas Fault in Southern California: Results from the Salton Seismic Imaging Project (SSIP) and strong ground motion expectations. *Bulletin of the Seismological Society of America, 107*(3), 1642–1662.

Fuis, G. S., Scheirer, D. S., Langenheim, V. E., & Kohler, M. D. (2012). A new perspective on the geometry of the San Andreas fault in southern California and its relationship to lithospheric structure. *Bulletin of the Seismological Society of America, 102*(1), 236–251.

Gentili, S., & Michelini, A. (2006). Automatic picking of P and S phases using a neural tree. *Journal of Seismology, 10,* 39–63.

Goebel, T. H. W., Hauksson, E., Shearer, P. M., & Ampuero, J. P. (2015). Stress-drop heterogeneity within tectonically complex regions: A case study of San Gorgonio Pass, southern California. *Geophysical Journal International, 202*(1), 514–528.

Gold, P. O., Behr, W. M., Rood, D., Sharp, W. D., Rockwell, T. K., Kendrick, K., et al. (2015). Holocene geologic slip rate for the Banning strand of the southern San Andreas Fault, southern California. *Journal of Geophysical Research, 120*(8), 5639–5663.

Guo, H., McGuire, J. J., & Zhang, H. (2018). *Imaging the subducted Gorda plate: Implications for the stress state and brittle-ductile transition of the Cascadia subduction zone.* In: American Geophysical Union Fall Meeting Abstract T13H-0334, Washington, DC.

Guo, H., & Zhang, H. (2017). Development of double-pair double difference earthquake location algorithm for improving earthquake locations. *Geophysical Journal International, 208,* 333–348.

Gutierrez, C., Bryant, W., Saucedo, G., & Wills, C. (2010). *Geologic map of California.* California: California Geological Survey.

Hamiel, Y., Liu, Y., Lyakhovsky, V., Ben-Zion, Y., & Lockner, D. (2004). A visco-elastic damage model with applications to stable and unstable fracturing. *Geophysical Journal International, 159,* 1155–1165. https://doi.org/10.1111/j.1365-246X.2004.02452.x.

Hardebeck, J. L., & Hauksson, E. (1999). Role of fluids in faulting inferred from stress field signatures. *Science, 285,* 236–239.

Harden, J. W., & Matti, J. C. (1989). Holocene and late Pleistocene slip rates on the San Andreas fault in Yucaipa, California, using displaced alluvial-fan deposits and soil chronology. *Geological Society of America Bulletin, 101,* 1107–1117.

Hauksson, E., Yang, W., & Shearer, P. M. (2012). Waveform relocated earthquake catalog for southern California (1981 to June 2011). *Bulletin of the Seismological Society of America, 102*(5), 2239–2244.

Hearn, T. M., & Clayton, R. W. (1986). Lateral velocity variations in southern California. I. Results for the upper crust from Pg waves. *Bulletin of the Seismological Society of America, 76*(2), 495–509.

Jones, L. M., Hutton, L. K., Given, D. D., & Allen, C. R. (1986). The North Palm Springs, California, earthquake sequence of July 1986. *Bulletin of the Seismological Society of America, 76,* 1830–1837.

Kendrick, K. J., Matti, J. C., & Mahan, S. A. (2015). Late Quaternary slip history of the Mill Creek strand of the San Andreas fault in San Gorgonio Pass, southern California: The role of a subsidiary left-lateral fault in strand switching. *Geological Society of America Bulletin, 127*(5–6), 825–849.

Langenheim, V. E., Jachens, R. C., Matti, J. C., Hauksson, E., Morton, D. M., & Christensen, A. (2005). Geophysical evidence for wedging in the San Gorgonio Pass structural knot, southern San Andreas fault zone, southern California. *Geological Society of America Bulletin, 117,* 1554–1572.

Langenheim, V. E., Jachens, R. C., Morton, D. M., Kistler, R. W., & Matti, J. C. (2004). Geophysical and isotopic mapping of preexisting crustal structures that influenced the location and development of the San Jacinto fault zone, southern California. *Geological Society of America Bulletin, 116,* 1143–1157.

Lee, E.-J., Chen, P., Jordan, T. H., Maechling, P. B., Denolle, M. A. M., & Beroza, G. (2014). Full-3-D tomography for crustal structure in Southern California based on the scattering-integral and the adjoint-wavefield methods. *Journal of Geophysical Research, 119*(8), 6421–6451.

Lévêque, J. J., Rivera, L., & Wittlinger, G. (1993). On the use of the checkerboard test to assess the resolution of tomographic inversions. *Geophysical Journal International, 115,* 313–318.

Lewis, M. A., & Ben-Zion, Y. (2010). Diversity of fault zone damage and trapping structures in the Parkfield section of the San Andreas Fault from comprehensive analysis of near fault seismograms. *Geophysical Journal International, 183*(3), 1579–1595.

Li, Y.-G., Aki, K., Adams, D., Hasemi, A., & Lee, W. H. K. (1994). Seismic guided waves trapped in the fault zone of the Landers, California, earthquake of 1992. *Journal of Geophysical Research, 99,* 11705–11722.

Lin, G., & Shearer, P. M. (2009). Evidence for water-filled cracks in earthquake source regions. *Geophysical Research Letters, 36,* L17315.

Lindsey, E. O., & Fialko, Y. (2013). Geodetic slip rates in the southern San Andreas Fault system: Effects of elastic heterogeneity and fault geometry. *Journal of Geophysical Research, 118,* 689–697.

Lockner, D. A., Byerlee, J. D., Kuksenko, V., Ponomarev, A., & Sidrin, A. (1992). Observations of quasistatic fault growth from acoustic emissions. *International Geophysics, 51,* 3–31.

Lozos, J. C. (2016). A case for historic joint rupture of the San Andreas and San Jacinto faults. *Science Advances, 2*(3), e1500621.

MacCarthy, J. K., Brochers, B., & Aster, R. C. (2011). Efficient stochastic estimation of the model resolution matrix diagonal and generalised cross-validation for large geophysical inverse problems. *Journal of Geophysical Research, 116,* B10304.

Magistrale, H., & Sanders, C. (1996). Evidence from precise earthquake hypocenters for segmentation of the San Andreas fault in San Gorgonio Pass. *Journal of Geophysical Research, 101,* 3031–3041.

Marliyani, G. I., Rockwell, T. K., Onderdonk, N. W., & McGill, S. F. (2013). Straightening of the northern San Jacinto Fault, California as seen in the fault-structure evolution of the San Jacinto Valley step-over. *Bulletin of the Seismological Society of America, 103*(1), 519–541.

Matti, J. C., Morton, D. M., & Cox, B. F. (1992). The San Andreas fault system in the vicinity of the central Transverse Ranges province, southern California, U.S. Geological Survey Open-File Report 92-354.

Mavko, G., Mukerji, T., & Dvorkin, J. (1998). *The rock physics handbook: Tools for seismic analysis in porous media* (p. 329). Cambridge: Cambridge University Press.

McGill, S., Owen, L., Weldon, R. J., & Kendrick, K. (2013). Latest Pleistocene and Holocene slip rate for the San Bernardino strand of the San Andreas Fault, Plunge Creek, Southern California: Implications for strain partitioning within the southern San Andreas Fault system for the last ~ 35 k.y. *Geological Society of America Bulletin, 125*(1/2), 48–72.

McGill, S. F., Spinler, J. C., McGill, J. D., Bennett, R. A., Floyd, M. A., Fryxell, J. E., et al. (2015). Kinematic modeling of fault slip rates using new geodetic velocities from a transect across the Pacific-North America plate boundary through the San Bernardino Mountains, California. *Journal of Geophysical Research, 120*(4), 2772–2793.

McGuire, J., & Ben-Zion, Y. (2005). High-resolution imaging of the Bear Valley section of the San Andreas Fault at seismogenic depths with fault-zone head waves and relocated seismicity. *Geophysical Journal International, 163,* 152–164.

Meade, B. J., & Hager, B. H. (2005). Block models of crustal motion in southern California constrained by GPS measurements. *Journal of Geophysical Research, 110,* B03403.

Mizuno, T., & Nishigami, K. (2006). Deep structure of the Nojima fault, southwest Japan, estimated from borehole observation of fault-zone trapped waves. *Tectonophysics, 417,* 231–247.

Najdahmadi, B., Bohnhoff, M., & Ben-Zion, Y. (2016). Bimaterial interfaces at the Karadere segment of the North Anatolian Fault, northwestern Turkey. *Journal of Geophysical Research, 121,* 931–950.

Nicholson, C. (1996). Seismic behavior of the southern San Andreas Fault Zone in the Northern Coachella Valley, California: Comparison of the 1948 and 1986 earthquake sequences. *Bulletin of the Seismological Society of America, 86,* 1331–1349.

Nicholson, C., Plesch, A., Shaw, J. H., & Marshall, S. T. (2018). Enhancements, updates, and improved access to the Community Fault Model. Presented at the 2018 Southern California Earthquake Center Annual Meeting.

Nippress, S. E. J., Rietbrock, A., & Heath, A. E. (2010). Optimized automatic pickers: Application to the ANCORP data set. *Geophysical Journal International, 181,* 911–925.

O'Connell, R. J., & Budiansky, B. (1974). Seismic velocities in dry and saturated cracked solids. *Journal of Geophysical Research, 79*(35), 5412–5426.

Onderdonk, N. W., McGill, S. F., & Rockwell, T. K. (2015). Short-term variations in slip rate and size of prehistoric earthquakes during the past 2000 years on the northern San Jacinto fault zone, a major plate-boundary structure in southern California. *Lithosphere, 7*(3), 211–234.

Oskin, M., Perg, L., Shelef, E., Strane, M., Gurney, E., Singer, B., et al. (2008). Elevated shear zone loading rate during an earthquake cluster in eastern California. *Geology, 36*(6), 507–510.

Paige, C. C., & Saunders, M. A. (1982). Algorithm 583; LSQR: Sparse linear equations and least squares problems. *ACM Transactions on Mathematical Software, 8*(2), 195–209.

Popp, T., & Kern, H. (1994). The influence of dry and water saturated cracks on seismic velocities of crustal rocks—A comparison of experimental data with theoretical model. *Surveys in Geophysics, 15,* 443–465.

Porter, K., Jones, L., Cox, D., Goltz, J., Hudnut, K., Mileti, D., et al. (2011). The ShakeOut Scenario: A hypothetical M_w7.8 earthquake on the Southern San Andreas fault. *Earthquake Spectra, 27*(2), 239–261.

Qin, L., Ben-Zion, Y., Qiu, H., Share, P.-E., Ross, Z. E., & Vernon, F. L. (2018). Internal structure of the San Jacinto fault zone in the trifurcation area southeast of Anza, California, from data of dense seismic arrays. *Geophysical Journal International, 213,* 98–114. https://doi.org/10.1093/gji/ggx540.

Qiu, H., Ben-Zion, Y., Ross, Z. E., Share, P.-E., & Vernon, F.L. (2017). Internal structure of the San Jacinto fault zone at Jackass Flat from data recorded by a dense linear array. *Geophysical Journal International*, *209*(3), 1369–1388.

Rawlinson, N., Fichtner, A., Sambridge, M., & Young, M. K. (2014). Seismic tomography and the assessment of uncertainty. *Advances in Geophysics, 55,* 1–76.

Rockwell, T. K., Dawson, T. E., Ben-Horin, J. Y., & Seitz, G. (2015). A 21-event, 4,000-year history of surface ruptures in the

Anza seismic gap, San Jacinto Fault, and implications for long-term earthquake production on a major plate boundary fault. *Pure and Applied Geophysics, 172*(5), 1143–1165.

Ron, H., Freund, R., Garfunkel, Z., & Nur, A. (1984). Block rotations by strike-slip faulting: Structural and paleomagnetic evidence. *Journal of Geophysical Research, 89*(B7), 6256–6270.

Ross, Z. E., & Ben-Zion, Y. (2014). Automatic picking of direct P, S seismic phases and fault zone head waves. *Geophysical Journal International, 199,* 368–381.

Ross, Z. E., Hauksson, E., & Ben-Zion, Y. (2017). Abundant off-fault seismicity and orthogonal structures in the San Jacinto fault zone. *Science Advances, 3,* e1601946. https://doi.org/10.1126/sciadv.1601946.

Ross, Z. E., White, M. C., Vernon, F. L., & Ben-Zion, Y. (2016). An improved algorithm for real-time S-wave picking with application to the (augmented) ANZA network in Southern California. *Bulletin of the Seismological Society of America, 106*(5), 2013–2022.

Sanders, C. O., & Kanamori, H. (1984). A seismotectonic analysis of the Anza seismic gap, San Jacinto fault zone, southern California. *Journal of Geophysical Research, 89*(B7), 5873–5890.

SCEDC. (2013). *Southern California earthquake data center.* California: Caltech.Dataset. https://doi.org/10.7909/C3WD3xH.

Share, P.-E., & Ben-Zion, Y. (2016). Bimaterial interfaces in the South San Andreas Fault with opposite velocity contrasts NW and SE from San Gorgonio Pass. *Geophysical Research Letters.* https://doi.org/10.1002/2016gl070774.

Share, P.-E., Ben-Zion, Y., Ross, Z. E., Qiu, H., & Vernon, F. (2017). Internal structure of the San Jacinto fault zone at Blackburn Saddle from seismic data of a linear array. *Geophysical Journal International, 210*(2), 819–832.

Sharp, R. V. (1967). San Jacinto Fault Zone in the Peninsular Ranges of Southern California. *Geological Society of America Bulletin, 78*(6), 705–730.

Shaw, J. H., Plesch, A., Tape, C., Suess, M. P., Jordan, T. H., Ely, G., et al. (2015). Unified structural representation of the southern California crust and upper mantle. *Earth and Planetary Science Letters, 415,* 1–15.

Shlomai, H., & Fineberg, J. (2016). The structure of slip-pulses and supershear ruptures driving slip in bimaterial friction. *Nature Communications, 7,* 11787.

Silver, L. T., & Chappell, B. W. (1987). The Peninsular Ranges batholith: An insight into the evolution of the Cordilleran batholiths of southwestern North America. *Transactions of Royal Society of Edinburgh: Earth Sciences, 79,* 105–121.

Spinler, J. C., Bennett, R. A., Anderson, M. L., McGill, S. F., Hreinsdottir, S., & McCallister, A. (2010). Present-day strain accumulation and slip rates associated with southern San Andreas and Eastern California shear zone faults. *Journal of Geophysical Research, 115,* B11407.

Tape, C., Liu, Q., Maggi, A., & Tromp, J. (2010). Seismic tomography of the southern California crust based on spectral-element and adjoint methods. *Geophysical Journal International, 180,* 433–462.

Tapponnier, P., Armijo, R., Manighetti, I., & Courtillot, V. (1990). Bookshelf faulting and horizontal block rotations between overlapping rifts in southern Afar. *Geophysical Research Letters, 17*(1), 1–4.

Thurber, C., & Eberhart-Phillips, D. (1999). Local earthquake tomography with flexible gridding. *Computers & Geosciences, 25*(7), 809–818.

Thurber, C., Zhang, H., Waldhauser, F., Hardebeck, J., Michael, A., & Eberhart-Phillips, D. (2006). Three-dimensional compressional wavespeed model, earthquake relocations, and focal mechanisms for the Parkfield, California, region. *Bulletin of the Seismological Society of America, 96,* S38–S49.

Trampert, J., Fichtner, A., & Ritsema, J. (2013). Resolution tests revisited: The power of random numbers. *Geophysical Journal International, 192,* 676–680.

Vernon, F., UC San Diego (1982). ANZA Regional Network. International Federation of Digital Seismograph Networks. Other/Seismic Network. https://doi.org/10.7914/sn/az.

Vernon, F., & Ben-Zion, Y. (2010). San Jacinto Fault Zone Experiment. International Federation of Digital Seismograph Networks. Other/Seismic Network. https://doi.org/10.7914/sn/yn_2010.

Waldhauser, F., & Ellsworth, W. L. (2000). A double-difference earthquake location algorithm: Method and application to the northern Hayward fault. *Bulletin of the Seismological Society of America, 90,* 1353–1368.

Weertman, J. (1980). Unstable slippage across a fault that separates elastic media of different elastic constants. *Journal of Geophysical Research, 85,* 1455–1461.

Wessel, P., Smith, W. H. F., Scharroo, R., Luis, J. F., & Wobbe, F. (2013). Generic mapping tools: Improved version released. *EOS Transactions, American Geophysical Union, 94,* 409–410.

Yang, W., Hauksson, E., & Shearer, P. (2012). Computing a large refined catalog of focal mechanisms for Southern California (1981–2010): Temporal stability of the style of faulting. *Bulletin of the Seismological Society of America, 102*(3), 1179–1194.

Yule, D. (2009). The enigmatic San Gorgonia Pass. *Geology, 37*(2), 191–192.

Yule, D., & Sieh, K. (2003). Complexities of the San Andreas fault near San Gorgonio Pass: Implications for large earthquakes. *Journal of Geophysical Research, 108,* B11.

Zhang, H., Nadeau, R. M., & Guo, H. (2017). Imaging the non-volcanic tremor zone beneath the San Andreas fault at Cholame, California using station-pair double-difference tomography. *Earth and Planetary Science Letters, 460,* 76–85.

Zhang, H., & Thurber, C. H. (2003). Double-difference tomography: The method and its application to the Hayward Fault, California. *Bulletin of the Seismological Society of America, 93*(5), 1875–1889.

Zhang, H., & Thurber, C. (2006). Development and applications of double-difference seismic tomography. *Pure and Applied Geophysics, 163*(2–3), 373–403.

Zhang, H., & Thurber, C. H. (2007). Estimating the model resolution matrix for large seismic tomography problems based on Lanczos bidiagonalization with partial reorthogonalization. *Geophysical Journal International, 170,* 337–345.

Zigone, D., Ben-Zion, Y., Campillo, M., & Roux, P. (2015). Seismic tomography of the Southern California plate boundary region from noise-based Rayleigh and Love waves. *Pure and Applied Geophysics, 172,* 1002–1032.

(Received May 10, 2018, revised October 26, 2018, accepted November 3, 2018, Published online November 21, 2018)

Pure Appl. Geophys. 176 (2019), 1145–1168

https://doi.org/10.1007/s00024-018-2055-y

Pure and Applied Geophysics

Spatiotemporal Variations of Stress and Strain Parameters in the San Jacinto Fault Zone

Niloufar Abolfathian,[1] Patricia Martínez-Garzón,[2] and Yehuda Ben-Zion[1]

Abstract—We analyze the background stress field around the San Jacinto fault zone (SJFZ) in Southern California with a refined inversion methodology using declustered focal mechanisms of background seismicity. Stress inversions applied to the entire fault zone, and three focus areas with high level of seismicity, provide three-dimensional distributions of the maximum horizontal compression direction (S_{Hmax}), principal stress plunges, and stress ratio $R = (\sigma_1 - \sigma_2)/(\sigma_1 - \sigma_3)$. The results are compared with coseismic strain parameters derived from direct summation of earthquake potencies and b values of frequency–size event statistics. The main stress regime of the SJFZ is strike-slip, although the northwest portion near Crafton Hills displays significant transtension. The S_{Hmax} orientation rotates clockwise with increasing depth, with the largest rotation (23°) observed near Crafton Hills. The principal stress plunges have large rotations below $\sim$ 9 km, near the depth section with highest seismicity rates and inferred brittle–ductile transition zone. The rotations produce significant deviations from Andersonian strike-slip faulting, likely generating the observed increased dip-slip faulting of relatively deep small events. The stress ratio parameters and b value results are consistent with increasing number of dip-slip faulting below $\sim$ 9 km. The derived coseismic strain parameters are in good agreement with the stress inversion results. No large-scale stress rotations are observed across the time of the 2010 M_{w} 7.2 El Mayor–Cucapah earthquake. The stress ratio near the Trifurcation area of the SJFZ changes after the El Mayor–Cucapah earthquake toward transpression, but this may have been produced by local $M > 5.4$ event or aseismic slip.

Key words: San Jacinto fault zone, stress inversion, coseismic strain, b value, spatiotemporal stress changes, depth variations of stress parameters.

1. Introduction

Crustal deformation has multiple components, including elastic strain that accumulates in interseismic periods between large events, inelastic strain increments associated with seismic failures of ongoing seismicity, aseismic slip on some fault sections, and viscoelastic failure near and below the brittle–ductile transition depth. Knowledge on the stress and strain distributions in the seismogenic crust is important for understanding the mechanics of faulting (e.g., Scholz 2002) and related slip patterns (Twiss and Unruh 1998). Occurrence of large earthquakes (e.g., $M > 6.5$) causes spatiotemporal stress perturbations on the scale of the seismogenic zone (e.g., Ben-Zion 2008; Hauksson et al. 2011; Meng and Peng 2014) that may lead to local rotations of the stress field (e.g., Hardebeck and Hauksson 2001; Wesson and Boyd 2007). With the ongoing densification of seismic instrumentation, larger seismicity catalogs and improved earthquake focal mechanisms are accessible and allow analyses of stress and strain parameters of fault zone regions at finer resolution.

The San Jacinto fault zone (SJFZ) is a major component of the plate boundary in Southern California. It is oriented subparallel to the San Andreas fault and is currently the most seismically active fault zone in Southern California (e.g., Hauksson et al. 2012; Ross et al. 2017a). The SJFZ accommodates a large proportion of the plate motion with slip rates of up to 21 mm year^{-1} (Fialko 2006). Paleoseismic and historic records indicate that the SJFZ is capable of large ($M_{\mathrm{W}} > 7.0$) earthquakes (e.g., Petersen and Wesnousky 1994; Rockwell et al. 2015), and has the potential to rupture the entire length of the fault zone in a single event (e.g., Salisbury et al. 2012; Onderdonk et al. 2013). The last through-going event probably occurred on 22 November 1800 (Salisbury et al. 2012),

Electronic supplementary material The online version of this article (https://doi.org/10.1007/s00024-018-2055-y) contains supplementary material, which is available to authorized users.

[1] University of Southern California, Los Angeles, USA. E-mail: niloufar.abolfathian@gmail.com

[2] GFZ German Research Centre for Geosciences, Potsdam, Germany.

and the average recurrence time for such events is estimated to be 257 ± 79 years (Rockwell et al. 2015). The SJFZ has complex geometry and high diversity of velocity structures and seismicity patterns (e.g., Allam et al. 2014; Cheng et al. 2018). During the era of highly dense seismic sensors (i.e., after year 2000), the SJFZ has not hosted any earthquake with $M > \sim 5.5$. However, the 2010 M_w 7.2 El Mayor–Cucapah (EMC) earthquake ruptured the southern continuation of the Elsinore Fault in Baja California and increased the seismicity rates across most of Southern California, including the SJFZ (Hauksson et al. 2011; Meng and Peng 2014).

In this study, we employed > 14,500 earthquake double-couple focal mechanisms (Yang et al. 2012) to derive high-resolution distributions of stress and coseismic strain parameters (i.e., orientation of principal axes and stress/strain ratios) around sections of the SJFZ. The stress inversions employ the improved methodology of Martínez-Garzón et al. (2016a) where (1) the initial seismicity catalog is declustered to minimize stress interaction between events, (2) an optimum number of events to be inverted is selected based on the uncertainties of the input data, and (3) the analyzed region is objectively discretized into small subdomains using an optimized *k*-means algorithm. The stress inversion is then refined by merging the damped least-squares algorithm (Hardebeck and Michael 2006) with the iterative stress inversion refinement based on the fault instability coefficient (Vavryčuk 2014). The inversion results provide the principal stress orientations and the stress ratio $R = (\sigma_1 - \sigma_2)/(\sigma_1 - \sigma_3)$. The elastic properties of rocks in a given spatiotemporal domain provide links between the stress and strain fields. Information on the inelastic strain tensor can be derived from calculating and stacking potency tensors of different earthquakes occurring within a region (e.g., Amelung and King 1997; Bailey et al. 2009, 2010). The advantage of seismic potency with respect to seismic moment is that the former does not depend on the shear modulus, which is poorly defined across faults and earthquake source volumes (Ben-Zion 2003).

Previous studies analyzed the stress field orientation in Southern California utilizing focal mechanisms (e.g., Jones 1988; Hardebeck and Hauksson 2001; Townend and Zoback 2001, 2004; Yang and Hauksson 2013). These studies analyzed relatively large regions with coarse spatial resolution. They also employed non-declustered seismicity catalogs, thus the recovered stress results represent a mixture of the background stress field and stress transfers generating aftershocks. The stress field orientation was found to be generally consistent with depth (e.g., Zoback 1992; Hardebeck and Hauksson 2001). However, rheological conditions may differ at some places, resulting in local stress variations with depth; For example, Bokelmann and Beroza (2000) observed rotations of focal mechanisms at the bottom of the seismogenic zone in Northern California, suggesting the existence of a weak lower crust layer. Hardebeck and Hauksson (2001) analyzed stress changes prior to and after the 1992 M_w 7.2 Landers earthquake in Southern California, and concluded that the stress field rotated near the main fault trace. Ross and Ben-Zion (2013) observed temporal variations of rotation angles between mechanisms of aftershocks of the 1992 Landers earthquake that may have contributed to the inferred rotation of the stress field. Bailey et al. (2009, 2010) analyzed coseismic strain fields around several major faults in Southern California, including the SJFZ. They concluded that properties of the coseismic strain field are controlled primarily by geometrical properties of the faults and plate motion.

In the current work, we derive stress and coseismic strain fields around the SJFZ at greater resolution than previous studies. We also estimate the brittle–ductile transition depth at different fault sections from depth variations of the *b* values of Gutenberg–Richter statistics (e.g., Amitrano 2003; Spada et al. 2013) and compare the results with derived parameters of the stress field. In addition, we investigate how the occurrence of the 2010 EMC earthquake in Baja California affected the background stress field in the SJFZ region ($\sim$ 200 km away). Our main goals are (1) to resolve the stress field orientation along and across the fault through the seismogenic thickness, (2) to validate the obtained stress results by investigating the consistency between deviatoric stress and strain tensors, and (3) evaluate the impact of a $M \approx 7$ regional earthquake at a large distance ($\sim$ 200 km) on the stress field parameters. The subsequent sections are organized as follows. Section 2 describes the utilized data, selected regions of interest, and catalog processing. Section 3

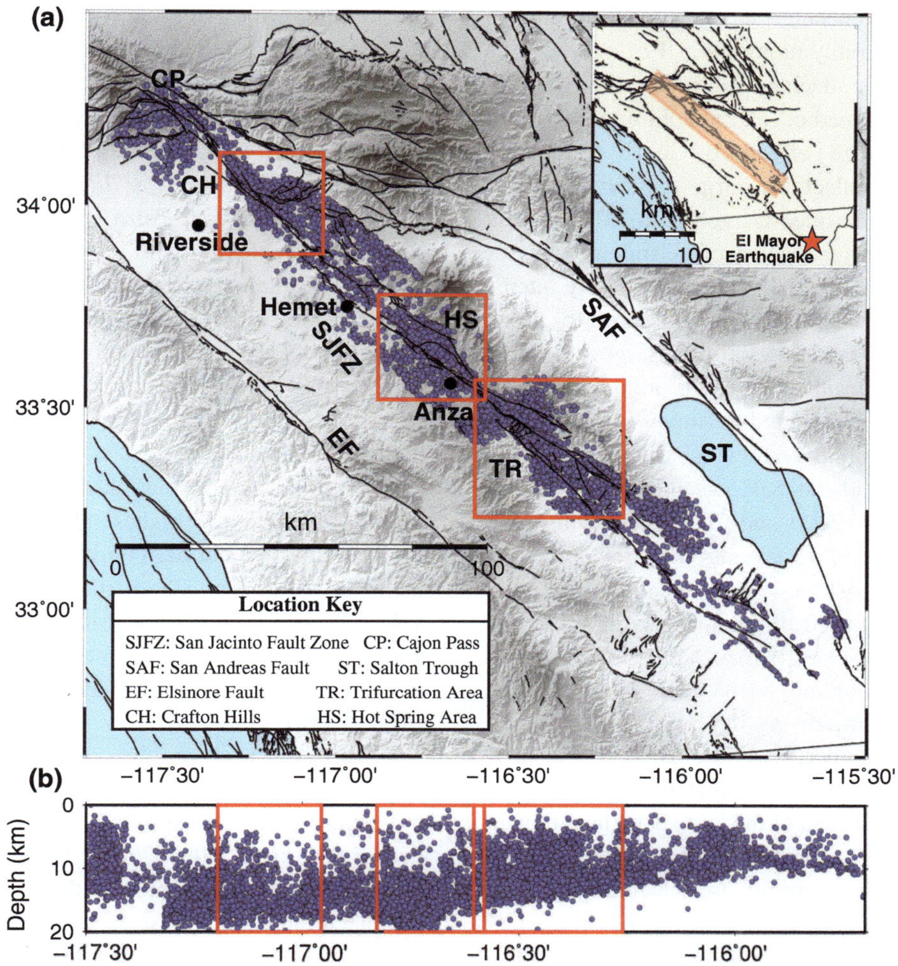

Figure 1

a Spatial distribution of seismicity in the San Jacinto fault zone (SJFZ) used in the study. Declustered earthquakes from 2000 to 2012 are shown by purple circles. Red rectangles denote three focus study areas: Crafton Hills (CH), Hot Springs (HS), and Trifurcation area (TR). Black lines indicate mapped surface traces of the SJFZ, San Andrea Fault (SAF), and Elsinore Fault (EF). The upper-right inset shows the study area at larger scale (Southern California). The red star marks the epicenter of the 2010 M_W 7.2 El Mayor–Cucapah Earthquake. **b** Distribution of seismicity with depth along the SJFZ. Red rectangles mark depth sections of the three focus areas

gives a summary of the methodologies employed to calculate the distribution of deviatoric stress and inelastic strain tensors. Section 4 describes the results obtained from the various performed analyses. Sections 5 and 6 provide a discussion of the results and the main conclusions of the study.

2. *Study Area and Data*

We utilize earthquake data recorded in years 2000–2012 in the region around the SJFZ in Southern California (Fig. 1). This time interval is selected because it is associated with relatively high density of stations and recorded earthquakes. Event hypocenters are taken from the relocated earthquake catalog of Hauksson et al. (2012), where 90 % of horizontal and vertical errors are smaller than 0.75 km and 1.25 km, respectively. Double-couple focal mechanisms with qualities ranging from A to D are taken from the catalog of Yang et al. (2012), including fault plane uncertainties between 5° and 55° with a median value of 31°. Larger background seismicity rates on the fault are found in the depth range 9–15 km, but the distribution of hypocenter locations and seismicity rates vary considerably along the SJFZ.

The catalog is declustered using the nearest-neighbor distance method of Zaliapin and Ben-Zion (2013) to minimize effects of local earthquake interactions and reflect the large-scale background stress field (e.g., Martínez-Garzón et al. 2016a). This is done by detecting clusters of seismicity without magnitude dependence (i.e., considering only interevent times and distances), using hypocentral locations for distances, and retaining only the largest event in each cluster. The declustered events are referred to in the text also as background seismicity. From more than 26,000 earthquakes with calculated focal mechanisms around the SJFZ in the initial catalog between the years 2000 and 2012, the declustered catalog contains 14,543 events with focal mechanisms. Figure 1 shows the locations of the declustered events used in the analyses of stress and strain fields along with the main faults in the area.

The thickness of the seismogenic crust is often defined from seismicity as the depth above which 90 % of the events occur. At the SJFZ, the seismogenic thickness decreases toward the southeast (Fig. 1b), likely because of the increasing heat flow in this direction (Doser and Kanamori 1986). Three regions with dense seismicity are selected (Fig. 1) as focus study areas: (1) Crafton Hills (CH), (2) Hot Springs (HS), and (3) Trifurcation area (TR). The three selected regions have different background seismicity rates, fault structural complexity, seismogenic thickness, and topography (Table 1). The CH region in the northwestern portion of the SJFZ contains the Crafton Hills fault zone. The main SJFZ is intersected in this region by orthogonal faults accommodating some normal faulting. It has a background seismicity rate of ~ 3.2 events per year per km along the fault and a seismogenic thickness of ~ 17.4 km. The HS region is located in the central part of the SJFZ and northwest of Anza. This region has the highest background seismicity rate of ~ 8.2 km^{-1} $year^{-1}$ among the study areas and a seismogenic thickness of ~ 17.5 km. The TR region located southeast of Anza displays the greatest structural complexity, with three main fault branches as well as smaller orthogonal faults and seismicity lineations (e.g., Ross et al. 2017a; Cheng et al. 2018). It has a background seismicity rate of 6.85 km^{-1} - $year^{-1}$ and seismogenic thickness of ~ 13.7 km (~ 4 km shallower than the other two areas). The topography of these regions is variable and includes areas adjacent to the San Jacinto and Santa Rosa Mountains, with the highest elevations at the Hot Springs fault (the northern fault strand in the HS) and the Buck Ridge fault (the northern fault strand in the TR). Inbal et al. (2017) suggested that the EMC event and the following M_w 5.4 Collins Valley earthquake in the TR area produced transient aseismic slip in the HS and TR regions.

First, we analyze the distribution of stress parameters along the SJFZ (Sect. 4.1). This is done by using 10,827 background events that occurred during the years 2000–2009 to suppress potential transient effects associated with the 2010 M_w 7.2 EMC earthquake in Baja California. Then, we describe the distribution of faulting styles, *b* values, and variations of stress and strain parameters through the seismogenic thickness using the same dataset (Sects. 4.2, 4.3, 4.4). Finally, we investigate spatiotemporal changes of stress parameters in relation to the EMC earthquake (Sect. 4.5). For this purpose, we select events from approximately 2 years prior to the EMC earthquake, from January 2008 to December 2009, and 2 years following the EMC event from

Table 1

Average seismic and stress parameters for the SJFZ, CH, HS, and TR areas, estimated based on the 2000–2009 declustered seismicity. The angle θ is the angle between S_{Hmax} and main fault strike direction (N55°W)

	Seismogenic thickness (km)	Seismicity rate (km^{-1} $year^{-1}$)	Stress ratio	S_{Hmax}	θ
SJFZ	16.7	~ 4	0.50 ± 0.10	5 ± 4.0°	60 ± 4.0°
CH	17.5	~ 3.2	0.20 ± 0.10	18 ± 5.5°	73 ± 5.5°
HS	17.4	~ 8.2	0.59 ± 0.08	10 ± 2.5°	65 ± 2.5°
TR	13.7	~ 6.85	0.48 ± 0.08	0 ± 3.5°	55 ± 3.5°

October 2010 to December 2012. These two data subsets have 2901 and 2429 declustered events with focal mechanisms, respectively.

3. Methodology

3.1. Stress Tensor Inversion of Focal Mechanisms

The formal stress inversion utilizes double-couple earthquake focal mechanisms to estimate the orientations of the three principal stresses σ_1, σ_2, and σ_3 (from most to least compressive) and the stress ratio, R, defined as

$$R = \frac{\sigma_1 - \sigma_2}{\sigma_1 - \sigma_3}. \quad (1)$$

The stress ratio, R, has values between 0 and 1. In strike-slip stress regime, larger stress ratios correspond to a stress regime closer to transpression (i.e., mixed strike-slip and reverse faulting), while smaller stress ratios represent a transition toward a transtensional stress field (i.e., mixed strike-slip and normal faulting).

The majority of stress inversion techniques are based on the following assumptions: (1) the stress field is homogeneous within a considered rock volume, (2) earthquakes occur on preexisting faults with varying orientations, and (3) slip on each fault occurs parallel to the direction of tangential traction (Wallace 1951; Bott 1959). This is equivalent to assuming that, at the time that an earthquake occurs, the stress and strain fields are consistent. The stress inversion further assumes that the seismicity is driven by the background tectonic stress and not by the stress interaction between the seismicity. For that reason, we decluster the seismicity catalog to only consider the background seismicity (see Martínez-Garzón et al. 2016a). Using the refined methodology of Martínez-Garzón et al. (2016a), we invert the focal mechanisms of the background seismicity for spatial and temporal variations of the stress field in the SJFZ region. The analysis employs the MSATSI software (Martínez-Garzón et al. 2014), which is an updated version from the SATSI algorithm (Michael 1984; Hardebeck and Michael 2006), applying a linear damped stress inversion to the earthquake focal mechanisms. Uncertainty assessments of 95 % confidence intervals are done by bootstrap resampling of the original set of focal mechanisms (Michael 1987). The resulting stress field orientation is classified according to the three Andersonian stress regimes—normal, strike-slip, and reverse—based on the trends and plunges of the principal stresses (Zoback 1992). From the orientation of the principal stress axes, the orientation of maximum horizontal compressional stress, S_{Hmax}, is computed following Lund and Townend (2007). The stress inversion selects one of the two available fault planes from each source mechanism solution. Considering the fault plane ambiguity, previous studies show that the choice of the main fault plane does not affect the estimation of the principal stress orientation (Michael 1984; Vavryčuk 2015), but can affect the resolution of the stress ratio. The linear inversion procedure of Michael (1984, 1987) selects one of the nodal planes randomly. Another approach is to use an iterative procedure. In each iteration, the stress field orientation is calculated and the fault plane with the largest instability coefficient I is selected for the next iteration (Vavryčuk 2011, 2014; Martínez-Garzón et al. 2016b). The parameter I is defined as

$$I = \frac{\tau - \mu(\sigma - 1)}{\mu + \sqrt{1 + \mu^2}}, \quad (2)$$

where μ is the apparent coefficient of friction and τ and σ are scaled shear and normal stresses, respectively. The values of I range between 0 and 1, indicating the faults least and most optimally oriented to failure within the given deviatoric stress field, respectively. When evaluating the fault instability, a grid search is applied over values of μ ranging between 0.2 and 0.8, and the value that produces the highest overall instability for the inverted data is selected (Vavryčuk 2014). As this is selected in relation to the inversions, we call it an apparent coefficient of friction.

Stress inversion methods cannot constrain the stress orientation of a region with a single focal mechanism (McKenzie 1969), and a minimum number of focal mechanisms is required to invert for stress parameters. Synthetic tests with focal mechanisms having realistic uncertainties are performed to estimate how many focal mechanisms are needed to retrieve the stress field orientation with an accuracy of ± 10° (Martínez-Garzón et al. 2016a). Based on these tests, ~ 60 events are the optimum in a strike-slip regime to converge on a stable stress tensor. In this analysis, the study volume is discretized into Voronoi cells containing the minimum number of events required for stable inversion. The Voronoi cells are implemented using the *k*-means technique (Hartigan and Wong 1979; Seber 1984; Martínez-Garzón et al. 2016a). The Voronoi cell sizes vary in relation to the density of seismicity and provide estimates for the spatial resolution of the inversion. In this study, the grid cells have dimensions in the range of ~ 1 to 5 km as an approximate radius of the grid cells. Therefore, the present study aims to resolve the stress field in the SJFZ region with a finer discretization than previous studies (e.g., Jones 1988; Hardebeck and Hauksson 2001; Yang and Hauksson 2013), which investigated spatial changes of stress parameters in Southern California with variable resolution between 5 and 20 km or larger.

3.2. Coseismic Strain Parameters from Focal Mechanisms

Unlike the estimation of the stress field, the estimation of the coseismic (i.e., inelastic) strain field does not require a formal inversion. The inelastic strain tensor in the faulting region defines the seismic potency density tensor per unit volume:

$$P_{ij} = \int_V d_{ij}^P \mathrm{d}V, \quad (3)$$

where d^P is the inelastic strain tensor, V is the source volume, and subscripts "i" and "j" denote the three Cartesian directions (Ben-Zion 2003, 2008). The tensor P_{ij} is directly related to the strike, dip, and rake of the source mechanisms of each event. The orientation of the principal strain axes, $\hat{d}_1 > \hat{d}_2 > \hat{d}_3$, can be calculated from the normalized sum of earthquake potencies, $\hat{P}_{ij}^{\mathrm{SM}}$, in each of the grid cells (Bailey et al. 2009) using

$$P_{ij}^{\mathrm{SM}} = \sum_{k=1}^{N} P_{ij}^{(k)}, \quad (4)$$

where N is the number of all earthquakes per grid cell and

$$\hat{P}_{ij}^{\mathrm{SM}} = \frac{P_{ij}^{\mathrm{SM}}}{\left\| P_{ij}^{\mathrm{SM}} \right\|}. \quad (5)$$

The use of normalized potencies equally distributes the effects of all events on the computed strain quantities and eliminates dominance of the largest events. The principal strain orientations are used to estimate a strain ratio parameter, D, defined as (Twiss and Unruh 1998)

$$D = \frac{\hat{d}_1 - \hat{d}_2}{\hat{d}_1 - \hat{d}_3}. \quad (6)$$

This parameter is a scalar invariant of the strain tensor related to the shape of the strain ellipsoid, and it is analogous to the stress ratio, R, defined in the previous section. The orientation of maximum horizontal compressional strain, D_{Hmax}, is estimated following Lund and Townend (2007).

The calculations of the potency and strain tensors require choosing one of the two possible fault planes in the earthquake focal mechanism. As done in the stress analysis, the strain tensor is calculated for each focal mechanism using the fault plane with the highest instability coefficient.

3.3. Estimation of Magnitude of Completeness and b Values

The *b* value of the frequency–magnitude distribution has often been used as an indicator of the differential stress level (Scholz 2015) as well as the fault damage (Amitrano 2003). Here, the magnitude of completeness M_c and *b* values of the Gutenberg–Richter statistics of earthquakes are estimated using the goodness of fit method (Wiemer and Wyss 2000). In each of the focus study areas, we calculate *b* value changes with depth (from 3 to 18 km) by considering the entire declustered seismicity catalog within the analyzed time period (2000–2009) and dividing the seismicity into moving windows of 3 km depth bins. We also group the declustered seismicity in different faulting styles (normal, reverse, and strike-slip) to investigate how the different faulting styles may affect the magnitude–frequency distribution. Only bins containing 60 or more earthquakes with

$M > M_c$, and b value error estimation of less than 0.1 are analyzed.

4. Results

Applying the refined methodologies for stress inversion, coseismic strain estimation, and b values, we obtain detailed information on the background stress and coseismic strain fields in the San Jacinto fault zone. In the following, we first analyze the distribution of stress parameters along the SJFZ. Then we describe the distribution of faulting styles, b values, and variations of stress and strain parameters for the three focused study regions (CH, HS, and TR; see Table 1) through depth. Finally, we investigate spatiotemporal changes of stress parameters in relation to the EMC earthquake.

4.1. Variations of Stress Parameters along the SJFZ

The declustered catalog of 10,827 events distributed along the entire SJFZ, having focal mechanisms and occurring between the years 2000 and 2009, is discretized into 111 groups of seismicity (grid cells) containing between 60 and 110 focal mechanisms in each group. Focal mechanisms from each grid cell are inverted for the stress field orientation and stress ratio, R, following the methodology discussed in Sect. 3.1.

Figure 2 shows the distribution of maximum horizontal compressional stress (S_{Hmax}) in the entire SJFZ and the focus areas CH, HS, and TR. As expected from the general tectonic setting and the previous larger-scale stress inversion study of Yang and Hauksson (2013), the main stress regime in the fault zone is strike-slip. Oblique faulting styles with a mixture of strike-slip and dip-slip deformation are prominent in the HS and CH, and in the northwest part of TR. The only region with a tendency towards transtension is CH, which includes subvolumes dominated by normal faulting. This can be explained in terms of extension associated with motion on both the SJFZ and the nearby San Andreas fault (Wesnousky 1986; Jennings 1994). Figure 3 presents inverted values of the stress ratio, R, along the different regions of the SJFZ. The stress ratio distributions are consistent with a main faulting style of strike-slip, superposed with some transtensional and transpressional components. In agreement with Fig. 2, the main recovered stress regime in the CH area is transtensional (strike-slip and normal faulting), while in HS area, the obtained oblique faulting represents a transpressional environment (strike-slip and reverse faulting).

The fine data discretization allows us to observe detailed changes of stress parameters along the fault zone. Table 1 lists the stress ratios and S_{Hmax} values averaged across the entire depth range for the three focused areas. From the northwest to southeast portions of the SJFZ, the median trend of S_{Hmax} rotates by $> 15°$. The angle between the S_{Hmax} and the main fault trace is larger in the CH area than in the other two regions, suggesting that the CH portion of the SJFZ could be mechanically weaker. In agreement with this, the estimated apparent friction coefficients (Eq. 2) generated by the iterative inversion (Vavryčuk 2014) are 0.45 for CH, 0.6 for HS, and 0.55 for TR areas.

We performed additional high-resolution inversions discretizing separately events that are within 1–2 km from the main faults, off-fault events with distance from the nearest major fault larger than 2 km, as well as events that are on the opposite sides of the SJFZ. These inversions show significant variations of stress ratios when a main fault is crossed (Fig. S1), but there are no significant changes between on- and off-fault events. Additional results for different depth sections are described in Sect. 4.3.

4.2. Seismicity Parameters and Faulting Styles in the Focused Study Areas

For b value estimation, the used seismicity is divided into moving windows of 3 km depth bins. This choice of depth bins is based on the distribution of events and the overall depth uncertainty ($\sim$ 1.25 km) of the hypocenter locations. Tables S1–S3 list the magnitude of completeness M_C and b values of various subsets of the employed declustered seismicity (different depth sections and different faulting styles) in the selected study regions. Higher b values represent relatively larger proportions of lower-magnitude events. The HS area

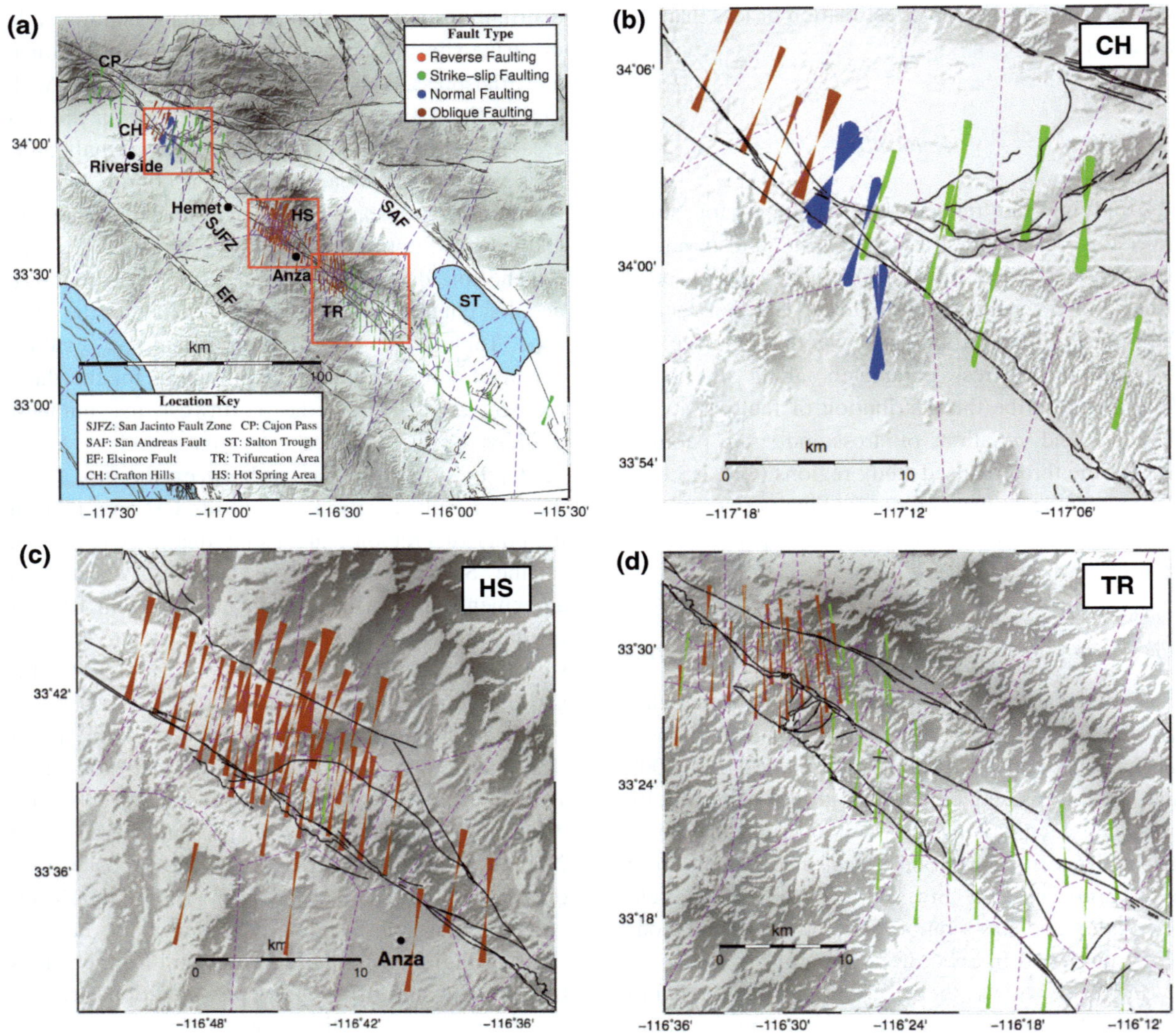

Figure 2

Maximum horizontal compressional stress orientations (S_{Hmax}) along the **a** SJFZ, **b** CH, **c** HS, and **d** TR areas. The S_{Hmax} orientations are color-coded with red, green, blue, and brown denoting reverse, strike-slip, normal, and oblique faulting, respectively. The variations in S_{Hmax} orientations show the uncertainty of 95 % confidence interval. Purple dashed lines mark the Voronoi cells used in the analysis of seismicity

displays the lowest average *b* value of ~ 0.72, while the CH and TR areas have average *b* values of ~ 0.95 and ~ 0.96.

Numerical and laboratory experiments have shown that the *b* value tends to decrease with increasing confining pressure (Amitrano 2003). Spada et al. (2013) inferred that *b* values in fault regions decrease in the seismogenic zone down to a given depth, considered to reflect the brittle–ductile transition, then increase below that depth. Figure 4 presents variations of *b* values with depth in the three focus study areas along the SJFZ. Among the three regions, only the TR area clearly displays a minimum *b* value at a given depth. In this area, the overall *b* value distribution and the one related to the strike-slip events reach a minimum value at 7–10 km (black and green line in Fig. 4c), suggesting that this may reflect the brittle–ductile transition depth. At CH, the *b* value from all events as well as the one calculated only from normal faulting have minima at 11–14 km

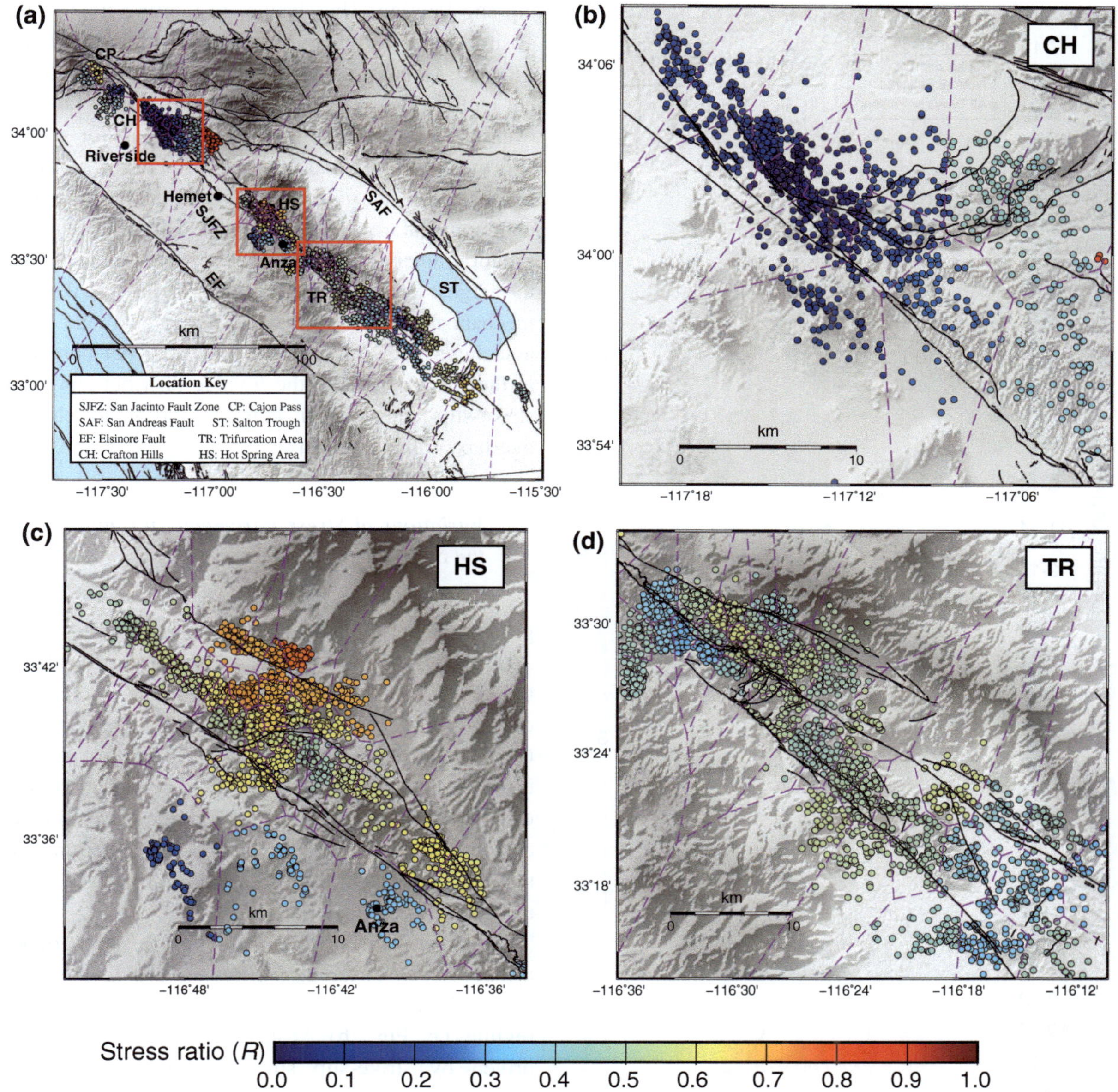

Figure 3

Seismicity along the **a** SJFZ, **b** CH, **c** HS, and **d** TR areas, color-coded with values of the stress ratio *R*. Purple dashed lines mark the used Voronoi cells. In a strike-slip faulting environment, *R* values around 0.5, 0, and 1 indicate pure strike-slip, transtension, and transpression, respectively

(black and blue line in Fig. 4a), but the uncertainties imply that this is not significant. At HS, there is not enough seismicity shallower than 9–12 km depth section. The overall *b* value shows two local minima at 10–13 km and at 12–15 km (black line in Fig. 4b). In principle, either of these two minima could be associated to the brittle–ductile transition.

The ternary plots (Frohlich 1992) in Fig. 5 show the distribution of events with different sizes and faulting types within 3 km depth bins through the seismogenic thickness. In all focus study areas,

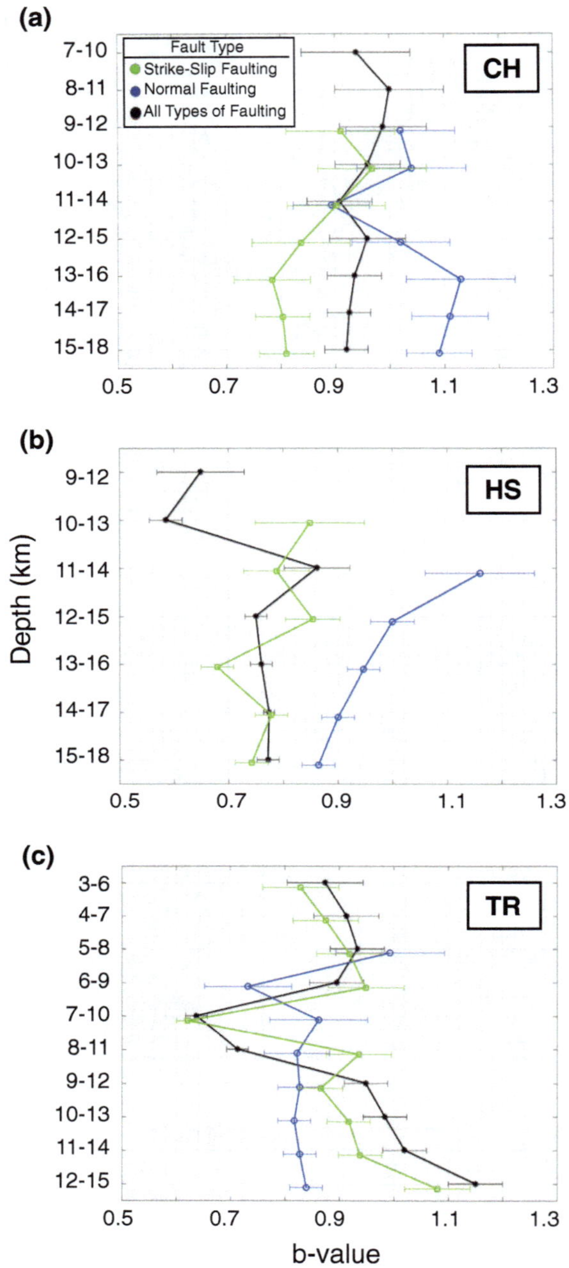

Figure 4
Variations of *b* values with depth in the **a** CH, **b** HS, and **c** TR areas. Green and blue lines display *b* values associated with strike-slip and normal faulting, respectively, while black line shows *b* value of all events

strike-slip faulting is seen to dominate in the shallow crust. However, in the HS area deeper than 9–12 km and the CH and TR regions deeper than 6–9 km, there is substantial increase in the relative amount of dip-slip faulting. The changes in faulting style occur at similar depths where *b* values show minima in the TR area. Comparing *b* values associated with different faulting types, *b* values related to normal faulting events are on average ~ 0.2 larger than for strike-slip events in CH and HS regions. Larger *b* values related to normal faulting have been observed to be a common feature of the Southern California catalog (Schorlemmer et al. 2005). These results indicate that the largest events in the CH and HS have strike-slip mechanisms, while normal faulting events tend to be associated with smaller events. This is not observed in the TR area, where the *b* value distributions from strike-slip and normal faulting appear to closely overlap (Tables S1–S3).

4.3. Variations of Stress Parameters through the Seismogenic Thickness

The dataset described in Sect. 4.1 (10,827 background events with focal mechanisms for the years 2000–2009) is divided into 3 km bins in the depth range of 3–18 km. As mentioned, the choices of depth bins are based on the distribution of events and the overall depth uncertainty (~ 1.25 km) of the resolved hypocenter locations. The data at the different depth bins are discretized into Voronoi cells and inverted for the orientation of the principal stresses, S_{Hmax}, and the stress ratio in the three focus areas (Figs. 6, 7, 8).

The variation of the stress regime with depth appears to be different for each of the focus study areas. The CH area displays an overall transtensional regime (i.e., Fig. 2b). At shallow depths, seismicity is driven by strike-slip stress regime, changing to normal and oblique faulting at deeper fault sections (Fig. 6). In the CH area, events in the depth bins of 9–12 km and 12–15 km appear to be mostly located off-fault. However, in the deepest bin of 15–18 km, earthquakes are localized within the main trace of the SJFZ. In contrast, in the HS area, the stress field orientation slightly changes towards transpression with depth (Fig. 7). This may be the result of higher topography related to the San Jacinto Mountains (Fialko et al. 2005). The event locations at 12–15 km depth in the HS area appear more localized than for the same depth range at CH. Within the HS,

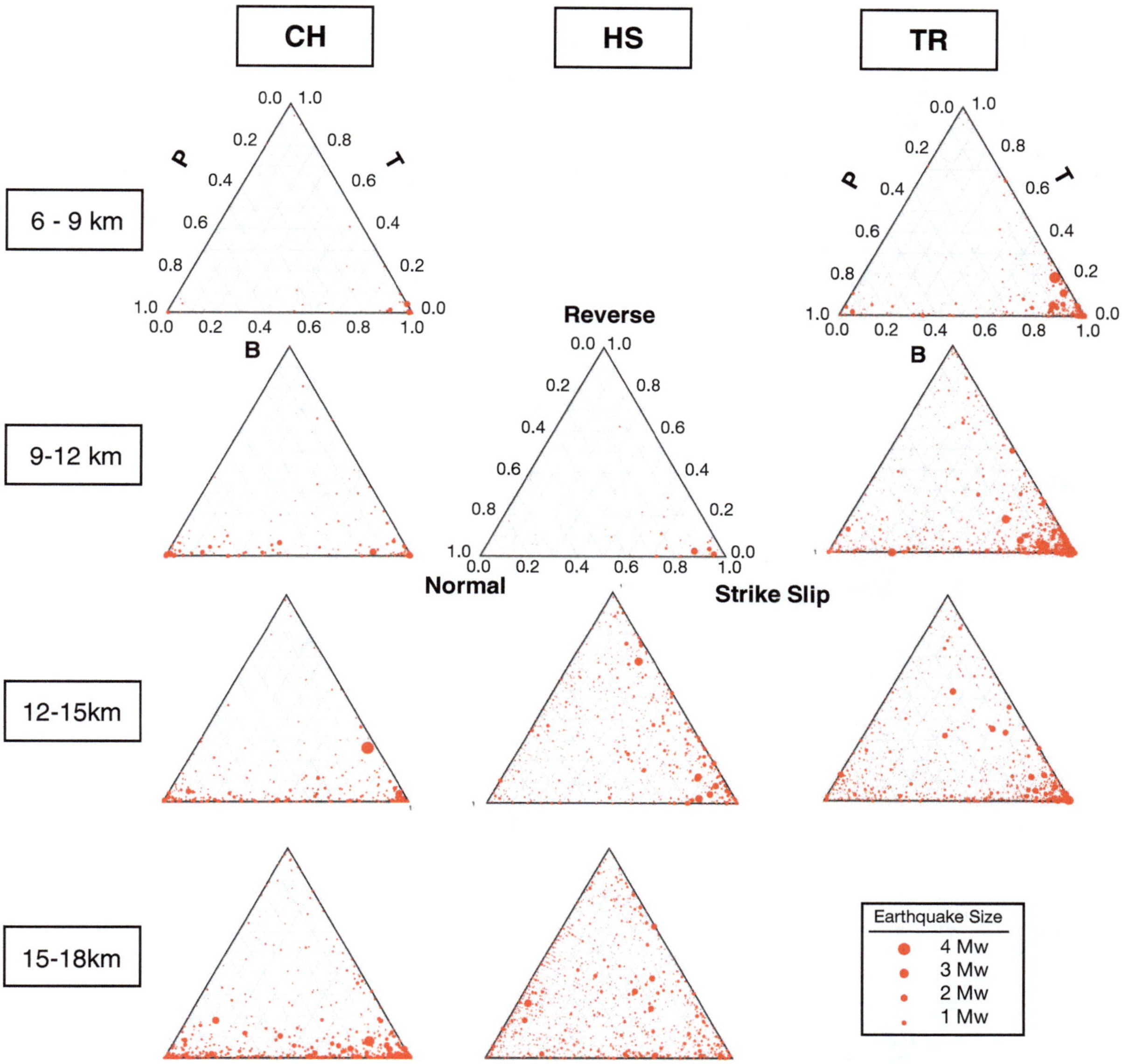

Figure 5
Ternary plots showing faulting type distribution versus depth for the three focus areas. The P, T, and B sides represent normalized plunge angles of the maximum, minimum, and intermediate compressional stresses, respectively. Red circles show events, with circle size denoting earthquake magnitude

seismicity in the 15–18 km section appears to be localized both on the Clark fault (main strand of the SJFZ) and also on Hot Springs fault in the northern section of HS region. In the TR area, strike-slip faulting is prevalent across the entire seismogenic zone. In the deepest section (12–15 km), the stress field includes a region with higher transpressional regime in the northern section adjacent to the Buck Ridge fault and the high topography associated with Santa Rosa Mountains. The seismicity distribution changes significantly with depth (Fig. 8). The seismicity in range 6–9 km covers both the mapped fault traces and the regions between (and also slightly outside) the main traces. With increasing depth, the

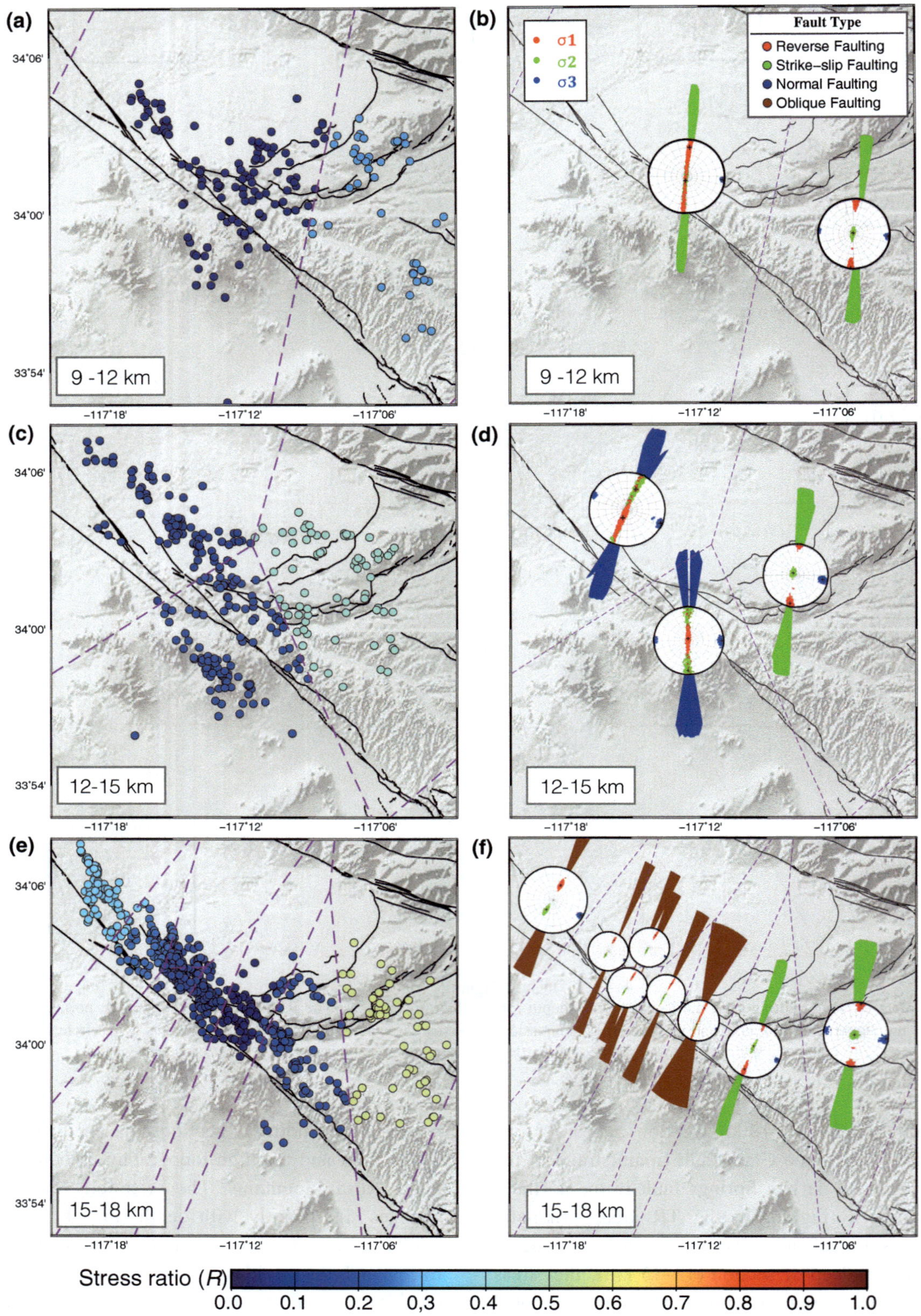
(a)
34°06'
34°00'
33°54'
9 -12 km
−117°18'
−117°12'
−117°06'
(b)
σ1
σ2
σ3
Fault Type
Reverse Faulting
Strike−slip Faulting
Normal Faulting
Oblique Faulting
9 -12 km
(c)
12-15 km
(d)
12-15 km
(e)
15-18 km
(f)
15-18 km
Stress ratio (R)
0.0
0.1
0.2
0,3
0.4
0.5
0.6
0.7
0.8
0.9
1.0

◀Figure 6
Distribution of stress parameters in the Crafton Hills area through depth sections of **a**, **b** 9–12 km, **c**, **d** 12–15 km, and **e**, **f** 15–18 km. Left plots (**a**, **c**, **e**) indicate the stress ratio distributions, while right plots (**b**, **d**, **f**) display the S_{Hmax} orientations (fan symbols) and the principal stress orientations (stereonets) in the Voronoi cells (purple dashed lines). The colors of fan symbols denote different fault types. The maximum, intermediate, and minimum principal stresses in the stereonets are indicated with red, green, and blue, respectively

seismicity becomes more localized. The majority of seismicity in the depth range 12–15 km is bounded by the main fault traces.

The event distributions with depth, median orientations of S_{Hmax}, principal stress plunges, and stress ratios in the three study areas are shown in Fig. 9. The CH and HS areas have maximum background seismicity in the depth range 15–18 km, while region TR with thinner seismogenic zone has maximum seismicity at a shallower depth range of 9–15 km. In all regions, the peak background seismicity is located near the bottom of the seismogenic zone (Fig. 9a). This can be explained by a strong gradient of slip and high stress concentration at this depth section, which, consequently, produce high rate of seismicity. A significant clockwise rotation in the S_{Hmax} orientation is observed in the entire SJFZ from the depth range 3–6 to 15–18 km, with an average rotation value of 13°. The largest S_{Hmax} rotation angle of about 23° is observed for the CH region, where the largest S_{Hmax} variation occurs close to the inferred brittle–ductile transition zone (below 11–14 km). At the TR region, a rotation of approximately 15° is observed and appears to be uniformly distributed across all the sampled depths. Finally, at the HS area, only a rotation of ~ 5° is visible below 9 km, which is comparable to the uncertainty of the estimated S_{Hmax} (Fig. 9b).

In all study regions, the stress at shallow depth sections follows Andersonian faulting showing a strike-slip regime with the intermediate principal stress axis oriented vertically and the other two horizontally (Fig. 9c). However, below the estimated brittle–ductile transition zone at each region, the results show mixed faulting regimes toward transtension or transpression. The principal stress plunges in all the areas rotate considerably below 9–12 km depth. This is in agreement with the observed increase in mixture of focal mechanisms in the ternary plots of Fig. 5. It is noteworthy that the grid cells in the eastern part of CH, containing only events off the SJFZ for all depths, do not show variations in either S_{Hmax} orientations or the principal stress plunges (Fig. 6).

The median stress ratios, R, in all regions also reflect variations of the stress field with depth. The stress ratio tends to first decrease with depth, reaching minima at 9–12 km, and increasing at deeper sections (Fig. 9d). The CH area has the lowest stress ratios ($R < 0.5$) among the focus study areas, consistent with a transtensional regime. The HS area has higher stress ratios than in the CH and TR regions, especially in the deepest section, consistent with a transpressional region. The TR area has stress ratios around 0.5, close to pure strike-slip faulting. The stress regime in TR changes only slightly with depth. The estimated instability coefficients of the fault planes employed in the stress inversions (Eq. 2) indicate that strike-slip ($I \geq 0.9$) and normal faulting events are better oriented for failure than the reverse faulting events.

4.4. *Variations of Coseismic Strain Parameters through the Seismogenic Zone*

The dataset and Voronoi cell discretization described in Sect. 4.3 are used to estimate the coseismic strain parameters: maximum horizontal compressional strain (D_{Hmax}), principal strain plunges, and strain ratio, D (Eq. 5). The derived strain parameters for equivalent depth bins as for the stress analysis are in good agreement with corresponding stress parameters in each region (Fig. S3). D_{Hmax} shows the same rotations with depth as S_{Hmax}, with the CH area displaying the largest rotation of ~ 24°, followed by the TR and HS areas with ~ 20° and ~ 8°, respectively. The principal strain plunges vary with depth in all three regions, changing from pure strike-slip in the shallow crust to oblique faulting for deeper sections. The CH area displays a large rotation of maximum and intermediate principal strain axes below 6–9 km. In the HS area, all three principal strain plunges rotate below 9–12 km. The

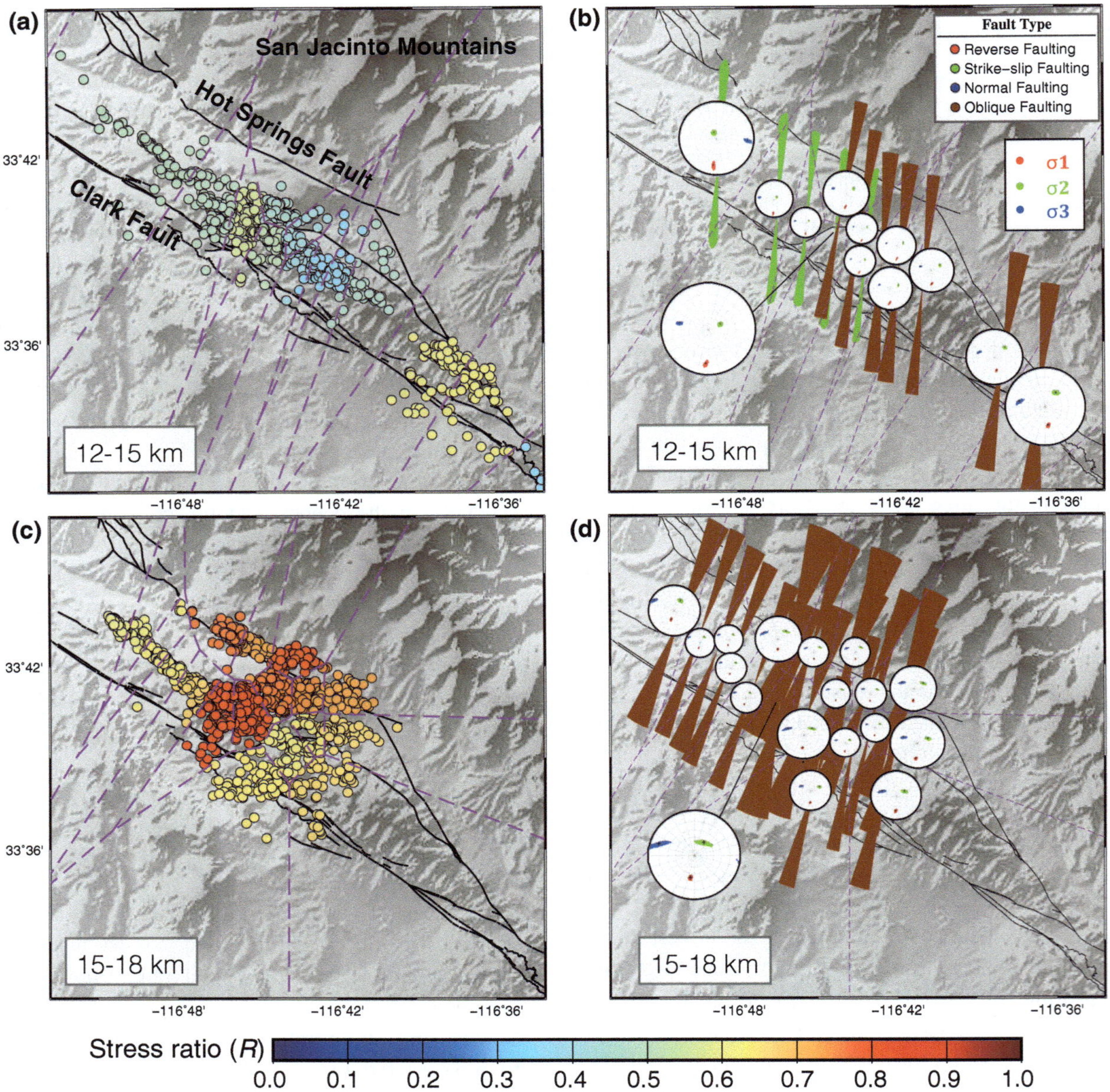

Figure 7
Same as Fig. 6 for the Hot Springs area

TR area has only a slight rotation in the plunges deeper than 9–12 km, but the variation is significant. The strain ratios show slight changes within seismogenic thickness, with changes mostly within the estimated uncertainty error and therefore not significant.

To quantitatively compare the stress and strain results for same regions, the median of the minimum rotation angle (i.e., the Kagan angle) between the stress and strain tensors and median differences in each corresponding stress and strain parameter are analyzed (Fig. 10). To be statistically comparable, we only compare results for depth sections including

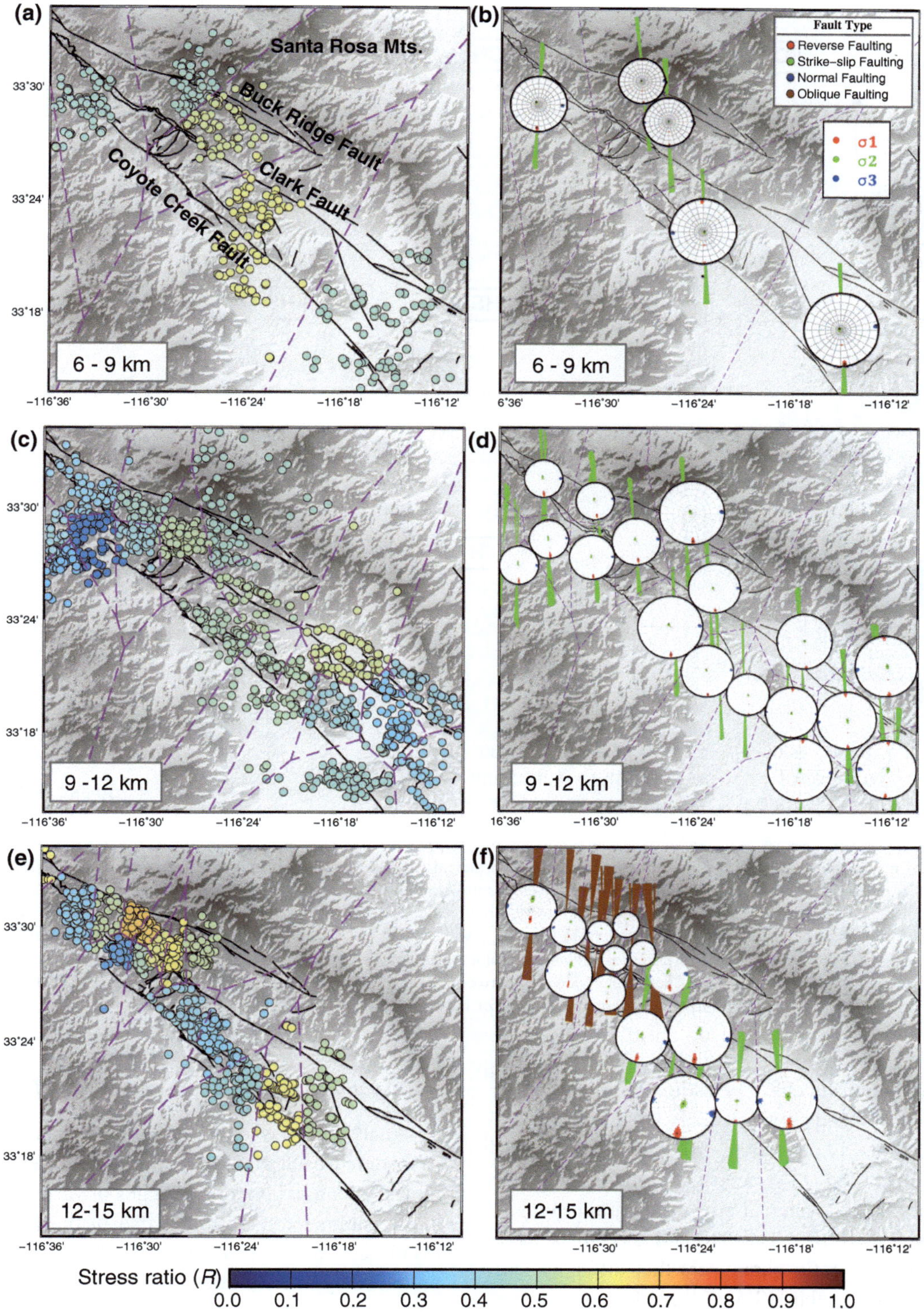

Figure 8
Same as Fig. 6 for the Trifurcation area

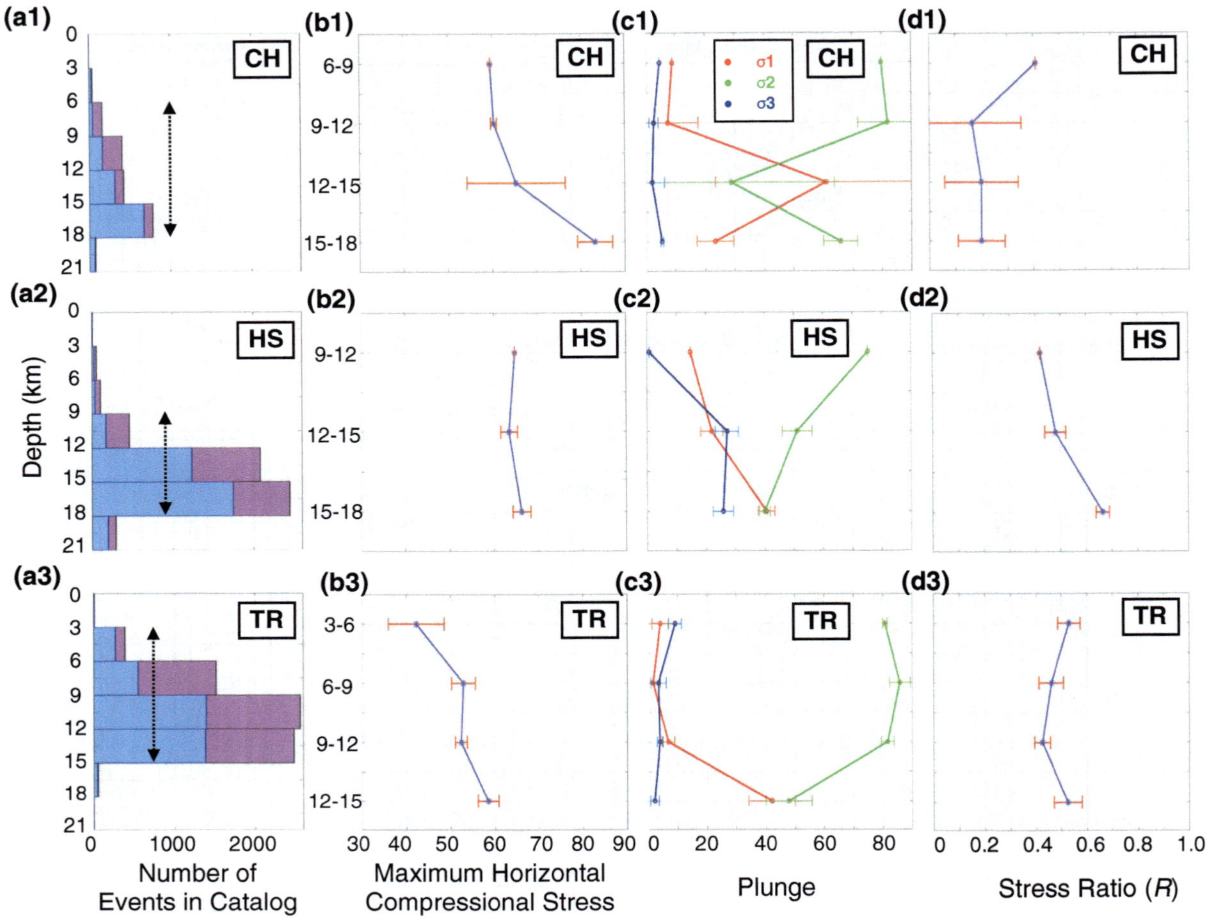

Figure 9

Variations of **a** seismicity rate, **b** S_{Hmax} orientation, **c** plunges of principal stress orientations, and **d** stress ratios through depth in the three focus areas for the period 2000–2009. The three rows display the results from the CH, HS, and TR areas, respectively. Purple and blue bars in the left panels show seismicity rates from the entire and declustered catalogs, respectively. Black arrows indicate the depth interval with enough data for analysis of the stress field orientation. In (**b**), S_{Hmax} values show the angle between the S_{Hmax} direction and the main fault strike direction of N55°W. In **c**, plunges of principal stresses are color coded with red, green, and blue showing maximum, intermediate, and minimum principal stresses, respectively. The median values of derived parameters from different grid cells within given depth intervals are shown in **b–d** with colored dots; error bars mark the 90 % confidence interval

more than two grid cells. The estimated Kagan angle between stress and strain tensors in all areas and depth sections is less than 20°. The differences between S_{Hmax} and D_{Hmax} are smaller than 5°, and the differences in the principal stress and strain plunges are smaller than 10°. The TR and CH areas have stress and strain ratio parameters that differ by less than 0.1, while in the HS area the difference is less than 0.2. In summary, the results indicate high consistency between the recovered stress and strain parameters in regions with sufficient seismicity.

4.5. Spatiotemporal Variations of Stress Parameters

Analyzing the entire catalog, rather than the declustered background seismicity as done here, can lead to apparent temporal changes of various inverted stress parameters across the time of the EMC event (Martínez-Garzón et al. 2016a). These changes are produced by internal stress variations associated with aftershocks, and they do not reflect changes in the background stress field operating on the study areas. To examine possible temporal changes of the

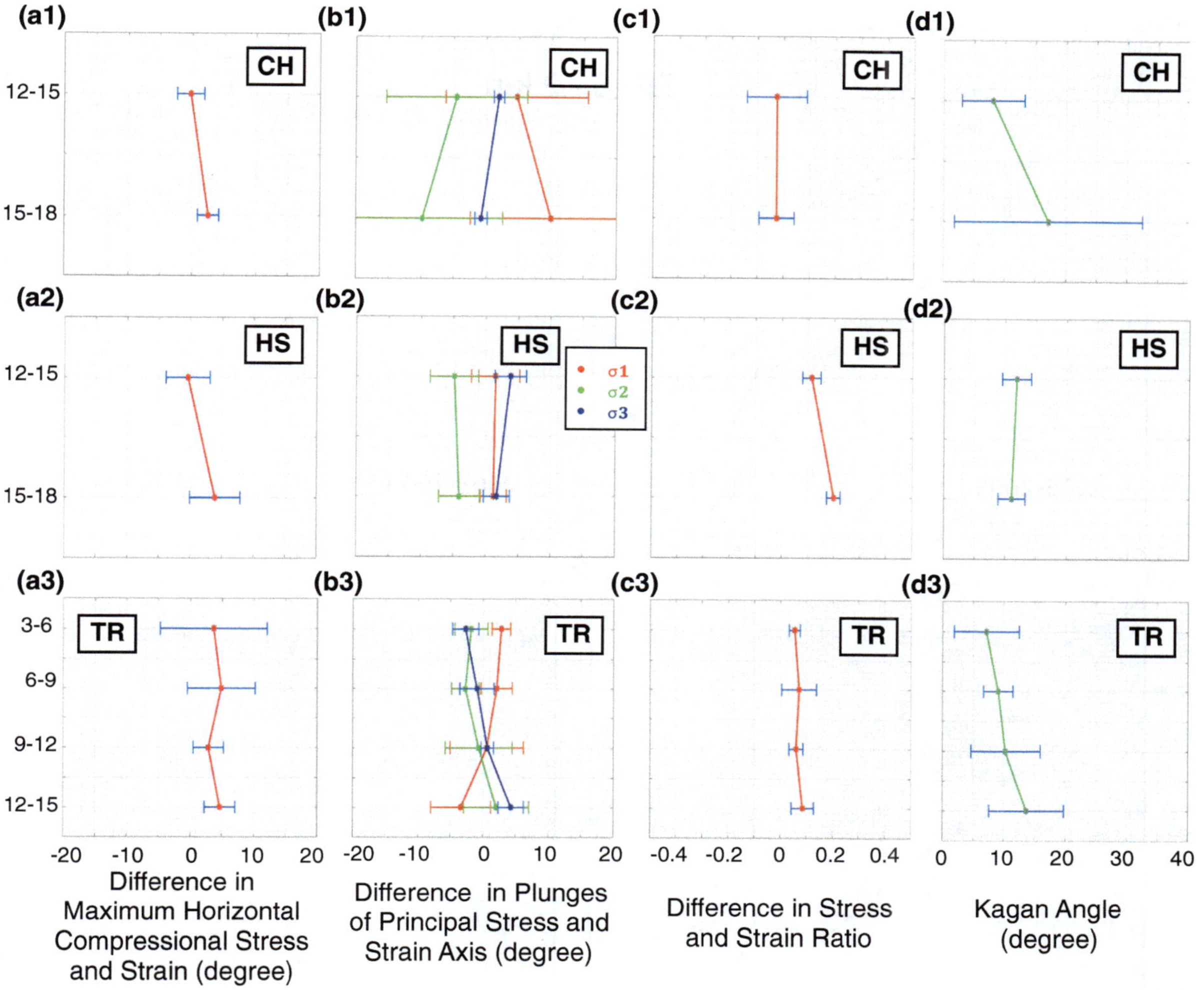

Figure 10
Variation of differences between **a** maximum horizontal compressional stress and strain, **b** plunges of principal stress and strain tensors, **c** stress and strain ratio, and **d** minimum rotation (Kagan) angle between the stress and strain tensors through the depth in the CH, HS, and TR areas. At all depth sections, median values are marked with circles and the error bars show 90 % confidence interval. Differences in the plunges of principal stresses in **b** are color coded as in Fig. 9

background stress in the TR, HS, and CH study regions, we use data from approximately 2 years before and after the 2010 M_w 7.2 EMC earthquake (see Sect. 2 for details). The datasets are divided in 3 km depth bins from 3 to 18 km depth (Figs. 11, S4–S10), discretized horizontally with Voronoi cells containing 60–110 events per cell, and inverted for stress parameters as discussed.

The obtained principal stress axes before and after the EMC earthquake have generally consistent orientations in all three focus areas and for all depth sections. The changes in S_{Hmax} after the EMC event (3°) are smaller than the uncertainties and hence insignificant (Figs. S11, S12). However, stress ratios show a significant increase after the EMC event at specific regions. The maximum variation of stress ratio is at 6–9 km depth in the TR area, increasing from a median value of 0.52 before the EMC event to 0.79 afterward. This changes the local stress regime from strike-slip toward transpressional (Figs. 11, 12). It should be noted that the largest event in the period 2008–2012, the M_w 5.43 Collins Valley earthquake

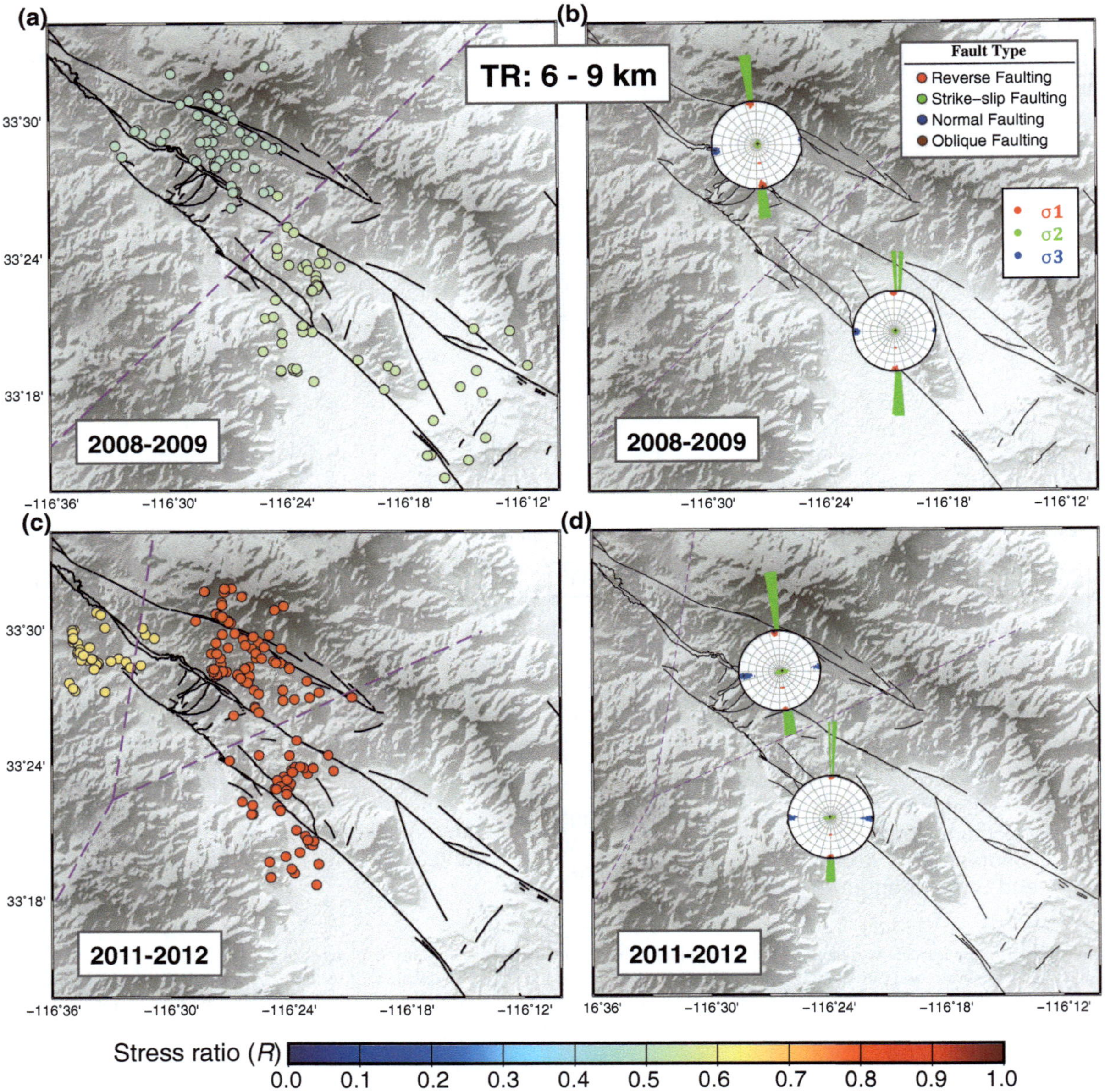

Figure 11
Temporal changes of stress ratios (left) and S_{Hmax} (right) in the TR area between 6 and 9 km depth using focal mechanisms from 2 years before the EMC earthquake (top panels) and 2 years after the event (bottom panels). The symbols and notations are the same as in Fig. 6

with a strike-slip focal mechanism, occurred in the TR area on 7 July 2010 at depth ~ 12 km. Hauksson et al. (2011) and Meng and Peng (2014) estimated a positive Coulomb stress change of ~ 10 kPa from the EMC earthquake near the TR area, and noted increase in seismicity along the SJFZ that is consistent with both dynamic and static stress increase from the EMC event. The observed stress variations in the TR can be produced in part by the positive Coulomb stress transfer from the EMC, along with changes generated by the M_w 5.4 Collins Valley earthquake and triggered moderate events and

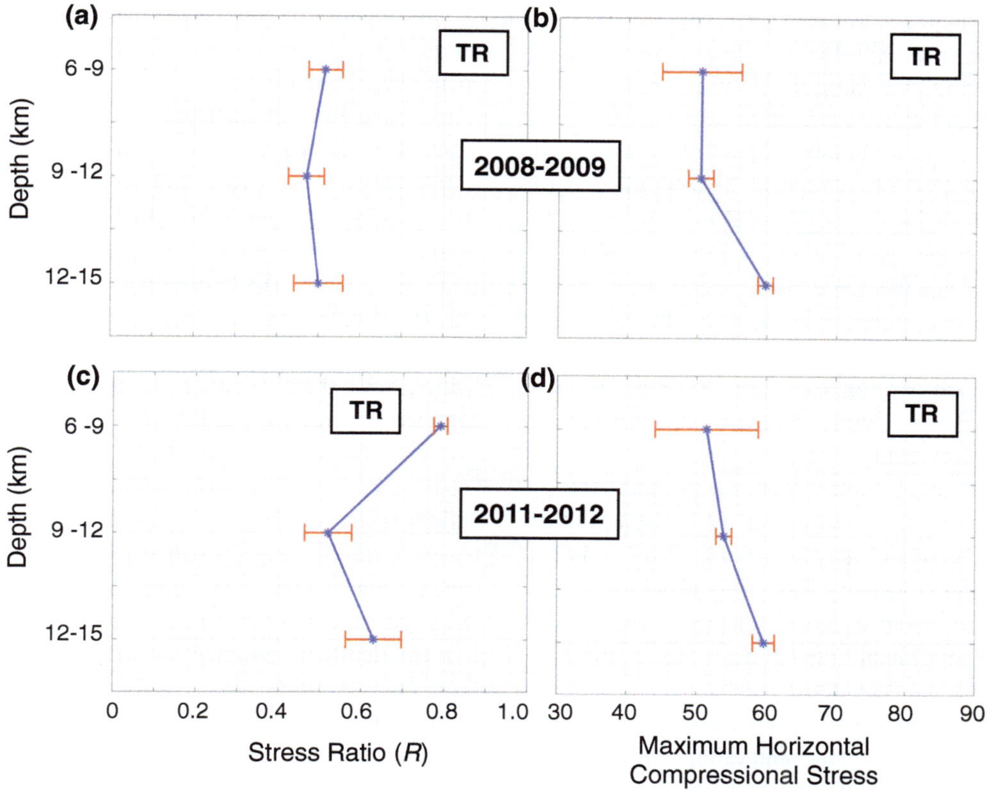

Figure 12
Variations of stress ratio (left) and S_{Hmax} (right) with depth in the TR area using data from 2 years before (top panels) and 2 year after (bottom panels) the EMC earthquake. Color circles and error bars denote the same as in Fig. 9

aseismic slip northwest of the EMC rupture (Ross et al. 2017b) and along the SJFZ (Inbal et al. 2017).

5. Discussion

This study aims to characterize spatiotemporal variations of crustal stress and strain in the San Jacinto fault zone region and to clarify the main sources of stress–strain variations. We examine data from the entire SJFZ and look in detail at three focus areas (CH, HS, and TR) associated with abundant seismicity (Fig. 1). The derived quantities include stress inversion results along the main fault and depth sections in the focus areas, direct analysis of coseismic strain (based on earthquake potency tensors), and calculations of frequency–size event statistics. In the following, we discuss some of the most important results and elaborate on their implications.

The finer selection of grid cell distribution, and improved data preprocessing (Martínez-Garzón et al. 2016a) lead to higher resolution of our results on stress parameters in the SJFZ than the earlier study of Yang and Hauksson (2013) in the same region. The results from Yang and Hauksson (2013) show primarily strike-slip regime along the SJFZ, with a few grid cells displaying tensional stress regime adjacent to the CH area. Our results indicate strike-slip as the main faulting type in the focus study areas, superposed with some variations. The stress ratios show a transtensional stress field near the CH area, which can be associated with an extensional region adjacent to the San Andreas and San Jacinto faults intersection. The stress ratios in the HS area indicate a transpressional stress environment, which may be the result of higher topography involving the San Jacinto Mountains in the northern section of the HS area closer to the Buck Ridge fault (Fialko et al. 2005).

The TR area is in general consistent with a strike-slip stress regime (Fig. 3; Table 1). The recovered S_{Hmax} orientation is mostly NNE–SSW (Fig. 2), in agreement with Yang and Hauksson (2013).

The apparent coefficient of friction is estimated for each of the analyzed regions, as a by-product of the stress tensor inversion. The CH area displays the smallest average apparent coefficient of friction ($\sim$ 0.45) among the three study areas. This region also displays the largest angle between the maximum horizontal compressional stress and the main fault trace. These observations suggest that the CH area may have a weaker crust than the other two focus study areas (Table 1).

The employed stress inversion methodology selects the fault plane best oriented for failure out of each focal mechanism (Vavryčuk 2014). Focal mechanism heterogeneity has been observed to be dominated by spatial or structural factors rather than by time or event magnitude (Bailey et al. 2010). Our analysis of the updated focal mechanisms and stress field with depth indicates increased proportion of dip-slip faulting with depth, suggesting larger focal mechanism heterogeneity in the deeper sections of the fault zones for all the focus areas (Fig. 5). The increased diversity with depth may be related to the changes of structural properties (e.g., Allam et al. 2014; Qiu et al. 2017; Qin et al. 2018), variations of rheology such as weak crustal layers (e.g., Bokelmann and Beroza 2000), or localized creep below the fault (e.g., Wdowinski 2009; Cooke and Beyer 2018).

The relation between the dominant fault strike direction, N55°W, and the asymmetry in the distribution of focal mechanism orientations is considered to correlate with the direction of the fault strike relative to the plate loading (Bailey et al. 2010). Based on this, when the plate motion direction deviates from a main fault strand, it can lead to second-order normal faulting (Bailey et al. 2010). Increased number of dip-slip faulting through depth includes a higher proportion of smaller normal events according to the *b* value analysis (Tables S1–S3). This can result from variations of the plate motion direction relative to the main SJFZ fault strike, and may hold for the CH area having significant larger *b* values associated with normal events compared with strike-slip events.

As mentioned, the three focus study areas show similar trends with depth for the estimated stress parameters, with different degrees of variation. The stress inversion results show a clear clockwise rotation of S_{Hmax} with increasing depth, increasing the angle between the S_{Hmax} and the main fault trace at the surface. At the same time, the stress regime appears to change with depth from more pure strike-slip to oblique (transtensional or transpressional) faulting depending on the region. All the observed rotations appear to be deeper than $\sim$ 9 km. Stress ratios have their minimum value at $\sim$ 9 km depth and slightly increase as one moves deeper (Fig. 9).

The average locking depth along the SJFZ is estimated from geodetic data and seismogenic thicknesses to be 11 ± 2 km and 17 ± 3 km, respectively. These provide two independent reference values associated with mechanical changes through depth (Wdowinski 2009). The $\sim$ 10 to 17 km depth is assumed to be the transition zone between the brittle crust to the ductile lower crust (Wdowinski 2009; Inbal et al. 2017). This depth section includes high seismic activity, but the elastic strain may be released both seismically and aseismically (Inbal et al. 2017). Proposed explanations for the mechanical changes in the transition zone include lower frictional strength at depth or higher pore pressure leading to a reduction in the effective normal stress acting on the fault (Wdowinski 2009). Further, rheology variations with depth may lead to decreasing horizontal shear traction at the brittle–ductile transition zone and related variations of focal mechanisms (Bokelmann and Beroza 2000). The stress field rotations and stress ratio increases resolved in this study appear to occur within the transition zone at $\sim$ 10 to 17 km depth, coinciding with the increased seismicity rates and mechanism heterogeneity.

To invert for the stress field orientation, we utilize focal mechanisms with qualities A to D representing fault plane uncertainties in the range 5–55° with a median value of 31°. To test the robustness of the results with respect to the focal mechanism uncertainties, the stress inversions were repeated selecting only focal mechanisms of quality A and B having $< 35°$ of fault plane uncertainty. The corresponding results confirm that the reported stress rotations and main changes with depth are robust and are not affected strongly by the uncertainty of the input data. However, due to the reduced number of available focal mechanisms, the resolution is lowered.

The coseismic strain distributions with depth for all focus areas show patterns highly consistent with the inverted stress field. The differences in principal stress and strain plunges, and maximum horizontal compressional stress and strain, are less than 10° in all cases. The difference in the stress ratio, R, and the analogous strain parameter, D, is less than 0.1, and the Kagan angle between the stress and strain tensors is less than 20° at all depth sections (Fig. 10), implying that the stress and strain orientations are nearly parallel in the study areas. One of the stress inversion assumptions is that the fault slip during an earthquake occurs in the direction of the shear traction vector (Wallace 1951; Bott 1959). This implies that, at the time of occurrence of an earthquake, the strain and stress fields are parallel. The general agreement between the obtained stress and strain results validates this assumption, and indicates that the velocity structure at the depth of the used seismicity is on the average approximately isotropic (e.g., Twiss and Unruh 1998). This also provides some validation of the stress inversion results, since the strain results reflect more directly the earthquake data with no additional assumptions.

Small secondary faults near the major faults in Southern California do not accommodate large displacements, but they are important for non-fault-parallel displacements related to transpressional and transtensional areas (Ghisetti 2000). Seismic sequences in Southern California have indicated slip heterogeneity and occurrence of dip-slip faulting on planes parallel to the major strike-slip fault (Ghisetti 2000). This is in accordance with our observations of increased dip-slip faulting and increased heterogeneity with depth in the SJFZ. Ghisetti (2000) found that slip along a large proportion of small-scale faults is compatible with S_{Hmax}, and inferred that the estimated stress and strain tensors are oriented coaxially based on shallow $M < 5$ earthquakes. Our results extend these conclusions to deeper sections of the SJFZ. The large events occur on the main faults and have mainly right-lateral strike-slip mechanisms, while the numerous off-fault smaller events with mixed faulting types accommodate geometrical variations of the faults in relation to the regional stress regime. These results are also consistent with observations of Bailey et al. (2010) and Cheng et al. (2018).

6. *Conclusions*

The analyses performed in this study allow us to obtain information on (1) the distribution of background stress field and coseismic strain in SJFZ regions with high resolution and (2) variations in the state of stress from shallow crust to the brittle–ductile transition region. Three focused areas, CH, HS, and TR, are analyzed in greater detail and compared. The main findings can be summarized as follows:

1. The b value analysis provides some constraints on the brittle–ductile transition depth in the analyzed areas. In all regions, the brittle crust extends down to 9–12 km and the seismicity in that depth section displays mainly strike-slip events. A change towards dip-slip is observed through the brittle–ductile transition zone to 15–18 km. This depth appears to be shallower at the TR than at the other two regions.
2. At least for the CH and TR areas, a significant clockwise rotation of S_{Hmax} with depth is observed. The angle between S_{Hmax} and the fault trace at the surface appears to increase, especially at the depth interval of the inferred brittle–ductile transition (10–17 km). This may indicate a progressive weakening of the fault with depth. This may also reflect partially changing dip of the main fault in the region (e.g., Ross et al. 2017a).
3. The distribution of stress and coseismic strain parameters appear to be largely consistent. This provides some validation of the stress inversion results with the direct input data and indicates that the velocity structure is on average approximately isotropic.
4. The stress parameters before and after the El Mayor–Cucapah earthquake do not show significant rotation of S_{Hmax}. The stress ratios move toward transpressional regime after the EMC

event in the TR area. However, an event with magnitude > 5.4 occurred in the TR area in 2010 adjacent to the maximum variations of the stress ratio. This event could have potentially affected the observed stress ratio (e.g., Martínez-Garzón et al. 2016a).

Acknowledgements

The study was supported by the Southern California Earthquake Center (based on NSF Cooperative Agreement EAR-1600087 and USGS Cooperative Agreement G17AC00047). P.M.-G. acknowledges funding from the Helmholtz Association under the Helmholtz Postdoc Programme and the Helmholtz Young Investigators Group SAIDAN. The manuscript benefited from useful comments by two anonymous referees.

Publisher's Note Springer Nature remains neutral with regard to jurisdictional claims in published maps and institutional affiliations.

References

Allam, A. A., Ben-Zion, Y., Kurzon, I., & Vernon, F. (2014). Seismic velocity structure in the Hot Springs and Trifurcation areas of the San Jacinto fault zone, California, from double-difference tomography. *Geophysical Journal International, 198*(2), 978–999.

Amelung, F., & King, G. (1997). Large-scale tectonic deformation inferred from small earthquakes. *Nature, 386*(6626), 702.

Amitrano, D. (2003). Brittle-ductile transition and associated seismicity: Experimental and numerical studies and relationship with the b value. *Journal of Geophysical Research Solid Earth, 108,* B1.

Bailey, I. W., Becker, T. W., & Ben-Zion, Y. (2009). Patterns of co-seismic strain computed from southern California focal mechanisms. *Geophysical Journal International, 177*(3), 1015–1036.

Bailey, I. W., Ben-Zion, Y., Becker, T. W., & Holschneider, M. (2010). Quantifying focal mechanism heterogeneity for fault zones in central and southern California. *Geophysical Journal International, 183*(1), 433–450.

Ben-Zion, Y. (2003). Appendix 2: Key Formulas in Earthquake Seismology. In W. H. K. Lee, H. Kanamori, P. C. Jennings, C. Kisslinger (Eds.) *International handbook of earthquake and engineering seismology, Part B,* 1857–1875.

Ben-Zion, Y. (2008). Collective behavior of earthquakes and faults: Continuum–discrete transitions, progressive evolutionary changes, and different dynamic regimes. *Reviews of Geophysics, 46*(4), RG4006. https://doi.org/10.1029/2008RG000260.

Bokelmann, G. H., & Beroza, G. C. (2000). Depth-dependent earthquake focal mechanism orientation: Evidence for a weak zone in the lower crust. *Journal of Geophysical Research Solid Earth, 105*(B9), 21683–21695.

Bott, M. H. P. (1959). The mechanics of oblique slip faulting. *Geological Magazine, 96*(02), 109–117.

Cheng, Y., Ross, Z. E., & Ben-Zion, Y. (2018). Diverse volumetric faulting patterns in the San Jacinto fault zone. *Journal of Geophysical Research, 123,* 5068–5081. https://doi.org/10.1029/2017JB015408.

Cooke, M. L., & Beyer, J. L. (2018). Off-fault focal mechanisms not representative of interseismic fault loading suggest deep creep on the Northern San Jacinto Fault. *Geophysical Research Letters, 45*(17), 8976–8984.

Doser, D. I., & Kanamori, H. (1986). Depth of seismicity in the Imperial Valley region (1977–1983) and its relationship to heat flow, crustal structure and the October 15, 1979, earthquake. *Journal of Geophysical Research Solid Earth, 91*(B1), 675–688.

Fialko, Y. (2006). Interseismic strain accumulation and the earthquake potential on the southern San Andreas fault system. *Nature, 441*(7096), 968–971.

Fialko, Y., Rivera, L., & Kanamori, H. (2005). Estimate of differential stress in the upper crust from variations in topography and strike along the San Andreas fault. *Geophysical Journal International, 160*(2), 527–532.

Frohlich, C. (1992). Triangle diagrams: Ternary graphs to display similarity and diversity of earthquake focal mechanisms. *Physics of the Earth and Planetary Interiors, 75*(1–3), 193–198.

Ghisetti, F. (2000). Slip partitioning and deformation cycles close to major faults in southern California: Evidence from small-scale faults. *Tectonics, 19*(1), 25–43.

Hardebeck, J. L., & Hauksson, E. (2001). Crustal stress field in southern California and its implications for fault mechanics. *Journal of Geophysical Research B, 106*(B10), 21859–21882.

Hardebeck, J. L., & Michael, A. J. (2006). Damped regional-scale stress inversions: Methodology and examples for southern California and the Coalinga aftershock sequence. *Journal of Geophysical Research Solid Earth, 111,* (B11).

Hartigan, J. A., & Wong, M. A. (1979). Algorithm AS 136: A k-means clustering algorithm. *Journal of the Royal Statistical Society Series C (Applied Statistics), 28*(1), 100–108.

Hauksson, E., Stock, J., Hutton, K., Yang, W., Vidal-Villegas, J. A., & Kanamori, H. (2011). The 2010 M w 7.2 El Mayor–Cucapah Earthquake Sequence, Baja California, Mexico and Southernmost California, USA: Active seismotectonics along the Mexican Pacific Margin. *Pure and Applied Geophysics, 168*(8–9), 1255–1277.

Hauksson, E., Yang, W., & Shearer, P. M. (2012). Waveform relocated earthquake catalog for southern California (1981 to June 2011). *Bulletin of the Seismological Society of America, 102*(5), 2239–2244.

Inbal, A., Ampuero, J. P., & Avouac, J. P. (2017). Locally and remotely triggered aseismic slip on the central San Jacinto Fault near Anza, CA, from joint inversion of seismicity and strainmeter data. *Journal of Geophysical Research Solid Earth, 122*(4), 3033–3061.

Jennings, C. W. (1994). Fault activity map of California and adjacent areas with location and ages of recent volcanic eruptions: California Division of Mines and Geology. *California Geologic Data Map Series, map,* (6).

Jones, L. M. (1988). Focal mechanisms and the state of stress on the San Andreas fault in southern California. *Journal of Geophysical Research Solid Earth, 93*(B8), 8869–8891.

Lund, B., & Townend, J. (2007). Calculating horizontal stress orientations with full or partial knowledge of the tectonic stress tensor. *Geophysical Journal International, 170*(3), 1328–1335.

Martínez-Garzón, P., Ben-Zion, Y., Abolfathian, N., Kwiatek, G., & Bohnhoff, M. (2016a). A refined methodology for stress inversions of earthquake focal mechanisms. *Journal of Geophysical Research, 121,* 8666–8687. https://doi.org/10.1002/2016JB013493.

Martínez-Garzón, P., Kwiatek, G., Bohnhoff, M., & Dresen, G. (2016b). Impact of fluid injection on fracture reactivation at The Geysers geothermal field. *Journal of Geophysical Research Solid Earth, 121*(10), 7432–7449.

Martínez-Garzón, P., Kwiatek, G., Ickrath, M., & Bohnhoff, M. (2014). MSATSI: A MATLAB package for stress inversion combining solid classic methodology, a new simplified user-handling, and a visualization tool. *Seismological Research Letters, 85*(4), 896–904.

McKenzie, D. P. (1969). The relation between fault plane solutions for earthquakes and the directions of the principal stresses. *Bulletin of the Seismological Society of America, 59*(2), 591–601.

Meng, X., & Peng, Z. (2014). Seismicity rate changes in the Salton Sea Geothermal Field and the San Jacinto Fault Zone after the 2010 *M*w 7.2 El Mayor–Cucapah earthquake. *Geophysical Journal International, 197*(3), 1750–1762.

Michael, A. J. (1984). Determination of stress from slip data: Faults and folds. *Journal of Geophysical Research Solid Earth, 89*(B13), 11517–11526.

Michael, A. J. (1987). Use of focal mechanisms to determine stress: A control study. *Journal of Geophysical Research Solid Earth, 92*(B1), 357–368.

Onderdonk, N., Rockwell, T., McGill, S., & Marliyani, G. (2013). Evidence for seven surface ruptures in the past 1600 years on the Claremont fault at Mystic Lake, northern San Jacinto fault zone, California. *Bulletin of the Seismological Society of America, 103*(1), 519–541.

Petersen, M. D., & Wesnousky, S. G. (1994). Fault slip rates and earthquake histories for active faults in southern California.

Qin, L., Ben-Zion, Y., Qiu, H., Share, P.-E., Ross, Z. E., & Vernon, F. L. (2018). Internal structure of the San Jacinto fault zone in the trifurcation area southeast of Anza, California, from data of dense seismic arrays. *Geophysical Journal International, 213,* 98–114. https://doi.org/10.1093/gji/ggx540.

Qiu, H., Ben-Zion, Y., Ross, Z. E., Share, P.-E., & Vernon, F. L. (2017). Internal structure of the San Jacinto fault zone at Jackass Flat from data recorded by a dense linear array. *Geophysical Journal International, 209,* 1369–1388. https://doi.org/10.1093/gji/ggx096.

Rockwell, T. K., Dawson, T. E., Ben-Horin, J. Y., & Seitz, G. (2015). A 21-event, 4,000-year history of surface ruptures in the Anza seismic gap, San Jacinto Fault, and implications for long-term earthquake production on a major plate boundary fault. *Pure and Applied Geophysics, 172*(5), 1143–1165.

Ross, Z. E., & Ben-Zion, Y. (2013). Spatio-temporal variations of double-couple aftershock mechanisms and possible volumetric earthquake strain. *Journal of Geophysical Research, 118,* 2347–2355. https://doi.org/10.1002/jgrb.50202.

Ross, Z. E., Hauksson, E., & Ben-Zion, Y. (2017a). Abundant off-fault seismicity and orthogonal structures in the San Jacinto fault zone. *Science Advances, 3*(3), e1601946.

Ross, Z. E., Rollins, C., Cochran, E. S., Hauksson, E., Avouac, J.-P., & Ben-Zion, Y. (2017b). Aftershocks driven by afterslip and fluid pressure sweeping through a fault-fracture mesh. *Geophysical Research Letters, 44,* 8260–8267. https://doi.org/10.1002/2017GL074634.

Salisbury, J. B., Rockwell, T. K., Middleton, T. J., & Hudnut, K. W. (2012). LiDAR and field observations of slip distribution for the most recent surface ruptures along the central San Jacinto fault. *Bulletin of the Seismological Society of America, 102*(2), 598–619.

Scholz, C. H. (2002). *The mechanics of earthquakes and faulting.* Cambridge: Cambridge University Press.

Scholz, C. H. (2015). On the stress dependence of the earthquake b value. *Geophysical Research Letters, 42*(5), 1399–1402.

Schorlemmer, D., Wiemer, S., & Wyss, M. (2005). Variations in earthquake-size distribution across different stress regimes. *Nature, 437*(7058), 539.

Seber, G. A. F. (1984). Multivariate analysis of variance and covariance. *Multivariate Observations, 1984,* 433–495.

Spada, M., Tormann, T., Wiemer, S., & Enescu, B. (2013). Generic dependence of the frequency size distribution of earthquakes on depth and its relation to the strength profile of the crust. *Geophysical Research Letters, 40*(4), 709–714.

Townend, J., & Zoback, M. D. (2001). Implications of earthquake focal mechanisms for the frictional strength of the San Andreas fault system. *Geological Society London Special Publications, 186*(1), 13–21.

Townend, J., & Zoback, M. D. (2004). Regional tectonic stress near the San Andreas fault in central and southern California. *Geophysical Research Letters, 31,* 15.

Twiss, R. J., & Unruh, J. R. (1998). Analysis of fault slip inversions: Do they constrain stress or strain rate? *Journal of Geophysical Research Solid Earth, 103*(B6), 12205–12222.

Vavryčuk, V. (2011). Principal earthquakes: Theory and observations from the 2008 West Bohemia swarm. *Earth and Planetary Science Letters, 305*(3–4), 290–296.

Vavryčuk, V. (2014). Iterative joint inversion for stress and fault orientations from focal mechanisms. *Geophysical Journal International, 199*(1), 69–77.

Vavryčuk, V. (2015). Earthquake mechanisms and stress field. In M. Beer, et al. (Eds.), *Encyclopedia of earthquake engineering* (pp. 728–746). Berlin: Springer.

Wallace, R. E. (1951). Geometry of shearing stress and relation to faulting. *The Journal of Geology, 59*(2), 118–130.

Wdowinski, S. (2009). Deep creep as a cause for the excess seismicity along the San Jacinto fault. *Nature Geoscience, 2*(12), 882–885.

Wesnousky, S. G. (1986). Earthquakes, Quaternary faults, and seismic hazard in California. *Journal of Geophysical Research Solid Earth, 91*(B12), 12587–12631.

Wesson, R. L., & Boyd, O. S. (2007). Stress before and after the 2002 Denali fault earthquake. *Geophysical Research Letters, 34,* 7.

Wiemer, S., & Wyss, M. (2000). Minimum magnitude of completeness in earthquake catalogs: Examples from Alaska, the western United States, and Japan. *Bulletin of the Seismological Society of America, 90*(4), 859–869.

Yang, W., & Hauksson, E. (2013). The tectonic crustal stress field and style of faulting along the Pacific North America Plate boundary in Southern California. *Geophysical Journal International, 194*(1), 100–117.

Yang, W., Hauksson, E., & Shearer, P. M. (2012). Computing a large refined catalog of focal mechanisms for southern California (1981–2010): Temporal stability of the style of faulting. *Bulletin of the Seismological Society of America, 102*(3), 1179–1194.

Zaliapin, I., & Ben-Zion, Y. (2013). Earthquake clusters in southern California I: Identification and stability. *Journal of Geophysical Research Solid Earth, 118*(6), 2847–2864.

Zoback, M. L. (1992). First- and second-order patterns of stress in the lithosphere: The World Stress Map Project. *Journal of Geophysical Research Solid Earth, 97*(B8), 11703–11728.

(Received June 2, 2018, revised November 2, 2018, accepted November 17, 2018, Published online November 27, 2018)

Pure Appl. Geophys. 176 (2019), 1169–1191

https://doi.org/10.1007/s00024-018-1988-5

Pure and Applied Geophysics

Structural Properties of the San Jacinto Fault Zone at Blackburn Saddle from Seismic Data of a Dense Linear Array

Pieter-Ewald Share,[1] Amir A. Allam,[2] Yehuda Ben-Zion,[1] Fan-Chi Lin,[2] and Frank L. Vernon[3]

Abstract—We image the San Jacinto fault zone at Blackburn Saddle using earthquake waveforms recorded by a ∼ 2-km across-fault linear array with 108 three-component sensors separated by ∼ 10–30 m. The length and spatiotemporal sampling of the array allow us to derive high-resolution information on the internal fault zone structure with spatial extent that can be merged with regional tomography models. Across-fault variations in polarization, amplitude, and arrival time of teleseismic *P* waves indicate abrupt changes in subsurface structure near the surface trace of the fault (sensor BS55) and ∼ 270 m to the northeast (sensor BS34). Analysis of fault zone head waves from local events reveals the existence of a deep bimaterial interface that extends from the array to at least 50 km southeast and has a section with > 10% velocity contrast. This analysis also corroborates the teleseismic results and indicates a broad damage zone primarily northeast of the fault bounded by a shallow bimaterial interface near BS34 that merges with the deep interface. Detection and waveform inversions of Love-type fault zone trapped waves generated by local events indicate a trapping structure within the broader damage zone with width of ∼ 150 m, velocity reduction of ∼ 55% from the surrounding rock and depth extent of ∼ 2 km. The performed analyses provide consistent results on the subsurface location of the main seismogenic fault and properties of a major bimaterial interface and damage structure. The imaged fault zone properties are consistent with preferred propagation direction of earthquake ruptures in the area to the northwest.

Key words: Earthquake dynamics, San Jacinto fault zone, high-resolution imaging, seismic array, body waves, interface waves, guided waves.

Electronic supplementary material The online version of this article (https://doi.org/10.1007/s00024-018-1988-5) contains supplementary material, which is available to authorized users.

[1] Department of Earth Sciences, University of Southern California, Los Angeles, CA 90089, USA. E-mail: pshare@usc.edu

[2] Department of Geology and Geophysics, The University of Utah, Salt Lake City, UT 84112, USA.

[3] Scripps Institution of Oceanography, University of California San Diego, La Jolla, CA 92093, USA.

1. Introduction

Clarifying the internal velocity structure of fault zones is important for many seismological and fault mechanics studies. A velocity contrast across a fault can affect the directivity, velocity and crack versus pulse style of earthquake ruptures (e.g., Weertman 1980; Andrews and Ben-Zion 1997; Shlomai and Fineberg 2016), as well as derived earthquake locations and focal mechanisms (e.g., Oppenheimer et al. 1988; McGuire and Ben-Zion 2005). Properties of the core damage zone around the fault contain information on processes and stress conditions operating during earthquake ruptures (e.g., Sibson 1989; Xu et al. 2012). Asymmetric damage zone with respect to the main fault may indicate a preferred directivity of large earthquake ruptures on that fault section (e.g., Ben-Zion and Shi 2005; Dor et al. 2006; DeDontney et al. 2011). Along-strike (dis)continuity of bimaterial fault interfaces can be relevant for the likely extent of earthquake ruptures (e.g., Dor et al. 2008; Share and Ben-Zion 2016).

A set of seismic arrays that cross the San Jacinto fault zone (SJFZ) in southern California at various locations (Fig. 1a) provides unique opportunities for detailed imaging of bimaterial fault interfaces and damage zones, in relation to geological (e.g., Sharp 1967; Rockwell et al. 2015) and regional tomographic (e.g., Allam and Ben-Zion 2012; Barak et al. 2015; Fang et al. 2016) results. Previous studies analyzed data recorded by different linear arrays with apertures of ∼ 200–450 m (e.g., Yang et al. 2014; Qiu et al. 2017) at various locations, and a dense rectangular array with a linear dimension of ∼ 600 m (Ben-Zion et al. 2015; Hillers et al. 2016; Qin et al. 2018) at the SGB site. In particular, Share et al.

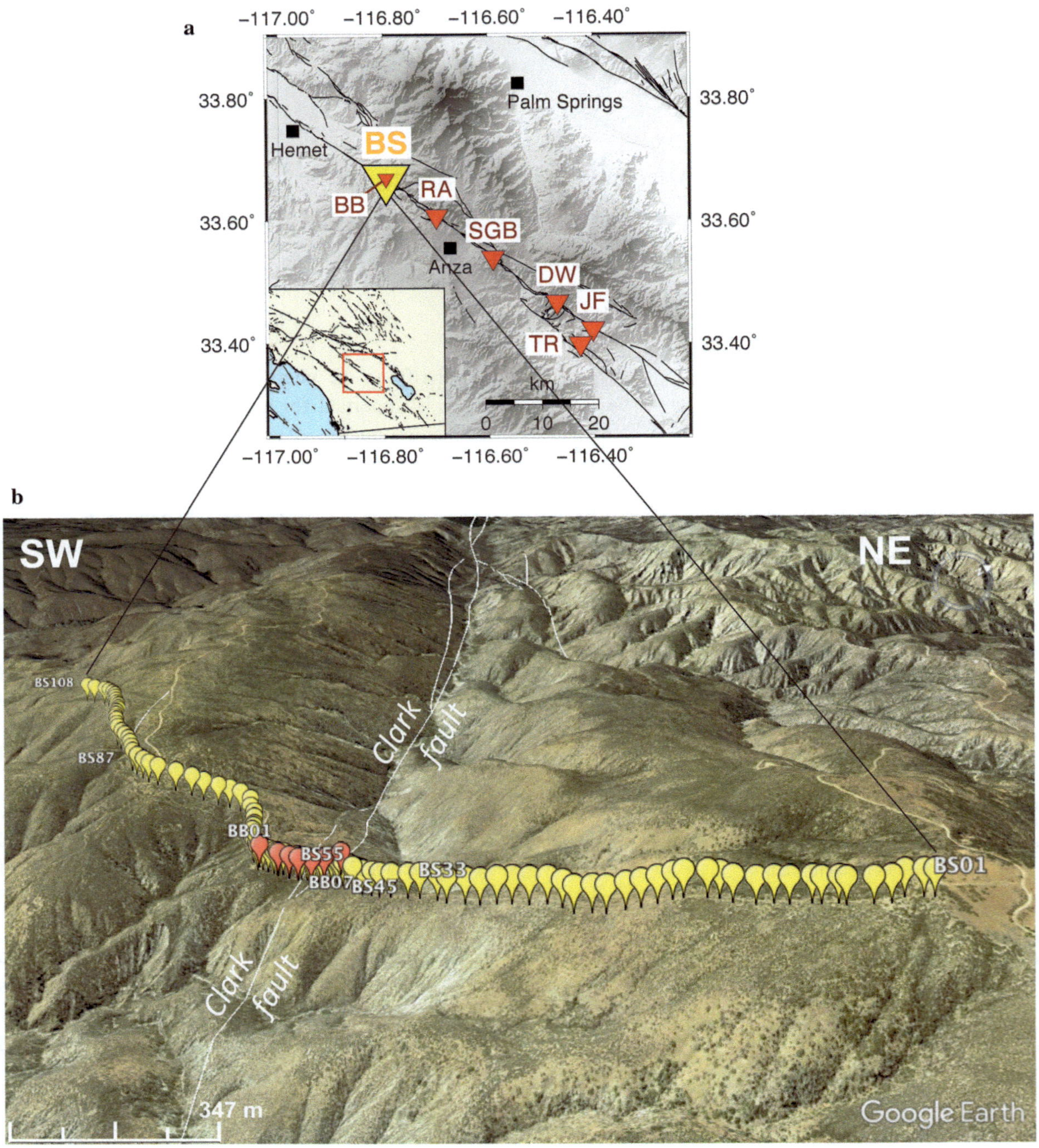

Figure 1
Dense linear arrays along the San Jacinto fault zone (SJFZ). **a** Locations of the Blackburn Saddle array analyzed in the present study (BS, large yellow triangle) and other arrays (BB, RA, SGB, DW, JF and TR, small red triangles) installed along the SJFZ. The towns of Hemet, Anza and Palm Springs (black squares) are shown for reference. **b** 108 nodes of the BS array (yellow icons) crossing the Clark fault (CF) surface trace. For reference, locations of the 7 BB stations are also shown (red icons)

(2017) analyzed data from local earthquakes and teleseismic events recorded by the BB linear array across the Clark branch (CF) of the SJFZ at Blackburn Saddle northwest of Anza, consisting of 7 sensors with ~ 200-m aperture. The site ruptured during two large earthquakes in the last 250 years, the M 7.2–7.5, 1800 (Salisbury et al. 2012) and the M 6.8, 1918 (Sanders and Kanamori 1984) events. The results indicated the presence of a > 100-m-wide asymmetric damage zone (most damage on northeast side of the CF), which terminates at a local bimaterial interface northeast of the array that is connected to a regional bimaterial interface at depth. Due to the short aperture of the array, the spatial extent of the damage zone, which can be wider than 1 km (e.g., Cochran et al. 2009), was not well constrained. Also, the location of the bimaterial interface at the surface was not observed but inferred from comparison with data recorded by stations of the regional seismic network.

To provide more complete information on the inner structure of the SJFZ at Blackburn Saddle over a much larger area, a new array (BS) extending ~ 1 km on each side of the CF was deployed in late 2015. The BS array had 108 densely spaced 3-component Zland nodes covering and extending well beyond the locations of the previous short BB array (Fig. 1b). In the present study, we analyze data recorded by this first-of-its-kind 3-component nodal array to derive a high-resolution structural model of the fault zone over a spatial extent that can be merged with the tomographic images in the region. The analyzed data include P and S waveforms generated by local earthquakes and P waveforms of teleseismic events. The employed seismic phases are sensitive to bimaterial interfaces, damage zone structure and the surrounding rocks, and can better be distinguished and analyzed using the large-aperture BS array spanning all these structures. Variations in teleseismic arrivals are used to determine large-scale structural changes across the fault zone including the extent of a broad damage zone at the site. Local arrivals are analyzed for the presence of waveform changes across the fault, fault zone head waves (FZHW) and fault zone trapped waves (FZTW). The fault zone phases are used to image properties of the fault bimaterial interface and the core inner damage zone that acts as a seismic trapping structure.

In the next section, we provide more details on the array stations and data. The analysis techniques and results are described in Sects. 3 and 4. Section 3 examines spatial variations of polarizations, amplitudes and delay times of teleseismic data across the array. Section 4 includes analysis of waveform changes across the array and analyses of FZHW and FZTW based on local earthquake waveforms. The results are summarized and discussed in relation to other studies of fault zones and earthquake physics in the final Sect. 5.

2. *Instrumentation and Data*

The BS array comprises 108 three-component 5-Hz geophones installed along an approximately fault-perpendicular line crossing the CF surface trace with 26 instruments placed off the main line. The array recorded data continuously at 1000 Hz sampling starting on 21 November 2015 for 35 days. Station BS55 was nearest to the CF trace and stations BS1 and BS108 were at the northeastern and southwestern ends, respectively (Fig. 1b). Along the line, stations were 10 m apart in a 400-m-wide zone centered on the surface trace and spaced ~ 30 m to the northeast and southwest of that zone. Only data from stations BS1 to BS108 were used in this study given our focus on across-fault structural variations. The instrument response was not removed from the waveforms as Zland nodes reliably record phases from ~ 0.1 Hz to well above local P wave bandwidth (Ringler et al. 2018). The slight reduction in amplitude of lower frequency teleseismic arrivals introduced by the instrument response does not influence our analysis results in Sect. 3, because we are interested in relative changes in waveform across the array and this amplitude reduction is equal for all stations.

The earthquake catalog of the Southern California Earthquake Data Center (SCEDC 2013) was used to extract P waveforms for all $M > 5$ teleseismic earthquakes (within 30°–100°) deeper than 500 km that occurred during the installation period. Four earthquakes matched these criteria but two of them occurred within 60 km and 5 min of each other, so the second of these was not considered. Waveforms for the 3 remaining events (Fig. 2a; Table 1) were extracted from continuous BS recordings using the P wave arrival time at the BS55 location at sea level predicted with TauP (Crotwell et al. 1999). After extraction, the mean and trend were removed and a

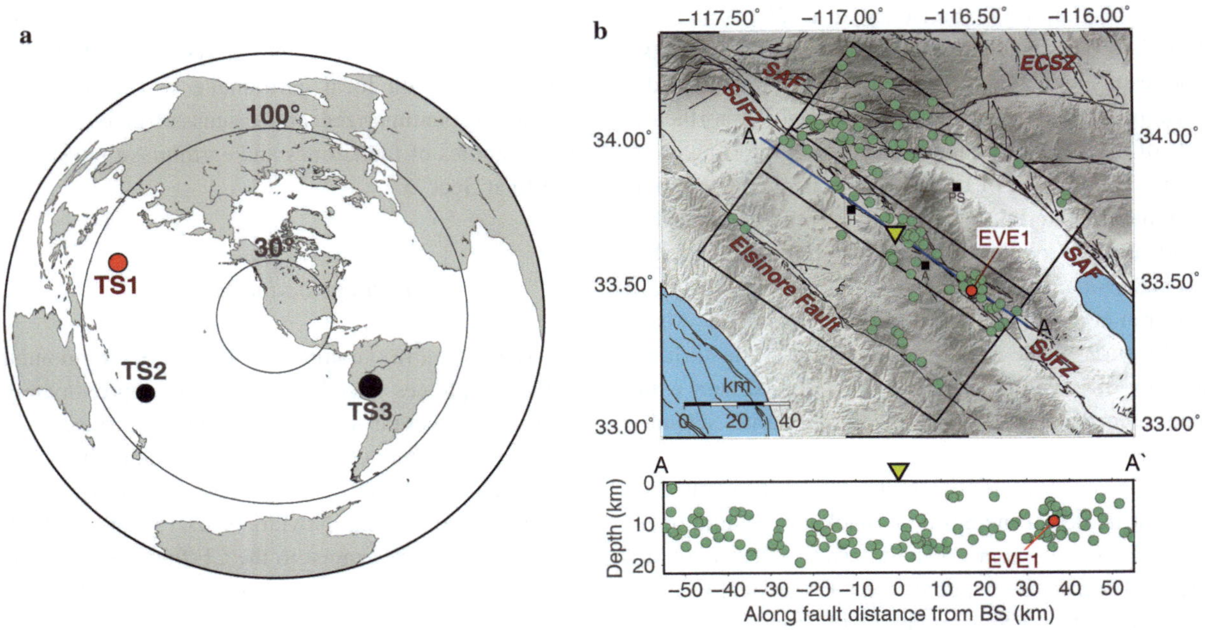

Figure 2

Data employed in the present study. **a** Locations of the 3 $M > 5$ teleseismic earthquakes used (TS1-TS3). Results of data analysis from an example event (TS1) are shown in Fig. 3. Waveforms from events TS2 and TS3 are given in Fig. S1. **b** Locations of $M_W > 1$ earthquakes (143 in total, circles) within a 110 km by 100 km region (big black box) centered on the BS array (yellow triangle) and aligned with the CF surface trace. The small black rectangle highlights earthquakes within 10 km of the CF. Waveforms from an example event (EVE1, red circle) are shown in Fig. 5a. H = Hemet; A = Anza; PS = Palm Springs; SAF = San Andreas fault; ECSZ = Eastern California Shear Zone. A depth section of events projected along the profile A–A′ is plotted at the bottom

low-pass Butterworth filter with corner frequency 2 Hz was applied.

The Hauksson et al. (2012) relocated catalog for Southern California (extended to later years) was used to extract waveforms from local events. In total, 143 $M_W > 1$ earthquakes located within a 110 km by 100 km box centered on the BS array and aligned with the CF trace occurred during the installation period (Fig. 2b). For these events, we used waveforms 5 s before and 30 s after the origin times reported in the catalog. A band-pass Butterworth filter between 2 and 20 Hz was applied to these waveforms after the initial removal of any trend from the data. As illustrated in Fig. 3a and later plots of waveforms, the long BS array provides a far better coverage of the fault zone than the previously analyzed data recorded by the short BB array.

3. *Teleseismic Earthquake Analyses*

The early *P* waves from teleseismic events are analyzed for changes in polarization, amplitude and arrival times. Polarization and amplitude will change in response to topographic and structural changes that have length scales comparable to the long

Table 1

Catalog of the teleseismic earthquakes analyzed

Event	SCEDC ID	Origin time UTC	Latitude	Longitude	Depth (km)	Magnitude
TS1	37268101	2015/11/24, 13:21:35.800	18.7906	145.3115	586.20	6.00
TS2	37268269	2015/12/06, 17:09:28.800	− 18.1903	− 178.6798	544.30	5.80
TS3	37268157	2015/11/24, 22:45:38.000	− 10.5484	− 70.9038	600.60	7.60

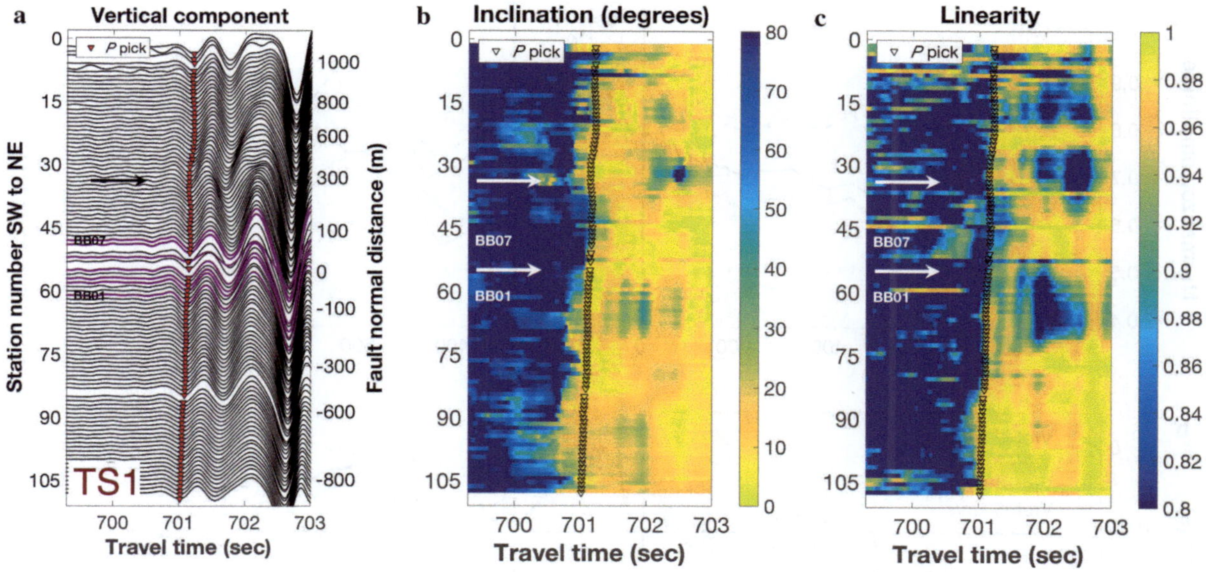

Figure 3
Analysis of waveforms recorded by the BS array from an example teleseismic event. **a** Early *P* waveforms (velocity seismograms) generated by event TS1. All seismograms are normalized by the same value. Noisy traces are removed. Red triangles denote manual *P* picks and the black arrow points to an abrupt change in waveform character across traces recorded near BS34. Waveforms recorded near BB array locations are colored purple with the southwestern most and northeastern most ones denoted by BB01 and BB07. **b**, **c** Inclination of the largest eigenvector and linearity calculated using a 0.75 s window. White arrows highlight abrupt changes in these parameters

wavelengths (> 1 km) of the teleseismic energy (Jepsen and Kennett 1990; Neuberg and Pointer 2000). Arrival times corrected for variations due to different propagation paths can be used to determine slowness changes at similar large length scales (e.g., Schmandt and Clayton 2013; Qiu et al. 2017). Here, such analyses are done to identify large-scale vertical (or near vertical) fault interfaces and regions of increased slowness potentially corresponding to fault damage zones.

3.1. Polarization and Amplitude Variations

For each of the 3 teleseismic events, we use the vertical component to manually pick the first coherent peaks (positive or negative) across the array and designate them *P* arrivals (*P* picks in Fig. 3a). We then visually inspect the waveforms 2 s after these picks for abrupt changes in character. As an example, for event TS1 (location in Fig. 2a), most rapid changes in arrival time and waveform are observed around station BS34 (270 m northeast of the CF trace, Fig. 3a). Similar changes in arrival time and waveform are found near the same station for the other 2 events (Fig. S1). A sharp change in arrival time and waveform character occurs near station BS55 (location of CF trace) for event TS3 (Fig. S1b).

Next, we apply polarization analysis (Jurkevics 1988) to all three components to quantify waveform changes recorded across the array. The analysis employs a moving time window of length 0.75 s (750 samples). For each window, the three components are combined in a 750 × 3 matrix and the covariance of that matrix is computed. The eigenvalues (λ_1, λ_2, λ_3) and eigenvectors ($\boldsymbol{u}_1$, $\boldsymbol{u}_2$, $\boldsymbol{u}_3$) of the covariance matrix give, respectively, the amplitudes and directions of the axes of the polarization ellipse. The largest eigenvalue λ_1 corresponds to the amplitude of maximum polarization with direction $\boldsymbol{u}_1$. For a *P* wave, direction $\boldsymbol{u}_1$ gives the inclination angle as

$$\mathrm{Inc} = \cos^{-1}(u_{11}). \tag{1}$$

From the eigenvalues, the degree of polarization/linearity can be calculated using

$$q = 1 - ((\lambda_2 + \lambda_3)/2\lambda_1), \tag{2}$$

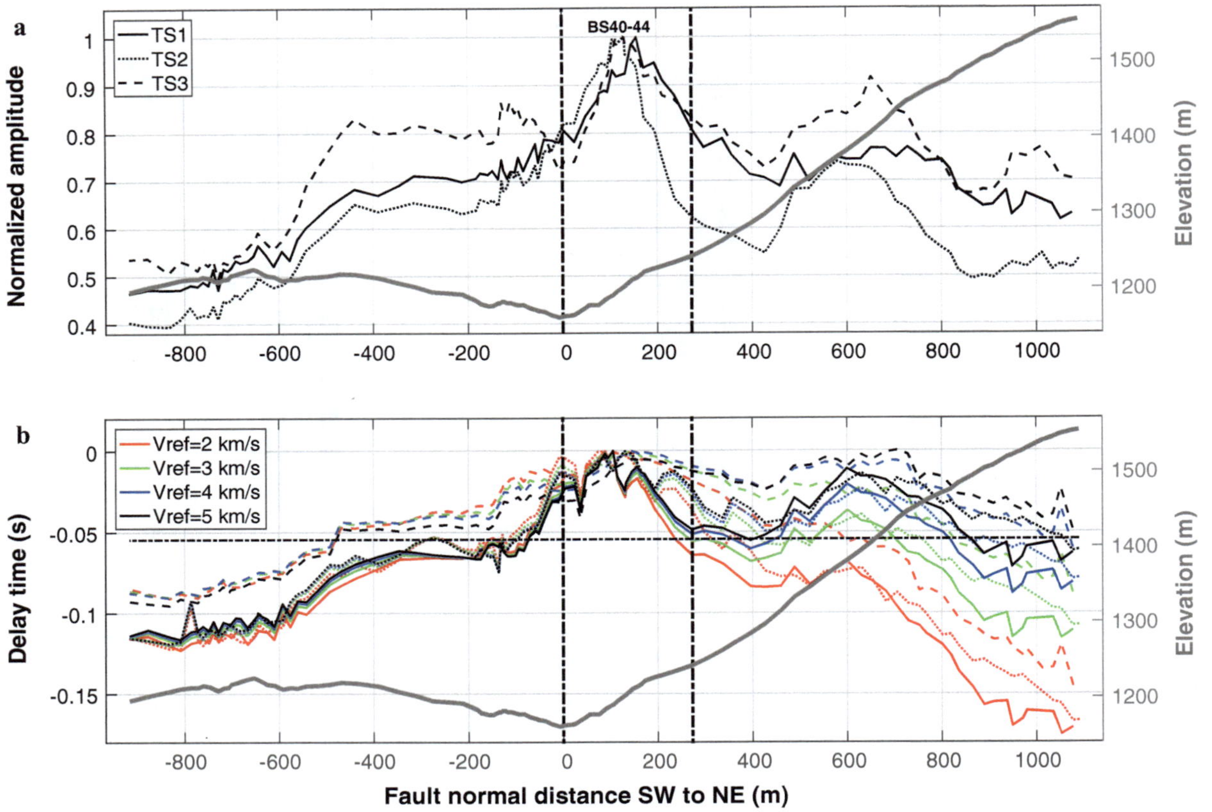

Figure 4

Amplitude and arrival time variations of teleseismic *P* waves across the array. **a** Maximum amplitudes within 1.5 s windows starting at the *P* picks for the 3 events (TS1—solid line; TS2—dotted line; TS3—dashed line, information in Table 1). Vertical black dot-dashed lines represent the locations of stations BS55 (left) and BS34 (right) and the horizontal black dot-dashed line depicts the median of the delay times shown. Gray line depicts changes in topography across the array. **b** Delay times for 3 events (same line styles as in **a**) and reference velocities of 2 km/s (red), 3 km/s (green), 4 km/s (blue) and 5 km/s (black) used during correction for topography

with q being 1 when the *P* wave is linearly polarized and 0 for perfectly uncorrelated motion. The inclination and linearity are calculated for each 0.75-s window using Eqs. (1) and (2) and are assigned to the time of the final sample in the moving window (time windows are trailing). Application to the example event TS1 indicates large changes in inclination (± 30°, Fig. 3b) and linearity (± 0.1, Fig. 3c) around station BS34. Large changes in inclination and linearity are also observed near BS55. Similar changes in inclination and linearity are detected at BS34 and BS55 for the other 2 events. Less prominent changes in inclination and linearity across the array (for example near BS20 and BS80) correspond to smaller scale structures in the area that are left unresolved.

The changes in arrival time and waveform character near BS34 correspond to lateral variations in large-scale subsurface structure rather than sudden changes in topography, since similar topographic gradient exists for stations BS1–55 (Fig. 4). The abrupt waveform change near station BS55 is correlated with the mapped surface trace of the CF at the site (Sharp 1967; Rockwell et al. 2015). Figure 4a displays the maximum amplitude within a 1.5-s window starting at each *P* pick for the 3 teleseismic events. A 1.5-s window is approximately equal to the largest period observed for the early *P* waves. The results show that stations BS40–44 (110–150 m northeast of the CF trace) consistently record the largest amplitude ground motion. The amplified motion may be caused by a core fault damage zone beneath stations BS40–44 (e.g., Ben-Zion and Aki 1990; Lewis et al. 2005). The smaller scale changes in the results of this section are likely related to secondary lateral variations of the fault zone structure and topography.

3.2. *Delay Time Analysis*

To focus on changes associated with the fault zone structure, arrival times are corrected in two steps. For a given event, the travel times from source to stations without taking elevation into account are predicted. Following this, travel times from sea level to the surface are computed using a homogeneous reference velocity. The total predicted arrival times are then subtracted from the picked times and the residual variations (delay times) between stations are used to infer subsurface slowness. Stations with larger delay after correction overlie regions of greater slowness.

Arrival times at sea level for different stations are approximated using TauP and a 1D velocity model that is a combination of the IASP91 model for the mantle (Kennett 1991) and the model of Hadley and Kanamori (1977) for the crust. Travel times from sea level to the surface are calculated using the ray parameter for each event obtained from TauP and reference velocities of 2 km/s, 3 km/s, 4 km/s and 5 km/s (Fig. 4b). After subtracting the calculated times from the picked times, each resulting delay time profile is normalized by subtracting the current maximum delay. This allows comparisons of profiles for different events and reference velocities. There is generally good agreement among the normalized profiles, although the large topographic gradient northeast of the CF produces a spread in delay time profiles for the different reference velocity corrections (Fig. 4b). Irrespective of the spread between the curves, maximum delay times are consistently observed for stations BS41–47 (90–140 m northeast of the CF trace). This suggests greater slowness beneath these stations, which agrees overall with the maximum amplitudes plotted in Fig. 4a and subsequent analysis of FZHW and FZTW in Sects. 4.2 and 4.3. Less prominent maxima in amplitude and delay time curves (for example 600–700 m northeast of CF trace, Fig. 4) again suggest minor damage structures or other heterogeneities. These features should be considered as potential targets for future geological/geophysical studies around the Blackburn Saddle site.

4. *Local Earthquake Analyses*

4.1. *Waveform Visual Inspection*

We perform a similar visual inspection of waveforms from local events for evidence of lateral changes in the subsurface structure beneath the array. The shorter wavelengths (higher frequencies) of local arrivals allow observation of large-scale variations across the fault at higher resolution compared to teleseismic arrivals.

Waveforms from local earthquakes (Fig. 2b) are first processed with an automatic algorithm (Ross and Ben-Zion 2014; Ross et al. 2016) to identify the onset times of *P* and *S* waves. We then visually inspect early *P* and *S* waves 0.5 s after these automatic picks for anomalous changes in character. Figure 5a shows vertical component waveforms from an example event EVE1 (location in Fig. 2b) and associated *P* and *S* picks. The *P* waveforms recorded southwest of station BS34 show a 4–5 factor increase in maximum amplitude compared to *P* waveforms recorded northeast of BS34. Similar to Sect. 3.1, this abrupt change in waveform character implies a lateral change in subsurface structure near BS34. Moreover, *P* peak amplitudes recorded southwest of BS34 are associated with phases (right orange line in Fig. 5b) arriving $\sim$ 0.2 s after the *P* first arrivals (left orange line in Fig. 5b). The *P* first arrivals recorded at these stations are potential FZHW, while the trailing large amplitude phases are direct *P* waves. This suggests that the structural change near BS34 is associated with a bimaterial interface. Following the *S* arrivals, stations BS41–45 record an anomalous large amplitude wave packet (orange box in Fig. 5c). These are potential FZTW associated with the core damage zone along the CF through Blackburn Saddle. Similar FZHW and FZTW are observed in waveforms from several other events and we analyze them in greater detail below.

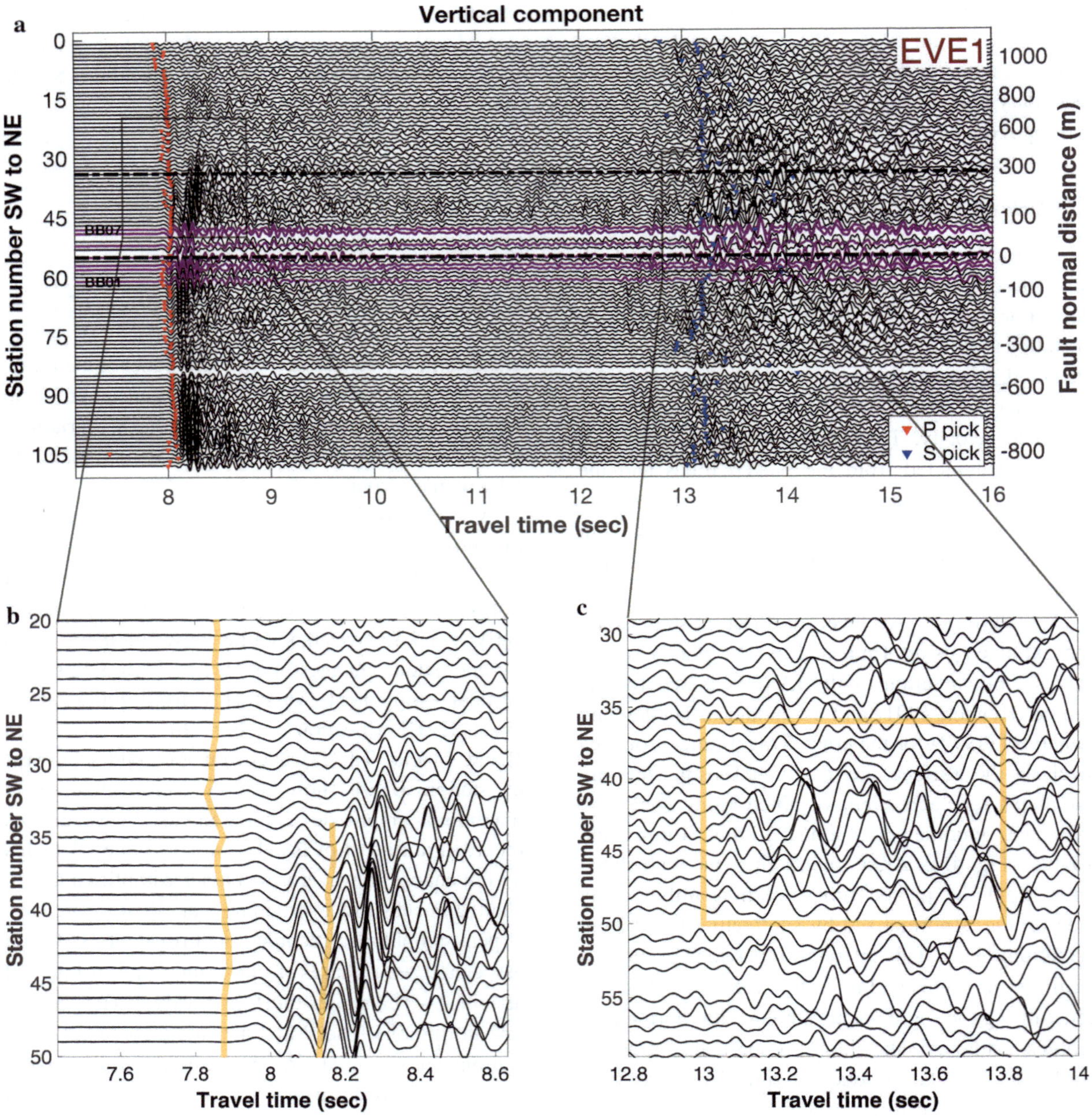

Figure 5
Waveforms (velocity seismograms) generated by local event EVE1 (M_W 1.9). **a** Waveforms recorded across the array with automatic *P* (red triangles) and *S* (blue triangles) picks highlighting the arrivals of these respective phases. All seismograms are normalized by the same value. Noisy traces are removed. The black dot-dashed lines show locations of the CF trace (bottom line) and another large-scale structure to the northeast of the CF (top line, see Sect. 3.1). Waveforms recorded near BB array locations are colored purple with the southwestern most and northeastern most ones denoted by BB01 and BB07. **b** Zoom in of early *P* arrivals at stations BS20–50. The leftmost orange line represents manually picked *P* first arrivals across these stations whereas the orange line to the right of it indicates the later arrival of large amplitude *P* phases at stations southwest of ~ BS34 only. **c** Zoom in of *S* arrivals at stations BS29–59. The orange square highlights an anomalous large amplitude wave packet recorded at stations BS41–45

4.2. Fault Zone Head Waves

4.2.1 Methodology

FZHW are critically refracted emergent phases that travel along a fault bimaterial interface with the velocity of the faster medium (Ben-Zion 1989). They arrive at stations in the slower medium before the impulsive direct *P* waves and are observed up to a fault-normal critical distance x_c given by

$$x_c = r \cdot \tan\left(\cos^{-1}(\alpha_s/\alpha_f)\right), \quad (3)$$

where r, α_s and α_f denote the propagation distance along the fault (both along-strike and up-dip direction) and average *P* wave velocities of the slower and faster media, respectively. The differential time Δt between FZHW and direct *P* wave for a station near the fault is given by

$$\Delta t = r/\alpha_f \cdot \eta/(1-\eta), \quad (4)$$

where $\eta = (\alpha_f - \alpha_s)/\alpha_f$ is the fractional velocity contrast across the fault (Ben-Zion 1989; Qiu et al. 2017). Because FZHW radiate from the fault and direct *P* waves propagate from the epicenter, FZHW have horizontal particle motion with a significant fault-normal component (Bulut et al. 2012; Allam et al. 2014; Share and Ben-Zion 2016) compared to that of direct *P* waves. For on-fault events, the FZHW and trailing direct *P* waves on the slow side of the fault have opposite first motion polarities (Ben-Zion 1989; Ben-Zion and Aki 1990).

4.2.2 Results

We identify FZHW with an automatic detector (Ross and Ben-Zion 2014) in conjunction with visual inspection and analysis using the aforementioned criteria. The detector flags *P* waveforms with an emergent phase followed by an impulsive arrival with a time separation between a minimum value (0.065 s representing the width of a narrow *P* wave wiggle) and a maximum value that depends on hypocentral distance. As an example, a maximum time separation of 0.8 s is allowed over a distance of 40 km using the default parameters of the Ross and Ben-Zion (2014) algorithm. The detector is applied similarly to previous linear SJFZ array studies (Qiu et al. 2017; Share et al. 2017) using a slightly lower STA/LTA trigger threshold due to the lower SNR of BS array waveforms compared to the other arrays.

Focusing first on the 75 events within 10 km of the CF trace (Fig. 6a), which have magnitudes M_W 1–3, the automatic detector flags 316 potential FZHW for the 108 stations analyzed. The picked FZHW are from events located both northwest and southeast of the array (Fig. 6a). Though all stations have potential detections, those southwest of BS34 have more than double detections compared to stations northeast of BS34 (Fig. 6b). Some of the automatic picks are expected to be false (Ross and Ben-Zion 2014), and some may be associated with the edge of the damage zone (Najdahmadi et al. 2016) or a basin (Yang et al. 2015; Qiu et al. 2017), rather than a deep fault bimaterial interface that is the main imaging target here.

Based on visual inspection of the 75 events generating candidate FZHW, we find that events with epicenters near the CF surface trace within the trifurcation area (having large along-fault propagation distance) produce the clearest head waves. The locations of two example events HW1 and HW2 are shown in the inset of Fig. 6a. These events generate emergent (lower frequency, Fig. S2a) low amplitude first *P* arrivals (candidate FZHW) at stations southwest of BS34 (Fig. 7a, 1st and 2nd row), followed by larger amplitude impulsive arrivals (direct *P* waves) 0.1–0.3 s later. The differential time between the identified FZHW and direct *P* arrivals is largest near BS34 and decreases towards BS108 in the southwest (Eq. 3). Only impulsive *P* arrivals are recorded at stations northeast of BS34. These observations imply that a bimaterial interface is present below the surface location of station BS34.

The candidate FZHW and trailing direct *P* waves from near-fault events are substantiated by comparing their move-outs across the array with those of nearby reference events that do not produce FZHW propagating at depth. In the presence of a bimaterial fault, a reference event located sufficiently far off-fault in the slower medium produces a first arriving direct *P* wavefront at locations on the slow side with azimuth pointing to the epicenter. In contrast, direct *P* waves from a fast side reference event and FZHW from an event near the interface are refracted along the fault; so their arrivals on the slow side have

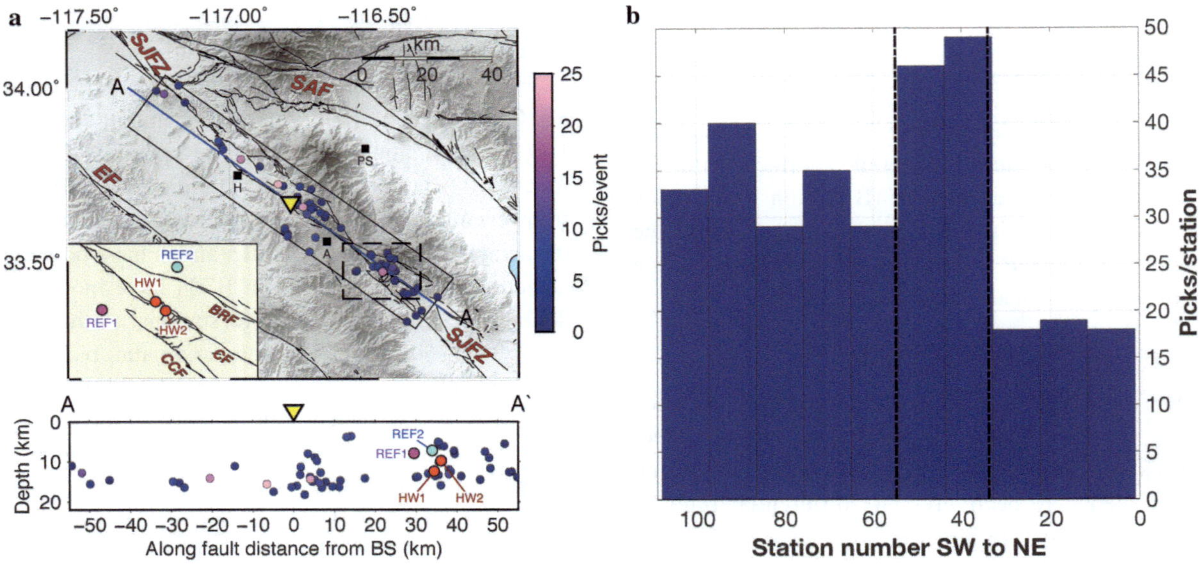

Figure 6

Identifying FZHW using an automatic detection algorithm. **a** Locations of events analyzed (circles inside solid black rectangle). Circle colors correspond to the number of stations with FZHW detections (blue = no detections, pink = most detections). The inset shows a zoom in of selected events located within the trifurcation area (dashed black box). FZHW are confirmed for events HW1 and HW2 near the CF (large red circles) using automatic detection and visual inspection (Fig. 7, 1st and 2nd rows). Reference events near the Coyote Creek (CCF) (REF1, magenta circle) and Buck Ridge (BRF) (REF2, cyan circle) faults that do not generate FZHW (Fig. 7, 3rd and 4th rows). A depth section of events projected along the profile A-A' is plotted at the bottom. **b** Histogram of number of FZHW detections per station for events in **a**. Dot-dashed lines represent the locations of stations BS55 (left) and BS34 (right)

azimuth with a larger fault-normal component compared to the source-receiver azimuth. If the fast side reference event is relatively close to a near-fault event, the move-out of the wavefront across locations on the slow side from the former will be similar to that of critically refracted FZHW generated by the latter. Likewise, move-outs of direct *P* wavefronts from slow side reference and near-fault events will be similar. We, therefore, identify FZHW and trailing direct *P* waves as arrivals with most similar move-outs to that of first *P* arrivals from fast side and slow side reference events, respectively. Comparing move-outs of reference and near-fault events in this manner to distinguish FZHW and direct *P* waves in a structurally complex fault zone with variable focal mechanisms is more robust (Najdahmadi et al. 2016; Share et al. 2017) than applying horizontal polarization analysis. The latter is useful primarily for simple faults and strike-slip mechanisms (e.g., Bulut et al. 2012; Allam et al. 2014; Li and Peng 2016).

We choose events near the Coyote Creek (REF1, Figs. 6a, inset, 7a, 3rd row) and Buck Ridge (REF2, Figs. 6a, inset, 7a, 4th row) faults as reference events on nominally slow and fast sides of the CF, respectively (Allam and Ben-Zion 2012). First arrival *P* waves from the reference events are impulsive and large in amplitude at all stations (Fig. S2b). Compared to REF1 (magenta line in Fig. 7a, 3rd row), REF2 has *P* arrival times that increase more with fault-normal distance from BS34 towards stations in the southwest (cyan line in Fig. 7a, 4th row). This implies that the wavefront from REF2 has the azimuth with larger fault-normal component on the nominally slow side of the fault. FZHW move-outs across stations southwest of BS34 from events HW1 and HW2 are highly correlated with the move-out of *P* arrivals from REF2 (cyan lines Fig. 7a, 1st and 2nd rows). At the same stations, direct *P* wave move-outs from events HW1 and HW2 are approximately equal to the move-out observed for REF1 (magenta lines Fig. 7a, 1st and 2nd rows). These comparisons show that the observed FZHW and direct *P* waves are radiated from the fault and the source, respectively. In addition, the comparisons reveal that the

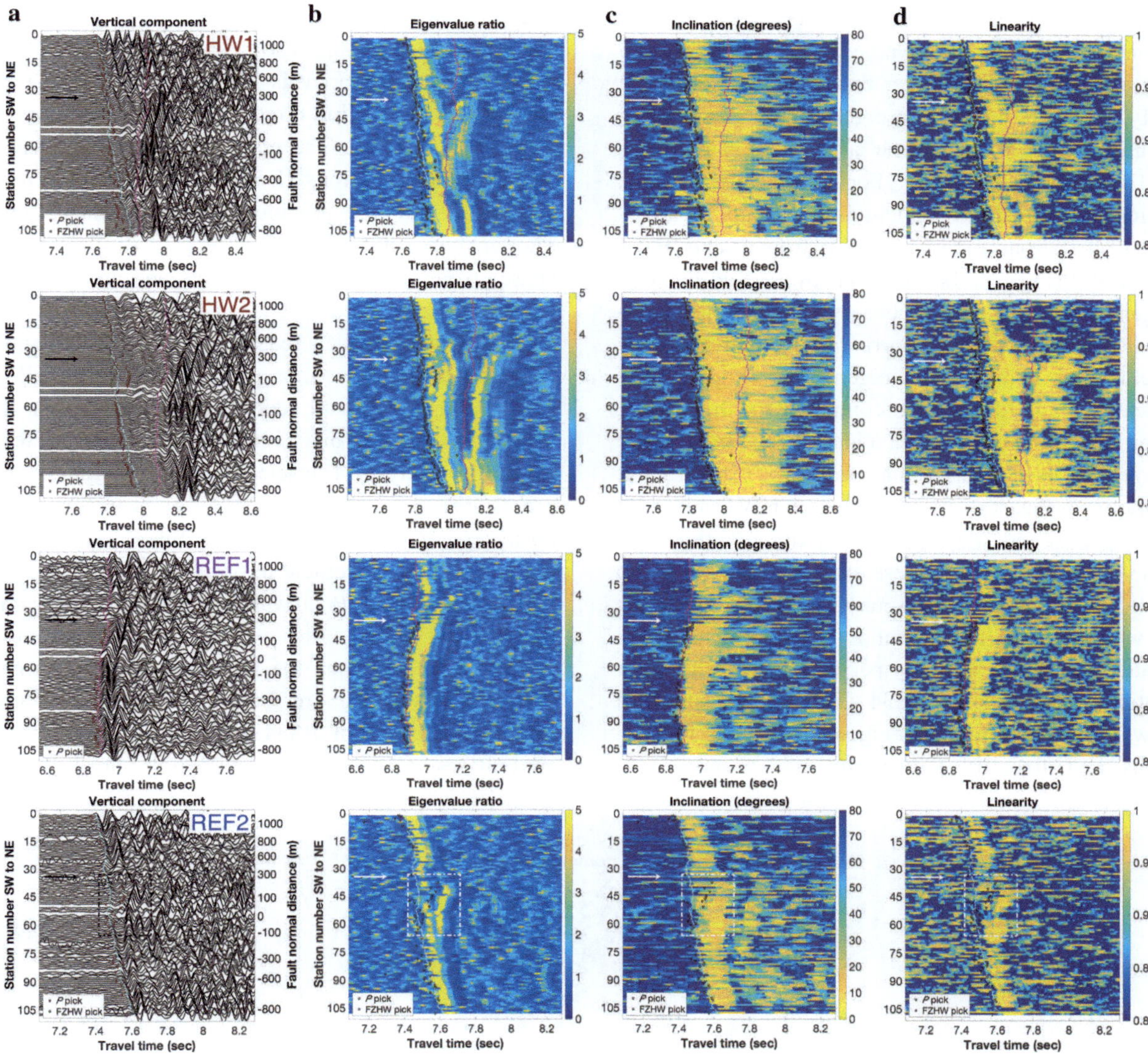

Figure 7

Observed waveforms and polarization analysis of FZHW. **a** *P* waveforms (velocity seismograms) generated by events HW1 (1st row), HW2 (2nd row), REF1 (3rd row) and REF2 (4th row). Seismograms are self-normalized and noisy traces are removed. Green squares and red triangles depict automatic FZHW and *P* picks, respectively. Waveforms recorded at locations southwest of BS34 (black arrows) from HW1 and HW2 contain FZHW while waveforms from REF1 and REF2 do not. Magenta line in row 3 and cyan line in row 4 represent manual picks for *P* first arrivals from events REF1 and REF2, respectively. Cyan and magenta lines in rows 1 and 2 are the same as in rows 3 and 4 but shifted in time to align with the *P* first arrival (candidate FZHW) and onset of the impulsive direct *P* wave at station BS55, respectively. **b–d** The ratio of the largest eigenvalue between neighboring windows, inclination of the largest eigenvector and linearity calculated within the leading window for events HW1 (1st row), HW2 (2nd row), REF1 (3rd row) and REF2 (4th row). Symbols and lines have the same meaning as in **a**. Dot-dashed box in 4th row depict waveforms with candidate local FZHW (see text for details)

differential time Δt between FZHW and direct *P* wavefronts is larger for event HW2 compared to HW1 (colored lines are farther apart for HW2 versus HW1). This observation and Eq. (4) imply a continuous deep bimaterial interface associated with the CF, since HW2 is located farther from the array than HW1 (38 km versus 36 km).

Waveform characteristics of FZHW are quantified using polarization analysis as described in Sect. 3.1. Eigenvalues are computed using consecutive sliding windows each 0.07 s long that overlap by 50%. In general, the largest eigenvalue of a window containing a FZHW will be larger than a signal containing only noise, and the largest eigenvalue corresponding to a direct *P* wave would be larger still (Allam et al. 2014; Share et al. 2017). For example, eigenvalue ratios computed between neighboring windows are > 5 at all stations for first arrivals from events HW1 and HW2 (Fig. 7b, 1st and 2nd rows). Additionally, eigenvalue ratios > 5 are observed at stations southwest of BS34 upon later arrivals of direct *P* waves. Inclination of the largest eigenvector (Fig. 7c) and linearity (Fig. 7d) are also computed for the leading time window. First arrivals appear as coherent < 15° anomalies in inclination that persist for 0.1–0.15 s at stations northeast of BS34 for events HW1 and HW2 (Fig. 7c, 1st and 2nd rows). For stations southwest of BS34 and same events, these anomalies persist for up to 0.3–0.4 s as they represent both FZHW and direct *P* waves. Coherent anomalies in linearity approximately equal to 1 similarly persist for longer times at stations southwest of BS34 (Fig. 7d, 1st and 2nd rows). Similar anomalous eigenvalue ratios, inclinations and linearity values are not observed for reference events REF1 and REF2 (Fig. 7b–d, 3rd and 4th rows). The polarization analysis indicates that FZHW have clear arrivals that stand out from the noise while the trailing direct *P* waves contain more energy than FZHW. The results are consistent with theoretical expectations (Ben-Zion 1989; Ben-Zion and Aki 1990) and previous observational studies. The analysis also shows, as expected, that both phases have near-vertical incidence and are linearly polarized.

Following the discussed analysis steps, 14 events near the trifurcation area generating clear FZHW propagating at depth are identified. In the case where events cluster, only one event per cluster is selected for further analysis. The selection process produces 5 events (Fig. 8a) with distinctly different hypocentral distances, and waveforms from these events are integrated to displacement prior to manual picking of FZHW and direct *P* arrivals. Using the picks and Eq. (4), we estimate the average velocity contrast across the CF between the BS array and generating events. FZHW picks are made on waveforms from stations southwest of BS34 where associated emergent phases begin to rise above the noise level (green

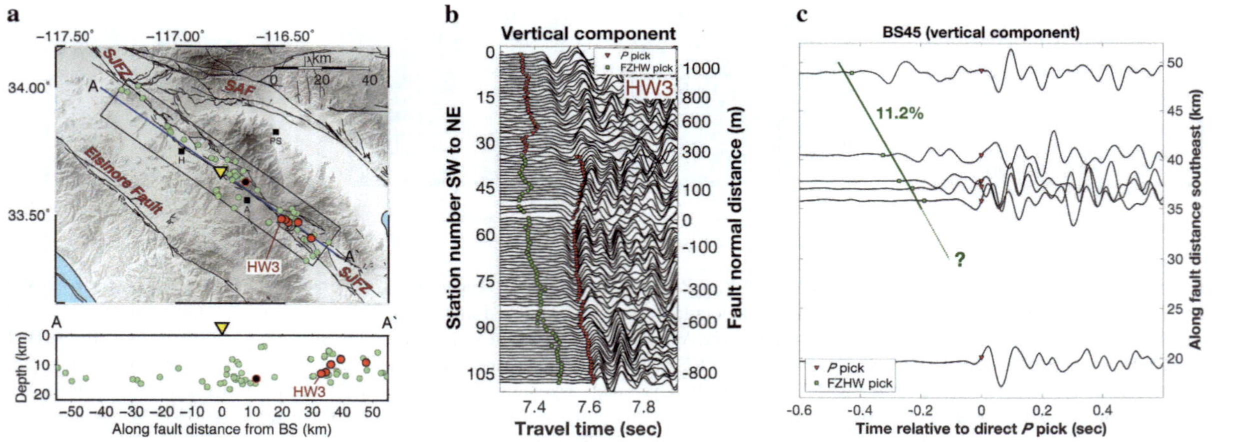

Figure 8

Average velocity contrast across the CF from FZHW move-out analysis. **a** Locations of 5 events (large red circles) used for the move-out analysis. The event closest to the array that generates FZHW is highlighted (HW3) and also shown is the location of an off-fault reference event (large red black-filled circle). A depth section of events projected along the profile A–A′ is plotted at the bottom. **b** *P* displacement seismograms from event HW3 and associated manual direct *P* (red triangles) and FZHW (green squares) picks (see text for picking procedure). Seismograms are self-normalized and noisy traces are removed. **c** Plot of displacement waveforms showing the linear move-out of FZHW (thick green solid line and squares) relative to direct *P* waves (red triangles) recorded at station BS45 for the 5 events in **a**. The move-out corresponds to a 11.2% average velocity contrast across that section of the CF. The thin green dotted line represents a speculative move-out in a region along the fault where there is no data (20–35 km). For comparison, the waveform from the reference event in **a** is also plotted at the bottom squares in Fig. 8b). Direct *P* waves are picked (red triangles in Fig. 8b) using both a comparison of horizontal particle motion of early *P* waves from reference event REF1 and waveforms containing FZHW (similar to Share et al. 2017) and the move-out comparison discussed earlier using REF1 (Fig. 7). The obtained average velocity contrast over the along-fault distance section of 35–50 km from the array is 11.2% (thick green solid line in Fig. 8c). This estimate assumes an average α_f (Eq. 4) of 6.1 km/s based on the tomographic results of Allam and Ben-Zion (2012). The calculation is made using waveforms from representative station BS45 but using other stations leads to similar average velocity contrasts. From the event locations in Fig. 8a, the 11.2% velocity contrast corresponds to the trifurcation area and a depth section around 10 km. The average velocity contrast in the shallow crust is likely to be larger, since the velocity contrast typically decreases with depth (Ben-Zion et al. 1992; Lewis et al. 2007). In the distance range 20–35 km from the array we have no data, but it is clear that the average velocity contrast decreases closer to the array since the move-out (green) line in Fig. 8c has *y*-intercept significantly larger than 0. This is consistent with the earlier results of Share et al. (2017) who determined an average velocity contrast of $\sim$ 3% along the CF closer to the (short BB) array.

The analysis of near-fault events also shows evidence for local FZHW propagating exclusively along the edge of the fault damage zone (e.g., Li and Peng 2016; Najdahmadi et al. 2016). Stations located within the low-velocity damage zone (Figs. 4, 5c) have the most automatic detections (Fig. 6b) with some detections associated with events that do not generate FZHW propagating at depth. For example, event REF2 (Fig. 7, 4th row) does not generate FZHW traveling at depth, but stations BS34–65 record emergent *P* first arrivals followed by large amplitude impulsive arrivals $\sim$ 0.1 s later (dot-dashed rectangle in Fig. 7, 4th row). The emergent arrivals are potential local FZHW refracting along the contact between the damage zone and host rock (at the faster speed of the latter), while the trailing impulsive phases are direct *P* waves propagating within the damage zone.

To explore this further, we apply the automated head wave picker to the 68 events > 10 km from the CF and combine the results of all events to search for local FZHW. Only near-fault events will generate FZHW propagating along a fault at depth. On the other hand, both near-fault and off-fault events can produce local FZHW since the waves from all events propagate through the damage zone before being recorded at surface stations above this zone. The automatic detector flags 357 potential local FZHW for the 68 off-fault events analyzed. As in Fig. 6b, stations within the damage zone have the most picks (Fig. 9a) that correspond to potential local FZHW. Stations northwest and southeast of the damage zone have similarly low numbers of detections (compare Figs. 6b and 9a), indicating (as expected) that none of the off-fault events generate FZHW propagating along the CF at depth.

Visual inspection and further analysis establish several clear examples of local FZHW at stations BS34–65 only. The locations of 15 such events and their respective waveforms recorded at representative stations within (BS43) and outside (BS25 and BS81) the damage zone are shown in Figs. 9b, c. The local FZHW are emergent phases (green line and squares) arriving before the direct *P* waves (red triangles) recorded at station BS43 (Fig. 9c, top). These phases are clearly not FZHW propagating along the CF at depth because an approximately constant Δt ($\sim$ 0.12 s) is observed with increasing hypocentral distance. If we assume that the local FZHW propagate near-vertically along the edge of the damage zone, Δt can be used to approximate the average *P* velocity contrast between damage zone and surrounding rock. Using Eq. (4), a depth extent of 2 km for the damage zone (see Sect. 4.3.2) and velocities of 2–5 km/s for the surrounding rock (similar to Sect. 3.2) give estimated velocity contrasts of $\sim$ 11–23%. This range is higher on average than the estimated contrast across the CF at depth as expected for a contrast between the damage zone and host rock.

4.3. Fault Zone Trapped Waves

4.3.1 Methodology

Spatially continuous low-velocity fault damage zones act as waveguides and generate constructive interference of *S*, *P* and noise phases that give rise to FZTW

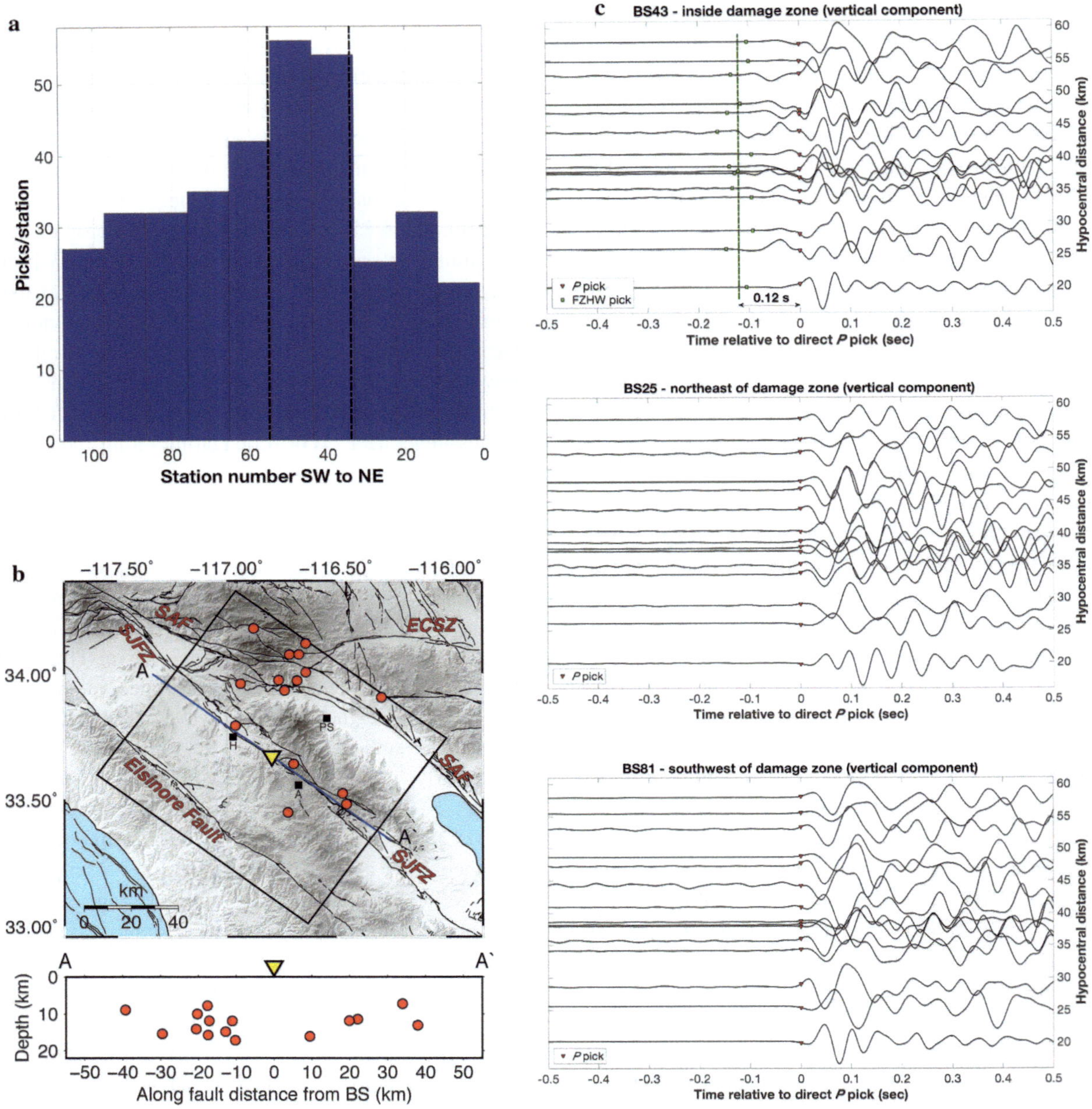

Figure 9

Analysis of local FZHW from off-fault events. **a** Histogram of number of FZHW detections per station for events > 10 km from CF (locations in Fig. 2b). Dot-dashed lines represent the locations of stations BS55 (left) and BS34 (right). **b** Locations of 15 example events (large red circles) that produce FZHW propagating exclusively along the damage zone edge. A depth section of events projected along the profile A–A′ is plotted at the bottom. **c** Self-normalized *P* velocity seismograms from events in **b** and recorded by stations within (BS43, top) and northeast (BS25, middle) and southwest (BS81, bottom) of the damage zone. Red triangles and green squares show manual direct *P* and local FZHW picks and the green line depicts the average early arrival time of the latter showing no apparent move-out

(e.g., Ben-Zion and Aki 1990; Hillers et al. 2014; Qin et al. 2018). These phases have been observed in various fault and geologic settings in California (Li et al. 1994; Lewis and Ben-Zion 2010) including the SJFZ using similar dense deployments (e.g., Qiu et al. 2017; Share et al. 2017) and elsewhere in the world

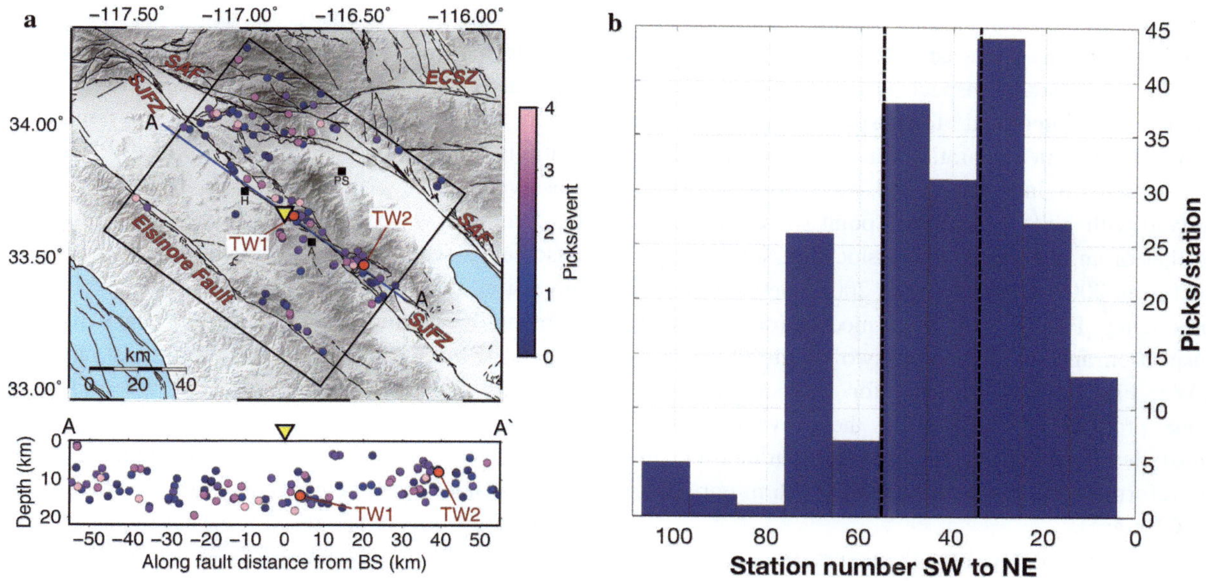

Figure 10
Identifying FZTW using an automatic detection algorithm. **a** Locations of the 143 events analyzed (circles inside black box). Circle colors correspond to the number of stations with FZTW detections (blue = no detections, pink = most detections). Waveforms from events TW1 (Fig. 11) and TW2 (Fig. S3) are inverted using a genetic algorithm approach to determine fault zone properties. A depth section of events projected along the profile A–A′ is plotted at the bottom. **b** Histogram of number of FZTW detections per station for events in **a**. Dot-dashed lines represent the locations of stations BS55 (left) and BS34 (right)

(e.g., Ben-Zion et al. 2003; Mizuno and Nishigami 2006; Eccles et al. 2015). FZTW appear on seismograms as high amplitude, long duration, anomalous frequency phases that follow direct arrivals and are observed only at stations that are within or near the trapping structure (e.g., Ben-Zion and Aki 1990; Li and Leary 1990; Lewis et al. 2005). Here, we focus on identifying and analyzing Love-type FZTW that follow the direct *S* wave (Ben-Zion 1998).

Similar to previous studies (Qiu et al. 2017; Share et al. 2017), we use an automated detection algorithm (Ross and Ben-Zion 2015) to flag candidate FZTW and visual inspection to confirm high-quality candidates. The detection algorithm flags a station(s), if the recorded *S* waveform(s) has(have) dominant period, wave energy, absolute peak amplitude to average amplitude ratio, and absolute peak delay (relative to *S* arrival) significantly larger than the median values of all stations analyzed. The computations are done on vertical and fault-parallel component velocity seismograms. After visual confirmation, waveforms from selected stations are inverted for parameters of the trapping structure using a genetic inversion algorithm with a forward kernel based on the analytical solution of Ben-Zion and Aki (1990) and Ben-Zion (1998). The topographic gradient to the northeast of the CF (Fig. 4) is not accounted for by the analytical solution; so, we expect an increased misfit during inversions related to stations atop this gradient compared to previous applications at other SJFZ arrays along flatter profiles (Qiu et al. 2017; Share et al. 2017; Qin et al. 2018).

4.3.2 Results

The automatic detection algorithm of Ross and Ben-Zion (2015) is applied on waveforms from local 143 earthquakes within a broad region that includes the SJFZ, San Andreas fault (SAF) and Elsinore fault (Fig. 10a). A search window for detecting FZTW is defined using the *S* picks described in Sect. 4.1. The automatic detector flags 194 potential FZTW from events throughout the region (Fig. 10a), with a maximum number of 4 picks made per event. The latter suggests that FZTW are most pronounced in waveforms recorded by ≥ 4 stations (as suggested in

Figs. 4, 5c). Approximately 80% of all detections are made for stations northeast of the CF trace (Fig. 10b).

We next visually inspect flagged FZTW. Seismograms are integrated to displacement prior to inspection and in preparation for waveform inversion. Subsequent to integration, the waveforms are convolved with $1/t^{1/2}$ to convert a point source response to that of an equivalent line dislocation source (e.g., Igel et al. 2002; Ben-Zion et al. 2003). Several events producing FZTW are confirmed through visual inspection and waveforms of two events (TW1 and TW2, Fig. 10a) are selected for inversion of fault zone properties. FZTW from these events appear most clearly (Figs. 11a, S3a) on fault-parallel rotated waveforms at stations around BS45 (106 m northeast of CF trace).

The forward model for the inversion consists of a fault zone layer sandwiched between two quarter spaces (Fig. 2 in Ben-Zion and Aki 1990). This basic model provides a useful modeling approach because it captures the key average geometrical and material properties affecting FZTW while accounting analytically for the strong trade-offs that exist between these parameters (Ben-Zion 1998). Various likely velocity gradients, internal scatterers and other small-scale heterogeneities in the trapping structure have small effects on FZTW that average out these heterogeneities (e.g., Igel et al. 1997; Jahnke et al. 2002). The inverted fault zone parameters are: (1–3) *S* velocities of the two quarter spaces (assumed different based on Sect. 4.2.2) and the fault zone layer, (4–5) width and *Q* value of the fault zone layer, (6) location of contact between the fault and left quarter space, and (7) propagation distance within the fault zone layer.

The genetic inversion algorithm maximizes the correlation between sets of synthetic seismograms calculated with the forward model and observed waveforms, while exploring systematically a large parameter space. This is accomplished by calculating fitness values associated with different sets of model parameters and migrating in the parameter space overall in the direction of larger fitness values. The fitness is defined as $(1 + C)/2$ where C is the correlation coefficient between synthetic and observed data. When C varies over the range -1 (perfect anti-correlation) to 1 (perfect correlation), the fitness value changes from 0 to 1. Only waveforms from stations BS34–58 (with noisy traces excluded) are inverted; this subset includes stations where FZTW are recorded and a sufficient number of stations to the northeast and southwest with no observed FZTW.

Synthetic waveforms (light blue lines in Fig. 11b) associated with the best-fit solution after 10,000 inversion iterations (black circles in Fig. 11c) are compared with recorded waveforms (black lines in Fig. 11b) from TW1. Summing the fitness values of the final 2000 inversion iterations (green dots in Fig. 11c) and normalizing the results to have unit sums give probability density functions for the various model parameters (curves in Fig. 11c). Corresponding inversion results for event TW2 are shown in Fig. S3. Because FZTW are generated by the same waveguide structure, inversions of waveforms associated with different high-quality candidates provide similar values for the most likely fault zone parameters. However, the parameters are subjected to significant trade-offs as discussed in previous studies (e.g., Ben-Zion 1998; Peng et al. 2003) and illustrated by the broad regions of high fitness values in Figs. 11c, S3c.

Based on the best-fit inversion results, the core damage zone is estimated to start at BS55 (main CF, Fig. 3b) and have width of ~ 150 m, *Q* value of ~ 40 and *S* velocity reduction of ~ 55% relative to the bounding rocks. Event TW1 is beneath the array (Fig. 10a), so energy from the event propagates almost exclusively in the vertical direction and the best-fitting propagation distance of ~ 2 km (Fig. 11c) reflects the depth extent of the core damage zone. On the other hand, event TW2 is located in the trifurcation area and FZTW generated by this event propagates a significant distance along-strike. Using for this event a propagation distance of 6 km (Fig. S3c) and a trapping structure depth of 2 km (Fig. 11c), the horizontal propagation distance within the waveguide from event TW2 is less than 6 km, considerably smaller than the epicentral distance of TW2 from the array (40 km, Fig. 10a). This indicates that the trapping structure does not extend continuously along-strike more than a few km, in agreement with inferences from other studies (e.g., Peng et al. 2003; Lewis and Ben-Zion 2010). For both

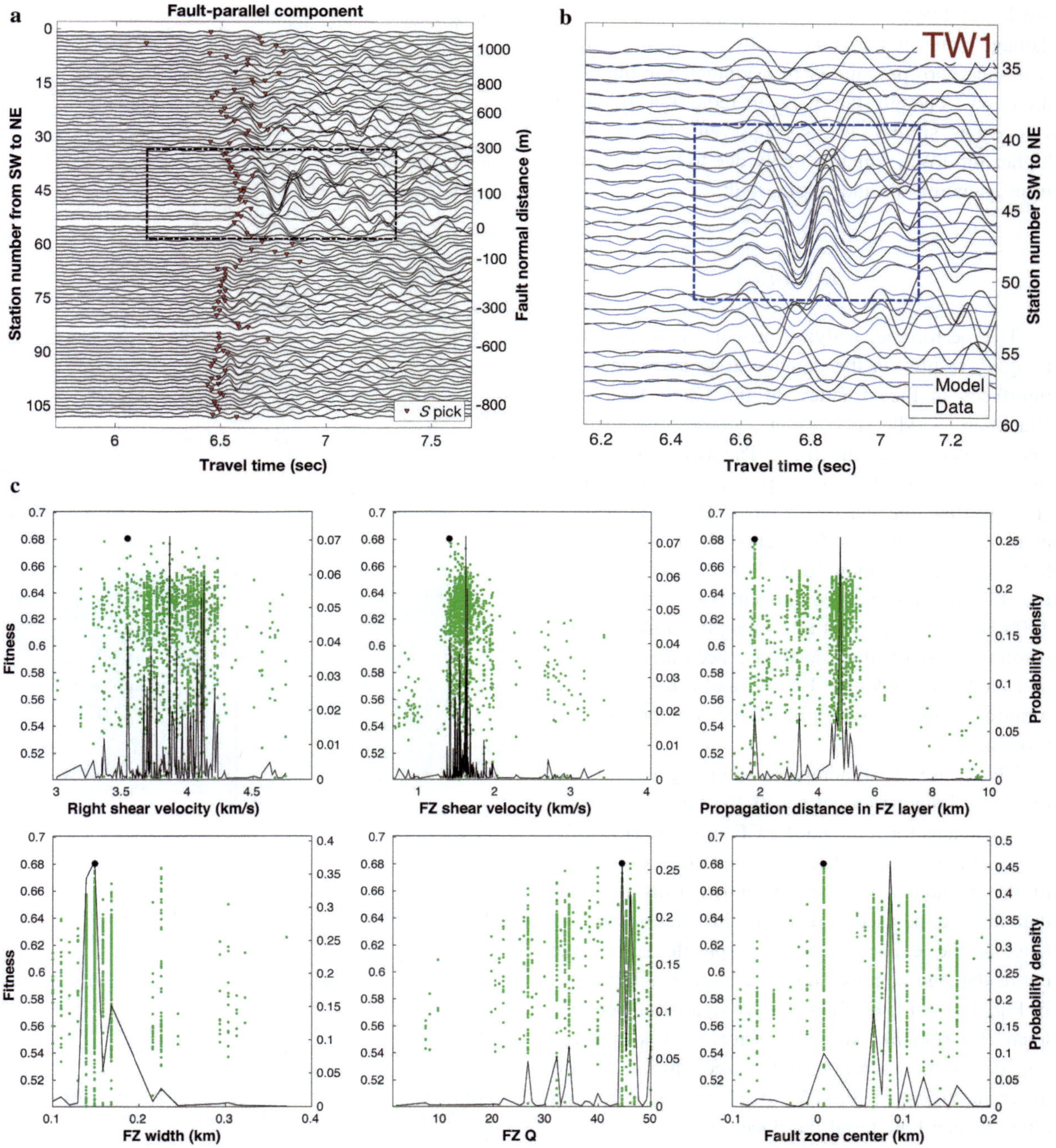

Figure 11
Fault model inversion results for event TW1. **a** Processed *S* waveforms (see text for processing steps) for the array with the dot-dashed box highlighting waveform analyzed during inversion. **b** Waveforms for stations BS34–58 (black traces) compared to synthetic waveforms (light blue traces) using fault parameters associated with the best-fit solution. The box shows the most pronounced FZTW. **c** Parameter space for the final 10 generations (2000 iterations). Green dots indicate fitness values for all solutions, black dots correspond to the best-fit solution and black lines are probability density curves of the inverted parameters

TW1 and TW2, the inversion yields an asymmetric damage zone centered northeast of the fault trace with near-zero probability of asymmetry in the other direction. Though the percent reduction in shear velocity is similar, the average velocity overall is higher for TW2, which is likely due to the difference in propagation distance along-fault.

5. Discussion

The performed analyses provide high-resolution imaging results on the damage zone and bimaterial interfaces in the structure of the Clark fault near Blackburn Saddle (BS) using a dense long-aperture linear array across the fault. The BS array has an order of magnitude larger aperture and number of stations than the earlier installed BB array (Share et al. 2017). Unlike other across-fault arrays (e.g., Schmandt and Clayton 2013; Ben-Zion et al. 2015), the BS array has three-component nodes recording at 1000-Hz sampling. This allows applications of high-resolution polarization analysis, accurate detection and modeling of *S*-type FZTW, and other tools that require dense arrays with three components of motion.

The denser spatiotemporal sampling over a significantly longer transect across the fault has several consequences for imaging the CF structure (Fig. 12). First, unlike the short BB array that covers only part of the damage zone, the longer BS array provides data that can be used to recover with teleseismic data (Fig. 4) the expected slowness profile across the entire damage zone (compare red and black profiles in Fig. 12a, top). Furthermore, comparisons between local *P* earthquake waveforms recorded within and outside the damage zone allow recognition of two different types of FZHW: head waves refracting along a deep bimaterial interface separating different crustal blocks across the CF (Fig. 12a, middle) and local FZHW propagating along the edge of the damage zone (Fig. 12a, bottom). The local FZHW were not analyzed in the earlier study using the BB array (Share et al. 2017) since that short array did not include stations outside the damage zone (Fig. 12b).

The results indicate that the main seismogenic CF at depth is close to its mapped surface trace (station

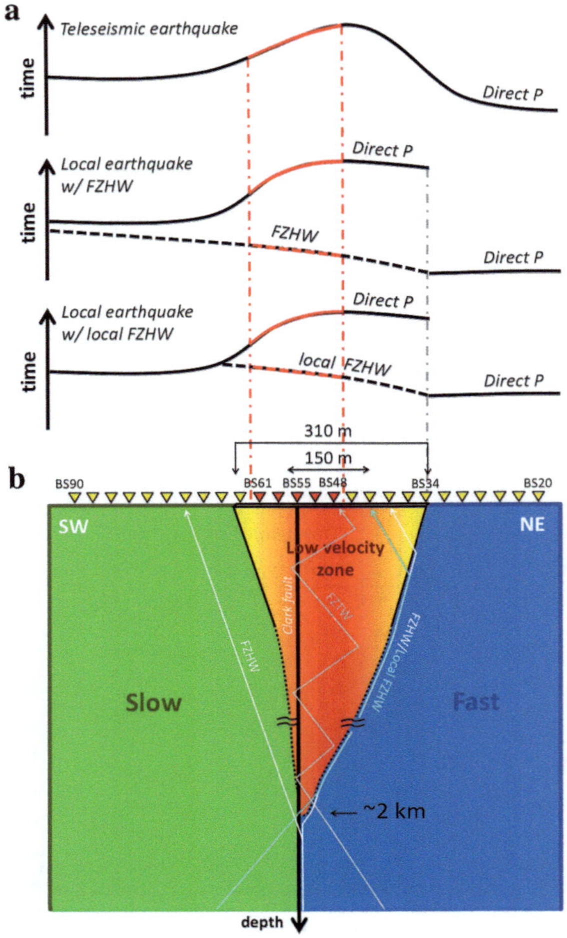

Figure 12
Schematic representation of arrival time variations of early *P* waves across the fault zone. **a** Expected variations in *P* first arrivals from teleseismic earthquakes (top), direct *P* and FZHW from on-fault local earthquakes (middle), and off-fault local earthquakes (bottom) given the fault model in **b**. The black lines represent profiles recorded by the BS array whereas the red profiles denote the limited profiles observed with the shorter BB array. The vertical gray dot-dashed line indicates abrupt changes in arrivals associated with a major bimaterial interface at the site and detected with the BS array only. **b** Simplified fault zone model suggested by analysis done in this work (not to scale) with nominally slow block on the southwest (green) separated from a fast block to the northeast (blue) by the main Clark fault (thick black line) and a low-velocity damage zone near the surface (warm colors with warmer representing greater damage). Fault zone parameters and example ray paths of several fault zone phases (white, gray and cyan lines) are indicated. Triangles on the surface show the BS array spanning the entire fault zone and red triangles mark the BB array covering only the southwestern side of the damage zone

BS55), and it manifests as abrupt changes in waveform character, inclination and linearity of teleseismic P waveforms (Figs. 3b, c, S1b). The large amplitude and delay time anomalies in teleseismic P arrivals (Fig. 4) and the presence of FZTW from local events (Figs. 5c, 11a, S3a) indicate an asymmetric damage zone (Fig. 12b) with a center 90–150 m (BS40–47) from the surface trace of the CF. The large-scale structural variation inferred from changes in waveforms recorded ~ 270 m northeast of the CF (BS34, Figs. 3, 5b, S1) is most likely the northeastern edge of this damage zone (Fig. 12b). The analysis of teleseismic P arrivals reveals potential minor faults and associated damage outside (> 300 m from CF) the broader damage zone. These features are unresolved in our analyses and could be targets for a future study in the area.

A combination of high-resolution polarization, across- and along-fault move-out analyses applied to waveforms of local events reveals FZHW at locations southwest of BS34 and not northeast of that station (Fig. 7). These results suggest the broader damage zone is bounded to the northeast by a sharp bimaterial interface separating structure to the northeast that is on average faster than the structure to the southwest (Fig. 12b). The existence and sense of velocity contrast are supported by tomographic models (Allam and Ben-Zion 2012) and regional geology (e.g., Sharp 1967; Gutierrez et al. 2010; Salisbury et al. 2012). The results also suggest that the bimaterial interface merges with the CF at depth and is continuous for at least 50 km to the southeast, as the observed FZHW are generated by events located within and beyond the trifurcation area (Fig. 8a). This implies the CF is continuous at depth through the Anza seismic gap and over a larger distance than previously estimated (Share et al. 2017), highlighting the potential for large earthquake rupture through the region as suggested by paleoseismic studies (Rockwell et al. 2015).

A P-wave velocity contrast in the range ~ 11–23% across the damage zone is calculated from the differential time between local FZHW propagating at the edge of the damage zone and direct P waves (Fig. 9c). This range of P velocity contrast corresponds to the S velocity contrast across the trapping structure derived from analysis of FZTW, assuming V_p/V_s ratios in the range 2.95–3.45 that may represent shallow damage rocks (e.g., O'Connell and Budiansky 1974; Mavko et al. 1998). Similar near-surface contrasts across the SJFZ are observed at a site ~ 50 km to the southeast (Qiu et al. 2017). An average velocity contrast of 11.2% across a deep section of the CF around the trifurcation area is calculated from the move-out between FZHW and direct P waves with along-fault distance (Fig. 8c). These values are comparable to contrasts across the SAF south of Hollister (McGuire and Ben-Zion 2005) and sections of the Hayward fault (Allam et al. 2014). A similar contrast of 9.2% is derived from velocities 1.5 km on either side of the trifurcation area CF up to 10-km depth (median depth of events generating FZHW) in the V_P model of Allam and Ben-Zion (2012). The smaller velocity contrast observed for the CF at depth closer to the array is, in turn, comparable to results for the SAF around San Gorgonio Pass (Share and Ben-Zion 2016) and North Anatolian fault (Najdahmadi et al. 2016). The observed FZHW indicate the existence of a continuous sharp bimaterial interface in the core structure of the CF, between the generating events and the BS array (Fig. 8a), despite structural discontinuities at the surface (Sanders and Magistrale 1997) and irrespective of (lateral and/or vertical) changes in the velocity structure.

Properties of the fault damage zone are obtained from delay time analysis and inversions of FZTW following S arrivals. The extent of the broader damage zone is inferred from the number of stations that record teleseismic P arrivals with above-average delay and local FZHW (Sects. 3.2, 4.2.2). Relative to the median delay (horizontal dot-dashed line Fig. 4b), stations BS35–61 consistently have larger delay times for all events and reference velocities considered. These stations, therefore, overlie a region of above-average slowness. The same set of stations also record local FZHW (Fig. 7, 4th row). This implies the damage zone edge(s) along which local FZHW propagate locate northeast of BS35 and southwest of BS61. We, therefore, estimate the width of the broader damage zone to be comparable to the 310-m-wide zone spanning these stations (Fig. 12b). Along the Parkfield section of the SAF, observations of FZHW similarly reveal the extent of a broad damage zone that is asymmetric with respect to the main fault trace (Li and Peng 2016). Inversions of waveforms with FZTW provide information on the fault zone trapping structure within this broader damage zone. The results indicate (Figs. 11c, S3c) a trapping structure that extends < 6 km along-strike and to a depth of 2 km based on the best-fit propagation distances from the FZTW inversions. The trapping structure is about 150 m wide and has S velocity reduction of 55% and Q value of about 40. These parameters are similar to properties of trapping structures along other sections of the SJFZ (Lewis et al. 2005; Qiu et al. 2017; Qin et al. 2018), SAF at Parkfield (Lewis and Ben-Zion 2010), Karadere branch of the North Anatolian fault (Ben-Zion et al. 2003) and other active strike-slip faults and rupture zones. The larger-aperture BS array (compared to BB, Figs. 3a, 5a) allows for better resolution of the trapping structure within the broader damage zone, and the properties estimated here (Figs. 11c, S3c) are, therefore, more accurate than those reported in Share et al. (2017).

The sense of velocity contrast across the CF at depth towards the southeast (Fig. 8), combined with results on bimaterial ruptures (e.g., Weertman 1980; Andrews and Ben-Zion 1997; Brietzke et al. 2009; Shlomai and Fineberg 2016), suggest that earthquakes in that fault section tend to propagate to the northwest. This is consistent with studies on directivity of small–moderate events along the CF between the trifurcation area and the BS array (Kurzon et al. 2014; Ross and Ben-Zion 2016). Persistent occurrence of bimaterial ruptures with preferred propagation direction is expected to produce more damage on the side with faster velocity at depth (Ben-Zion and Shi 2005). This agrees with the geological mapping of Dor et al. (2006) near Anza, and the observed damage asymmetry across the CF to the northeast (Figs. 4, 5, 9, 11) summarized in Fig. 12b. Similar asymmetric damage zones are also associated with CF at sites southeast of Blackburn Saddle (Qiu et al. 2017; Qin et al. 2018). A recent example of both rupture directivity and damage asymmetry across the CF is provided by the June 2016 M_W 5.2 Borrego Springs earthquake, where rupture initiated in the trifurcation area, propagated to the northwest and generated significantly more aftershocks (damage) northeast of the fault (Ross et al. 2017a, b).

The results of this work are obtained by a set of separate analyses that examine different phases with different techniques and largely ignore the topography (other than travel time corrections). A more complete analysis that merges the detailed fault zone structure within a larger-scale tomographic model and accounts for topography may be done with 3D numerical simulations of the type explored by Allam et al. (2015). This approach is computationally very demanding but is currently feasible. Additional information about the fault zone structure may be obtained by analyzing the ambient seismic noise recorded by the dense linear array (e.g., Hillers et al. 2014). These studies may be the subject of future work.

Acknowledgements

We thank Marianne Karplus and Jerry Schuster for providing geophones for the experiment, and Hsin-Hua Huang, Elizabeth Berg, Yadong Wang, Scott Palmer, Kathleen Ritterbush, Jon Gonzalez, Jerry Schuster, Robert Zinke and Cooper W. Harris for assistance during the array deployment. The research was supported by the King Abdullah University of Science and Technology (Award OCRF-2014-CRG3-2300), the National Science Foundation (Grants CyberSEES-1442665 and EAR-1620601), and the Department of Energy (Awards DESC0016520 and DE-SC0016527). The data collected with the BS array are stored at the University of Utah Seismograph Stations and are available upon request. Most maps are made using the Generic Mapping Tools (Wessel et al. 2013). The manuscript benefitted from useful comments by two anonymous referees.

References

Allam, A., & Ben-Zion, Y. (2012). Seismic velocity structures in the southern California plate boundary environment from double difference tomography. *Geophysical Journal International, 190*, 1181–1196.

Allam, A. A., Ben-Zion, Y., & Peng, Z. (2014). Seismic imaging of a bimaterial interface along the Hayward fault, CA, with fault zone head waves and Direct P arrivals. *Pure and Applied Geophysics, 171*(11), 2993–3011.

Allam, A. A., Tape, C., & Ben-Zion, Y. (2015). Finite-frequency sensitivity kernels of seismic waves to fault zone structures. *Geophysical Journal International, 203,* 2032–2048. https://doi.org/10.1093/gji/ggv413.

Andrews, D. J., & Ben-Zion, Y. (1997). Wrinkle-like slip pulse on a fault between different materials. *Journal of Geophysical Research, 102,* 553–571.

Barak, S., Klemperer, S. L., & Lawrence, J. F. (2015). San Andreas fault dip, peninsular ranges mafic lower crust and partial melt in the Salton Trough, Southern California, from ambient–noise tomography. *Geochemistry, Geophysics, Geosystems.* https://doi.org/10.1002/2015GC005970.

Ben-Zion, Y. (1989). The response of two joined quarter spaces to SH line sources located at the material discontinuity interface. *Geophysical Journal International, 98,* 213–222.

Ben-Zion, Y. (1998). Properties of seismic fault zone waves and their utility for imaging low-velocity structures. *Journal of Geophysical Research, 103,* 12567–12585.

Ben-Zion, Y., & Aki, K. (1990). Seismic radiation from an SH line source in a laterally heterogeneous planar fault zone. *Bulletin of the Seismological Society of America, 80,* 971–994.

Ben-Zion, Y., Katz, S., & Leary, P. (1992). Joint inversion of fault zone head waves and direct P arrivals for crustal structure near major faults. *Journal of Geophysical Research, 97,* 1943–1951.

Ben-Zion, Y., Peng, Z., Okaya, D., Seeber, L., Armbruster, J. G., Ozer, N., et al. (2003). A shallow fault-zone structure illuminated by trapped waves in the Karadere–Duzce branch of the North Anatolian Fault, western Turkey. *Geophysical Journal International, 152*(3), 699–717.

Ben-Zion, Y., & Shi, Z. (2005). Dynamic rupture on a material interface with spontaneous generation of plastic strain in the bulk. *Earth and Planetary Science Letters, 236*(1), 486–496.

Ben-Zion, Y., Vernon, F., Ozakin, Y., Zigone, D., Ross, Z., Meng, H., et al. (2015). Basic data features and results from a spatially dense seismic array on the San Jacinto fault zone. *Geophysical Journal International, 202,* 370–380.

Brietzke, G. B., Cochard, A., & Igel, H. (2009). Importance of bimaterial interfaces for earthquake dynamics and strong ground motion. *Geophysical Journal International, 178,* 921–938.

Bulut, F., Ben-Zion, Y., & Bonhoff, M. (2012). Evidence for a bimaterial interface along the Mudurnu segment of the North Anatolian Fault Zone from polarization analysis of P waves. *Earth and Planetary Science Letters, 327–328,* 17–22.

Cochran, E. S., Li, Y.-G., Shearer, P. M., Barbot, S., Fialko, Y., & Vidale, J. E. (2009). Seismic and geodetic evidence for extensive, long-lived fault damage zones. *Geology, 37*(4), 315–318.

Crotwell, H. P., Owens, T. J., & Ritsema, J. (1999). The TauP Toolkit: Flexible seismic travel-time and ray-path utilities. *Seismological Research Letters, 70*(2), 154–160.

DeDontney, N., Templeton-Barrett, E. L., Rice, J. R., & Dmowska, R. (2011). Influence of plastic deformation on bimaterial fault rupture directivity. *Journal of Geophysical Research, 116,* B10312. https://doi.org/10.1029/2011JB008417.

Dor, O., Rockwell, T. K., & Ben-Zion, Y. (2006). Geological observations of damage asymmetry in the structure of the San Jacinto, San Andreas and Punchbowl Faults in Southern California: A possible indicator for preferred rupture propagation direction. *Pure and Applied Geophysics, 163,* 301–349.

Dor, O., Yildirim, C., Rockwell, T. K., Ben-Zion, Y., Emre, O., Sisk, M., et al. (2008). Geological and geomorphologic asymmetry across the rupture zones of the 1943 and 1944 earthquakes on the North Anatolian Fault: Possible signals for preferred earthquake propagation Direction. *Geophysical Journal International, 173,* 483–504.

Eccles, J. D., Gulley, A. K., Malin, P. E., Boese, C. M., Townend, J., & Sutherland, R. (2015). Fault Zone Guided Wave generation on the locked, late interseismic Alpine Fault, New Zealand. *Geophysical Research Letters, 42*(14), 5736–5743.

Fang, H., Zhang, H., Yao, H., Allam, A., Zigone, D., Ben-Zion, Y., et al. (2016). A new three-dimensional joint inversion algorithm of body-wave and surface-wave data and its application to the Southern California Plate Boundary Region. *Journal of Geophysical Research, 121*(5), 3557–3569.

Gutierrez, C., Bryant, W., Saucedo, W.G., & Wills, C. (2010). Geologic map of California, California Geological Survey.

Hadley, D., & Kanamori, H. (1977). Seismic structure of the Transverse Ranges, California. *Geological Society of America Bulletin, 88,* 1469–1478.

Hauksson, E., Yang, W., & Shearer, P. M. (2012). Waveform relocated earthquake catalog for southern California (1981 to June 2011). *Bulletin of the Seismological Society of America, 102*(5), 2239–2244.

Hillers, G., Campillo, M., Ben-Zion, Y., & Roux, P. (2014). Seismic fault zone trapped noise. *Journal of Geophysical Research, 119*(7), 5786–5799.

Hillers, G., Roux, P., Campillo, M., & Ben-Zion, Y. (2016). Focal spot imaging based on zero lag cross correlation amplitude fields: Application to dense array data at the San Jacinto fault zone. *Journal of Geophysical Research.* https://doi.org/10.1002/2016JB013014.

Igel, H., Ben-Zion, Y., & Leary, P. C. (1997). Simulation of *SH*- and *P-SV*-wave propagation in fault zones. *Geophysical Journal International, 128*(3), 533–546.

Igel, H., Jahnke, G., & Ben-Zion, Y. (2002). Numerical simulation of fault zone guided waves: Accuracy and 3-D effects. *Pure and Applied Geophysics, 159,* 2083–2967.

Jahnke, G., Igel, H., & Ben-Zion, Y. (2002). Three-dimensional calculations of fault-zone-guided waves in various irregular structures. *Geophysical Journal International, 151*(2), 416–426.

Jepsen, D. C., & Kennett, B. L. N. (1990). Three component analysis of regional seismograms. *Bulletin of the Seismological Society of America, 80,* 2032–2052.

Jurkevics, A. (1988). Polarization analysis of three-component array data. *Bulletin of the Seismological Society of America, 78*(5), 1725–1743.

Kennett, L. N. (1991). IASPEI 1991 Seismological Tables. *Bibliotech, 167.*

Kurzon, I., Vernon, F. L., Ben-Zion, Y., & Atkinson, G. (2014). Ground motion prediction equations in the San Jacinto Fault Zone—Significant effects of rupture directivity and fault zone amplification. *Pure and Applied Geophysics, 171,* 3045–3081.

Lewis, M. A., & Ben-Zion, Y. (2010). Diversity of fault zone damage and trapping structures in the Parkfield section of the San Andreas Fault from comprehensive analysis of near fault seismograms. *Geophysical Journal International, 183*(3), 1579–1595.

Lewis, M. A., Ben-Zion, Y., & McGuire, J. (2007). Imaging the deep structure of the San Andreas Fault south of Hollister with joint analysis of fault-zone head and Direct P arrivals. *Geophysical Journal International, 169,* 1028–1042.

Lewis, M. A., Peng, Z., Ben-Zion, Y., & Vernon, F. L. (2005). Shallow seismic trapping structure in the San Jacinto fault zone near Anza, California. *Geophysical Journal International, 162,* 867–881.

Li, Y.-G., Aki, K., Adams, D., Hasemi, A., & Lee, W. H. K. (1994). Seismic guided waves trapped in the fault zone of the Landers, California, earthquake of 1992. *Journal of Geophysical Research, 99,* 11705–11722.

Li, Y.-G., & Leary, P. C. (1990). Fault zone trapped seismic waves. *Bulletin of the Seismological Society of America, 80*(5), 1245–1271.

Li, Z., & Peng, Z. (2016). Automatic identification of fault zone head waves and direct *P* waves and its application in the Parkfield section of the San Andreas Fault. *Geophysical Journal International, 205,* 1326–1341.

Mavko, G., Mukerji, T., & Dvorkin, J. (1998). *The rock physics handbook: Tools for seismic analysis in porous media* (p. 329). Cambridge, UK: Cambridge University Press.

McGuire, J., & Ben-Zion, Y. (2005). High-resolution imaging of the Bear Valley section of the San Andreas Fault at seismogenic depths with fault-zone head waves and relocated seismicity. *Geophysical Journal International, 163,* 152–164.

Mizuno, T., & Nishigami, K. (2006). Deep structure of the Nojima fault, southwest Japan, estimated from borehole observation of fault-zone trapped waves. *Tectonophysics, 417,* 231–247.

Najdahmadi, B., Bohnhoff, M., & Ben-Zion, Y. (2016). Bimaterial interfaces at the Karadere segment of the North Anatolian Fault, northwestern Turkey. *Journal of Geophysical Research, 121,* 931–950.

Neuberg, J., & Pointer, T. (2000). Effects of volcano topography on seismic broad-band waveforms. *Geophysical Journal International, 143,* 239–248.

O'Connell, R. J., & Budiansky, B. (1974). Seismic velocities in dry and saturated cracked solids. *Journal of Geophysical Research, 79*(35), 5412–5426.

Oppenheimer, D. H., Reasenberg, P. A., & Simpson, R. W. (1988). Fault plane solutions for the 1984 Morgan Hill, California, earthquake sequence: evidence for the state of stress on the Calaveras fault. *Journal of Geophysical Research, 143,* 239–248.

Peng, Z., Ben-Zion, Y., Michael, A. J., & Zhu, L. (2003). Quantitative analysis of seismic trapped waves in the rupture zone of the 1992 Landers, California earthquake: Evidence for a shallow trapping structure. *Geophysical Journal International, 155,* 1021–1041.

Qin, L., Ben-Zion, Y., Qiu, H., Share, P.-E., Ross, Z. E., & Vernon, F. (2018). Internal structure of the San Jacinto fault zone in the trifurcation area southeast of Anza, California, from data of dense seismic arrays. *Geophysical Journal International, 213,* 98–114. https://doi.org/10.1093/gji/ggx540.

Qiu, H., Ben-Zion, Y., Ross, Z. E., Share, P.-E., & Vernon, F. (2017). Internal structure of the San Jacinto fault zone at Jackass Flat from data recorded by a dense linear array. *Geophysical Journal International, 209*(3), 1369–1388.

Ringler, A. T., Anthony, R. E., Karplus, M. S., Holland, A. A., & Wilson, D. C. (2018). Laboratory tests of three Z-land Fairfield Nodal 5-Hz, three component sensors. *Seismological Research Letters*. https://doi.org/10.1785/0220170236.

Rockwell, T. K., Dawson, T. E., Young-Ben Horton, J., & Seitz, G. (2015). A 21 event, 4000-year history of surface ruptures in the Anza Seismic Gap, San Jacinto Fault and implications for long-term earthquake production on a major plate boundary fault. *Pure and Applied Geophysics, 172*(5), 1143–1165.

Ross, Z. E., & Ben-Zion, Y. (2014). Automatic picking of Direct P, S seismic phases and fault zone head waves. *Geophysical Journal International, 199,* 368–381.

Ross, Z. E., & Ben-Zion, Y. (2015). An algorithm for automated identification of fault zone trapped waves. *Geophysical Journal International, 202,* 933–942.

Ross, Z. E., & Ben-Zion, Y. (2016). Toward reliable automated estimates of earthquake source properties from body wave spectra. *Journal of Geophysical Research*. https://doi.org/10.1002/2016JB013003.

Ross, Z. E., Hauksson, E., & Ben-Zion, Y. (2017a). Abundant off-fault seismicity and orthogonal structures in the San Jacinto fault zone. *Science Advances, 3,* e1601946.

Ross, Z. E., Kanamori, H., & Hauksson, E. (2017b). Anomalously large complete stress drop during the 2016 M_W 5.2 Borrego Springs earthquake inferred by waveform modeling and near-source aftershock deficit. *Geophysical Research Letters, 44,* 5994–6001.

Ross, Z. E., White, M. C., Vernon, F. L., & Ben-Zion, Y. (2016). An improved algorithm for real-time S-wave picking with application to the (augmented) ANZA network in Southern California. *Bulletin of the Seismological Society of America, 106*(5), 2013–2022.

Salisbury, J. B., Rockwell, T. K., Middleton, T. J., & Hudnut, K. W. (2012). LiDAR and field observations of slip distribution for the most recent surface ruptures along the central San Jacinto Fault. *Bulletin of the Seismological Society of America, 102*(2), 598–619.

Sanders, C. O., & Kanamori, H. (1984). A seismotectonic analysis of the Anza seismic gap, San Jacinto fault zone, southern California. *Journal of Geophysical Research, 89*(B7), 5873–5890.

Sanders, C., & Magistrale, H. (1997). Segmentation of the northern San Jacinto fault zone, southern California. *Journal of Geophysical Research, 102*(B12), 27453–27467.

SCEDC. (2013). *Southern California Earthquake Data Center*. Caltech: Dataset. https://doi.org/10.7909/C3WD3xH.

Schmandt, B., & Clayton, R. W. (2013). Analysis of teleseismic P waves with a 5200-station array in Long Beach, California: Evidence for an abrupt boundary to Inner Borderland rifting. *Journal of Geophysical Research, 118,* 1–19.

Share, P.-E., & Ben-Zion, Y. (2016). Bimaterial interfaces in the South San Andreas Fault with opposite velocity contrasts NW and SE from San Gorgonio Pass. *Geophysical Research Letters*. https://doi.org/10.1002/2016GL070774.

Share, P.-E., Ben-Zion, Y., Ross, Z. E., Qiu, H., & Vernon, F. (2017). Internal structure of the San Jacinto fault zone at Blackburn Saddle from seismic data of a linear array. *Geophysical Journal International, 210*(2), 819–832.

Sharp, R. V. (1967). San Jacinto fault zone in the peninsular ranges of Southern California. *Geological Society of America Bulletin, 78*(6), 705–730.

Shlomai, H., & Fineberg, J. (2016). The structure of slip-pulses and supershear ruptures driving slip in bimaterial friction. *Nature Communications, 7,* 11787.

Sibson, R. H. (1989). Earthquake faulting as a structural process. *Journal of Structural Geology, 11,* 1–14.

Weertman, J. (1980). Unstable slippage across a fault that separates elastic media of different elastic constants. *Journal of Geophysical Research, 85,* 1455–1461.

Wessel, P., Smith, W. H. F., Scharroo, R., Luis, J. F., & Wobbe, F. (2013). Generic mapping tools: improved version released. *EOS Transactions of the American Geophysical Union, 94,* 409–410.

Xu, S., Ben-Zion, Y., & Ampuero, J. P. (2012). Properties of inelastic yielding zones generated by in-plane dynamic ruptures-II. Detailed parameter-space study. *Geophysical Journal International, 191*(3), 1343–1360.

Yang, H., Li, Z., Peng, Z., Ben-Zion, Y., & Vernon, F. (2014). Low velocity zones along the San Jacinto Fault, Southern California, from body waves recorded in dense linear arrays. *Journal of Geophysical Research, 119,* 8976–8990.

Yang, W., Peng, Z., Wang, B., Li, Z., & Yuan, S. (2015). Velocity contrast along the rupture zone of the 2010 Mw6.9 Yushu, China, earthquake from fault zone head waves. *Earth and Planetary Science Letters, 416,* 91–97.

(Received May 27, 2018, revised August 27, 2018, accepted September 3, 2018, Published online September 18, 2018)

Pure Appl. Geophys. 176 (2019), 1193–1206

https://doi.org/10.1007/s00024-018-2041-4

Pure and Applied Geophysics

Shallow Crustal Structure of the Tanlu Fault Zone Near Chao Lake in Eastern China by Direct Surface Wave Tomography from Local Dense Array Ambient Noise Analysis

Ning Gu,[1] Kangdong Wang,[1,2] Ji Gao,[1] Ning Ding,[3] Huajian Yao,[1,4] and Haijiang Zhang[1]

Abstract—Ambient noise tomography (ANT) has been used successfully to image shallow earth structure. Here we perform ANT on a local dense seismic array around the Tanlu fault zone (TFZ) to the southeast of Hefei City, Anhui Province in eastern China. The array consists of 53 stations with average spacing close to 5 km. Cross-correlations of vertical-component ambient noise data of different station pairs are computed in 1-h segments and stacked over 1 month from 17 March to 26 April 2017. Clear fundamental-mode Rayleigh waves are observed between 0.2 and 5 s period. We then use the direct surface-wave tomographic method with period-dependent ray tracing to invert group and phase dispersion travel-time data simultaneously for three-dimensional (3D) shear-wave velocity (V_s) structure. The V_s model shows clear correlation with the known geologic features. The TFZ is associated with a high-velocity anomaly zone in the shallow crust, corresponding to metamorphic rocks due to magma intrusion. Low-velocity anomaly zones are mainly located to the west of the TFZ, caused by thick sedimentary layers in the Hefei Basin. Our study shows that, with ambient noise data recorded on a dense array and an advanced surface-wave inversion method, we can image detailed structure around the fault zone.

Key words: Tanlu fault zone, magma intrusion, ambient noise tomography, dense seismic array, direct surface-wave tomography.

1. Introduction

Ambient noise tomography (ANT) has emerged as one of the powerful geophysical methods to study Earth structures at different scales over the last ~ 15 years. It is capable of turning seismic "noise" into useful signals to retrieve surface waves between stations based on cross-correlation and stacking analysis of ambient noise. The retrieved surface waves can then be used to invert high-resolution images of structures in the crust and upper mantle. Since the first ambient noise tomographic images of Rayleigh wave group velocity for southern California (Sabra et al. 2005; Shapiro et al. 2005), this method has been applied successfully around the world to image high-resolution crustal structures, including at regional scales in the Tibetan Plateau (Yao et al. 2006, 2008), the western USA (Moschetti et al. 2007; Lin et al. 2008; Yang et al. 2008b), the Iberian Peninsula (Villasenor et al. 2007), Southern Africa (Yang et al. 2008a), Southern Korea (Cho et al. 2007), Japan (Nishida et al. 2008), and New Zealand (Lin et al. 2007) and at continental scales across Europe (Yang et al. 2007), Australia (Saygin and Kennett 2010), China (Zheng et al. 2008), and the USA (Bensen et al. 2008; Liang and Langston 2008; Lin et al. 2008) as well as other places.

Recently, ANT has been combined with local dense seismic arrays to image high-resolution structures of the shallow crust (e.g. Lin et al. 2013; Roux et al. 2016; Liu and Ben-Zion 2017). This method is not limited by the temporal and spatial distribution of seismic events and has thus become an important method for studying shallow structures (Li et al. 2016; Zhang et al. 2016; Liu et al. 2018).

In this study, we aim to apply the ANT method to image the detailed structure of the Tanlu fault zone (TFZ) in the shallow crust based on a local dense array, which is located near Chao Lake, to the southeast of Hefei City, eastern China. The TFZ

[1] Laboratory of Seismology and Physics of Earth's Interior, School of Earth and Space Sciences, University of Science and Technology of China, Hefei 230026, Anhui, China. E-mail: zhang11@ustc.edu.cn
[2] Geological Exploration Technology Institute of Anhui Province, Hefei, Anhui, China.
[3] Anhui Institute of Geological Survey, Hefei, Anhui, China.
[4] National Geophysical Observatory at Mengcheng, University of Science and Technology of China, Hefei 230026, Anhui, China.

strikes for a length of 2400 km of sinistral displacement in China, being one of the largest continental strike-slip faults in the world. It is regarded as an important mid-Cenozoic magmatic activity belt, metallogenic belt, as well as seismic belt in eastern China. According to structural and geochronological studies, the TFZ can be subdivided into southern (in Jiangsu and Anhui Provinces), middle (from Shandong Province to Shenyang in Liaoning Province), and northern (from Shenyang in Liaoning Province to Zhaoxing in Heilongjiang Province) segments (Deng et al. 2013; Zhang et al. 2007). The northern segment is closely related to the subduction of the western Pacific Plate (Xu and Zhu 1994; Zhang et al. 2007; Zhao et al. 2016), while the middle and southern segments are related to the subduction of the South China Plate to the North China Plate (Okay and Celal Sengör 1992; Yin and Nie 1993; Li 1994; Gilder et al. 1999; Wang et al. 2003; Zhu et al. 2009; Zhao et al. 2016). Overall, the middle and southern segments of the TFZ show the characteristics of vertical strike-slip fault.

Due to its complicated crustal structure, strong tectonic activity, and seismic activity, many studies have focused on the southern segment of the TFZ, including seismic tomography studies for determining crustal velocity structure (Ma et al. 1991; Dong et al. 1998; Yang et al. 1999a, b; Liu et al. 2009, 2015; Huang et al. 2011), fault activity studies for determining how the activity of the fault varies along the strike (Shi et al. 2003; Xu et al. 2014), electromagnetic studies for determining electrical structure (Xiao et al. 2009; Zhang et al. 2010), and integrated geophysical studies (Wang et al. 1995; Hao et al. 2004). However, the detailed structure of the southern segment of TFZ in the shallow crust is still scarcely studied.

In this study, we present the detailed structure of the TFZ near Chao Lake to the southeast of Hefei, eastern China, by applying a new direct surface-wave tomography method using the surface-wave dispersion data extracted from continuously recorded ambient noise on a local dense array. Compared with the traditional ANT method, the direct surface-wave tomography method considers ray-bending effects on surface-wave tomography in the shallow complex medium. This is the first time that a local dense seismic array has been used to characterize the shallow structure of the TFZ.

2. Ambient Noise Data Analysis

We deployed 53 short-period (5 s–100 Hz) QS-5A seismic stations around the TFZ to the southeast of Hefei, the capital city of Anhui Province. The temporary array continuously recorded the data for about 1 month from 17 March to 26 April 2017. The seismic array spans an area of about 30 km $\times$ 40 km with average station spacing of about 5 km (Fig. 1). The vertical component data were used for ANT in this study.

Following the standard ambient noise processing procedure (Bensen et al. 2007), the vertical component data were first decimated at a sampling rate of 50 Hz. We then removed the mean and trend of the data, and band-pass filtered the data in the frequency band of 0.2–10 s. To suppress the effects of earthquakes, we also performed spectral whitening and multiple-band temporal normalization on the data (Zhang et al. 2018). After these steps, we performed time-domain ambient noise cross-correlation between two stations for each hourly data with lag time from -60 to 60 s. All the hourly cross-correlation functions (CCFs) from each station pair were stacked by using normalized linear stacking method. The time-domain empirical Green's function was then derived from the time derivative of the CCFs (Sabra et al. 2005; Yao et al. 2006). We can observe the emergence of surface-wave signals (Fig. 2a) from cross-correlation functions with signal-to-noise ratio (SNR) greater than five in the 0.2–5 s period band. Here, the SNR is defined as the ratio of the maximum amplitude of the signal window of CCFs (-30 to $\sim$ 30 s) and the average absolute amplitude of the noise window (-60 to $\sim$ -30 s and 30 to $\sim$ 60 s). To check the azimuthal distribution of ambient noise sources, we compute the relative amplitude ratios of the positive- and negative-time CCFs for all station pairs. Most values of relative amplitude ratios are between 0.5 to 1.5 (Fig. 2b), which means there are small amplitude differences between positive- and negative-time CCFs. Thus, we can deduce that the noise source distribution does not have significant

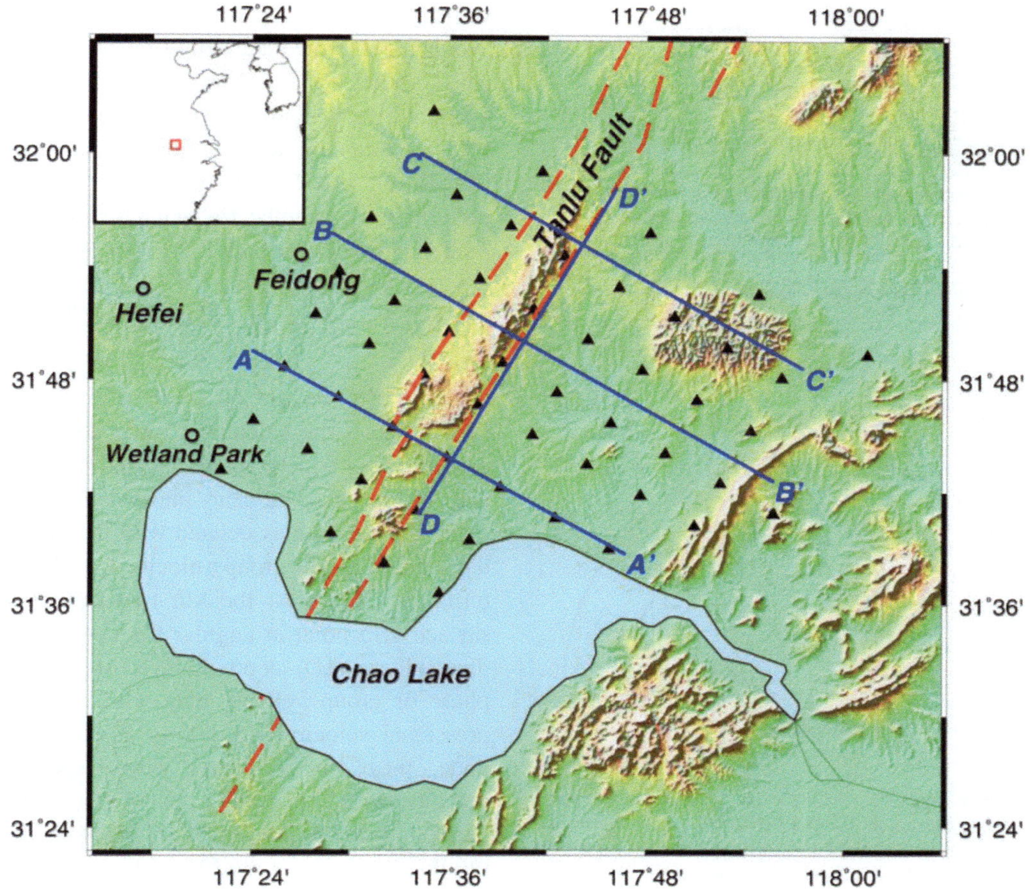

Figure 1
Distribution of seismic array stations (black triangles) and TFZ (red dashed lines) near Chao Lake located to the southeast of Hefei City. Red square in the upper-left inset shows the location of the survey. The blue lines marked by *AA'*, *BB'*, *CC'*, and *DD'* represent the profiles shown in Fig. 11

azimuthal variations, which will not cause large bias in dispersion measurements (Yao and Van der Hilst 2009). For this reason, we can expect the noise sources to be generally distributed randomly. After stacking the positive-time and negative-time parts linearly, we used the image analysis method by Yao et al. (2006, 2011) to extract the fundamental mode Rayleigh wave group and phase velocity dispersion curves.

Figure 3a and b show all the group and phase velocity dispersion curves for the period range of 0.2–5 s for the fundamental mode Rayleigh waves, respectively. The quantity of dispersion data decreases with increasing period. Above the 5 s period, no measurements can be made due to relatively short interstation distance. The velocity variation in the study region appears to be very large, as inferred from the dispersion curves; For example, at the period of 0.5 s, group velocity varies from 0.7 to 3 km/s and phase velocity changes from 0.8 to 2.9 km/s. In total, we obtained 482 dispersion data, including 246 group velocity measurements and 236 phase velocity measurements.

3. Direct Surface Wave Tomography of 3D V_s Model

We used a direct surface-wave tomography method (DSurfTomo) to simultaneously invert group and phase velocity dispersion data for the 3D shear

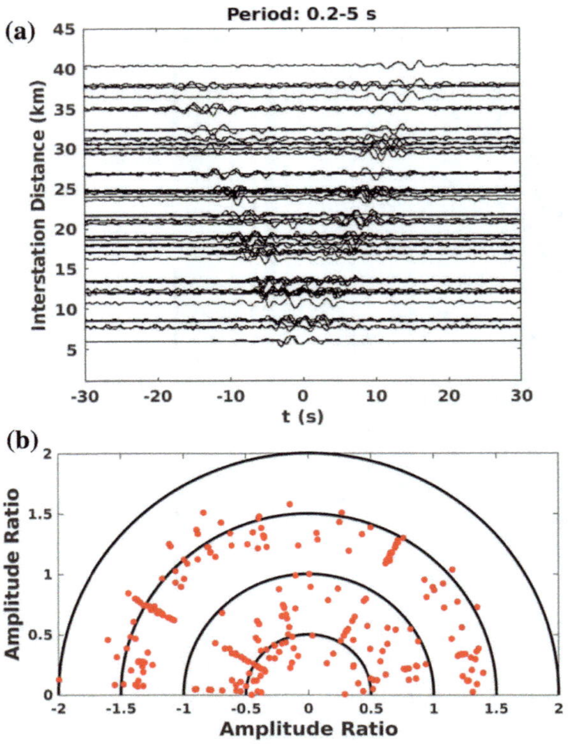

Figure 2
a Cross-correlation functions (CCFs) with signal-to-noise ratio (SNR) greater than five in the 0.2–5 s period band obtained from normalized linear stacking method. **b** Azimuthal distribution of the amplitude ratio (red dots) between the maximum envelope amplitudes of the positive-time and negative-time parts of CCFs in **a**. The azimuth of each red dot gives the azimuth direction of each station pair. Some relative amplitude ratios are linearly distributed, caused by the linear distribution of stations (Fig. 1)

velocity structure (Fang et al. 2015). This method avoids the intermediate step of inversion for group or phase velocity maps. Importantly, it considers ray-bending effects of surface waves due to complicated velocity structure based on the fast-marching method (Rawlinson and Sambridge 2004). The travel-time perturbation at angular frequency ω with respect to a reference model for path **i** is given by

$$\delta t_i(\omega) = t_i^{\mathrm{obs}} - t_i(\omega) \approx -\sum_{k=1}^{K} v_{ik} \frac{\delta C_k(\omega)}{C_k^2(\omega)}, \qquad (1)$$

where $t_i^{\mathrm{obs}}(\omega)$ is the observed surface-wave travel time, $t_i(\omega)$ is the calculated travel time from a reference model which can be updated in the inversion, and v_{ik} are the bilinear interpolation coefficients along the ray path associated with the **i**th travel-time data, the phase or group velocity $C_k(\omega)$ and its perturbation $\delta C_k(\omega)$ at the **k**th two-dimensional (2D) surface grid point at angular frequency ω. For one-dimensional (1D) depth kernels of Rayleigh wave phase or group velocity data with respect to compressional velocity and density at each surface grid node, they are converted with respect to shear velocity with the scaling factor $R_\alpha(z_j)$ and $R_\rho(z_j)$ derived from the empirical relationship of Brocher (2005) (see Fang et al. 2015). Thus, Eq. (1) can be rewritten as (Fang et al. 2015)

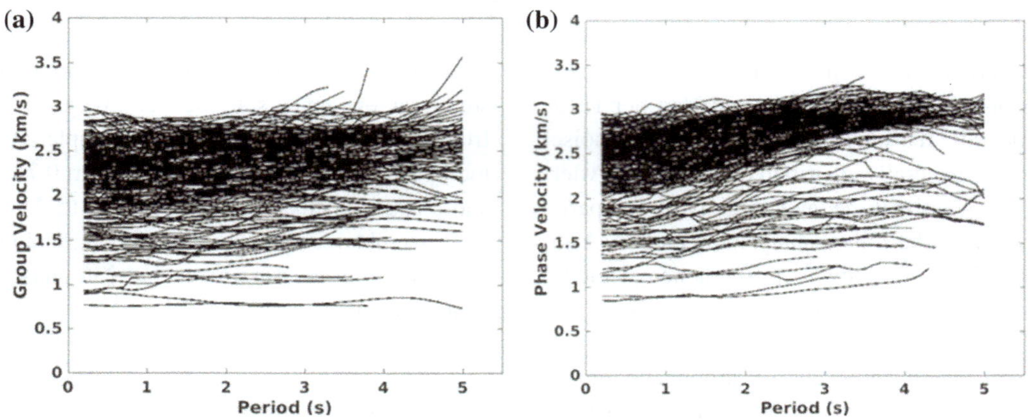

Figure 3
a Group velocity and **b** phase velocity dispersion curves in the 0.2–5 s period band

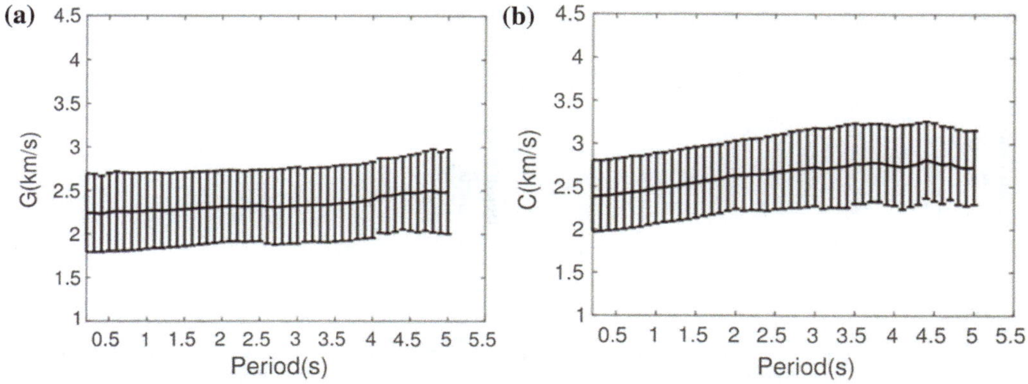

Figure 4
Average **a** group dispersion curve and **b** phase dispersion curve. Black bars represent standard deviations

$$\delta t_i(\omega) = \sum_{k=1}^{K}\left(-\frac{v_{ik}}{C_k^2}\right)$$
$$\sum_{j=1}^{J}\left[R'_\alpha(z_j)\frac{\partial C_k}{\partial \alpha_k(z_j)} + R'_\rho(z_j)\frac{\partial C_k}{\partial \rho_k(z_j)} + \frac{\partial C_k}{\partial \beta_k(z_j)}\right]\Bigg|_{\theta_k}$$
$$\delta\beta_k(z_j) = \sum_{l=1}^{M} G_{il}m_l, \tag{2}$$

where θ_k represents the 1D reference model at the **k**th surface grid point on the surface and $\alpha_k(z_j)$, $\beta_k(z_j)$, and $\rho_k(z_j)$ are the compressional velocity, shear velocity, and mass density at the jth node in the depth direction, respectively, and $M = KJ$, which is the total number of grid points of the 3D model.

The direct surface-wave tomography Eq. (2) can be further formulated into matrix form as

$$\mathbf{d} = \mathbf{Gm}, \tag{3}$$

where **d** is the surface-wave travel-time residual vector for all ray paths at different periods, **G** is the data sensitivity matrix, and **m** is the model parameter vector. To make the inversion stable, both damping and first-order smoothing regularizations are applied.

4. Inversion Details and Model Resolution Tests

In general, the selection of the initial model is important for linearized inversion. To determine an optimal 1D shear-wave velocity model, we used the average group and phase velocity dispersion curves for the study region (Fig. 4). An average 1D model is then determined by using the average dispersion curves, which is used as the initial model for the direct 3D surface-wave tomography (Fig. 5).

We first check the depth sensitivities of the surface-wave phase velocity and group velocity at different periods with respect to the compressional velocity, shear velocity, and density. At a given frequency f, the depth sensitivity kernels of phase velocity can be written as

$$\frac{\delta C(f)}{C(f)} = \int_0^R \left\{K_\alpha(f)\frac{\delta\alpha}{\alpha} + K_\beta(f) + K_\rho(f)\frac{\delta\rho}{\alpha\rho}\right\}\mathrm{d}z, \tag{4}$$

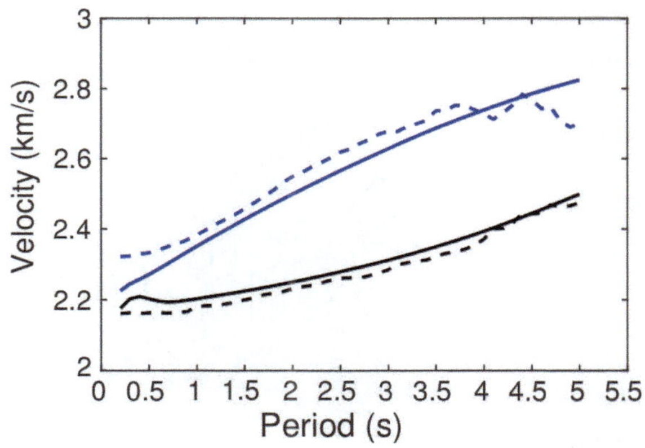

Figure 5
Observed and synthetic group and phase velocity dispersion curves. The blue dashed line represents the average phase dispersion curve, and the blue solid line represents the synthetic phase dispersion curve. The black dashed and solid line represent the average and the synthetic group dispersion curves, respectively

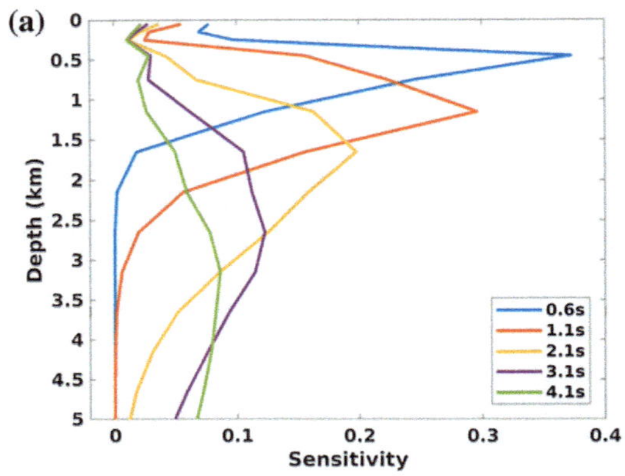

Figure 6
Depth sensitivity kernels for **a** phase velocity and **b** group velocity at five periods (0.6 s, 1.1 s, 2.1 s, 3.1 s, and 4.1 s)

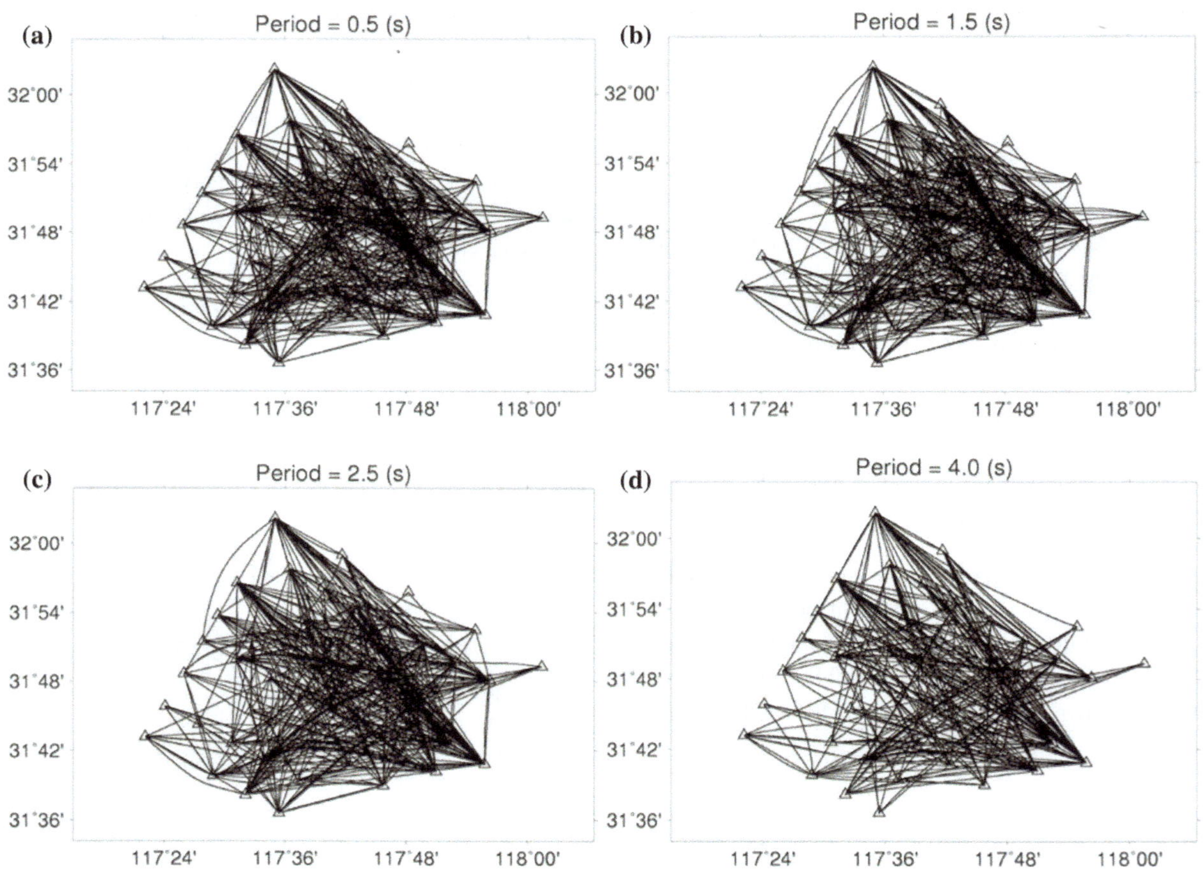

Figure 7
Ray-path coverage for four selected periods: **a** 0.5 s, **b** 1.5 s, **c** 2.5 s, and **d** 4.0 s. Triangles represent stations, and black lines represent ray paths

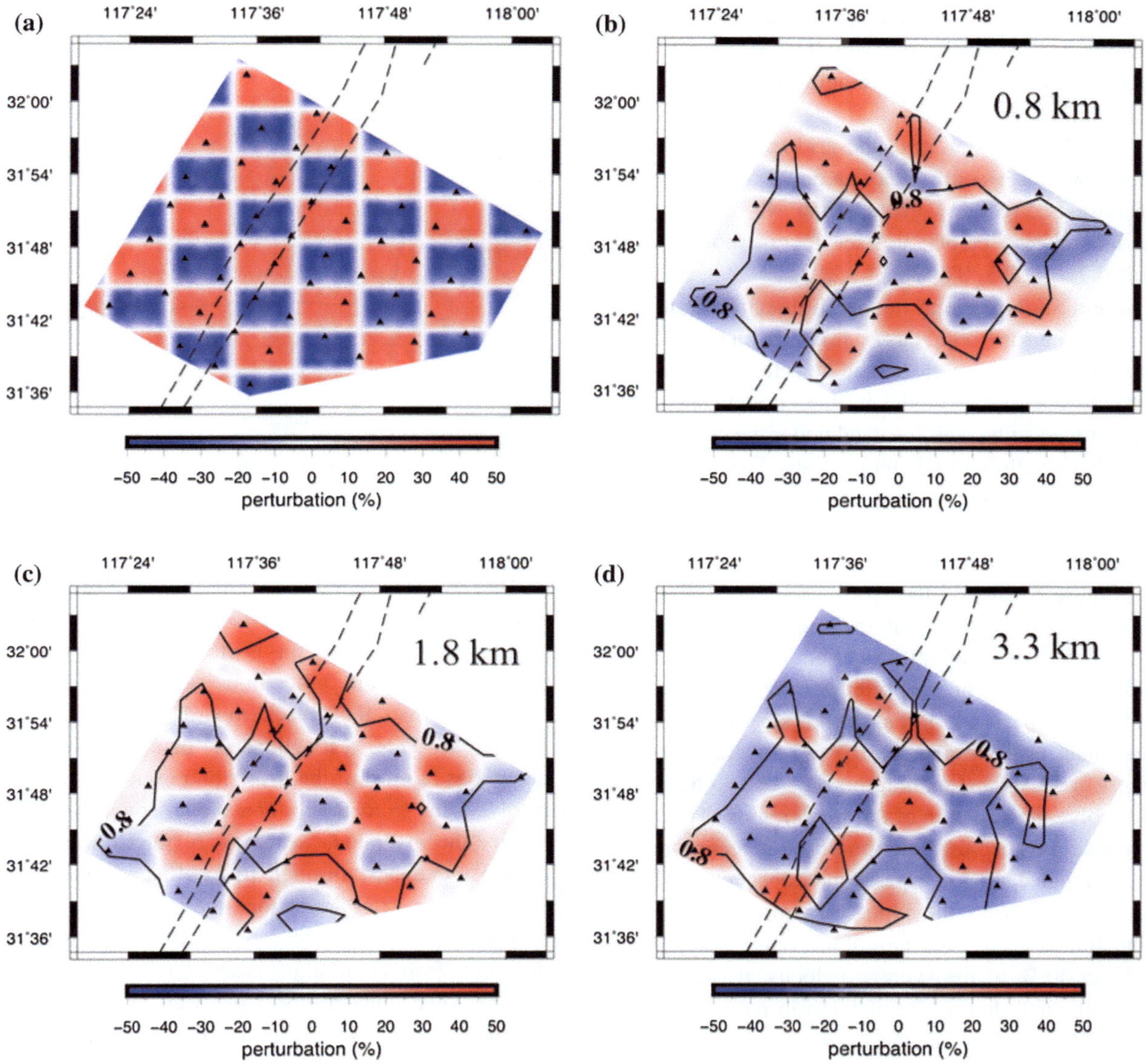

Figure 8
Checkerboard resolution tests of the inversion. **a** The input checkerboard model and **b–d** the recovered checkerboard models at depths of 0.8, 1.8, and 3.3 km, respectively. The black dashed lines represent TFZ. The black contour shows the semblance value of 0.8 between true and recovered checkerboard models

where z represents the depth, C is the Rayleigh wave group or phase velocity, and $K_\alpha(f)$, $K_\beta(f)$, and $K_\rho(f)$ are the depth sensitivity kernels for Rayleigh wave group or phase velocity to α, β, and ρ, respectively. Figure 6a, b show the depth sensitivity kernels for Rayleigh wave group and phase velocities at five periods (0.6, 1.1, 2.1, 3.1, and 4.1 s), using the average 1D velocity model obtained for the study area. It can be seen that, at the period of 0.6 s, the data have some sensitivity to the structure shallower than 400 m depth. At about 3.1 s, our dispersion data can have good sensitivity to structures around 3 km in depth. It can be seen that the surface-wave dispersion data between 0.2 and 5 s have the ability to resolve at depth down to 5 km. It can also be seen that the sensitivity kernels for the phase velocity and group velocity are complementary at the same frequency. As a result, we simultaneously use both phase velocity and group velocity dispersion curves for joint inversion.

For the inversion grid, the grid spacing is 0.015° along the north–south direction and 0.025° along the east–west direction, thus giving 40 × 40 = 1600 horizontal grid points. We set 15 grid points (at 0, 0.1, 0.2, 0.3, 0.5, 0.8, 1.3, 1.8, 2.3, 2.8, 3.3, 3.8, 4.3, 4.8, and 5.3 km) along the depth direction from surface to 5.3 km, with larger grid spacing at deeper region. The inversion grid is selected by trial and error to ensure the model is fine enough but at the same time can be well resolved under the current data distribution through resolution tests. To evaluate the model resolution, we first check the surface-wave path coverage at different periods and then conduct checkerboard resolution test.

Figure 7 shows the path coverage of phase velocity measurements at the four selected periods (0.5, 1.5, 2.5, and 4.0 s) based on the final 3D velocity model. The path number decreases from 235 at the 0.5 s period to 154 at the 4.0 s period. It can be seen that, at these periods, the ray coverage is relatively good, indicating that this dataset has the ability to resolve the structure of the study region. To further evaluate the spatial resolution of the inversion model at different depths, we performed checkerboard resolution tests with the input anomalies following a sine function, and the amplitude range varies from −50 to 50 % of the average velocity at each depth. Each anomaly has a size of 5 km in the north–south direction and 6 km in the east–west direction. Figure 8 shows the recovered checkerboard models at different depths with the same path distribution as the real data. It can be seen that checkerboard models are recovered well in the central part of the study region where the ray coverage is dense (Fig. 8).

To quantitatively measure the similarity between input and recovered checkerboard models, we calculate the semblance (S) between two models (Neidell and Taner 1971):

$$S = \frac{(\mathrm{d}V_\mathrm{r} + \mathrm{d}V_\mathrm{t})^2}{2(\mathrm{d}V_\mathrm{r}^2 + \mathrm{d}V_\mathrm{t}^2)}, \quad (5)$$

where $\mathrm{d}V_\mathrm{t}$ and $\mathrm{d}V_\mathrm{r}$ are V_s perturbations of true and recovered checkerboard models, respectively.

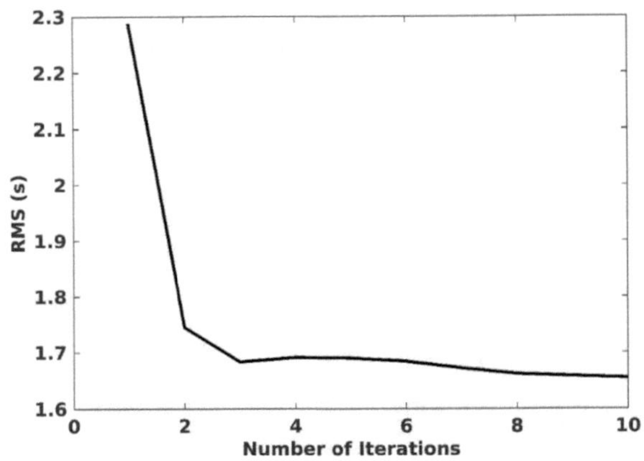

Figure 9
Variation of the RMS value of surface-wave travel-time residuals through the iterations

S denotes the similarity between recovered and true models, being closer to 1 if the two models are similar. We incorporate the contour value of 0.8 for semblance into the recovered checkerboard models (Fig. 8). It shows that the central part of the study region is within the 0.8 contour value of semblance, indicating that the checkerboard model is well recovered. This test suggests that the velocity anomalies in most of the inversion region can be well resolved due to relatively dense data coverage.

For the inversion, we carefully selected regularization parameters through trade-off analysis (Aster et al. 2013). After inversion, the root-mean-square (RMS) value of surface-wave travel-time residuals decreases from about 2.3 s to about 1.65 s (Fig. 9). It can be seen that the RMS value of travel-time residuals decreases quickly in the first three iterations. After that, the RMS residual value decreases slowly and converges after the 10th iteration.

5. Results and Discussion

Figure 10 shows the V_s model at different depths from the joint inversion of group and phase velocity dispersion data. Figure 11 shows the cross-sections of

Figure 10 ▶
Map view of the V_s model at depths of 0.2, 0.5, 1.3, 1.8, 2.3, 2.8, 3.3, and 3.8 km. Black triangles represent stations. Dashed line shows the TFZ

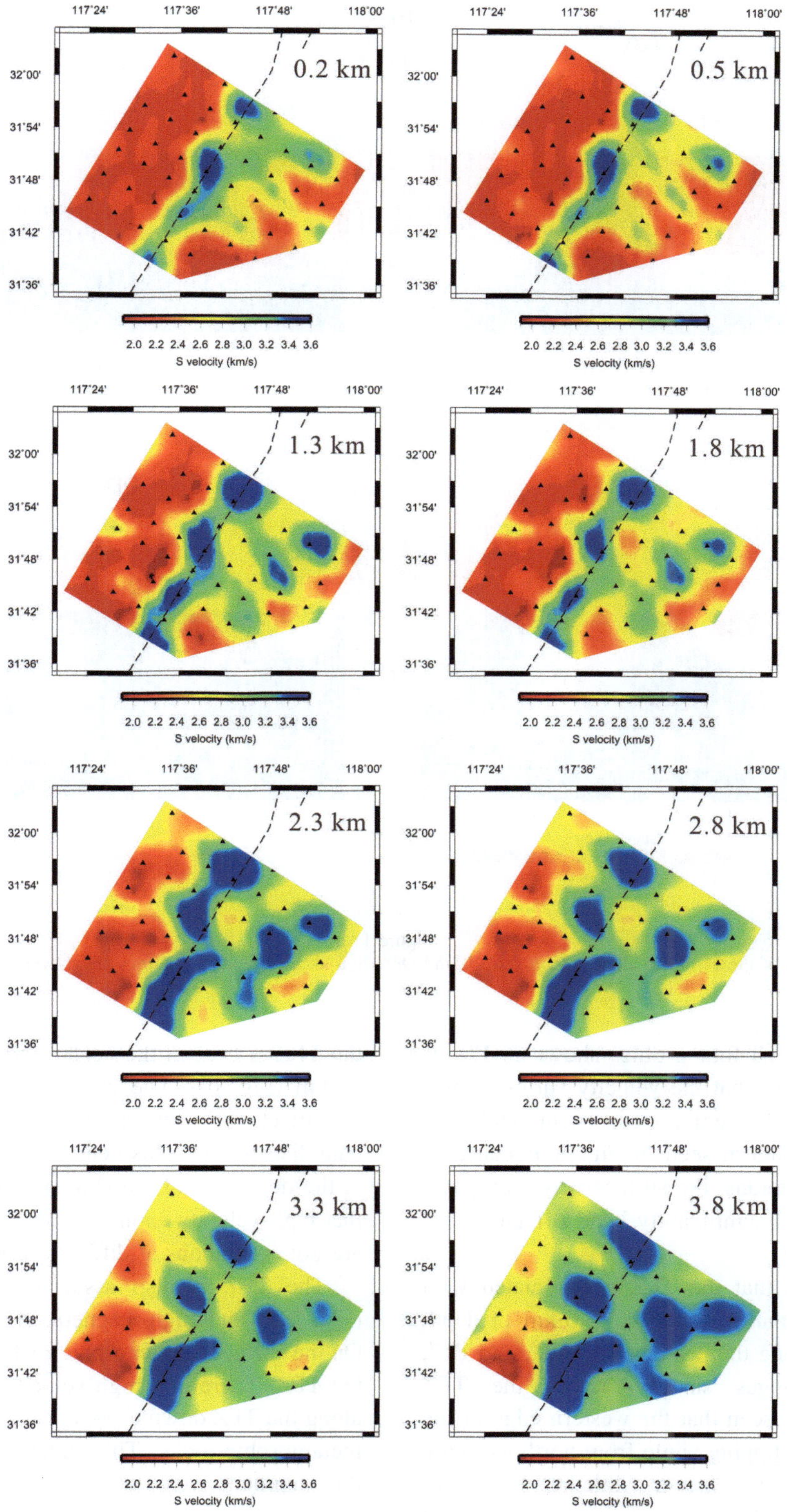
117°24'
117°36'
117°48'
118°00'
32°00'
31°54'
31°48'
31°42'
31°36'
0.2 km
0.5 km
1.3 km
1.8 km
2.3 km
2.8 km
3.3 km
3.8 km
2.0 2.2 2.4 2.6 2.8 3.0 3.2 3.4 3.6
S velocity (km/s)

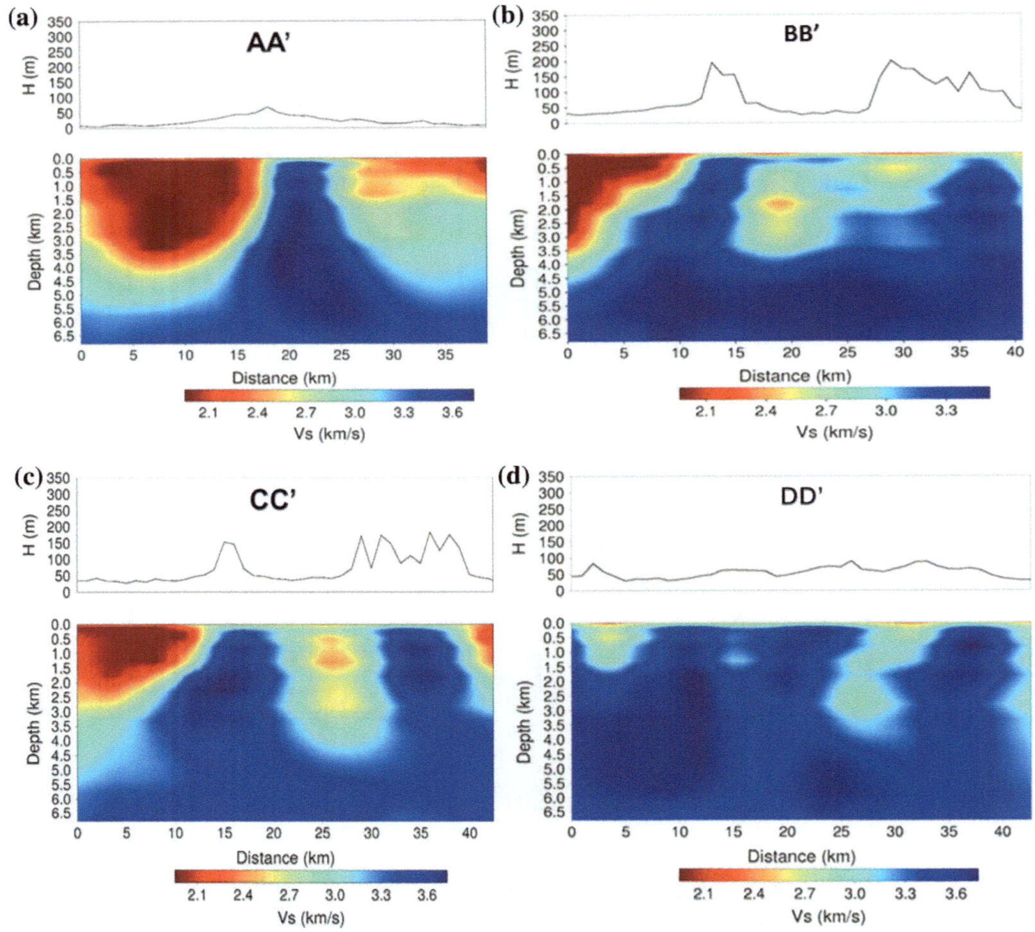

Figure 11
Cross-sections of the V_s model along *AA*′, *BB*′, *CC*′, and *DD*′. In **a–d**, topography variations are shown along the profile

the V_s model through four profiles shown in Fig. 1. The V_s model shows that the velocity varies significantly in the study region, consistent with the dispersion curve pattern seen in Fig. 3. Because of large velocity variations, the surface-wave ray paths at different periods significantly bend off the great-circle paths (Fig. 7).

It can be seen that the TFZ is associated with high-velocity anomalies down to ~ 6 km in depth (Figs. 10, 11). Along the profiles *AA*′, *BB*′, and *CC*′, the velocity changes sharply across the TFZ (Fig. 11). It can be seen that the western edge of the TFZ has different dipping angle from north to south. In the southern part it is steeper, while in the northern part it is gradual. In the map view, it can be seen that the TFZ is bent in the center and the offset is about 10 km (Fig. 10). The checkerboard test shows that the offset of 10 km can be resolved by the current data distribution. This bent shape can be clearly seen at the depth of 2.3 km (Fig. 10). It can also be seen that high-velocity anomalies associated with the TFZ are not continuous at different depths.

The TFZ is also associated with high-gravity (Fig. 12a) and high-magnetic (Fig. 12b) anomalies. On the surface, metamorphic rocks are exposed along the TFZ. Therefore, high-velocity anomalies imaged along the TFZ down to ~ 6 km can be interpreted as metamorphic rocks. This indicates that the TFZ in this segment may serve as channels for magma intrusion. According to Niu et al. (2005), large basalt

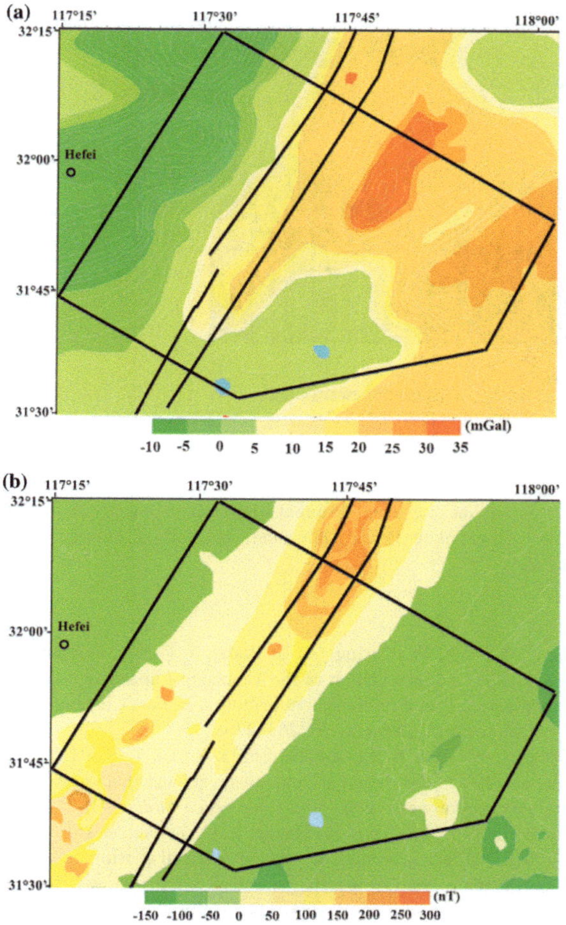

Figure 12
a Bouguer gravity and **b** aeromagnetic anomalies. Black lines represent the TFZ. Black box shows the study area

eruptions occurred in the Cenozoic era during the evolution of the TFZ. It was thought that the TFZ could cut into the upper mantle during the strong Pleistocene extension, and thus could constitute the channel for the basalt eruption (Niu et al. 2005; Zhu et al. 2001a, b). Our ambient noise tomography V_s model in the shallow crust for the first time provides direct evidence supporting this hypothesis.

To the northwest of the TFZ, low velocity anomalies reach down to ~ 3.5 km, especially at the southwest corner of the study region (Figs. 10, 11). They correspond to sedimentary layers in the Hefei basin. Evidently, the low velocity anomaly zone goes deeper in the southwest corner of the study region (Fig. 10), which is close to Chao Lake and is associated with the Hefei Binhu wetland park. The low velocity anomalies to the west of the TFZ are associated with low gravity anomalies. To the east of the TFZ, it is mainly composed of old strata of the Qingbaikou, Sinian, and Cambrian (Zhu et al. 2001a, b; Niu et al. 2005; Wang et al. 2016). The shear-wave velocity varies from ~ 2.5 to 3.0 km/s. About 20 km farther from the TFZ, there is an obvious high-velocity anomaly zone starting from depth of 1.5 km. It also corresponds to high-gravity anomalies but normal magnetic anomalies (Fig. 12). This indicates that this high-velocity anomaly zone may have different origins from the high-velocity zone along the TFZ.

We also compared our velocity model with a known geological profile along profile *BB'* (Fig. 13). Although the overall features are similar between the geophysical and geological profiles, there are apparent inconsistencies between them; For example, our high-resolution velocity result shows that the western edge of the TFZ dips gently, while the geological mapping gives a sharper edge. This indicates that ambient noise tomography using a local dense array can be very useful for depicting an accurate geological profile.

6. *Conclusions*

We applied direct surface-wave tomography using 1 month of ambient noise data recorded by a local dense array to investigate the structure of TFZ to the southeast of Hefei City, China. We are able to extract Rayleigh wave group and phase velocity dispersion curves for periods of 0.2–5 s. A high-resolution 3D shear-wave velocity structure shows high-velocity anomalies beneath the TFZ, which is consistent with locally high-gravity and high-magnetic anomalies. They are associated with metamorphosed rocks exposed on the surface, likely caused by magma intrusion via the TFZ. The Hefei basin is characterized by low velocity anomalies down to ~ 3.5 km due to thick sedimentary layers. Overall, our results agree well with the local geology and can provide strong constraints on geological modeling in this region.

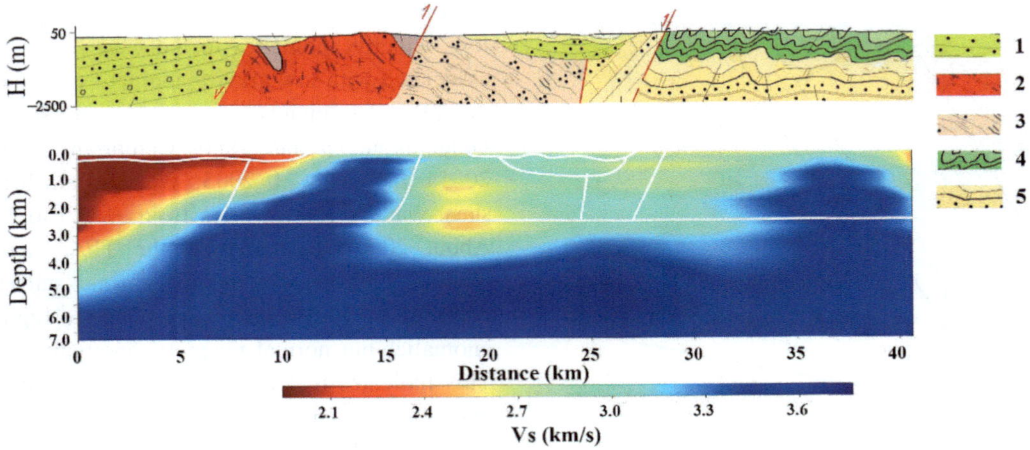

Figure 13
Local geological section along the same *S* wave velocity profile *BB′* in Fig. 9. White lines represent different stratigraphic units by geological mapping. 1 = Upper Cretaceous sandstone; 2 = Paleoproterozoic gneiss; 3 = Qingbaikou Period limestone; 4 = Early Cambrian shale; 5 = Sinian limestone

Acknowledgements

Most figures made with Generic Mapping Tools (GMT) (Wessel and Smith 1995). The authors would like to thank two anonymous reviewers, Cheng Li, and Qiushi Zhai for their constructive comments on this paper. This research is supported by the Department of Land and Resources of Anhui Province (grant no. 2015-g-35). H.Y. is partially supported by the National Natural Science Foundation of China (grant no. 41790464).

References

Aster, R. C., Borchers, B., & Thurber, C. H. (2013). *Parameter estimation and inverse problems* (2nd ed.). Academic Press.

Bensen, G. D., Ritzwoller, M. H., Barmin, M. P., Levshin, A. L., Lin, F., Moschetti, M. P., et al. (2007). Processing seismic ambient noise data to obtain reliable broad-band surface wave dispersion measurements. *Geophysical Journal International, 169*(3), 1239–1260.

Bensen, G. D., Ritzwoller, M. H., & Shapiro, N. M. (2008). Broadband ambient noise surface wave tomography across the United States. *Journal of Geophysical Research: Solid Earth, 113*, B05306. https://doi.org/10.1029/2007JB005248.

Brocher, T. M. (2005). Empirical relations between elastic wave-speeds and density in the Earth's crust. *Bulletin of the Seismological Society of America, 95*(6), 2081–2092.

Cho, K. H., Herrmann, R. B., Ammon, C. J., & Lee, K. (2007). Imaging the upper crust of the Korean Peninsula by surface-wave tomography. *Bulletin of the Seismological Society of America, 97*(1B), 198–207.

Deng, Y., Fan, W., Zhang, Z., & Badal, J. (2013). Geophysical evidence on segmentation of the Tancheng-Lujiang fault and its implications on the lithosphere evolution in East China. *Journal of Asian Earth Sciences, 78*, 263–276.

Dong, S. W., Wu, X. Z., Gao, R., Lu, D. Y., Li, Y. K., He, Y. Q., et al. (1998). On the crust velocity levels and dynamics of the Dabieshan orogenic belt. *Acta Geophysica Sinica, 41,* 361–369.

Fang, H., Yao, H., Zhang, H., Huang, Y. C., & van der Hilst, R. D. (2015). Direct inversion of surface wave dispersion for three-dimensional shallow crustal structure based on ray tracing: Methodology and application. *Geophysical Journal International, 201*(3), 1251–1263.

Gilder, S. A., Leloup, P. H., Courtillot, V., Chen, Y., Coe, R. S., Zhao, X., et al. (1999). Tectonic evolution of the Tancheng-Lujiang (Tan-Lu) fault via middle Triassic to Early Cenozoic paleomagnetic data. *Journal of Geophysical Research: Solid Earth, 104*(B7), 15365–15390.

Hao, T. Y., Suh, M., & Liu, J. H. (2004). Deep structure and boundary belt position between Sino-Korean and Yangtze blocks in Yellow Sea. *Earth Science Frontiers, 11,* 51–62.

Huang, Y., Li, Q. H., Zhang, Y. S., Sun, Y. J., Bi, X. M., Jin, S. M., et al. (2011). Crustal velocity structure beneath the Shandong-Jiangsu-Anhui segment of the Tancheng-Lujiang fault zone and adjacent areas. *Chinese Journal of Geophysics (in Chinese), 54*(10), 2549–2559. https://doi.org/10.3969/j.issn.0001-5733.2011.10.012.

Li, Z.-X. (1994). Collision between the North and South China blocks: A crustal-detachment model for suturing in the region east of the Tanlu fault. *Geology, 22*(8), 739–742.

Li, C., Yao, H., Fang, H., Huang, X., Wan, K., Zhang, H., et al. (2016). 3D near-surface shear-wave velocity structure from ambient-noise tomography and borehole data in the Hefei urban area, China. *Seismological Research Letters, 87*(4), 882–892.

Liang, C., & Langston, C. A. (2008). Ambient seismic noise tomography and structure of eastern North America. *Journal of*

Geophysical Research: Solid Earth, 113, B03309. https://doi.org/10.1029/2007JB005350.

Lin, F. C., Moschetti, M. P., & Ritzwoller, M. H. (2008). Surface wave tomography of the western United States from ambient seismic noise: Rayleigh and Love wave phase velocity maps. *Geophysical Journal International, 173*(1), 281–298.

Lin, F. C., Ritzwoller, M. H., Townend, J., Bannister, S., & Savage, M. K. (2007). Ambient noise Rayleigh wave tomography of New Zealand. *Geophysical Journal International, 170*(2), 649–666.

Lin, F. C., Li, D., Clayton, R. W., & Hollis, D. (2013). High-resolution 3D shallow crustal structure in Long Beach, California: application of ambient noise tomography on a dense seismic arrayNoise tomography with a dense array. *Geophysics, 78*(4), Q45–Q56.

Liu, X., & Ben-Zion, Y. (2017). Analysis of non-diffuse characteristics of the seismic noise field in southern California based on correlations of neighbouring frequencies. *Geophysical Journal International, 212*(2), 798–806.

Liu, Y., Wang, Q., Jiang, M., & Wang, Y. J. (2009). The receiver function image of the deep structure in Sulu orogenic belt. *Acta Petrologica Sinica (in Chinese), 25*(7), 1658–1662.

Liu, Y., Zhang, H., Fang, H., Yao, H., & Gao, J. (2018). Ambient noise tomography of three-dimensional near-surface shear-wave velocity structure around the hydraulic fracturing site using surface microseismic monitoring array. *Journal of Applied Geophysics, 159,* 209–217.

Liu, B., Zhu, G., Zhang, M. J., Gu, C. C., Song, L. H., & Liu, S. (2015). Features and genesis of active faults in the Anhui segment of the Tan-Lu fault zone. *Chinese Journal of Geology (Scientia Geologica Sinica), 2,* 017.

Ma, X. Y., Liu, C. Q., & Liu, G. D. (1991). Xiangshui (Jiangsu Province) to Mandal (Nei Monggol) geoscience transect. *Acta Geologica Sinica, 3,* 199–215.

Moschetti, M. P., Ritzwoller, M. H., & Shapiro, N. M. (2007). Surface wave tomography of the western United States from ambient seismic noise: Rayleigh wave group velocity maps. *Geochemistry, Geophysics, Geosystems, 8*, Q08010. https://doi.org/10.1029/2007GC001655.

Neidell, N. S., & Taner, M. T. (1971). Semblance and other coherency measures for multichannel data. *Geophysics, 36*(3), 482–497.

Nishida, K., Kawakatsu, H., & Obara, K. (2008). Three-dimensional crustal wave velocity structure in Japan using microseismic data recorded by Hi-net tiltmeters. *Journal of Geophysical Research: Solid Earth, 113*, B10302. https://doi.org/10.1029/2007JB005395.

Niu, M. L., Zhu, G., Liu, G. S., Song, C. Z., & Wang, D. (2005). Cenozoic volcanic activities and deep processes in the middle-south sector of the Tan-Lu fault zone. *Chinese Journal of Geology, 40,* 390–403.

Okay, A. I., & Celal Şengör, A. M. (1992). Evidence for intracontinental thrust-related exhumation of the ultra-high-pressure rocks in China. *Geology, 20*(5), 411–414.

Rawlinson, N., & Sambridge, M. (2004). Wave front evolution in strongly heterogeneous layered media using the fast marching method. *Geophysical Journal International, 156*(3), 631–647.

Roux, P., Moreau, L., Lecointre, A., Hillers, G., Campillo, M., Ben-Zion, Y., et al. (2016). A methodological approach towards high-resolution surface wave imaging of the San Jacinto Fault Zone using ambient-noise recordings at a spatially dense array. *Geophysical Journal International, 206*(2), 980–992.

Sabra, K. G., Gerstoft, P., Roux, P., Kuperman, W. A., & Fehler, M. C. (2005). Surface wave tomography from microseisms in Southern California. *Geophysical Research Letters, 32*, L14311. https://doi.org/10.1029/2005GL023155.

Saygin, E., & Kennett, B. L. (2010). Ambient seismic noise tomography of Australian continent. *Tectonophysics, 481*(1–4), 116–125.

Shapiro, N. M., Campillo, M., Stehly, L., & Ritzwoller, M. H. (2005). High-resolution surface-wave tomography from ambient seismic noise. *Science, 307*(5715), 1615–1618.

Shi, W., Zhang, Y. Q., & Dong, S. W. (2003). Quaternary activity and segmentation behavior of the middle portion of the Tan-Lu fault zone. *Acta Geoscientia Sinica, 24*(1; ISSU 73), 11–18.

Villasenor, A., Yang, Y., Ritzwoller, M. H., & Gallart, J. (2007). Ambient noise surface wave tomography of the Iberian Peninsula: Implications for shallow seismic structure. *Geophysical Research Letters, 34*, L11304. https://doi.org/10.1029/2007GL030164.

Wang, L., Li, C., Shi, Y., & Wang, Y. (1995). Distributions of geotemperature and terrestrial heat flow density in lower Yangtze area. *Chinese Journal of Geophysics, 38*(04), 469–476.

Wang, E., Meng, Q., Burchfiel, B. C., & Zhang, G. (2003). Mesozoic large-scale lateral extrusion, rotation, and uplift of the Tongbai-Dabie Shan belt in east China. *Geology, 31*(4), 307–310.

Wang, W., Song, C., Ren, S., Li, J., Zhang, Y., Wang, P., et al. (2016). PT conditions and zircon U-Pb analysis of the Taoyuan ductile shear zone in Tan-Lu fault zone. *Acta Petrologica Sinica, 3,* 011.

Wessel, P., & Smith, W. H. (1995). New version of the generic mapping tools. *Eos, Transactions American Geophysical Union, 76*(33), 329–329.

Xiao, Q., Zhao, G., Wang, J., Zhan, Y., Chen, X., Tang, J., et al. (2009). Deep electrical structure of the Sulu orogen and neighboring areas. *Science in China, Series D: Earth Sciences, 52*(3), 420–430.

Xu, R., Chen, S. J., & Wang, F. (2014). Study on tectonic stress field in Yishu fault zone and its adjacent areas. *China Earthquake Engineering Journal (in Chinese), 36*(1), 158–169.

Xu, J. W., & Zhu, G. (1994). Tectonic models of the Tan-Lu fault zone, eastern China. *International Geology Review, 36*(8), 771–784.

Yang, W. C., Chen, Z. Y., Chen, G. J., Hu, Z. Y., & Bai, J. (1999a). Geophysical investigations in northern Sulu UHPM belt, Part I: Deep seismic reflection. *Chinese Journal of Geophysics-Chinese Edition, 42*(1), 41–52.

Yang, W. C., Hu, Z. Y., Cheng, Z. Y., Ni, C. C., Fang, H., & Bai, J. (1999b). Long profile of geophysical investigation from Tanchen to Lianshui, East-Central China. *Chinese Journal of Geophysics-Chinese Edition, 42*(2), 206–217.

Yang, Y., Li, A., & Ritzwoller, M. H. (2008a). Crustal and uppermost mantle structure in southern Africa revealed from ambient noise and teleseismic tomography. *Geophysical Journal International, 174*(1), 235–248.

Yang, Y., Ritzwoller, M. H., Levshin, A. L., & Shapiro, N. M. (2007). Ambient noise Rayleigh wave tomography across Europe. *Geophysical Journal International, 168*(1), 259–274.

Yang, Y., Ritzwoller, M. H., Lin, F. C., Moschetti, M. P., & Shapiro, N. M. (2008b). Structure of the crust and uppermost mantle beneath the western United States revealed by ambient noise and earthquake tomography. *Journal of Geophysical Research: Solid Earth, Earth,* 113(B12).

Yao, H., Beghein, C., & Van Der Hilst, R. D. (2008). Surface wave array tomography in SE Tibet from ambient seismic noise and two-station analysis—II. Crustal and upper-mantle structure. *Geophysical Journal International, 173*(1), 205–219.

Yao, H., Gouedard, P., Collins, J. A., McGuire, J. J., & van der Hilst, R. D. (2011). Structure of young East Pacific Rise lithosphere from ambient noise correlation analysis of fundamental- and higher-mode Scholte–Rayleigh waves. *Comptes Rendus Geoscience, 343*(8–9), 571–583.

Yao, H., & Van Der Hilst, R. D. (2009). Analysis of ambient noise energy distribution and phase velocity bias in ambient noise tomography, with application to SE Tibet. *Geophysical Journal International, 179*(2), 1113–1132.

Yao, H., van Der Hilst, R. D., & De Hoop, M. V. (2006). Surface-wave array tomography in SE Tibet from ambient seismic noise and two-station analysis—I. Phase velocity maps. *Geophysical Journal International, 166*(2), 732–744.

Yin, A., & Nie, S. (1993). An indentation model for the North and South China collision and the development of the Tan-Lu and Honam fault systems, eastern Asia. *Tectonics, 12*(4), 801–813.

Zhang, B. L., Li, Z. W., Bao, F., et al. (2016). Shallow shear-wave velocity structures under the Weishan volcanic cone in Wudalianchi volcano field by microtremor survey. *Chinese Journal of Geophysics (in Chinese), 59*(10), 3662–3673. https://doi.org/10.6038/cjg20161013.

Zhang, P., Wang, L. S., Zhong, K., & Ding, Z. Y. (2007). Research on the segmentation of Tancheng-Lujiang fault zone. *Geological Review, 53*(5), 586–591.

Zhang, Y., Yao, H., Yang, H. Y., Cai, H. T., Fang, H., Xu, J., et al. (2018). 3-D crustal shear-wave velocity structure of the Taiwan Strait and Fujian, SE China, revealed by ambient noise tomography. *Journal of Geophysical Research: Solid Earth, 123*(9), 8016–8031.

Zhang, J. H., Zhao, G. Z., Xiao, Q. B., & Tang, J. (2010). Analysis of electric structure of the central Tan-Lu fault zone (Yi-Shu fault zone, 36 N) and seismogenic condition. *Chinese Journal of Geophysics (in Chinese), 53*(3), 605–611.

Zhao, T., Zhu, G., Lin, S., & Wang, H. (2016). Indentation-induced tearing of a subducting continent: evidence from the Tan-Lu Fault Zone, East China. *Earth-Science Reviews, 152,* 14–36.

Zheng, S., Sun, X., Song, X., Yang, Y., & Ritzwoller, M. H. (2008). Surface wave tomography of China from ambient seismic noise correlation. *Geochemistry, Geophysics, Geosystems, 9*, Q05020. https://doi.org/10.1029/2008GC001981.

Zhu, G., Liu, G. S., Niu, M. L., Xie, C. L., Wang, Y. S., & Xiang, B. (2009). Syn-collisional transform faulting of the Tan-Lu fault zone, East China. *International Journal of Earth Sciences, 98*(1), 135–155.

Zhu, G., Song, C., Wang, D., Liu, G., & Xu, J. (2001a). Studies on 40 Ar/39 Ar thermochronology of strike-slip time of the Tan-Lu fault zone and their tectonic implications. *Science in China, Series D: Earth Sciences, 44*(11), 1002–1009.

Zhu, G., Wang, D. X., Liu, G. S., Song, C. Z., Xu, J. W., & Niu, M. L. (2001b). Extensional activities along the Tan-Lu fault zone and its geodynamic setting. *Chinese Journal of Geology, 36*(3), 269–278.

(Received June 25, 2018, revised October 17, 2018, accepted November 3, 2018, Published online November 21, 2018)

Pure Appl. Geophys. 176 (2019), 1207–1224

https://doi.org/10.1007/s00024-018-2050-3

Pure and Applied Geophysics

Coseismic Radiation of the 2008 M_w 7.9 Wenchuan Earthquake and Its Relationship to Fault Complexities

Taizi Huang[1] and Huajian Yao[1,2]

Abstract—The long rupture length and complex fault geometries of the 2008 M_w 7.9 Wenchuan earthquake offer a great opportunity to investigate the relationship between coseismic radiation and fault complexities. In this study, we first improved the compressive-sensing back-projection method (CS-BP) with window time adjustment and parallel computation to enhance its spatial–temporal resolution and computational efficiency. Then we applied the CS-BP to invert for the coseismic radiation of the Wenchuan earthquake in multiple frequency bands. Our results show that the coseismic radiation of the Wenchuan earthquake is frequency-dependent and strongly associated with the fault geometries: low-frequency radiation sources are concentrated on smooth fault segments with large coseismic slip, while higher-frequency sources are mainly related to fault complexities (e.g., Xiaoyudong fault, Gaochuan fault-bend, Leigu fault-bend, and Nanba fault step-over). An acceleration of rupture speed from ~ 1.5 to ~ 3.0 km/s is also observed the near Xiaoyudong fault, where strong high-frequency energies were radiated. Finally, we discuss the role of rupture speed in mediating the relationship between coseismic radiation and fault complexities. On one hand, sudden changes of rupture speeds contribute to strong high-frequency radiation. On the other hand, fault complexities can also strongly affect the rupture speed: a sudden break of a large asperity is able to propel the acceleration, while rupture front may decelerate when it jumps across geometric barriers.

Key words: Wenchuan earthquake, rupture, compressive sensing, coseismic radiation, fault complexity.

1. Introduction

Great earthquakes commonly cause grievous fatalities and massive financial losses for human beings, so understanding the physics of their rupture processes is crucial to mitigate risks of earthquake hazards. Even though it is still challenging to simulate the real dynamic rupture process of a natural earthquake, much effort has been devoted to imaging the slip history and seismic radiation in order to better describe the kinematic rupture process. Implementation of a back-projection method serves this purpose.

The classic back-projection method (Ishii et al. 2005; Kruger and Ohrnberger 2005; Wang and Mori 2011; Zhang and Ge 2010) and the high-resolution MUSIC back-projection method (Meng et al. 2011) back-project teleseismic waveforms of dense arrays to the source region to obtain the spatial–temporal distribution of subevents (local energy bursts), by which the rupture process can be effectively depicted. Because this slant-stacking or time-reversal process requires no Green's functions, it is very efficient and fast to obtain the extent and duration of great earthquakes. Yao et al. (2011) extended the back-projection method into the compressive-sensing back-projection method (CS-BP) based on the assumption of sparse distribution of seismic radiation sources. The CS-BP is essentially a frequency-domain back-projection method, while it has advantages over the classic back-projection methods mainly in several aspects: a higher spatial resolution, ability to resolve multiple sources, and better capability to resolve frequency-dependent seismic radiation sources (Yin and Yao 2016). With the application of back-projection, many studies revealed that seismic energies of different frequency bands are radiated from varying locations of the source region (Allmann and Shearer 2007; Yao et al. 2011). Low-frequency radiation is mainly located where large slip occurs (e.g., Wang and Mori 2011; Yao et al. 2011, 2013; Yin and Yao 2016); higher-frequency energies are

[1] Laboratory of Seismology and Physics of Earth's Interior, School of Earth and Space Sciences, University of Science and Technology of China, Hefei 230026, China. E-mail: hjyao@ustc.edu.cn

[2] National Geophysical Observatory at Mengcheng, University of Science and Technology of China, Anhui, China.

more concentrated near the edges of large slip patches (Uchide et al. 2013; Yin et al. 2018) or small slip area (Meng et al. 2011; Yao et al. 2011, 2013). Furthermore, ruptures at fault complexities such as the step-over typically emit a large amount of high-frequency energies (Okuwaki and Yagi 2018; Uchide et al. 2013; Zhang and Ge 2010). In-depth research on the relationship between high-frequency radiation and fault complexities can help to constrain the dynamic processes of earthquake rupture.

On May 12, 2008, the M_w 7.9 Wenchuan earthquake struck northwest of the Sichuan province. Figure 1 shows the source region. The event is located at the eastern margin of the Tibetan Plateau, which has experienced strong crustal shortening and uplift of the Longmen Shan range (Hubbard and Shaw 2009), generating two parallel northeast-trending faults named the Beichuan Fault (BCF) and the Pengguan Fault (PGF). Despite different fault models and a combination of different datasets, including geodetic, teleseismic, and strong-motion data, slip distribution resolved by many studies shared a similar pattern (Fielding et al. 2013; Hao et al. 2009; Hartzell et al. 2013; Shen et al. 2009; Tong et al. 2010). This event propagated almost unilaterally northeastwards and ruptured BCF and PGF simultaneously. During the rupture, the fault type transforms from dominantly thrusting to purely strike-slip. Two maximum slip patches are located near Yingxiu and Beichuan, and another relatively smaller slip region is near Nanba. Back-projection images (Xu et al. 2009a; Zhang and Ge 2010) also revealed the unilateral rupture propagation, which estimated a rupture length of over 290 km and a rupture velocity of 2.8–3.0 km/s, which is consistent with slip inversion results.

Field surveys, fault geometry inversion, and aftershock relocation results all revealed the complexity of this rupture process (Liu-Zeng et al. 2009; Shen et al. 2009; Xu et al. 2009b; Yi et al. 2012). The 240-km-long surface rupture of BCF is composed of piece-wise subfaults characterized by many fault complexities including the secondary fault, fault-bend, and step-over. In the southern part, the northwest-trending Xiaoyudong fault (XDF) intersects the two major faults perpendicularly with a surface rupture of ~ 8 km (Chang et al. 2012; Liu-Zeng et al. 2012). The dominant strike-slip aftershocks around the area extends further northwest, indicating the strain partition during the coseismic rupture (Chang et al. 2012). In the middle part, the right-stepping dextral Gaochuan fault-bend and the left-stepping Leigu fault-bend near Beichuan county link the south and north of BCF. The transtension environment has generated several normal-fault aftershocks (Fig. 1). In the northern part, the Nanba step-over is located near Nanba county, where the dipping angle changes abruptly (Shen et al. 2009; Wan et al. 2017). The complex fault geometries of the Wenchuan earthquake offer a great opportunity to investigate their relationship with coseismic radiation. Okuwaki and Yagi (2018) found that the high-frequency radiation of the Wenchuan earthquake is associated with those geometric complexities acting as barriers.

In this paper, we improved the compressive-sensing back-projection method and applied it to the inversion of the spatial–temporal distribution of coseismic radiation sources in different frequency bands of the Wenchuan earthquake. By comparing our results with the fault geometry given by previous studies, we discussed the spatial correlation and physical mechanisms between the fault complexities and the spatial distribution of high-frequency coseismic radiation.

2. *Data and Methodology*

We downloaded the first 150-s vertical-component direct P waveforms (velocity seismograms) with the epicentral distance range of 66°–82° and azimuthal range of 23°–34° from the stations in Alaska through the Incorporated Research Institutions for Seismology–Data Management Center (IRIS–DMC). First, we removed the instrumental response, trend, and mean of waveforms, and deleted waveforms with signal–noise-ratios (SNRs) less than 10. Next, the first 10-s P waveforms of all seismograms were aligned by the multichannel cross-correlation (Vandecar and Crosson 1990; Ishii et al. 2005) to suppress the 3-D path effect of structural heterogeneities, and we only kept seismograms whose cross-correlation coefficients are larger than 0.7. Finally, signals were

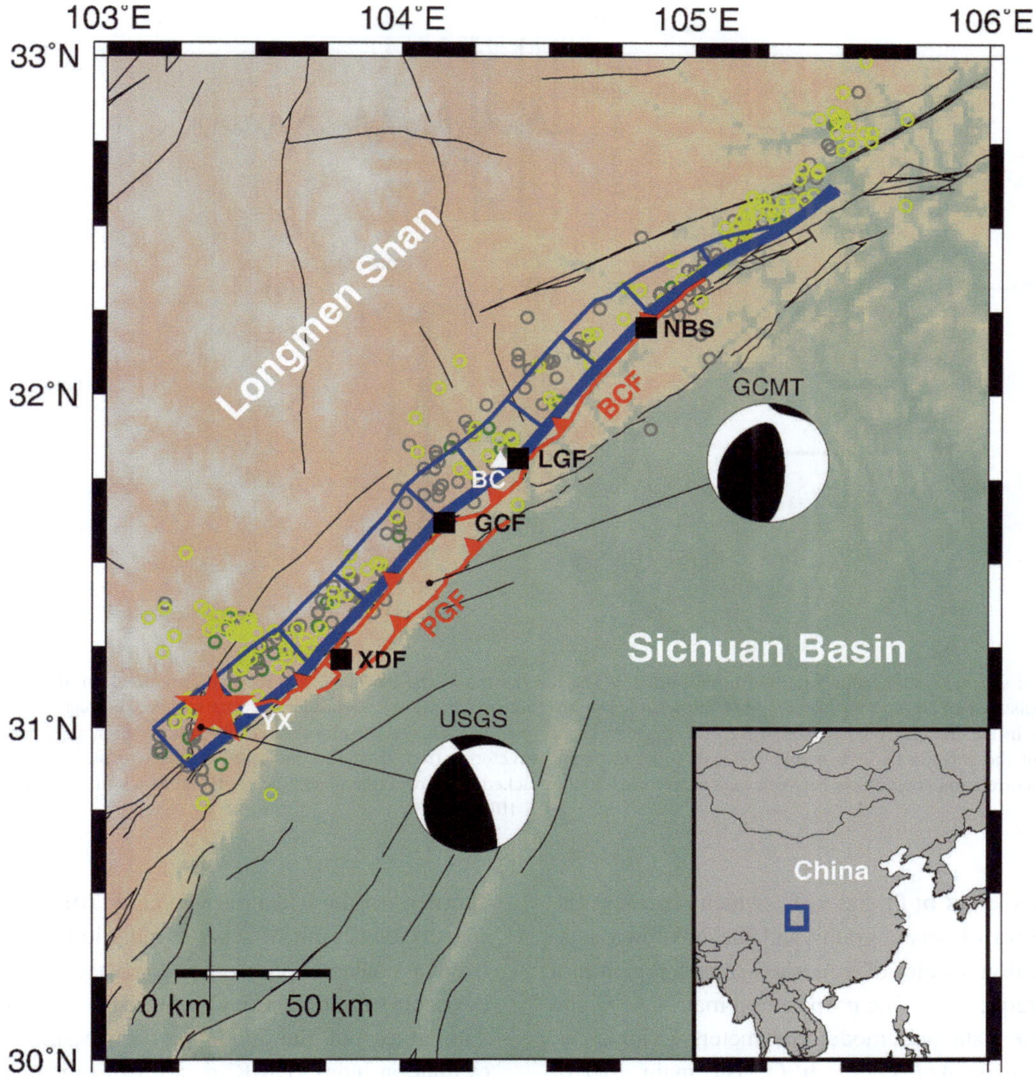

Figure 1

Map view of the source region of the 2008 M_w 7.9 Wenchuan earthquake. The red star is the location of epicenter. The blue boxes show the piece-wise fault planes obtained by Shen et al. (2009), and the thicker blue lines indicate the shallowest edges of fault segments. The colored circles show the location of aftershocks within 3 months of the main shock from a relocated aftershock catalog (Yi et al. 2012). Their colors in green, grey, and yellow represent normal, thrust, and strike-slip focal mechanisms of aftershocks, respectively. Black lines are regional faults. Red lines show surface traces of the coseismic rupture from Xu et al. (2009b), and the red triangles on them indicate the dipping direction. Two beach balls show the first-motion focal mechanism solution obtained from the United States Geological Survey (USGS) and the CMT solution by the Global CMT project (GCMT), respectively. The white triangles point out Yingxiu town and Beichuan county. Black squares denote four major fault complexities: Xiaoyudong fault (XDF), Gaochuan fault-bend (GCF), Leigu fault-bend (LGF), and Nanba step-over (NBS). BCF and PGF stand for the Beichuan fault and Pengguan fault, respectively. The blue box in the inset map shows the location of the source region

resampled to 10 Hz and filtered into the frequency band of 0.05–4 Hz. Waveforms of 72 stations meet our criteria above, and the station distribution is shown in Fig. 2.

The compressive-sensing back-projection method (CS-BP) aims to obtain the spatial–temporal distribution of coseismic radiation of large earthquakes at various frequencies. It involves constructing two

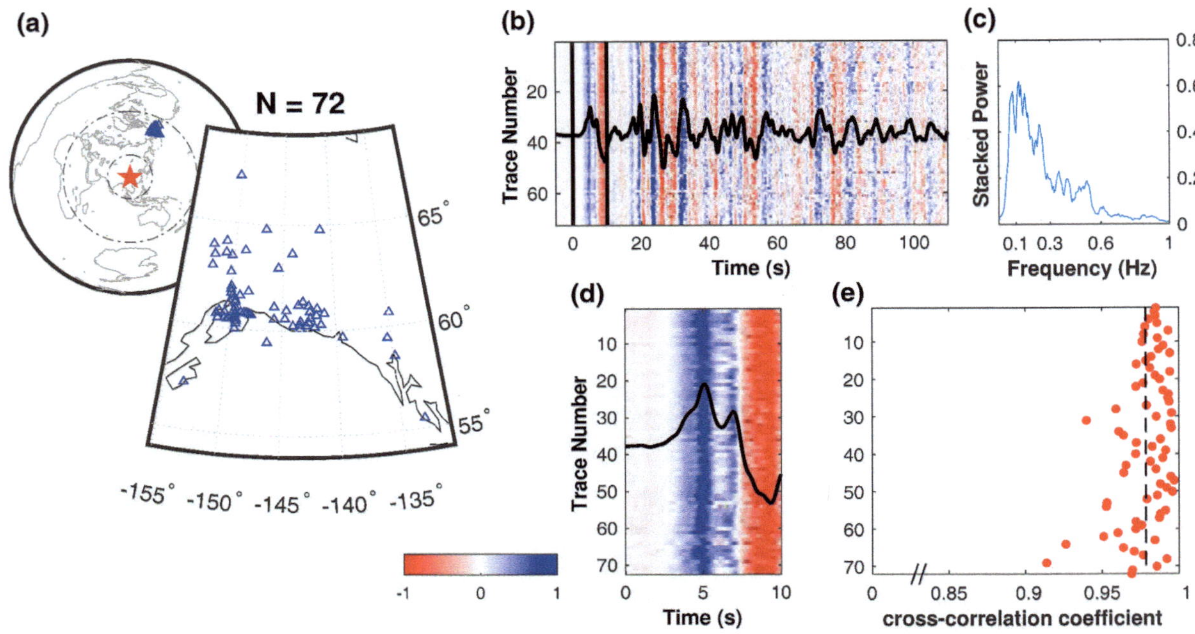

Figure 2
Overview of seismic stations and data used in this study. **a** The red star marks the epicenter, and the two grey dot-dash lines represent the epicentral distance of 30° and 90°. Seventy-two teleseismic stations (blue triangles) in the Alaska region are selected. **b** Image of the aligned P waveforms in the frequency band 0.05–4 Hz. The thick black curve is the stacked waveform from all stations after alignment. **c** Power spectrum of the waveforms in **b** within 1 Hz. **d** Zoom-in of the waveforms from 0 to 10 s indicated by the two black vertical lines in **b**. **e** Cross-correlation coefficients between each waveform and the stacked waveform (the black trace in **b** and **d**) within the time window of [− 5 10] s

complex vectors $\mathbf{b}(f)$ and $\mathbf{x}(f)$, which represent the data spectra of seismograms and the unknown seismic radiation spectra of a meshed source region at a certain frequency f. One propagation matrix $\mathbf{A}(f)$ can bridge the data and model parameters (Yao et al. 2011), that is, $\mathbf{A}(f)\mathbf{x}(f) = \mathbf{b}(f)$. The main goal of CS-BP is to invert for the L1-norm model constrained optimization problem:

$$\hat{\mathbf{x}}(f) = \arg\min_{x} \|\mathbf{A}(f)\mathbf{x}(f) - \mathbf{b}(f)\|_1 + \lambda\|\mathbf{x}(f)\|_1 \tag{1}$$

where λ is the damping factor regulating data misfit and the L1-norm of the model vector ($\|\mathbf{x}(f)\|_1$). Here we also chose the L1-norm for the data misfit term because it is more resistant to data outliers (Yin and Yao 2016). Following Yin and Yao (2016), we defined $\lambda = rN$ where N is the number of stations, then we selected λ by plotting an the L-curve of the data misfit $\|\mathbf{Ax} - \mathbf{b}\|_1$ versus the model constraint $\|\mathbf{x}\|_1$ with logarithmically spaced r and picking the "corner" of the L-curve (Yao et al. 2011). Under the sparsity constraint of seismic radiation $\mathbf{x}(f)$, we utilized a convex optimization package CVX (http://cvxr.com) to solve Eq. (1). In order to evaluate the data fitness of the model $\mathbf{x}(f)$, we introduced an evaluation index (DMR: data misfit ratio) written as:

$$\mathrm{DMR} = \frac{\|\mathbf{A}(f)\mathbf{x}(f) - \mathbf{b}(f)\|_2}{\|\mathbf{b}(f)\|_2} \tag{2}$$

By this definition, the larger the DMR is, the poorer the spectra data are fitted. The choices of thresholds are empirical. We typically increase the DMR gradually and find an appropriate value that results in a clear radiation pattern without many scattered radiation sources due to inversion instability and the effect of noise.

We applied a sliding time-window technique to study the time evolution of seismic radiation subevents (Yao et al. 2011). For each time window, $\mathbf{b}(f)$ is the Fourier transform of the normalized seismograms at f with a dimension of N (number of stations), and $\mathbf{x}(f)$ is the representation of source spectra of that time window with a dimension of M (number of source grids). If we only consider the phase shift of P wave while it is propagating, we can thus write:

$$\mathbf{A}_{\mathrm{mn}}(f) = \mathrm{e}^{-2j\pi f\left(\tau_{\mathrm{mn}} - t_n^0\right)} \tag{3}$$

where τ_{mn} is the travel time from a source grid m to station n, and t_{n}^0 is the travel time from the epicenter to station n. It is the time difference $\left(\tau_{\mathrm{mn}} - t_n^0\right)$ rather than travel time τ_{mn} in Eq. (3) because seismograms have already been aligned with respect to the epicenter.

Large earthquakes usually rupture hundreds of kilometers long, so teleseismic waveforms are almost aligned perfectly at the first break while they become slant as subevents move far away from the epicenter, especially for an array with a large aperture. Slant waveforms of the later arrived P phase are truncated by the sliding time window with a tapered boxcar shape, which leads to inaccurate subevent locations after inversion (Yin et al. 2018). To address this problem, we proposed to set the sliding window with window time adjustment (WTA) utilizing reference rupture fronts calculated at every time step, which is generally similar as the approach taken by Yin et al. (2018) in the frequency domain and Fang et al. (2018) by in the wavelet domain. If we suppose that the reference rupture velocity is a constant v, where the Wenchuan earthquake roughly meets this condition, the coordination of the reference rupture front in the time window $[t_1, t_2]$ is thus written as:

$$\begin{cases} x_1 = x_0 + vt_1 * \sin(\phi_s) \\ y_1 = y_0 + vt_1 * \cos(\phi_s) \end{cases} \tag{4}$$

where x_0 and y_0 give the coordinates of the epicenter, and ϕ_s is the strike angle of the rupture direction. Then the corresponding start and end time of the time window $[t_1, t_2]$ of each aligned trace can be modified as:

$$\begin{cases} t_1' = t_1 - \left(t_n^1 - t_n^0\right) + \delta t \\ t_2' = t_2 - \left(t_n^1 - t_n^0\right) + \delta t \\ \delta t = \mathrm{median}\left(t_n^1 - t_n^0\right) \mathrm{for} n \in [1, N] \end{cases} \tag{5}$$

in which t_n^1 is the travel time from the rupture front (x_1, y_1) to station n and the δt is the median value to offset the significant time shift caused by $\left(t_i^1 - t_i^0\right)$ due to the change of the reference point from the epicenter to (x_1, y_1). Meanwhile, the propagation matrix $\mathbf{A}$ should be updated to $\mathbf{A}'_{\mathrm{mn}}(f) = \exp\{-2j\pi f\left(\tau_{\mathrm{mn}} - t_n^1\right)\}$. In Eq. (5), we can see that the adjusted time window of every station is slightly different due to the calibration and they are used to cut waveform data and invert for the source radiation model. Finally, for a subevent resolved in the modified time window $\left[t_1', t_2'\right]$, its source time should be corrected following the equation from Yao et al. (2011):

$$t_{\mathrm{real}} = (t_1 + t_2)/2 + \mathrm{median}\left(t_n^1 - t_n^{\mathrm{sub}}\right) \tag{6}$$

where t_n^{sub} is the travel time from the obtained subevent to the nth station. The error of t_{real} comes from two parts: first, we choose the center of the modified time window as the base time, but theoretically, it could be any time in the time window $\left[t_1', t_2'\right]$; the second part is the time correction error transferred by the location error of subevents. Since the seismic radiation result obtained from each time window is independent, we have increased the computational efficiency through parallel computation and the total running time of this study has been significantly reduced with multiple cores.

In this study, the hypocenter was set at (31.064° N, 103.372° E) with a depth of 7.6 km and the strike angle in Eq. (4) was set as 51° (data from IRIS). Previous back-projection results (Xu et al. 2009a; Zhang and Ge 2010) indicated that the rupture extended over 300 km northeastwards, so we chose a long and narrow inversion region (Fig. 3) whose length is 320 km and width is 100 km, and we meshed it with a grid spacing of 5 km. All theoretical travel times were calculated by interpolating the travel time table based on the IASPI 91 model (Kennett and Engdahl 1991). The constant velocity v of rupture front was set as 2.5 km/s. Seismograms are bandpassed into three frequency bands: 0.1–0.3 Hz as the low-frequency band, 0.3–0.6 Hz as the middle-frequency band, and 0.6–1.0 Hz as the high-frequency band. After bandpass filtering, the initial 5 s of P waves of each frequency band are aligned again using the waveform cross-correlation. The length of sliding time window is 14 s, 10 s, and 8 s for the low-, middle-, and high-frequency bands, respectively,

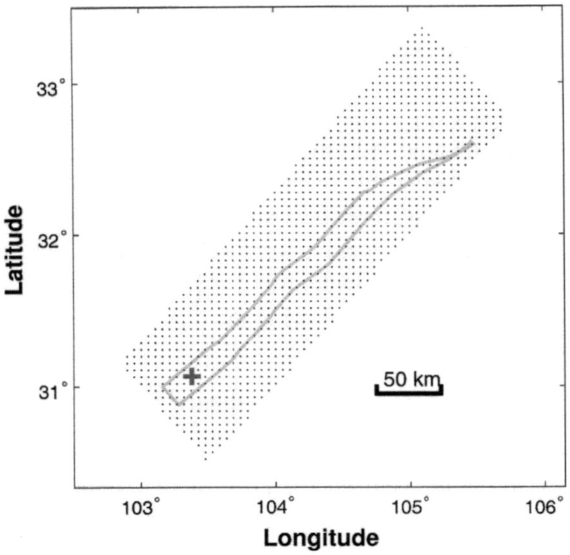

Figure 3
The distribution of source grids at the hypocentral depth of the Wenchuan earthquake (7.6 km) used for the CS-BP inversion. Each black dot denotes the center of a grid. The total amount of grids is 1244 and the grid spacing is 5 km. The magenta cross marks the epicenter and the grey solid line denote the fault planes obtained from Shen et al. (2009)

and the window slides every 2 s through the total 120 s time range.

3. Resolution and Synthetic Tests

We have performed three experiments to validate our seismic radiation inversion results. The first is relocation of nearby moderate magnitude earthquakes (Fan et al. 2016) utilizing the obtained time shifts to align the waveforms from the Wenchuan earthquake. CS-BP hypothesizes that structural heterogeneity of different positions in the source region is similar, and we can mitigate the 3-D path effect by aligning the beginning part of P waveforms concerning the epicenter. The time shift of each waveform is the time for the alignment. Meanwhile, moderate earthquakes can be treated as point sources with small rupture extents and a minor frequency-dependent pattern. So, we can test this hypothesis by relocating moderate events in the source region using time shifts of the main shock and compare our results of earthquake locations with the catalogue locations. Hence, we selected four moderate earthquakes ($M_w < 6.0$) along the rupture faults of the Wenchuan earthquake and downloaded seismic records of the same stations used in the main shock. Since we removed low-quality data with low SNRs, the number of available stations is fewer for those smaller magnitude events (Table 1). Moreover, we decreased the threshold of SNR for selecting waveforms for the 06/29/2009 M_w 5.3 earthquake because the quality of its waveforms is poorer. Time shifts of stations were calculated by cross-correlation of the beginning part of P waveforms of the main shock, and we added them to the corresponding waveforms of all four earthquakes. Then we applied CS-BP to obtain seismic radiations (Fig. 4) of all four events in three frequency bands 0.1–0.3 Hz, 0.3–0.6 Hz and 0.6–1.0 Hz. Deviations relative to their locations Yi et al. (2012) are listed in Table 2. In the low-frequency band, locations are all well resolved except the 06/29/2009 event, whose peak energy is located at ∼ 40 km northeast of its epicenter. This is possibly due to its smallest waveform amplitudes and the poor SNRs. The middle-frequency band results are all well resolved with energy peaks near the epicenters with higher resolution and the smallest mean distance deviation (16.3 km). The 06/29/2009 event that has a large location deviation in the low-frequency band only has ∼ 12 km deviation in this band. For the high-frequency band, the location accuracy for two events (05/17/2008, 06/29/2009) appears very good as their deviations are ∼ 5–6 km. However, the large location deviation of the 08/01/2008 event could possibly result from contamination of waveforms from some other high-frequency sources or noise. Considering the location errors of these events in the catalogue (mostly less than or around 20 km), as well as lateral heterogeneities in the rupture region, we think CS-BP can locate seismic radiation sources of the main shock with good spatial accuracy.

The second synthetic experiment is designed to verify the sliding window with WTA as shown in Eq. (5). We set up four point sources at the distances of 0, 45, 90, and 200 km with respect to the epicenter along the strike of rupture and the source times at 0 s, 25 s, 50 s, and 80 s. It simulates a rupture that propagates unilaterally northeastwards with inconstant speeds, as the rupture speed is 1.75 km/s during

Table 1

Selected four moderate magnitude aftershocks for the robustness tests of CS-BP

Date	Latitude (°)/Longitude (°)/Depth (km)	M_w	SNR	Number of stations
5/17/2008	32.25/104.9/9	5.8	10	46
7/23/2008	32.83/105.5/7	5.5	10	37
8/1/2008	32.09/104.64/10	5.7	10	52
6/29/2009	31.455/103.95/14	5.3	3	39

The information of aftershocks (the first 3 columns) is from Yi et al. (2012). The fourth column gives the minimum SNR of P waveform that is required for station selection. The number of stations that were finally selected is given in the fifth column

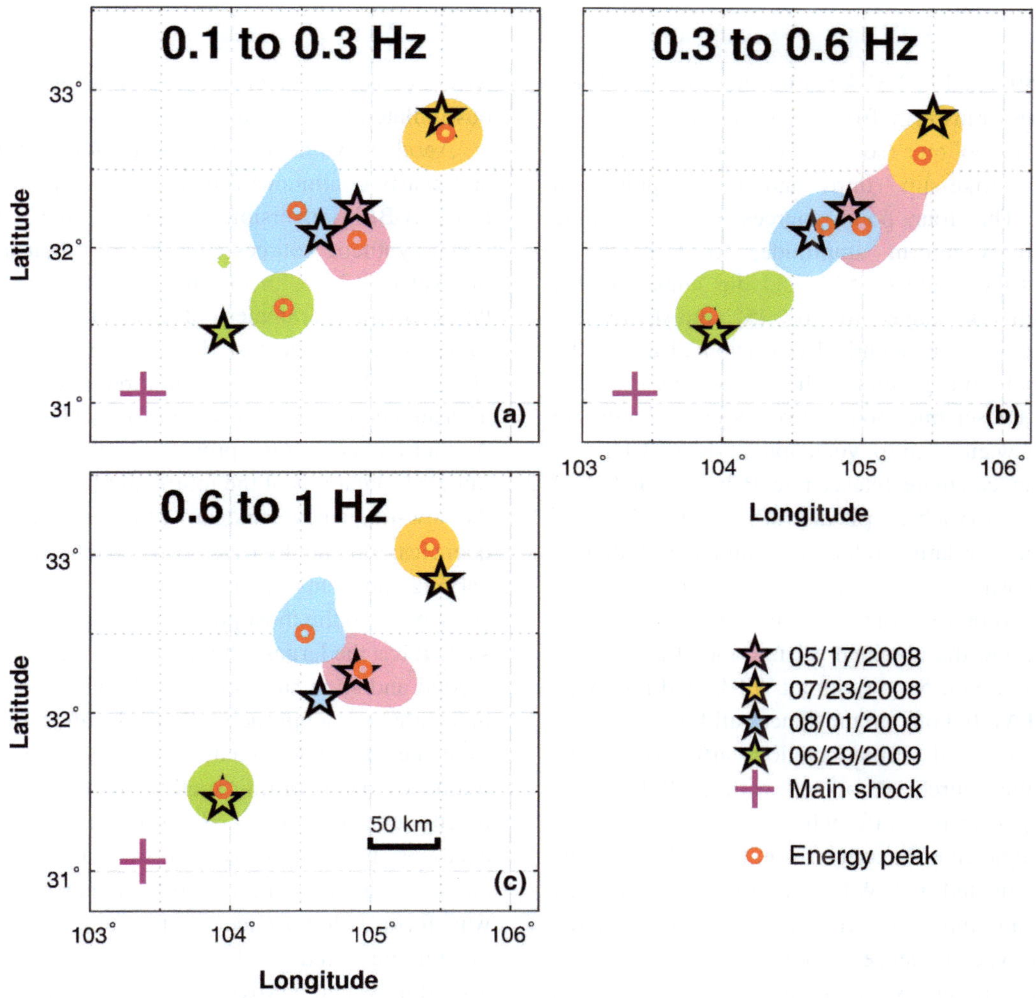

Figure 4
Coseismic radiation regions (colored patches) from CS-BP of four moderate aftershocks (Table 1) in the frequency bands of 0.1–0.3 Hz (**a**), 0.3–0.6 Hz (**b**) and 0.6–1.0 Hz (**c**) within the source region. The colored patches obtained from the inversion represent 80% energy release and the red circles point out the energy peak of every patch. The colored star gives the location of the corresponding aftershock from Yi et al. (2012). The epicenter of the Wenchuan earthquake is marked by the magenta cross

Table 2

Deviation distance between each aftershock location in the catalog (Table 1, Yi et al. 2012) and the corresponding positions of our relocated energy peaks in three frequency bands

Date	Distance		
	LF (km)	MF (km)	HF (km)
05/17/2008	21.8	15.1	5.4
07/23/2008	11.6	27.4	24.8
08/01/2008	22.3	10.3	46.7
06/29/2009	43.1	12.5	6.5
Mean	24.7	16.3	20.8

The last row gives the mean deviation distance of the four events in each frequency band. LF: low-frequency band (0.3–0.1 Hz); MF: middle-frequency band (0.3–0.6 Hz); HF: high-frequency band (0.6–1 Hz)

the first 50 s and is 3.67 km/s for the last 50 s. Their fault types transform from thrusting (first pulse in Fig. 5a), to oblique (second pulse in Fig. 5a), and lastly to strike-slip (third and fourth pulses in Fig. 5a). The four point sources share the same maximum waveform amplitude, focal depth of 10 km, strike angle of 51°, and dip angle of 50°, while their rake angles are 90°, 45°, and 0°, respectively. Stations are identically distributed as for the main shock. We generated the waveforms (Fig. 5a) using a teleseismic body-wave synthetic program (http://www.eic.eri.u-tokyo.ac.jp/etal./KIKUCHI/), which can calculate teleseismic P waves including depth-phases (such as pP and sP). A 10% level of random noise relative to the maximum amplitude was added to waveforms. The constant velocity of the reference rupture front for inversion was set as 2.5 km/s, and the obtained seismic radiation sources are shown in Fig. 5b, d (without WTA) and Fig. 5c, e (with WTA). In both circumstances all four subevents are well resolved without obvious artifacts. Nevertheless, the sources recovered without WTA have much larger spatial shifts when they are more away from the epicenter. In comparison, those shifts can be greatly mitigated with WTA, and the recovered peak energies are more concentrated to the true source locations. The influence of depth phases that might introduce artifacts is minor and it doesn't have a significant effect on the location accuracy both in time and space. Since the synthetic rupture speed is not constant, it also implies that our WTA method with a constant reference speed is adaptable to various unilateral rupture modes.

Another synthetic test with a more complicated and nearly continuous rupture mode is performed to test CS-BP's inversion capability. We generated seven synthetic sources (Fig. 6a) at a constant time interval of 13 s with different amplitudes and fault types. Source parameters of each fault type are the same as the second test (Fig. 5). The noise level is also 10%. Radiation sources in three frequency bands after inversion with WTA are displayed in Fig. 6c–e. A continuous rupture process is observed and the mingled radiations of the fourth and fifth subevents in the low-frequency band are separate at higher frequencies. Our method can also reflect amplitudes of varying subevents qualitatively. The fourth subevent contributes to the most proportion of radiation energy since it has the largest amplitude. The radiation of the second and sixth subevents is relatively weak due to their smallest amplitudes. Some location deviations exist, i.e. the sixth radiation sources in the middle-frequency band shifting southwestwards, leaving the nearby area blank (Fig. 6d, g). It could be the influence of the neighboring fifth subevents whose amplitude is the highest. The dominant frequency of waveforms is low frequency (Fig. 6b) so noise levels of middle- and high-frequency bands increase accordingly and DMRs also increase as frequency increases. Overall, our CS-BP can well image complex continuous ruptures with varying amplitudes and fault types.

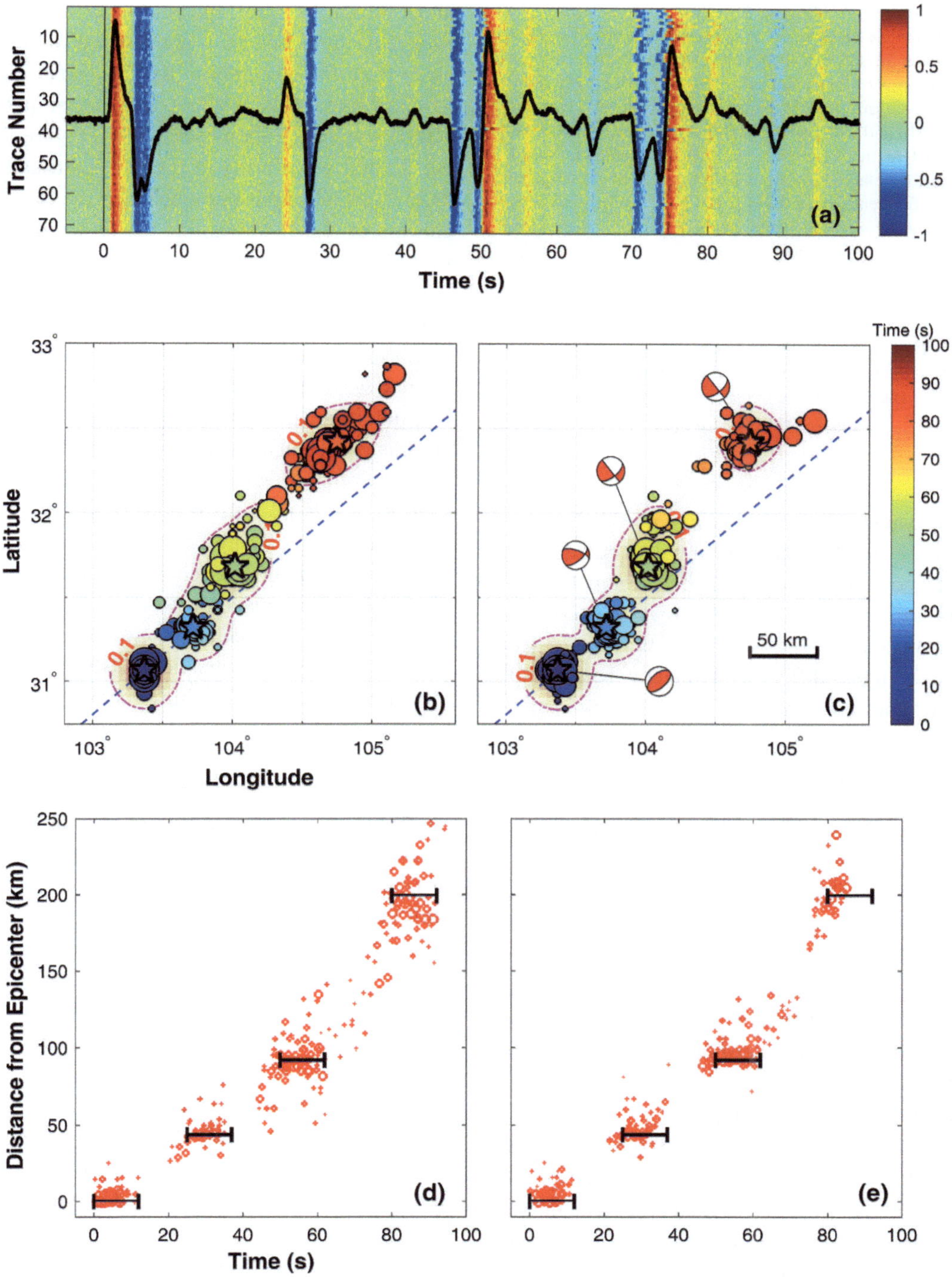

4. Results

In this study, the DMR (Eq. 2) threshold was set as 0.2, 0.4 and 0.7 for the low-, middle- and high-frequency bands, respectively. The threshold values increase as the bandpassed waveforms are less coherent at higher frequencies. Radiation sources in the time window whose misfit is larger than the specified threshold will not appear in the result figures. It should be noted that because CS-BP uses

◀Figure 5
Synthetic tests for the verification of the window time adjustment (WTA) approach in CS-BP. Top panel: image of the aligned synthetic waveforms of four point sources. The black curves show the stacked waveform. Middle panel: comparison of inversion results without (**b**) and with (**c**) WTA. Colored circles, whose areas are proportional to their energies, indicate the spatial–temporal distribution of the obtained radiation sources at all single frequencies within the frequency band 0.1–1.0 Hz. Four synthetic sources (colored stars) are connected by beach balls indicate their fault types. Colors of both circles and stars denote their source time. The blue dashed line marks the strike direction. The magenta contour line shows the 10% level of peak seismic radiation energy. Background brown color shows the total source power obtained from CS-BP. Bottom panel: the distance-time plot of recovered subevents (red dots) without (**d**) and with (**e**) WTA. The black horizontal bars indicate the source times and locations of synthetic sources

teleseismic waveforms and thus has no depth resolution, depths of all subevents are considered the same as that of the hypocenter, i.e., 7.6 km for the Wenchuan earthquake. Figure 7 shows the spatial–temporal distribution of subevents of the Wenchuan earthquake in three different frequency bands and Fig. 8 shows the time evolution of the imaged sources in three frequency bands.

In the low-frequency (LF) (0.1–0.3 Hz) band (Fig. 7a), seismic radiation propagated unilaterally northeastwards from the epicenter. A distinctive dense subevent assembly is located ~ 50 km from the epicenter near XDF. Other sources within the source time of 20–50 s are concentrated on the southern BCF between XDF and Beichuan county where the fault geometry is smooth. We did not observe significant and reliable LF radiation sources after ~ 50 s in the northeastern fault segment, which has much smaller slip compared to the southwestern fault segment (Shen et al. 2009).

The middle-frequency (MF) (0.3–0.6 Hz) band results are shown in Fig. 7b. Because of the good spatial resolution and waveform quality, results in the MF band are best resolved. There are four distinct subevent clusters along the BCF. Compared with the other three, the first energy peak that marked the earthquake initiation is relatively small. Notably, the last three subevents are all located on fault complexities indicated by thick dash lines (Fig. 7a-c). The second energy bursts at around 20 s are located about 30 km from the epicenter near XDF. The third subevents at about 50 s are located about 100 km away from the epicenter at the Gaochuan fault-bend. The last subevents at about 80 s are situated at the Nanba step-over (Shen et al. 2009). Figure 9b shows the filtered waveforms in the MF band and there is nearly perfect consistency between the four signified clusters of signal pulses and the four obtained subevent clusters, indicating the robustness of our results.

Figure 7c shows the high-frequency (HF) (0.6–1.0 Hz) results. Because of better theoretical spatial resolution, but less spatial coherency of teleseismic waveforms, the realistic HF spatial resolution is worse than that of the MF band. There are six clusters of HF subevents and likewise, they also show good correlation with fault complexities. The first energy peak is near the epicenter and has the strongest radiation compared to other subevents in this frequency band due to excellent waveform coherency of the initial P waveforms (Fig. 9c). The second subevents at 20 s, such as the circumstance of the LF and MF bands, are located near XDF at around 30 km from the epicenter. The third subevent is near GCF and the fourth subevent resides right on LGF. These two subevents have the similar source times but are apart from 40 km, possibly due to the inaccuracy of the source times and location uncertainties in the CS inversion (Fig. 6e). The fifth subevent (orange circle in Fig. 7c) is neither on rupture faults nor near any fault complexity, so it could be due to inaccuracy of the subevent location or some triggered event dominated by high-frequency waveforms. The last subevent is located 250 km away from the epicenter at ~ 90 s, marking the termination of the rupture.

Time slices of the imaged sources (Fig. 8) reveal the different radiation pattern of the lower frequency band (< 0.3 Hz) and higher frequency band (> 0.3 Hz). As in the observations from Okuwaki and Yagi (2018), LF subevents appear only in the first half rupture zone and concentrate on smooth faults. HF subevents, however, are mainly clustered and reside near or on fault complexities. Also, HF sources appear to move faster than MF and LF ones. So, migration of HF subevents provides us a direct view on the propagation of the rupture front (Yin and Yao 2016).

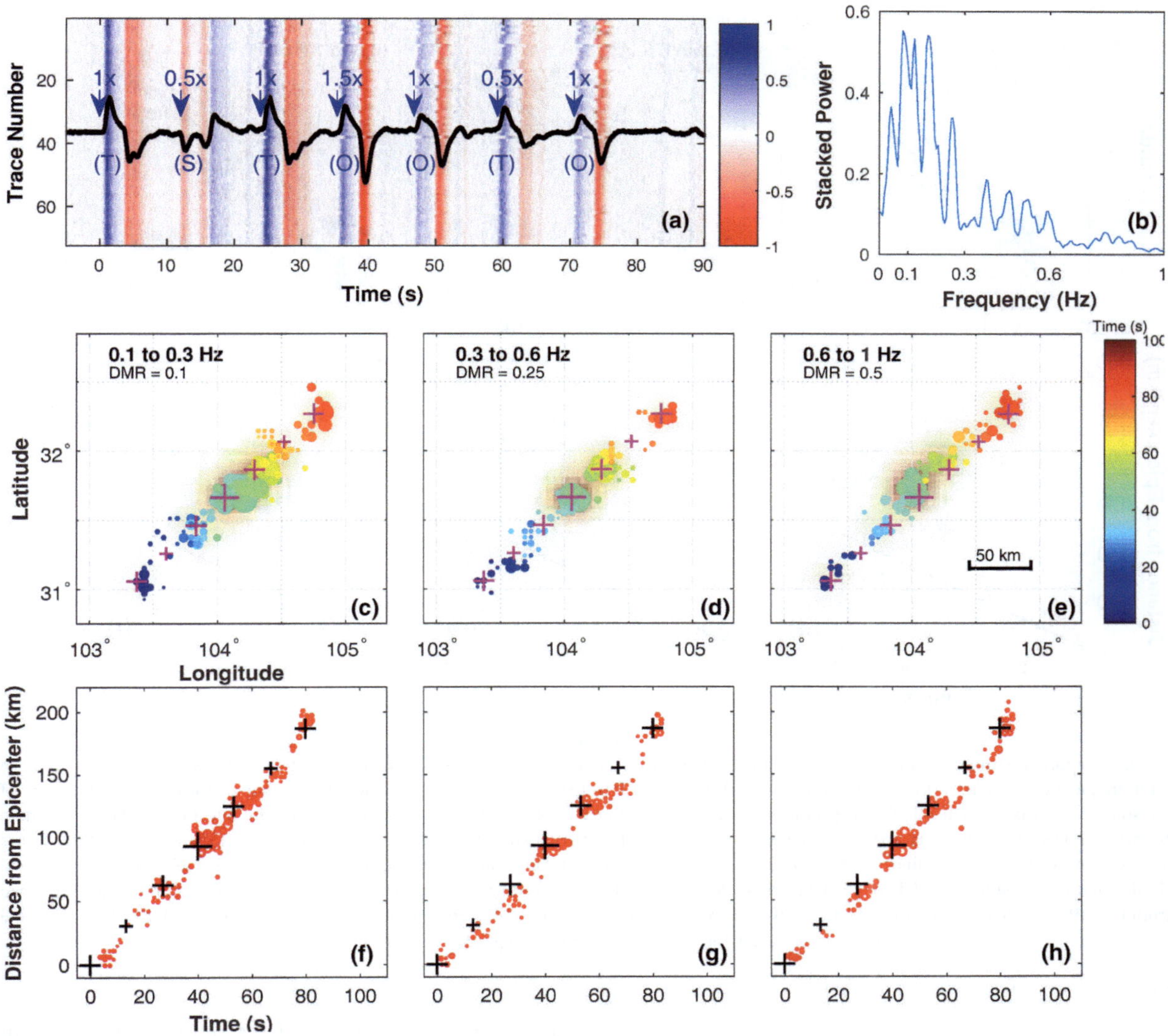

Figure 6
Synthetic test to evaluate the capability of inverting for a complex continuous rupture. **a** Image of aligned synthetic seismograms. Blue texts above the stacked waveform (thick black line) denote the relative amplitudes of point sources during the rupture, whose fault types are marked by blue texts in brackets: T for thrust fault, S for strike-slip fault and O for oblique fault. **b** Power spectrum of the stacked waveform. **c–e** Spatial–temporal distribution of the obtained radiation sources at all single frequencies within each frequency band with WTA. Background brown color shows the total source power. **f–h** Time-distance plots of radiation sources along the strike direction corresponding to the sources in **c–e**, respectively. Magenta (**c–e**) and black (**f–h**) crosses represent the exact locations of synthetic sources whose amplitudes are marked by three different sizes

We have also observed variations in rupture speeds. As is shown in the distance-time plot (Fig. 7d–f), the primary speed during the first 20 s is about 1.5 km/s. Then, the rupture speeds up to around 3.0 km/s and keeps propagating unilaterally northeastwards at a roughly constant speed until its termination. The last subevent as observed in the HF band is at a distance of 250 km and source time of 90 s. Based on this, the average rupture speed of the main shock is about 2.8 km/s, which is consistent with previous estimations using back-projection (Xu et al. 2009a; Zhang and Ge 2010).

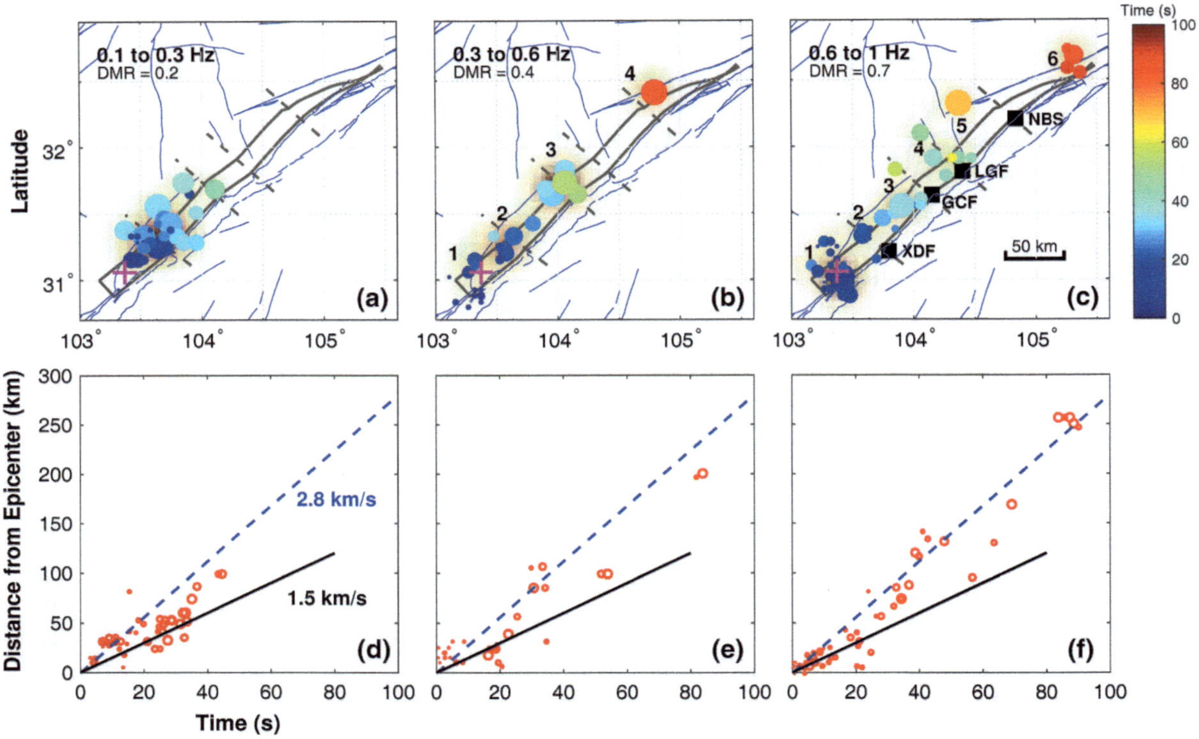

Figure 7

Spatial-temporal distribution of coseismic radiation sources (**a–c**) and distance-time plots of subevents (**d–f**) of the Wenchuan earthquake in three frequency bands: **a**, **d** 0.1–0.3 Hz, **b**, **e** 0.3–0.6 Hz, and **c**, **f** 0.6–1.0 Hz. The area of the colored circle stands for its radiation energy of the source and it is normalized with respect to the largest source in each frequency band. Numbers (1–4) point out four subevent clusters in **b** and numbers (1–6) give six subevent clusters in **c**. Grey solid lines in (**a–c**) depict the fault geometries (Shen et al. 2009), whose intersections with grey dash lines mark four fault complexities: Xiaoyudong fault (XDF), Gaochuan fault-bend (GCF), Leigu fault-bend (LGF), and Nanba step-over (NBS). Magenta cross denotes the epicenter. Background brown color shows the total source power obtained from CS-BP. Blue lines are regional faults. In **d–f** the blue dash line indicates the average rupture speed of 2.8 km/s, and the black solid line indicates the primary slow rupture speed of 1.5 km/s

5. Discussion

5.1. Robustness of Our CS-BP Results

We carried out some further synthetic tests to examine the robustness of our inversion results. We first tried to check whether the sliding WTA is effective. Figure 10a, b display the comparison of subevents inversion results with and without WTA in the middle-frequency band. It is reasonable to see that the overall patterns of the two figures have not changed much. However, with WTA the second and third radiation clusters are more concentrated, and seismic radiation of the fourth subevent is larger. This confirms that WTA based on a constant rupture front can resolve subevents better and increase spatial resolution. We also tested the influence of different DMRs on the results (Fig. 11). As DMR increases, more scattered subevents appear in the inversion results. However, increasing the threshold only leads to retaining some weaker energy subevents obtained in less-fitted sliding time windows, which does not influence our conclusions.

Many studies have applied the back-projection method to track the rupture process. Xu et al. (2009a) backprojected teleseismic P waveforms from the global network. Their results showed three energy peaks at 0, 23 and 57 s after the origin time and an average rupture speed of 2.8 km/s, which is consistent with our estimation. H. Zhang and Ge (2010) chose 16 stations from the Australian subarray and

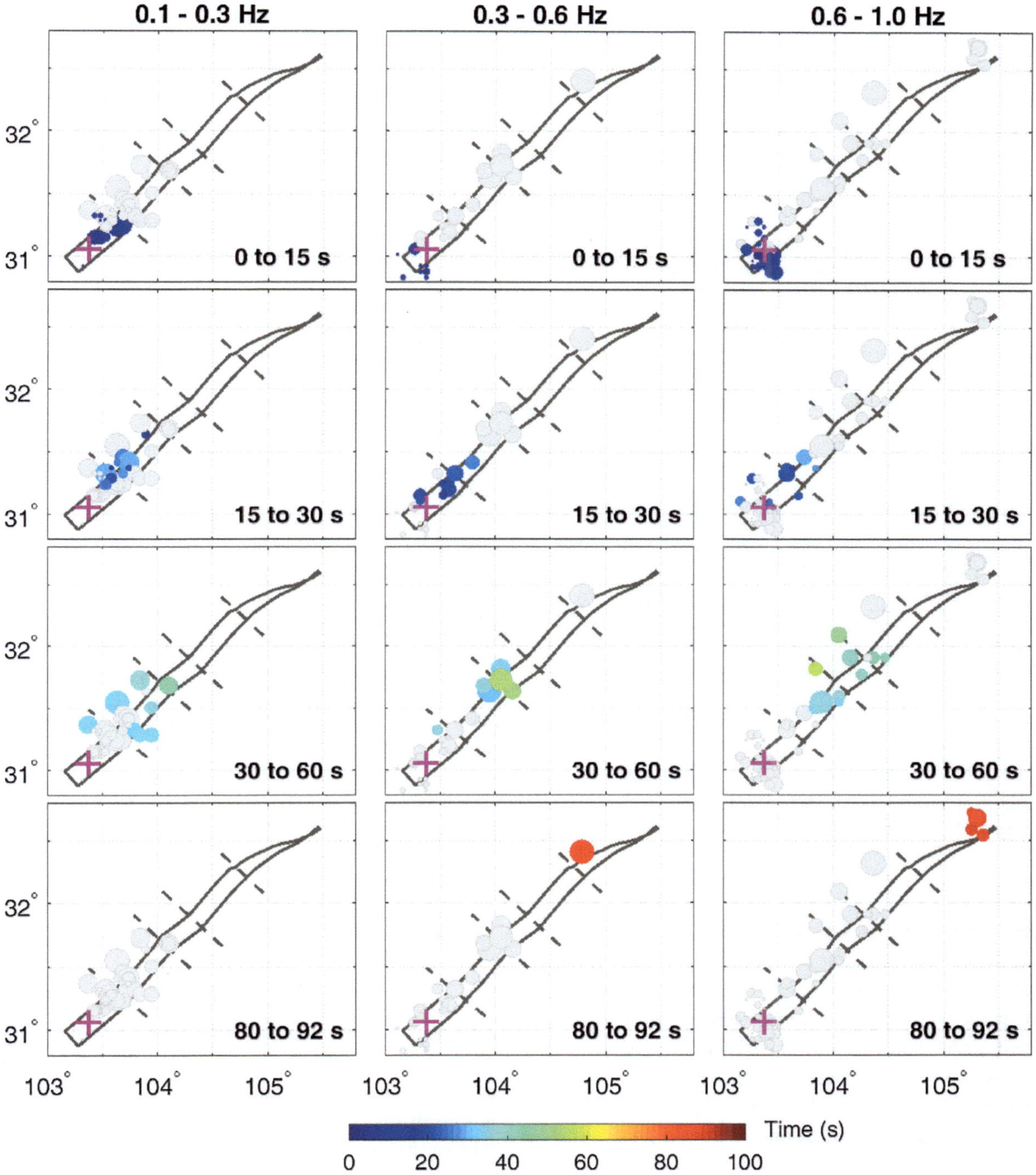

Figure 8
Time evolution of the imaged sources in three frequency bands within different time intervals as indicated in each plot. Imaged sources within the time interval have bright colors or otherwise are grey. Other symbols are the same as Fig. 7a

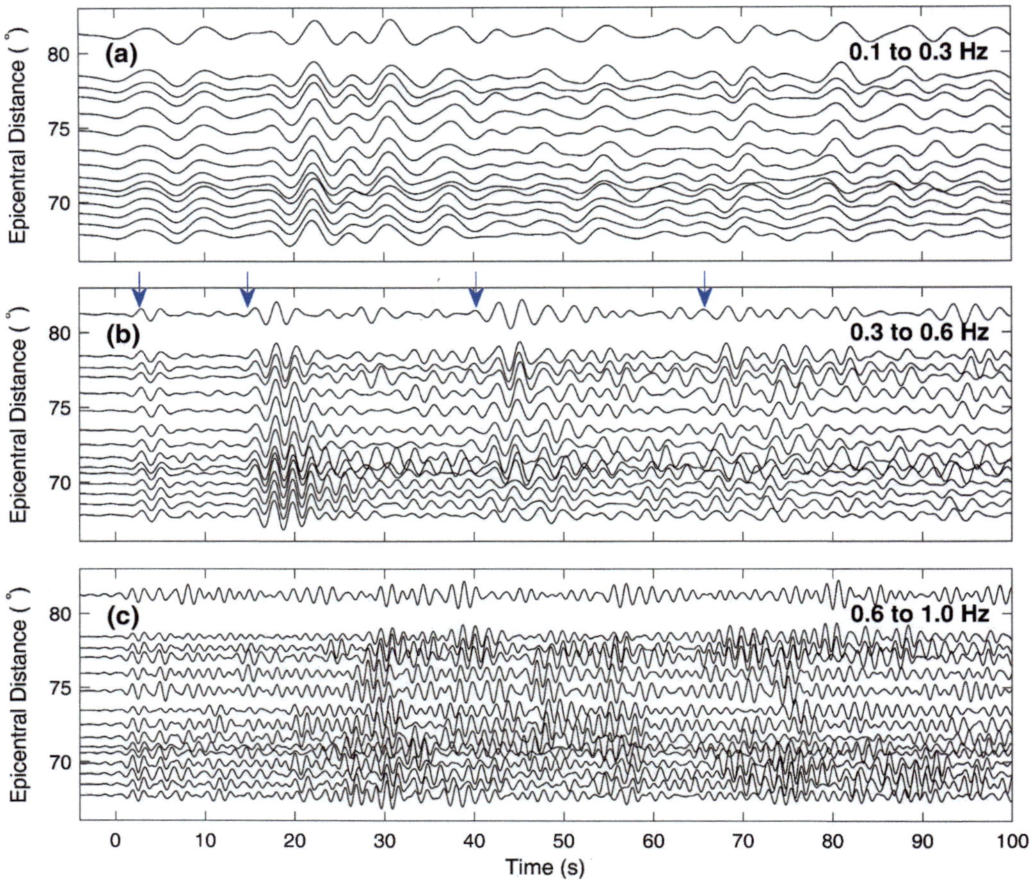

Figure 9
Bandpassed waveforms of the 2008 Wenchuan main shock for 15 selected stations in three frequency bands: **a** 0.1–0.3 Hz, **b** 0.3–0.6 Hz, and **c** 0.6–1.0 Hz. The blue arrows in **b** denote four clusters of signal pulses, which are of good consistency with the four obtained subevent clusters in Fig. 7b

used the relative back-projection method. One major difference between their and our results is the rupture speed. Their distance and time relationship of subevents is stable and almost linear, but we identified a change in rupture speeds near the epicenter that is supported by other studies (Wen et al. 2012; Zhang et al. 2012). In order to investigate the location of high-frequency radiation, Okuwaki and Yagi (2018) applied a hybrid back-projection method to the GSN stations. Similar to our results, they found that high-frequency radiation sources to the southwestern part of the main rupture are closely associated with fault complexities.

5.2. Coseismic Radiation and its Relationship with Fault Complexities

The spatial–temporal distribution of the obtained subevents offers detailed information about where and when seismic energies radiate, and we have confirmed that higher-frequency (> 0.3 Hz) radiation is closely associated with fault complexities. Such relationship with fault complexities is mediated by the rupture speed. Previous analytical studies have revealed that sudden stop at the end of rupture (Madariaga 1977) or abrupt rupture speed changes (Bernard and Madariaga 1984) would generate sharp high frequency waves. A good example is the strong high-frequency radiation near XDF, where there experienced an acceleration of rupture velocity

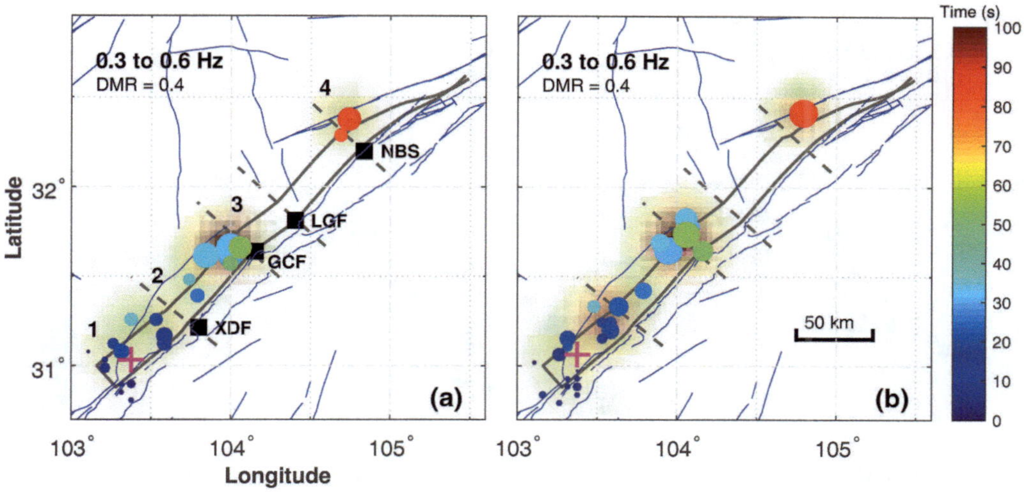

Figure 10
Comparison of the inversion results in the frequency band of 0.3–0.6 Hz without **a** and with **b** WTA. Numbers (1–4) point out four subevent clusters. Other symbols are the same as Fig. 7a

from ~ 1.5 to ~ 3.0 km/s. The initial slower speed is probably due to the resistance of a large asperity (Zhang et al. 2012). Then, the increasing loading stress overcomes the resistance and causes a sudden break of the asperity, which accelerates the rupture front. This acceleration was also discovered by other slip inversion results. At the same location, Zhang et al. (2012) resolved a sudden acceleration 25 s after the earthquake initiation utilizing strong-motion data. Wen et al. (2012) found a rapid increase of rupture speed from 1.7 to 3.3 km/s based on empirical Green's functions.

Dynamic rupture simulations also observed large amount of high-frequency seismic radiation at the fault segments such as a step-over or kink (Adda-Bedia and Madariaga 2008; Hu et al. 2018). One example is that the HF subevents reside right on the Gaochuan and Leigu fault-bend (Fig. 7b, c). The abrupt change of strike direction at the fault-bend would suddenly change the rupture speed, which contributes to the strong high-frequency radiation (Madariaga 1977). The Nanba step-over shares the same feature. The dip angle of fault planes near the step-over has changed abruptly from ~ 51° to ~ 70° (Wan et al. 2017). Thus, the rupture speed would change dramatically when the rupture jumps across the step-over, leading to the burst of high-frequency energy.

We observed that the lower- and higher-frequency radiation sources are complementary to each other. The LF subevents are mainly distributed between XDF and Beichuan county where the fault is relatively smooth. The HF subevents, however, appear on fault complexities or edges. As seen from the subevent distance-time plot (Fig. 7d, e), this complementarity is quite clear. Such a frequency-dependent radiation pattern of the Wenchuan earthquake, which was also revealed by Okuwaki and Yagi (2018), is similar to results from other inter- and intra-plate earthquakes (Wang and Mori 2011; Yao et al. 2011; Uchide et al. 2013) for which seismic energies in different frequency bands originate from different locations of faults. We speculate that when the rupture front crosses a linear and long fault plane such as the segment between XDF and Beichuan county at a relatively stable speed, it radiates mainly LF energies. However, when the rupture propagates through fault complexities, it generates many HF energies due to significant variations in rupture speeds. In summary, fault complexities can control acceleration and deceleration of the rupture speed, which leads to generation of HF seismic radiation. A sudden break of a large asperity is able to propel the

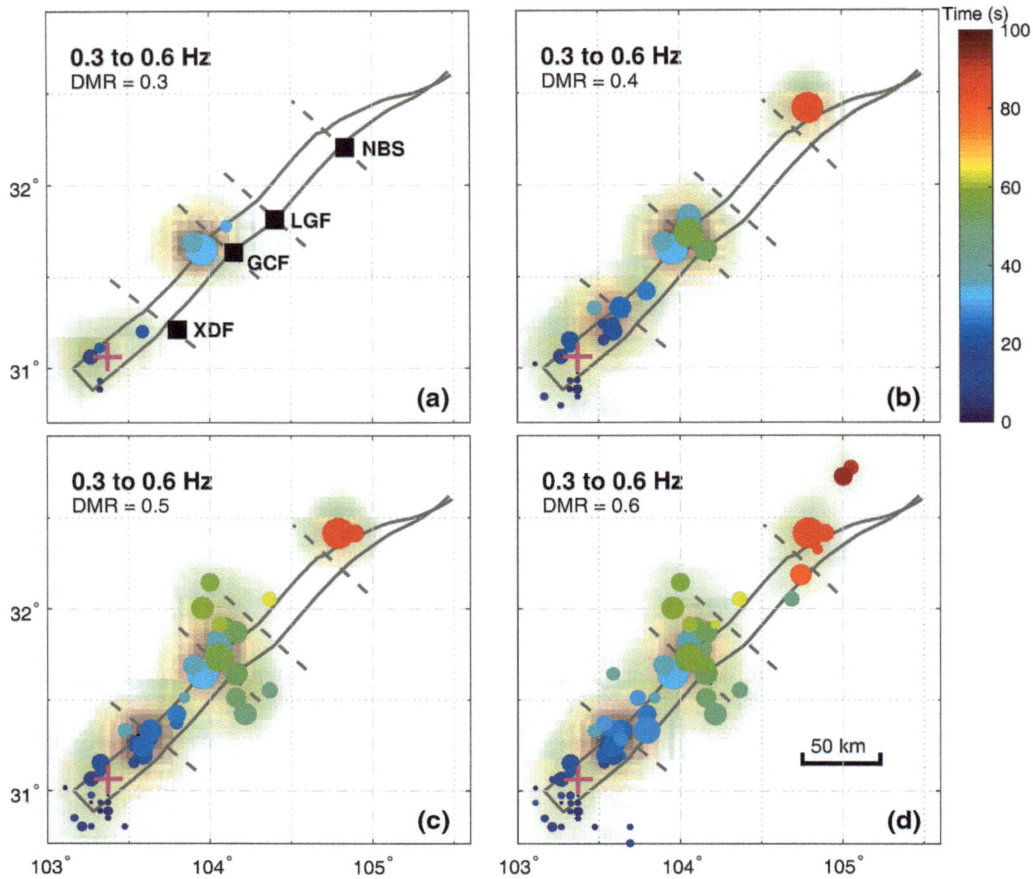

Figure 11
Influence of using different data misfit ratio (DMR) thresholds on the inversion results in the middle-frequency band as indicated. Other symbols are the same as Fig. 7a

rupture front and cause rupture acceleration. Barriers including step-overs and fault-bends, however, may slow the rupture front when it jumps across the step-over or changes its propagation direction.

6. Conclusions

We applied the improved CS-BP with a window time adjustment scheme to image the coseismic radiation of the 2008 M_w 7.9 Wenchuan earthquake in three different frequency bands. We have designed several experiments to verify the robustness of the method. Our results show that the rupture propagated unilaterally northeastwards with a total length of over 250 km. The average rupture speed is about 2.8 km/s and an acceleration from $\sim$ 1.5 to $\sim$ 3.0 km/s at $\sim$ 20 s near Xiaoyudong fault is observed in our subevent results. The coseismic radiation of the Wenchuan earthquake is frequency-dependent and strongly associated with fault geometries: lower-frequency (0.3–0.1 Hz) sources are concentrated on smooth fault segments between Xiaoyudong fault and Gaochuan fault-bend; higher-frequency (0.3–1 Hz) sources are mainly located on fault complexities including Xiaoyudong fault, Gaochuan fault-bend, Leigu fault-bend, and Nanba step-over. The close relationship between high-frequency radiation and fault complexities, to our analysis, is probably mediated by the rupture speed. Our findings reveal the different rupture behaviors between low- and high-frequency waves, which help to better constrain the rupture dynamics of Wenchuan earthquake.

Acknowledgements

We appreciate the constructive comments from two anonymous reviewers, which help to improve the manuscript significantly. We acknowledge the Incorporated Research Institutions for Seismology–Data Management Center (IRIS-DMC) for providing the waveform data used in this study. We appreciate Guixi Yi for providing the information of aftershocks. This work is benefitted from discussion with Yehuda Ben-Zion. Some figures are plotted using GMT (the Generic Mapping Tools, http://gmt.soest.hawaii.edu). This work is supported by the National Natural Science Foundation of China (grant no. 41374055) and China Earthquake Science Experiment Project, China Earthquake Administration (Project Code 2017CESE0101).

References

Adda-Bedia, M., & Madariaga, R. (2008). Seismic Radiation from a Kink on an Antiplane Fault. *Bulletin of the Seismological Society of America, 98*(5), 2291–2302. https://doi.org/10.1785/0120080003.

Allmann, B. P., & Shearer, P. M. (2007). A high-frequency secondary event during the 2004 Parkfield earthquake. *Science, 318*(5854), 1279–1283. https://doi.org/10.1126/science.1146537.

Bernard, P., & Madariaga, R. (1984). A new asymptotic method for the modeling of near-field accelerograms. *Bulletin of the Seismological Society of America, 74*(2), 539–557.

Chang, C.-P., Chen, G.-H., Xu, X.-W., Yuan, R.-M., Kuo, Y.-T., & Chen, W.-S. (2012). Influence of the pre-existing Xiaoyudong salient in surface rupture distribution of the M_w 7.9 Wenchuan earthquake, China. *Tectonophysics, 530–531,* 240–250. https://doi.org/10.1016/j.tecto.2011.12.038.

Fan, W., Shearer, P. M., Ji, C., & Bassett, D. (2016). Multiple branching rupture of the 2009 Tonga-Samoa earthquake. *Journal of Geophysical Research: Solid Earth, 121*(8), 5809–5827. https://doi.org/10.1002/2016jb012945.

Fang, H., Yao, H., & Zhang, H. (2018). Earthquake rupture imaging with the wavelet domain compressive sensing: methodology and application to the 2011 Tohoku earthquake. *Geophysical Journal International, 215*(3), 2060–2070. https://doi.org/10.1093/gji/ggy384.

Fielding, E. J., Sladen, A., Li, Z., Avouac, J.-P., Bürgmann, R., & Ryder, I. (2013). Kinematic fault slip evolution source models of the 2008 M7.9 Wenchuan earthquake in China from SAR interferometry, GPS and teleseismic analysis and implications for Longmen Shan tectonics. *Geophysical Journal International, 194*(2), 1138–1166. https://doi.org/10.1093/gji/ggt155.

Hao, K. X., Si, H., Fujiwara, H., & Ozawa, T. (2009). Coseismic surface-ruptures and crustal deformations of the 2008 Wenchuan earthquake M_w 7.9 China. *Geophysical Research Letters*. https://doi.org/10.1029/2009gl037971.

Hartzell, S., Mendoza, C., Ramirez-Guzman, L., Zeng, Y., & Mooney, W. (2013). Rupture History of the 2008 M_w 7.9 Wenchuan, China, Earthquake: Evaluation of Separate and Joint Inversions of Geodetic, Teleseismic, and Strong-Motion Data. *Bulletin of the Seismological Society of America, 103*(1), 353–370. https://doi.org/10.1785/0120120108.

Hu, F., Wen, J., & Chen, X. (2018). High frequency near-field ground motion excited by strike-slip step overs. *Journal of Geophysical Research: Solid Earth, 123*(3), 2303–2317. https://doi.org/10.1002/2017jb015027.

Hubbard, J., & Shaw, J. H. (2009). Uplift of the Longmen Shan and Tibetan plateau, and the 2008 Wenchuan (M = 7.9) earthquake. *Nature, 458*(7235), 194–197. https://doi.org/10.1038/nature07837.

Ishii, M., Shearer, P. M., Houston, H., & Vidale, J. E. (2005). Extent, duration and speed of the 2004 Sumatra-Andaman earthquake imaged by the Hi-Net array. *Nature, 435*(7044), 933–936. https://doi.org/10.1038/nature03675.

Kennett, B. L. N., & Engdahl, E. R. (1991). Traveltimes for global earthquake location and phase identification. *Geophysical Journal International, 105*(2), 429–465. https://doi.org/10.1111/j.1365-246X.1991.tb06724.x.

Kruger, F., & Ohrnberger, M. (2005). Tracking the rupture of the M_w = 9.3 Sumatra earthquake over 1,150 km at teleseismic distance. *Nature, 435*(7044), 937–939. https://doi.org/10.1038/nature03696.

Liu-Zeng, J., Sun, J., Wang, P., Hudnut, K. W., Ji, C., Zhang, Z., et al. (2012). Surface ruptures on the transverse Xiaoyudong fault: A significant segment boundary breached during the 2008 Wenchuan earthquake, China. *Tectonophysics, 580,* 218–241. https://doi.org/10.1016/j.tecto.2012.09.024.

Liu-Zeng, J., Zhang, Z., Wen, L., Tapponnier, P., Sun, J., Xing, X., et al. (2009). Co-seismic ruptures of the 12 May 2008, M-s 8.0 Wenchuan earthquake, Sichuan: East-west crustal shortening on oblique, parallel thrusts along the eastern edge of Tibet. *Earth and Planetary Science Letters, 286*(3–4), 355–370. https://doi.org/10.1016/j.epsl.2009.07.017.

Madariaga, R. (1977). High-frequency radiation from crack (Stress Drop) models of earthquake faulting. *Geophysical Journal of the Royal Astronomical Society, 51*(3), 625–651. https://doi.org/10.1111/j.1365-246X.1977.tb04211.x.

Meng, L., Inbal, A., & Ampuero, J.-P. (2011). A window into the complexity of the dynamic rupture of the 2011 M_w 9 Tohoku-Oki earthquake. *Geophysical Research Letters, 38*(7), n/a-n/a. https://doi.org/10.1029/2011gl048118.

Okuwaki, R., & Yagi, Y. (2018). Role of geometric barriers in irregular-rupture evolution during the 2008 Wenchuan earthquake. *Geophysical Journal International, 212*(3), 1657–1664. https://doi.org/10.1093/gji/ggx502.

Shen, Z.-K., Sun, J., Zhang, P., Wan, Y., Wang, M., Bürgmann, R., et al. (2009). Slip maxima at fault junctions and rupturing of barriers during the 2008 Wenchuan earthquake. *Nature Geoscience, 2*(10), 718–724. https://doi.org/10.1038/ngeo636.

Tong, X., Sandwell, D. T., & Fialko, Y. (2010). Coseismic slip model of the 2008 Wenchuan earthquake derived from joint inversion of interferometric synthetic aperture radar, GPS, and field data. *Journal of Geophysical Research*. https://doi.org/10.1029/2009jb006625.

Uchide, T., Yao, H., & Shearer, P. M. (2013). Spatio-temporal distribution of fault slip and high-frequency radiation of the 2010 El Mayor-Cucapah, Mexico earthquake. *Journal of Geophysical Research: Solid Earth, 118*(4), 1546–1555. https://doi.org/10.1002/jgrb.50144.

Vandecar, J. C., & Crosson, R. S. (1990). Determination of teleseismic relative phase arrival times using multi-channel cross-correlation and least-squares. *Bulletin of the Seismological Society of America, 80*(1), 150–169.

Wan, Y., Shen, Z.-K., Bürgmann, R., Sun, J., & Wang, M. (2017). Fault geometry and slip distribution of the 2008 M_w 7.9 Wenchuan, China earthquake, inferred from GPS and InSAR measurements. *Geophysical Journal International, 208*(2), 748–766. https://doi.org/10.1093/gji/ggw421.

Wang, D., & Mori, J. (2011). Frequency-dependent energy radiation and fault coupling for the 2010 M_w 8.8 Maule, Chile, and 2011 M_w 9.0 Tohoku, Japan, earthquakes. *Geophysical Research Letters*. https://doi.org/10.1029/2011gl049652.

Wen, Y. Y., Ma, K. F., & Oglesby, D. D. (2012). Variations in rupture speed, slip amplitude and slip direction during the 2008 M_w 7.9 Wenchuan Earthquake. *Geophysical Journal International, 190*(1), 379–390. https://doi.org/10.1111/j.1365-246x.2012.05476.x.

Xu, Y., Koper, K. D., Sufri, O., Zhu, L., & Hutko, A. R. (2009a). Rupture imaging of the M_w 7.9 12 May 2008 Wenchuan earthquake from back projection of teleseismicPwaves. *Geochemistry, Geophysics, Geosystems*. https://doi.org/10.1029/2008gc002335.

Xu, X., Wen, X., Yu, G., Chen, G., Klinger, Y., Hubbard, J., et al. (2009b). Coseismic reverse- and oblique-slip surface faulting generated by the 2008 M_w 7.9 Wenchuan earthquake. *China. Geology, 37*(6), 515–518. https://doi.org/10.1130/g25462a.1.

Yao, H., Gerstoft, P., Shearer, P. M., & Mecklenbräuker, C. (2011). Compressive sensing of the Tohoku-Oki M_w 9.0 earthquake: Frequency-dependent rupture modes. *Geophysical Research Letters, 38*(20), n/a-n/a. https://doi.org/10.1029/2011gl049223.

Yao, H., Shearer, P. M., & Gerstoft, P. (2013). Compressive sensing of frequency-dependent seismic radiation from subduction zone megathrust ruptures. *Proceedings of the National academy of Sciences of the United States of America, 110*(12), 4512–4517. https://doi.org/10.1073/pnas.1212790110.

Yi, G., Long, F., & Zhang, Z. W. (2012). Spatial and temporal variation of focal mechanisms for aftershocks of the 2008 M(s) 8.0 Wenchuan earthquake. *Chinese Journal of Geophysics-Chinese Edition, 55*(4), 1213–1227. https://doi.org/10.6038/j.issn.0001-5733.2012.04.017.

Yin, J., Denolle, M. A., & Yao, H. (2018). Spatial and Temporal Evolution of Earthquake Dynamics: Case Study of the M_w 8.3 Illapel Earthquake, Chile. *Journal of Geophysical Research: Solid Earth, 123*(1), 344–367. https://doi.org/10.1002/2017jb014265.

Yin, J., & Yao, H. (2016). Rupture and frequency-dependent seismic radiation of the 2012 M_w 8.6 Sumatra strike-slip earthquake. *Geophysical Journal International, 205*(3), 1682–1693. https://doi.org/10.1093/gji/ggw105.

Zhang, H., & Ge, Z. (2010). Tracking the rupture of the 2008 Wenchuan earthquake by using the relative back-projection method. *Bulletin of the Seismological Society of America, 100*(5B), 2551–2560. https://doi.org/10.1785/0120090243.

Zhang, G., Vallée, M., Shan, X., & Delouis, B. (2012). Evidence of sudden rupture of a large asperity during the 2008 M_w 7.9 Wenchuan earthquake based on strong motion analysis. *Geophysical Research Letters*. https://doi.org/10.1029/2012gl052516.

(Received July 14, 2018, revised October 21, 2018, accepted November 15, 2018, Published online November 23, 2018)

Pure Appl. Geophys. 176 (2019), 1225–1241
© 2018 Springer Nature Switzerland AG
https://doi.org/10.1007/s00024-018-2067-7

Pure and Applied Geophysics

Seismicity Analysis of the 2016 Ms5.0 Yunlong Earthquake, Yunnan, China and Its Tectonic Implications

Jinzhong Jiang,[1] Jiao Li,[1] and Hong Fu[1]

Abstract—We relocate earthquakes occurring 1 week before and 1 month after the 2016 Yunlong Ms5.0 earthquake by simultaneously using P (and S) travel times and waveform cross-correlation data in the double-difference relocation algorithm. We then use the well-relocated earthquakes as template events and scan through the continuous waveforms to search for and locate weak events by using the match and locate method. A total of 4602 events are detected by using 660 templates, which is ~ 3 times the number of events listed in the Yunnan Seismic Networks catalogue. Our refined catalogue reveals three stages of seismic activity and energy release processes during the 2016 Yunlong Ms5.0 earthquake sequence. We also refine the focal depths and mechanisms of $12M \geq 3.0$ earthquakes using the broadband waveform modeling method. Both the relocated hypocenters and focal mechanisms of the $12M \geq 3.0$ earthquakes delineate fault strike and dip angles of ~ 200° and ~ 75°, respectively, indicating that the seismogenic fault could be a NNE-striking blind or subfault located between the Weixi-Qiaohou Fault and Lancangjiang Fault. Our results could contribute to better understanding of local seismic hazard evaluations in the vicinity of Yunlong area of Yunnan, China.

1. Introduction

The Yunnan Province of China, located at the southeastern margin of the Tibetan Plateau, is one of the most active tectonic regions in Mainland China (Fig. 1a). Three major fault systems have developed in Yunnan as a result of the collision between the Indian and Eurasian Plates, including the N–S–NNW-striking Xiaojiang Fault system, the NW-striking Red River fault system, and the N–S-striking Nujiang fault system (Fig. 1a), which separate the South China Block into several subblocks (Deng et al. 2007; Lei et al. 2011). Moreover, abundant smaller fault systems (e.g. Lancangjiang Fault and Weixi-Qiaohou Fault in Fig. 1a) are also well developed in this area due to the complex tectonic setting. Yunnan Province also experiences a high level of seismic hazard, with more than $300M \geq 5$ and $10M \geq 7$ earthquakes occurring within the area during the 20th century (Huang et al. 2007).

On 18 May 2016, a moderate earthquake (Ms5.0), the Yunlong earthquake, occurred at a depth of about 15 km (local catalogue) in the northwest of Yunnan Province. As a consequence, the Yunnan Earthquake Agency deployed three temporary seismic stations immediately after the mainshock (~ 12 h after the origin time of the mainshock) to monitor the aftershocks. Combined with the permanent stations in this area (more than 20 broadband seismic stations are available within a 200 km radius of the mainshock epicenter), the Yunnan Seismic Networks (YSN) recorded more than 1800 earthquakes with magnitude from − 0.5 to 4.6 (Fig. 1a). Based on the YSN catalogue, the epicenter of the mainshock is located between two mapped faults (Fig. 1a), the NW-striking Weixi-Qiaohou and Lancangjiang Faults. However, the locations of the $M \geq 3.0$ earthquakes show that the seismogenic fault could be a NE-striking fault (Fig. 1b). As the earthquake locations in the YSN catalogue are routinely processed by using an absolute location method with a one-dimensional (1D) regional velocity model only, it is difficult to determine the seismogenic fault accurately due to the large location uncertainties in the YSN catalogue.

Estimation of accurate source parameters (e.g., focal locations and mechanisms) of aftershocks is becoming an essential tool for understanding the seismotectonics of active faulting (e.g., Ben-Zion

Electronic supplementary material The online version of this article (https://doi.org/10.1007/s00024-018-2067-7) contains supplementary material, which is available to authorized users.

[1] Yunnan Earthquake Agency, Kunming, China. E-mail: jz_jiang@foxmail.com

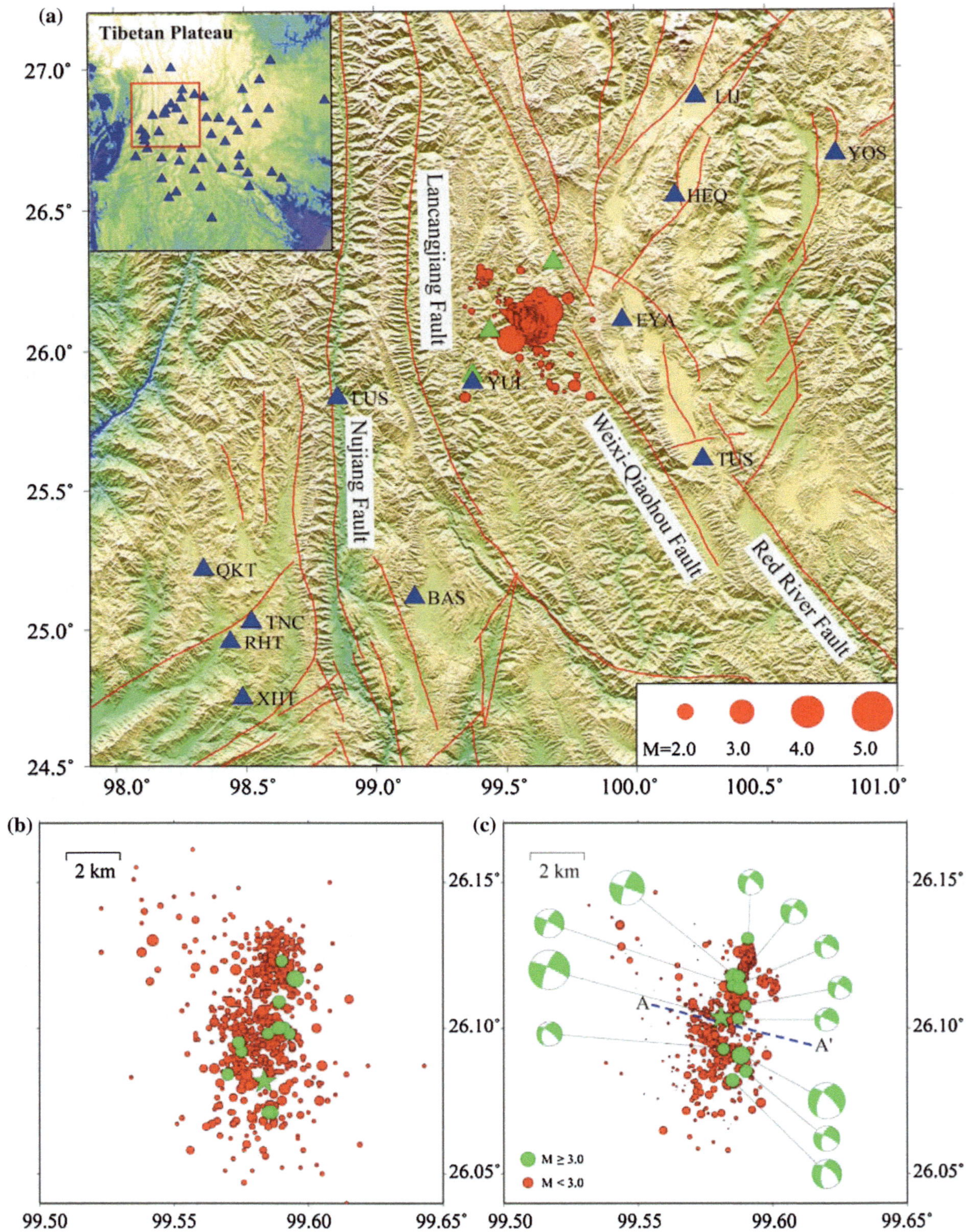
(a)
Tibetan Plateau
Lancangjiang Fault
Nujiang Fault
Weixi-Qiaohou Fault
Red River Fault
LIJ
YOS
HEQ
EYA
YUL
LUS
TUS
QKT
BAS
TNC
RHT
XIIT
M=2.0
3.0
4.0
5.0
27.0°
26.5°
26.0°
25.5°
25.0°
24.5°
98.0°
98.5°
99.0°
99.5°
100.0°
100.5°
101.0°
(b)
2 km
99.50
99.55
99.60
99.65°
26.15°
26.10°
26.05°
(c)
2 km
A
A'
M ≥ 3.0
M < 3.0
99.50
99.55
99.60
99.65°
26.15°
26.10°
26.05°

◀Figure 1
Map view of hypocenters of Yunlong Ms5.0 earthquake sequence. **a** Topographic map showing the Yunlong and adjacent regions in northwestern Yunnan, China. Blue triangles indicate seismic stations of the Yunnan Seismic Networks (YSN); three green triangles indicate temporary seismic stations deployed after the mainshock. Red circles indicate the earthquakes of the 2016 Yunlong Ms5.0 sequence listed in the YSN catalogue. Red lines are major active faults in this area (from Deng et al. 2007), where the Red River, Weixi-Qiaohou, Lancangjiang, and Nujiang Faults are highlighted. **b** Enlarged map view of hypocenters in Fig. 1a; green circles and the star indicate the $M \geq 3.0$ earthquakes and mainshock, respectively. **c** Enlarged map view of hypocenters in Fig. 1b after DD relocation, where beach balls indicate focal mechanisms of $M \geq 3.0$ earthquakes from gCAP inversion

et al. 2003; Ben-Zion 2008; Yang et al. 2009, 2014; Fang et al. 2015; Wang et al. 2017, 2018); For example, source parameters can provide crucial information to infer the geometry of existing or blind faults, and furthermore help to understand the seismogenic model and regional stress field (Mandal and Horton 2001; Luo et al. 2015; Wang et al. 2017, 2018; Xie et al. 2018). In addition, detailed investigations on the spatial and temporal evolution of earthquake sequences usually include statistics on foreshocks and aftershocks in the time, space, and magnitude domain (Ogata 1988; Vere-Jones et al. 2005; Huang et al. 2016). Different laboratory experiments have found that the *b*-value in the Gutenberg–Richter (G–R, 1944) law varies systematically for different types of faulting and stress status of rocks (Scholz 1968; Schorlemmer et al. 2005). Another classical statistical model, the epidemic-type aftershock sequence (ETAS) model for quantifying time-varying seismicity rates in a source region, is usually employed to understand the whole process of seismic activity and predict the expected occurrence rate of earthquakes in normal sequences (Ogata 1988, 1992; Lei et al. 2008, 2013; Omi et al. 2014; Ogata and Tsuruoka 2016; Kumazawa et al. 2017). Thus, a more complete earthquake catalogue and detailed investigation on the spatial and temporal evolution are important for understanding the physical model of earthquakes to potentially mitigate seismic hazards and risk (Peng and Zhao 2009; Yang et al. 2009; Zhang and Wen 2015; Wu et al. 2017).

In this study, we employ the double-difference (DD) location method combined with waveform cross-correlation (WCC) analysis to improve the earthquake locations (Waldhauser and Ellsworth 2000). To obtain a more complete earthquake catalogue, we use the waveforms of earthquakes listed in the manually picked catalogue as templates to detect weaker events following the mainshock. We also apply the generalized cut-and-paste (gCAP, Zhu and Helmberger 1996; Zhu and Ben-Zion 2013) method to investigate the focal mechanisms and focal depth of $M \geq 3.0$ earthquakes. The refined source parameters (locations and focal mechanisms) for both the mainshock and aftershocks indicate that the seismogenic fault could be a NNE-striking blind or subfault, which could contribute to improve future local earthquake hazard evaluations. In this study, we also investigate the seismic activity in the G–R law and ETAS model based on our refined catalogue to constrain the stress status in the source region, as well as to discriminate distinct seismic activities and energy release processes after the Yunlong mainshock.

2. *Earthquake Detection and Relocation*

The seismic data used in the analysis are selected within the time window from 7 days prior to 30 days after the Yunlong mainshock (between 11 May 2016, 16:00:00 and 18 June 2016, 16:00:00 UTC). During this period, more than 20 broadband seismic stations (blue triangles in Fig. 1a) were available within a 200 km radius of the mainshock epicenter. In addition, three short-period seismometers (green triangles in Fig. 1a) were deployed immediately after the mainshock (about 12 h after the origin time of the mainshock) within a 20 km radius of the mainshock to monitor the aftershocks.

We first employ waveform cross-correlation (WCC) analysis and the double-difference (DD) relocation method (Waldhauser and Ellsworth 2000) to refine the locations of more than 1800 events in the local YSN catalogue from 11 May to 18 June 2016. We select 1544 well-recorded earthquakes and apply the WCC method to calculate the relative arrival times between earthquakes for both P- and S-waves. To ensure high signal-to-noise ratio (SNR) in the

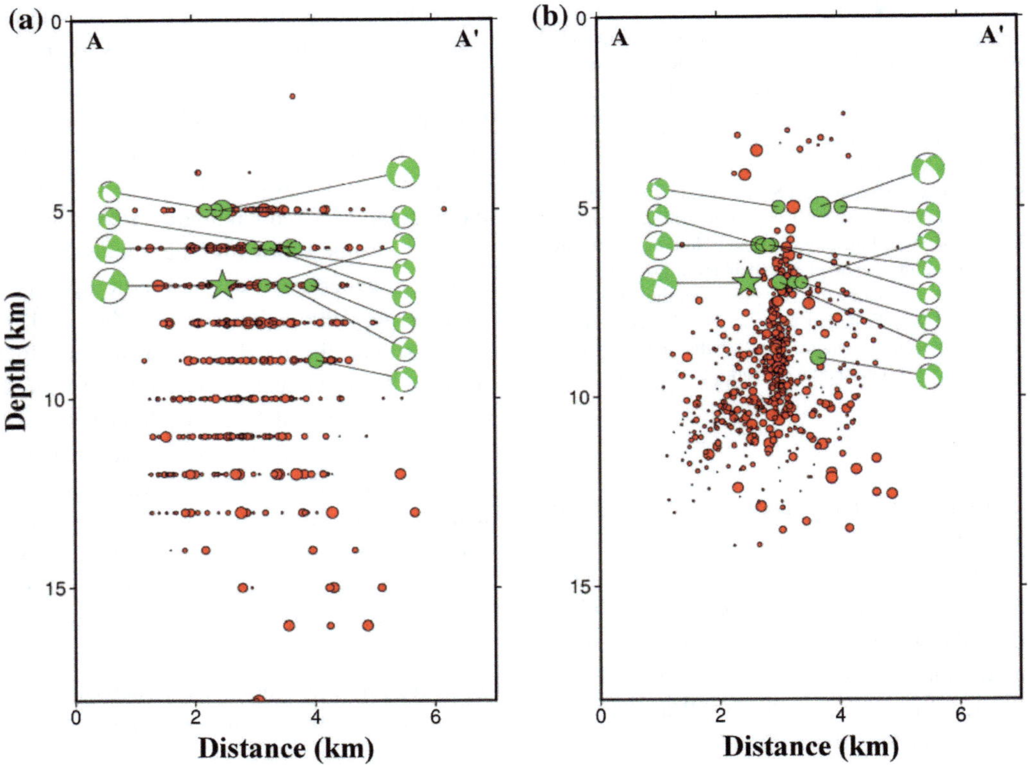

Figure 2
Cross-section view of hypocenters **a** before and **b** after DD relocation along profile AA′ in Fig. 1c. The green star indicates the Ms5.0 mainshock; green circles and beach balls indicate $M \geq 3.0$ earthquakes and focal mechanisms from the gCAP inversion, respectively

correlation window and enhance the robustness of the WCC measurements, we apply a bandpass filter between 1 and 10 Hz on the waveforms before the WCC measurement. The length of the time window is 2.0 s (0.5 s before and 1.5 s after the onset of P- and S-phase, respectively). We obtained 54,866 P- and 48,304 S-phase differential times with correlation coefficient (CC) larger than 0.7 from event pairs recorded by more than three common stations. The DD method accepts both catalogue and WCC differential travel times; however, the DD inversion could be instable when an event pair has both catalogue (phase-picking) and WCC differential travel times at the same station (Wang et al. 2015; Jiang et al. 2016). Therefore, we only preserve the WCC differential times of an event pair and remove the corresponding catalog differential times before DD relocation.

We apply different weighting for P and S wave arrivals (1.0 for P-phase and 0.5 for S-phase) in 15 iterative inversions. We obtain DD locations of 1432 earthquakes, which is ~ 93 % of the total number of earthquake events listed in the YSN catalogue. We also estimate the DD location uncertainties by means of the bootstrapping method according to the P- and S-phase picking errors from the WCC analysis, resulting in location errors of ~ 0.7 km and ~ 1.5 km in horizontal and depth direction, respectively. After DD location, the relocated hypocenters (with more than 10 WCC differential times used in the DD location for each event) clearly distribute along a NNE direction, which is consistent with the strike of the mainshock focal mechanism (Fig. 1c). In addition, more than 90 % of earthquakes are located in the 5–12 km depth range and the hypocenters delineate a ~ 75° steep dip fault plane, which is

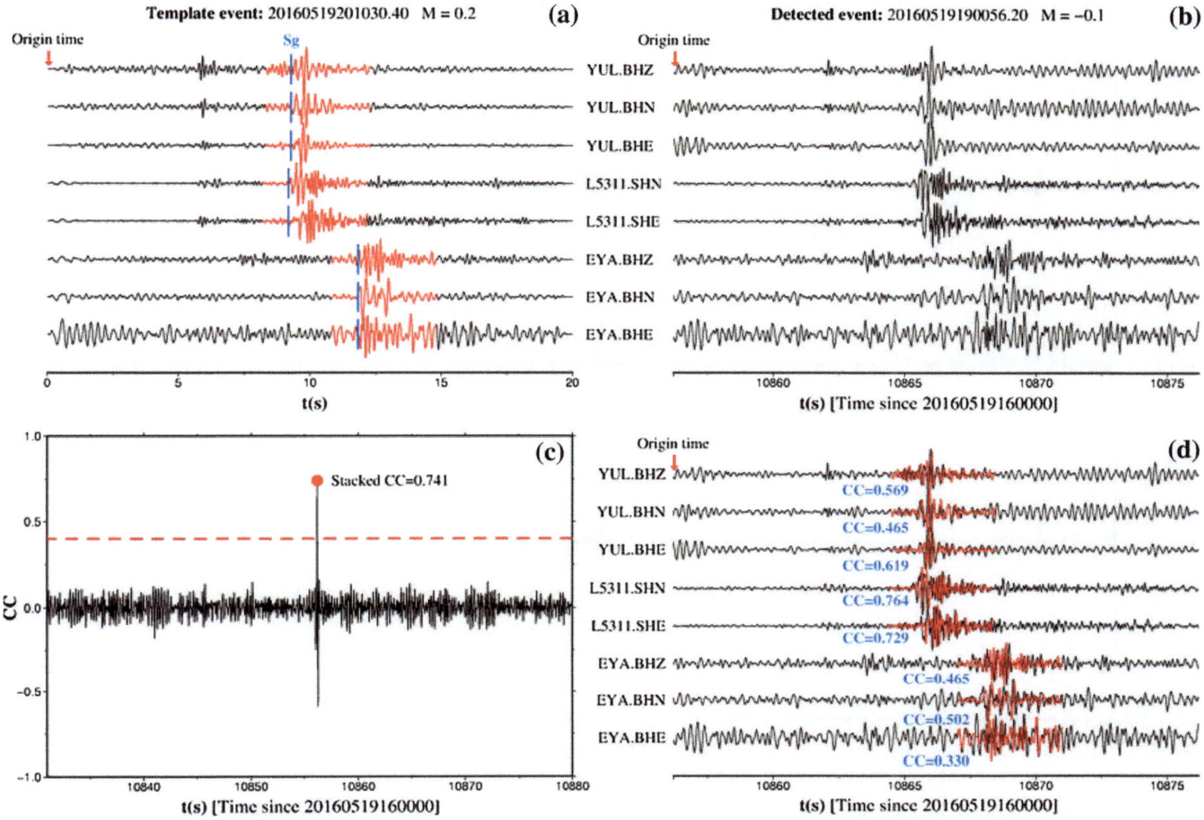

Figure 3
A sample of M&L detection. **a** Template seismograms of an *M*0.2 event. The red waveforms indicate the templates used in M&L detection. **b** Waveforms recorded by the same stations for a detected *M*-0.1 event. **c** The maximum stacked CC value in this detection. **d** Waveform cross-correlation of the template (red) and detected (black) event. The CC values are marked correspondingly

consistent with the dip angle of the mainshock focal mechanism (Fig. 2b).

To detect more weaker aftershocks, we use the match and locate method (M&L, Zhang and Wen 2015) in this study. The M&L method detects small events by stacking cross-correlograms between waveforms of the template events and potential small event signals in the continuous waveforms, then searches for an optimal location of the detected event around the template (Zhang and Wen 2015). Compared with traditional template matching methods, the M&L method extends event detection to a lower magnitude level and improves the capability for detecting small events. To ensure the detection efficiency and accuracy, we select 660 well-relocated earthquakes (with more than 20 WCC differential times used in the DD location for each event) as the template events. We then apply a bandpass filter of 2–8 Hz to both the 1-day continuous waveforms and template to improve the SNR. The P-phase usually has week amplitude, which results in low SNR. Thus, we also include the S-phase with stronger amplitude in the template seismogram segments to run the cross-correlation with the continuous data stream at each station and data component. We calculate the mean CC and SNR of the stacked cross-correlograms and empirically set the thresholds as 0.40 for CC and 9.0 for SNR to detect weaker events (for definition of thresholds, see Zhang and Wen 2015). The location with the maximum mean CC value corresponds to the location of the detected event. Figure 3 shows an example of an event (*M*-0.1) detected using the waveforms of an *M*0.2 template event.

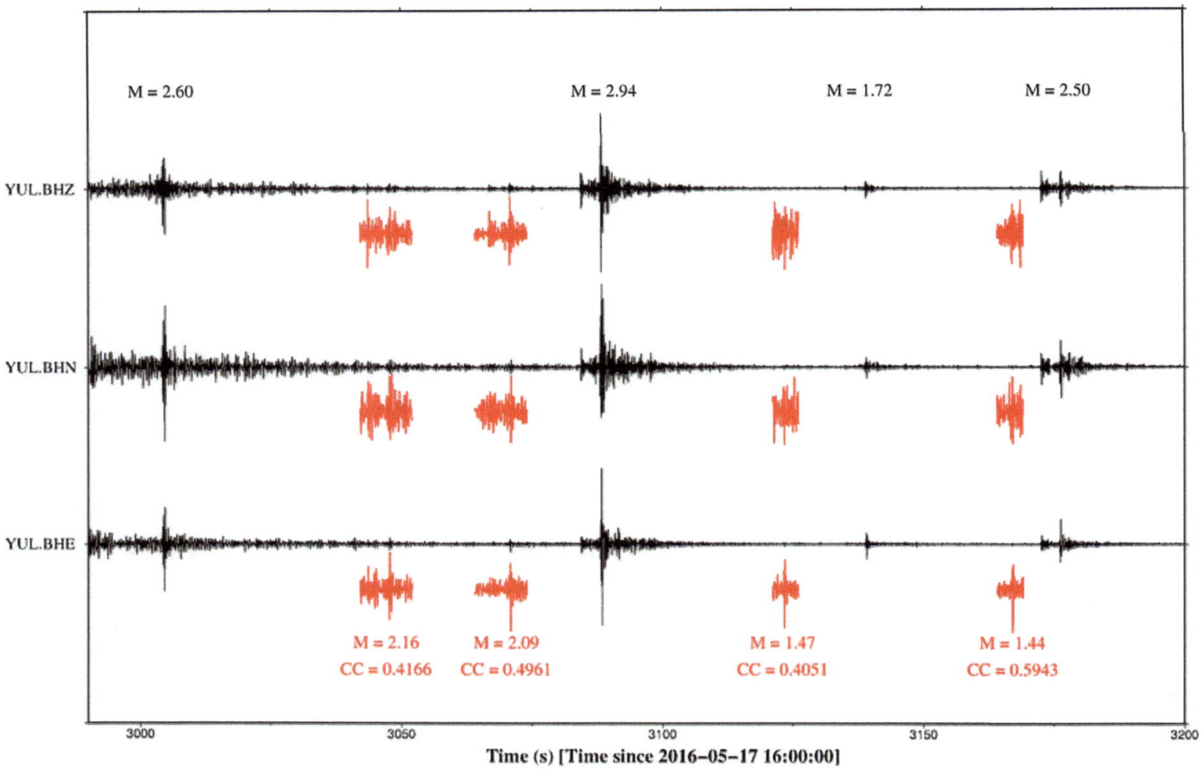

Figure 4
Waveforms in the ~ 300 s following the Ms5.0 mainshock record by the YUL station. Magnitude values (*M*) of four apparent aftershocks listed in the YSN catalogue are marked in black. The red waveform sections indicate four detected events not listed in the YSN catalogue, with the *M* and cross-correlation (CC) values marked in red. The section length is 10 s for the first two events, but 5 s for the last two events due to the smaller *M* and lower SNR

After the M&L detection and location, we found 4602 events, which is about seven times the number of selected templates (660) and three times the number of events listed in the YSN catalogue (1544). Note also that we detect more than 96 % of the events listed in the YSN catalogue while using only 660 template events. This indicates that we selected suitable thresholds. Figure 4 shows four events detected in this study within 300 s after the origin time of the Ms5.0 mainshock, while these earthquakes are not listed in the YSN catalogue. We also analyze the seismic activities in the time and magnitude domain for the two different catalogues. The magnitude of completeness (M_c) decreases from 0.6 for the YSN catalogue to 0.0 for our refined catalogue obtained using the M&L method, and the magnitude of most detected events is smaller than 0.5 (Fig. 5a). The daily cumulative number of earthquakes shows that most of the detected events occurred within 1 week after the mainshock, and we can clearly observe three stages of seismic activity and energy release in our refined catalogue (Fig. 5b).

3. Statistics on the Yunlong Ms5.0 Sequence

The G–R law (Gutenberg and Richter 1944) describes the magnitude and total number of earthquakes as follows:

Figure 5 ▶
Comparison of the number of events versus magnitude (**a**) and time (**b**) between the M&L catalogue (blue squares) and YSN catalogue (red circles)

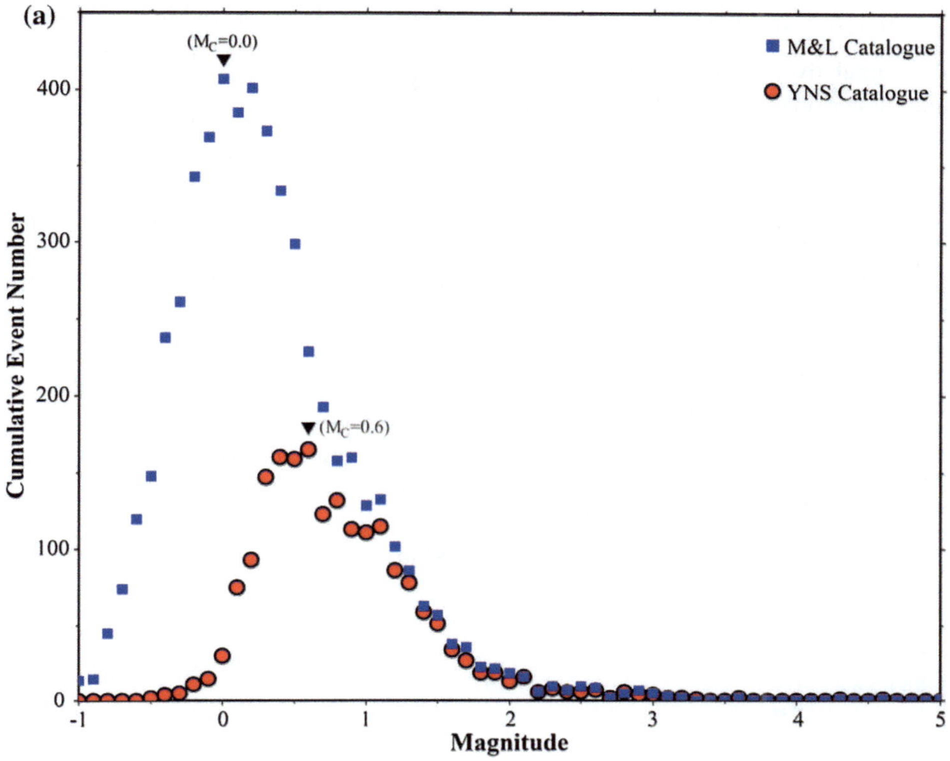
(a)
(M_C=0.0)
(M_C=0.6)
M&L Catalogue
YNS Catalogue
Cumulative Event Number
Magnitude

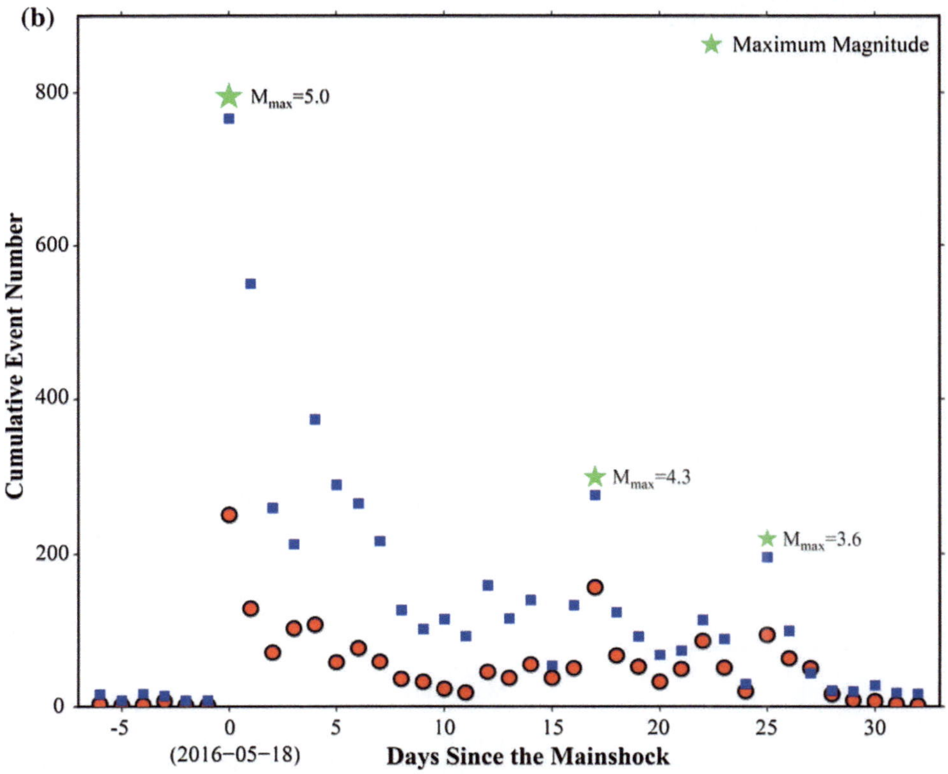
(b)
Maximum Magnitude
M_{max}=5.0
M_{max}=4.3
M_{max}=3.6
Cumulative Event Number
(2016–05–18)
Days Since the Mainshock

$$\log_{10} N = a - bM, \tag{1}$$

where N is the cumulative number of earthquakes with magnitude larger than or equal to M. The parameter a characterizes the seismic activity, or earthquake productivity, while b describes the relative ratio between large and small events. Moreover, the b-value is found to vary for different types of faulting (Schorlemmer et al. 2005) and stress status of rocks (Scholz 1968) in rock deformation experiments. Thus, the b-value is helpful for understanding the stress evolution in the source region before and after a mainshock (Zhang and Zhou 2016; Liu and Pei 2017).

In this study, we estimate the b-value for the Yunlong Ms5.0 sequence by using the YSN catalogue and M&L catalogue, respectively. The completeness magnitude, M_c, is crucial in b-value estimation. Thus, we set it as 0.6 for the YSN catalogue and 0.0 for the M&L catalogue, then employ the least-squares method to obtain the b-values. The b-value calculated from the two different catalogues is 0.763 ± 0.103 and 0.728 ± 0.102, respectively (Fig. 6), in general consistent with b-values estimated in the northwestern Yunnan area (Li et al. 2010; Zhang and Zhou 2016), but smaller than the global average value ($\sim$ 1.0, Schorlemmer et al. 2005). The relatively low b-value indicates relatively high accumulated stress and high seismicity activity in the vicinity of the source region.

The epidemic-type aftershock sequence model (ETAS) (Ogata 1988, 1992; Lei et al. 2008; Kawamura et al. 2014; Ogata and Tsuruoka 2016; Kumazawa et al. 2017) is a point process model for

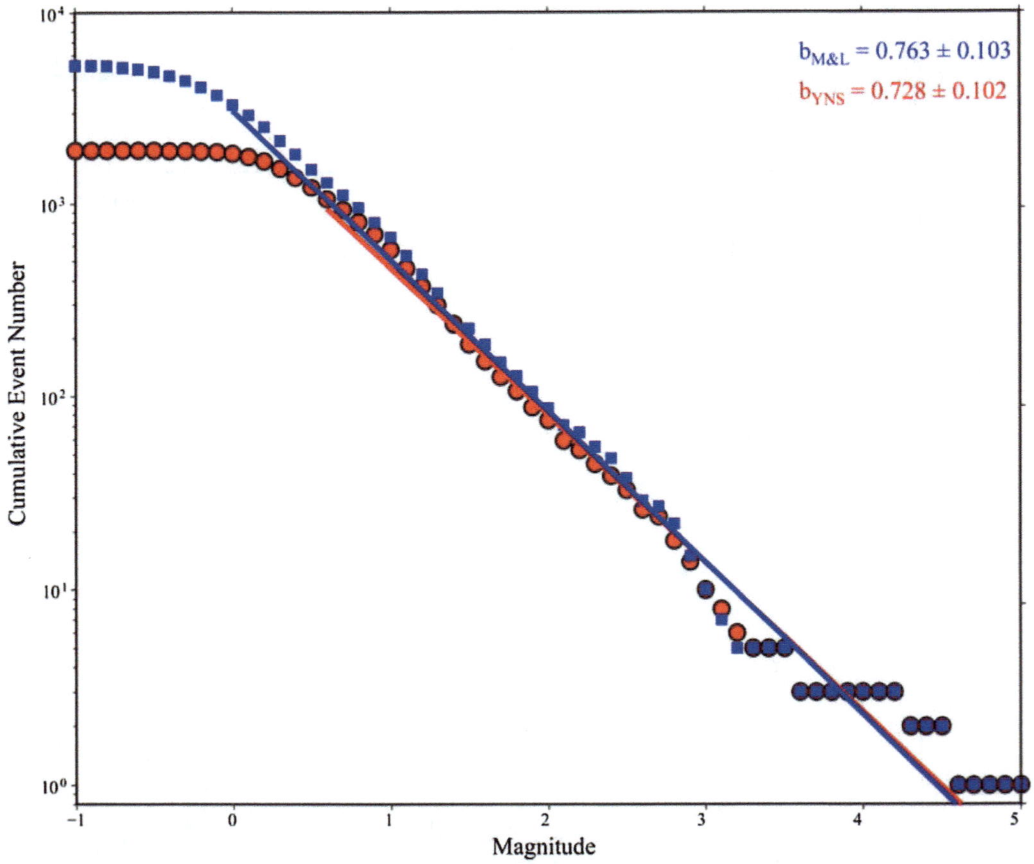

Figure 6
b-Values and fit errors estimated from the M&L catalogue (blue squares) and YSN catalogue (red circles)

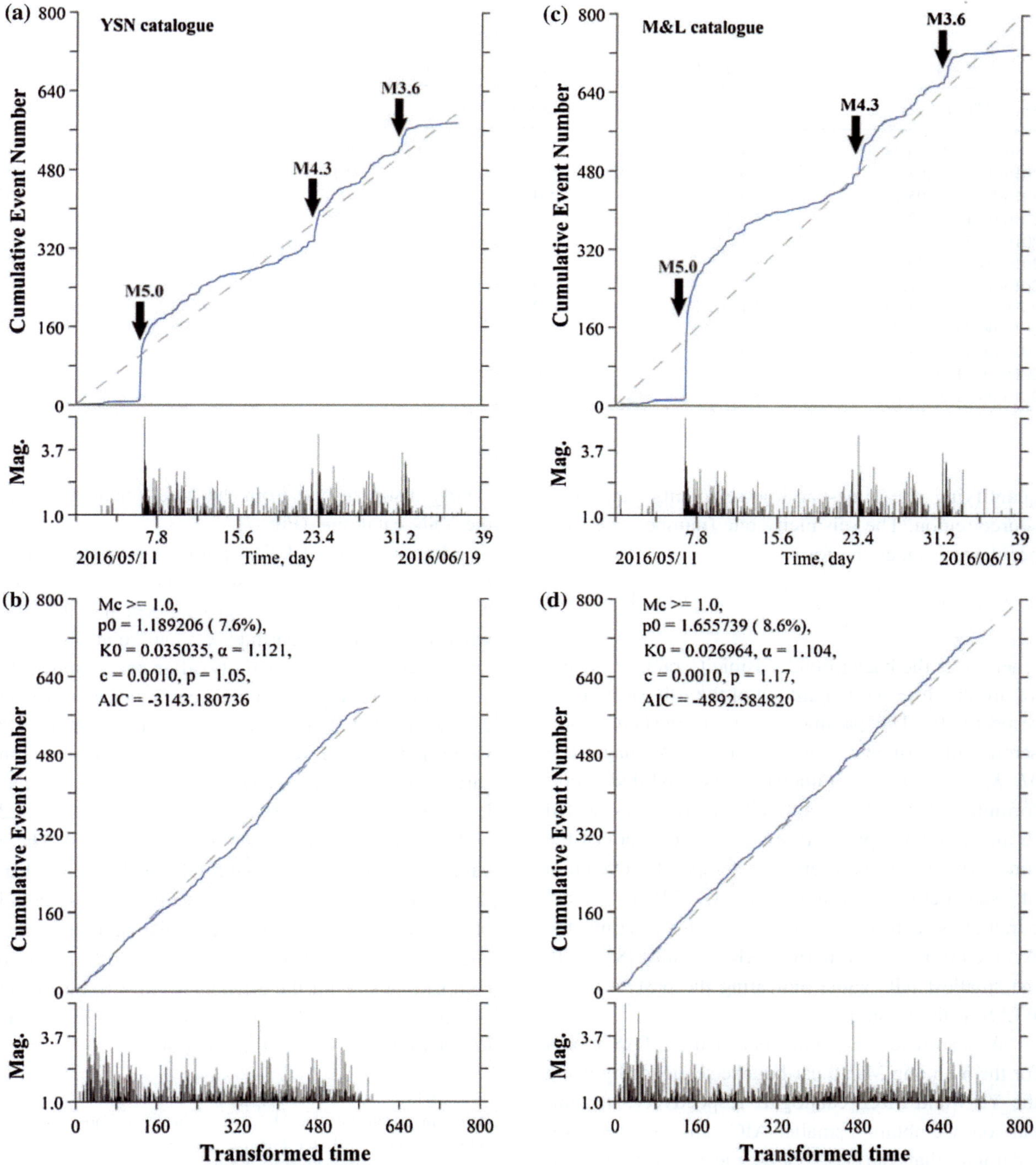

Figure 7
Epidemic-type aftershock sequence (ETAS) modeling for the YSN catalogue (**a**, **b**) and M&L catalogue (**c**, **d**). Black arrows indicate the occurrence day for three earthquakes. The transformed time is the synthetic cumulative number of earthquakes predicted by the ETAS model. Solid and dashed lines show the observed and synthetic fitting lines of the ETAS model, respectively. The ETAS model parameters (P_0, K_0, α, c, p) are shown in **b** and **d**

Table 1

Double-couple solutions and moment magnitude for $M \geq 3.0$ earthquakes

Time (UTC)	Mag. (Mw)	Dep. (km)	Strike	Dip	Rake
2016-05-17 16:00:48	4.8	7.0	201°	75°	− 12°
2016-05-17 17:03:51	3.6	9.0	191°	56°	− 40°
2016-05-17 17:05:11	4.5	5.0	207°	54°	− 26°
2016-05-17 17:29:18	3.3	5.0	198°	40°	− 25°
2016-05-17 17:49:46	3.4	5.0	197°	55°	− 15°
2016-05-17 18:45:06	3.2	7.0	193°	56°	− 24°
2016-05-17 19:47:31	3.2	7.0	200°	60°	− 17°
2016-06-03 10:01:53	4.1	6.0	199°	75°	− 1°
2016-06-04 21:32:27	3.3	6.0	195°	56°	− 12°
2016-06-11 12:41:25	3.6	7.0	206°	75°	− 8°
2016-06-11 19:41:24	3.3	6.0	204°	69°	− 24°
2016-06-11 20:55:06	3.3	6.0	204°	59°	− 24°

quantifying the time-varying seismicity rate in a source region. The seismicity rate (number of events per day), λ, takes the form

$$\lambda(t) = p_0 + \sum_{\{i:t_i<t\}} K_0 e^{\alpha(M_i - M_c)}(t - t_i + c)^{-p}, \quad (2)$$

where p_0 is the background seismicity rate, and t_i and M_i are the time and magnitude of the i-th earthquake, respectively. The parameter α is a constant, representing the efficiency of earthquakes of magnitude M_i. K_0, c, and p are parameters in the modified Omori formula (Utsu and Ogata 1995), where K_0 is related to the aftershock productivity, c is used to adjust the time axis, and p represents the temporal attenuation of seismicity (Kawamura et al. 2014). Model parameters can be estimated by minimizing the Akaike information criterion (AIC, Ogata 1988), with the smallest AIC value indicating the best fit of the ETAS to the seismic data.

We estimate the parameters of the ETAS model for the Yunlong Ms5.0 earthquake sequence by using the YSN and M&L catalogue, respectively. In comparison, we obtain a smaller AIC value for the M&L catalogue than the YSN catalogue when the parameter M_c is 1.0 and c is fixed as 0.001 (Fig. 7), which indicates that the M&L catalogue fits the ETAS model better than the YSN catalogue. We also test different models and obtain similar parameters (P_0, K_0, α, c, p). The three stages of earthquake occurrence and energy release processes can be more clearly distinguished from the M&L catalogue than the YSN catalogue (Fig. 7).

Moreover, the fit ETAS parameter P_0 shows that 8.6 % of the earthquakes listed in the M&L catalogue can be attributed to background seismic activity, while the rest of the earthquakes can be attributed to Omori-type triggered activity after the mainshock. The α value for typical tectonic earthquakes is around 2 (Ogata 1992; Lei et al. 2008), while a smaller α value indicates a relatively weak dependence on earthquake magnitude, suggesting that static triggering should have a relatively weak impact compared with tectonic seismic activity (Lei et al. 2013). The α value we obtain here is around 1.1, much smaller than the value of 2, indicating that tectonic stress could play a dominant role in the occurrence of aftershocks. High tectonic stress in this region also explains a comparable size for the aftershock (Ms4.6) occurring $\sim$ 20 min after the Ms5.0 mainshock, as well as the low b-value ($\sim$ 0.73) in this region.

Figure 8 ▶

gCAP inversion for the Ms5.0 mainshock. **a** Three different velocity models used and tested in this inversion, with blue, red, and green solid (dashed) lines, indicate the P (S) velocity models from Luo et al. (2014, YN01), Wu et al. (2004), and Zhu et al. (2016, YN03), respectively. **b** Normalized fit error versus different focal depths. **c** Comparison between observed (black) and synthetic (red) seismograms. Numbers below station codes are epicentral distances in kilometers. The first number below each seismogram is the time shift between data and synthetics, and the second number is the cross-correlation (CC) value as a percentage. Small red triangles in the beach ball indicate stations with different azimuths

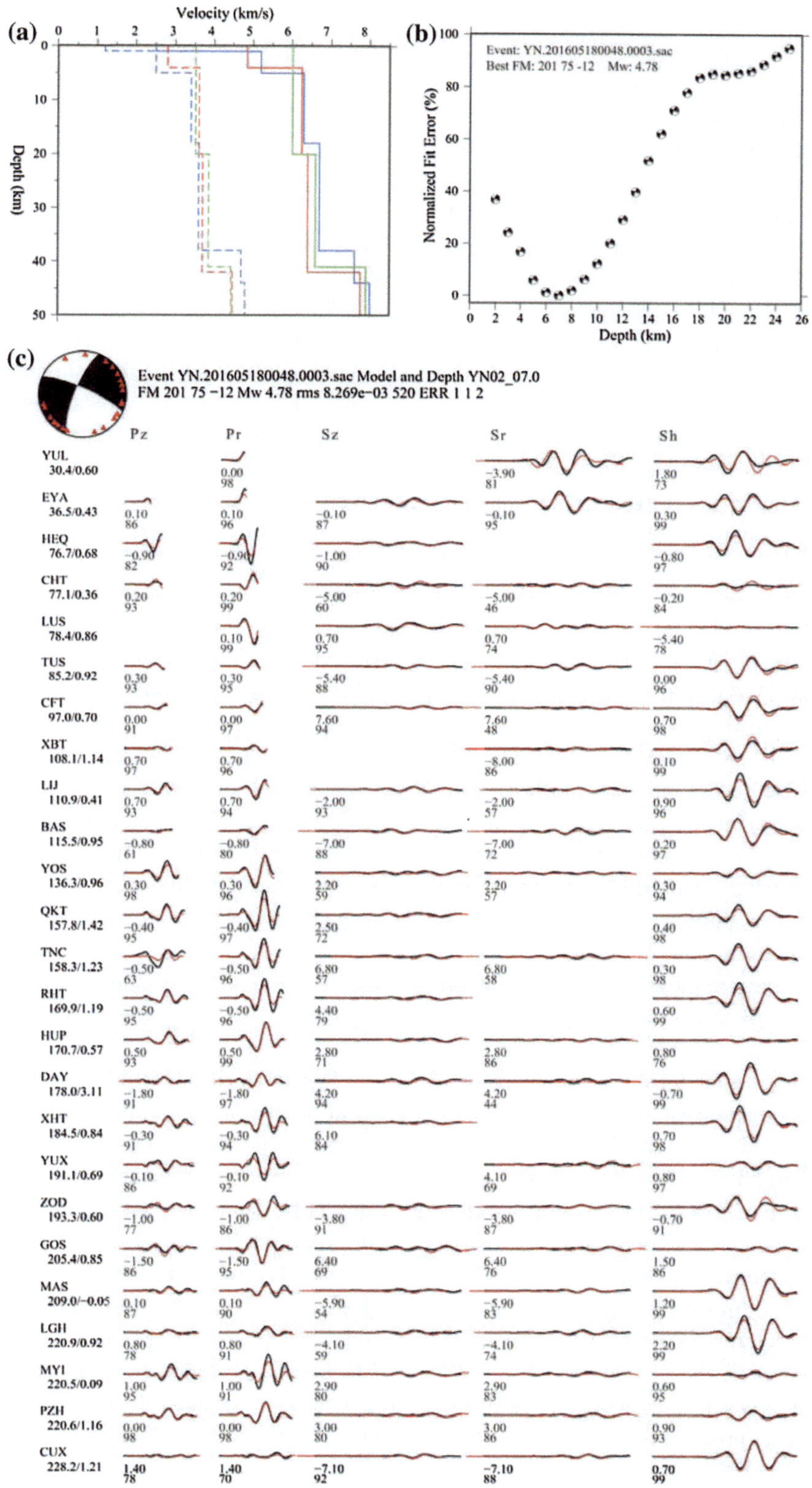
(a)
Velocity (km/s)
Depth (km)
(b)
Event: YN.201605180048.0003.sac
Best FM: 201 75 -12 Mw: 4.78
Normalized Fit Error (%)
Depth (km)
(c)
Event YN.201605180048.0003.sac Model and Depth YN02_07.0
FM 201 75 -12 Mw 4.78 rms 8.269e-03 520 ERR 1 1 2
Pz
Pr
Sz
Sr
Sh
YUL 30.4/0.60
EYA 36.5/0.43
HEQ 76.7/0.68
CHT 77.1/0.36
LUS 78.4/0.86
TUS 85.2/0.92
CFT 97.0/0.70
XBT 108.1/1.14
LIJ 110.9/0.41
BAS 115.5/0.95
YOS 136.3/0.96
QKT 157.8/1.42
TNC 158.3/1.23
RHT 169.9/1.19
HUP 170.7/0.57
DAY 178.0/3.11
XHT 184.5/0.84
YUX 191.1/0.69
ZOD 193.3/0.60
GOS 205.4/0.85
MAS 209.0/-0.05
LGH 220.9/0.92
MYI 220.5/0.09
PZH 220.6/1.16
CUX 228.2/1.21

4. Determining Focal Mechanisms for $M \geq 3.0$ Earthquakes

We also determine the double-couple solutions (strike, dip, and rake) and moment magnitude (Mw) for $12M \geq 3.0$ earthquakes (including the mainshock, listed in Table 1) by using the generalized cut-and-paste (gCAP) method, in which the waveform inversion is used for both the Pnl and S (or surface) waves to constrain the source parameters (Zhu and Helmberger 1996; Zhu and Ben-Zion 2013; Han et al. 2014). We collect local high-quality broadband seismograms (at epicentral distances of less than 300 km) recorded by the YSN stations, and calculate the corresponding Green's functions by using the frequency–wavenumber (F–K) technique (Zhu and Rivera 2002) based on three different regional 1D velocity models (Wu et al. 2004; Luo et al. 2014; Zhu et al. 2016; Fig. 8a).

In our inversion, the Pnl waves (with maximum time window length of 30 s) are bandpass filtered between 0.05 and 0.2 Hz, while the S (or surface) waves (with maximum time window length of 60 s) are bandpass filtered between 0.05 and 0.1 Hz. We then invert the best double-couple solutions by minimizing the waveform misfits between the observed waveform segments and synthetics. Figure 8b, c shows the double-couple solutions and the best waveform fits for the Yunlong Ms5.0 ($\sim$ Mw4.8) mainshock. With the minimum misfit errors between observed and synthetic seismograms achieved at the depth of 7 km, the strike, dip, and rake of the focal mechanisms are 201°/75°/−12°, and the auxiliary nodal plane solutions are 294°/78°/−165°, respectively. Our results are consistent with the double-couple solutions provided by the Institute of Geophysics, China Earthquake Administration (http://www.cea-igp.ac.cn/tpxw/274495.html). We also test the effects of different 1D velocity models on the inversion. For the mainshock, similar double-couple solutions are obtained from three different velocity models (Wu et al. 2004; Luo et al. 2014; Zhu et al. 2016; Fig. 8a). However, the velocity model of Wu et al. (2004, model YN02) gives the best waveform fits with smallest misfit errors (Fig. 8b, Fig. S1). Thus we only employ the YN02 velocity model of Wu et al. (2004) to constrain the focal mechanisms for the $12M \geq 3.0$ earthquakes.

Figure 9 ►
gCAP inversion for a Mw3.2 earthquake occurring on 17 May 2016, 18:45:06 UTC. Comparison between observed (black) and synthetic (red) seismograms at different stations. Other symbols the same as defined in Fig. 8c

We list the double-couple solutions of another 11 aftershocks in Table 1, which are similar to double-couple solutions of the mainshock, especially for the Mw > 3.5 aftershocks. We also show gCAP results for a smaller Mw3.1 earthquake occurring at 18:45:06 UTC, 17 May 2016 in Fig. 9, and gCAP inversion results for other $3.0 \leq M \leq 3.5$ earthquakes in Figs. S2–S7. Note that we fail to obtain focal mechanisms for two aftershocks occurring at 17:06:31 UTC, 2016-05-17 (M3.3) and 17:08:45 UTC, 2016-05-17 (M3.0), which are early aftershocks occurring immediately after the larger event at 17:05:11 UTC, 2016-05-17 (Mw4.5). This is because the long-period seismic signals were contaminated by surface waves or coda from the previous large event. Comparison of the focal mechanisms of Ms5.0 mainshock and aftershocks and the DD locations (Figs. 1c, 2b) indicates that the strike and dip of the seismogenic fault are $\sim$ 200° and $\sim$ 75°, respectively.

5. Discussion

Typically, it is difficult to determine accurate focal depths for earthquakes due to the strong trade-off between the origin time and depth in travel-time-based location approaches. The focal depths listed in the YSN catalogue, which are obtained from the absolute earthquake location method by using a simple 1D velocity model, show large scatter (Figs. 2a, 10). In addition, the reported focal depths for some $M \geq 3.0$ earthquakes (including the mainshock) are around 15–20 km (Fig. 10). Thus, it is difficult to connect these large events with their smaller aftershocks located at around 5–10 km in depth. The waveform modeling method, including depth–phase modeling and CAP inversion, has been

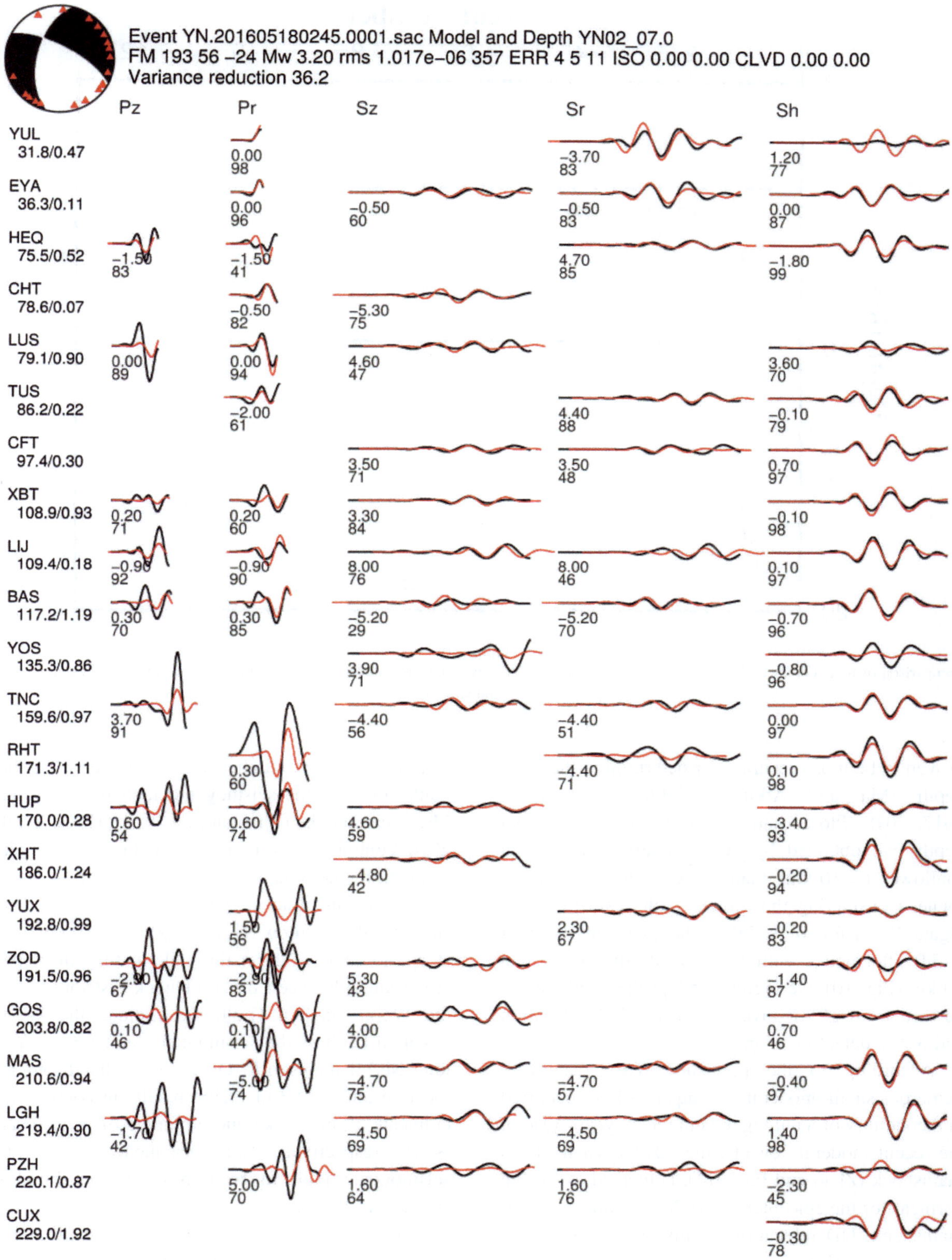
Event YN.201605180245.0001.sac Model and Depth YN02_07.0
FM 193 56 −24 Mw 3.20 rms 1.017e−06 357 ERR 4 5 11 ISO 0.00 0.00 CLVD 0.00 0.00
Variance reduction 36.2
Pz
Pr
Sz
Sr
Sh
YUL 31.8/0.47
EYA 36.3/0.11
HEQ 75.5/0.52
CHT 78.6/0.07
LUS 79.1/0.90
TUS 86.2/0.22
CFT 97.4/0.30
XBT 108.9/0.93
LIJ 109.4/0.18
BAS 117.2/1.19
YOS 135.3/0.86
TNC 159.6/0.97
RHT 171.3/1.11
HUP 170.0/0.28
XHT 186.0/1.24
YUX 192.8/0.99
ZOD 191.5/0.96
GOS 203.8/0.82
MAS 210.6/0.94
LGH 219.4/0.90
PZH 220.1/0.87
CUX 229.0/1.92

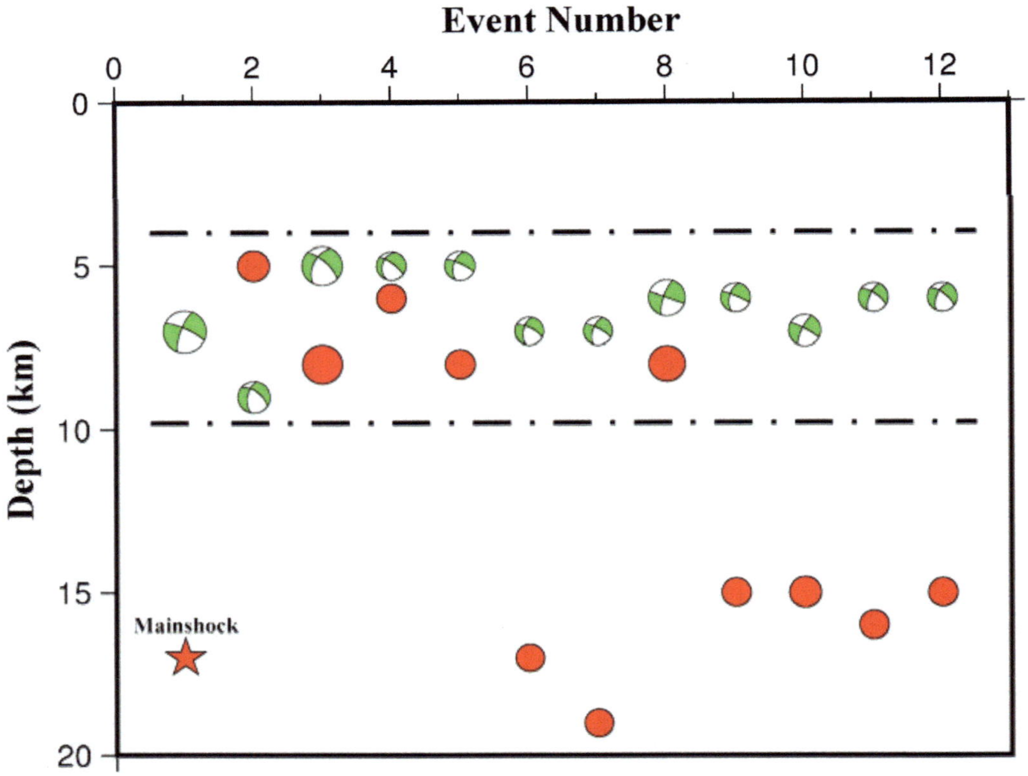

Figure 10
Comparison of focal depths of $12 M \geq 3.0$ earthquakes from gCAP inversion (green beach balls) and those listed in the YSN catalogue (red star and circles)

proven to be more accurate for determining the focal depth (Ma and Atkinson 2006; Wang et al. 2017, 2018; Florez and Prieto 2017). The centroid depths we obtained by gCAP inversion are much shallower (< 10 km) than those in the YSN catalogue. Considering the limited fault scale (< 3 km, Ogata 1988) for $M \approx 5.0$ earthquakes and other $M \geq 3.0$ aftershocks with focal depth shallower than 10 km (Fig. 10), we prefer to adopt the depth values derived from gCAP inversion in the DD relocation and M&L detection (Fig. 2).

According to the published YSN catalogue, numerous small–moderate earthquakes have occurred in the vicinity of Yunlong area in recent years. One of the recent moderate events, the 2013 Eryuan Ms5.5 mainshock (Zhao and Fu 2014), is located ~ 30 km northeast of the epicenter of the 2016 Yunlong Ms5.0 mainshock. DD relocation results and focal mechanism investigations for $2.9 \leq M \leq 4.5$ aftershocks as well as field survey indicate that the Weixi-Qiaohou Fault could be the seismogenic fault of the Eryuan Ms5.5 earthquake (Zhao and Fu 2014). However, the 2016 Yunlong Ms5.0 mainshock is located ~ 20 km away from the Weixi-Qiaohou Fault and ~ 30 km away from the Lancangjiang Fault (Fig. 1a). Considering the shallow focal depths (< 10 km), the steep dip angle of the mainshock, and the discrepant strike angle between the aftershock distribution and the Weixi-Qiaohou Fault, we exclude the Weixi-Qiaohou Fault as the seismogenic fault for the Yunlong Ms5.0 mainshock. On the other hand, our DD location and gCAP inversion results in combination delineate a fault strike and dip angle of ~ 200° and ~ 75°, respectively. As a consequence, we infer the seismogenic fault to be a NNE-striking blind or subfault in this area.

From the DD location results (Fig. 1c), the NNE section of the seismogenic fault displays a much simpler and more uniform geometry than the SSW section, which could indicate a much more complex and dispersive secondary fault system in the SSW section. Moreover, the DD-relocated hypocenters of early aftershocks define a $\sim$ 10 km^2 aftershock zone, which could approximate the subsurface rupture area of the Yunlong Ms5.0 earthquake, which is consistent with the empirical relationship between moment magnitude and subsurface rupture area (Wells and Coppersmith 1994).

Numerical simulations of strike-slip faulting show that the seismogenic width (depth) plays an important role in rupture propagation. All ruptures are self-arresting when the seismogenic width of faults is smaller than a critical value (e.g. 10 km); in other words, the final earthquake magnitude may be limited due to subcritical seismogenic width (Weng and Yang 2017). Our refined source depths from DD location and gCAP inversion show that the seismogenic width of the newly identified fault is $\sim$ 7 km (Fig. 2b), suggesting that the narrow seismogenic width may have significant impacts on rupture propagation and may limit the maximum magnitude of future earthquakes caused by this fault.

6. Conclusions

We obtain a more complete and accurate earthquake catalogue for the 2016 Yunlong Ms5.0 earthquake sequence by using the M&L and DD relocation methods, as well as the focal mechanisms for 12$M \geq$ 3.0 earthquakes by the gCAP method. Our earthquake location and focal mechanism results reveal a NNE-striking blind or subfault in this area, accounting for the occurrence of the Yunlong Ms5.0 sequence. Based on our refined catalogue, we find three obvious stages of seismic activity and energy release processes in the Yunlong Ms5.0 earthquake sequence. Statistical results indicate that tectonic stress could play a dominant role in the Yunlong Ms5.0 sequence, rather than static triggering. Our results could contribute to improve future local earthquake hazard evaluations in the vicinity of Yunlong area in Yunnan, China.

Acknowledgements

This research was supported by the Earthquake Scientific Research of Yunnan Earthquake Agency (grants 201601, 2018ZX01, and C3-201706). Seismic waveform and catalogue data were obtained from Yunnan Seismic Networks. We are grateful to Dr. Miao Zhang for providing the M&L program package, as well as Professor Xinlin Lei for teaching and discussing in the ETAS analysis module. Seismic Analysis Code (SAC) and Generic Mapping Tools (GMT) were used for basic data processing and figure development. We thank Dr. Xin Wang for help in paper writing and revision. We also thank the editor, Dr. Haijiang Zhang, and two anonymous reviewers for constructive comments and suggestions that significantly improved the quality of this paper.

References

Ben-Zion, Y. (2008). Collective behavior of earthquakes and faults: Continuum-discrete transitions, progressive evolutionary changes, and different dynamic regimes. *Reviews of Geophysics, 46*, RG4006. https://doi.org/10.1029/2008RG000260.

Ben-Zion, Y., Peng, Z., Okaya, D., Seeber, L., Armbruster, J. G., Ozer, N., et al. (2003). A shallow fault-zone structure illuminated by trapped waves in the Karadere–Duzce branch of the North Anatolian Fault, western Turkey. *Geophysical Journal International, 152*(3), 699–717.

Deng, Q., et al. (2007). *Map of active tectonics in China*. Beijing: Seismological Press. **(in Chinese)**.

Fang, L., Wu, J., Wang, W., Du, W., Su, J., Wang, C., et al. (2015). Aftershock observation and analysis of the 2013 Ms 7.0 Lushan earthquake. *Seismological Research Letters, 86*(4), 1135–1142.

Florez, M. A., & Prieto, G. A. (2017). Precise relative earthquake depth determination using array processing techniques. *Journal of Geophysical Research: Solid Earth, 122*, 4559–4571. https://doi.org/10.1002/2017jb014132.

Gutenberg, B., & Richter, C. F. (1944). Frequency of earthquakes in California. *Bulletin of the Seismological Society of America, 34*(4), 185–188.

Han, L., Zeng, X., Jiang, C., Ni, S., Zhang, H., & Long, F. (2014). Focal mechanisms of the 2013 Mw6.6 Lushan, China earthquake and high-resolution aftershock relocations. *Seismological Research Letters, 85*(1), 8–14.

Huang, Q., Gerstenberger, M., & Zhuang, J. (2016). Current challenges in statistical seismology. *Pure and Applied Geophysics, 173*(1), 1–3.

Huang, F., Qin, J., Li, Z., & Wu, C. (2007). Subarea characteristics of earthquake types in Yunnan. *Acta Seismologica Sinica (in Chinese), 29*(2), 142–150.

Jiang, J., Fu, H., & Chen, Q.-F. (2016). Characteristics of seismicity of the Xiaowan reservoir in tectonic active area from double-difference relocation analysis. *Chinese Journal of Geophysics, 59*(5), 605–621.

Kawamura, M., Chen, C. C., & Wu, Y. M. (2014). Seismicity change revealed by ETAS, PI, and Z-value methods: A case study of the 2013 Nantou, Taiwan earthquake. *Tectonophysics, 634,* 139–155.

Kumazawa, T., Ogata, Y., & Tsuruoka, H. (2017). Measuring seismicity diversity and anomalies using point process models: Case studies before and after the 2016 Kumamoto earthquakes in Kyushu, Japan. *Earth, Planets and Space, 69*(1), 169.

Lei, X., Ma, S., Chen, W., Pang, C., Zeng, J., & Jiang, B. (2013). A detailed view of the injection-induced seismicity in a natural gas reservoir in Zigong, southwestern Sichuan Basin, China. *Journal of Geophysical Research: Solid Earth, 118*(8), 4296–4311.

Lei, X., Xie, C., & Fu, B. (2011). Remotely triggered seismicity in Yunnan, southwestern China, following the 2004 Mw9.3 Sumatra earthquake. *Journal of Geophysical Research: Solid Earth, 116,* B08303. https://doi.org/10.1029/2011jb008245.

Lei, X., Yu, G., Ma, S., Wen, X., & Wang, Q. (2008). Earthquakes induced by water injection at $\sim$ 3 km depth within the Rongchang gas field, Chongqing, China. *Journal of Geophysical Research: Solid Earth, 113,* B10310. https://doi.org/10.1029/2008JB005604.

Li, T., Wu, X. P., & Wang, Y. (2010). Study on *b*-value distribution in Yunnan based on regional spatial scanning. *Journal of Yunnan University (in Chinese), 32*(6), 670–675.

Liu, Y., & Pei, S. (2017). Temporal and spatial variation of *b*-value before and after Wenchuan earthquake and its tectonic implication. *Chinese Journal of Geophysics, 60*(6), 2104–2112. https://doi.org/10.6038/cjg20170607 **(in Chinese)**.

Luo, Y., Zhao, L., Zeng, X., & Gao, Y. (2015). Focal mechanisms of the Lushan earthquake sequence and spatial variation of the stress field. *Science China Earth Sciences, 58*(7), 1148–1158. https://doi.org/10.1007/s11430-014-5017-y.

Luo, J., Zhao, C., & Zhou, L. (2014). Characteristics of focal mechanisms and stress field of the Chuan-Dian rhombic block and its adjacent regions. *Seismology and Geology (in Chinese), 36*(2), 405–421.

Ma, S., & Atkinson, G. M. (2006). Focal depths for small to moderate earthquakes (mN $\geq$ 2.8) in Western Quebec, Southern Ontario, and Northern New York. *Bulletin of the Seismological Society of America, 96*(2), 609–623.

Mandal, P., & Horton, S. (2007). Relocation of aftershocks, focal mechanisms and stress inversion: Implications toward the seismo-tectonics of the causative fault zone of Mw7.6 2001 Bhuj earthquake (India). *Tectonophysics, 429*(1-2), 61–78.

Ogata, Y. (1988). Statistical models for earthquake occurrences and residual analysis for point processes. *Journal of the American Statistical Association, 83*(401), 9–27.

Ogata, Y. (1992). Detection of precursory relative quiescence before great earthquakes through a statistical model. *Journal of Geophysical Research: Solid Earth, 97*(B13), 19845–19871.

Ogata, Y., & Tsuruoka, H. (2016). Statistical monitoring of aftershock sequences: A case study of the 2015 Mw7.8 Gorkha, Nepal, earthquake. *Earth, Planets and Space, 68*(1), 44.

Omi, T., Ogata, Y., Hirata, Y., & Aihara, K. (2014). Estimating the ETAS model from an early aftershock sequence. *Geophysical Research Letters, 41*(3), 850–857. https://doi.org/10.1002/2013GL058958.

Peng, Z., & Zhao, P. (2009). Migration of early aftershocks following the 2004 Parkfield earthquake. *Nature Geoscience, 2,* 877–881. https://doi.org/10.1038/ngeo697.

Scholz, C. H. (1968). The frequency-magnitude relation of micro fracturing in rock and its relation to earthquakes. *Bulletin of the Seismological Society of America, 58*(1), 399–415.

Schorlemmer, D., Wiemer, S., & Wyss, M. (2005). Variations in earthquake-size distribution across different stress regimes. *Nature, 437*(7058), 539.

Utsu, T., & Ogata, Y. (1995). The centenary of the Omori formula for a decay law of aftershock activity. *Journal of Physics of the Earth, 43*(1), 1–33.

Vere-Jones, D., Ben-Zion, Y., & Zúñiga, R. (2005). Statistical seismology. *Pure and Applied Geophysics, 162*(6–7), 1023–1026.

Waldhauser, F., & Ellsworth, W. L. (2000). A double difference earthquake location algorithm method and application to the northern Hayward fault, California. *Bulletin of the Seismological Society of America, 96*(6), 1353–1368.

Wang, X., Bradley, K. E., Wei, S., & Wu, W. (2018). Active backstop faults in the Mentawai region of Sumatra, Indonesia, revealed by teleseismic broadband waveform modeling. *Earth and Planetary Science Letters, 483,* 29–38.

Wang, X., Wei, S., & Wu, W. (2017). Double-ramp on the Main Himalayan Thrust revealed by broadband waveform modeling of the 2015 Gorkha earthquake sequence. *Earth and Planetary Science Letters, 473,* 83–93. https://doi.org/10.1016/j.epsl.2017.05.032.

Wang, Q., Zhu, L., Su, Y., & Wang, G. (2015). Double-difference relocation of the 7 September 2012 Yiliang earthquake and its aftershock sequence. *Chinese Journal of Geophysics (in Chinese), 58*(9), 3205–3221. https://doi.org/10.6038/cjg20150916.

Wells, D., & Coppersmith, K. (1994). New empirical relationships among magnitude, rupture length, rupture width, rupture area, and surface displacement. *Bulletin of the Seismological Society of America, 84*(4), 974–1002.

Weng, H., & Yang, H. (2017). Seismogenic width controls aspect ratios of earthquake ruptures. *Geophysical Research Letters, 44*(6), 2725–2732. https://doi.org/10.1002/2016gl072168.

Wu, J., Ming, Y., & Wang, C. (2004). Source mechanism of small-to-moderate earthquakes and tectonic stress field in Yunnan province. *Acta Seismologica Sinica (in Chinese), 26*(5), 457–465.

Wu, J., Yao, D., Meng, X., Peng, Z., Su, J., & Long, F. (2017). Spatial-temporal evolutions of early aftershocks following the 2013 Mw 6.6 Lushan earthquake in Sichuan, China. *Journal of Geophysical Research: Solid Earth, 122,* 2873–2889. https://doi.org/10.1002/2016jb013706.

Xie, Z., Zheng, Y., Yao, H., Fang, L., Zhang, Y., Liu, C., et al. (2018). Preliminary analysis on the source properties and seismogenic structure of the 2017 Ms7.0 Jiuzhaigou Earthquake. *Science China Earth Sciences, 61*(3), 339–352. https://doi.org/10.1007/s11430-017-9161-y.

Yang, H., Li, Z., Peng, Z., Ben-Zion, Y., & Vernon, F. (2014). Low-velocity zones along the San Jacinto Fault, Southern

California, from body waves recorded in dense linear arrays. *Journal of Geophysical Research: Solid Earth, 119*(12), 8976–8990.

Yang, H., Zhu, L., & Chu, R. (2009). Fault-plane determination of the 18 April 2008 Mount Carmel, Illinois, earthquake by detecting and relocating aftershocks. *Bulletin of the Seismological Society of America, 99*(6), 3413–3420. https://doi.org/10.1785/0120090038.

Zhang, M., & Wen, L. (2015). An effective method for small event detection: Match and locate (M&L). *Geophysical Journal International, 200*(3), 1523–1537.

Zhang, S., & Zhou, S. (2016). Spatial and temporal variation of *b*-values in southwest China. *Pure and Applied Geophysics, 173*(1), 85–96.

Zhao, X., & Fu, H. (2014). Seismogenic structure identification of the 2013 Eryuan Ms5.5 and Ms5.0 earthquake sequence. *Acta Seismologica Sinica (in Chinese), 36*(4), 640–650. https://doi.org/10.3969/j.issn.0253-3782.2014.04.010.

Zhu, L., & Ben-Zion, Y. (2013). Parametrization of general seismic potency and moment tensors for source inversion of seismic waveform data. *Geophysical Journal International, 194*(2), 839–843.

Zhu, L., & Helmberger, D. V. (1996). Advancement in source estimation techniques using broadband regional seismograms. *Bulletin of the Seismological Society of America, 86*(5), 1634–1641.

Zhu, L., & Rivera, L. A. (2002). A note on the dynamic and static displacements from a point source in multilayered media. *Geophysical Journal International, 148*(3), 619–627.

Zhu, Y., Song, X., Liu, S., et al. (2016). *Research on earthquake reference velocity structure of China*. Beijing: Seismological Press. **(in Chinese)**.

(Received May 2, 2018, revised November 19, 2018, accepted November 28, 2018, Published online December 7, 2018)

Pure Appl. Geophys. 176 (2019), 1243–1254

https://doi.org/10.1007/s00024-018-2014-7

Pure and Applied Geophysics

4-D Imaging of Subsurface Changes with Coda Waves: Numerical Studies of 3-D Combined Sensitivity Kernels and Applications to the M_w 7.9, 2008 Wenchuan Earthquake

Anne Obermann,[1] Thomas Planès,[2] Eric Larose,[3] and Michel Campillo[3]

Abstract—In the context of seismic monitoring, recent studies made successful use of seismic coda waves to image the lateral extent of medium changes. Locating the depth of the changes, however, remains a challenge. We use multiply scattered body- and surface-wave 3-D combined sensitivity kernels to address this problem. We show that we can locate velocity perturbations at depth with numerical data from elastic wave-field simulations in 3-D heterogeneous media. This procedure is then applied to assess the extent of the crustal damage due to the M_w 7.9, 2008 Wenchuan earthquake. We discuss the potential and limitations of our approach to retrieve the depth information of temporal changes occurring in heterogeneous structures.

1. Introduction

The detection of temporal variations in the coda of repeating earthquakes or ambient seismic noise correlations has been successfully applied in different areas in seismology. This method allows monitoring of temporal changes occurring in various settings, such as oil and gas reservoirs (Meunier et al. 2001), CO_2 and geothermal fluid injection studies (Ugalde et al. 2013; Obermann et al. 2015; Hillers et al. 2015), volcanoes (Poupinet et al. 1996; Grêt et al. 2005; Sens-Schönfelder and Wegler 2006a), and fault zones (Poupinet et al. 1984; Schaff and Beroza 2004; Peng and Ben-Zion 2006; Wu et al. 2009; Takagi et al. 2012; Roux and Ben-Zion 2014).

Not only the temporal evolution of the changes is of interest, but also their spatial distribution within the medium, as well as their physical origin (subject to interpretation). Retrieving the spatial distribution of the changes using coda waves is not a straightforward problem given the complexity of the multiply scattered wave paths. This is why coda waves measurements were initially limited to detection purposes only. However, Pacheco and Snieder (2005) showed that the phase shift of coda waves induced by a localized velocity change could be modelled using a 'statistical' sensitivity kernel. This kernel describes the proportion of scattered wave energy that interacts, on average, with the velocity change. Larose et al. (2010) and Rossetto et al. (2011) showed that a similar sensitivity kernel could describe the coda wave decoherence induced by a local scattering change (or structural change). The similarities and differences between these kernels are described in Planès et al. (2014) and Margerin et al. (2016). To retrieve the spatial distribution of the changes, a linear inverse problem can be solved. This was applied to numerical simulations (Planès 2013; Planès et al. 2015; Kanu and Snieder 2015b), locating precursors of volcanic eruptions (Obermann et al. 2013a; Lesage et al. 2014; Sanchez et al. 2018), assessing fault zone damages (Froment 2011; Obermann et al. 2014a), subsurface changes due to high-pressure fluid injections (Obermann et al. 2015; Hillers et al. 2015), and crack/fissure growth in concrete blocks (Larose et al. 2010, 2015; Zhang et al. 2016).

These kernels are based on the description of the wave propagation intensity in the scattering medium. In the aforementioned applications, the coda waves are assumed to be composed of a single propagation mode, either surface waves or body waves. This assumption allows use of analytical expressions of the multiply scattered intensity propagator. Hence, either a 2-D sensitivity kernel describing the 2-D surface wave intensity propagation, or a 3-D

[1] Swiss Seismological Service, ETH Zurich, Zurich, Switzerland. E-mail: anne.obermann@sed.ethz.ch
[2] Université de Genève, Geneva, Switzerland.
[3] ISTerre, CNRS, Université de Grenoble, Grenoble, France.

sensitivity kernel describing the 3-D body-wave intensity propagation is used. However, coda waves are composed of different propagation modes, namely, *P*, *S* and surfaces waves, in proportions that depend on the degree of heterogeneity, coda lapse time, source mechanism, etc. (Perton et al. 2009; Margerin et al. 1999; Hennino et al. 2001; Obermann et al. 2013b). Building sensitivity kernels that describe the full complexity of the coda wave composition is not an easy task as the elastic radiative transfer equation does not even have an analytical solution for the half space. An interesting approach has been proposed by Kanu and Snieder (2015a), based on 2-D numerical simulations of the acoustic and elastic wave fields in heterogeneous media. This approach consists of computing the intensity propagator through an ensemble average of the envelope of the coda wave field obtained in different realisations of the disorder. This approach has the strong advantage of intrinsically taking into account all the complexity of the wave field (mode conversions, inhomogeneous velocity or scattering, topography, etc.). However, the computational cost of the 2-D acoustic simulations is high and the convergence to a stable solution slow. The approach has yet to be extended to 3-D and to an elastic wave propagation.

Another approach based on the linear combination of surface- and body-wave sensitivity kernels was proposed by Obermann et al. (2016). Obermann et al. (2016) have shown that a linear combination of the 2-D (surface wave) and 3-D (body wave) kernel describes well the sensitivity of the multiply scattered waves as a function of depth and lapse time in the coda. The downside is that this method (at the moment) can only describe the subsurface with uniform velocity and heterogeneity.

In the present paper, we show with wave-field simulations in 3-D multiply scattering media that a combination of body- and surface-wave 3-D sensitivity kernels (combined kernels) can indeed be used to constrain the depth location of medium changes. We first describe the combined kernels (Sect. 2.1) and the numerical wave-field simulations (Sect. 2.2). Then, imaging results obtained from numerical simulations with local velocity perturbations at various depths are shown (Sect. 3) and the potential and limitations of our approach are discussed. We then apply the method to the M_w 7.9, 2008 Wenchuan earthquake, to assess the extent of the crustal damage (Sect. 4).

2. Methods

2.1. Multiply Scattered Body- and Surface-Wave 3-D Combined Sensitivity Kernels

Pacheco and Snieder (2005) first described the sensitivity of acoustic coda waves to a velocity perturbation in a multiply scattering medium. The velocity change in the medium induces a relative phase shift in the coda waves that can be measured in the time or spectral domain as $\varepsilon(t) = \frac{-\delta t}{t}$, where $\delta t(t)$ is the phase shift and t is the lapse time in the coda (Sens-Schönfelder and Wegler 2006a). Pacheco and Snieder (2005) showed that the sensitivity of the coda depends on the position of the change relative to the source and receiver, the lapse time in the coda, and the strength of the velocity change. They introduced the following sensitivity kernel K to model the induced relative phase shifts ε at a lapse time t in the coda:

$$\varepsilon(t) = \frac{\delta V}{t}\frac{\mathrm{d}v}{v}K, \qquad (1)$$

where δV is the perturbation volume and $\frac{\mathrm{d}v}{v}$ the actual local velocity change. The sensitivity kernel K is defined as follows:

$$K(\boldsymbol{S},\boldsymbol{R},\boldsymbol{r}_0,t) = \frac{\int_0^t p(\boldsymbol{S},\boldsymbol{r}_0,u)p(\boldsymbol{r}_0,\boldsymbol{R},t-u)\mathrm{d}u}{p(\boldsymbol{S},\boldsymbol{R},t)}, \qquad (2)$$

where $\boldsymbol{S}$ and $\boldsymbol{R}$ are the positions of the source and the receiver, respectively, $\boldsymbol{r}_0$ is the position of the local velocity variation, and t is the centre of the time interval in the coda where the phase shifts are evaluated. Here, $p(\boldsymbol{a},\boldsymbol{b},t)$ is the intensity propagator of the scattered wave field from $\boldsymbol{a}$ to $\boldsymbol{b}$ at time t. So far, most studies used the analytical solution of the acoustic isotropic radiative transfer equation (Paasschens 1997) or the solution of the diffusion equation for p. As discussed in the introduction, in more general and realistic scattering media (elastic, non-diffusive, anisotropically scattering), the intensity propagators would have to be computed numerically.

This same kernel was later shown to also describe the decorrelation of coda waves induced by a scattering perturbation (Larose et al. 2010; Rossetto et al. 2011; Planès et al. 2014). Note that this kernel is only valid in the diffusion regime, i.e. when the heterogeneity level is high enough to render the wave energy flux isotropic at the local variation position r_0. Outside the diffusion regime, a propagator including the direction of the flux needs to be used (i.e. the specific intensity, see Margerin et al. 2016).

Obermann et al. (2016) proposed to obtain 3-D combined sensitivity kernels that can describe the complexity of the coda, as a linear combination of 2-D surface ($\varepsilon^{\mathrm{Surf}}$) and 3-D body wave ($\varepsilon^{\mathrm{Body}}$) sensitivities:

$$\varepsilon(r_0,t) = \alpha(t)\varepsilon^{\mathrm{Surf}}(r_0,t) + (1-\alpha(t))\varepsilon^{\mathrm{Body}}(r_0,t). \tag{3}$$

Here, $\varepsilon(r_0,t)$ is the relative phase shift of coda waves measured at time t in the coda for a change/variation at position r_0 in space. $\varepsilon^{\mathrm{Surf}}$ is the contribution of surface waves:

$$\varepsilon^{\mathrm{Surf}}(r_0,t) = \frac{\delta V}{t}\frac{\mathrm{d}v}{v}K^{\mathrm{Surf}}(x_0,y_0,t)\Gamma(z_0), \tag{4}$$

and $\varepsilon^{\mathrm{Body}}$ is the contribution of body waves:

$$\varepsilon^{\mathrm{Body}}(r_0,t) = \frac{\delta V}{t}\frac{\mathrm{d}v}{v}K^{\mathrm{Body}}(r_0,t). \tag{5}$$

In Eq. 4, $\varepsilon^{\mathrm{Surf}}$ is computed with a 2-D sensitivity kernel to describe lateral variations of the sensitivity of surface waves in the horizontal plane:

$$K^{\mathrm{Surf}}(x_0,y_0,t) = \frac{\int P^{2D}(S,x_0,y_0,u)P^{2D}(x_0,y_0,R,t-u)\mathrm{d}u}{P^{2D}(S,R,t)}, \tag{6}$$

and multiplied by a 1-D depth sensitivity profile $\Gamma(z_0)$ corresponding to the sensitivity of the fundamental mode of Rayleigh waves.

$\varepsilon^{\mathrm{Body}}$ in Eq. 5 is computed using a 3-D kernel describing the sensitivity of an effective scattered body-wave mode:

$$K^{\mathrm{Body}}(r_0,t) = \frac{\int P^{3D}(S,r_0,u)P^{3D}(r_0,R,t-u)\mathrm{d}u}{P^{3D}(S,R,t)}. \tag{7}$$

Following Planès et al. (2014), we use the 2-D and 3-D radiative transfer solution for isotropic scattering (Paasschens 1997) to model the intensity transport of the surface (P^{2D}) and body waves (P^{3D}), respectively. With this formulation, we do not consider P- to S-mode conversions but describe P and S waves as a single effective mode with an effective speed c_{E}. We thus make the assumption of the diffusive regime (isotropic wave energy flux) implying the equipartition of the energy of different states of P and S waves in the coda (Weaver 1982, 1985). In a Poisson half-space, the theoretical energy ratio of S/P is 7.19 at the surface and 10.39 at depth (absence of surface waves; Hennino et al. 2001; Tregourès and van Tiggelen 2002). Obermann et al. (2016) observed an average ratio of $\langle S^2\rangle/\langle P^2\rangle = 9 \pm 2$ for the simulations we are using here, implying a strong dominance of S waves over P waves. Using this ratio, the effective travel time of the waves t_{E} can be determined in a statistical way, as a linear combination of the time that the waves spend in each mode:

$$t_{\mathrm{E}} = \frac{P}{P+S}t_{\mathrm{P}} + \frac{S}{S+P}t_{\mathrm{S}}. \tag{8}$$

The proportion of P-wave energy can be expressed as $\frac{P}{P+S} = \frac{1}{1+x}$ and the proportion of S-wave energy can be expressed as $\frac{S}{P+S} = \frac{x}{1+x}$, with $x = S/P = 9$. Substituting travel times with velocities yields the effective velocity c_{E}:

$$\frac{1}{c_{\mathrm{E}}} = \frac{0.1}{V_{\mathrm{S}}} + \frac{0.9}{V_{\mathrm{P}}}. \tag{9}$$

For the settings in these numerical simulations, $c_{\mathrm{E}} \approx 3.9$ km/s. The contribution of surface and body waves at different positions r_0 and at each lapse time t in the coda is modelled with the lapse time-dependent partition coefficient α. In this way, we take into account surface body wave conversions, as we allow the partition ratio to vary along time in the coda. Numerical simulations (Obermann et al. 2013b, 2016) have shown that the partition coefficient follows a universal behaviour with respect to the coda lapse time expressed in terms of the mean free time $t^{\star} = \frac{\ell^{\star}}{c_{\mathrm{E}}}$. The mean free time depends on the degree of medium heterogeneity measured through the transport mean free path $\ell^{\star}$ and the energy velocity c_{E}. Roughly speaking, the transport mean free path corresponds to the propagation distance after which the energy flux of the multiply scattered waves becomes isotropic. Obermann et al. (2016) observed that the surface waves dominate the depth sensitivity for about eight mean free times, while the body waves dominate the depth sensitivity at later times.

The advantage of this approach of combining surface- and body-wave sensitivity is that no numerical wave-field simulation is required to construct the kernels. These kernels can, however, only represent the sensitivity in media described by a uniform velocity c_E and a uniform transport mean free path $\ell^\star$. Additionally, because of the approximation of the diffusive regime, these kernels might not realistically describe the coda sensitivity in low scattering regimes.

In Fig. 1, slices through the 3-D combined kernels are shown; two horizontal slices at different depth [0.5 km (a), 1.5 km (b)] and a vertical slice (c) along the position indicated with the red line in (a). We observe a sensitivity concentration along the source/receiver line as well as a large sensitivity at shallow depths (c).

2.2. Numerical Wave-Field Simulations

We perform numerical simulations of wave-field propagation in 3-D heterogeneous elastic media without intrinsic attenuation. We use SPECFEM3D for the simulations, which is based on a continuous Galerkin spectral element method (Komatitsch and Tromp 2002; Peter et al. 2011). The model size is $10 \times 10 \times 6\,\mathrm{km}^3$ in the x, y and z directions, respectively. The grid spacing is $\delta x = \delta y = \delta z = 50\,\mathrm{m}$. The record length is 16 s with a time step of 10^{-4} s. For the heterogeneous medium, we use a von Karman distribution with a correlation length of $a = 300\,\mathrm{m}$, which is in the order of one P wavelength ($\lambda_P = 325\,\mathrm{m}$), and consider velocity fluctuations of $\sigma = 20\%$ around a background P-wave velocity of $v_P = 6500\,\mathrm{m/s}$. The S-wave velocity depends on the P-wave velocity with $v_S = v_P/\sqrt{3}$ ($\lambda_s = 190\,\mathrm{m}$). We use an isotropic source with a central frequency of 20 Hz at nine different positions (all combinations of $x = 1, 5, 9\,\mathrm{km}$ and $y = 1, 5, 9\,\mathrm{km}$) in the first grid point below the surface ($z = 50\,\mathrm{m}$). Receivers that measure vertical displacement are placed in a regular grid on the free surface with 100-m spacing. An example of this setup is shown in Fig. 2, with some selected receivers. The simulations follow the setup of Obermann et al. (2016), who determined the transport mean free path for these settings as $\ell^\star = 1840\,\mathrm{m}$ from the intensity of the recorded signals that was fitted with the diffusion equation (Margerin et al. 1999). Please note that the different wavelengths of P and S waves cause a different sampling of the scattering pattern and result in different mean free paths for P and S waves. Obermann et al. (2016) did not separate both wave types, but considered the effective scattering mean free path for the wave field as a whole, which is associated with the propagation speed of the energy of the mixture of the P and S waves; the effective velocity c_E.

For field data, the degree of heterogeneity can often be estimated; for instance, via the decay of the coda envelope (Margerin et al. 1999; Nakahara and Carcolé 2010), phase statistics (Obermann et al.

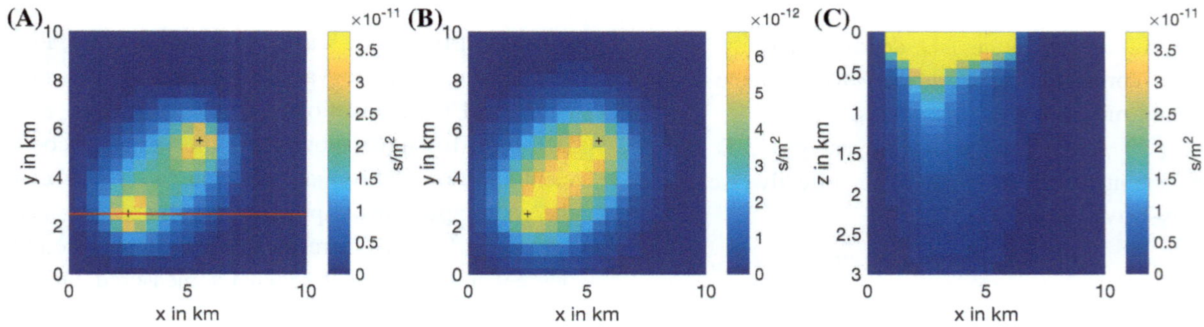

Figure 1
3-D combined kernels. Horizontal slices at depths of 0.5 km (**a**) and 1.5 km (**b**) and a vertical slice (**c**) along the red line indicated in **a**. The crosses mark the source and receiver positions

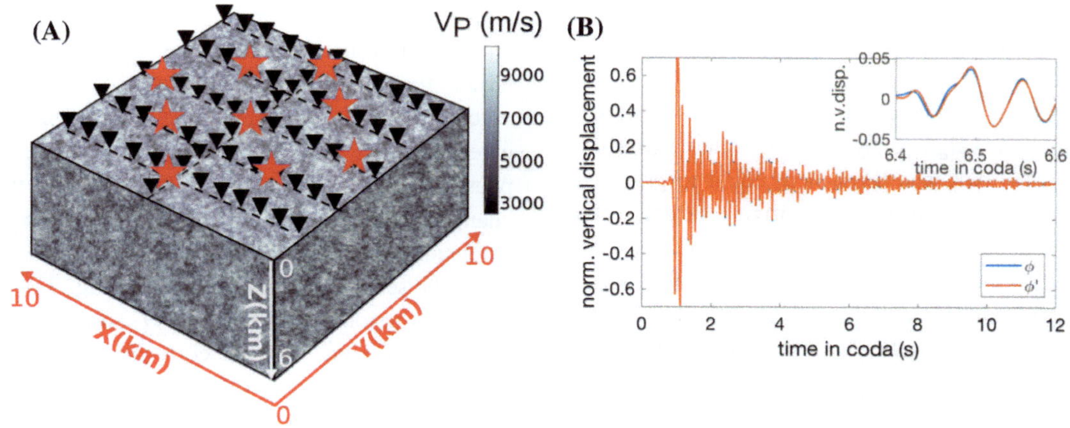

Figure 2
a Heterogeneous model $(x, y) = 10\,\mathrm{km}$, $z = 6\,\mathrm{km}$. The red stars mark the source positions and the black inverted triangles mark some exemplary receiver positions. **b** Synthetic seismograms recorded without (ϕ, blue) and with a velocity perturbation (ϕ', red) at a 1.5-km depth in a medium with 20% velocity fluctuation

2014b), or stochastic medium characterization (Nakata and Beroza 2015).

For each source setup, the wave field is first simulated in an unperturbed medium. Then, a velocity reduction of 500 m/s ($\approx 8\%$) within a cube of 500-m side and length is introduced at $(x, y) = 6\,\mathrm{km}$ and $z = 0.5, 1,$ and 2 km, respectively. The relative phase shifts between both simulations are measured at various lapse times in the coda (2–10 s) using the stretching technique (Lobkis and Weaver 2003; Sens-Schönfelder and Wegler 2006a). The stretching technique is applied in narrow moving time windows of 1 s. These time windows are long enough to estimate a stable waveform distortion at each lapse time. With lapse time, the diffusive halo of the coda waves changes and different parts of the medium are sampled, which affects the measured relative phase shift. The relative phase shift is constant with coda lapse time only when the velocity change is global (i.e. uniformly affecting the medium); if the medium change is local, the relative phase shift with lapse time is non-linear (Obermann et al. 2013b).

2.3. *Linear Least-Square Inversion Procedure*

To estimate the spatial distribution of the velocity changes, we pose the direct problem $d_i = G_{ij} m_j$, where $d_i (i = 1 \ldots n)$ is a vector containing the relative phase shifts that we measured between the n combinations of receivers and different sources. G is a matrix for which each component $G_{ij} = \frac{\Delta V}{t} K_{ij}$ corresponds to the combined sensitivity kernel for the receiver/source pair i in cell j evaluated at time t in the coda and weighted by the volume of the cells Δv and the lapse time t in the coda. m is a vector, for which each component m_j contains the actual relative velocity changes that we estimate for each pixel j (unitless).

To determine m, we use the regularized linear least-square method as formulated by Tarantola and Valette (1982):

$$\boldsymbol{m} = \boldsymbol{m_0} + \boldsymbol{C}_{\mathrm{m}} \boldsymbol{G}^t (\boldsymbol{G} \boldsymbol{C}_{\mathrm{m}} \boldsymbol{G}^t + \boldsymbol{C_d})^{-1} (\boldsymbol{d} - \boldsymbol{G} \boldsymbol{m_0}), \quad (10)$$

where $\boldsymbol{m_0}$ is the initial model, a zero vector in our case, as we do not possess any a priori information about the expected changes. $\boldsymbol{C_d}$ is the diagonal covariance matrix for the data and contains the variances of the data $\sigma_{d,i}^2$ estimated following Weaver et al. (2011). $\boldsymbol{C_m}$ is the covariance matrix for the model. To limit the number of independent parameters, we introduce a spatial smoothing obtained from the correlation of neighbouring cells in the covariance matrix of the model $\boldsymbol{C}_{\mathrm{m}}$ via an exponential function. The elements (i, j) of the else-wise diagonal matrix are given by:

$$C_{\mathrm{m}}(i,j) = \left(\sigma_{\mathrm{m}}\frac{\lambda_0}{\lambda}\right)^2 \exp\left(-\frac{\delta_{i,j}}{\lambda}\right), \qquad (11)$$

where $\delta_{i,j}$ is the distance between two cells i and j. The matrix is weighted by $(\sigma_{\mathrm{m}}\frac{\lambda_0}{\lambda})^2$ where λ_0 is the scaling factor that was taken here equal to grid discretisation of the combined sensitivity kernels (500 m) and λ is the correlation length that defines the length over which the parameters are correlated. σ_{m} is the a priori standard deviation of the model. λ and σ_{m} are determined using the L curve criterion (Hansen 1992).

We perform this inversion jointly for different times ($t = 2, 3,$ and 4 s) in the coda with the corresponding universal partition ratios ($\alpha = 0.80, 0.75, 0.69$) (Obermann et al. 2014a) of surface- and body-wave sensitivity. The idea behind the joint inversion is that the diffusive halo as well as the contribution of body and surface waves and hence the region sampled by the waves, change with lapse time. A joint inversion of different lapse times will hence help to constrain the depth position of the perturbation. At later times > 4 s, because the energy fills the volume of our model, the imprint of the change is homogeneously spread over a very large area, with little spatial contrast and, thus, no imaging potential with the full space kernels we are using.

3. *Results from the Numerical Wave-Field Simulations*

In Fig. 3, we show cuts of the inverted 3-D velocity change model trough the vertical plane (at $y = 5$ km) corresponding to perturbations introduced at depth of 0.5 km ($1.5\lambda_{\mathrm{P}}$, $2.6\lambda_{\mathrm{S}}$), 1 km ($3\lambda_{\mathrm{P}}$, $5\lambda_{\mathrm{S}}$) and 2 km ($6\lambda_{\mathrm{P}}$, $10.5\lambda_s$). We compare inversions performed with the 3-D combined kernels (first column) with kernels describing body-wave propagation only (second column) and with kernels describing surface wave propagation only (third column). The blue square indicates the location of the cubic velocity perturbation that we want to image.

For all depth positions, the inversions using 3-D combined kernels resolve the spatial locations of the velocity perturbation reasonably well (Fig. 3, first column). For better visualisation, the colour scale is adjusted to the min./max. values of the retrieved changes. Because we introduce smoothing to regularize the linear inversion method, the imaged velocity changes are extended in space. The retrieved extended but weak velocity changes produce similar phase shifts than the true small but strong velocity changes. Integrating the changes over their spatial extension leads to similar values.

In the second and third column of Fig. 3, we focus on the inversion results using body- and surface-wave sensitivities alone, which corresponds to partition ratios of $\alpha = 0$ and $\alpha = 1$, respectively. While the inversions provide an accurate location in the horizontal plane, except for the inversion using the surface-wave-only kernels at largest depth (lowest subplot in the third column in Fig. 3), these kernels cannot provide a depth estimate of the changes. In our simulations, a significant part of the energy is contained in the surface waves ($\alpha > 0.6$ at $t > 2s$). The body-wave-only kernel can hence not accurately describe the coda-wave sensitivity. Concerning the surface-wave-only kernel, there are several reasons to explain the bad performance. The shallowest perturbation, at 0.5-km depth, is already at $2.5\lambda_{\mathrm{S}}$, which is deep for the surface-wave kernel to probe. We also only use results from a single frequency (source with a central frequency at 20 Hz) to describe the surface-wave kernel. Surface waves are dispersive and their sensitivity to changes at depth depends on the period. Shorter periods with short wavelengths are sensitive to shallower changes, while longer periods with longer wavelengths, sample increasing depth ranges (Rivet et al. 2011). Taking a frequency analysis into account, the performance of the surface-wave-only kernel might be improved in settings that are dominated by surface waves and help to locate changes at shallow depth ($<1\lambda_{\mathrm{S}}$).

In the following, we discuss limitations of the method focussing on physical, as well as practical issues.

Let us focus on the *source/receiver configuration*. In Fig. 4a, b we show horizontal slices through the relative velocity change model close to the surface resulting from a central, cubic velocity perturbation at 2-km depth (blue square). For the simulations in Fig. 4b, we used eight sources distributed along a circle (white crosses). For the simulations in Fig. 4a,

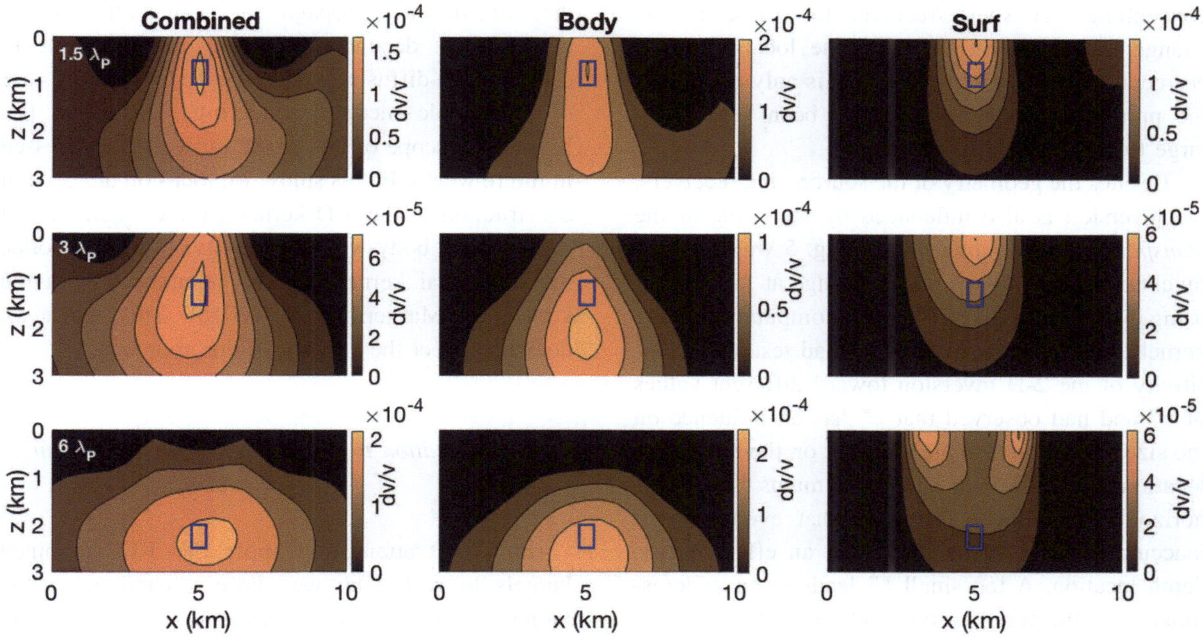

Figure 3
Vertical cuts through the relative velocity change model along $y = 5$ km. Cubic velocity perturbations (blue square) are placed at depths of 0.5 km (first row), 1 km (second row) and 2 km (third row). The inversions were performed using the 3-D combined kernels (first column), 3-D kernels describing body-wave propagation only (second column) and 2-D kernels describing surface-wave propagation only (third column)

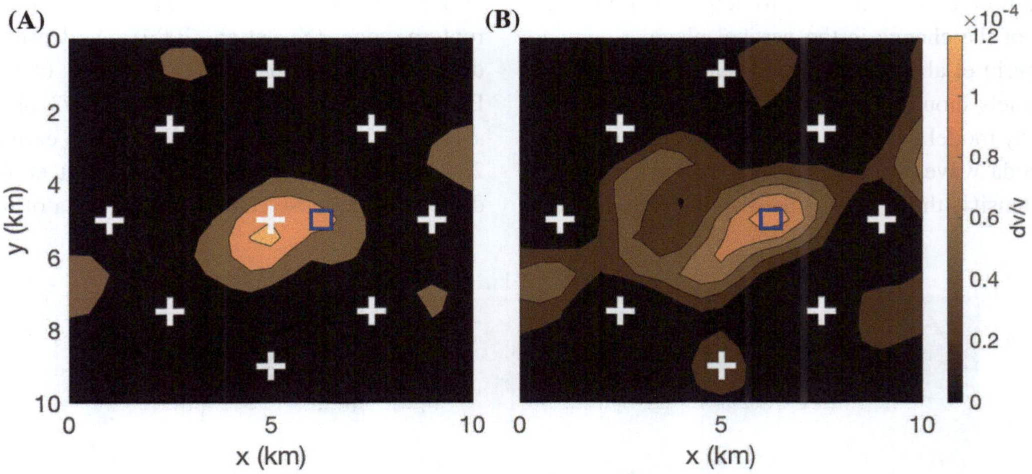

Figure 4
Horizontal cuts through the relative velocity change model close to the surface for a cubic velocity change at 2-km depth. The projection of the location of the velocity change is shown with the blue square. For the simulations, nine (**a**) and eight (**b**) sources are used as indicated with the white crosses

we added a ninth source in the centre that is much closer to the location of the velocity perturbation compared to the other sources. Consequently, we observe that the location of the velocity perturbation, as seen by the inversion, is shifted towards the position of the central source, where the sensitivity concentrates. A source/receiver too close to the change has a tendency to shift the localisation. A correct estimation of the sensitivity is only possible if the propagation distance/time has been sufficiently large to enter the diffusive regime.

Besides the geometry of the sources and receivers, the inversion is also influenced by the value of the *transport mean free path* $\ell^{\star}$. In Fig. 5 we repeat an inversion with a cubic velocity change at 1-km depth using different values of $\ell^{\star}$ in the computation of the kernels. Obermann et al. (2013a) had tested the sensitivity of the 2-D inversion toward different values of $\ell^{\star}$ and had observed that $\ell^{\star}$ has an influence on the size of the affected area but not on the horizontal location itself. While in 3-D, this remains true for the horizontal location, we observe that even a slight inaccuracy in the value of $\ell^{\star}$ has an effect on the depth location. A too small $\ell^{\star}$ leads to an underestimation of the perturbation depth, while the opposite leads to an overestimation. The exact value for this set of simulations is $\ell^{\star} = 1.84\,\text{km}$. $\ell^{\star}$ controls the shape of the diffusive halo, including in the vertical direction. On the horizontal plane, the effect is less important as the errors are compensated for by the distribution of the receivers. In our configuration, there are no receivers at depth to help constraint the position of the change in the vertical plane.

Margerin et al. (2016) pointed out that the sensitivity kernels should be based on the *specific intensity* to properly model the effect of a velocity perturbation on the coda waves. Contrary to the (angularly averaged) intensity, the specific intensity depends on the direction of wave propagation. Taking into account this angular dependency would be important to describe non-diffusive regimes, especially in the case of anisotropic medium scattering. This issue lies beyond the scope of this paper and will be addressed in future works. In this study, we focus on the general construction of the 3-D kernel via a combination of surface- and body-wave sensitivity. In future works, the individual kernels should be updated with the kernels of Margerin et al. (2016). This, however, does not affect the validity of our approach.

4. Application to the M_w 7.9, 2008 Wenchuan Earthquake

In a first attempt to apply the 3-D combined kernels to real data, we chose a passive seismic monitoring application. Obermann et al. (2014a) had previously processed data from a dense network in China that was in operation during the M_w 7.9, 2008 Wenchuan earthquake that struck the Longmen Shan region along the Eastern margin of the Tibetan Plateau (Burchfiel et al. 2008; Zhang et al. 2010).

The earthquake occurred at a depth of about 19 km (Chen et al. 2009) and affected a 240-km-long rupture zone (Xu et al. 2009). A dense network deployed by the Institute of Geology of the China Earthquake Administration covered 2/3 of the fault system activated during the Wenchuan earthquake in 2008. Chen et al. (2010) and Froment et al. (2013) detected time-lapse changes in the coda of the cross-

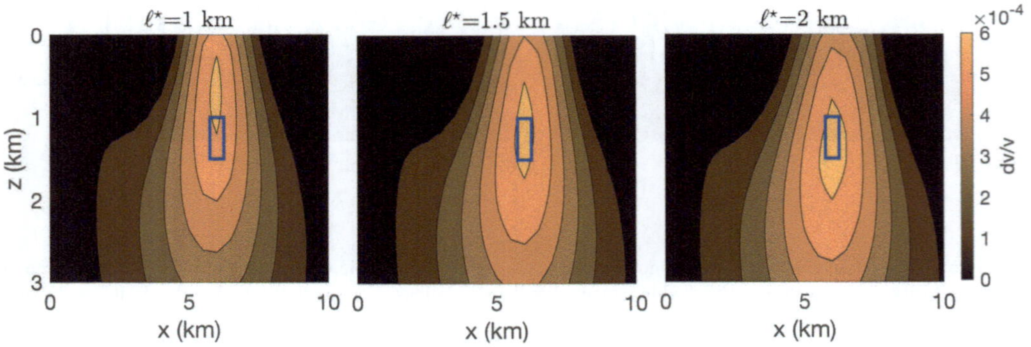

Figure 5
Vertical cuts through the relative velocity change model along $x = 6$ km for a cubic velocity perturbation at 1 km. For the inversions, we used 3-D combined sensitivity kernels with different scattering mean free paths. The exact value for this set of simulations is $\ell^{\star} = 1.84\,\text{km}$

correlations from 156 stations covering the fault system within two different period bands, in which the surface waves are sensitive to shallower (1–3 s) and larger depths (12–20 s). In the period band of 12–20 s, Froment et al. (2013) and Obermann et al. (2014a) could observe a strong post-seismic signature of the middle crust to the earthquake, and also detect a clear seasonal signature of the velocity changes within the Sichuan basin that they related to a loading effect due to the heavy rainfalls during the monsoon season. Obermann et al. (2014a) studied the changes in the 12–20-s period band at different lapse times. The post-seismic change is more prominent at later lapse times, indicating a greater depth, while the change due to the monsoon rain is more prominent at earlier times, indicating a shallower depth. In this region, surface waves in the period band from 12–20 s are estimated to be sensitive to a depth range of 5–40 km (Froment et al. 2013), leaving a wide range for the depth of a relatively 'shallow' or 'deep' change. Obermann et al. (2014a) used 2-D probabilistic sensitivity kernels to locate the apparent velocity changes at different lapse times on the horizontal plane (Fig. 6a, b). In this previous work, the loading effect of the monsoon rain (a) was constrained to the Sichuan basin and could be clearly separated from the post-seismic signature (b).

In Fig. 6c, d, we show a vertical cross section through the model of velocity perturbation that we obtain from an inversion with the 3-D combined kernels along the red line indicated in Fig. 6a, b. To compute the diffusivity for the kernels, we used an effective velocity of 5 km/s and a transport mean free path of $\ell^{\star} = 500$ km, which corresponds to the values used for the 2-D inversion by Obermann et al. (2014a). The exact value for $\ell^{\star}$ in this region in China is not known, but our estimate lies well within the broad range of observed values in these period bands for scattering around the globe (Sato and Nohechi 2001; Lee et al. 2003; Hillers et al. 2013; Sens-Schönfelder and Wegler 2006b).

From the results in Fig. 6c, d, we see that the inversion with the 3-D combined kernels allows us to constrain the depth of the changes. The peak velocity reduction in the Sichuan basin is constrained to the upper 15 km of the crust. This strengthens our interpretation that the velocity change is due to a loading effect of the rain and not related to water infiltration, which cannot reach such depths. Loading can significantly increase the pore pressure (Bettinelli et al. 2008) and reduce the velocity (Dvorkin et al. 1999; Carcione and Tinivella 2001).

The maximal velocity reduction in the month following the Wenchuan earthquake is constrained to 20–30 km of depth. As pointed out by Froment et al. (2013) and Obermann et al. (2014a), no co-seismic velocity reduction in this period range (12–20 s) could be observed. Given the depth range, the observed changes might have been caused by earthquake-related processes at depth (after-slip, viscoelastic response, fluid transfer, etc.). Specifically, the slip of a deep decollement beneath Eastern Tibet as an aftereffect of the Wenchuan earthquake has been proposed by Fielding et al. (2013) and Obermann et al. (2014a).

5. Conclusion

We have shown that it is possible to constrain the depth of medium changes in multiply scattered media with an imaging procedure that involves a linear combination of body- and surface-wave sensitivity kernels. We have seen that the accuracy of the location results depends on the geometry of the sources and receivers, and the accurate characterisation of the subsurface scattering. A limitation for real data applications is that the scattering properties of the subsurface are often poorly constrained. We show a promising first application of the kernels to locate velocity changes at depth that were obtained from the correlations of ambient seismic noise data covering the region affected by the 2008 Wenchuan earthquake. Constraining the depth location of the changes can significantly improve our understanding of the nature of the medium variations revealed by seismic monitoring. As standard methods are not effective to provide images of the texture of the rock at depth (e.g. small-scale heterogeneities, cracks, voids), we think that new imaging approaches that make use of multiply scattered and sensitive coda waves can help to make progress in this direction. Following the presented approach, an improvement should be done on the level of the individual kernels (2-D surface and

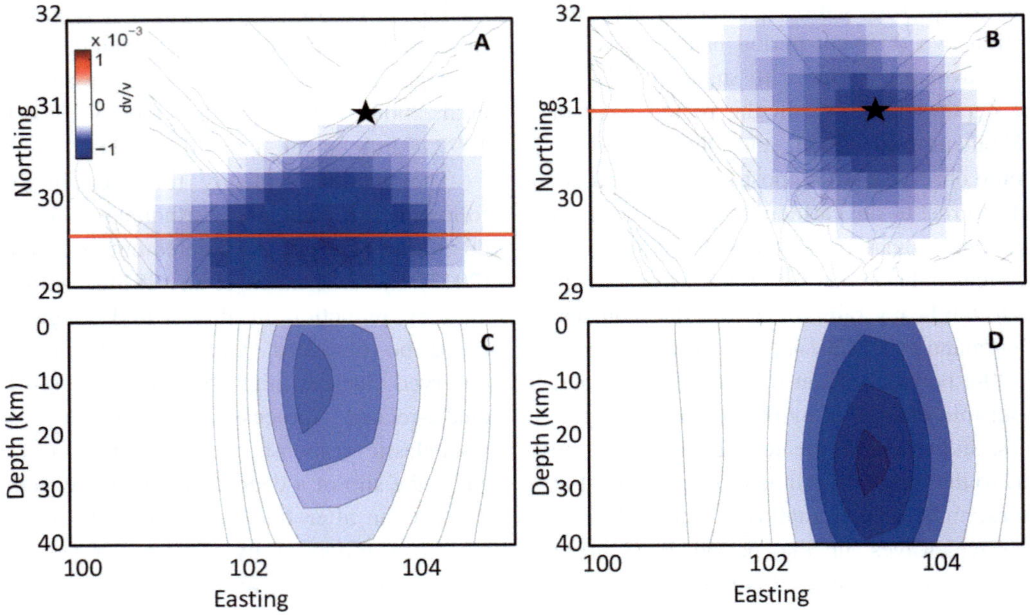

Figure 6
Inversion results on the horizontal plane of the apparent velocity model in the Sichuan region after the M_w 7.9, 2008 Wenchuan earthquake (star) at 50 s in the coda **a** featuring the loading effect due to the monsoon rain and at 90 s in the coda **b** featuring the post-seismic effects (modified from Obermann et al. (2014a)). The vertical cuts (**c**, **d**) along the red line indicated in **a** and **b** were computed with the 3-D combined kernels

3-D body) to account for the directionality of the energy flux and address non-diffusive regimes.

Acknowledgements

The research leading to these results received funding from the European Community Seventh Framework Programme under grant agreement no. 608553 (Image) and the Swiss Federal Office of Energy with the project GEOBEST. The authors also want to thank the Leibniz-Rechenzentrum (LRZ) and Jens Oeser for access and support in using the SUPER-MUC system at the LRZ. The work has benefited from discussions within the European Cooperation in Science and Technology (COST) action ES1401-TIDES, supported by COST. MC acknowledges the support of the European Research Council (ERC) under agreement no. 742335 (F-Image). The authors wish to thank Gregor Hillers and an anonymous reviewer for their very constructive and detailed comments that helped to significantly improve the manuscript, as well as the editor Yehuda Ben-Zion for his helpful suggestions.

References

Bettinelli, P., Avouac, J. P., Flouzat, M., Bollinger, L., Ramillien, G., Rajaure, S., et al. (2008). Seasonal variations of seismicity and geodetic strain in the Himalaya induced by surface hydrology. *Earth and Planetary Science Letters*, *266*, 332–344.

Burchfiel, B. C., Royden, L. H., van der Hilst, R. D., Chen, Z., King, R. W., Li, C., et al. (2008). A geological and geophysical context for the Wenchuan earthquake of 12 May 2008, Sichuan, People's Republic of China. *GSA Today*, *18*(7), 5.

Carcione, J. M., & Tinivella, U. (2001). The seismic response to overpressure: A modelling study based on laboratory, well and seismic data. *Geophysical Prospecting*, *49*, 523–539.

Chen, J. H., Froment, B., Liu, Q. Y., & Campillo, M. (2010). Distribution of seismic wave speed changes associated with the 12 May 2008 M_w 7.9 Wenchuan earthquake. *Geophysical Research Letters*, *37*(L18302), 1–4.

Chen, J. H., Liu, Q. Y., Li, S. C., Guo, B., Wang, J., & Qi, S. H. (2009). Seismotectonic study by relocation of the Wenchuan M(S)8.0 earthquake sequence. *Chinese Journal of Geophysics-Chinese Edition*, *52*(2), 390–397.

Dvorkin, J., Mavko, G., & Nur, A. (1999). Overpressure detection from compressional- and shear-wave data. *Geophysical Research Letters*, *26*(22), 3417–3420.

Fielding, E. J., Sladen, A., Li, Z., Avouac, J.-P., Bürgmann, R., & Ryder, I. (2013). Kinematic fault slip evolution source models of the 2008 M7. 9 Wenchuan earthquake in China from SAR interferometry, GPS and teleseismic analysis and implications for Longmen Shan tectonics. *Geophysical journal international, 194*(2), 1138–1166.

Froment, B. (2011). *Utilisation du bruit sismique ambiant dans le suivi temporel de structures géologiques*. PhD thesis, Université de Grenoble.

Froment, B., Campillo, M., Chen, J. H., & Liu, Q. Y. (2013). Deformation at depth associated with the May 12, 2008 M_w 7.9 Wenchuan earthquake from seismic ambient noise monitoring. *Geophysical Research Letters, 40*, 78–82.

Grêt, A., Snieder, R., Aster, R. C., & Kyle, P. R. (2005). Monitoring rapid temporal change in a volcano with coda wave interferometry. *Geophysical Research Letters, 32*(6), L06304.

Hansen, P. C. (1992). Analysis of discrete ill-posed problems by means of the L-curve. *SIAM Review, 34*(4), 561–580.

Hennino, R., Tregoures, N., Shapiro, N. M., Margerin, L., Campillo, M., van Tiggelen, B. A., et al. (2001). Observation of equipartition of seismic waves. *Physical Review Letters, 86*(15), 3447–3450.

Hillers, G., Ben-Zion, Y., Landes, M., & Campillo, M. (2013). Interaction of microseisms with crustal heterogeneity: A case study from the San Jacinto fault zone area. *Geochemistry, Geophysics, Geosystems, 14*(7), 2182–2197.

Hillers, G., Husen, S., Obermann, A., Planès, T., Campillo, M., & Larose, E. (2015). Noise-based monitoring and imaging of aseismic transient deformation induced by the 2006 Basel reservoir stimulation. *Geophysics, 80*(4), KS51–KS68.

Kanu, C., & Snieder, R. (2015a). Numerical computation of the sensitivity kernel for monitoring weak changes with multiply scattered acoustic waves. *Geophysical Journal International, 203*(3), 1923–1936.

Kanu, C., & Snieder, R. (2015b). Time-lapse imaging of a localized weak change with multiply scattered waves using numerical-based sensitivity kernel. *Journal of Geophysical Research: Solid Earth, 120*(8), 5595–5605.

Komatitsch, D., & Tromp, J. (2002). Spectral-element simulations of global seismic wave propagation II. Three-dimensional models, oceans, rotation and self-gravitation. *Geophysical Journal International, 150*(1), 303–318.

Larose, E., Obermann, A., Digulescu, A., Planès, T., Chaix, J. F., Mazerolle, F., et al. (2015). Locating and characterizing a crack in concrete with LOCADIFF: A four-point bending test. *The Journal of the Acoustical Society of America, 138*(1), 232–241.

Larose, E., Planès, T., Rossetto, V., & Margerin, L. (2010). Locating a small change in a multiple scattering environment. *Applied Physics Letters, 96*(204101), 1–3.

Lee, W. S., Sato, H., & Lee, K. (2003). Estimation of S-wave scattering coefficient in the mantle from envelope characteristics before and after the ScS arrival. *Geophysical Research Letters, 30*(24), 2248–2252.

Lesage, P., Reyes-Dávila, G., & Arámbula-Mendoza, R. (2014). Large tectonic earthquakes induce sharp temporary decreases in seismic velocity in Volcán de Colima, Mexico. *Journal of Geophysical Research: Solid Earth, 119*(5), 4360–4376.

Lobkis, O. I., & Weaver, R. L. (2003). Coda-wave interferometry in finite solids: Recovery of P-to-S conversion rates in an elastodynamic billiard. *Physical Review Letters, 90*(254302), 1–4.

Margerin, L., Campillo, M., Shapiro, N. M., & van Tiggelen, B. (1999). Residence time of diffuse waves in the crust and the physical interpretation of coda Q. Application to seismograms recorded in Mexico. *Geophysical Journal International, 138*, 343–352.

Margerin, L., Planès, T., Mayor, J., & Calvet, M. (2016). Sensitivity kernels for coda-wave interferometry and scattering tomography: Theory and numerical evaluation in two-dimensional anisotropically scattering media. *Geophysical Journal International, 204*(1), 650–666.

Meunier, J., Huguet, F., & Meynier, P. (2001). Reservoir monitoring using permanent sources and vertical receiver antennae The Céré-la-Ronde case study. *The Leading Edge, 20*(6), 622–629.

Nakahara, H., & Carcolé, E. (2010). Maximum-likelihood method for estimating coda Q and the Nakagami-m parameter. *Bulletin of the Seismological Society of America, 100*(6), 3174–3182.

Nakata, N., & Beroza, G. C. (2015). Stochastic characterization of mesoscale seismic velocity heterogeneity in Long Beach, California. *Geophysical Journal International, 203*(3), 2049–2054.

Obermann, A., Froment, B., Campillo, M., Larose, E., Planès, T., Valette, B., et al. (2014). Seismic noise correlations to image structural and mechanical changes associated with the M_w 7.9 2008 Wenchuan earthquake. *Journal of Geophysical Research, 119*(4), 3155–3168.

Obermann, A., Kraft, T., Larose, E., Wiemer, S., et al. (2015). Potential of ambient seismic noise techniques to monitor the St. Gallen geothermal site (Switzerland). *Journal of Geophysical Research, 120*(6), 4301–4316.

Obermann, A., Larose, E., Margerin, L., & Rossetto, V. (2014b). Measuring the scattering mean free path of Rayleigh waves on a volcano from spatial phase decoherence. *Geophysical Journal International, 197*, 435–442.

Obermann, A., Planès, T., Hadziioannou, C., & Campillo, M. (2016). Lapse-time-dependent coda-wave depth sensitivity to local velocity perturbations in 3-D heterogeneous elastic media. *Geophysical Journal International, 207*(1), 59–66.

Obermann, A., Planès, T., Larose, E., & Campillo, M. (2013a). Imaging pre- and co-eruptive structural and mechanical changes on a volcano with ambient seismic noise. *Journal of Geophysical Research, 118*(12), 6285–6294.

Obermann, A., Planès, T., Larose, E., Sens-Schönfelder, C., & Campillo, M. (2013b). Depth sensitivity of seismic coda waves to velocity perturbations in an elastic heterogeneous medium. *Geophysical Journal International, 194*(1), 372–382.

Paasschens, J. C. J. (1997). Solution of the time-dependent Boltzmann equation. *Physical Review E, 56*(1), 1135.

Pacheco, C., & Snieder, R. (2005). Time-lapse travel time change of multiply scattered acoustic waves. *The Journal of the Acoustical Society of America, 118*, 1300–1310.

Peng, Z., & Ben-Zion, Y. (2006). Temporal changes of shallow seismic velocity around the Karadere–Düzce branch of the north Anatolian fault and strong ground motion. *Pure and Applied Geophysics, 163*(2–3), 567–600.

Perton, M., Sánchez-Sesma, F., Rodríguez-Castellanos, A., Campillo, M., & Weaver, R. (2009). Two perspectives on equipartition in diffuse elastic fields in three dimensions. *The Journal of the Acoustical Society of America, 126*(3), 1125–1130.

Peter, D., Komatitsch, D., Luo, Y., Martin, R., Le Goff, N., Casarotti, E., et al. (2011). Forward and adjoint simulations of

seismic wave propagation on fully unstructured hexahedral meshes. *Geophysical Journal International, 186*(2), 721–739.
Planès, T. (2013). *Imagerie de changements locaux en régime de diffusion multiple*. PhD thesis, Université de Grenoble.
Planès, T., Larose, E., Margerin, L., Rossetto, V., & Sens-Schönfelder, C. (2014). Decorrelation and phase-shift of coda waves induced by local changes: Multiple scattering approach and numerical validation. *Waves in Random and Complex Media*, *2*(24), 1–27.
Planès, T., Larose, E., Rossetto, V., & Margerin, L. (2015). Imaging multiple local changes in heterogeneous media with diffuse waves. *The Journal of the Acoustical Society of America*, *2*(137), 660–667.
Poupinet, G., Ellsworth, W. L., & Frechet, J. (1984). Monitoring velocity variations in the crust using earthquake doublets: An application to the Calaveras fault. *California. Journal of Geophysical Research, 89*(B7), 5719–5731.
Poupinet, G., Ratdomopurbo, A., & Coutant, O. (1996). On the use of earthquake multiplets to study fractures and the temporal evolution of an active volcano. *Annals of Geophysics, 39*, 253–264.
Rivet, D., Campillo, M., Shapiro, N. M., Cruz-Atienza, V., Radiguet, M., Cotte, N., et al. (2011). Seismic evidence of nonlinear crustal deformation during a large slow slip event in Mexico. *Geophysical Research Letters, 38*(L08308), 1–5.
Rossetto, V., Margerin, L., Planès, T., & Larose, E. (2011). Locating a weak change using diffuse waves: Theoretical approach and inversion procedure. *Journal of Applied Physics, 109*(034903), 1–11.
Roux, P., & Ben-Zion, Y. (2014). Monitoring fault zone environments with correlations of earthquake waveforms. *Geophysical Journal International, 196*(2), 1073–1081.
Sanchez, P., Obermann, A., Schimmel, M., et al. (2018). Detecting and locating precursory signals during the 2011 El Hierro, Canary Islands, submarine eruption. *Geophysical Research Letters*. https://doi.org/10.1029/2018GL079550.
Sato, H., & Nohechi, M. (2001). Envelope formation of long-period Rayleigh waves in vertical component seismograms: Single isotropic scattering model. *Journal of Geophysical Research: Solid Earth, 106*(B4), 6589–6594.
Schaff, D. P., & Beroza, G. C. (2004). Coseismic and postseismic velocity changes measured by repeating earthquakes. *Journal of Geophysical Research, 109*(B10302), 1–14.
Sens-Schönfelder, C., & Wegler, U. (2006a). Passive image interferometry and seasonal variations of seismic velocities at Merapi Volcano. *Indonesia. Geophysical Research Letters, 33*(21), L21302.
Sens-Schönfelder, C., & Wegler, U. (2006b). Radiative transfer theory for estimation of the seismic moment. *Geophysical Journal International, 167*(3), 1363–1372.
Takagi, R., Okada, T., Nakahara, H., Umino, N., & Hasegawa, A. (2012). Coseismic velocity change in and around the focal region of the 2008 Iwate-Miyagi Nairiku earthquake. *Journal of Geophysical Research, 117*(B06315), 1–19.
Tarantola, A., & Valette, B. (1982). Generalized nonlinear inverse problems solved using the least squares criterion. *Reviews of Geophysics, 20*(2), 219–232.
Tregourès, N. P., & van Tiggelen, B. A. (2002). Generalized diffusion equation for multiply scattered elastic waves. *Waves in Random Media, 12*, 21–38.
Ugalde, A., Villaseñor, A., Gaite, B., Casquero, S., Martí, D., Calahorrano, A., et al. (2013). Passive seismic monitoring of an experimental CO_2 geological storage site in Hontomín (Northern Spain). *Seismological Research Letters, 84*(1), 75–84.
Weaver, R. L. (1982). On diffuse waves in solid media. *The Journal of the Acoustical Society of America, 71*, 1608–1609.
Weaver, R. L. (1985). Diffuse elastic waves at a free surface. *The Journal of the Acoustical Society of America, 78*, 131–136.
Weaver, R. L., Hadziioannou, C., Larose, E., & Campillo, M. (2011). On the precision of noise correlation interferometry. *Geophysical Journal International, 185*(3), 1384–1392.
Wu, C., Peng, Z., & Ben-Zion, Y. (2009). Non-linearity and temporal changes of fault zone site response associated with strong ground motion. *Geophysical Journal International, 176*(1), 265–278.
Xu, X. W., Wen, X. Z., Yu, G. H., Chen, G. H., Klinger, Y., Hubbard, J., et al. (2009). Coseismic reverse- and oblique-slip surface faulting generated by the 2008 M_w 7.9 Wenchuan earthquake, China. *Geology, 37*(6), 515–518.
Zhang, P. Z., Wen, X. Z., Shen, Z. K., & Chen, J. H. (2010). Oblique, high-angle, listric-reverse faulting and associated development of strain: The Wenchuan earthquake of May 12, 2008, Sichuan, China. *Annual Review of Earth and Planetary Sciences, 38*, 353–382.
Zhang, Y., Planès, T., Larose, E., Obermann, A., Rospars, C., & Moreau, G. (2016). Diffuse ultrasound monitoring of stress and damage development on a 15-ton concrete beam. *The Journal of the Acoustical Society of America, 139*(4), 1691–1701.

(Received May 2, 2018, revised September 30, 2018, accepted October 4, 2018, Published online October 15, 2018)

Pure Appl. Geophys. 176 (2019), 1255–1267
© 2018 Springer Nature Switzerland AG
https://doi.org/10.1007/s00024-018-2075-7

Pure and Applied Geophysics

Significant Effects of Shallow Seismic and Stress Properties on Phase Velocities of Rayleigh Waves Up to 20 s

CHENHAO YANG,[1] GUOLIANG LI,[1,2] FENGLIN NIU,[1,2] and YEHUDA BEN-ZION[3]

Abstract—Seismic data are increasingly used to monitor subsurface velocity changes associated with tectonic and environmental processes that occur in different depth sections. To clarify the differences between effects associated with shallow and deep changes of properties, we conduct numerical experiments using simple layered models that include low velocities, low attenuation coefficients and stress-sensitivity of cracked rocks in the shallow crust. We find significant phase-velocity drops in the period range of 5–20 s when large structural changes occur in the top 1–3 km. The apparent velocity changes (δv/v) measured from the first part of the synthetic Rayleigh waves with a cross-correlation based technique show significant velocity drops in the period bands of 5–10 s and 10–20 s that are consistent with reported values of changes at seismogenic depth. The results highlight the importance of accounting for low velocities, attenuation coefficients and stress-sensitivity of parameters in the top 1–3 km in studies aiming to determine the source region of temporal changes. Analyses using different frequency ranges and calculations of apparent delay times over multiple period bands are essential for resolving the depth range of temporal changes of properties. For temporal changes occurring at seismogenic depths, the measured δv/v values at 5–10 s are significantly larger than those of 10–20 s, which are not observed for shallow changes.

Key words: Time-lapse seismic imaging, temporal changes in seismic structure, coseismic stress drop, shallow structural damage, Rayleigh wave phase velocity, apparent velocity change.

1. Introduction

Recent advances in seismic data acquisition and computational power have led to significant progress in time-lapse (4-D) seismic imaging. A growing number of studies report on temporal changes of seismic velocities associated with the occurrence of earthquakes (e.g., Schaff and Beroza 2004; Peng and Ben-Zion 2006; Sawazaki et al. 2009), volcanic deformation (e.g., Sens-Schönfelder and Wegler, 2006; Brenguier et al. 2008), and various non-tectonic processes such as precipitation, thermoelastic strain and changes of water level in reservoirs (e.g., Meier et al. 2010; Chen et al. 2014; Hillers et al. 2015; Lecocq et al. 2017). The techniques used to monitor temporal changes of seismic properties are associated with different types of data and analysis time steps. Studies using earthquake waveforms are associated mainly with body and coda waves, and have time resolution that can approach in cases involving aftershock sequences of minutes to seconds. Several such studies reported on co-seismic temporal changes of velocities with amplitudes of a few percent to tens of percent (e.g., Rubinstein and Beroza 2005; Wu et al. 2009; Zhao and Peng 2009; Nakata and Snieder 2011; Roux and Ben-Zion 2014), and concluded from borehole observations and modeling that the changes occur primarily in the top few hundred meters of the crust.

Using the ambient seismic noise to monitor temporal changes has become a significant technique with many applications (e.g., Brenguier et al. 2008, 2014; Cheng et al. 2010; Sens-Schönfelder and Larose 2010; Obermann et al. 2014). These studies use typically Rayleigh waves and time steps of days or more, since stacking is required to increase the signal-to-noise ratio of the waves that are extracted from the noise correlations. The amplitudes of temporal changes reported by noise-based studies are typically a fraction of a percent, and the depth section

[1] Department of Earth, Environmental and Planetary Sciences, Rice University, 6100 Main St, Houston, TX 77005, USA. E-mail: niu@rice.edu
[2] State Key Laboratory of Petroleum Resource and Prospecting, and Unconventional Gas Institute, China University of Petroleum at Beijing, Beijing 102249, China.
[3] Department of Earth Sciences, University of Southern California, Los Angeles, CA 90089-0740, USA.

sustaining the changes is inferred from the used frequency band. For example, Cheng et al. (2010) observed ~ 0.4% temporal changes of seismic velocities after the 2008 M_w 7.9 Wenchuan earthquake in China. The analysis employed 10–25 s Rayleigh waves, leading to inferred depth sustaining the changes up to 25 km. Froment et al. (2013) observed similar changes following the M_w 7.9 Wenchuan event using both 1–3 s and 12–20 s Rayleigh waves, and interpreted the results as reflecting deformation that extends to seismogenic depth. Rivet et al. (2011) found ~ 0.1% temporal changes of seismic velocities following a slow slip event in the Guerrero region, Mexico, with periods ranging from 4 s to over 25 s. The changes were largest for 7–17 s Rayleigh waves, leading to inferred depth range of 5–20 km. Brenguier et al. (2014) observed ~ 0.01–0.1% velocity reductions across Japan following the 2011 M_w 9.0 Tohoku-Oki earthquake, with the most significant changes beneath Quaternary volcanoes. The analysis used Rayleigh waves with frequencies around 0.1–0.9 Hz, and the changes were attributed to a depth range between the surface and 10 km.

Clearly, resolving the depth range sustaining temporal changes of seismic properties requires analysis over different frequency ranges. Obermann et al. (2013) investigated with numerical experiments the depth sensitivity of high-frequency surface and coda waves to velocity perturbations, and found that the shallow structure can have a significant influence on the wavefield. In the present study we analyze with numerical simulations the sensitivity of intermediate-frequency Rayleigh waves to velocity, attenuation and stress changes at various depth sections of crustal structures. The analysis includes computing apparent velocity changes δv/v with a technique based on cross-correlation of the first part of Rayleigh waves in different frequency bands. Significant changes of both phase and apparent velocities at periods up to 20 s are found in some cases with changes of properties limited to the top 1–3 km (i.e., above the seismogenic zone). Comparisons of the relative phase and apparent velocity changes observed using different frequency ranges can help to resolve the depth section where the temporal changes occur.

2. Numerical Modeling

2.1. Model Settings

We first set a reference 1-D layered model for the numerical simulations with constant mass density, seismic velocity and attenuation quality factor in each layer. As many studies of temporal changes are based on seismic stations located in sedimentary basins, we include 3.5 km thick sediments in the model. The model consists of four sedimentary layers, two crystalline crustal layers and an underlying half-space substrate (Fig. 1; Table 1). We refer to this model as the original model in the rest of the paper. To investigate how structural and stress perturbations

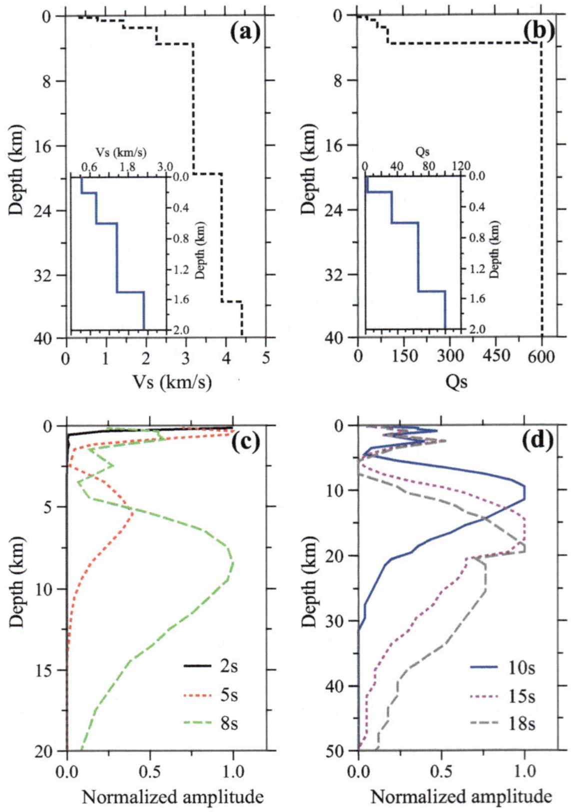

Figure 1
S-wave velocity and Qs of the original model are shown in dashed lines in **a**, **b**, respectively. Blue insets show the top 2 km of the velocity and attenuation structure. **c** Phase velocity sensitivity kernels of 2 s (black solid line), 5 s (red dotted line), and 8 s (green dashed line) Rayleigh waves are plotted in the depth range of 0–20 km. **d** is similar to **c** except for 10 s (blue solid line), 15 s (purple dotted line), and 18 s (grey dashed line) Rayleigh waves, and the depth range, which is 0–50 km

Table 1

Parameters of the initial model for case 1

H^a (km)	α (km/s)	β (km/s)	ρ (g/cm^3)	Qs
0.2	1.50	0.35	1.93	3.0
0.4	2.22	0.80	2.12	33.0
0.9	2.88	1.44	2.25	66.0
2.0	4.57	2.28	2.52	100.0
16.0	5.80	3.20	2.60	600.0
16.0	6.80	3.90	2.90	600.0
80.0	8.02	4.40	3.38	600.0

[a]Thickness of each layer

affect the velocity of Rayleigh waves at the surface, we generate 12 models by changing some parts of the original model. We divide these cases into four categories summarized in Table 2. Categories 1 and 2 with eight models are designed to investigate effects associated with changes occurring in sediments shallower than 1.5 km, model 9 in category 3 is used to probe effects involving perturbations at seismogenic depth, and five additional models in category 4 to investigate stress-induced velocity changes.

The four cases in category 1 incorporate changes of velocity and density, whereas those in category 2 have additional changes of attenuation. A 30% reduction of seismic velocity in the shallow sediments is assumed in models 1–8, in agreement with large co-seismic drops observed by studies using earthquake waveforms and very short time sampling (e.g., Karabulut and Bouchon 2007; Wu et al. 2009). A 5% velocity reduction over the depth section 5–10 km in model 9 is an upper bound to the inferred observational values. We expect that large coseismic stress drops within seismogenic zones can cause significant changes in seismic velocity. Velocity drops in models 10 and 11 are computed with a velocity-stress sensitivity coefficient ($\delta \ln V/\delta p$) based on laboratory experiments of Nur and Simmons

Table 2

Parameters of the initial model for case 1

Cat.	Model	Dep. range (km)		Vel. change (%)		Q change (%)		δv/v (%)	
		Top	Bottom	Top	Bottom	Top	Bottom	5–10 s	10–20 s
1	1	0.00	0.60	30.00	30.00	0.0	0.0	− 0.32	− 0.51
	2	0.00	0.60	30.00	0.00	0.0	0.0	− 0.42	− 0.49
	3	0.00	1.50	30.00	30.00	0.0	0.0	− 1.34	− 1.27
	4	0.00	1.50	30.00	0.00	0.0	0.0	− 0.69	− 0.73
2	5	0.00	0.60	30.00	30.00	30.0	30.0	− 0.31	− 0.50
	6	0.00	0.60	30.00	0.00	30.0	0.0	− 0.41	− 0.49
	7	0.00	1.50	30.00	30.00	30.0	30.0	− 1.34	− 1.27
	8	0.00	1.50	30.00	0.00	30.0	0.0	− 0.68	− 0.73
3	9	5.00	10.00	5.0	5.00	0.0	0.0	− 2.98	− 1.84
4	10[a]	0.00	3.34	9.40	0.06	0.0	0.0	− 0.05	− 0.03
	11[a]	0.00	3.34	28.19	0.18	0.0	0.0	− 0.08	− 0.09
	12	0.00	3.00	30.0	5.00	0.0	0.0	− 2.01	− 1.63
	13	0.00	3.00	10.0	5.00	0.0	0.0	− 1.04	− 0.85
	14	0.00	3.00	5.0	5.00	0.0	0.0	− 0.74	− 0.64

[a]Velocity drops of models 10 and 11 are computed with laboratory velocity-stress sensitivity data of Nur and Simmons (1969) listed in Table 3 with a stress drop of 10 MPa and 30 MPa, respectively

Table 3

Velocity–stress sensitivity of Nur and Simmons (1969)

Pressure (Kbar)	Depth (km)	dlnV/dp (10^{-8} Pa^{-1})
0.017	0.06	9.399
0.039	0.13	6.285
0.059	0.20	5.047
0.077	0.26	1.183
0.142	0.47	0.710
0.195	0.65	0.543
0.250	0.83	0.473
0.288	0.96	0.304
0.346	1.15	0.205
0.416	1.39	0.167
0.555	1.85	0.128
0.683	2.28	0.063
1.001	3.34	0.063

(1969) using stress drops of 10 MPa and 30 MPa, respectively. Model 12 has a velocity-stress sensitivity based on field observations (Niu et al. 2008, and references therein) and a stress drop of 10 MPa. Details of δlnV/δp are shown in Table 3.

The physical basis for the stress sensitivity of seismic velocity in the crust is the presence of cracks in the rocks. Their effect on seismic velocity on the one hand, and their sensitivity to the ambient stress field on the other, means that seismic velocities in cracked rocks are generally stress sensitive (e.g., Lyakhovsky et al. 1997; Hamiel et al. 2009) and depend on the presence of fluids. The measured stress sensitivity of dry rock samples in the lab is of the order of 10^{-7} Pa^{-1} at atmospheric pressure and decreases rapidly to 10^{-10}–10^{-9} Pa^{-1} at a pressure of $\sim$ 100 MPa. With a stress drop of 10 MPa, model 10 has a velocity reduction of $\sim$ 10% at the surface that decreases rapidly to $\sim$ 0.06% around 3 km depth. These changes are tripled in model 11, which has a stress drop of 30 MPa. Field measurements of δlnV/δp are more scattered than laboratory results and range from 10^{-9} to 10^{-6} Pa^{-1} near the surface (e.g., Yamamura et al. 2003, Silver et al. 2007). Niu et al. (2008) measured velocity perturbations at a depth of $\sim$ 1 km in the SAFOD drill site near Parkfield, California, and found stress sensitivity of $\sim$ 10^{-7} Pa^{-1}, which is approximately two orders of magnitude larger than the results of Nur and Simmons (1969) for corresponding confining pressure. These studies suggest that the in situ stress sensitivity at depth could be larger than lab measurements due to pervasive heterogeneities and fluids. Therefore, we construct model 12 that has a velocity drop of 30% at the surface and decreases linearly to 5% at 3 km depth, considering the field measurements of the δlnV/δp and depth variable stress changes. For comparison, we also construct two more models (model 13 and model 14), which have a velocity drop of 10% and 5% at the surface and change linearly to 5% at 3 km depth.

For each of the 15 models (one original + 14 perturbed models), we compute the phase velocity in the period range of 2 to 20 s, the corresponding sensitivity kernels of the phase velocities, and the Rayleigh wave Green's functions. A cross-correlation based method is used to estimate the apparent velocity changes in the period bands of 5–10 s and 10–20 s for the 15 perturbed models.

2.2. *Calculation of Rayleigh Wave Phase Velocities*

The Thomson-Haskell propagator matrix is used to compute Rayleigh waves in a 1-D structure consisting of a stack of constant velocity layers on top of a half space. The displacement-stress field observed at the surface is linked to those at the top of the half space by the product of the propagator matrices of each layer. The stress-free boundary condition is applied to the surface, and displacements are set to zero at the bottom of the half space, which is implemented by allowing only downgoing P and SV waves within the half spaces. This results in the following equation:

$$\begin{pmatrix} `p_{n+1}(z_n) \\ `s_{n+1}(z_n) \\ 0 \\ 0 \end{pmatrix} = \begin{pmatrix} B_{11} & B_{12} & B_{13} & B_{14} \\ B_{21} & B_{22} & B_{23} & B_{24} \\ B_{31} & B_{32} & B_{33} & B_{34} \\ B_{41} & B_{42} & B_{43} & B_{44} \end{pmatrix} \begin{pmatrix} r_1(z_0) \\ r_2(z_0) \\ 0 \\ 0 \end{pmatrix}. \quad (1)$$

Here, $r_1(z_0)$ and $r_2(z_0)$ are vertical and radial displacements at the surface, and $`p_{n+1}$ and $`s_{n+1}$ are the amplitudes of the downgoing P- and SV-waves. The B matrix is the product of the propagator matrices, which can be computed from the P, S wave velocities and density for given angular frequency ω and wavenumber k. The eigenvalue k and eigenvector

$((r_1(z_0),\ r_1(z_0))^T$ can be determined from the zero-determinant of the submatrix B:

$$\begin{vmatrix} B_{31} & B_{32} \\ B_{41} & B_{42} \end{vmatrix} = 0. \tag{2}$$

To find the root of this determinant for a given ω, a series of trial values of k are tested until (2) is met. This method provides the dispersion curves of a given layered elastic model.

To compute the phase velocities in an anelastic medium with attenuation coefficients Qp and Qs, one can simply replace the frequency independent elastic P- and S-wave velocities (α_e and β_e) with frequency dependent complex velocities (α_a and β_a) in the propagator matrix of each layer (Liu et al. 1976):

$$\begin{aligned} \alpha_e \to \alpha_a &= \alpha(\omega)\left(1 + \frac{i}{2Q_p(\omega)}\right) \\ &= \alpha(\omega_0)\left(1 + \frac{1}{\pi Q_p}\log\frac{\omega}{\omega_0} + \frac{i}{2Q_p}\right) \\ \beta_e \to \beta_a &= \beta(\omega)\left(1 + \frac{i}{2Q_s(\omega)}\right) \\ &= \beta(\omega_0)\left(1 + \frac{1}{\pi Q_s}\log\frac{\omega}{\omega_0} + \frac{i}{2Q_s}\right) \end{aligned}. \tag{3}$$

Here, ω_0 is a reference angular frequency set to 2π, equivalent to a reference frequency of 1 Hz.

The anelastic phase velocities can also be computed from perturbation theory. If we treat the complex velocities as adding to the elastic velocities with the following small perturbations:

$$\begin{aligned} \delta\alpha = \alpha_a - \alpha(\omega_0) &= \frac{\alpha(\omega_0)}{Q_p}\left(\frac{1}{\pi}\ln\frac{\omega}{\omega_0} + \frac{i}{2}\right) \\ \delta\beta = \beta_a - \beta(\omega_0) &= \frac{\beta(\omega_0)}{Q_s}\left(\frac{1}{\pi}\ln\frac{\omega}{\omega_0} + \frac{i}{2}\right) \end{aligned}, \tag{4}$$

then the phase velocities in anelastic media can be written as:

$$c_a(\omega) = c_e(\omega) + \delta c(\omega). \tag{5}$$

Here, the phase-velocity perturbations, $\delta c(\omega)$, can be computed from

$$\delta c(\omega) = \left(\frac{1}{\pi}\ln\frac{\omega}{\omega_0} + \frac{i}{2}\right) \cdot \int_0^\infty \left[\frac{\alpha(z)}{Q_p(z)}\frac{\partial c}{\partial \alpha} + \frac{\beta(z)}{Q_s(z)}\frac{\partial c}{\partial \beta}\right] \mathrm{d}z\,. \tag{6}$$

Since $\delta c(\omega)$ are complex functions, the phase velocities in anelastic media have complex values. In the results section we take the real parts as the phase velocities.

2.3. *Calculation of Phase Velocity Sensitivity Kernels*

In order to compute the sensitivity kernel of a Rayleigh wave phase velocity with respect to S-wave velocity at a fixed frequency ω_0, we first increase the S velocity of a given model by 1% at depth z, and then calculate the phase velocity difference between the given and perturbed models. The ratio of the relative phase velocity change and the S velocity perturbation, $d\ln c(\omega_0)/d\ln\beta(z)$, is the sensitivity at depth z. Varying z over the depth range provides the phase velocity sensitivity kernel at frequency ω_0.

2.4. *Green's functions of the fundamental Rayleigh wave*

A wave packet with a given spectral density $A(\omega)$ and initial phase $\phi_0(\omega)$ composed of a single mode can be expressed as:

$$f(r,t) = \frac{1}{2\pi\sqrt{r}}\int_{-\infty}^{+\infty} A(\omega)\mathrm{e}^{-i\omega\left(t-\frac{r}{c_a(\omega)}\right)+i\phi_0(\omega)} d\omega. \tag{7}$$

Here, $r^{-1/2}$ accounts for geometrical spreading and $c_a(\omega)$ is the complex phase velocity in anelastic media. For an impulse source, $A(\omega) = 1$ and $\phi_0(\omega) = 0$. Thus the Green's functions of the fundamental mode Rayleigh wave at epicentral distance r can be computed by:

$$f(r,t) = \frac{1}{2\pi\sqrt{r}}\int_{-\infty}^{+\infty} \mathrm{e}^{-i\omega\left(t-\frac{r}{c_a(\omega)}\right)} d\omega. \tag{8}$$

2.5. Calculation of Apparent Velocity Change $\delta v/v$

We first bandpass-filter the Green's functions of the original model and the 12 perturbed models in two period bands, 5–10 s and 10–20 s. In each period band, we use the original Green's function as the reference trace and compute the cross-correlation functions of the Green's functions of the 12 models. Each cross-correlation is computed in time domain and a cosine function fitting is employed to obtain sub-sample delay times τ (Cespedes et al. 1995). We choose the first 10 s of the Rayleigh wave time window to highlight apparent velocity changes of the long-wavelength and, therefore, deeply penetrated surface waves. Finally, we divide the measured delay times by the arrival times T of the central point of the time window to obtain the apparent velocity drops, i.e., $\delta v/v = -\tau/T$.

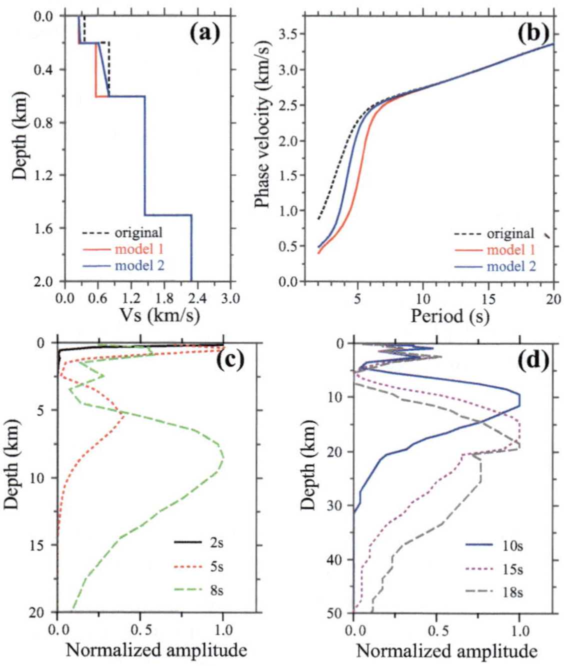

Figure 2
a S-wave velocity of model 1 (red solid line) and model 2 (blue solid line) are shown together with the original model (black dashed line). Only the top 2 km of the models is shown here. Model 1 has a constant 30% velocity drop in the top two layers from the origin model, while model 2 has a velocity drop of 30% at the surface, which decays gradually to zero at 0.6 km deep. **b** The phase velocity dispersion curves of the three model in the period range of 2–20 s. **c**, **d** Are similar to **c**, **d** in Fig. 1, except that they are computed with model 1

3. Results

Figure 1c, d display, respectively, the sensitivity kernels of the original model at periods 2 s, 5 s, 8 s, and 10 s, 15 s, 18 s. Note that different depth ranges are used in plotting the kernels of the two groups. All the kernels indicate that short to medium periods Rayleigh waves have significant sensitivity to structures shallower than 5 km. Therefore, large velocity perturbations at shallow depth may lead to significant changes of phase velocities at these periods. We illustrate this below by the 13 models with shallow structural changes that are limited to the top 3 km.

In model 1, we impose 30% drops of the P- and S-wave velocities as well as mass density, but not the quality factors. The reductions are limited to the two top sedimentary layers with a thickness of 0.6 km (Table 2, red solid line in Fig. 2a). Related milder changes are incorporated in model 2, where the reductions decrease linearly from 30% at the surface to 0% at the bottom of the second sedimentary layer (Table 2, blue solid line in Fig. 2a). The phase velocities of these two models are shown in Fig. 2b

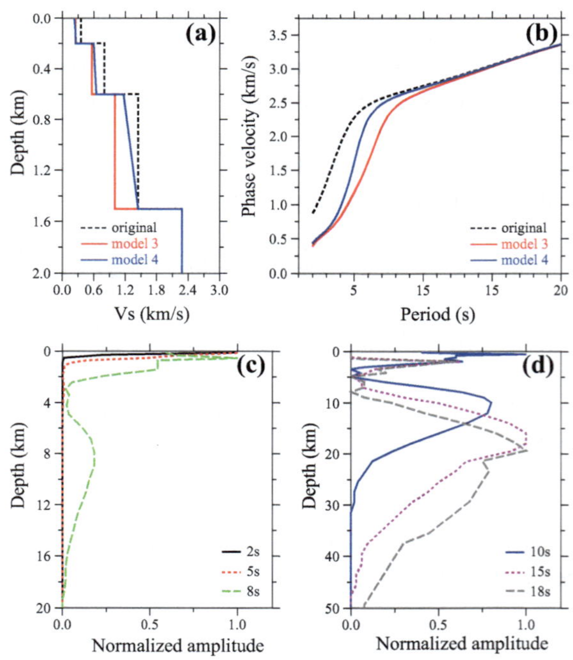

Figure 3
Results from model 3 and model 4. Same as Fig. 2, except that model 3 and model 4 are shown here, and the phase velocity kernels are computed with model 3

by red and blue solid lines, together with those from the original model (black dashed line). It is clear that the phase velocities calculated in the period range of 2-8 s of the two models are significantly lower than those computed from the original model. Figure 2c, d present the sensitivity kernels for models 1 and 2. The results are generally similar to those of the original model.

In model 3 we extend the 30% drops to the third sedimentary layer extending to 1.5 km below the surface, and in model 4 the reduction of properties decays from 30% at the top to 0% at 1.5 km depth. The two models and their corresponding phase velocities are shown (along with the dispersion curve of the original model) in Fig. 3a, b, respectively. Because of the additional velocity and density perturbation in the third sedimentary layer, the phase-velocity reduction of these two models is more significant and the effects on the dispersion curves extend to almost 12 s. The sensitive depth ranges of the three short periods (2 s, 5 s, and 8 s) also shift significantly to shallower values (Fig. 3c).

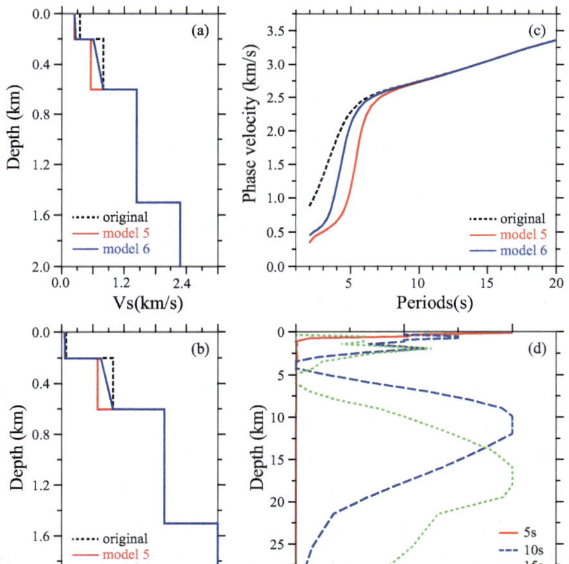

Figure 4
S-wave velocity and Qs of model 5 and model 6 are shown with the original model in **a**, **b**, respectively. Only the top 2 km of the models are shown here. **c** The phase velocity dispersion curves of the three models in the period range of 2–20 s. **d** Phase velocity sensitivity kernels of 5 s (red solid line), 10 s (blue dashed line), and 15 s (green dotted line) Rayleigh waves computed from model 5 are plotted in the depth range of 0–30 km

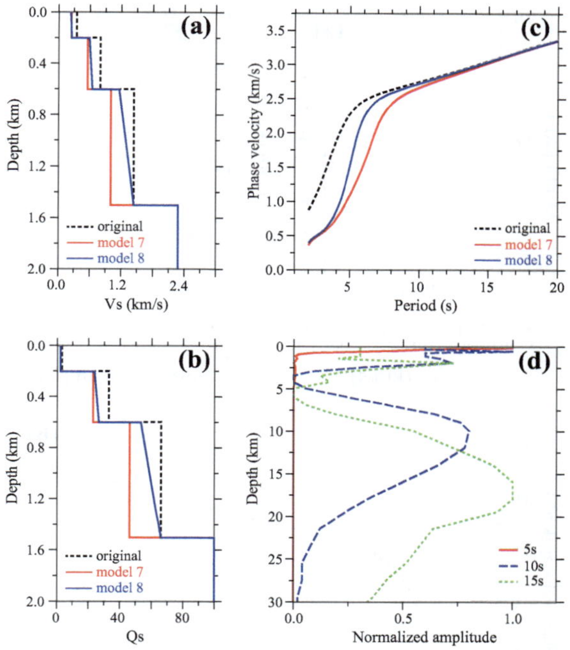

Figure 5
Results from model 7 and model 8. Same as Fig. 4, except that model 7 and model 8 are shown here, and the phase velocity kernels are computed with model 7

The four models in category 2 have the same velocity and density drops as in category 1, while also incorporating similar reduction of the S-wave quality factor. The four models and their corresponding phase velocities are shown in Figs. 4 and 5, respectively. All four models show significant phase velocity drops compared to the original model. The amount of observed drops are similar to their counterpart models in the category 1, suggesting that attenuation change plays a less important role than velocity and density perturbations in affecting properties of the Rayleigh waves. However, a changing attenuation quality factor can affect inferences on velocity structure by modifying the observable range of periods.

To investigate how velocity and density drops at seismogenic depth influence Rayleigh wave speed, we use model 9 with a 5% drop in P- and S-wave velocities and mass density in the depth range 5-10 km (Table 2, Fig. 6a). The attenuation structure is kept unchanged. The corresponding phase velocity

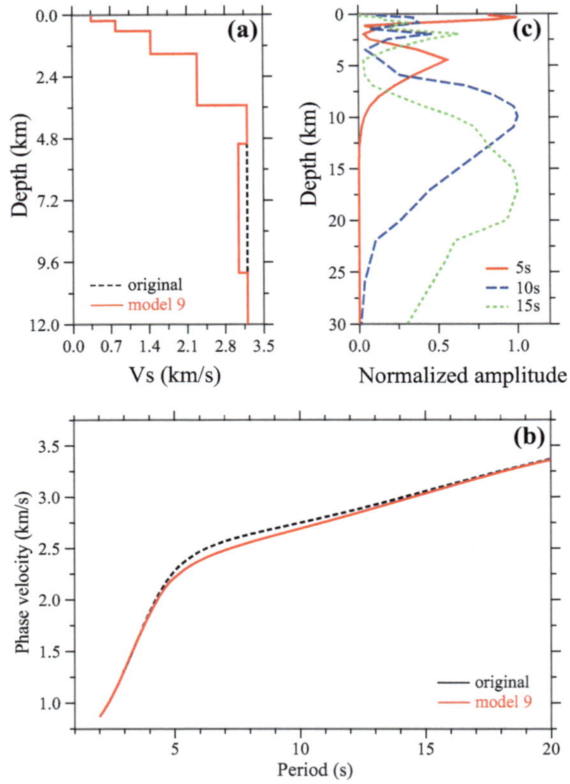

Figure 6
a S-wave velocity of model 9 (red solid line) is shown together with the original model (black dashed line). The model has a constant velocity drop of 5% in the depth range of 5–10 km. **b** Phase velocities of the model 9 and the original model are plotted as functions of period from 2 to 20 s. **c** Phase velocity sensitivity kernel of 5 s (red solid line), 10 s (blue dashed line), and 15 s (green dotted line) Rayleigh waves computed from model 9 (red solid line) are plotted in the depth range of 0–30 km

dispersion curve is shown in a red solid line in Fig. 6b, together with the original dispersion curve (black dashed line). The results show phase-velocity drops of a few percent in the period range ~ 4–15 s (Fig. 6b). The sensitivity kernel of the 8 s phase velocity also changes slightly (Fig. 6c).

The five stress-related models in category 4 are associated with a velocity drop that extends to 3 km deep (Fig. 7a, b). This is considerably deeper than in the first eight models but still shallower than the seismogenic depth range in model 9. The attenuation structure of the five models is not changed. The calculated phase velocities of models 11 and 12 are shown in Fig. 7c with blue and red solid lines, respectively, along with the original dispersion curve

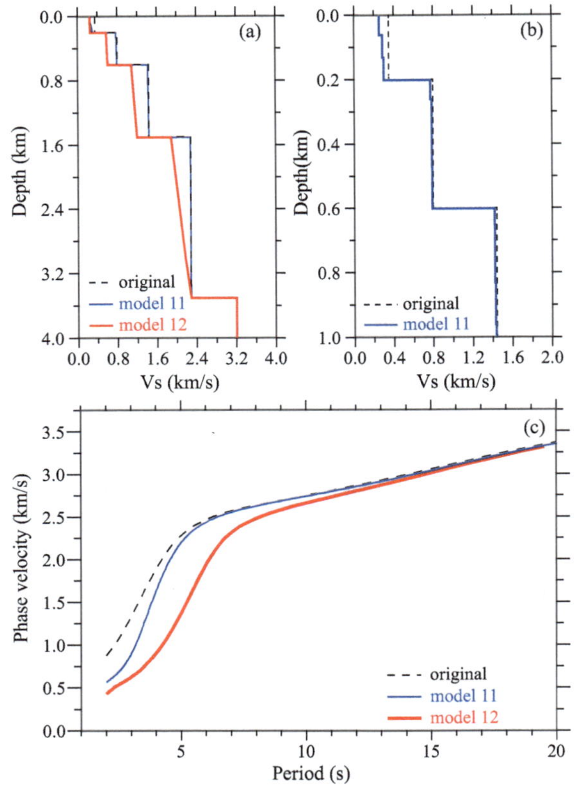

Figure 7
a S-wave velocities of model 11 (blue solid line) and 12 (red solid line) are shown together with the original model (black dashed line). The three models have the same velocity structure below ~ 3 km. **b** Shows a comparison of model 11 and the original model of the top 1 km where most of the velocity changes occur. **c** Phase velocities of model 11 (blue solid line) and 12 (red solid line) are plotted as functions of period from 2 to 20 s. The phase velocity dispersion curve of the original model (black dashed line) is also shown for comparison

(black dashed line). The differences in phase velocity between the two stress-related models and the original model are quite large at short periods and decrease gradually with increasing periods.

For comparison, we further show the calculated percentile phase velocity drops of the 14 models in Fig. 8. The left panels show changes of the 14 models in the period range of 2–20 s, while the right panels focus on changes in the period band of 15–20 s. Model 7 and model 12 in category 4 exhibit a phase velocity drop greater than 0.5% at a period of 20 s. This is comparable to the inferred co-seismic velocity changes reported for the 2004 M_w 6.6 Mid-Niigata earthquake (Wegler and Sens-Schönfelder 2007) and

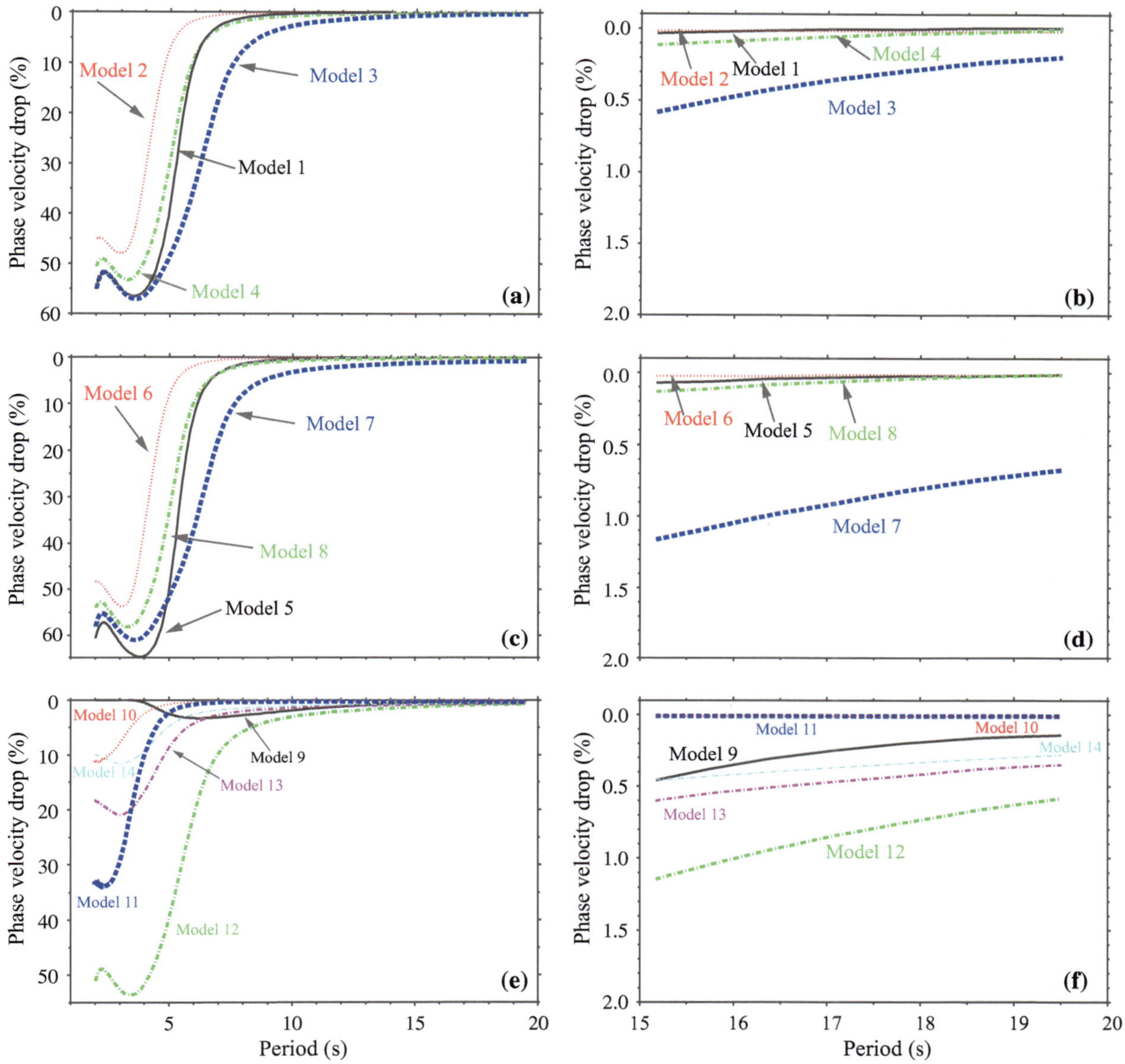

Figure 8
Phase velocity drops of models 1–4 are shown in the period band of 2–20 s (**a**) and 15–20 s (**b**), respectively. **c**, **d** Are similar to **a**, **b** except for models 5–8. **e**, **f** Are similar to **a**, **b** except for models 9–14

the 2008 M_w 7.9 Wenchuan earthquake (Cheng et al. 2010). Among the 14 perturbed models, only model 9 with a temporal change occurring at seismogenic depths shows a phase velocity reduction peaked at a period greater than 5 s. Model 9 is also featured by a slower decrease against increasing periods as compared to other models.

Figure 9 shows the filtered Green's functions of the 14 models in two period bands (5–10 s and 10–20 s) and measured apparent velocity changes, which are also listed in Table 2. Among the 14 examined cases, the measured apparent velocity changes of the first two stress-related models (models 10 and 11) are less than 0.1%. Seven models (models 1, 2, 4, 5, 6, 8, 14) show an apparent velocity drop in

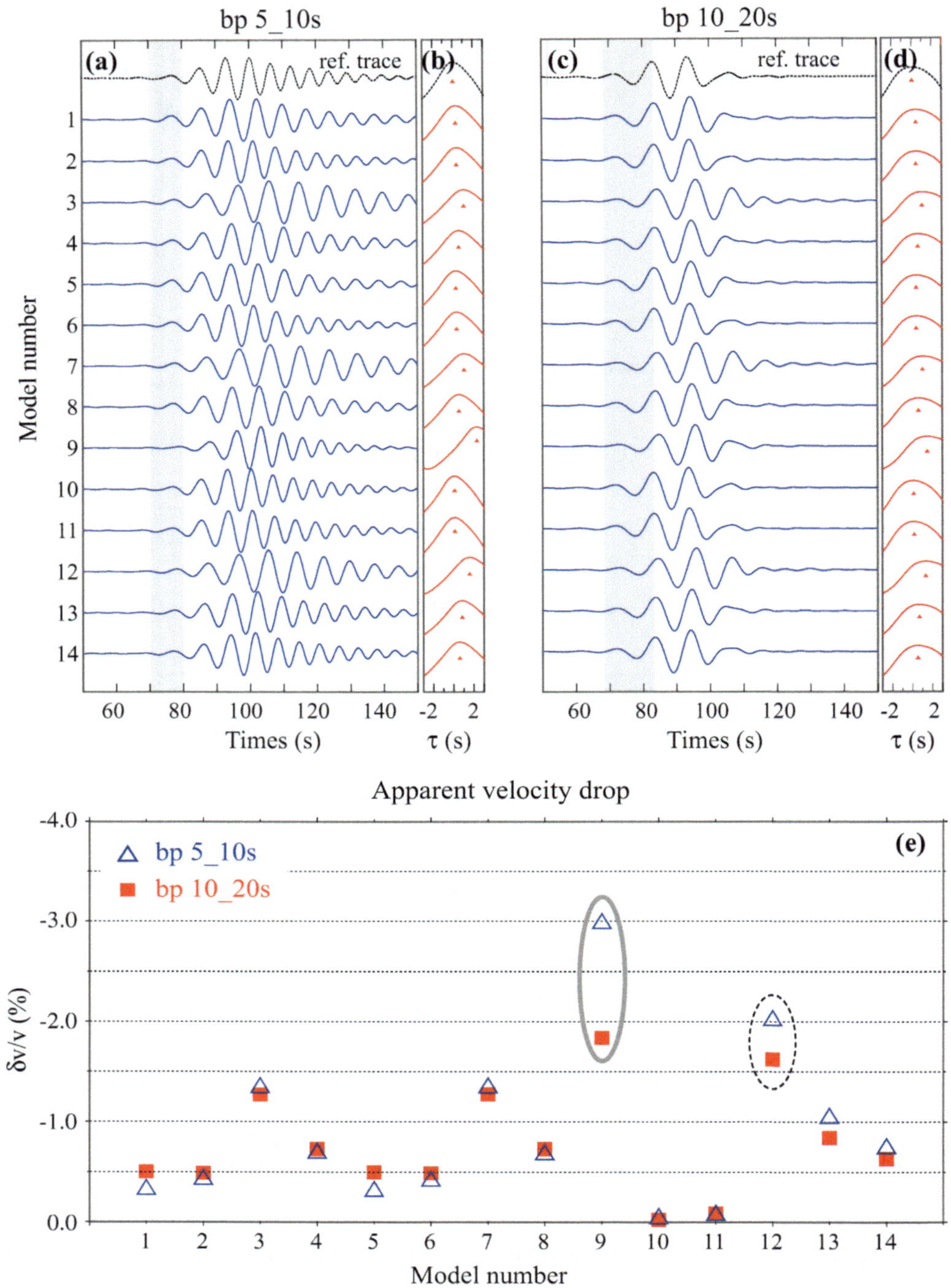

Figure 9
Rayleigh wave Green's functions computed from the original model and the 14 models with various types of structural changes are shown in grey and blue, respectively. The data are filtered in the period band from 5 to 10 s. **b** Shows the cross-correlation (cc) functions calculated between the Green's function of each model and the grey reference trace. The red dots indicate the peak of the cc functions. **c**, **d** Are similar to **a**, **b** except that the filtered period band is from 10 to 20 s. **e** The measured apparent velocity changes of the 14 models are shown in blue open triangles and red solid squares for the period bands of 5–10 s and 10–20 s, respectively. Note the large differences in the measurements of models 9 and 12, which are marked with grey solid and black dashed ellipses, respectively

the range of 0.3-1.0%. Models 3, 7, 9, 12 and 13 exhibit a velocity change of more than 1% (Fig. 9e; Table 2). The apparent velocity drops measured at the two period bands are generally comparable, other than for models 9 and 12 (the solid and dashed line ellipses in Fig. 9e; Table 2). The apparent velocity drops of the two models measured at the shorter period band 5–10 s (open triangles in Fig. 9e) are significantly larger than those measured at the longer period band 10–20 s (solid squares in Fig. 9e). It seems that the amount of difference in the measured apparent velocity drops between the two period bands is mostly sensitive to changes at deep parts of the media.

4. *Discussion and Conclusions*

We conduct numerical simulations to investigate how shallow structural and stress properties affect Rayleigh wave phase velocities in the period range 5–20 s, and to clarify the types of analysis needed to separate temporal changes limited to the shallow crust from those occurring in seismogenic zones. Calculations of sensitivity kernels and measurements of apparent velocity changes of Rayleigh waves in different frequency ranges show clearly the following results: (1) Phase velocities and sensitivity kernels of Rayleigh waves at periods up to 20 s can be affected significantly by the velocity structure in the top few kilometers of the crust. (2) Large velocity drops at shallow depth that may be caused by strong shaking of large earthquakes, with values consistent with those observed in some studies (e.g., Wu et al. 2009; Bonilla et al. 2018), can significantly affect Rayleigh wave phase velocities at periods up to 20 s. (3) Large velocity drops within the shallow sediments can produce $\sim$ 0.5% apparent velocity changes even in analysis of temporal changes using 10–20 s Rayleigh wave data. (4) Stress changes in the top 3 km of the crust can produce considerable velocity changes that have significant effects on 10–20 s Rayleigh wave data.

The results highlight the importance of accounting for low velocities, attenuation coefficients and stress changes in the shallow crust in analyses of sensitivity kernels of Rayleigh waves, and in efforts to determine the depth range sustaining temporal changes of seismic properties. Comparisons of the relative amplitudes of apparent velocity changes observed using different frequency ranges (e.g., Rivet et al. 2011; Liu et al. 2014; Obermann et al. 2014; Wu et al. 2016) are essential for separating effects limited to the very shallow structure from those occurring at greater depth. The apparent velocity changes measured from the first part of Rayleigh waves at the 10–20 s period band are significantly lower than those of the 5–10 s band for model 9 with changes at seismogenic depth (solid ellipse in Fig. 9e). This is in contrast to models 1–8 with velocity drops in the top 0.6 and 1.5 km. However, model 12 with stress-related coseismic velocity changes in the top $\sim$ 3 km of the crust produces apparent velocity changes that are comparable to model 9 with velocity drop in the depth range 5–10 km (dashed ellipse in Fig. 9e).

The results of the present paper can be strengthen by performing similar calculations in 3-D models that incorporate lateral variations of properties in addition to the depth-dependent changes assumed here. Simulations using 3-D velocity models with the presence of randomly distributed scatterers are particularly important, as many lag time measurements are based on the coda wavefield (e.g., Brenguier et al. 2008, 2014; Obermann et al. 2014, 2016, 2018). Such calculations are computationally very intensive but should be attempted in follow-up works.

Acknowledgements

All figures are made using the Generic Mapping Tools (www.gmt.soest.hawaii.edu; Wessel et al. 2013). We thank Dr. Anne Obermann and another anonymous reviewer for their constructive comments and suggestions, which significantly improved the quality of this paper. C.Y. and F.N. were supported by NSF (Grant EAR-1251667) and G.L. was funded by NSFc (Grant 41630209). Y.B.Z. acknowledges support by the Wiess visiting professorship program of Rice University and the Department of Energy (award DESC0016520).

Publisher's Note Springer Nature remains neutral with regard to jurisdictional claims in published maps and institutional affiliations.

References

Bonilla, L. F., Guéguen, P., Ben-Zion, Y. (2018). Monitoring co-seismic temporal changes of shallow material during strong ground motion with interferometry and autocorrelation, Bulletin of the Seismological Society of America, in press.

Brenguier, F., Campillo, M., Takeda, T., Aoki, Y., Shapiro, N. M., Briand, X., et al. (2014). Mapping pressurized volcanic fluids from induced crustal seismic velocity drops. *Science, 345,* 80–82.

Brenguier, F., Shapiro, N. M., Campillo, M., Ferrazzini, V., Duputel, Z., Coutant, O., et al. (2008). Towards forecasting volcanic eruptions using seismic noise. *Nature Geoscience, 1,* 126–130. https://doi.org/10.1038/ngeo104.

Cespedes, I., Huang, Y., Ophir, J., & Spratt, S. (1995). Methods for estimation of sub-sample time delays of digitized echo signals. *Ultrasonic Imaging, 17,* 142–171.

Chen, H., Ge, H., & Niu, F. (2014). Semiannual velocity variations around the 2008 M_w 7.9 Wenchuan Earthquake fault zone revealed by ambient noise and ACROSS active source data. *Earthquake Science*. https://doi.org/10.1007/s11589-014-0089-5.

Cheng, X., Niu, F., & Wang, B. (2010). Coseismic velocity change in the rupture zone of the 2008 M_w 7.9 Wenchuan earthquake observed from ambient seismic noise data. *Bulletin of the Seismological Society of America, 100,* 2539–2550. https://doi.org/10.1785/0120090329.

Froment, B. M., Campillo, M., Chen, J. H., & Liu, Q. Y. (2013). Deformation at depth associated with the 12 May 2008 M_w 7.9 Wenchuan earthquake from seismic ambient noise monitoring. *Geophysical Research Letters, 40,* 78–82. https://doi.org/10.1029/2012GL053995.

Hamiel, Y., Lyakhovsky, V., Stanchits, S., Dresen, G., & Ben-Zion, Y. (2009). Brittle deformation and damage-induced seismic wave anisotropy in rocks. *Geophysical Journal International, 178,* 901–909. https://doi.org/10.1111/j.1365-246X.2009.04200.x.

Hillers, G., Ben-Zion, Y., Campillo, M., & Zigone, D. (2015). Seasonal variations of seismic velocities in the San Jacinto Fault area observed with ambient seismic noise. *Geophysical Journal International, 202,* 920–932. https://doi.org/10.1093/gji/ggv151.

Karabulut, H., & Bouchon, M. (2007). Spatial variability and non-linearity of strong ground motion near a fault. *Geophysical Journal International, 45,* 22–33. https://doi.org/10.1111/j.1365-246x.2007.03406.x.

Lecocq, T., Longuevergne, L., Pedersen, H. A., Brenguier, F., & Stammler, K. (2017). Monitoring ground water storage at mesoscale using seismic noise: 30 years of continuous observation and thermo-elastic and hydrological modeling. *Scientific Reports*. https://doi.org/10.1038/s41598-017-14468-9.

Liu, H.-P., Anderson, D. L., & Kanamori, H. (1976). Velocity dispersion due to anelasticity; implications for seismology and mantle composition. *Geophysical Journal International, 47*(1), 41–58.

Liu, Z., Huang, J., Peng, Z., & Su, J. (2014). Seismic velocity changes in the epicentral region of the 2008 Wenchuan earthquake measured from three-component ambient noise correlation techniques. *Geophysical Research Letters, 41,* 37–42. https://doi.org/10.1002/2013GL058682.

Lyakhovsky, V., Ben-Zion, Y., & Agnon, A. (1997). Distributed damage, faulting, and friction. *Journal of Geophysical Research, 102,* 27635–27649.

Meier, U., Shapiro, N. M., & Brenguier, F. (2010). Detecting seasonal variations in seismic velocities within Los Angeles basin from correlations of ambient seismic noise. *Geophysical Journal International, 181,* 985–996. https://doi.org/10.1111/j.1365-246X.2010.04550.x.

Nakata, N., & Snieder, R. (2011). Near-surface weakening in Japan after the 2011 Tohoku-Oki earthquake. *Geophysical Research Letters, 38,* L17302.

Niu, F., Silver, P., Daley, T., Cheng, X., & Majer, E. (2008). Preseismic velocity changes observed from active source monitoring at the Parkfield SAFOD drill site. *Nature, 454,* 204–208.

Nur, A., & Simmons, G. (1969). The effect of saturation on velocity in low porosity rocks. *Earth and Planetary Science Letters, 7,* 183–193.

Obermann, A., Froment, B., Campillo, M., Larose, E., Planes, T., Valette, B., et al. (2014). Noise correlations to image structural and mechanical changes associated with the M_w 7.9 2008 Wenchuan earthquake. *Journal of Geophysical Research, 119,* 3155–3168.

Obermann, A., Planes, T., Hadziioannou, C., & Campillo, M. (2016). Lapse-time-dependent coda-wave depth sensitivity to local velocity perturbations in 3-d heterogeneous elastic media. *Geophysical Journal International, 207,* 59–66.

Obermann, A., Planes, T., Hadziioannou, C., & Campillo, M. (2018). 4-D imaging of subsurface changes with coda waves: numerical studies of sensitivity kernels and applications to the M_w 7.9, 2008 Wenchuan earthquake, Pure and Applied Geophysics, in press.

Obermann, A., Planes, T., Larose, E., & Sens-Schönfelder, C. (2013). Depth sensitivity of seismic coda waves to velocity perturbations in an elastic heterogeneous medium. *Geophysical Journal International, 194,* 372–382.

Peng, Z., & Ben-Zion, Y. (2006). Temporal changes of shallow seismic velocity around the Karadere-Duzce branch of the north Anatolian fault and strong ground motion. *Pure and Applied Geophysics, 163,* 567–599. https://doi.org/10.1007/s00024-005-0034-6.

Rivet, D., Campillo, M., Shapiro, N. M., Cruz-Atienza, V., Radiguet, M., Cotte, N., et al. (2011). Seismic evidence of nonlinear crustal deformation during a large slow slip event in Mexico. *Geophysical Research Letters, 38,* L08308. https://doi.org/10.1029/2011GL047151.

Roux, P., & Ben-Zion, Y. (2014). Monitoring fault zone environments with correlations of earthquake waveforms. *Geophysical Journal International, 196,* 1073–1081. https://doi.org/10.1093/gji/ggt441.

Rubinstein, J. L., & Beroza, G. C. (2005). Depth constraints on nonlinear strong ground motion from the 2004 Parkfield earthquake. *Geophysical Research Letters, 32,* L14313. https://doi.org/10.1029/2005GL023189.

Sawazaki, K., Sato, H., Nakahara, H., & Nishimura, T. (2009). Time-lapse changes of seismic velocity in the shallow ground caused by strong ground motion shock of the 2000 Western-Tottori Earthquake, Japan, as revealed from coda deconvolution analysis. *Bulletin of the Seismological Society of America, 99,* 352–366.

Schaff, D. P., & Beroza, G. C. (2004). Coseismic and postseismic velocity changes measured by repeating earthquakes. *Journal of Geophysical Research, 109,* B10302. https://doi.org/10.1029/2004JB003011.

Sens-Schönfelder, C., & Larose, E. (2010). Lunar noise correlation, imaging and monitoring. *Earthquake Science, 23,* 519–530.

Sens-Schönfelder, C., & Wegler, U. (2006). Passive image interferometry and seasonal variations of seismic velocities at Merapi volcano, Indonesia. *Geophysical Research Letters, 33,* L21302.

Silver, P. G., Daley, T. M., Niu, F., & Majer, E. L. (2007). Active source monitoring of cross-well seismic travel time for stress-induced changes. *Bulletin of the Seismological Society of America, 97,* 281–293.

Wegler, U., & Sens-Schönfelder, C. (2007). Fault zone monitoring with passive image interferometry. *Geophysical Journal International, 168,* 1029–1033.

Wessel, P., Smith, W. H. F., Scharroo, R., Luis, J. F., & Wobbe, F. (2013). Generic mapping tools: improved version released. *Eos, Transactions American Geophysical Union, 94*(45), 409–410. https://doi.org/10.1002/2013EO450001.

Wu, C., Delorey, A., Brenguier, F., Hadziioannou, C., Daub, E. G., & Johnson, P. (2016). Constraining depth range of S wave velocity decrease after large earthquakes near Parkfield, California. *Geophysical Research Letters, 43,* 6129–6136. https://doi.org/10.1002/2016GL069145.

Wu, C., Peng, Z., & Ben-Zion, Y. (2009). Non-linearity and temporal changes of fault zone site response associated with strong ground motion. *Geophysical Journal International, 176,* 265–278. https://doi.org/10.1111/j.1365-246X.2008.04005.x.

Yamamura, K., Sano, O., Utada, H., Takei, Y., Nakao, S., & Fukao, Y. (2003). Long-term observation of in situ seismic velocity and attenuation. *Journal of Geophysical Research: Solid Earth.* https://doi.org/10.1029/2002jb002005.

Zhao, P., & Peng, Z. (2009). Depth extent of damage zones around the central Calaveras fault from waveform analysis of repeating earthquakes. *Geophysical Journal International, 179,* 1817–1830. https://doi.org/10.1111/j.1365-246X.2009.04385.x.

(Received June 14, 2018, revised November 25, 2018, accepted December 5, 2018, Published online December 17, 2018)

Pure Appl. Geophys. 176 (2019), 1269–1291

https://doi.org/10.1007/s00024-018-1971-1

Pure and Applied Geophysics

Stress Inversion of Regional Seismicity in the Sea of Marmara Region, Turkey

CHRISTOPHER WOLLIN,[1] MARCO BOHNHOFF,[1] VÁCLAV VAVRYČUK,[2] and PATRICIA MARTÍNEZ-GARZÓN[1]

Abstract—In this study we derive the stress tensor and its local variations throughout the Marmara region, Turkey. Based on a recently compiled 10-year earthquake catalogue, we directly invert first-motion polarity data and quantify confidence intervals for the principal stress orientations. We find a combined strike-slip and normal faulting stress field for the Marmara region generally reflecting the overall transtensional setting. However, the results clearly show moderate local variations of the stress field. The largest (σ_1) and intermediate (σ_2) principal stresses show an average regional trend of N125°E and locally varying plunges. The least principal stress (σ_3) is well resolved in its confidence interval and consistent throughout the region with an average trend of $\sim$ N35°E and a subhorizontal plunge. The eastern Sea of Marmara shows local stress field orientations with pronounced strike-slip (northern part) and normal faulting (southern part) components. Along the central Marmara region, normal faulting tends to dominate, while a well resolved strike-slip stress regime is found in the western Sea of Marmara region. Regarding the faulting mechanism of an earthquake with magnitude up to 7.4 which is expected in this area in direct vicinity of the Istanbul metropolitan region, our results imply that neither strike-slip nor normal faulting kinematics can be excluded.

1. Introduction

The vast majority of earthquakes are caused by failure of critically stressed faults within the seismogenic layer of the earth's crust and along active tectonic plate boundaries (e.g. Wallace 1951; Bott 1959; McKenzie 1969; Vavryčuk 2015; Hardebeck and Okada 2018). Whether a particular fault is going to rupture or not is controlled by the fault's orientation with respect to the present stress field orientation thereby defining the preferred faulting mechanism to be expected in a future earthquake (Vavryčuk 2011). The Sea of Marmara region in northwestern Turkey is located along the plate-bounding North Anatolian Fault Zone (NAFZ) at the transition zone between the pure right-lateral part of the NAFZ to the east and the north–south extensional Aegean region to its southwest. This setting results in a first-order transtensional tectonic regime (McClusky et al. 2000; Le Pichon et al. 2015; Bohnhoff et al. 2016b) and in the opening of the Sea of Marmara as a pull-apart structure (e.g. Armijo et al. 1999, 2005; Le Pichon et al. 2001). The submarine Marmara segment of the NAFZ currently represents a seismic gap capable of generating a major ($M > 7$) earthquake in the next decades (Parsons 2004; Bohnhoff et al. 2013; Murru et al. 2016). This translates into significant hazard and risk for the neighboring Istanbul metropolitan area with its 15+ million inhabitants (Bąs and Yăgci 2008). Much progress has been achieved in characterizing the Marmara Section of the NAFZ with regard to potential nucleation points and rupture propagation (e.g. Karabulut et al. 2011; Hergert et al. 2011; Bohnhoff et al. 2013, 2017b). However, in order to quantify the earthquake and potential tsunami risk for the region (e.g. Yalçıner et al. 2002; Hébert et al. 2005; Latcharote et al. 2016), a key role is taken by the expected type of faulting. For its characterization, knowledge of the local and regional stress field orientation is crucial.

In this paper we study the stress field orientation in the broader Marmara region based on the recently published earthquake catalogue of Wollin et al. (2018) which covers the time period of 2006–2016. Since the catalogue does not allow to determine a sufficiently high number of single-event focal mechanisms due to absence of moderate-size ($M > 3$) events, we calculate the stress tensor by a

Electronic supplementary material The online version of this article (https://doi.org/10.1007/s00024-018-1971-1) contains supplementary material, which is available to authorized users.

[1] German Research Center for Geosciences, Section 4.2: "Geomechanics and Rheology", Telegrafenberg, 14473 Potsdam, Germany. E-mail: wollin@gfz-potsdam.de

[2] Institute of Geophysics of the Czech Academy of Science, Bocni II/1401, 14100 Praha 4, Czech Republic.

direct inversion of first-motion P-wave polarities of seismicity clusters throughout the study region. The resulting uncertainties of the resolved stress are discussed and expressed as intervals or areas of confidence of the marginal probability density functions (PDFs). The reliability of our results is quantified by synthetic modeling. We discuss which type of faulting can be expected for the pending Marmara earthquake and its potential consequences for seismic risk scenarios.

2. *Tectonic Setting of the Marmara Region in NW Turkey*

The North Anatolian Fault Zone (NAFZ) is one of the largest plate bounding transform faults (Ketin 1948). It extends for some 1200 km between the Karliova triple junction in Eastern Anatolia and the Gulf of Saros in the northern Aegean (e.g. Barka 1992; Şengör et al. 2005; Bohnhoff et al. 2016b) separating the Anatolian and Eurasian plates. Wedged by the northward moving Arabian plate in the east and accelerated by the pull of the southward rollback of the Hellenic subduction zone in the west, the Anatolian plate performs a westward motion in combination with counter-clockwise rotation with respect to Eurasia resulting in a right-lateral strike-slip system along most of its extent (Flerit et al. 2004; Bohnhoff et al. 2005; Reilinger et al. 2006; Bulut et al. 2012). GPS data suggest an average slip rate along the NAFZ of 20–25 mm/year, where values increase towards the west (Reilinger et al. 2006). The pure strike-slip system along the eastern and central NAFZ turns into a transtensional setting in NW Turkey, where the opening of the Sea of Marmara as a large pull-apart structure is a result of the rollback of the Hellenic subduction zone that started a few million years ago, overprinting earlier tectonic units of the region (e.g. Armijo et al. 1999, 2005; Le Pichon et al. 2001; McClusky et al. 2000; Şengör et al. 2005; Le Pichon et al. 2015). This tectonically young development still evolves and leads to the progression of the transtensional system towards the east. Younger and thus less well developed pull-apart structures are currently progressing further east below the eastern tip of the Sea of Marmara (Cinarcik Basin) (Le Pichon et al. 2001; Karabulut et al. 2002; Acarel et al. 2014) and also extend further on land along the 1999 Izmit rupture (Akyazi Plain) (Bohnhoff et al. 2006; Bulut et al. 2007; Najdahmadi et al. 2016, 2018).

The NAFZ forms a single main fault strand along most of its part between eastern Anatolia and the Bolu region until approximately 150 km east of the city of Izmit where it splits into two major branches, the northern and the southern NAFZ strand (Le Pichon et al. 2014). Below the Sea of Marmara most of the NAFZs deformation is accommodated by the northern strand (Reilinger et al. 2006; Hergert and Heidbach 2010; Ergintav et al. 2014), also referred to as the 'Marmara Section' (Fig. 1, Wollin et al. 2018). To the east and west, the Marmara Section is bound by the three most recent major regional earthquakes ($M_w > 7$): the 1912 Mürefte–Ganos event in the west (Ambraseys 1970; Janssen et al. 2009) and the 1999 Izmit and Düzce events in the east (Tibi et al. 2001; Pinar et al. 2001; Barka et al. 2002; Bohnhoff et al. 2016a) (Fig. 1). The last major earthquake along the Marmara Section occurred in 1766 (*M*7.4, Ambraseys 1970). Given the average recurrence time of 200–250 years this fault is currently in the final phase of its seismic cycle (Parsons 2004; Bohnhoff et al. 2016b) and thus believed to represent a seismic gap whose activation in a $M_w \geq 7.0$ earthquake threatens the metropolitan area of Istanbul. The two end-members for the potential faulting style of this earthquake discussed in the literature are: (1) a single through-going strike-slip event activating the entire so-called "Main Marmara Fault" between the 1912 and 1999 ruptures (Le Pichon et al. 1999, 2001), or (2) an activation of several smaller en-echelon normal faults (Armijo et al. 2002, 2005). While the strike-slip scenario implies intense ground-shaking in succession to a single event with a magnitude of up to 7.4, the normal faulting events could cause local tsunamis as documented in the historic past (Yalçıner et al. 2002).

First information on the stress field orientation in the broader Marmara region was provided by Kiratzi (2002) who inverted focal mechanisms of 12 large regional earthquakes recorded during the second half of the twentieth century, most of which occurred on the mainland surrounding the Sea of Marmara. The

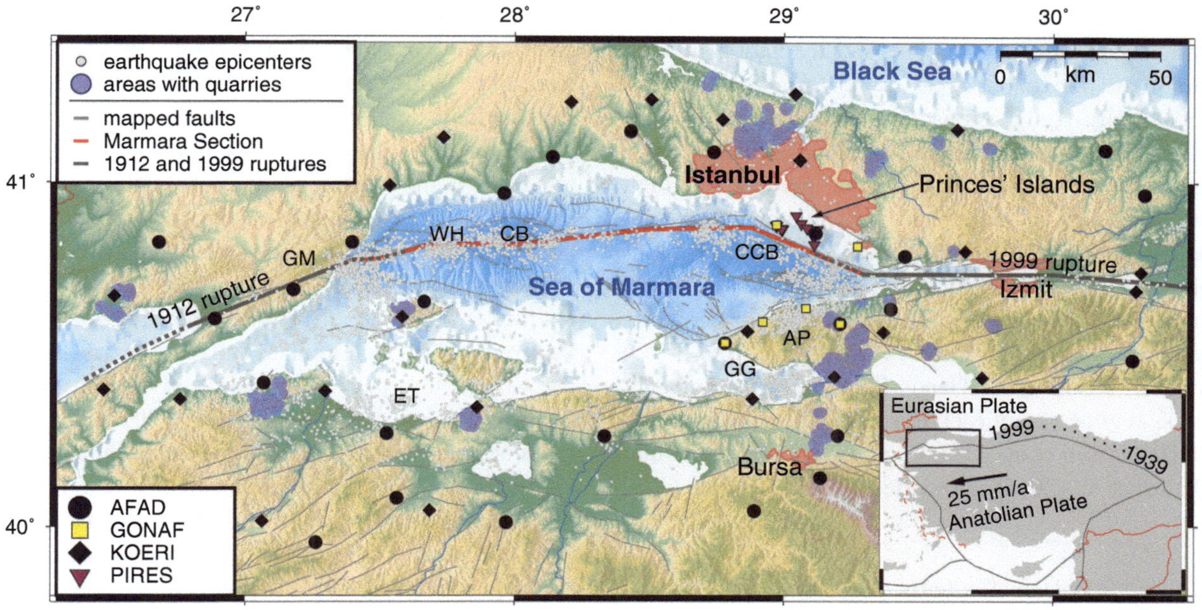

Figure 1
The Sea of Marmara Region with the offshore Marmara Section of the North Anatolian Fault Zone (NAFZ) delineated in red. In its continuation lie the ruptures of the 1912 Ganos and 1999 Izmit earthquakes (delineated in black). Location of topographic features are indicated with capital letters: Western High (WH), Central Basin (CB), Central High (CH), Çınarcık Basin (CCB), Armutlu Peninsula (AP) all after (Bécel et al. 2009), Gulf of Gemlik (GG), Erdek Tombolo (ET), Ganos Mountains (GM) all after (Şengör et al. 2014). Permanent seismic stations are marked by different symbols according to their associated network. The bathymetry, submarine faults and onshore faults are after Le Pichon et al. (2001), Armijo et al. (2005) and the Turkey General Directorate of Mineral Research and Exploration (pers. comm.), respectively. The extent of larger cities (Istanbul, Izmit, Bursa) is indicated by pink areas. The inset in the lower right shows the area of study in the broader tectonic regime where the movement of the Anatolian Plate is given with respect to stable Eurasia. 6812 epicenters of selected seismicity are shown. Events within blue areas are likely to be quarry blasts (after Wollin et al. 2018)

results show a first-order strike-slip stress field with an oblique component. The largest horizontal stress (S_{Hmax}) has a WNW–ESE orientation and a transpressional signature. However, also $M > 6$ normal faulting earthquakes such as the 1963 Armutlu earthquake below the eastern Sea of Marmara are well documented (Bulut and Aktar 2007). Örgülü (2011) used a refined data set of numerous smaller events and showed that a strike-slip regime can be separated from a normal-faulting regime in the eastern Marmara region between the Princes' Islands Segment in the north and the neighboring Armutlu Peninsula in the south (see Fig. 1 for location). This transition is accompanied by a rotation of S_{Hmax} from NW–SE to WNW–ESE and correlates well with the results for the fast polarization direction obtained from the shear-wave-splitting analysis (Eken et al. 2013). Here the polarization is sub-parallel to the maximum shortening direction at seismic stations in the Princes' Islands and rather diffuse for those in the Armutlu Peninsula. Similar stress orientations were found by Pinar et al. (2003) who document a pure strike-slip regime with an intermediate shape ratio $R = 0.5$ for the entire eastern Sea of Marmara, where the overlap of the confidence areas of the smallest and the intermediate principal stresses, σ_2 and σ_3, indicate a transpressional stress regime. This is in contrast to the results of Örgülü (2011), who document an overlap of the confidence areas of the largest principal stresses, σ_1 and σ_2, thus indicating transtension. At the westernmost tip of the Princes' Islands Segment, Armijo et al. (2002) report on en-echelon normal fault scarps apparently produced by composite normal and strike-slip faulting. In the northern Armutlu Peninsula, normal-faulting with a more pronounced strike-slip component at the south is confirmed by Kinscher et al. (2013). For the Princes' Islands Segment, Armutlu Peninsula and the

Gulf of Gemlik areas, Öztürk et al. (2015) find a normal-faulting stress regime and a rotation of the minimum horizontal stress S_{Hmin} from SW–NE to SSW–NNE, along a trajectory from north to south. Interestingly, several reverse mechanisms are reported further west. Strike-slip and normal faulting events generally dominate the offshore segment below the NAFZ, which evidently lacks near-fault observations (Pinar et al. 2003; Örgülü 2011; Wollin et al. 2018). Below the western Sea of Marmara, where the Marmara Section changes strike from E–W to ESE–WNW at the Ganos-Bend, a transpressional stress regime (Armijo et al. 1999, 2002; Pinar et al. 2003; Janssen et al. 2009; Örgülü 2011) or strike slip Öztürk et al. (2015) is reported in the literature.

3. Data Synthesis and Methods

3.1. Earthquake Catalog and Determination of First-Motion Polarities

Recently, Wollin et al. (2018) combined seismic recordings (2006–2016) from the major permanent networks in the region (black dots in Fig. 1), including the two national networks (AFAD and KOERI), to ensure the best possible azimuthal coverage. This is crucial since the largest portion of the target fault lies below the Sea of Marmara, thus preventing the deployment of on-land near-fault seismic stations needed for the determination of reliable focal mechanisms of small to moderate earthquakes. To compile a waveform data base and subsequent hypocenter catalogue, Wollin et al. (2018) re-picked the entire data set employing a novel automatized picking scheme. The new Marmara hypocenter catalogue reports 6812 reliable hypocenters down to magnitude $M_w = 0.7$ for the time period 2006–2016, all of which are based on at least seven picks (including P- and S-picks) and have a maximum azimuthal gap $GAP \leq 270°$ as well as an error ellipse area of less than $A_{ell} \leq 256\,km^2$ (Wollin et al. 2018). In this catalogue, each P-phase onset was assigned a first-motion polarity, if established with a sufficient confidence. Otherwise, the first motion was rated neutral. The signal-to-noise ratio (SNR) was used as a measure for the first-motion uncertainty and Fig. 2 shows the fraction of the P-wave onsets and first-motion polarities as a function of the SNR threshold. With the SNR threshold equaling 4, slightly more than half of the picks were assigned the first motion (blue distribution in Fig. 2a). In order to obtain a more homogeneous distribution of the first-motion measurements on the focal sphere, we later consider phase onsets measured within an epicentral distance of ≤ 60 km. In this way, we retained $\sim 50\%$ of all picks, amongst which the fraction of well determined first motions is higher ($\sim 40\%$ with respect to all picks, red distribution in Fig. 2a). Nevertheless, Wollin et al. (2018) use 758 reference picks on 72 manually picked earthquakes to calculate the fraction of automatically determined first motions being equal or different to the manually derived values. The result of this analysis yields an error rate with respect to all measurements of $\sim 10\%$, i.e. on average $\sim 10\%$ of all up- and down-polarities are erroneously flipped. Figure 3a, b show exemplary **U**p and **D**own first motions (i.e. compressional and dilatational or positive '+' and negative '−', respectively) of 75 P-wave onsets each, which were recorded from earthquakes of M_w between 1.7 and 3.8 at epicentral distances between 5 and 210 km.

In this study, we utilize P-wave polarities of 6812 events. In general, the number of measured phase arrivals and first-motion polarities per event positively correlates with the moment magnitude (Fig. 4a, b). The bulk of earthquakes have a magnitude M_w in the range between 1.7 and 2.3 and are located with approximately 20 (P- and S-) phase readings on average. However, these events yield usually only 5 high-quality first-motion polarities. This number is too small to calculate stable single-event focal mechanisms, but may be utilized to determine the stress tensor when directly inverting the combined P-wave polarities from many collocated earthquakes.

3.2. Clustering of Seismicity and Selection of Polarity Data

The results from the stress-tensor inversion strongly depend on the grouping of the underlying seismicity when investigating the spatial (or

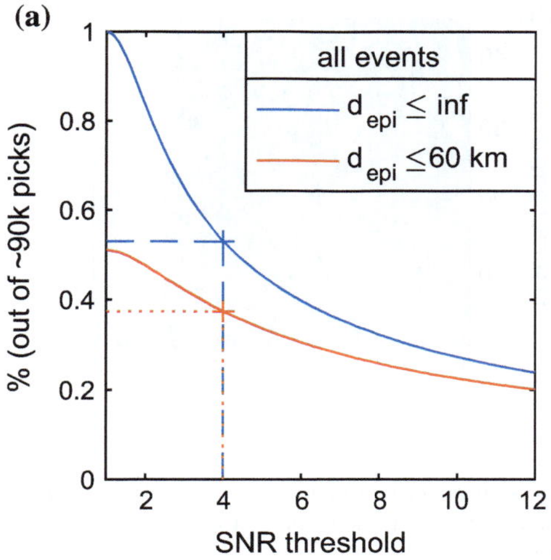

Figure 2

Fraction of $\sim 90,000$ picked P-wave onsets as a function of the threshold for the signal-to-noise ratio (SNR). Picks with SNR ≥ 4 were assigned an up or down first motion. **a** Onsets of all earthquakes in the catalog, **b** only those associated with a cluster for which the stress tensor was calculated. Both figures differentiate between all onsets and those measured within an epicentral distance of $d \leq 60$ km

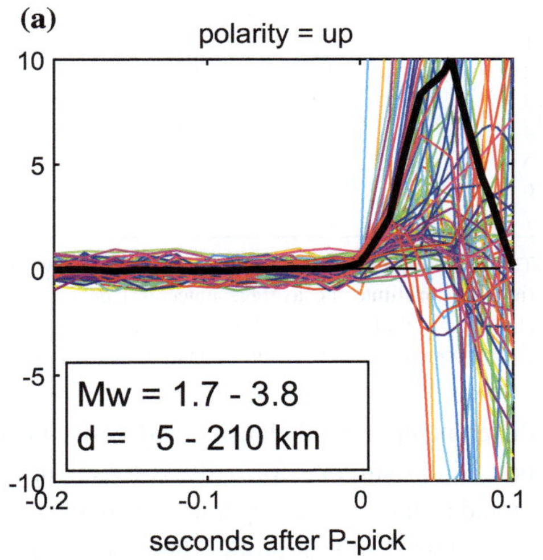

Figure 3

Two sets of 75 exemplary P-wave onsets of events with M_w between 1.7 and 3.8 at epicentral distances d between 5 and 210 km with **a** down (negative) and **b** up (positive) first motions. The wavelets are normalized to the largest absolute value within 1 s before the P-wave onset. The black graph shows the averaged wavelet

temporal) variations of the stress tensor (e.g. Hardebeck and Hauksson 1999; Townend and Zoback 2001; Hardebeck and Michael 2006; Martínez-Garzón et al. 2013, 2016a; Ickrath et al. 2015; Öztürk et al. 2015). In this study, we aim at investigating the stress field orientation along local tectonic features of particular interest as well as in areas with a pronounced spatially clustered seismic activity. We used rectangular polygons to select groups of earthquakes along fault segments and defined seismicity clusters with the 'Density-based spatial clustering of applications with noise' (DBSCAN) method (Ester

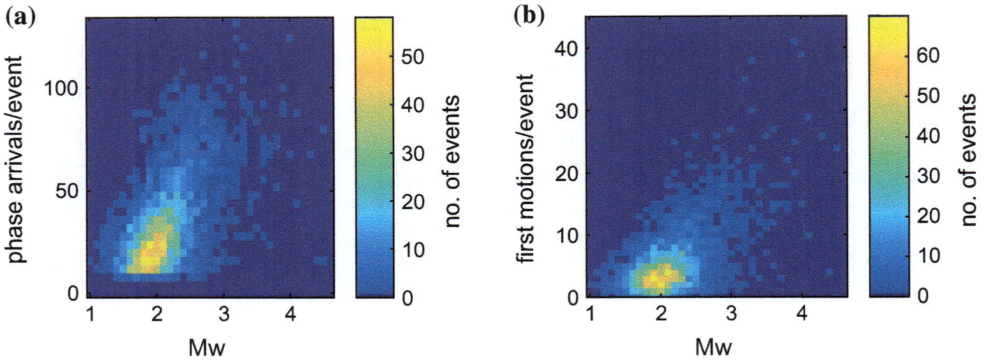

Figure 4

Density plots of the number of earthquakes as a function of two parameters: **a** the moment magnitude M_w versus the number of phase arrivals (P- and S-waves) used per earthquake location, and **b** the moment magnitude M_w versus the number of acceptable "Up" or "Down" first-motion polarities per earthquake

et al. 1996). This method essentially groups earthquakes according to the density of their occurrence, which is expressed by two parameters, N_c and r_c, the minimum number of elements and the maximum distance between elements, respectively. DBSCAN may identify arbitrarily shaped clusters and neither requires the a-priori estimation of the number of expected clusters nor some arbitrary initial clustering (*k*-means, Han et al. 2011) nor a termination condition (hierarchical clustering, Han et al. 2011). We use the inter-epicentral distance as a metric and choose seven sets of N_c and r_c with decreasing density $\rho_c = \frac{r_c}{N_c}$, thus obtaining 207 nested clusters. The cloud-shaped boundary of a seismicity cluster may be defined by the union of the r_c neighborhoods of all events associated with the respective cluster. For all cluster densities we require a minimum cluster size of $N_c = 10$ events. However, clusters of only 10 events are too small for any accurate stress inversion. After the density based clustering we thus add events located in the wider neighborhood of a cluster defined by $r_c + r_b$ (for numeric values of the parameters see Table 1). Some clusters contain a similar subset of events and we only kept the largest representative and only those clusters which contain between 100 and 1000 events (see Appendix 1 for more details regarding the usage of DBSCAN in this study and examples of the cloud-shaped cluster boundaries in dependence of different parameter sets in Fig. 14). In this way we obtain a set of 20 nested and overlapping clusters representing earthquake agglomerations of different epicentral density and distributional scale, potentially suitable for a stress-tensor inversion. We include three seismogenic volumes by defining longitudinal and latitudinal intervals delineating crucial portions of the Marmara Section, from east to west, namely the western Basins, the central Marmara Section and the Princes' Islands Segment.

Table 1

DBSCAN parameters used in this study to cluster seismicity in the Sea of Marmara Region

#DBSCAN	N_c	r_c	ρ_c	r_b
1	10	0.8	1.99	1.6
2	10	1.0	1.59	1.0
3	10	1.3	1.22	1.0
4	10	1.5	1.06	1.5
5	10	1.7	0.94	1.3
6	10	2.5	0.64	0.5
7	10	4.0	0.40	0.0

The minimum number of events in a cluster, N_c, within a radius r_c (in km) constitute the average epicentral density in a seismicity cluster, ρ_c. Radius r_b is added to r_c (in km) to "blow up" the cluster boundary and to include some of the surrounding seismicity

The stress-tensor inversion using the first-motion polarities requires a uniform sampling of the focal sphere. Refracted P-wave rays are emitted within narrow angle intervals and they tend to form dense concentric rings when mapped onto the focal sphere. Therefore, we predominantly used upward radiated (e.g. direct) rays for the inversion. For a source at 12 km depth, which is an average depth of the seismicity in the Marmara region, the velocity model of Karabulut et al.

(2011) predicts direct rays up to an epicentral distance of 50 km. Thus, we only processed the first-motion polarities measured along the ray paths covering epicentral distances of less than 60 km and emerging from events in the depth interval of 8–16 km. This procedure significantly decreases the available first-motion measurements (Fig. 2b) but produces more reliable results for the resolved stress tensor.

In order to avoid oversampling of rays measured by densely collocated seismic stations, we assigned the corresponding first motion to one station, if more than 3/4 of the first motions measured at the neighboring stations have the same sign. In the eastern Sea of Marmara, this affected the PIRES (Bohnhoff et al. 2013) and the GONAF (Bohnhoff et al. 2017a) networks, which consist of several dense near-fault seismograph arrays. The PIRES network includes local five-station arrays on the islands of Yassiada and Sivriada in direct vicinity to the Princes' Islands Segment as part of the Marmara Section (Fig. 1). Furthermore, the GONAF vertical arrays are distributed throughout the eastern Sea of Marmara region in near-shore locations or on the Princes' Islands (Sivriada and Büyakada) (Fig. 1). We processed earthquake clusters only if they contained more than 250 reliable first-motion polarities. In this way, we remained with 18 seismicity clusters which consisted of 37–310 earthquakes and yielded between 256 and 3118 first-motion polarities.

3.3. *Inversion for Stress from First-Motion Polarities*

The local stress tensor is inverted from data of the spatial seismicity clusters and the selected fault branches using software package MOTSI (First **MOT**ion Polarity **S**tress **I**nversion) of Abers and Gephart (2001). The stress field is described by the normalized deviatoric stress tensor and the MOTSI performs a non-linear inversion with two nested grid searches. The outer search loops over the parameter space of the principal stress axis orientations and the shape ratio

$$R = \frac{\sigma_2 - \sigma_1}{\sigma_3 - \sigma_1}, \tag{1}$$

where $\sigma_1 \geq \sigma_2 \geq \sigma_3$ are the principal stress magnitudes and their positive values mean compression (Zoback 2007). For every possible stress tensor, Abers and Gephart (2001) calculate a set of "stress consistent focal mechanisms" and select the best synthetic focal mechanisms fitting the first-motion polarities of an individual event. The fit of observations to each imposed stress field is expressed by a score, which translates into marginal PDFs for the stress field orientations and the shape ratio. The method is not capable of determining absolute stress values.

The advantage of the approach proposed by Abers and Gephart (2001) is that it circumvents uncertainties, which are usually inherent in the determination of focal mechanisms and which are typically not less than 10° regarding the orientation of either of the fault planes (e.g. Bohnhoff et al. 2004). Focal mechanisms are particularly uncertain for $M < 3.5$ earthquakes, which constitute the bulk of events used in this study (Fig. 4). Thus, the approach allows to include individual events with only few reliable first-motion polarities that would not lead to a reliable single-event focal mechanism otherwise. The possibility of determining the stress field from such events with $M \leq 3$ was confirmed by, e.g. Robinson and McGinty (2000) and Ickrath et al. (2015).

We investigate how well the input data match the theoretical prediction of a synthetic first-motion distribution consistent with the stress tensor retrieved during the inversion. Uncertainties in the stress field parameters, which are expressed as marginal PDFs, may stem from: (1) insufficient observational constraints, for instance in the case of a limited directional diversity of recorded ray coverage, (2) inaccurate first-motion measurements, (3) deviations from the assumption of stress-field homogeneity within the considered cluster area, or (4) a non-negligible stress interaction between the events. Under the premise that MOTSI maximizes the fit of the searched stress-field parameters to the underlying first-motion data-set, we may quantify the adequacy of our input data with respect to the combined effects of (2)–(4).

Theoretical first-motion distributions were calculated following an approach of Vavryčuk (2011). The approach can be summarized as follows: whereas Abers and Gephart (2001) consider all possible plane

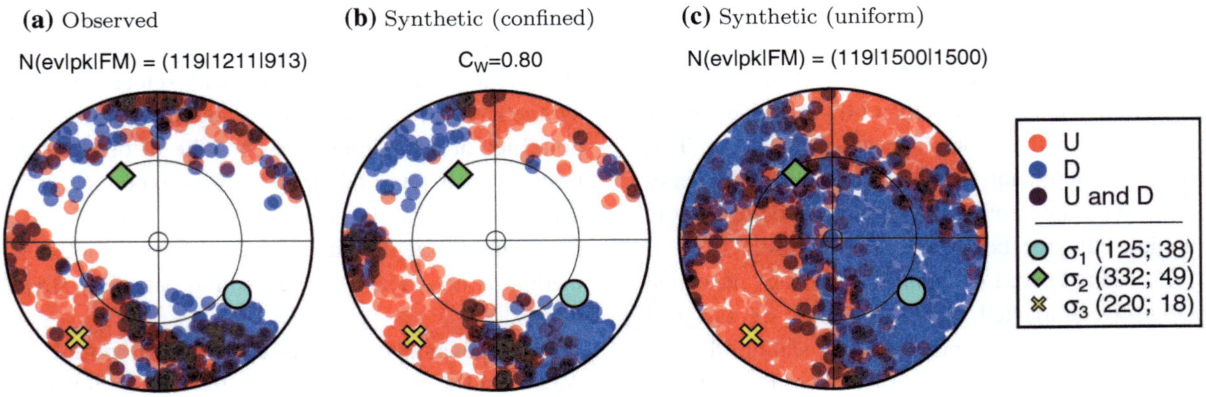

Figure 5

Distributions of up and down first motions on the lower hemisphere in equal-area projection. **a** 913 first motions out of 1211 picked P-wave arrivals observed along the Princes' Islands Segment for a total of 119 seismic events. **b**, **c** Synthetically calculated first motion distributions consistent with the stress tensor, which was obtained from the inversion of the data shown in **a**. Azimuths and takeoff-angles in **b**, **c** are confined to those observed (913 in total) and uniformly distributed (1500 in total), respectively

orientations as potentially activated, Vavryčuk (2011) restricts the variety of potentially activated faults to orientations satisfying the Mohr–Coulomb failure criterion and assumes activation of a representative set of such critically stressed faults (also see Vavryčuk 2014; Martínez-Garzón et al. 2016b). The first-motion polarities along the combined rays of all sources yield polarity distributions which are characterized by three types of areas: the first and second areas display only positive and negative first motions, respectively, and the third area shows an overlap of both (Fig. 5c). This overlap is due to the first motions belonging to seismic events activated with a variety of different focal mechanisms. The details as well as examples of first-motion distributions in dependence of the shape ratio R are presented in Appendix 2.

We obtained synthetic first-motion distributions by calculating polarities along measured ray paths with respect to the stress field retrieved by MOTSI. A measured distribution and the corresponding synthetic first-motion distribution are shown in Fig. 5a, b. For the quantitative comparison of both distributions, we perform a weighted cross-correlation (see Appendix 3), where the correlated vectors are composed of the differences of fractions of up and down polarities (Eqs. 4, 5). A difference of fractions of up and down polarities was evaluated along uniformly distributed directions, where each difference of fractions may take any value between -1 and 1. Compared to a normal cross-correlation, the weighted cross-correlation C_W (Eq. 6) puts more emphasis on areas of the focal sphere, which are more densely populated by measured first motions, thus reducing the uncertainty of the corresponding difference of fractions. A correlation with a random first-motion distribution throughout the focal sphere would yield $C_W = 0$. Areas of two first-motion distributions with opposing differences of fractions of up and down first motions yield negative contributions, such that $C_W < 0$ express dissimilarity. Values of C_W larger than 0.5 display a significant similarity.

4. Results and Discussion

We calculated the deviatoric stress tensor for a number of spatial seismicity clusters following the previously described methodology. In the following, we present and discuss the results for the individual study areas. First, we analyze the Princes' Islands Segment, for which we provide all methodological details. Then, we present the results for all other analyzed regions.

4.1. The Princes' Islands Segment

Seismicity along the Princes' Islands Segment is distributed along a ~ 50 km long and ~ 10 km wide stripe (Fig. 6) parallel and south of the fault escarpment, which strikes WNW–ESE. The seafloor

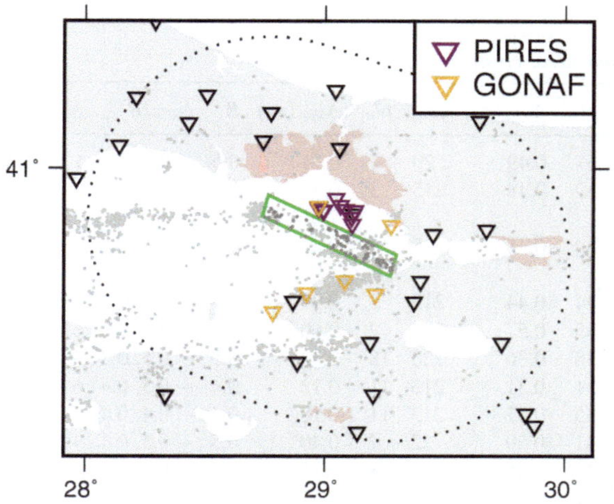

Figure 6
Map of the Princes' Islands Segment with the investigated volume delineated by a colored boundary. The contained seismicity used for the inversion is plotted in darker gray. First motions were only considered if measured at a station within 60 km epicentral distance. The corresponding stations (triangles) are situated within the dotted boundary. The permanent seismic networks with densely spaced stations, PIRES and GONAF, are marked by colored triangles

expression of this fault segment is situated only ~ 5 km south of the Princes' Islands (Figs. 1, 6; Bohnhoff et al. 2013; Wollin et al. 2018). This fault covers a prominent part of the Marmara seismic gap as it is situated in close proximity to Istanbul. It has been found to host a locked fault patch and thus represents a potential nucleation point of the pending Marmara earthquake (Bohnhoff et al. 2013; Ergintav et al. 2014; Wollin et al. 2018). The stress tensor inversion for this area was based on a total of 913 first-motion polarities of 119 local earthquakes spread along the fault. Due to the seismic gap, the epicentral distribution displays large earthquake densities at the northwestern and southeastern tip of the segment. For this reason, the DBSCAN either associates the tips to different clusters or merges them along with seismicity agglomerations further to the south-east. In order to jointly investigate the stress field along the entire Princes' Islands Segment, we select seismicity located within a polygon of the above given measurements and recorded by ~ 30 stations (Fig. 5a).

The orientations of the largest (σ_1) and smallest (σ_3) principal stress axes are given by the global maxima of the marginal PDFs obtained with MOTSI and read $(\mathrm{az}, \mathrm{pl})_{\sigma_1} = (125°, 38°)$ and $(\mathrm{az}, \mathrm{pl})_{\sigma_3} = (220°, 18°)$ (Fig. 7), thus constraining $(\mathrm{az}, \mathrm{pl})_{\sigma_2} = (332°, 49°)$. The area delineated by the 95% confidence limit of the σ_3-axis orientation, $A_{95\%}(\sigma_3)$, is small and bounded to an approximately circular patch showing that the σ_3-axis is well resolved. By contrast, the orientation of the σ_1-axis is less well resolved as indicated by the shape and size of the corresponding area of confidence, $A_{95\%}(\sigma_1)$ (for the size, also see Table 2). The trend is well bound along

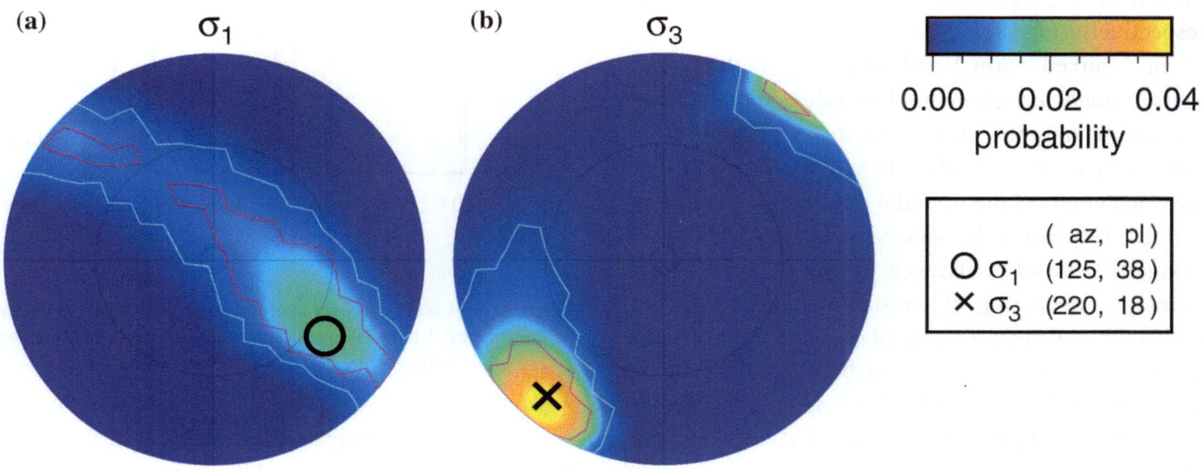

Figure 7
Marginal probability distributions of orientations of the σ_1 and σ_3 principal stress axis on the lower hemisphere in equal area projection calculated for seismicity at the Princes' Islands Segment (Fig. 6). Contour lines delineate the 68 and 95% confidence limits where the color scheme is as in Fig. 8

Table 2

Stress-tensor results for individual clusters

#	Abrev.	Name	N_{events}	N_{Picks}	N_{FMs}	C_W	az_{σ_1}	pl_{σ_1}	$A_{95\%}(\sigma_1)$	az_{σ_3}	pl_{σ_3}	$A_{95\%}(\sigma_3)$	R	$\Delta_{95\%}(R)_{l,r}$	#DBSCAN
1	**WH**	Western High	64	538	405	0.88	134	67	0.49	29	2	0.34	0.5	– 0.4, 0.4	1
2	**WC1**	Western High to Central Basin 1	72	602	453	0.82	126	59	0.48	215	3	0.33	0.5	– 0.4, 0.4	6
3	**WCw**	Western High to Central Basin wide	79	657	493	0.86	126	59	0.48	215	3	0.33	0.5	– 0.4, 0.4	NaN
4	**CMS**	Central Marmara Section	138	1263	965	0.78	141	44	0.44	215	3	0.29	0.5	– 0.4, 0.4	NaN
5	**CCB**	Cinarcik Basin	52	550	435	0.67	326	35	0.57	223	2	0.60	0.5	– 0.4, 0.5	6
6	**PIS**	Princes' Islands Segment	119	1211	913	0.81	125	38	0.36	220	18	0.22	0.5	– 0.5, 0.3	NaN
7	**EM**	Eastern-Marmara	310	3118	2484	0.76	122	31	0.31	213	11	0.12	0.4	– 0.3, 0.4	6
8	**YT**	Yalova-Tuzla	155	1584	1223	0.85	114	35	0.37	213	11	0.18	0.4	– 0.4, 0.4	5
9	**YC**	Yalova Coast	43	412	326	0.86	122	31	0.50	22	7	0.40	0.5	– 0.5, 0.4	4
10	**YTm**	Yalova-Tuzla (medium)	56	554	392	0.84	117	43	0.45	211	18	0.37	0.5	– 0.5, 0.4	4
11	**YTs**	Yalova-Tuzla (small)	39	389	282	0.79	121	51	0.49	211	18	0.42	0.5	– 0.5, 0.4	3
12	**AP2**	Armutlu 2	104	1007	839	0.86	306	67	0.41	14	12	0.32	0.5	– 0.4, 0.4	2
13	**AP1**	Armutlu 1	79	753	627	0.83	328	65	0.45	213	11	0.33	0.5	– 0.4, 0.4	1
14	**GGm**	Golf of Gemlik (medium)	37	310	256	0.61	121	51	0.48	198	9	0.45	0.5	– 0.5, 0.4	6
15	**GGl**	Gulf of Gemlik (large)	42	351	285	0.72	121	51	0.49	200	1	0.42	0.5	– 0.4, 0.4	7
16	**ET**	Erdek Tombolo	74	679	499	0.91	305	14	0.44	37	5	0.39	0.5	– 0.4, 0.5	7
17	**G2MI**	Ganos to Marmara Island	102	871	682	0.79	128	8	0.31	232	24	0.40	0.5	– 0.3, 0.4	4
18	**GF3**	Ganos Fault 3	86	741	583	0.76	135	14	0.32	232	24	0.40	0.5	– 0.3, 0.4	3

The rows are sorted according to the insets' appearance in clockwise sense in Fig. 12, beginning with the inset in the upper left corner. Azimuths (*az*) and plunges (*pl*) are given in degrees (°). $A_{95\%}$ is the fraction of the area on the unit sphere that the 95% confidence limits of the stress axis σ_1 and σ_3 occupy, and $\Delta_{95\%,l/r}$ are the left and right limits of the 95% confidence interval of the shape ratio *R*. The column #DBSCAN identifies the parameters used during the density-based clustering to define the local selection of seismic events (Table 1). Manual selection of the seismicity is marked by NaN

(N125° ± 10°E), but the plunge is more uncertain. The marginal PDF for the shape ratio has a maximum at $R = 0.5$. The 95% confidence intervals stretch widely being skewed towards lower values, namely the left and right confidence limit reach zero and 0.8, respectively (Fig. 8).

In a mixed normal and strike-slip faulting, i.e. transtensional, environment, low values of *R* indicate similar magnitudes of σ_1 and σ_2. Vanishing *R* is equivalent to $\sigma_1 = \sigma_2$ (Eq. 1) and would mean that the orientations of the σ_1 and σ_2 axes are not uniquely defined. Consequently, values of $R < 0.5$ lead to an elongated area of confidence for the σ_1-axis (which develops into a ring of constant probability in the limit of $R = 0$) and a value of $R = 0.5$ should yield circular areas of confidence for both σ_3 and σ_1 axis. The elongated marginal PDF of σ_1 might also be caused by missing observations along a ring-like area of the focal sphere. However, the trend of the σ_1-axis marginal PDF and the ring-like observational gap are noticeably different and the latter might thus only impede the proper numeric constraint of *R*.

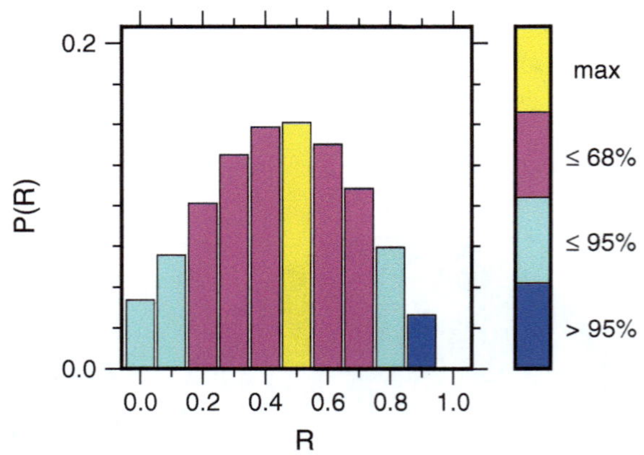

Figure 8
Marginal probability distribution of the shape ratio *R* for seismicity at the Princes' Islands Segment (Fig. 6). Bar colors indicate the extent of the 68th and 95th percentiles

In order to investigate the resolution of the stress tensor recovered for the Princes' Islands Segment, we performed additional numerical tests. We perturbed the stress-axes orientations and the shape ratios and calculated synthetic first-motion distributions (Fig. 9).

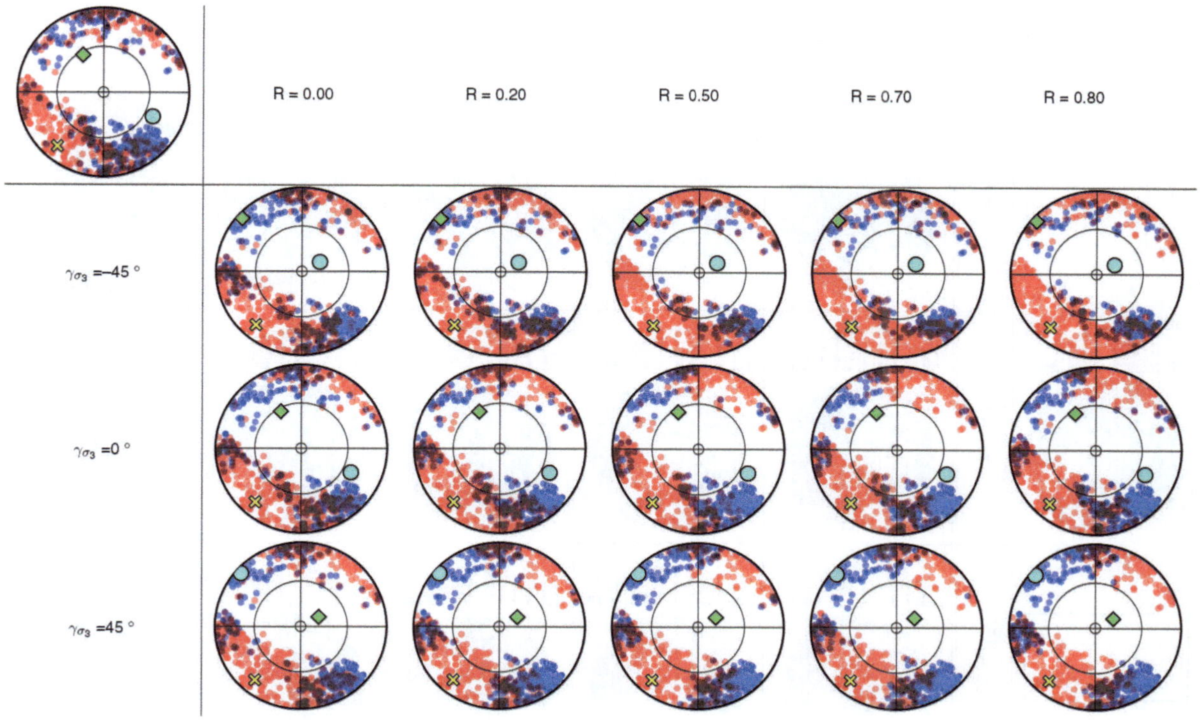

Figure 9
Simulated first-motion distributions representing different stress tensors with the same ray coverage as for observations. The stress axes are marked as in Fig. 5. Columns and rows show different shape ratios and different stress-axes orientations rotated round the σ_3-axis by the angle γ_{σ_3}. The measured first-motion distribution and the stress axes retrieved from measurements are shown in the upper left panel

These were again inverted using the MOTSI-software package. The simulated stress tensors had shape-ratios varying from $R = 0$ to $R = 0.8$ (corresponding to the 68 and 95% confidence intervals of the marginal probability distribution of R shown in Fig. 8) and stress axes which were rotated by 45° around the well resolved and thus fixed σ_3-axis. The first motions were evaluated along the observed rays and, before the inversion, 10% were randomly selected and their polarities reversed.

The results are summarized in Fig. 10 and show that the direction of the σ_3 is accurately retrieved in all cases. The direction of the σ_1 is well retrieved in all cases except for the $R = 0$ when the σ_1 and σ_2 have the same values and their directions cannot, in principal, be determined uniquely. However, the correct value of the shape ratio R is not retrieved in most cases. The histograms display a stable pattern with its maximum around $R = 0.5$ irrespective of the imposed value of R. This is an indication that the histograms are misleading and we will not present them when studying other fault segments in the Sea of Marmara. As seen from Fig. 10, some indication about the value of R is provided by the shape of the confidence areas of the σ_1-axis. This is expressed in a smeared and ring-like shaped 68% confidence area as for $R = 0$ and nearly horizontal σ_1.

Between the synthetic first-motion distribution obtained on the bases of the stress-inversion result and the corresponding measured distribution, the weighted cross-correlation coefficient reads $C_W = 0.81$ expressing a high similarity. The remaining differences may have three reasons: (1) we expected 10% of false polarities (Sect. 3.1). In a first-motion distribution composed of many different seismic events, false polarities can be identified in areas that are otherwise strictly dominated by the opposite first motion, e.g. a single down first motion was falsely determined in the south–east near zero plunge (Fig. 5a). (2) Systematic differences may be attributed to a violation of the stress homogeneity. (3) The shape ratio R was badly resolved by the inversion.

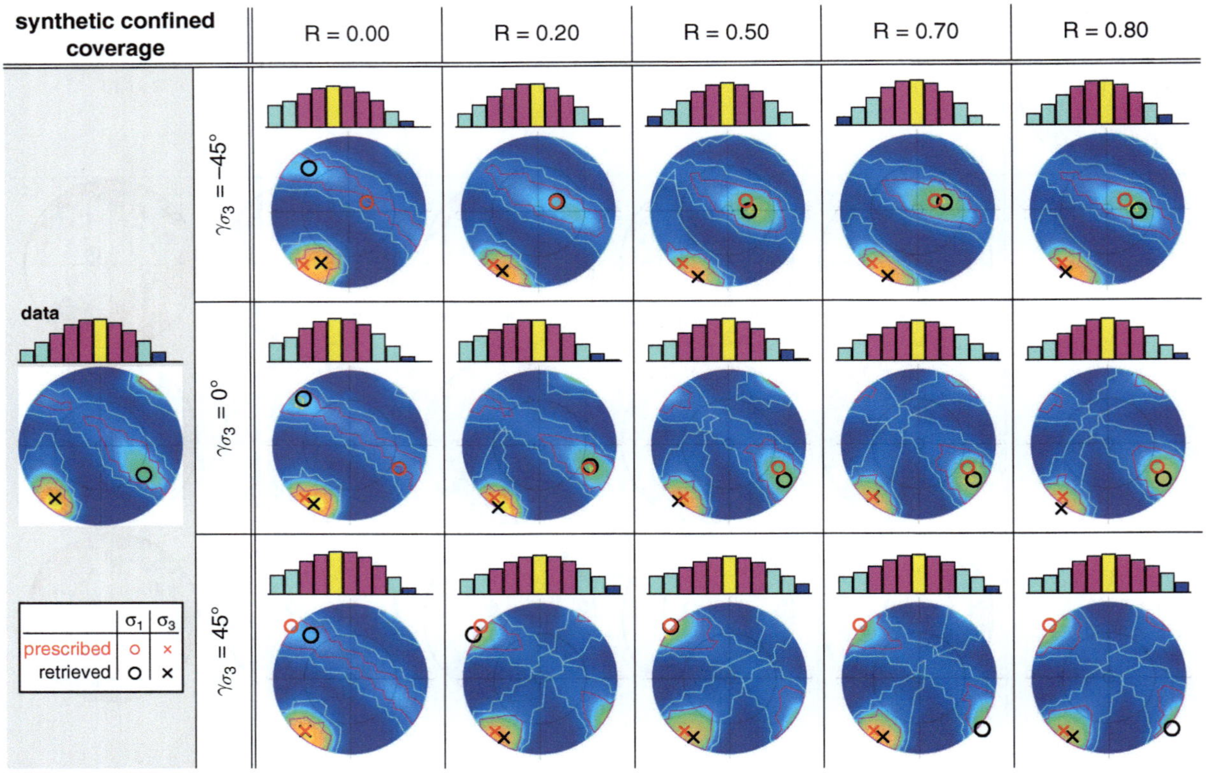

Figure 10

Stress-tensor inversions of the simulated first-motion distributions shown in Fig. 9 where the marginal PDFs of σ_1 and σ_3 are combined. The prescribed and retrieved stress axes are indicated by red and black symbols, respectively. The stress tensor inverted from the measured data is shown in the left column

Figure 15 in Appendix 2 shows that the largest change in the difference of fractions of up and down first motions occurs in the neighborhood of the piercing point of the σ_2-axis. For the first-motion distribution inverted for the Princes' Islands Segment, this area lies near the observational gap described above and thus might explain the poorly resolved shape ratio. However, a low shape ratio also provokes a broad overlap of areas with up and down first motions as observed here.

In summary, the synthetic and measured first-motion distributions display a large similarity along the Princes Islands Segment. This verifies the inversion results for the stress-axis orientations, particularly for the well constrained σ_3-axis. The differences between the measured and simulated first-motion distributions are rather systematic and cannot be explained by erroneous polarities. They more likely originate in an inaccurate recovery of the shape ratio R or an inhomogeneous stress field. The σ_1-axis displays a strike-slip orientation with a slight normal component and its smeared 68% confidence area indicates a transtensional stress regime.

These results imply that neither a pure strike-slip nor pure normal faulting can be predicted for the pending activation of the Princes' Islands Segment. They are in agreement with the stress tensor orientation of this region directly after the 1999 Izmit earthquake (Bohnhoff et al. 2006). The stress field orientation recovered along the Princes Islands Segment is generally congruent with earlier findings by Örgülü (2011) who also report transtension[1] and a sub-horizontal NE–SW orientation of the σ_1-axis including a minor normal component. The NW–SE striking σ_3-axis reported here is also in accordance with observations by Öztürk et al. (2015). Their results for the σ_1-axis, however, show a vertical

[1] Using (a different definition than us) $R = \frac{\sigma_2 - \sigma_3}{\sigma_1 - \sigma_2}$, Örgülü (2011) observes $R \approx 0.7$ for the entire Sea of Marmara.

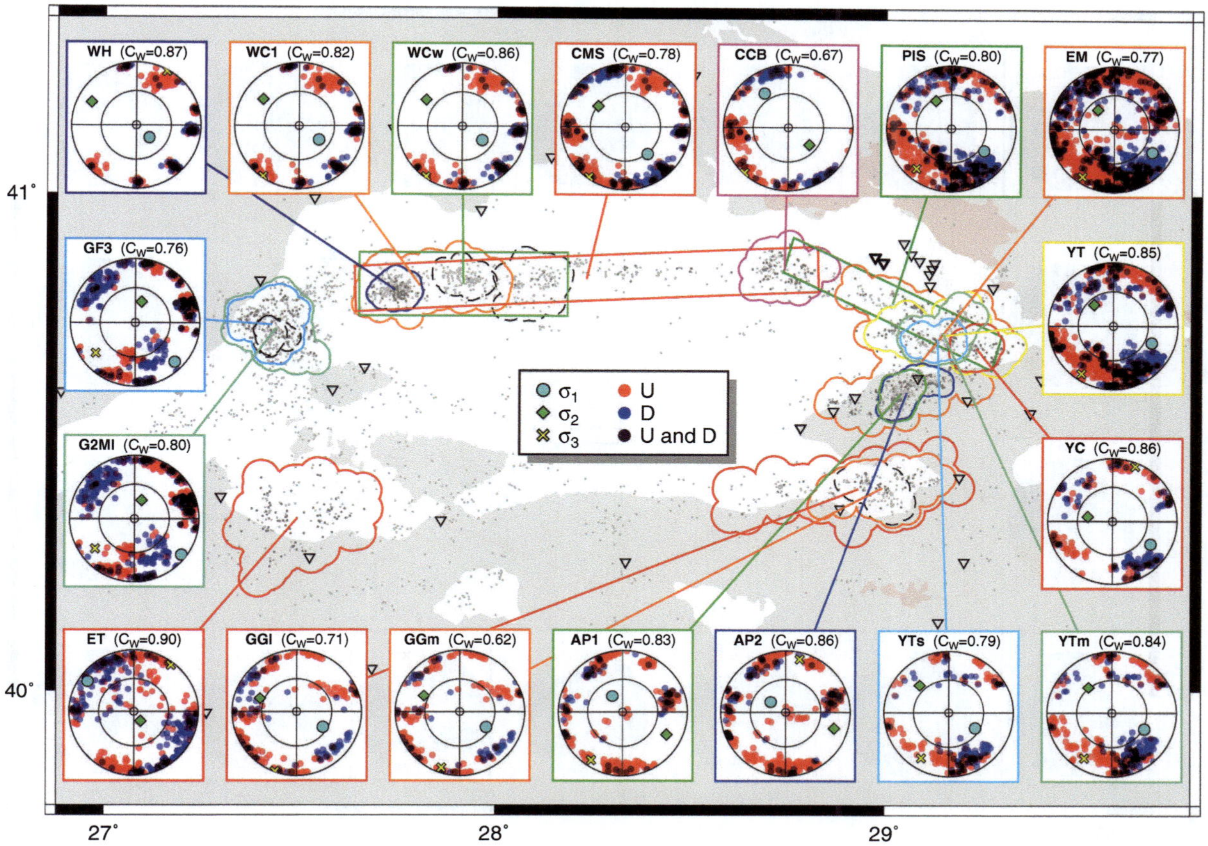

Figure 11

The insets show the measured first-motion distributions of local seismicity clusters in the Sea of Marmara. The clusters are delineated in the color of the insets' frame. The title of each inset indicates the abbreviation of the cluster name in bold letters (see Table 2) and the weighted cross-correlation coefficient with the corresponding synthetic first-motion distribution, C_W

plunge, which supports our findings of it being weakly constrained not only due to a systematic observational gap, but also due to transtension.

4.2. Stress Field in the Sea of Marmara

We found a regionally consistent well-constrained SW–NE striking σ_3-axis for all local seismicity clusters in the study area. All deviations of σ_3-axis orientations lie within the uncertainties delineated by the boundary of the area of 68% confidence. This is in correspondence with a transtensional regional stress regime as indicated by the GPS-derived velocity field (Reilinger et al. 2006) and geological findings describing the Sea of Marmara region as a pull-apart structure (Armijo et al. 2005; Acarel et al. 2014). Except for the westernmost areas of the region, the orientation of the σ_1-axis is consistent in its trend of about N125°E but generally less well constrained in its plunge, the 95% confidence interval of its marginal PDF forming a great circle in the plane spanned with the σ_2-axis. This indicates that σ_1 and σ_2 do not differ substantially in their magnitudes. Unambiguous interpretation is aggravated by observational gaps in the first-motion distributions foremost preventing the numerical constrain of the shape ratio R, which is thus not further interpreted (also see Sect. 4.1).

However, we obtain a significant correlation between the measured and the corresponding synthetic first-motion distributions, C_W ranging between 0.62 and 0.91, and we conclude that the first-motion

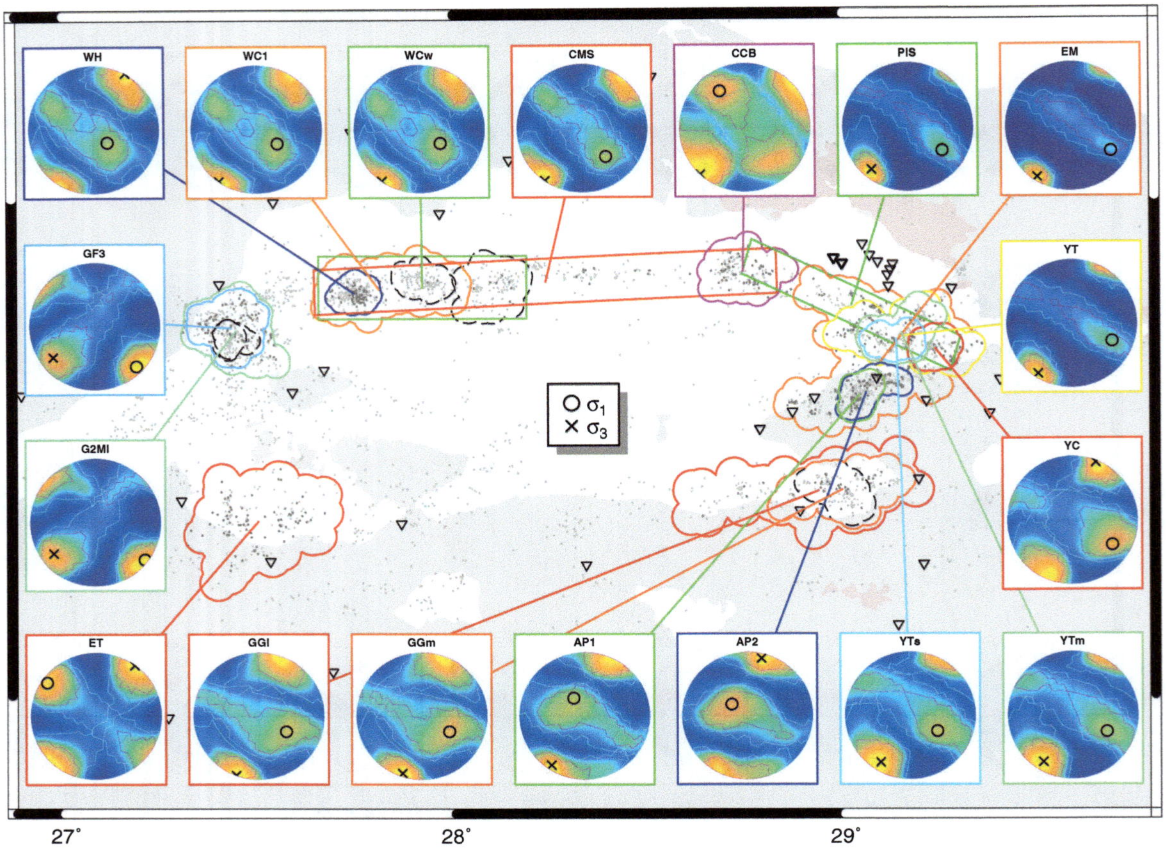

Figure 12
The insets show the combined marginal PDFs for the σ_1- and σ_3-axis orientations as obtained from the stress inversion of local seismicity clusters in the Sea of Marmara. The frame of each inset is colored as the corresponding polygon delineating the seismicity used for an inversion. The corresponding first-motion distributions are shown in Fig. 11 and the numeric results are summarized in Table 2

measurements (1) were reliably determined, (2) fulfill the assumption of stress-field homogeneity to a large degree and (3) are representative for the calculated stress tensor within the areas of confidence. All measured first-motion distributions and the corresponding stress-inversion results are summarized in Figs. 11 and 12, respectively. Results for individual seismicity clusters are shown in Figs. 16–33 of Appendix 4 (Suplementary Material) and numeric results are listed in Table 2 where the inversions can be identified by the abbreviation of the seismicity clusters' name in bold letters. For each stress tensor, we also calculated the 95% confidence intervals of S_{Hmax} following Lund and Townend (2007). The results are summarized in Fig. 13 where the different faulting regimes are categorized after Zoback (1992).

4.2.1 Eastern Sea of Marmara Region

Compared to the Princes' Islands Segment (**PIS**, dark green), the σ_1- and σ_3-axis in the eastern Sea of Marmara between Yalova and Tuzla (**YT**, yellow) are slightly rotated counter-clockwise. This corresponds to the changing strike of the main fault branch further east, where it rotates from ESE–WNW to E–W. The Yalova-Tuzla cluster (**YT**, yellow) consists of several sub-clusters, of medium and small size (**YTm** and **YTs**), as well as a cluster at the coast of Yalova (**YC**). The stress-inversion results yield larger areas of 95% confidence for the principal stress axes ($A_{95\%}(\sigma_1)$ and $A_{95\%}(\sigma_3)$ in Table 2), which is likely to stem from fewer observations per cluster. The results for these sub-clusters are congruent with those obtained for the parent and the Princes' Islands cluster (**YT** and **PIS**).

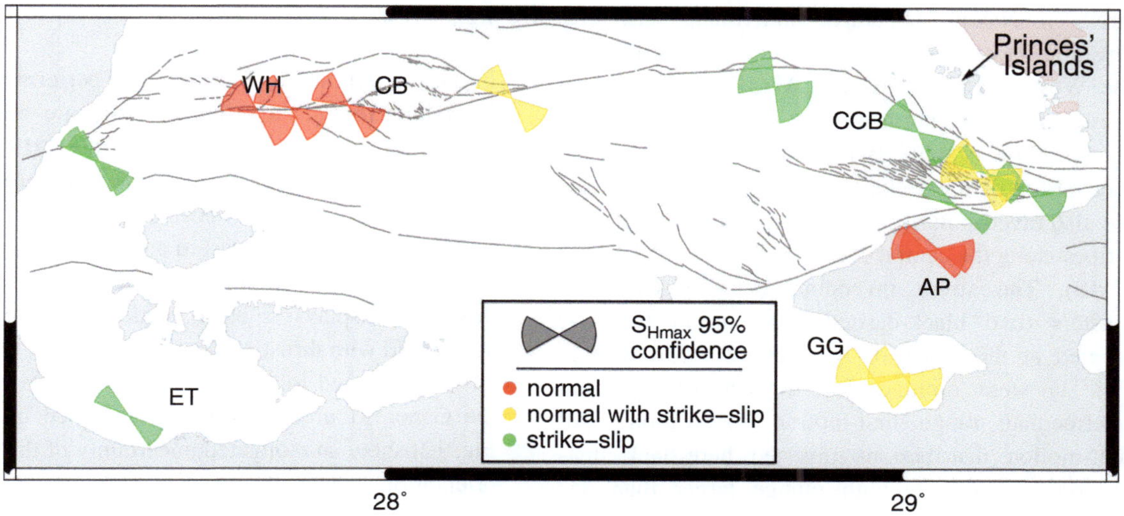

Figure 13
95% confidence intervals of S_{Hmax} for all derived stress tensors. As in Fig. 12, the wedges are situated at the centroids of the corresponding clusters. Different faulting regimes are color coded and the regional abbreviations are as in Fig. 1

Stress tensors from seismicity clusters located on the Armutlu Pensinsula (**AP1** and **AP2**) indicate a predominant normal faulting stress regime with strike-slip components. This is in agreement with earlier studies observing a substantial amount of normal faulting events (Örgülü 2011; Kinscher et al. 2013; Öztürk et al. 2015). However, a substantial number of positive first-motion polarities around the pole (Figs. 27a–28a in Appendix 4) indicate that this region shows a spatially heterogeneous tectonic setting observed also in focal mechanisms of the 1999 Izmit aftershocks (Bohnhoff et al. 2006). In general, the weighted cross-correlation between the distributions of observed and synthetic polarities yields a high similarity coefficient ($C_W = 0.86$).

A joint inversion of the polarities from seismicity in the entire eastern Sea of Marmara (**EM**, orange) results in a stress tensor congruent with the results of the sub-clusters. The large number and the spread of seismicity over a large area leads to a dense first-motion distribution covering large parts of the focal sphere. Whilst the σ_3-axis orientation is the best resolved in the entire study, the confidence interval of the σ_1-axis delineates a ring-like area crossing the pole and striking ENE–WSW. Despite large uncertainties in the inversion results for the individual small sub-clusters, the results indicate the decomposition of the eastern Sea of Marmara into two stress regimes, one in the north tending towards strike-slip with a normal faulting component and one in the south tending towards normal faulting, respectively (Fig. 13). The orientations of the least compressive stress axes (σ_3) are the same for both stress regimes.

The two seismicity clusters in the Gulf of Gemlik (**GGm** and **GGl**) south of the Armutlu Peninsula have comparatively fewer observations. The results for both marginal PDFs for the principal stress orientations are very similar and congruent with the values obtained for the offshore clusters north of the Armutlu Peninsula. The stress field orientation derived for the Gulf of Gemlik is dominantly NE–SW extensional strike-slip faulting with a significant normal component (Fig. 13). The extended confidence intervals show, however, a larger uncertainty. Still, the measured first-motion distribution is representative for the recovered stress field as it shows significant similarity with the corresponding synthetic first-motion distribution, i.e. $C_W = 0.62$ for the medium and even $C_W = 0.74$ for the larger cluster.

4.2.2 Central Marmara Section

The central Marmara Section is subject to large observational gaps. Hence, we can only perform stress inversions for three individual seismicity clusters. Two clusters are located in proximity to the Western High (**WH** and **WC1**), and the third cluster under the north-western part of the Cinarcik Basin at the Istanbul Bend (**CCB**). To resolve the local stress field orientation in the best possible way, we also invert polarities along two manually selected stripes along the central Marmara Section (**WCw** and **CMS**). The stress inversion for two seismicity clusters (two black-dashed polygons in Fig. 12) located at the edges of the Central Basin (CB in Fig. 1) were omitted due to lack of sufficient intermediate range first-motion observations. All first-motion distributions inverted here lack first-motion observations with plunge larger than 45° (Figs. 16a–20a in Appendix 4).

The resolved stress-field orientation along the central Marmara Section indicates a first-order NE–SW normal faulting stress regime with a well-constrained σ_3-axis and a bimodal marginal PDF for the σ_1-axis where the two maxima approximately enclose an angle of 90°. No pure strike-slip component is observed while the oblique plunge of σ_1 indicates a dominantly normal faulting regime (Fig. 13). Below the northwestern part of the Cinarcik Basin at the Istanbul Bend the stress inversion displays large uncertainties. The maximum of the marginal PDF for the σ_3-axis orientation is in agreement with that of the neighboring central Marmara Section and the Princes Islands Segment. However, the 68% confidence area stretches over almost 40°. The marginal PDF of the σ_1-axis is bimodal indicating a trend of about N125°E. Although the measured and synthetic first-motion distributions display a significant similarity, the weighted cross-correlation coefficient $C_W = 0.67$ is low and comparable to the C_W-values observed in the Gulf of Gemlik where the stress inversion was performed with only half as many first-motion observations. A possible explanation for the badly constrained stress field in the northwestern Cinarcik Basin is a prevalent stress heterogeneity at the Istanbul Bend where the fault trace changes strike abruptly and reverse faulting has been observed (Pinar et al. 2003).

4.2.3 Western Sea of Marmara

In the western Sea of Marmara, we performed stress inversions for: (1) a prominent offshore seismicity cluster at the eastern termination of the 1912 Ganos-rupture, and (2) scattered seismic activity around the Erdek–Tombolo region (between GM and TB as well as ET in Fig. 1). In contrast to all other stress regimes presented in the eastern and central Marmara region, this area displays a clear predominant strike-slip stress field with thrust components. Further, the stress tensors retrieved for two seismicity clusters offshore the Ganos Mountains (**GF3** is contained in **G2MI**, Fig. 12) show an elongated uncertainty of the σ_3-axis orientation which indicates a stress field with a transpressional component. While the obtained stress field clearly separates from all other areas of the region, it is in agreement with previous studies reporting transpression at this part of the Marmara Section (Armijo et al. 1999, 2002; Pinar et al. 2003; Janssen et al. 2009; Örgülü 2011). Generally, the distributions of measured and synthetic first motions show a high similarity with $C_W \approx 0.78$ (Table 2), supporting the reliability of our results.

The stress regime in the Erdek Tombolo (**GF3** is contained in **G2MI**) region along the southwestern shore of the Sea of Marmara is well resolved despite the fact that seismicity in this region is less spatially clustered than in all other regions considered here. This probably indicates a homogeneous stress field. The stress orientation is similar to that in the Tekirdak–Ganos region further to the north showing a predominant strike-slip stress regime. The 68% confidence areas of both σ_1- and σ_3-axis form circular regions.

5. Conclusions

We determine the stress field in the Sea of Marmara region along the main branch of the North Anatolian Fault Zone, where a major ($M > 7$) earthquake is pending in direct vicinity to the Istanbul population center. The stress field was calculated by inverting first-motion polarities of earthquakes reported in a recently published 10-year earthquake catalogue and we tested the reliability of our results by comparing measured and synthetic first-motion distributions.

Our results show a combined strike-slip and normal faulting regime reflecting the overall transtensional setting of the region. Whereas the orientation of the least compressive principal stress (σ_3) is well resolved and almost constant throughout the region with an average trend of $\sim$N35°E and a subhorizontal plunge, the largest (σ_1) and intermediate (σ_2) principal stresses show a similar trend of N125°E on average but locally varying plunges. Local variations of the stress field orientation are found throughout the region. Synthetic tests show that the shape ratio R (also called the relative stress magnitude) cannot be well constrained. However, a transtensional stress regime might be indicated by observations of elongated confidence intervals of the σ_1- and σ_2-axis.

In the eastern Sea of Marmara, the Armutlu Peninsula displays a normal faulting regime. To its north and south, around the Princes' Islands Segment and the Gulf of Gemlik, strike-slip regimes with minor or more pronounced normal faulting components prevail. Along the central Marmara Section, normal faulting tends to dominate while a strike-slip stress regime is found in the western Sea of Marmara region. Regarding the pending major Marmara earthquake, the results indicate that neither a pure strike-slip nor a pure normal faulting mechanism is to be expected. Rather, a combination of both faulting styles needs to be considered with implications to local hazard and subsequent risk. It is conceivable that several en-echelon normal faulting earthquakes in contrast to an earthquake with a significant strike-slip component, would expose the near coastal areas to a tsunami and areas further inland to substantial ground motions, respectively. Our results could be used to calculate the traction of individual faults whose strike and dip are known, to further constrain their potential faulting style.

Acknowledgements

We acknowledge funding from the Helmholtz Association in the frame of the Young Investigators Group SO-053 and Helmholtz Postdoc Programme, and from the German Research Foundation DFG under Grant Bo1877/6. The work was partially supported by the Grant Agency of the Czech Republic, Grant 16-19751J. The maps and lower hemisphere projections were plotted with the Generic Mapping Tool (GMT) (Wessel et al. 2013). The automatic picker was implemented with MATLAB®. We thank two anonymous reviewers for their constructive remarks that substantially improved our manuscript.

Appendix 1: Density-Based Clustering

Given a set of elements for which a metric is defined, Density-based spatial clustering of applications with noise (DBSCAN) (Ester et al. 1996) uses two parameters, (1) N_c, the minimum number of elements, and (2) r_c, the maximum distance between elements, to define a cluster-density and to group the set of elements. DBSCAN categorizes the set of elements into three groups: (1) core-points define a cluster of elements where the sets of core-points of different clusters are disjunct; (2) border-points may belong to several clusters and (3) noise are outliers that are not associable to any cluster. The algorithm is deterministic regarding the separation of core-points and a boundary including core- and border-points may be defined by the union of the r_c neighborhoods of the core-points. We "blew up" the set of events associated to a cluster by calculating the union of $r_c + r_b$ neighborhoods of its core- and border-points. We added all events within this boundary to a cluster, condoning the intersection of clusters. An example of how these three parameters may define clusters of different size and density is given in Fig. 14 for four sets of N_c, r_c and r_b (the parameters used in this study are summarized in Table 1). Despite being calculated with different sets of parameters, some of the "blown up" clusters consist of almost the same subset of earthquakes. We define the similarity of two subsets

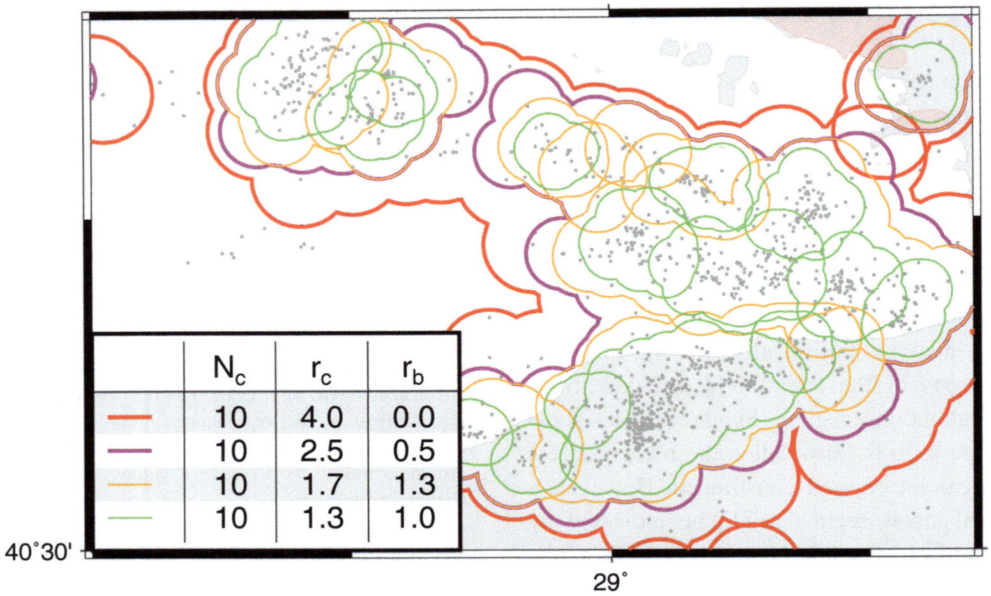

Figure 14
Results of density-based clustering (DBSCAN) applied to the seismicity in the eastern Sea of Marmara with four different sets of parameters N_c, r_c and r_b emphasizing clusters of increasing density. N_c is the minimum number of earthquakes within an epicentral distance r_c, and r_b is an additional radius that defines the cluster boundary

as the fraction of the number of earthquakes in their intersection over those in their union. If this fraction is larger than 0.9, the two subsets are considered similar and only the larger one is kept as a representative.

Appendix 2: Synthetic First-Motion Polarity Distributions

We apply the theory on the mechanics of faulting as presented in Vavryčuk (2011, 2015) (see also Martínez-Garzón et al. 2016b) to derive distributions of first-motion polarities on the focal sphere from stress field parameters.

Given the orientation and shape ratio of the principal stresses, the shear and normal stresses acting on an arbitrary plane are functions of its orientation. In the Mohr–Circle diagram, every pair of shear and normal stress represents a possible fault orientation. For activated faults, the shear stress should exceed the critical threshold t_c given by the Mohr–Coulomb failure-criterion

$$t_c = C + \mu(\sigma_n - p), \tag{2}$$

where σ_n denotes the normal stress, i.e. the normal component of the traction $\boldsymbol{t}$. Here compressional stress has positive sign (convention in rock mechanics Shearer 1999; Zoback 2007) and the parameters in this equation are dimensionless, because we normalize the stress tensor by fixing the stress magnitudes to the values $\sigma_1 = 2$ and $\sigma_3 = 0$ (e.g. Vavryčuk 2015; Martínez-Garzón et al. 2016b). The friction $\mu = 0.6$ is chosen to reflect commonly accepted values (Zoback 2007; Byerlee 1978; Vavryčuk 2015; Martínez-Garzón et al. 2016b) and the cohesion $C = 0.2$ and the pore pressure $p = 0.5$ are adapted such that the Mohr–Coulomb failure-criterion well penetrates into the outer Mohr–Circle of normalized stress magnitudes. The slip vector is parallel to the shear stress and together with the fault normal constrains a focal mechanism. The first-motion polarity of a P-wave along some ray $\boldsymbol{x}$ was then calculated after Shearer (1999),

$$f(\boldsymbol{x}) = \mathrm{sign}(\boldsymbol{x} \cdot \dot{\boldsymbol{M}} \cdot \boldsymbol{x}), \tag{3}$$

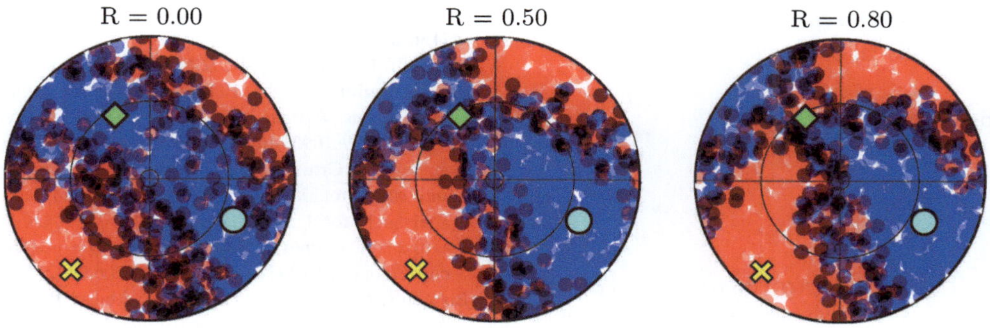

Figure 15
Synthetic first-motion distributions for three different shape ratios (Eq. 1). Symbols are as in Fig. 5

where $\dot{\boldsymbol{M}}$ is the time derivative of the moment tensor and $f = \pm 1$ is equivalent to an up and down first motion, respectively.

Figure 15 shows three exemplary synthetic first-motion distributions calculated for the same stress-axis orientations but for three different values of the shape ratio R (Eq. 1) where the first motions were evaluated along uniformly distributed rays. The distribution for $R = 0$ is rotationally symetric around the σ_3 axis (yellow 'x') and represents a pure transtensional stress regime. With increasing R the neighborhood of the σ_2-axis is increasingly populated with positive first motions until, towards a transpressional stress regime at $R = 0.8$, they begin to dominate there. At $R = 1.0$, the rotational symmetry would be established around the σ_1-axis.

Appendix 3: Weighted Cross-Correlation of First-Motion Distributions

Let us consider a circular patch P of the focal sphere which is centered around the piercing point of some unit vector $\boldsymbol{r}$. We then search the rays with an up polarity piercing the focal sphere within this patch and define their total number as $N_\mathrm{u}(P)$. Accordingly, the number of down first motions piercing the focal sphere through the patch P is given by $N_\mathrm{d}(P)$. The difference of fractions of first-motion polarities within P is then given by

$$\varrho(P) = \frac{N_\mathrm{u}(P) - N_\mathrm{d}(P)}{N_\mathrm{u}(P) + N_\mathrm{d}(P)}, \tag{4}$$

which is not defined for patches without any piercing rays. If the nonempty patch P contains, for instance,

1. only up-polarites, i.e. $N_\mathrm{d}(P) = 0$, then $\varrho(P) = 1$,
2. only down-polarites, i.e. $N_\mathrm{u}(P) = 0$, then $\varrho(P) = -1$,
3. as many up- as down-polarities, then $\varrho(P) = 0$.

For a set of M equally sized, non-overlapping and non-empty patches P_m, i.e. patches pierced by at least one ray, where $m = 1, \ldots, M$, we obtain a vector of differences of fractions

$$\boldsymbol{\varrho} = (\varrho_1, \ldots, \varrho_M) \,. \tag{5}$$

The weighted cross-correlation coefficient between two first-motion distributions represented by the vectors $\boldsymbol{\varrho}_1$ and $\boldsymbol{\varrho}_2$ finally reads

$$C_\mathrm{W}(\boldsymbol{\varrho}_1, \boldsymbol{\varrho}_2) = \frac{\sum_{m=1}^{M} \varrho_{1m} \cdot \varrho_{2m} \cdot N(P_m)}{\sqrt{\sum_{m=1}^{M} \varrho_{1m}{}^2 \cdot N(P_m)}\sqrt{\sum_{m=1}^{M} \varrho_{2m}{}^2 \cdot N(P_m)}}, \tag{6}$$

where $N(P_m) = N_u(P) + N_d(P)$ represent the total number of first-motion polarities (up or down) measured within the mth patch P_m such that their sum $\sum_{m=1}^{M} N(P_m)$ equals the total number of rays or polarities constituting a first-motion distribution. C_W thus emphasizes patches containing many rays for which a better estimate of the difference of fractions of up and down first motions ϱ_m can be assumed.

Appendix 4: Stress-Tensor Inversion Results for Local Seismicity Clusters

See Supplementary Material.

References

Abers, G. A., & Gephart, J. W. (2001). Direct inversion of earthquake first motions for both the stress tensor and focal mechanisms and application to Southern California. *Journal of Geophysical Research Solid Earth, 106*(B11), 26523–26540. https://doi.org/10.1029/2001JB000437 **(ISSN 2156-2202)**.

Acarel, D., Bulut, F., Bohnhoff, M., & Kartal, R. (2014). Coseismic velocity change associated with the 2011 Van Earthquake (M7.1): Crustal response to a major event. *Geophysical Research Letters, 41*(13), 4519–4526. https://doi.org/10.1002/2014GL060624**(ISSN 1944-8007)**.

Ambraseys, N. N. (1970). Some characteristic features of the Anatolian fault zone. *Tectonophysics, 9*(2–3), 143–165. https://doi.org/10.1016/0040-1951(70)90014-4**(ISSN 0040-1951)**.

Armijo, R., Meyer, B., Hubert, A., & Barka, A. (1999). Westward propagation of the North Anatolian fault into the northern Aegean: Timing and kinematics. *Geology, 27*(3), 267–270.

Armijo, R., Meyer, B., Navarro, S., King, G., & Barka, A. (2002). Asymmetric slip partitioning in the Sea of Marmara pull-apart: A clue to propagation processes of the North Anatolian Fault? *Terra Nova, 14*(2), 80–86. https://doi.org/10.1046/j.1365-3121.2002.00397.x**(ISSN 1365-3121)**.

Armijo, R., Pondard, N., Meyer, B., Uçarkus, G., de Lépinay, B. M., Malavieille, J., et al. (2005). Submarine fault scarps in the sea of Marmara Pull-Apart (North Anatolian Fault): Implications for seismic hazard in Istanbul. *Geochemistry Geophysics Geosystems, 6*, 6. https://doi.org/10.1029/2004GC000896**(ISSN 1525-2027)**.

Bąs, M., & Yăgci, M. (2008). Disaster prevention mitigation studies for a safer urban life measures for preventing disaster of Istanbul Metropolitan Municipality. In *Megacity Istanbul Megacity Project Reports*. Istanbul, Turkey: Municipality Disaster Management Center (AKOM). http://www.cedim.de/download/Megacity_Istanbul_Final_Report2008.pdf.

Barka, A. (1992). The north Anatolian fault zone. *Annales Tectonicae, 6*, 164–195.

Barka, A., Akyüz, H. S., Altunel, E., Sunal, G., Çakir, Z., & Dikbas, A. (2002). The surface rupture and slip distribution of the 17 August 1999 İzmit Earthquake (M 7.4), North Anatolian Fault. *Bulletin of the Seismological Society of America, 92*(1), 43. https://doi.org/10.1785/0120000841.

Bécel, A., Laigle, M., de Voogd, B., Hirn, A., Taymaz, T., & Galvé, A. (2009). Moho, crustal architecture and deep deformation under the North Marmara Trough, from the SEISMARMARA Leg 1 offshore-onshore reflection-refraction survey. *Tectonophysics, 467*(1–4), 1–21. https://doi.org/10.1016/j.tecto.2008.10.022. **(ISSN 0040-1951)**.

Bohnhoff, M., Baisch, S., & Harjes, H.-P. (2004). Fault mechanisms of induced seismicity at the superdeep German Continental Deep Drilling Program (KTB) borehole and their relation to fault structure and stress field. *Journal of Geophysical Research Solid Earth, 109*, B2. https://doi.org/10.1029/2003JB002528**(ISSN 2156-2202)**.

Bohnhoff, M., Harjes, H.-P., & Meier, T. (2005). Deformation and stress regimes in the Hellenic subduction zone from focal Mechanisms. *Journal of Seismology, 9*(3), 341–366. https://doi.org/10.1007/s10950-005-8720-5**(ISSN 1573-157X)**.

Bohnhoff, M., Grosser, H., & Dresen, G. (2006). Strain partitioning and stress rotation at the North Anatolian fault zone from aftershock focal mechanisms of the 1999 Izmit Mw = 7.4 earthquake. *Geophysical Journal International, 166*(1), 373–385. https://doi.org/10.1111/j.1365-246X.2006.03027.x.

Bohnhoff, M., Bulut, F., Dresen, G., Malin, P. E., Eken, T., & Aktar, M. (2013). An earthquake gap south of Istanbul. *Nature Communications*,. https://doi.org/10.1038/ncomms2999.

Bohnhoff, M., Ickrath, M., & Dresen, G. (2016a). Seismicity distribution in conjunction with spatiotemporal variations of coseismic slip and postseismic creep along the combined 1999 Izmit-Düzce rupture. *Tectonophysics, 686*, 132–145. https://doi.org/10.1016/j.tecto.2016.07.029.

Bohnhoff, M., Martínez-Garzón, P., Bulut, F., Stierle, E., & Ben-Zion, Y. (2016b). Maximum earthquake magnitudes along different sections of the North Anatolian fault zone. *Tectonophysics, 674*, 147–165. https://doi.org/10.1016/j.tecto.2016.02.028**(ISSN 0040-1951)**.

Bohnhoff, M., Dresen, G., Ceken, U., Kadirioglu, F. T., Kartal, R. F., & Kilic, T. (2017). GONAF–the borehole geophysical observatory at the North Anatolian Fault in the eastern Sea of Marmara. *Scientific Drilling, 22*, 19–28. https://doi.org/10.5194/sd-22-19-2017.

Bohnhoff, M., Wollin, C., Domigall, D., Küperkoch, L., Martínez-Garzón, P., Kwiatek, G., et al. (2017b). Repeating Marmara Sea earthquakes: Indication for fault creep. *Geophysical Journal International, 210*(1), 332–339. https://doi.org/10.1093/gji/ggx169.

Bott, M. H. P. (1959). The mechanics of oblique slip faulting. *Geological Magazine, 96*(2):109–117. http://geolmag.geoscienceworld.org/content/96/2/109**(ISSN 0016-7568)**.

Bulut, F., & Aktar, M. (2007). Accurate relocation of İzmit Earthquake (Mw = 7.4, 1999) Aftershocks in Çınarcık Basin using double difference method. *Geophysical Research Letters, 34*(10), 2007. https://doi.org/10.1029/2007GL029611.

Bulut, F., Bohnhoff, M., Aktar, M., & Dresen, G. (2007). Characterization of aftershock-fault plane orientations of the 1999 İzmit (Turkey) earthquake using high-resolution aftershock locations. *Geophysical Research Letters, 34*, 20. https://doi.org/10.1029/2007GL031154**(ISSN 1944-8007)**.

Bulut, F., Bohnhoff, M., Eken, T., Janssen, C., Kılıç, T., & Dresen, G. (2012). The East Anatolian Fault Zone: Seismotectonic setting and spatiotemporal characteristics of seismicity based on precise earthquake locations. *Journal of Geophysical Research Solid Earth, 117*, B7. https://doi.org/10.1029/2011JB008966**(ISSN 2156-2202)**.

Byerlee, J. (1978). Friction of rocks. *Pure and Applied Geophysics, 116*(4), 615–626. https://doi.org/10.1007/BF00876528**(ISSN 1420-9136)**.

Eken, T., Bohnhoff, M., Bulut, F., Can, B., & Aktar, M. (2013). Crustal anisotropy in the eastern sea of Marmara Region in Northwestern Turkey. *Bulletin of the Seismological Society of America, 103*(2A), 911–924. https://doi.org/10.1785/0120120156**(ISSN 0037-1106)**.

Ergintav, S., Reilinger, R. E., Çakmak, R., Floyd, M., Cakir, Z., Doğan, U., et al. (2014). Istanbul's earthquake hot spots:

Geodetic constraints on strain accumulation along faults in the Marmara Seismic Gap. *Geophysical Research Letters, 41*(16), 5783–5788. https://doi.org/10.1002/2014GL060985(**ISSN 1944-8007**).

Ester, M., Kriegel, H.-P., Sander, J., & Xu, X. (1996). A density-based algorithm for discovering clusters in large spatial databases with noise. In *Proceedings of the Second International Conference on Knowledge Discovery and Data Mining, KDD'96* (pp. 226–231). New York: AAAI Press.

Flerit, F., Armijo, R., King, G., & Meyer, B. (2004). The mechanical interaction between the propagating North Anatolian Fault and the back-arc extension in the Aegean. *Earth and Planetary Science Letters, 224*(3), 347–362. https://doi.org/10.1016/j.epsl.2004.05.028(**ISSN 0012-821X**).

Han, J., Pei, J., & Kamber, M. (2011). Data mining: Concepts and techniques. In *The Morgan Kaufmann series in data management systems*. Oxford: Elsevier Science.

Hardebeck, J. L., & Hauksson, E. (1999). Role of fluids in faulting inferred from stress field signatures. *Science, 285*(5425), 236–239. https://doi.org/10.1126/science.285.5425.236(**ISSN 0036-8075**).

Hardebeck, J. L., & Michael, A. J. (2006). Damped regional-scale stress inversions: Methodology and examples for southern California and the Coalinga aftershock sequence. *Journal of Geophysical Research Solid Earth, 111*, B11. https://doi.org/10.1029/2005JB004144(**ISSN 2156-2202**).

Hardebeck, J. L., & Okada, T. (2018). Temporal stress changes caused by earthquakes: A review. *Journal of Geophysical Research Solid Earth,*. https://doi.org/10.1002/2017JB014617(**ISSN 2169-9356**).

Hébert, H., Schindelé, F., Altinok, Y., Alpar, B., & Gazioglu, C. (2005). Tsunami Hazard in the Marmara Sea (Turkey): A numerical approach to discuss active faulting and impact on the Istanbul Coastal Areas. Tsunamis in the Pacific, in the Atlantic and in Europe: relevant cases and hazard evaluation. *Marine Geology, 215*(1), 23–43. https://doi.org/10.1016/j.margeo.2004.11.006(**ISSN 0025-3227**).

Hergert, T., & Heidbach, O. (2010). Slip-rate variability and distributed deformation in the Marmara Sea fault system. *Nature Geoscience, 3*(2), 132–135.

Hergert, T., Heidbach, O., Bécel, A., & Laigle, M. (2011). Geomechanical model of the Marmara Sea Region–I. 3-D contemporary kinematics. *Geophysical Journal International, 185*(3), 1073–1089. https://doi.org/10.1111/j.1365-246X.2011.04991.x.

Ickrath, M., Bohnhoff, M., Dresen, G., Martínez-Garzón, P., Bulut, F., & Kwiatek, G. (2015). Detailed analysis of spatiotemporal variations of the stress field orientation along the Izmit-Düzce rupture in NW Turkey from inversion of first-motion polarity data. *Geophysical Journal International, 202*(3), 2120–2120. https://doi.org/10.1093/gji/ggv273.

Janssen, C., Bohnhoff, M., Vapnik, Y., Görgün, E., Bulut, F., Plessen, B., et al. (2009). Tectonic evolution of the Ganos segment of the North Anatolian Fault (NW Turkey). *Journal of Structural Geology, 31*(1), 11–28. https://doi.org/10.1016/j.jsg.2008.09.010(**ISSN 0191-8141**).

Karabulut, H., Bouin, M.-P., Bouchon, M., Dietrich, M., Cornou, C., & Aktar, M. (2002). The seismicity in the Eastern Marmara Sea after the 17 August 1999 İzmit earthquake. *Bulletin of the Seismological Society of America, 92*(1), 387–393. https://doi.org/10.1785/0120000820(**ISSN 0037-1106**).

Karabulut, H., Schmittbuhl, J., Özalaybey, S., Lengliné, O., Kömeç-Mutlu, A., Durand, V., et al. (2011). Evolution of the seismicity in the eastern Marmara Sea a decade before and after the 17 August 1999 Izmit earthquake. *Tectonophysics, 510*(1–2), 17–27. https://doi.org/10.1016/j.tecto.2011.07.009(**ISSN 0040-1951**).

Ketin, I. (1948). Über die tektonisch-mechanischen Folgerungen aus den großen anatolischen Erdbeben des letzten Dezenniums. *Geologische Rundschau, 36*(1), 77–83. https://doi.org/10.1007/BF01791916(**ISSN 1432-1149**).

Kinscher, J., Krüger, F., Woith, H., Lühr, B., Hintersberger, E., & Irmak, T. (2013). Seismotectonics of the Armutlu Peninsula (Marmara Sea, NW Turkey) from geological field observation and regional moment tensor inversion. *Tectonophysics, 608*, 980–995. https://doi.org/10.1016/j.tecto.2013.07.016.

Kiratzi, A. A. (2002). Stress tensor inversions along the westernmost North Anatolian Fault Zone and its continuation into the North Aegean Sea. *Geophysical Journal International, 151*(2), 360–376. https://doi.org/10.1046/j.1365-246X.2002.01753.x.

Latcharote, P., Suppasri, A., Imamura, F., Aytore, B., & Yalciner, A. C. (2016). Possible worst-case tsunami scenarios around the Marmara Sea from combined earthquake and landslide sources. *Pure and Applied Geophysics, 173*(12), 3823–3846. https://doi.org/10.1007/s00024-016-1411-z(**ISSN 1420-9136**).

Le Pichon, X., Şengör, A., & Taymaz, T. (1999). *The Marmara fault and the future Istanbul earthquake* (pp. 41–54). Istanbul: Istanbul Technical University.

Le Pichon, X., Şengör, A., Demirbăg, E., Rangin, C., Imren, C., Armijo, R., et al. (2001). The active Main Marmara Fault. *Earth and Planetary Science Letters, 192*(4), 595–616. https://doi.org/10.1016/S0012-821X(01)00449-6(**ISSN 0012-821X**).

Le Pichon, X., İmren, C., Rangin, C., Şengör, A., & Siyako, M. (2014). The South Marmara Fault. *International Journal of Earth Sciences, 103*(1), 219–231. https://doi.org/10.1007/s00531-013-0950-0(**ISSN 1437-3254**).

Le Pichon, X., Şengör, A., Kende, J., İmren, C., Henry, P., Grall, C., et al. (2015). Propagation of a strike slip plate boundary within an extensional environment: The westward propagation of the North Anatolian Fault. *Canadian Journal of Earth Sciences,*. https://doi.org/10.1139/cjes-2015-0129.

Lund, B., & Townend, J. (2007). Calculating horizontal stress orientations with full or partial knowledge of the tectonic stress tensor. *Geophysical Journal International, 170*(3), 1328–1335. https://doi.org/10.1111/j.1365-246X.2007.03468.x.

Martínez-Garzón, P., Bohnhoff, M., Kwiatek, G., & Dresen, G. (2013). Stress tensor changes related to fluid injection at The Geysers geothermal field. *California. Geophysical Research Letters, 40*(11), 2596–2601. https://doi.org/10.1002/grl.50438(**ISSN 1944-8007**).

Martínez-Garzón, P., Ben-Zion, Y., Abolfathian, N., Kwiatek, G.,& Bohnhoff, M. (2016a). A refined methodology for stress inversions of earthquake focal mechanisms: Refined stress inversion methodology. *Journal of Geophysical Research Solid Earth.*

Martínez-Garzón, P., Vavryčuk, V., Kwiatek, G., & Bohnhoff, M. (2016b). Sensitivity of stress inversion of focal mechanisms to pore pressure changes. *Geophysical Research Letters, 43*(16), 8441–8450. https://doi.org/10.1002/2016GL070145(ISSN 1944-8007)(**ISSN 1944-8007**).

McClusky, S., Balassanian, S., Barka, A., Demir, C., Ergintav, S., Georgiev, I., et al. (2000). Global positioning system constraints on plate kinematics and dynamics in the eastern Mediterranean and Caucasus. *Journal of Geophysical Research: Solid Earth, 105*(B3), 5695–5719. https://doi.org/10.1029/1999JB900351**(ISSN 2156-2202)**.

McKenzie, D. P. (1969). The relation between fault plane solutions for earthquakes and the directions of the principal stresses. *Bulletin of the Seismological Society of America, 59*(2):591–601. http://bssa.geoscienceworld.org/content/59/2/591**(ISSN 0037-1106)**

Murru, M., Akinci, A., Falcone, G., Pucci, S., Console, R., & Parsons, T. (2016). M $\geq$ 7 earthquake rupture forecast and time-dependent probability for the sea of Marmara region, Turkey. *Journal of Geophysical Research Solid Earth, 121*(4), 2679–2707. https://doi.org/10.1002/2015JB012595**(ISSN 2169-9356)**.

Najdahmadi, B., Bohnhoff, M., & Ben-Zion, Y. (2016). Bimaterial interfaces at the Karadere segment of the North Anatolian Fault, Northwestern Turkey. *Journal of Geophysical Research Solid Earth, 121*(2), 931–950. https://doi.org/10.1002/2015JB012601**(ISSN 2169-9356)**.

Najdahmadi, B., Hrubcová, P., Vavryčuk, V., & Bohnhoff, M. (2018). Imaging the Mudurnu segment of the North Anatolian fault zone from waveforms of small earthquakes. *Journal of Geophysical Research Solid Earth, 123*(1), 493–512. https://doi.org/10.1002/2017JB015198.

Örgülü, G. (2011). Seismicity and source parameters for small-scale earthquakes along the splays of the North Anatolian Fault (NAF) in the Marmara Sea. *Geophysical Journal International, 184*(1), 385–404. https://doi.org/10.1111/j.1365-246X.2010.04844.x.

Öztürk, Y. K., Özal, N. M., & Özbakir, A. D. (2015). States of local stresses in the Sea of Marmara through the analysis of large numbers of small earthquakes. *Tectonophysics, 665*, 37–57. https://doi.org/10.1016/j.tecto.2015.09.027**(ISSN 0040-1951)**.

Parsons, T. (2004). Recalculated probability of M7 earthquakes beneath the Sea of Marmara, Turkey. *Journal of Geophysical Research Solid Earth, 109*, B5. https://doi.org/10.1029/2003JB002667**(ISSN 2156-2202)**.

Pinar, A., Honkura, Y., & Kuge, K. (2001). Seismic activity triggered by the 1999 Izmit earthquake and its implications for the assessment of future seismic risk. *Geophysical Journal International, 146*(1), F1–F7. https://doi.org/10.1046/j.0956-540x.2001.01476.x**(ISSN 1365-246X)**.

Pinar, A., Kuge, K., & Honkura, Y. (2003). Moment tensor inversion of recent small to moderate sized earthquakes: Implications for seismic hazard and active tectonics beneath the Sea of Marmara. *Geophysical Journal International, 153*(1), 133–145. https://doi.org/10.1046/j.1365-246X.2003.01897.x.

Reilinger, R., McClusky, S., Vernant, P., Lawrence, S., Ergintav, S., Cakmak, R., et al. (2006). GPS constraints on continental deformation in the Africa-Arabia-Eurasia continental collision zone and implications for the dynamics of plate interactions. *Journal of Geophysical Research Solid Earth, 111*, B5a. https://doi.org/10.1029/2005JB004051**(ISSN 2156-2202)**.

Robinson, R., & McGinty, P. J. (2000). The enigma of the Arthur's Pass, New Zealand, earthquake: 2. The aftershock distribution and its relation to regional and induced stress fields. *Journal of Geophysical Research: Solid Earth, 105*(B7), 16139–16150. https://doi.org/10.1029/2000JB900012 **(ISSN 2156-2202)**.

Şengör, A. C., Tüysüz, O., İmren, C., Sakınç, M., Eyidoğan, H., & Görür, N. (2005). The North Anatolian fault: A new look. *Annual Review of Earth and Planetary Sciences, 33*(1), 37–112. https://doi.org/10.1146/annurev.earth.32.101802.120415.

Şengör, A. C., Grall, C., İmren, C., Le Pichon, X., Görür, N., Henry, P., et al. (2014). The geometry of the North Anatolian transform fault in the Sea of Marmara and its temporal evolution: Implications for the development of intracontinental transform faults. *Canadian Journal of Earth Sciences, 51*(3), 222–242. https://doi.org/10.1139/cjes-2013-0160**(ISSN 0008-4077)**.

Shearer, P. M. (1999). *Introduction to seismology*. Cambridge: Cambridge University Press.

Tibi, R., Bock, G., Xia, Y., Baumbach, M., Grosser, H., & Milkereit, C. (2001). Rupture processes of the 1999 August 17 Izmit and November 12 Düzce (Turkey) earthquakes. *Geophysical Journal International, 144*(2), F1–F7. https://doi.org/10.1046/j.1365-246x.2001.00360.x. **(ISSN 1365-246X)**.

Townend, J., & Zoback, M. D. (2001). Implications of earthquake focal mechanisms for the frictional strength of the San Andreas fault system. *Geological Society London Special Publications, 186*(1), 13–21. https://doi.org/10.1144/GSL.SP.2001.186.01.02**(ISSN 0305-8719)**.

Vavryčuk, V. (2014). Iterative joint inversion for stress and fault orientations from focal mechanisms. *Geophysical Journal International, 199*(1), 69–77. https://doi.org/10.1093/gji/ggu224.

Vavryčuk, V. (2011). Principal earthquakes: Theory and observations from the 2008 West Bohemia swarm. *Earth and Planetary Science Letters, 305*(3), 290–296. https://doi.org/10.1016/j.epsl.2011.03.002**(ISSN 0012-821X)**.

Vavryčuk, V. (2015). Earthquake mechanisms and stress field. In M. Beer, I. A. Kougioumtzoglou, E. Patelli, & S.-K. Au (Eds.), *Encyclopedia of earthquake engineering* (pp. 728–746). Berlin: Springer. https://doi.org/10.1007/978-3-642-36197-5_295-1 **(ISBN 978-3-642-35344-4)**.

Wallace, R. E. (1951). Geometry of shearing stress and relation to faulting. *The Journal of Geology, 59*(2), 118–130. https://doi.org/10.1086/625831.

Wessel, P., Smith, W. H. F., Scharroo, R., Luis, J., & Wobbe, F. (2013). Generic mapping tools: Improved version released. *Eos, Transactions American Geophysical Union, 94*(45), 409–410. https://doi.org/10.1002/2013EO450001**(ISSN 2324-9250)**.

Wollin, C., Bohnhoff, M., Martínez-Garzón, P., Küperkoch, L., & Raub, C., (2018). A unified earthquake catalogue for the Sea of Marmara Region, Turkey, based on automatized phase picking and travel-time inversion: Seismotectonic implications. *Tectonophysics*,. https://doi.org/10.1016/j.tecto.2018.05.020 **(ISSN 0040-1951)**.

Yalçıner, A. C., Alpar, B., Altınok, Y., Özbay, İ., & Imamura, F. (2002). Tsunamis in the Sea of Marmara: Historical documents for the past. *Models for the future. Marine Geology*, *190*(1), 445–463. https://doi.org/10.1016/S0025-3227(02)00358-4(**ISSN 0025-3227**).

Zoback, M. D. (2007). *Reservoir geomechanics*. Cambridge: Cambridge University Press.

Zoback, M. L. (1992). First- and second-order patterns of stress in the lithosphere: The world stress map project. *Journal of Geophysical Research Solid Earth*, *97*(B8), 11703–11728. https://doi.org/10.1029/92JB00132(**ISSN 2156-2202**).

(Received May 9, 2018, revised August 1, 2018, accepted August 3, 2018, Published online September 3, 2018)

Pure Appl. Geophys. 176 (2019), 1293–1321

https://doi.org/10.1007/s00024-018-2027-2

Pure and Applied Geophysics

Quantifying the Partition Between Seismic and Aseismic Deformation Along Creeping and Locked Sections of the North Anatolian Fault, Turkey

Maor Kaduri,[1] Jean-Pierre Gratier,[1] Cécile Lasserre,[2] Ziyadin Çakir,[3] and François Renard[1,4]

Abstract—Shallow aseismic creep is a key deformation component along plate boundaries that contributes to the energy budget during the seismic cycle. Several major active continental faults show spatial alternation of creeping and locked sections. The present study focuses on the evaluation of the aseismic part of the total displacement along the North Anatolian Fault in Turkey. Detailed microstructural analyses of finite strain were performed using various methods, based on change of length or angle, on six representative samples collected over 32 outcrops along locked and creeping sections of the fault. Chemical analyses were used to map mineral composition of fault rocks and to calculate relative volume changes associated with creep. The relationship between finite strain and volume change allowed quantifying the evolution of the penetrative pressure solution cleavage mechanism of creep. In volcanic and analogous creeping rocks, finite strain measurements revealed two spatial scales of strain that correspond to the alternation of two types of shear zones, with cleavages either oblique or sub-parallel to the fault displacement. Using geodetic and geologic data, cumulative aseismic displacement was calculated in the range 9–49% of the total 80-km displacement in the creping sections and was negligible in locked sections. The large uncertainty in the kilometer-width creeping sections was related to the difficulty of quantifying the high strain values associated with high shear displacement and for which measurement uncertainties are large. A promising way to improve such quantification would be to develop reliable statistical analysis of cleavage orientation in the field.

1. Introduction

Aseismic creep has been extensively documented along several active faults worldwide (Chen and Bürgmann 2017; Harris 2017; Bürgmann 2018). Active fault creep processes develop either as transient (mostly post-seismic) or permanent sliding. The post-seismic creep rate can evolve through time with various patterns: with an exponential or a power law decay, down to zero or to a residual constant creep velocity (Çakir et al. 2005). In some cases, shallow creep can accommodate the whole tectonic loading as, for example, along the permanent creeping section of the San Andreas fault (Savage and Burford 1973). In cases where shallow creep only partly releases the tectonic loading, as is the case along the North Anatolian, the Hayward or the Longitudinal Valley faults in Turkey, California and Taiwan, respectively (Çakir et al. 2005, Graymer et al. 2005, Thomas et al. 2014), major earthquakes may still occur at depth and propagate toward the surface.

These contrasting creep behavior patterns may have different implications for seismic hazard assessment. It is therefore crucial to understand the spatio-temporal characteristics of creep and its mechanisms (Bürgmann 2018). Rock types can partly control fault creep behavior. For example, observations have shown that surface creep rates along the North Anatolian Fault are almost null along fault sections that are rich in massive limestones, but become significant along sections that host various kinds of volcanic rocks that were softened by progressive deformation processes (Kaduri et al. 2017). However, other important characteristics of creep, such as creep deformation duration during a single seismic cycle and over several seismic cycles, as well as the aseismic part of the total displacement during geologic fault history, remain to be determined more accurately.

[1] Université Grenoble Alpes, Université Savoie Mont Blanc, CNRS, IRD, IFSTTAR, ISTerre, 38000 Grenoble, France. E-mail: francois.renard@geo.uio.no

[2] Université de Lyon, UCBL, ENSL, CNRS, LGL-TPE, 69622 Villeurbanne, France.

[3] Department of Geology, Istanbul Technical University, Istanbul, Turkey.

[4] The Njord Centre, PGP, Department of Geosciences, University of Oslo, Box 1048, 0316 Blindern, Oslo, Norway.

To make progress in understanding creep mechanisms, the displacement associated with creep processes must be evaluated. On one hand, the total displacement over the lifetime of a fault accommodated by both creep and earthquakes can be evaluated by measuring the offset of a number of geologic markers (Emre et al. 2013). On the other hand, the displacement associated with creep processes can be deduced only from geodetic or geologic strain measurements, quantifying the short-term interseismic strain and the long-term finite strain, respectively. However, surface geodetic measurements include both the irreversible (ductile creep) and reversible (elastic) parts of the deformation. The distinction between creep and tectonic loading is generally made through simple modeling by considering an elastic half-space surrounding a fault plane (Savage and Burford 1973; Okada 1985). In addition, geodetic measurements are representative of slip only in the recent past, covering a few years to several decades. Extending geodetic displacement measurements to the long term (geologic times of several millions years) must include an assumption on the variations of the displacement rate with time, which is almost impossible to estimate, especially in areas associating seismic and aseismic processes. Consequently, to evaluate only the irreversible part of the large finite strain associated with the creep processes, complementary detailed finite strain measurements in the gouge and damage zone rocks of fault zones are required.

The aim of the present study is to quantify the displacement related to aseismic irreversible creep processes in both creeping and apparently presently locked sections along the North Anatolian Fault in Turkey. To reach this aim, one needs to measure the associated geologic finite strain in both the gouge and damage zone rocks (Fig. 1). To calculate this displacement, two types of data are required: the finite strain values and width of the zone of the associated shear deformation (Ramsay 1980; Ramsay and Graham 1970). However, the measurement and interpretation of such data are rather complex because strain and fault width values evolve with time and along the fault strike during deformation since the beginning of the formation of the fault to the present day. Strike-slip creeping faults accumulate strain in the upper crust by plastic and viscoelastic mechanisms involving both shear deformation and mass transfer, developing what is known as compaction or dilatant shear zones (Ramsay 1980; Ramsay and Graham 1970). In such zones, the mineral composition and rock fabric change because of fracturing, mass transfer, fluid circulation, chemical alteration and metamorphism (Gratier et al. 2011, 2013, Stewart et al. 2000; Imber et al. 2001; Jefferies et al. 2006; Collettini et al. 2009). Moreover, clay gouge comprising low-friction minerals such as montmorillonite and saponite also develop (Carpenter et al. 2016; Kaduri et al. 2017; Lockner et al. 2011; Samuelson and Spiers 2012). In addition, fault growth is not linear with displacement. Power law relationships between maximum displacement and fault width are often reported (Pennacchioni 2005; Scholz 2002). At the outcrop scale, irreversible shear strain gradients generally decrease to zero away from the fault over distances of several centimeters to several kilometers and sometimes over distances of up to 25 km (Mavko 1981). The lateral evolution of such strain profiles is often non-linear (Pennacchioni 2005). The change in shear zone width in time and space thus depends on the deformation mechanisms that combine simple shear, pure shear and volume change and that determine the displacement behavior (Bos and Spiers 2002; van der Pluijm and Marshak 2010; Fossen and Cavalcante 2017). Such heterogeneous shear zones can be classified into three categories based on their width variations: increasing, decreasing or constant in time (Hull 1988). The shape of displacement profiles across faults depends on whether the deformation process is strain hardening (e.g., deformation diffuses into the host rock) or strain softening (e.g., deformation is localized in narrow zones) (Vitale and Mazzoli 2008). In particular, strain-softening processes involving simple shear with volume loss are consistent with clay gouge formation (Kaduri et al. 2017).

Measuring and interpreting finite strain and the width of the associated creeping shear from geodetic and geologic observations are challenging tasks, especially when deformation involves very high strain values. Despite considerable uncertainties associated with both types of observations, the data presented in this study of the North Anatolian fault can be used to characterize: (1) the spatial correlation

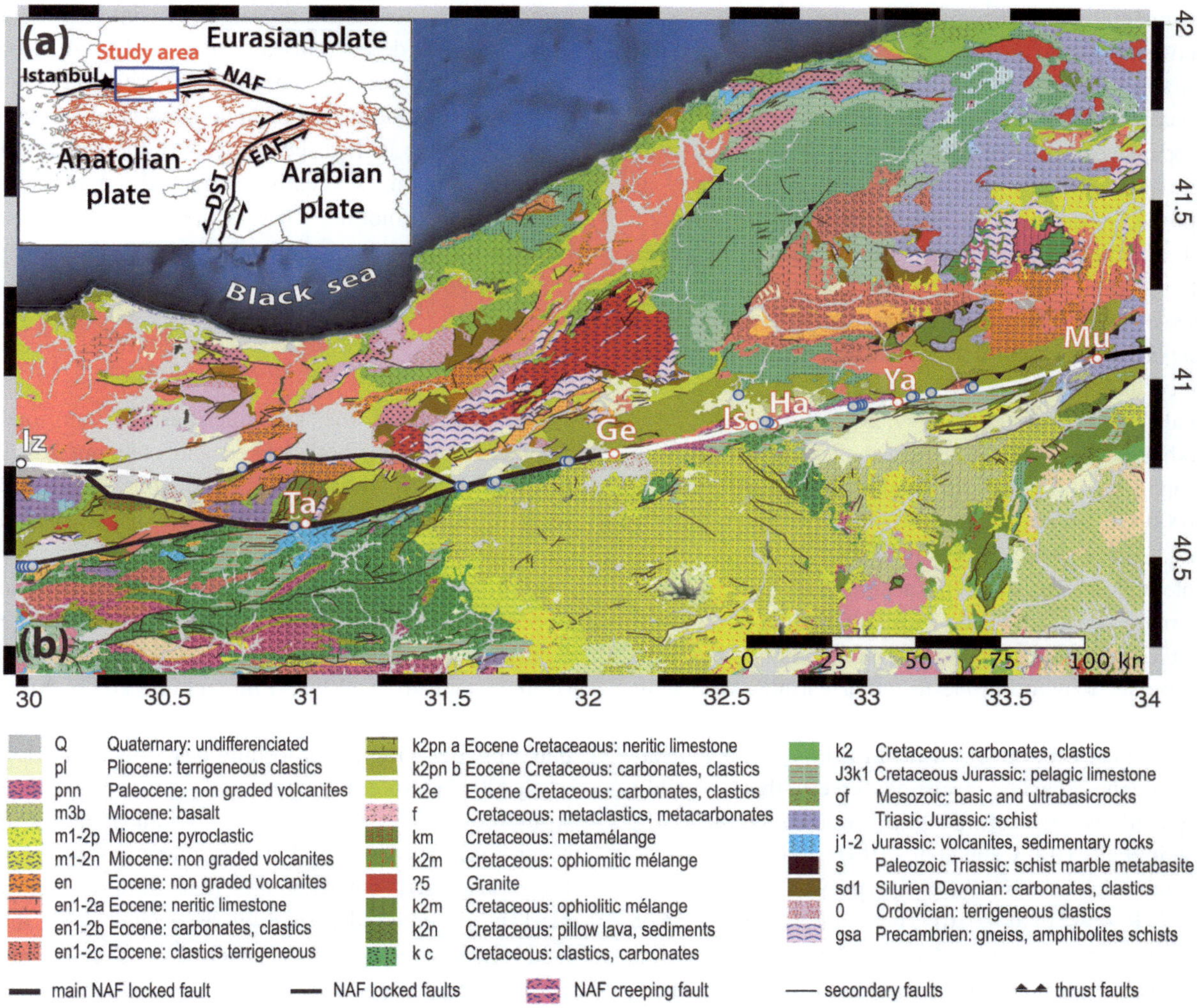

Figure 1

Structural and geologic map of the study area along the North Anatolian Fault (NAF), modified from Kaduri et al. (2017). **a** Tectonic setting of study area, located east of Istanbul (blue rectangle with NAF in red), adapted from Emre et al. (2013). *EAF* East Anatolian Fault; *DST* Dead Sea Transform fault. **b** Geologic map of the study area with outcrop locations indicated as blue or red circles for the 32 investigated outcrops. Red circles with names locate the six representative outcrops with strain markers described in the present study: *Ta* Taşkesti (40°34′54.00″N/31°1′60.00″E); *Ge* Gerede (40°47′3.00″N/32°6′30.00″E); *Is* Ismetpasa (40°51′55.00″N/32°35′41.00″E); *Ha* Hamamli (40°52′12.19″N/32°39′8.78″E) *Ya* Yazioren (40°56′1.59″N/33°6′11.09″E); *Mu* Mülayim (41°3′5.28″N/33°48′17.61″E). The presently creeping segments of Izmit = Iz (west) and Ismetpasa (east) are marked in white (Çakir et al. 2005; Cetin et al. 2014). The locked segments are marked in black. The regional geologic map is extracted from an interactive map of the Turkish geologic survey (Akbas et al. 2016) available at http://yerbilimleri.mta.gov.tr/anasayfa.aspx. The legend gives the lithology of the formations that are cut by the studied North Anatolian Fault section

between the creep-related strain measured by geodetic methods and that measured by geologic methods, (2) the control of rock lithology on aseismic creep mechanisms and temporal evolution, and (3) the aseismic part of the total displacement on the fault.

2. *Measuring Creep-Related Deformation from Fault Zone Outcrops: Setting and Approach*

2.1. *Seismotectonic Setting*

The North Anatolian Fault (NAF) in Turkey is a right-lateral strike-slip plate boundary. This major continental fault accommodates the relative motion between the Anatolian plate to the south and the Eurasian plate to the north. The NAF formed approximately 13–11 Ma ago in the east and slowly propagated westward (Şengör et al. 2004). The long-term geologic rate is 20 ± 8 mm/year (Şengör et al. 2004), consistent with short-term geodetic rates of 24 ± 2 mm/year measured from GPS data (Reilinger et al. 2006) and 25 ± 1.5 mm/year measured from InSAR data (Cetin et al. 2014). The total cumulative displacement along the studied NAF segment (Fig. 1) since its initiation is on the order of 80 km (Janssen et al. 1997; Armijo et al. 1999, 2000; Emre et al. 2013; Akbas et al. 2016). A sequence of large earthquakes propagated from east to west and ruptured over ~ 1000 km of the fault during the twentieth century (e.g., Emre et al. 2016), providing evidence of the potential of the fault to accumulate large seismic displacements. However, a striking feature of the NAF is the existence of two slip modes along the fault. Some sections are prone to major earthquakes (Stein et al. 1997) and remain locked in between earthquakes, while other sections, well-identified by geodesy measurements, display aseismic creep that possibly initiated as post-seismic slip (Kaneko et al. 2013; Çakir et al. 2014; Cetin et al. 2014) and may behave as a succession of transient creep episodes (Bilham et al. 2016; Rousset et al. 2016). The relative contribution of seismic and aseismic displacement in the 80-km cumulated displacement remains to be evaluated, which is the aim of the present study.

At least two sections of the inland part of the North Anatolian Fault have been identified in recent years as aseismic creeping fault sections (Ismetpasa and Izmit sections, Fig. 1). Based on an InSAR study of Envisat satellite data along descending orbits, Cetin et al. (2014) showed that the Ismetpasa creeping section runs from longitude 32.2° to 34.0°, with a maximum horizontal surface creep rate of ~ 20 mm/year, located 30 km east of Ismetpasa, and a creeping depth of ~ 5 km. These results were revised from a new Envisat InSAR analysis including both descending and ascending data, which helped to obtain a better separation of the vertical and horizontal creep components. The Ismetpasa creeping section was found to be longer (running from 31.0° to 35.0°), with two peaks of the maximum horizontal creep rate up to ~ 14 mm/year, and creep extending down to 9 km depth (Hussain et al. 2016). A clear correlation between the shallow creep rate and near-surface fault lithology has been demonstrated (Kaduri et al. 2017): the locked section west of the Ismetpasa creeping section is mostly composed of massive limestone and dolomite whereas the creeping section is composed of volcanic and ophiolite rocks including clay-rich gouges (Fig. 1).

2.2. *Fault Sampling and Strategy Analysis*

We calculated the aseismic part of the total displacement along both locked and creeping zones of the North Anatolian Fault. The finite strain and creeping width for various lithologies were evaluated by investigating 32 outcrops over a 400-km-long section between the city of Izmit to the west and the eastern end of the Ismetpasa creeping section to the east (Fig. 1). Satellite images and geologic maps (Herece and Akay 2003, Emre et al. 2013, 2016, Akbaş et al. 2016) were used to identify all accessible outcrops following the procedure described in Kaduri et al. (2017). No continuous outcrop was found that would cover the whole kilometer width of the NAF shear zone. The links and common features between these dispersed outcrops are that they all show markers of creep deformation associated with the NAF shear zone as either penetrative cleavage (in volcanic or analogous rocks) or stylolites (in limestones). Therefore, they all attest to the finite aseismic strain, which was associated with the large strike-slip 80-km displacement of the NAF shear zone during geologic times. It must be noted that geologic evidence of such creep deformation was found away from the present-day creeping fault, sometimes at more than 1-km distance. This observation indicates that the NAF was not a single localized fault during geologic times but a kilometer-width shear zone that includes several seismic faults and large associated zones of aseismic deformation, an observation consistent with geologic maps (Herece and Akay 2003; Emre et al. 2013, 2016; Akbaş et al. 2016). Building on our first study on the implication of fault rock transformation on aseismic creep (Kaduri et al. 2017), we focus in this following study on the calculation of the total cumulative aseismic displacement from microstructural observations. The main challenge was not only to find outcrops but also to find samples that allow calculating strain values. From the 32 visited outcrops (7 outcrops in the locked segments, 25 outcrops in the creeping segment), 90 oriented hand samples were collected and analyzed in the laboratory, and 130 thin polished sections of rock samples were prepared for mineralogic, geochemical and microstructural analyses. All outcrops and samples were chosen to be representative of the wide NAF shear zone and the geologic deformation associated with the NAF displacement. However, estimating strain values was not possible for all these outcrops and samples. Six representative outcrops with strain markers are presented here, one from the locked section and five from the creeping section (Fig. 1). They represent the various aspects of the strain measurement methods and give a representative focus on the evolution from the initial rock to damaged rock and gouges to characterize the strain related to aseismic creep.

Strain was measured at various scales from thin section to outcrop scales. Micro- to meso-structural strain measurements were performed on thin sections from the millimeter to decimeter scales. Chemical maps were acquired using either an X-ray fluorescence spectrometer (XRF, Eagle III) (mm-cm) or Wavelength Dispersive Spectroscopy on an Electron Probe Micro Analyzer (WDS-EPMA) (microns-mm) (Kaduri et al. 2017). Image processing techniques based on MATLAB codes were implemented to calculate internal finite strain from particle distributions or deformation of geologic objects. Meso- to macro-structural strain measurements at the meter to kilometer scales were derived from field observations and measurements in quarries with outcrops > 100 m wide, located within the NAF shear zone, of kilometer width. Due to different deformation markers, two distinct approaches to creeping (volcanic and analogous rocks) and locked (limestone) sections were used to evaluate the strain values. The techniques are briefly described below including a discussion on the uncertainty of the measurements. These techniques rely on the classical strain analysis procedure, well described, for example, in Ramsay (1967, 1980).

3. Strain Measurements Methodology

3.1. Strain Measurements Along the Creeping Section (Volcanic and Analogous Rocks)

The main parameters of the shear zone are defined in Fig. 2a: the kinematic axes *a* (direction of shear), *ab* (shear plane) and *ac* (displacement plane) and the corresponding coordinate axes *x*, *y* and *z*.

3.1.1 Boudinage and Folding Analysis

Strain values were extracted from XRF chemical maps relying on the difference of composition of the boudins and the veins relative to the matrix. The quadratic extension/contraction λ is defined as $\sqrt{\lambda} = l_f/l_i$ where l_f is the final length of a deformed object, and l_i is its initial length. The quadratic extension is calculated from boudinage or fractured structures: (1) the initial length (l_i) is measured by adding the length of each boudin, and (2) the final length (l_f) is the length from the first to the last boudin (Fig. 2b). The quadratic contractional strain is estimated from folded veins with constant length during deformation by measuring (1) the final length l_f of the vein and (2) its initial length l_i as the length of an arc along a deformed object using $l_i = \sum_{i=1}^{n} ds_i = \int_0^{l_0} dx\sqrt{1 + (dy/dx)^2}$, where ds_i is the length of the section between two successive points and can be written as $ds_i = \sqrt{(x_i - x_{i-1})^2 + (y_i - y_{i-1})^2}$ (Fig. 2b).

Uncertainty is linked to the ratio between the resolution of the image and measured lengths. For the boudinage, the total error is the sum of the errors on the measurements of each boudin element that gives the initial length plus the error on the final length. For the folding, it is the sum of the error on the initial and final length. In the following, uncertainties are given with the results and are shown as error bars in the figures.

3.1.2 Volume Change Due to Mass Transfer

The relative mass change was estimated by comparing the composition of insoluble and soluble minerals between protected (as undeformed as

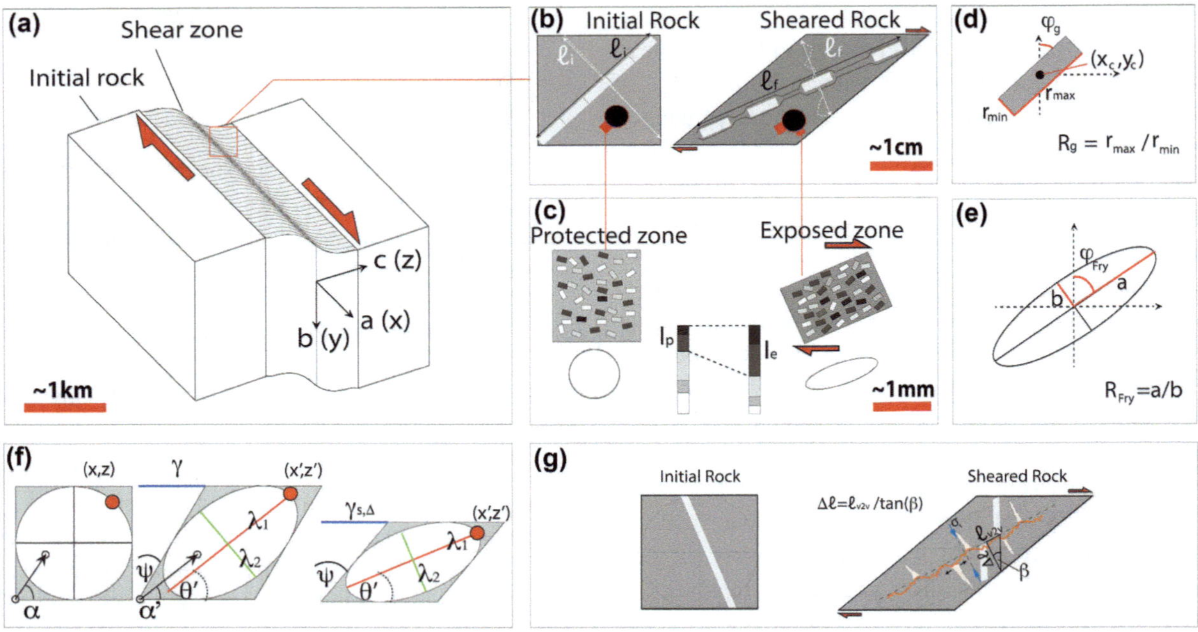

Figure 2

Strain measurements in shear zones. **a** Kinematic axes with *a* the direction of shear, *ab* the shear plane and *ac* the displacement plane and corresponding coordinate axes *x*, *y*, *z*. **b** Example of a boudinage strain marker at meso-scale. **c** Magnification of protected and exposed zones with their mineral composition and strain ellipse. **d** Definition of measured parameters of a single grain: major and minor axes r_{min}, r_{max} aspect ratio R_g and angle φ_g and the location of the center of the grain (x_c, y_c). **e** Definition of a strain ellipse using the Fry method with major and minor axes *a*, *b* and the aspect ratio between them $R_{Fry} = a/b$ with the angle φ_{Fry}. **f** Schematic deformation in simple shear coupled to volume change with distortion of a unit circle into an ellipse, adapted from (Ramsay 1967, 1980). **g** Contraction evaluated from shifted veins along the tectonic stylolite

possible) and exposed (deformed) zones at the micro- to meso-scales within a given shear zone (Fig. 2c). The relative mass change is $dm/m = I_p/I_e - 1$, where I_p and I_e are the concentration of all insoluble minerals in the protected and exposed zones, respectively (Gratier et al. 2013). If the rock density does not change, which is the case in the studied samples, the mass change is equivalent to the volume change. In plane strain, the product of the two principal strain values is related to the volumetric change as follows:

$$\sqrt{\lambda_1 \lambda_2} = 1 + \Delta, \tag{1}$$

where Δ is the change in volume (or surface area) (Ramsay 1967). Because this measurement is based on mineral maps, the accuracy depends on the dimension and the resolution of these maps. Then, the relative mass change depends on the amount of the insoluble minerals in the protected and exposed zones. An error, defined as a percentage of the calculated volume change or strain, is given in the following. In general, it is $< 2\%$.

3.1.3 Grain Geometry and Orientation Analysis

Mineral phases were identified on the chemical maps and the following geometrical properties of individual grains were measured: the minor and major axes of each grain (r_{min}, r_{max}), grain orientation φ_g defined clockwise relative to the north and coordinates of the center of each grain (x_c, y_c) (Fig. 2d). In such cases, the standard deviation of the data is calculated and used to estimate the uncertainty.

3.1.4 Fry Method

This method is also known as the center-to-center method (Fry 1979, Genier and Epard 2007). The center of every strain marker (e.g., oxide grains) is plotted with respect to the position of all other markers. A plot is built by locating one marker at the origin and by plotting as dots the positions of all other markers. Then, another marker is located at the origin and the positions of all other markers are plotted until all markers have been considered. The result is a cloud of points that contains an empty space (i.e., void) at the origin and that represents the strain ellipse. Strain is quantified using (1) the aspect ratio R_{Fry} between the elongation axis (a) and the contraction axis (b) of the strain ellipse and (2) the angle between the x-axis and the elongation axis also defined as φ_{Fry}. This method is similar to an autocorrelation function, which can be used for the same type of object (Heilbronner 2002; Heilbronner and Barrett 2014). The normalized center-to-center method (Erslev 1988; Erslev and Ge 1990) improves the Fry method by taking the grain size into account and was used in the present study. The strain ellipse parameters $(R_{\mathrm{Fry}}, \varphi_{\mathrm{Fry}})$ were measured by superimposing all minerals together (Fig. 2e). The actual values of the principal strain axes were then obtained using $R_{\mathrm{Fry}} = \sqrt{\lambda_1}/\sqrt{\lambda_2}$ with $\sqrt{\lambda_1} = \sqrt{R_{\mathrm{Fry}}(1+\Delta)}$ and $\sqrt{\lambda_2} = \sqrt{(1+\Delta)/R_{\mathrm{Fry}}}$. Uncertainty derives from fitting an ellipse to the points along the rim of the void to extract $(R_{\mathrm{Fry}}, \varphi_{\mathrm{Fry}})$ values. For the best fitting, higher weights are given to points that are close to the rim while none of the clustered points remain out of the calculation as suggested by Mulchrone (2013). The angle of the ellipse was fixed using the main orientation of the grains. We evaluate the accuracy of this fitting procedure to be $\pm$ 5%.

3.1.5 Cleavage Angle Evolution Within the Shear Zone

Evolution of the cleavage orientation within the shear zone was used to evaluate the displacement (Ramsay 1980; Ramsay and Graham 1970) when integrating the volume change (Fossen and Tikoff 1993) at the millimeter to decimeter scales on thin sections and up to the macro-scale at the meter to hectometer scales. In a shear plane, passive line markers (e.g., cleavage or vein) originally making an angle α with the shear zone walls (Fig. 2f) are sheared to make a new angle α' (Ramsay 1980) such that

$$\cot(\alpha') = \frac{\cot(\alpha) + \gamma}{1 + \Delta} \tag{2}$$

High γ values are very sensitive to a' angle field measurements: a' values of 1°, 0.5° or 0.1° lead to γ values of 57, 114 or 572, respectively (Eq. 2). Consequently, estimating γ from a' is extremely challenging for high strain values because of the difficulty of measuring a' with sufficient accuracy when it is $< 1°$. In the following, the uncertainty of this technique is given as the standard deviation of the calculated strains.

3.2. Strain Measurements Along the Locked Sections (Limestone)

The strain contraction component was calculated by two techniques. The first involves identifying veins shifted by stylolites (Gratier et al. 2013) and using $\Delta l = l_{v2v}/\tan(\beta)$, where Δl is the thickness of the layer dissolved by the stylolite, l_{v2v} is the distance between the vein walls, and β is the angle between the direction of the vein and the stylolite peaks (Fig. 2g). The initial length of the vein is then obtained by adding the shifted length to the final length of the shifted vein $l_i = l_f + \Delta l$, and the quadratic contraction can be calculated. The second technique involves using the maximum amplitude of the stylolite peaks (Toussaint et al. 2018) and applying: $\sqrt{\lambda_2}_{\mathrm{stylolite}}(\varphi) = \frac{L_{\mathrm{Tot}}(\varphi)}{L_{\mathrm{Tot}}(\varphi) + N_{\mathrm{Stylolite}}(\varphi)\cdot\langle Amp_{max}(\varphi)\rangle}$, where $\sqrt{\lambda_2}_{\mathrm{stylolite}}$ is the quadratic contraction due to stylolite in their peak direction φ, $N_{\mathrm{stylolite}}$ is the number of stylolites in direction φ, $\langle Amp_{max}\rangle$ is the average maximum peak height in direction φ, and L_{Tot} is the total length of the sample in direction φ. The strain extension component is measured perpendicular to the vein direction. The final length is defined from side to side, and the initial length is the spacing between the veins. Finally, the contraction and extension components are combined and plotted in polar coordinates to define the strain ellipse. Uncertainty is linked to the ratio between resolution of the image and measured lengths. The total error is the sum of the errors on the measurements of shifted veins or stylolites size and on the final length and is given with the results.

3.3. Relation Between Shear Displacement and Finite Strain Values

For simple shear followed by volume change (Fig. 2f), the deformation matrix transforms the undeformed vector (x, z) (initial state) to a new position after deformation (deformed state) (x', z'): $\begin{pmatrix} x' \\ z' \end{pmatrix} = \begin{pmatrix} 1 & \gamma_{s,\Delta} \\ 0 & 1+\Delta \end{pmatrix} \cdot \begin{pmatrix} x \\ z \end{pmatrix}$ where $\gamma_{s,\Delta}$ is the shear strain due to shear and dilation (Fossen and Tikoff 1993), and the principal strains are:

$$\lambda_{1,2} = \frac{1}{2}\left(1+\gamma^2+(1+\Delta)^2 \pm\sqrt{\left[1+\gamma^2+(1+\Delta)^2\right]^2-4(1+\Delta)^2}\right), \tag{3}$$

where Δ is the volume change. Both the aspect ratio between the principal strain axes $R(\gamma,\Delta) = \sqrt{\lambda_1/\lambda_2}$ and $\theta'(\gamma,\Delta)$ are nonlinear functions of γ and Δ (i.e., surfaces in a three-dimensional coordinate system). Such functions are known as $R-\theta'$ diagrams when plotted together (Fossen and Tikoff 1993). θ' is the angle between the quadratic extension λ_1 and shear direction. The total shear displacement $d_{\hat{a}}$ is obtained by integrating the shear strain along the z-axis (Ramsay 1980) (Fig. 2)

$$d_{\hat{a}} = \int_0^w \gamma \cdot \mathrm{d}z \tag{4}$$

where w is the width of the shear zone, and γ is the shear strain.

4. Results of Finite Strain Measurements at Selected Sites

The structural strain markers differed from one outcrop to another although all were consistent with dextral slip on the North Anatolian Fault. Therefore, the strain values were extracted using different methods, depending on the particular type of strain markers along the shear zones. For five selected sites (Taşkesti, Hamamli, Ismetpasa, Yazioren and Mülayim, Fig. 1), strain measurements at the thin-section scale are shown in Figs. 3, 4, 5, 6 and 7. For each outcrop, the mineral maps, mineral content histograms, Fry and rose diagrams have a similar color codes. For the last selected site, Gerede (Fig. 1), strain measurements were integrated at the outcrop and regional scales (Fig. 8). Finally, all strain measurements on the six outcrops are integrated in Fig. 9.

Figure 3 ►
Taşkesti outcrop. **a** Limestone hand sample in a horizontal displacement plane (*ac*, Fig. 2a) in the Taşkesti outcrop deformed by tectonic stylolites associated with two families of veins. Locations of the two thin sections Ta1a1 and Ta1a2 are indicated. A third thin-section, Ta1d, is not shown here. **b** Digitized tectonic stylolites and veins. **c** Rose diagrams representing the orientations of the tectonic stylolite surfaces (red) and peaks (gray) and the first and second generations of veins. **d** Strain ellipse based on stylolite and vein data giving contraction and extension values (black arrows), respectively. The locations of individual veins v1–v8 are given in **b**

4.1. Taşkesti Outcrop (Locked Section)

The Taşkesti outcrop (Fig. 1) exposes the fault zone in massive limestone formations, which are deformed by tectonic stylolites associated with veins as seen in a hand sample (Fig. 3). This sample is representative of the seven visited outcrops in limestones, which all show several relatively narrow, meter-wide, shear zones with numerous stylolites. The cumulated width of all these narrow shear zones is a few meters at maximum for the whole NAF width. Six thin sections in three orthogonal planes were cut from the Taşkesti sample. Two types of tectonic stylolites can be distinguished from the different color of their insoluble residues (either red or black) (Fig. 3a). Several generations of veins that cross cut each other can be seen. Some are sub-parallel to the shear fault plane; others are oblique to it. In places, some of the veins are dissolved and shifted perpendicularly to the stylolite plane, indicating the shortening direction. In other places, some of the horizontal veins are stylolitized, while some vertical stylolites are filled with calcite (Fig. 3b). Figure 3c shows the orientation of maximum amplitude peak directions and stylolite surfaces that are oriented in various directions, with the highest density in the range N70–N90. It also shows the various orientations of the oblique-to-the-fault veins, the most frequent one in the direction N110, and

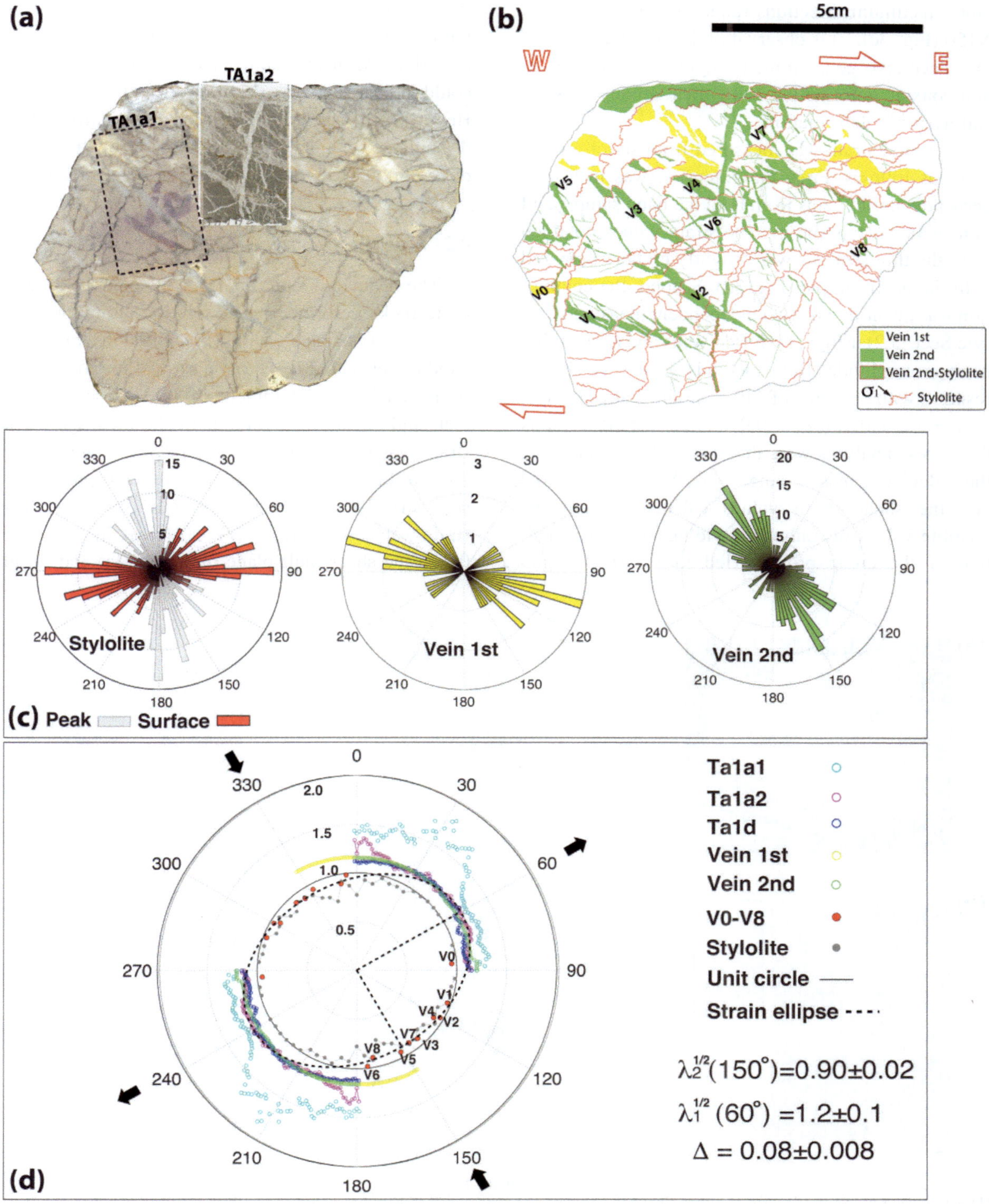
(a)
TA1a2
TA1a1
(b)
5cm
W
E
V0
V1
V2
V3
V4
V5
V6
V7
V8
Vein 1st
Vein 2nd
Vein 2nd-Stylolite
σ1
Stylolite
Stylolite
Vein 1st
Vein 2nd
(c) Peak
Surface
(d)
Ta1a1
Ta1a2
Ta1d
Vein 1st
Vein 2nd
V0-V8
Stylolite
Unit circle
Strain ellipse
$\lambda_2^{1/2}(150°)=0.90\pm0.02$
$\lambda_1^{1/2}(60°)=1.2\pm0.1$
$\Delta = 0.08\pm0.008$

other maximum directions range between N120 and N150 (Fig. 3c). This observation is well indicative of the successive generations of stylolites and veins in non-coaxial deformation with alternatively seismic and aseismic deformation.

The total strain was measured from stylolites and veins as follows. First, the extension was measured perpendicular to veins from both the hand sample and thin sections (Fig. 3a, b). The strain values obtained from the thin sections were slightly higher compared with the hand sample (Fig. 3d). Second, the contraction was quantified by two independent measurements (see Sect. 3.2) using: (1) shifted veins and (2) stylolite average amplitude. Both methods gave very similar results with a fit of the strain ellipse with $\sqrt{\lambda_1} = 1.2 \pm 0.1$, $\sqrt{\lambda_2} = 0.9 \pm 0.02$. The main results from these analyses were that, at the decimeter scale, the calculated mass change was $\Delta = 0.08 \pm 0.008$, and the shear strain values calculated from the stylolite values (as only the stylolites are representative of the creep process) led to $\gamma = 0.20 \pm 0.05$ (Eq. 3). It must be noted that, assuming that the shear fault corresponds to the shear direction, the angle of the ellipse with the fault plane, which is about 30°, could indicate a higher shear value of about $\gamma = 1.15$. However, in such semi-brittle deformation, it is difficult to evaluate the orientation of the shear direction, which is not always parallel to the fault.

4.2. Hamamli Outcrop (Creeping Section)

The Hamamli outcrop (Fig. 1) exposes serpentinite rocks deformed by dissolution along cleavage surfaces associated with precipitation in veins (Kaduri et al. 2017). In places, the strain was measured by the alignment of iron oxide minerals that agglomerate sub-parallel to the cleavages. The strain was measured along two planes: a shear plane on the fault (*ab*) and a displacement plane (*ac*) (Fig. 2a). The results are presented in Fig. 4. The strain values are $R_{Fry} = 2.3 \pm 0.1$ and $R_{Fry} = 3.2 \pm 0.2$ in the shear fault plane (*ab*) (Fig. 4a) and in the

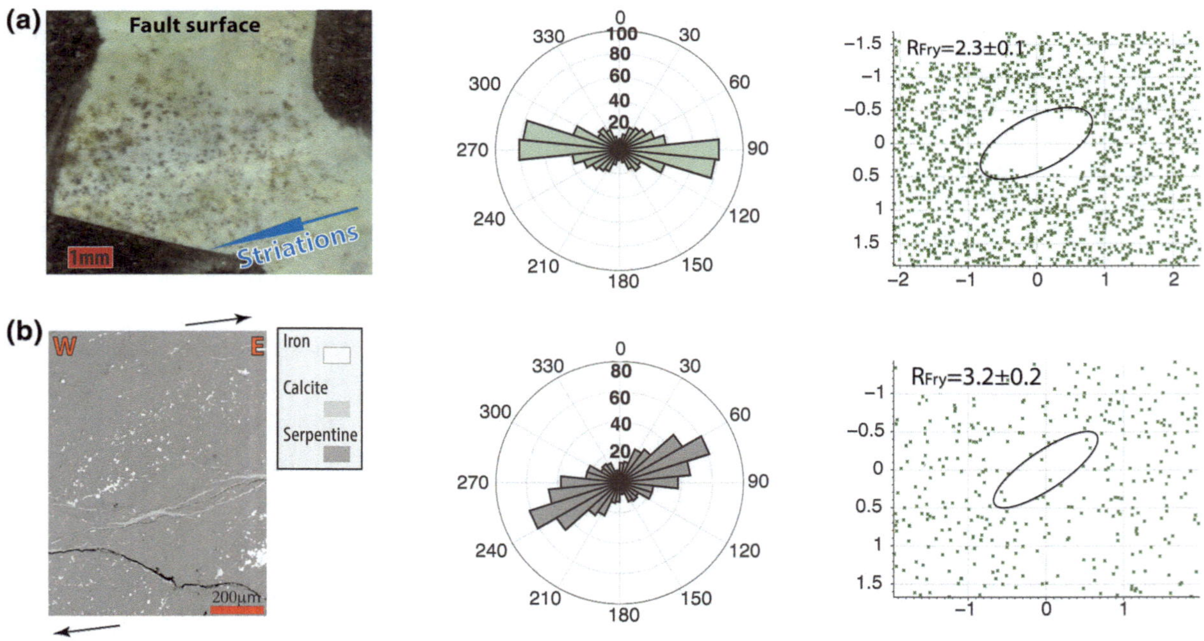

Figure 4
Hamamli outcrop. Serpentine rocks show cleavage-foliation and anisotropic clustering of iron oxide minerals in the cleavage. These microstructures were used to evaluate the strain values in two directions. **a** Strain analysis of a fault surface in vertical shear plane (*ab*), with the rose diagram of Fe oxides, and plots of the normalized Fry using Fe oxides. **b** Strain analysis with the same two methods in a horizontal displacement plane (*ac*). The shear arrows are indicative of the sense of shear and do not show an accurate orientation of the shear direction, which is not available on the outcrop where the sample was collected

displacement plane (*ac*) (Fig. 4b), respectively. Rose diagrams of the iron oxides showed alignments in the strain elongation direction in the displacement plane (Fig. 4b) but were slightly oblique to the strain elongation in the shear plane (Fig. 4a). This result may be due to the effect of episodic oblique displacements along the fault surface. Because it was not possible to find an area that was less deformed than the studied samples, the mass change could not be calculated. A maximum shear strain value in the displacement plane $\gamma = 1.3 \pm 0.1$ was derived from Eq. 3 (Fig. 2f).

4.3. *Ismetpasa Outcrop (Creeping Section)*

The Ismetpasa outcrop (Fig. 1) exposes the fault within volcanic rocks near their boundary with sandstone-shale units. At this site, the deformation in the gouge is associated with tectonic layering leading to more or less fine-grained foliated to granular rocks (*F*-gouge to *G*-gouge; Fig. 5a). The foliations are deformed around a rigid clast of the residual domain of the initial (undeformed) rock. Figure 5a shows a digitized horizontal thin section, parallel to the displacement plane (*ac*) of the shear zone, with four XRF maps of silicon on which the Fry method analysis gave four R_{Fry} values ranging from 2.7 ± 0.1 to 7.5 ± 0.4 at the centimeter scale (Fig. 5b). It was also possible to calculate the strain distribution at the millimeter scale within this thin section from six EPMA maps of mineral content acquired in exposed (the most deformed areas) and protected zones (the less deformed area, which is the clast of undeformed initial rock; Fig. 5a). Some of the maps were masked to obtain a more accurate evaluation of the localized strain and mineral composition in zones of disaggregated but relatively undeformed large quartzite clasts (map 3, Fig. 5a). The relative mass change was calculated excluding magnesite and calcite from the calculation considering that these minerals sealed the porosity of the rock more recently in the deformation process and at least after the process of massive deformation considered here (Kaduri et al. 2017). There was a clear decrease in the soluble mineral content from the protected (undeformed) zone to the exposed zones (Fig. 5c). Then, in parallel, the strain was measured on these EPMA maps using the Fry method and R_{Fry} values for a few representative minerals (albite, anorthite, orthoclase, quartz, smectite). These values were plotted as a function of the relative mass change and showed a near-linear trend (Fig. 5d). The values of R_{Fry} converted to quadratic contraction/extension $\sqrt{\lambda_{1,2}}$ using the Δ values (Eq. 1) showed the actual successive change in deformation as a function of the relative mass change (Fig. 5e). Maximum shear strain values are given by maps 2 and 1 with $\gamma = 1.2 \pm 0.1$ and $1.3 \pm 0.1$with $R_{\mathrm{Fry}} = 7.3$ and 4.5 and with $\Delta = -0.66$ and -0.3, respectively. The orientation of the grains in the exposed zones was most often aligned with the local foliation. On the contrary, the protected (near undeformed) zones showed random grain distribution with negligible preferred orientation (Fig. 5f). It must be noted that the estimated grain orientations and the Fry plots are not parallel, because the calculations are not performed at the same scale and because the deformation is heterogeneous around the clast. The Fry plots are obtained from large XRD maps that represent a sampling of the whole thin section (area XRF I, II, III, IV) whereas the grain orientation is obtained from local EPMA chemical maps and is located in zones where the orientation of the deformation is quite the same (area map 1, 2, 3b).

4.4. *Yazioren Outcrop (Creeping Section)*

The Yazioren outcrop (Fig. 1) exposes a shear zone in a mélange of carbonate and volcanic rocks with tectonic layering (rich in alumino-silicates) and carbonate boudinage. A horizontal thin section parallel to the displacement plane (*ac*) of the shear zone was imaged by XRF to segment and measure the strain markers. Figure 6a shows the calcium XRF map, which emphasizes carbonate boudinage. The strain measurement included two steps. The first step was to measure the quadratic extension $\sqrt{\lambda_1}$ along the preserved boudinage (strain markers 1–8 in Fig. 6a), which varied in the range 1.4 ± 0.1–4.1 ± 0.2. The second step was to measure the deformation recorded in the deformed matrix located near the boudins using the Fry method on maps of elements acquired from EPMA measurements (location m1–m13 in the Al map, Fig. 6b). The R_{Fry} values varied from 1.0 ± 0.05

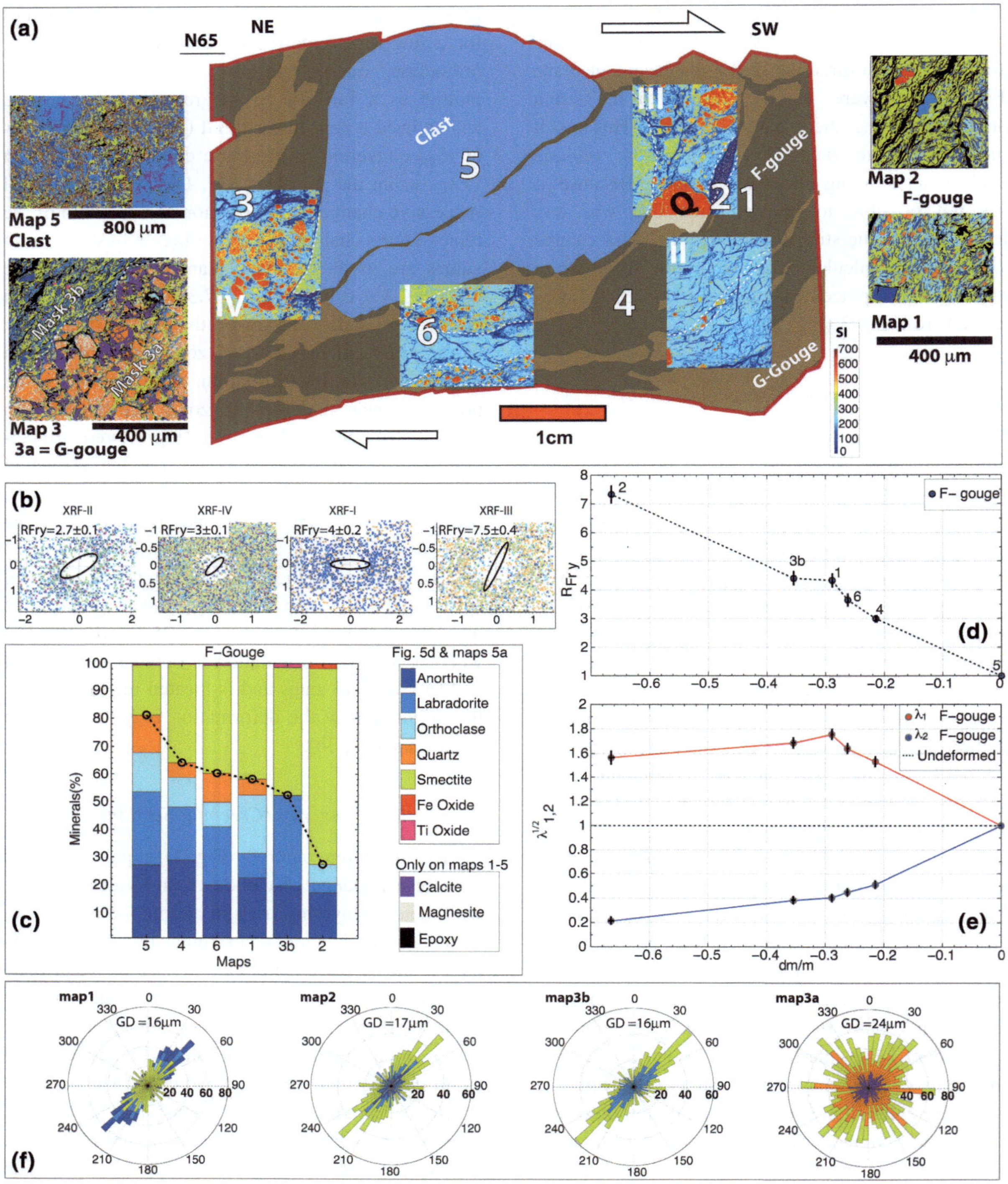

to 4.5 ± 0.2. The plot of these two types of measurement showed a good correlation with a fitted linear regression coefficient $R^2 = 0.89$ (Fig. 6c). The relative mass change was calculated using the EPMA maps by comparing exposed and protected zones. Figure 6d shows an example of these two types of

◀Figure 5
Ismetpasa outcrop. **a** Thin section in horizontal displacement plane (*ac*) with two types of gouge layers: foliated and granular gouges (*F*-gouge and *G*-gouge) around an undeformed clast of initial rock, with four XRF maps of silicon (I–VI, centimeter-scale) and six detailed EPMA maps of minerals located by numbers (1–6, millimeter scale); each corresponds to a different mineral, given in the legend in **c**. **b** Fry method based on the four XRF large maps (I–IV) with their location above and using the minerals in the legend in **c**. **c** Mineral content of the mineral maps. **d** Strain measurements using the Fry method and R_{Fry} on these detailed maps plotted as a function of the relative mass change (uncertainty is lower than the symbol size). The reference (protected zones) is map 5 (clast). **e** Plot of the calculated quadratic contraction/extension as a function of the relative mass change. **f** Rose diagrams of grain orientations from mineral maps with mean grain diameters (GD). The color code is the same as in **c**. The shear arrows are indicative of the sense of shear and do not show an accurate orientation of the shear direction, which is not available on the outcrop where the sample was collected

zones: an exposed zone (map 13) and a protected zone (pressure shadow contoured with a dashed line in map 2) around a calcite clast. From the protected to the more exposed zones, the mineral assemblage showed a reduction in soluble mineral content and a relative increase in insoluble mineral content (Fig. 6e). It is thus possible to calculate the mass change relative to the protected zone (map 2) (Fig. 6f). There is a near-linear trend between these strain values evaluated by two different methods, $\sqrt{\lambda_1}$ and R_{Fry}, and the mass change (Fig. 6f). Note that such mass change values were only minimum values since the protected zone was not an initial state but rather a less deformed state as attested by the preferred orientation of the grains (Fig. 6g). Based on these measurements, boudinage values associated with mean $\Delta = -0.1$ led to strain ratios ranging from 2 to 19 (Eq. 1) and gave a shear strain in the range $\gamma = 1.3 \pm 0.1$ to 4.0 ± 0.2 while the Fry method gave a range $\gamma = 0.5 \pm 0.05$ to 1.5 ± 0.1 (Eq. 3). These values may be compared with the value of $\gamma = 1.3 \pm 0.1$ obtained by the Fry method at the decimeter scale with $R_{\mathrm{Fry}} = 2.8 \pm 0.1$ (Fig. 6h) and considering a mean value $\Delta = 0.1$. The variation of γ values at the millimetric scale indicated the development of parallel-to-the-cleavage differentiated layers with various associated shear values.

4.5. Mülayim Outcrop (Creeping Section)

In this outcrop (see Fig. 1), the fault crosscuts a block of ophiolitic mélange embedded into schist, marble and metabasite units. Measurements were performed in the schist, which constitutes the main fraction of the rock body. Deformation was measured based on the Fry method at different scales also with the use of folded carbonate veins and the statistical analysis of grain clusters. A horizontal thin section parallel to the displacement plane (*ac*) revealed a shear zone with cleavage foliation sub-parallel to the fault and crosscut by carbonate veins in all directions (Fig. 7a). The cleavage can be seen on the XRF aluminum map (Fig. 7b). Parallel-to-the-cleavage and oblique-to-the-cleavage veins registered only a small part of the finite extension $\sqrt{\lambda_1}(0°) = 1.1 \pm 0.02$, and $\sqrt{\lambda_1}(90°) = 1.2 \pm 0.02$, respectively. Contraction perpendicular to the foliation was measured from folded veins giving $\sqrt{\lambda_2} = 0.45 \pm 0.01$ from the decimeter-scale thin section (Fig. 7a) to millimeter-size EPMA map (Fig. 7d) indicating a minimum shear strain $\gamma = 1.8 \pm 0.1$. The deformation evaluated from the Fry method using XRF aluminum maps had a relatively high value $R_{Fry} = 8.5 \pm 0.4$ (Fig. 7c), corresponding to a shear strain $\gamma = 2.6 \pm 0.2$. No volume change can be calculated here because of the lack of clear evidence of undeformed (or at least less deformed) protected zones, despite cleavage foliation being visible with alternating phyllosilicate-rich and albite-quartz-rich tectonic layers indicating significant pressure solution mass transfer (Fig. 7f). Nonetheless, the deformation was highly heterogeneous. Using EPMA maps at the millimeter size, some strain variations can be measured when using the Fry method in three different zones: the hinge (*H*) and limb (*L*) of an arcuate structure and an intermediate area (*I*) (Fig. 7d) with R_{Fry} ranging from 1.4 ± 0.07 to 4.8 ± 0.2 (Fig. 7e). Such strain values, lower at millimeter size than at the decimeter scale, imply the existence of some highly localized zones with very-high-strain values at the sub-grain size. Within these zones, phyllosilicates were oriented sub-parallel to the shear zone with embedded residues of feldspar minerals and Fe oxides (Fig. 7f), and such

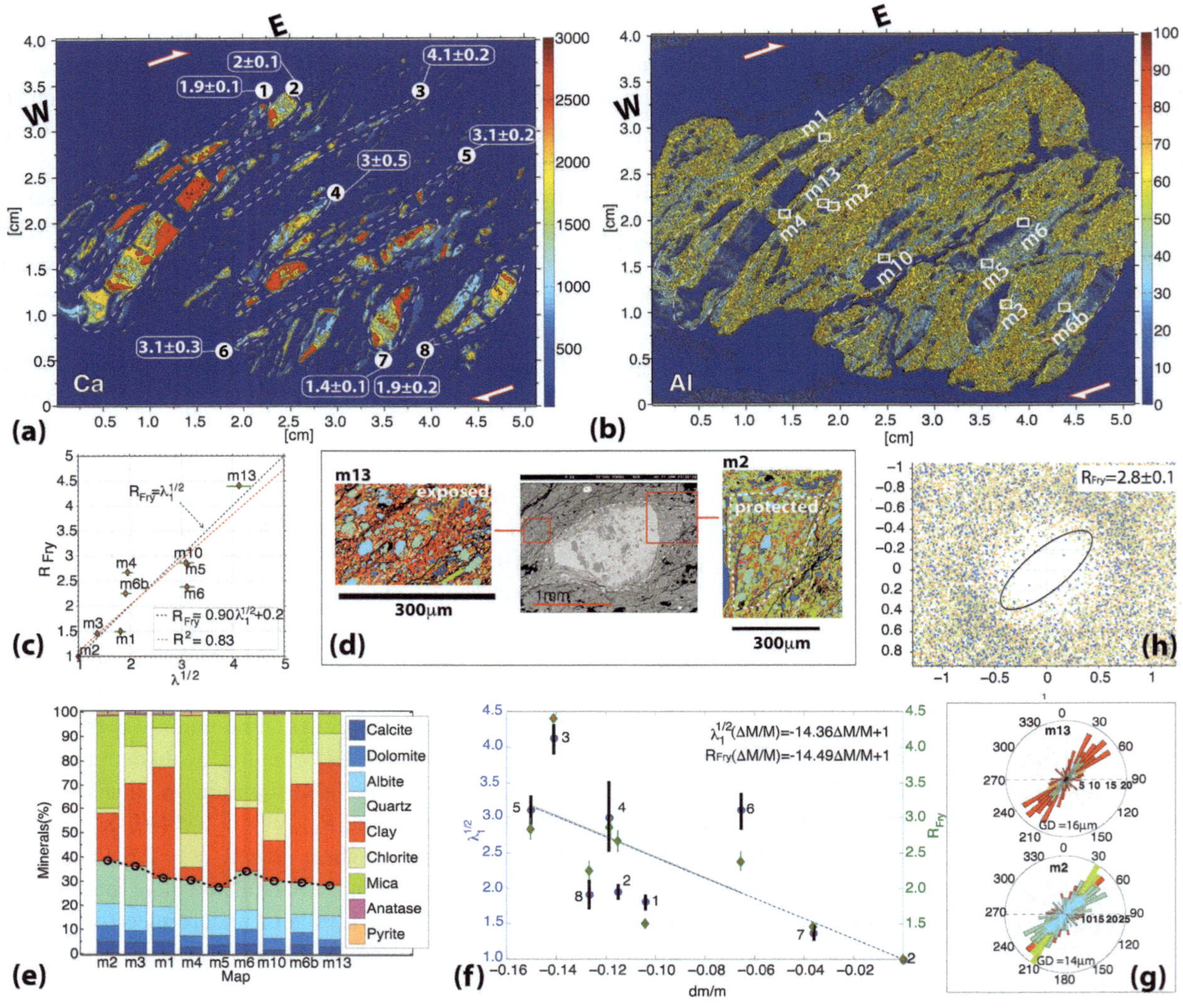

Figure 6

Yazioren outcrop. **a** Calcium map measured by XRF showing boudinage with variation in extensional strain in the horizontal displacement plane (*ac*). **b** Aluminum map measured by XRF showing the foliation with the locations of the EPMA mineral maps. In **a**, **b** color bars indicate relative content. **c** Correlation between the $\sqrt{\lambda_1}$ extension of the boudinage and R_{Fry} of the nearby matrix (with map numbers). **d** SEM-BSE image and EPMA mineral maps, showing protected (right) and exposed (left) zones. The color coding for the minerals is given in the legend of **e**. **e** Comparative mineral compositions; the dashed line separates soluble (mobile) and insoluble (non-mobile) minerals. **f** Extension values $\sqrt{\lambda_1}$ (from boudinage with numbers) and R_{Fry} (from EPMA mineral maps) versus the relative mass change (uncertainty is lower than the symbol size). **g** Rose diagrams of grain orientations from EPMA mineral maps with mean grain diameters (GD) and orientation with respect to the North. **h** Fry diagram at centimeter scale on the whole thin section from XRF data using the minerals in the legend in **e**. The shear arrows are indicative of the sense of shear and do not show an accurate orientation of the shear direction, which is not available on the outcrop where the sample was collected

fine-grained thin structures were probably not taken into account well enough by the Fry method.

This problem was investigated by a statistical analysis of the Al grain clusters. More precisely, relying on the detailed observations of the grain orientation and associated Fry ellipses (Fig. 7d–f), the distribution of the orientation of elongated grain clusters containing aluminum was fitted with two normal probability density functions. At the decimeter scale (Fig. 7b), the cleavage foliation is overall

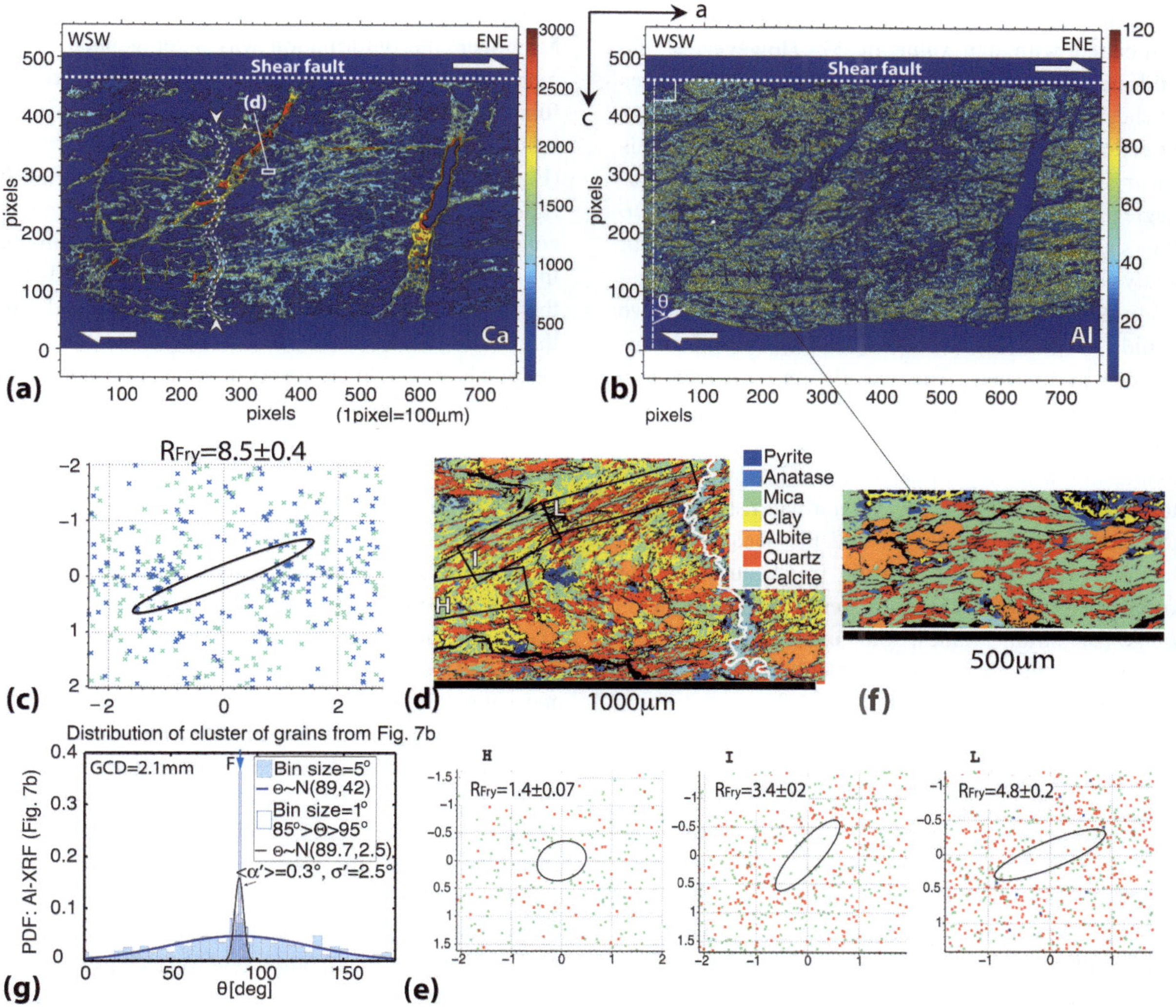

Figure 7

Mülayim outcrop. **a** Calcium map in the horizontal displacement plane (*ac*) measured by XRF with a folded calcite vein (ptygmatic vein) shown between arrows along dashed white lines. **b** Aluminum map in the same sample indicates sub-parallel cleavage foliation rich in phyllosilicates. In **a**, **b**, color bars indicate relative content. **c** Fry diagram based on XRF maps (decimeter scale, **a**, **b**) using the minerals in the legend in **d**: anatase and mica. **d** Mineral maps with fold separated by masks in the hinge, the intermediate area and the limb with different cleavage angles (locations are given in **a**. **e** Fry diagram based on the selected zones of the mineral map with increasing values of R_{Fry} from hinge to limb (millimeter scale) using the minerals in the legend in **d**. **f** Mineral map in a highly localized deformation zone sub-parallel to the shear displacement micas and embedded fine-grained quartz. **g** Orientation distribution of clusters of grains containing Al (**b**) fitted by two normal probability density functions (PDF), with the mean diameter of the grain clusters containing aluminum (GCD). Parameters of the two normal distributions are $\langle\alpha'\rangle$ the mean angle with the fault (F) and $\langle\sigma'\rangle$ the standard deviation: $\langle\alpha'\rangle = 1°$ and 0.3° and $\sigma' = 42°$ and 2.5° for the wide and narrow distributions, respectively

sub-parallel to the fault at the boundary of the sample. In this ductile context, this fault is presumably parallel to the shear direction and, by measuring the angle θ between the line normal to the fault and the long axis of elongated grain clusters, the amplitude of shear strain is evaluated: the angle between the long axis of elongated grains (assimilated to the cleavage plane) and shear direction (a) is α' when $\alpha = 45°$ (initial cleavage) in Eq. 2. This calculation corresponds to the wider normal distribution in Fig. 7g that gives an angle $\alpha' = 1°$ associated with a γ value of 57. However, at the millimeter scale (Fig. 7d, f), we described two types of shear bands: those with grains oblique to the shear direction and those with grains sub-parallel to this shear direction. To better constrain the value of the α' angle in the shear bands with grains sub-parallel to the shear direction, we chose to concentrate our analysis on these bands (with angles ranging from $-5°$ to $+5°$ with the shear direction). Doing so, we avoid the possible perturbing effect of the grains oblique to the shear. This calculation corresponds to the narrower normal distribution in Fig. 7g that gives the angle $\alpha' = 0.3°$. This value of the α' angle is associated with a γ value of 190. However, uncertainty on the angle between the normal distribution of the elongated grains cluster and the shear displacement is probably $> 1°$, leading to γ values ranging from 40 to > 200. The coexistence of such a wide range of strain values, the uncertainty of their evaluation and their use to calculate the aseismic displacement is discussed in Sect. 5.2.

4.6. *Gerede Outcrop (Creeping Section)*

The Gerede outcrop (Figs. 1, 8b) is an abandoned quarry at the western end of the Ismetpasa creeping segment (Hussain et al. 2016; Cetin et al. 2014). It is located 1 km south of the present-day active fault zone with several other major strike-slip faults in between mostly buried under Quaternary sediments (Kaduri et al. 2017). The quarry presents a continuous 150-m-wide outcrop (Fig. 8a) that reveals a network of 5–30-cm-thick clay-rich soft gouges with parallel-to-the-fault cleavage foliation (Fig. 8a, c, d). These faults cross the volcanic units of the Galatia massif (Adiyaman et al. 2001), which are a mélange of dacite, andesite and trachy-basalts (Wilson et al. 1997). These different units are difficult to distinguish. In the gouges, they have been transformed by deformation coupled to low temperature metamorphism (Kaduri et al. 2017). In the damage zones, they have been highly damaged with fractured blocks and dense networks of vertical veins in all directions (Fig. 8e). Such sealed fractures were related to episodic inflow events of carbonate-rich fluids that were associated with the successive earthquakes that affected this section of the NAF (Kaduri et al. 2017). Moreover, the deformation was heterogeneous with more or less rigid fractured blocks surrounded by foliated zones with intense deformation displaying a braided fault pattern at all scales from the thin section (Fig. 5a) to regional scale (Fig. 8a). The size of such rigid blocks decreased drastically in gouges and in zones with cleavage sub-parallel to the shear fault but the boundaries of such zones were undulated and their widths varied along the strike (Fig. 8a–d). In thin sections, only some islands of volcanic rocks can be seen between the networks of veins. Consequently, it was not possible to evaluate the strain and shear values using particle distribution as in the other studied outcrops.

However, it was possible to evaluate the shear values from the angles between the cleavage and the shear fault planes, similar to what was done at the microstructural scale for the Mülayim outcrop (Fig. 7g). Such angles can be measured either directly in the field or in thin sections. It was not possible to measure the variations of these angles along a continuous path over the entire 150 m of the outcrop. However, the damage zones appeared to be more layered in areas near the gouges than away from them (Fig. 8c, d). Thin sections showed that, in these areas near the gouges, the cleavage orientation was sub-parallel to the shear fault plane (Fig. 8f–h), whereas away from the gouge the cleavage was oblique to the fault (Fig. 8e), as in other outcrops (Figs. 5f, 6g). Such change of cleavage orientation being rather sharp, it was possible to evaluate the width of the shear zones characterized by the parallel-to-the-fault cleavage with an uncertainty of about 10% in the field. Such shear zones include all the gouges plus damage zones around them that have the same structural aspect as that seen on thin sections (Fig. 8f, g). These shear zones are highlighted in near-transparent white in Fig. 8a–d, and a cumulative width of 6.0 ± 0.6 m was found along the 150 m width of the quarry. The ratio between this cumulative width of shear zones with parallel-to-the-fault cleavage and the total 150-m width of the outcrop may be evaluated and is equal to 0.04 ± 0.004. This ratio can be considered a representative value at a more regional scale (100–1000 m). The shear strain values associated with cleavage sub-parallel to the

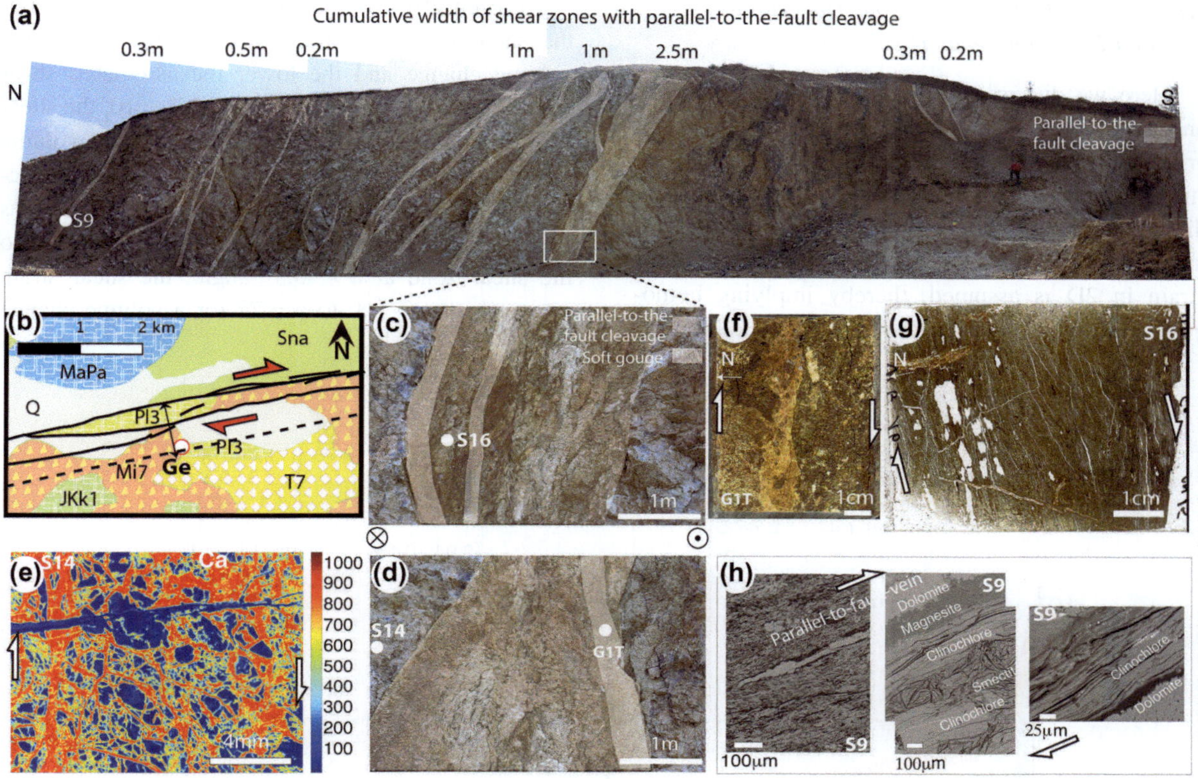

Figure 8

Gerede outcrop. **a** Photograph of the 150-m-wide outcrop showing the fault network and the location and widths of zones (shaded in white) with parallel-to-the-fault cleavage foliation. **b** Geologic map with the trace of the main fault in the outcrop (dashed line), *Q* Quaternary, *Pl3* Pleistocene conglomerate sandstone, *M7* Miocene agglomerate tuff, *T7* Eocene andesite basalt, *MaPa* Cretaceous limestone, Sna Cretaceous limestone, *JKk1* Jurassic limestone. **c** Zone with parallel-to-the-fault cleavage and soft gouge in the upper part of the quarry: see location in **a**. **d** Zone with parallel-to-the-fault cleavage and soft gouge in the lower part of the quarry (not seen in a). **e** Element map of Ca using XRF with typical network of calcite-dolomite veins parallel and perpendicular to the cleavage. The blue perpendicular-to-the-fault vein is a late magnesite vein. **f** Parallel-to-the-fault cleavage in a thin section of soft gouge. **h** Parallel-to-the-fault cleavage in a thin section near the gouge (see location in **c**). **g** Parallel-to-the-fault cleavage in the SEM image with dolomite and magnesite veins parallel to the fault and cleavage foliation (layers of clinochlore and smectite)

fault were not easy to evaluate and will be discussed in Sect. 5.2. These strain values and the width of the shear zones with fault parallel cleavage were used to estimate the total fault displacement accommodated by creep during the geologic times in Sect. 5.3.

5. Discussion

Several parameters are needed to calculate the displacement associated with the creep processes in the NAF shear zone: the quadratic extension and contraction strain values, the volume change, that allow calculating the shear strain when data of all outcrops are combined (Fig. 9) and the width of the shear zone that allows calculating the shear displacement (Eq. 4). The evaluation of these parameters is discussed below, as well as their uncertainty. In Sect. 5.1, we discuss the volume change in the creeping sections and its effect on shear displacement. In Sect. 5.2, we discuss the strain and width evaluation. We evaluate the total displacement in locked sections. We discuss the effect of heterogeneous deformation in creeping sections, and we propose a model of two normal shear strain distributions. Finally, the modeling of the total aseismic displacement along the creeping sections is discussed in Sect. 5.3.

5.1. *Volume change Evaluation: Uncertainty and Effect on Shear Displacement*

Here we discuss the relationship between strain and mass-volume change because volume change is an important parameter of the relation between total strain and shear strain (Eq. 3). These relations are discussed based on a "strain-ratio/shear-strain/volume-change" diagram (Fig. 9). A horizontal plane strain in 2D is assumed, thereby implying homogeneity along the y-axis (Ramsay 1980; Heilbronner and Barrett 2014; Fossen and Cavalcante 2017). For the same measured strain ratio, R, the higher the volume change and the lower the associated shear strain are (Fig. 9). For example, for $R = 10$, the shear strain ranges from $\gamma = 2.9$ to $\gamma = 2.0$ for a volume change ranging from $\Delta = 0$ to $\Delta = -0.5$, respectively (Fig. 9). For a higher strain ratio, such as $R = 10{,}000$, when cleavage becomes sub-parallel to the shear zone at a $< 0.5°$ angle, the shear strain ranges from $\gamma = 100$ to $\gamma = 72$ for a volume change

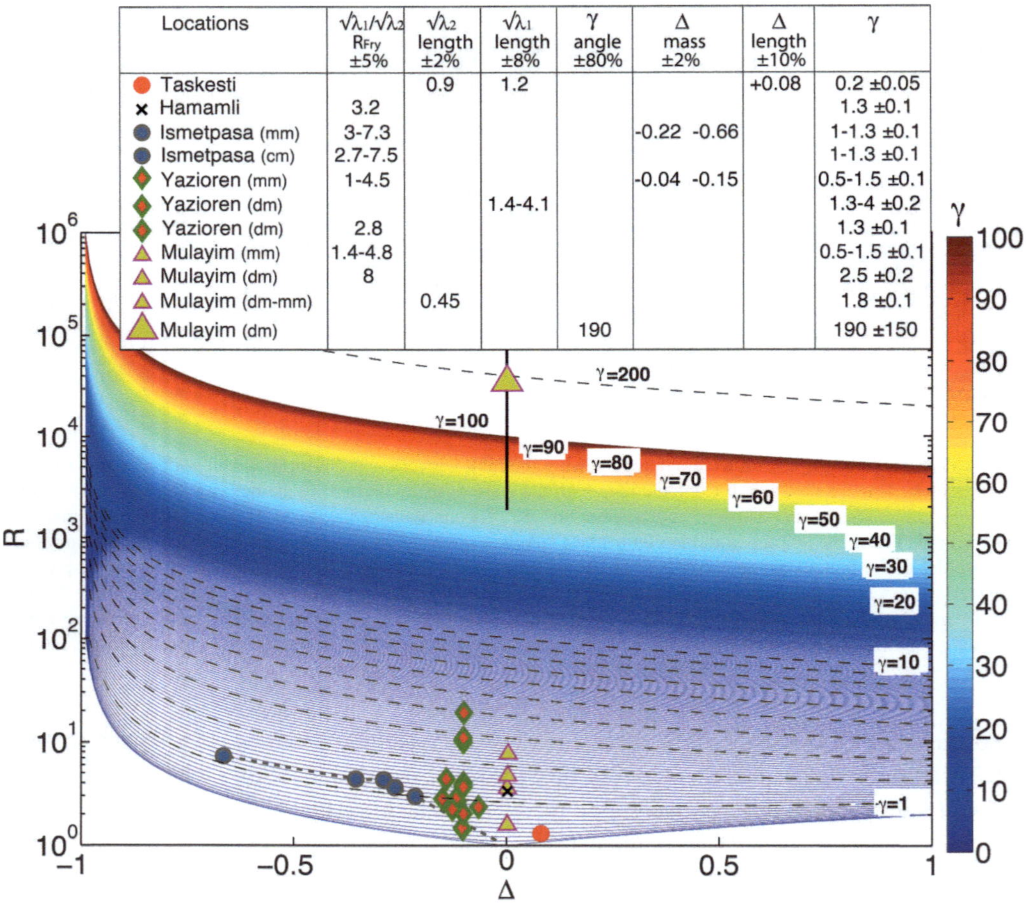

Locations	$\sqrt{\lambda_1}/\sqrt{\lambda_2}$ R_{Fry} ±5%	$\sqrt{\lambda_2}$ length ±2%	$\sqrt{\lambda_1}$ length ±8%	γ angle ±80%	Δ mass ±2%	Δ length ±10%	γ
Taskesti		0.9	1.2			+0.08	0.2 ±0.05
Hamamli	3.2						1.3 ±0.1
Ismetpasa (mm)	3-7.3				-0.22 -0.66		1-1.3 ±0.1
Ismetpasa (cm)	2.7-7.5						1-1.3 ±0.1
Yazioren (mm)	1-4.5				-0.04 -0.15		0.5-1.5 ±0.1
Yazioren (dm)			1.4-4.1				1.3-4 ±0.2
Yazioren (dm)	2.8						1.3 ±0.1
Mulayim (mm)	1.4-4.8						0.5-1.5 ±0.1
Mulayim (dm)	8						2.5 ±0.2
Mulayim (dm-mm)		0.45					1.8 ±0.1
Mulayim (dm)				190			190 ±150

Figure 9

Plot of finite strain and mass volume change raw measurements from various microstructural methods in the displacement plane (*ac*). The principal strain axis aspect ratio (R) is plotted versus the relative mass and volume change (Δ) and the finite shear strain (lines of various colors from $\gamma = 0$—blue- to $\gamma = 100$—dark red). Highly contrasted shear strain values γ are found at the Mülayim site, depending on the measurement methods, ranging from 0.5–2.5 to 40–> 200 using either the Fry method (small triangles) or the angle between the cleavage and shear zone (large triangle), respectively. Other symbols refer to sites in Fig. 3 (Taşkesti), Fig. 4 (Hamamli), Fig. 5 (Ismetpasa), Fig. 6 (Yazioren) and Fig. 7 (Mülayim). The size of the symbols is larger than the uncertainty except for the high shear strain value deduced from angle measurement where error bars are indicated. The table at the top of the figure summarizes the strain values at different sites and for different methods: R_{Fry}, measures of length or angle changes, measures of volume changes from chemical content or length changes with the mean percentage uncertainty and calculated shear strain values with specific uncertainty. In all cases only the range of values is given for simplicity ranging from $\Delta = 0$ to $\Delta = -0.5$, respectively (Fig. 9). Despite such variations of strain for such large variation from $\Delta = 0$ to $\Delta = -0.5$, it does not change the order of magnitude of the shear strain values. Notably, the calculated mass change is a relative value. It is obtained by comparing a deformed "exposed" area with a "protected" area that is either undeformed (in the best case) or is just less deformed. This is why we also evaluated the deformation of the protected zone from grain geometry and orientation analysis. Moreover, as the rock density does not significantly change, mass change is equivalent to volume change.

As previously discussed in Sect. 2.2, the six samples presented here are representative of all visited sites along the NAF shear zone (Fig. 1): Taşkesty in limestone, Hamamli in serpentine and the four other sites (Ismetpasa, Yazioren, Mülayim, Gerede) in various rocks with volcanic origin.

All the sites in limestone show several narrow parallel-to-the-fault shear zones of decimeter width with numerous stylolites. The Taşkesti sampling is well representative of all these shear zones for which the deformation near the fault measured at decimeter scale shows no significant volume change with $\Delta = 0.08$ (Fig. 3). Observations (Fig. 3) always show clear evidence of alternating seismic (fractures) and aseismic (stylolites) processes (Kaduri et al. 2017). However, the aseismic part remains very low, with mean quadratic values of $\sqrt{\lambda_1} = 1.2 \pm 0.1$ and $\sqrt{\lambda_2} = 0.9 \pm 0.02$ leading to a low shear strain value $\gamma = 0.2 \pm 0.05$, a value not significantly affected by uncertainties on volume change.

At the Hamamli serpentine site, it was not possible to calculate a volume change as no area less deformed than any other was found. This difficulty is a general problem of volume change calculation in serpentine where heterogeneities are rare (Andreani et al. 2005). In Mülayim, the same absence of a protected zone prevented us from evaluating the volume change. The corresponding strain values at these sites are then plotted in Fig. 9 assuming no volume change.

At the other sites, volume change was measured by comparative chemical analyses (Figs. 5, 6) at the millimeter scale. These calculations assume that in the deformed areas the soluble minerals dissolve and are removed. A minor part of these soluble minerals, as some feldspars, were transformed into newly crystallized phyllosilicate minerals rather than removed, however, this does not significantly alter the calculations (Kaduri et al. 2017). There is a clear correlation between strain and volume change indicating that the same pressure solution mechanism is responsible for the deformation with the development of parallel-to-the-cleavage tectonic layering. When the protected zone is in an undeformed state, as in Ismetpasa (Fig. 5), the volume change varies from one layer to another from 0.22 ± 0.005 to more than 0.66 ± 0.01. When the protected zone is slightly deformed, as in Yazioren (Fig. 6), the variations are lower and the uncertainty is much higher. This pattern of successive perpendicular-to-the-compressive-stress layers with various compositions is typical of a pressure solution self-organized process that has been reproduced experimentally (Gratier et al. 2015). This observation was made in the 25 investigated outcrops in volcanic and analogous rocks of the NAF creeping shear zone. It indicates a very slow ductile and aseismic deformation (Gratier et al. 2013). To extrapolate this quantitative result to the whole shear zone, some complementary geologic observations are needed. No traces of recrystallization of the soluble minerals (quartz and feldspar) are found in the shear zone, so it is likely that the whole shear zone was developed with a decrease of volume at least during the early time of its development. However, later in the NAF development, numerous carbonate veins crosscut some of the damage zone as seen in Gerede (Fig. 8e). As such carbonates need to come from outside since they were not present in the initial rocks, they may have contributed to a local increase of volume. Consequently, to evaluate shear strain values representative of the entire NAF shear zone in volcanic and analogous rocks (Fig. 9), we used a median regional value of the volume change with a large uncertainty, $\Delta = -0.20 \pm 0.15$.

Finally, the possibility of much higher volume changes that we may have missed must be discussed because large volume change values below $\Delta = -0.8$ have a significant effect on the calculation of shear strain (Fig. 9). A common observation is that volume change associated with pressure solution is limited to about $\Delta = -0.7$. This is due to several effects, especially the progressive decrease of the soluble mineral content and progressive isolation in the soft matrix of insoluble species that render its stress-driven dissolution increasingly difficult (Gratier et al. 2013). Moreover, there is another effect in shear zones that limits the volume change, linked to the rotation of the cleavage. The pressure solution initiates along the solution cleavage planes near rigid objects (Fig. 5a) and develops progressively further away at 45° from the shearing direction at the beginning of the shearing process. At this stage, the shearing displacement contributes to the tectonic layering process with a volume decrease. However, at a later stage with the rotation of the cleavage that becomes sub-parallel to the shear zone, the displacement is decreasingly dependent on the volume reduction perpendicular to the cleavage. When the cleavage is sub-parallel to the shear zone (Fig. 7f), only the effect of the perpendicular-to-the-shear zone compaction contributes to the dissolution. At this stage, large sliding along the parallel-to-the-fault cleavage must switch to grain boundary sliding. This mechanism can always be a pressure solution process at grain size scale in order to accommodate relative grain displacement (Ashby and Verrall 1973). However, the transfer distance is reduced to the size of the asperities along the grain boundaries. Consequently, the strain rate of such viscous behavior increases significantly for a given stress state (Gratier et al. 2013). This effect contributes to the strain localization parallel to the shear zones. The system can then evolve to a combined frictional and viscous flow behavior (Bos and Spiers 2002) or can depend only on the friction of individual minerals and on the geometry of their assemblage if mass transfer is no longer possible (Lockner et al. 2011; Sone et al. 2012; Carpenter et al. 2016). In all cases, this effect increases the segregation process between shear zones with cleavage oblique to the shear displacement and shear zones with cleavage sub-parallel to this displacement (Fig. 10a). The implication for aseismic displacement estimate is discussed in the next section.

5.2. *Strain and Width Evaluation: Uncertainty, Heterogeneity and Combination of Two Normal Shear Strain Distributions in Creeping Sections*

Uncertainties on strain measurements are discussed in the methodologic Sect. 3 and are given for each measured value in Sect. 4 and Fig. 9. However, such uncertainties are valid for homogeneous deformation. The effect of the heterogeneity of the deformation and the validity of extrapolating measurements from thin sections to the entire shear zone are discussed below and integrated in a normal distribution model of the shear strain, which is used to calculate the aseismic displacement in the next Sect. 5.3.

Several independent strain measurements on several different outcrops show the same trend: data are clustered around the same values in Fig. 9. However, observation shows variation in measured strain values: the ratio between the principal strain axes (R) ranges from close to 1 in the massive limestone of Taşkesti (Fig. 3) to 8.5 (Mülayim, Fig. 7) and 7.5–19 (Ismetpasa, Fig. 4) in volcanic and analogous rocks. The contrasting behavior of the limestone (locked sections) versus the volcanic rocks (creeping sections) has been discussed in another study (Kaduri et al. 2017) and is only briefly summarized here. In the locked sections, using the low shear strain value of $\gamma = 0.2 \pm 0.05$ for the ductile shear zones and their cumulative width of a few meters leads to calculating a total displacement of few meters, indicating that the creep-related displacement in locked sections is almost negligible. This effect is explained by the contrasting behavior of monomineralic (limestone) rocks and polymineralic (volcanic) rocks. Diffusion at the contact surface between two identical grain minerals as in limestone, whose boundaries are most often healed, is much slower than diffusion at the contact surface between minerals of a different nature as in volcanic rocks, where the fluid phase boundary is more continuous (Zubtsov et al. 2004). The difference in behavior is even greater if one of the minerals is a phyllosilicate, which prevents healing (Bos and Spiers 2000; Zubtsov et al. 2004) or activates dissolution (Renard et al. 2001). This concept that creep deformation is controlled by lithology is supported by a wide range of

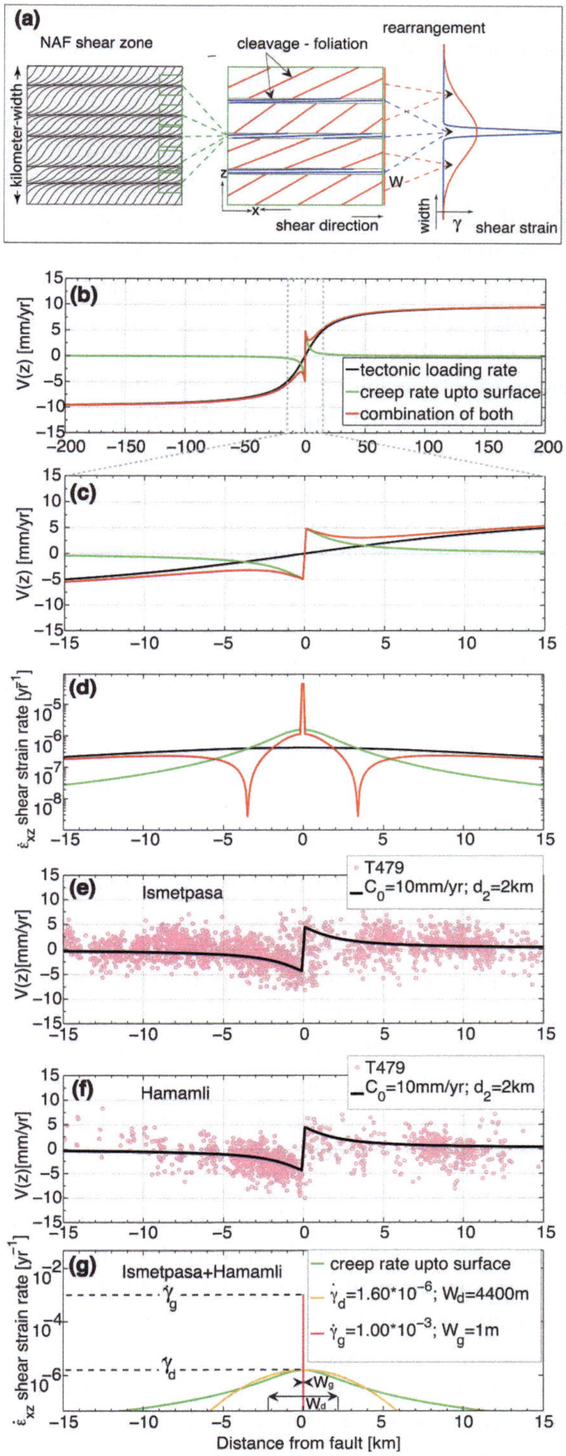

◀Figure 10
Aseismic strain along the North Anatolian Fault. **a** Schematic representation of the strain distribution in the creeping zone, which is made up of parallel zones of various strain values representative of Fig. 8a, b from left to right, respectively, and that are represented on the right by two normal (i.e., Gaussian) distributions of the shear strain. **b** Horizontal surface velocity as a function of distance from the fault (black profile is for a fault locked at the surface and creeping at depth; green profile is for a fault creeping at the surface and locked at depth; red profile is for a combination of both). **c** Zoom on **b**. **d** Surface strain rate as a function of distance from the fault, derived from **c**. **e**, **f** Geodetic horizontal creep rate profiles perpendicular to the North Anatolian Fault across Ismetpasa and Hamamli outcrops, respectively; the data (dots) and model (black lines) are taken from Cetin et al. (2014). **g** Strain rate profiles derived from creep rate models in **e**, **f**. These profiles are fitted with the superposition of two normal distributions to model the shear strains and the widths of the gouge and the damage zones for Ismetpasa-Hamamli outcrops

observations in several major creeping faults (Graymer et al. 2005; Sone et al. 2012; Richard et al. 2014; Thomas et al. 2014; Kaduri et al. 2017).

However, in the creeping sections in volcanic or analogous rocks even at the same site the strain values vary widely, for example, from $R = 1.4$ to 8.5 in Mülayim from the Fry method. This strain ratio is even much greater, up to $R = 10000$, when derived from the angles of the cleavage with the shear zone as in Mülayim (Figs. 7g, 9). Such a heterogeneous pattern is compatible with the development of the pressure solution layering process discussed in the preceding section (Sect. 5.1). Not only the volume change but also the strain values vary from one layer to another in a shear zone. This pattern is schematized in Fig. 10a. On the left panel, a sketch of the whole kilometer-width NAF shear zone is presented with several faults and parallel-to-the-displacement shear zones (as in Fig. 8b). In the middle panel, the hectometer scale is displayed with alternating shear zones with cleavage oblique or sub-parallel to the shear displacement (as in Fig. 8a). Finally, on the right panel, we propose a combination of two normal (i.e., Gaussian) shear strain distributions at the scale of the entire NAF shear zone that can be used to evaluate the total aseismic displacement. This model combines (1) a first normal distribution that represents the low strain and wide shear zones with cleavage oblique to the fault zone, with γ_{wsz} and W_{wsz} as maximum shear strain and cumulative width characteristics, respectively, and (2) a second normal distribution that represents the high strain and narrow shear zones with cleavage sub-parallel to the fault zone, with γ_{nsz} and W_{nsz} as maximum shear strain and cumulative width characteristics. We discuss below how strain measurements on thin sections are extended to the regional scale with the associated uncertainties and evaluation of the cumulative width of these two types of shear zones.

5.2.1 Shear Zones with Cleavage Oblique to the Shear Displacement

The width of the shear zone W_{wsz} is taken as $2\sigma_{\mathrm{wsz}}$, where σ_{wsz} is the standard deviation of the normal distribution. There are few possible measurements of the width of this large zone of shearing. Evidence of several parallel faults indicating kilometer width of deformation is seen on all the geologic maps (Herece and Akay 2003, Emre et al. 2013, 2016, Akbaş et al. 2016) or in Gerede (Fig. 8b), where the presently active fault is located 1 km away from the studied creeping zone. The width W_{wsz} is thus considered to be at least 1 km wide. Another kilometer-width value can be obtained from the modeling of geodetic data (see Sect. 5.3).

Uncertainty about strain measurements assuming homogeneous deformation is rather low (see table in Fig. 9). However, heterogeneity is everywhere, amplifying the uncertainty. The normal distribution model implies choosing the highest value of the shear strain γ_{wsz} that is representative of the zones with oblique cleavage (Fig. 10a). For a kilometer-width normal distribution, samples for such a γ value representing some meters in size seem realistic. We did not perform strain measurement at this scale but will consider that measurements on large thin sections at the decimeter scale are representative of this meter scale. The higher strain values at thin section scale are from Mülayim and Yazioren sites. Use of the Fry method gives a strain ratio $R = 8.5 \pm 0.4$ in Mülayim leading to $\gamma = 2.6 \pm 0.2$ (Fig. 7c). This value is a minimum because of the limitation of the method to evaluate high strain values because of the difficulty of measuring a very elongated elliptical void in a cloud of points. In Yazioren, it is possible to compare two different methods: boudinage measurements give shear strain values in the range $\gamma = 1.3 \pm 0.1$ to 4.0 ± 0.2, whereas the Fry method gives $\gamma = 1.3 \pm 0.1$ (Fig. 6). The two methods do not rely on the same markers, and the heterogeneity of the deformation with undeformed truncated blocks and highly localized deformation is probably not well accounted for by the Fry method. The boudinage measurement is also limited because with large extension it is difficult to link the truncated blocks. Moreover, the boudinage measurement distributions should also be discussed. As for extension deduced from truncated belemnites (Ramsay 1967), one can either consider that the truncated objects were heterogeneously fragmented in homogeneous deformation, and then the maximum extension values are the most representative of the whole deformation, or consider that the deformation was heterogeneous and a mean value is best representative of the whole deformation (Gratier et al. 2013). Here, as we clearly see heterogeneous shear deformation at all scales, it seems more realistic to take a mean value of the boudinage measurement of $\sqrt{\lambda_1} = 2.6$, leading to a shear strain of about $\gamma = 2.4$ (with $\Delta = -0.1$), close to the value $\gamma = 2.6 \pm 0.2$ (with $\Delta = 0$) measured in Mülayim. These values must be mitigated by the effect of the volume change ($\Delta = -0.2 \pm 0.15$) as discussed before (Sect. 5.1), leading to a range of shear strain of $\gamma_{\mathrm{wsz}} = 1.9 - 2.7$ that will be used in Sect. 5.3.

5.2.2 Shear Zones with Cleavage Sub-parallel to the Shear Displacement

Zones of cleavage sub-parallel to the shear displacement are revealed at thin section scale by two types of observations. The first one results from the statistical analysis of the orientation of elongated cluster of grains, for which the long axis can be assimilated to cleavage planes, which shows normal distribution with very low angles with the shear plane ($\alpha' = 0.3°$); see Fig. 7g. Another key observation is that the shear strain values deduced from particles and length change measurements are sometime higher at the decimeter than millimeter scale (Fig. 7). This observation implies the existence of some highly localized zones with very high strain values associated with cleavage sub-parallel to shear displacement (Fig. 7f) that cannot be captured by such measurements. At the regional scale, as shown in the Gerede outcrop, there is clear evidence of zones with cleavage sub-parallel to the main shear zone displacement among zones with cleavage oblique to the shear displacement (Fig. 8). These intensively deformed zones include gouges but also some parts of the damage zones that show cleavage planes sub-parallel to the shear zone (Fig. 8f, g). The ratio between the widths of shear zones with parallel-to-the-fault cleavage and the total width has been evaluated in the range 0.04 ± 0.004. This value can be considered as a representative value at a regional scale (100–1000 m). This ratio is used to evaluate the cumulative width of the narrow zones, for example, $W_{nsz} = 40 \pm 4$ m for a total width of 1 km.

Regarding the reliability of the strain measurements to evaluate the maximum value of the normal distribution of the shear strain values, γ_{nsz}, as before, for a kilometer-width normal distribution samples representing some meters in size seem realistic. We used measurements on large thin sections at the decimeter scale as representative of meter size. The statistical analysis made in the Mülayim thin section gives a value of $\gamma_{nsz} = 190$ associated with an angle $a' = 0.3°$ with the shear displacement. However, uncertainty about the angle between the normal distribution of the elongated grain cluster and the shear displacement is probably $> 1°$, leading to a γ_{nsz} value ranging from 40 to > 200 (see Sect. 4.5). Using such large uncertainty in the displacement calculation is very difficult, whereas choosing some arbitrary values does not seem realistic either. Moreover, we must also integrate the fact that we were not able to evaluate the volume change in this thin section because of the lack of protected zones representative of the initial state, despite observations of mass transfer, so we took $\Delta = 0$. However, according to other measurements as in Fig. 5d, it is likely that the volume change in the shear zones with cleavage sub-parallel to the displacement must be very large, up to $\Delta = -0.6$. With such large volume change, the shear strain value of 190 decreases down to near 100 (Fig. 9). Therefore, we finally choose $\gamma_{nsz} = 100$ as a conservative representative order of magnitude. Then, after integrating all the other parameters with lower uncertainty in the modeling to calculate the total displacement, we will be able to evaluate more precisely the possible range of the shear strain values γ_{nsz} in the zones with cleavage sub-parallel to the shear displacement.

5.3. *Calculation of the Aseismic Part of the Total Displacement Along Creeping Sections of the North Anatolian Fault*

In this section, the relative contribution of aseismic displacement to the total cumulative displacement, along creeping sections of the NAF, is estimated. At the continental scale, rocks deform by both reversible deformation (elastic tectonic loading) and irreversible deformation (ductile or brittle) that involves plastic dissipative processes and depends on the system size (Karmakar et al. 2010). We propose using a combination of geodetic and geologic data to evaluate the width of the creeping zone, which we combine with shear strain measurements at the outcrop scale to estimate the aseismic part of the total displacement on the fault (Ramsay 1980) (Eq. 4).

5.3.1 *Modeling Creep Behavior at the Present Time*

Geodetic data used in this section are horizontal surface velocity maps across the NAF taken from the InSAR time series analysis of Cetin et al. (2014). These maps are based on Envisat satellite data acquired between 2003 and 2010. Horizontal velocity profiles perpendicular to the fault were extracted from these maps and modeled using:

$$V(z) = \frac{S}{\pi}\arctan\left(\frac{z}{d_1}\right) + C\left(\frac{1}{\pi}\arctan\left(\frac{z}{d_2}\right) - H(z)\right), \tag{5}$$

where z is the distance perpendicular to the fault. Equation (5) is derived from the classical model of a dislocation into an elastic space of Weertman and Weertman (1964) and Savage and Burford (1973). This equation models surface horizontal velocities V across a strike slip fault locked from the surface to a locking depth d_1 and sliding at the tectonic loading rate S below (first term of the equation). It was modified to take into account creep at velocity C from the surface down to a depth d_2 (second term of the equation), as in Hussain et al. (2016). This second term of the equation including the Heaviside step function H thus represents the expected velocity step across fault due to shallow creep.

The corresponding shear strain rate $\dot{\varepsilon}_{xz}$ along the profile is the spatial derivative of the velocity profile:

$$\dot{\varepsilon}_{xz} = \frac{\partial V(z)}{\partial z} = \frac{S}{\pi} \cdot \frac{d_1}{d_1^2 + z^2} + C\left(\frac{1}{\pi} \cdot \frac{d_2}{d_2^2 + z^2} - \delta(z)\right), \tag{6}$$

where x is the distance along the fault and z along the width of the shear zone (see the coordinate system in Fig. 2a).

Figure 10b, c shows the theoretical surface velocity profiles expected across a fault that: (1) slides at a rate of 20 mm/year < 15 km (tectonic loading) and is locked above, (2) creeps between the surface and a depth of 2 km at a rate of 10 mm/year and is locked below between 2 and 15 km and (3) combines both slip processes. These values correspond to the averaged values of the tectonic loading rate (20 mm/year), locking depth (15 km), creep rate (10 mm/year) and depth of creep (2 km), at the longitudes of Ismetpasa and Hamamli sampling sites along the creeping section of the NAF. The first two parameters (tectonic loading rate and locking depth) are based on GPS measurements (Reilinger et al. 2006). The last two (rate and depth of creep) are from the creep rate distribution model shown in Fig. 9 of Cetin et al. (2014) (from inversion of InSAR data). The corresponding theoretical strain rate profiles are also derived in Fig. 10d.

Figure 10e, f shows two horizontal velocity profiles derived from Cetin et al. (2014), perpendicular to the fault (30 km-wide and 240 km-long) and intersecting it at the Ismetpasa and Hamamli sites. Data are corrected from the tectonic loading at depth, from (Reilinger et al. 2006), so that the profiles correspond only to the shallow creep signal. Superimposed on these profiles are models from Cetin et al. (2014), with similar parameters as for Fig. 10b, c [black lines in Fig. 10e, f are computed from the right-hand term of Eq. (5) considering only shallow creep]. The creep rate is considered to be uniform from the surface to a given depth (10 mm/year down to 2 km depth). Given that the tectonic loading rate is about 25 mm/year on average for the entire North Anatolian Fault (Reilinger et al. 2006; Cetin et al. 2014) and 20 mm/year near Ismetpasa (Cetin et al. 2014), the present-day shallow creep accommodates about 40–50% of the tectonic loading.

These models on Fig. 10e, f are derived to obtain shear strain rate profiles (Fig. 10g) at the sampling sites in order to:

1. Evaluate the relevance of the proposed combination of two normal (i.e., Gaussian) shear strain distributions (Fig. 10a).
2. Calculate the width of the wide shear zone in the past assuming that it keeps the same width (at least the same order of magnitude) throughout the deformation process.

We propose to fit the geodetic strain rate model of Fig. 10g (green curve) at Ismetpasa and Hamamli with a superposition of two normal distributions representing the cumulative width and shear strain associated with both types of shear zones (e.g., damage and gouge zones in the present time). The shear components can be approximated with normal distributions $\gamma \sim N(0, \sigma^2)$ so that:

$$\gamma(z) = \sum_k \gamma_{0,k} \cdot \mathrm{e}^{-z^2/2\sigma_k^2}, \tag{7}$$

where γ is the shear strain, and index k refers to the type of shear zone. The width of the shear zone is considered to be $W_k = 2\sigma_\mathrm{k}$. Postulating that the shear strain is the only time-dependent variable $\gamma(t)$, the shear strain rate $\dot{\gamma}$ obtained by deriving Eq. (7) over time is $\dot{\gamma}(z) = \sum_k \dot{\gamma}_{0,k} \cdot \mathrm{e}^{-z^2/2\sigma_k^2}$.

Considering that strain rate and width values are independent for the damage zone and the gouge zone, the strain rate and width of the damage zone are first inverted: $\dot{\gamma}_d = 1.6 \times 10^{-6}$ year^{-1}, and $W_\mathrm{d} = 4.4$ km, respectively (Fig. 10g, orange normal distribution). This value of several kilometers for the width of the damage zone is consistent with the kilometer scale fault zone width observed in the field (Fig. 8b). Owing to the low resolution (pixel size 20 m east–west, 4 m north–south) and high noise level (Fig. 10e, f) of InSAR satellite data, it is not possible to provide a strain rate model in the fault near-field that is sufficiently precise to estimate the width of the gouge zone properly. This width is therefore taken as $W_\mathrm{g} = 1$ m, based on field measurements from Kaduri et al. (2017). The strain rate of the gouge zone is finally estimated as follows. The cumulative displacement rate along a fault, $\dot{D}$, is given by the integration of the two normal distributions over a z-

axis perpendicular to the fault plane, in a range $(-\infty, +\infty)$ (placing Eq. 7 in Eq. 4):

$$\dot{D} = \dot{\gamma}_g \times \int_{-\infty}^{\infty} e^{-z^2/2\sigma_g^2} dz + \dot{\gamma}_d \times \int_{-\infty}^{\infty} e^{-z^2/2\sigma_d^2} dz = \dot{\gamma}_g \times \sigma_g \times \sqrt{2\pi} + \dot{\gamma}_d \times \sigma_d \times \sqrt{2\pi} \quad (8)$$

Assuming that $\dot{D}$ is 10 mm/year based on the InSAR analysis of Cetin et al. (2014), (Fig. 10c, d), and given the values of $\dot{\gamma}_d$ and W_d obtained above, it is estimated that $\dot{\gamma}_g = 0.89 \times 10^{-3}$ year^{-1} from Eq. (8), with $W_k = 2\sigma_k$. It is worth noting that this partition of the displacement rate between damage zone and gouge is strongly dependent on the imposed width of the gouge, and significant detailed improvements of this partition would rely on a better spatial definition of present-day geodetic measurements. However, this formalism has the merit of representing in a consistent framework both structural geologic and geodesic data, which have different spatial and temporal scales.

5.3.2 *Modeling the Creeping Behavior in the Past Since the Beginning of Fault Sliding*

As discussed in Sect. 5.2, two types of strain distributions must be taken into account when calculating the displacement associated with creep processes since the beginning of fault sliding (Fig. 10a). First, a normal distribution represents the low strain and wide shear zones with cleavage oblique to the fault zone, with γ_{wsz} and W_{wsz} as maximum shear strain and width characteristics, respectively. Second, another normal distribution represents the high strain and narrow shear zones with cleavage sub-parallel to the fault zone, with γ_{nsz} and W_{nsz} as maximum shear strain and width characteristics, respectively. Based on geodetic data, the width of the wide shear zone W_{wsz} is evaluated to be 4.4 km (Fig. 10g). As discussed in Sect. 5.2, geologic observations near Gerede lead to a minimum width of the fault zone of 1 km (Fig. 8b). Strain measurements on field samples give shear strain values ranging from $\gamma_{wsz} = 1.9$ to 2.7 when integrating the uncertainty on strain measurements and volume change at the scale of the NAF shear zone (see Sect. 5.2). For the narrow shear zones, as discussed in Sect. 5.2, the width of the narrow zones can vary from 0.036 × 1000 m = 36 m to 0.044 × 4400 m = 194 m, and a conservative value of $\gamma_{nsz} = 100$ is used for the shear strain (see Sect. 5.2). All these strain and width values can be used to evaluate the limits on the total displacement associated with aseismic creep by integrating the shear strain values of the two normal distributions across the fault. The maximum and minimum aseismic displacement is calculated using Eq. (8) with γ instead of $\dot{\gamma}$. The maximum displacement results from the combination of the maximum displacement associated with the low strain and wide shear zones (using the maximum values $\gamma_{wsz} = 2.7$and $W_{wsz} =$ 4.4 km, corresponding to σ_{wsz} = 2.2 km), which is 15 km, and of the maximum displacement associated with the high strain and narrow shear zone (maximum values $\gamma_{nsz} = 100$ and W_{nsz} = 194 m (corresponding to σ_{wsz} = 97 m), which is 24 km, leading to a total maximum displacement of 39 km. Similarly, the minimum displacement is estimated using the minimum values of γ_{wsz} and W_{wsz} for the low strain and wide shear zone ($\gamma_{wsz} = 1.9$; W_{wsz} = 1 km), giving 2.4 km of displacement, and the minimum values of γ_{nsz} and W_{nsz} for the high strain and narrow shear zone ($\gamma_{nsz} = 100$; W_{nsz} = 36 m), giving 4.5 km of displacement, leading to a total minimum displacement of 6.9 km. The estimated total aseismic displacement thus ranges between 9 and 49%, of the total shallow displacement of 80 km during geologic times.

The conservative value $\gamma_{nsz} = 100$ can now be discussed in relation to geologic observations. Using shear strain values of $\gamma_{nsz} = 200$ for the zone with cleavage sub-parallel to the shear displacement would mean that nearly 80% of the total displacement was accommodated by aseismic creep, which seems a maximum since we see evidence of earthquakes all along the NAF deformed zone. Using shear strain values of γ_{nsz} less that 70 would mean that the zones with cleavage sub-parallel to the shear zone accommodated less displacement than the wide zones, which seems rather unrealistic. Consequently, these two shear strain values may be considered the upper and lower bound of the shear strain values in zones with cleavage sub-parallel to the shear displacement.

From a more general point of view, this type of calculation reveals the difficulty in quantifying large displacements along regional shear zones. Measurements based on length changes are limited to shear strain below 2.5–4, depending on the methods. Measurements based on angle changes are much more efficient, but they need to be done with reliable statistical analysis that still needs to be developed on the field with numerous cleavage measurements, which was not possible to do here because of the quality of the outcrops.

6. Conclusion

The quantitative evaluation of the aseismic part of the total displacement along the North Anatolian Fault provides methodologic and rheologic insights on strain measurements and fault mechanisms. From a methodologic point of view, it has been shown that the measurement of finite strain in shear zones, which can reach very large strain values, is strongly dependent on the methodology. Methods based on the analysis of particle distribution, such as the Fry method, or length change, such as boudinage measurements, do not allow retrieving shear strains > 2.5–4, respectively. Conversely, using the variations in cleavage angle with the shear zone, very large strain values up to 100 or more can be estimated, with great uncertainty about this value. However, in both cases, evaluation of volume change is important and must be integrated in the calculation.

The relationship between strain and volume change shows that the penetrative cleavage observed in the whole NAF shear zone affecting volcanic or analogous rocks is a pressure solution cleavage that evolves from oblique to sub-parallel to the shear displacement. A combination of two normal (i.e., Gaussian) shear strain distributions is proposed to represent this pattern of deformation in the kilometer-width creeping zones. These zones are partitioned into parallel deformation zones with contrasting shear strain characteristics: (1) wide (10–100 m) shear zones with cleavage oblique to the shear displacement and (2) narrow (0.2–2 m) shear zones with cleavage planes sub-parallel to the shear displacement, with sub-grain shear localization.

Finally, finite shear strain characteristics evaluated from geologic analyses were used in parallel with geodetic data to calculate the total cumulative shallow aseismic displacement that has been accommodated by creep. By modeling the present-day creep behavior, it is possible to propose a model with the two types of shear zones and evaluate the maximum width of the creeping zone. By modeling the creeping behavior since the beginning of fault sliding, it is possible to calculate an aseismic part of the total 80-km displacement in the range 9–49% in the creeping (volcanic and analogous rocks) zone and a negligible displacement in the locked (limestone) zones, showing the crucial effect of the lithology in the seismic versus aseismic partition.

The large uncertainty in the evaluation of the NAF aseismic displacement is partly related to the difficulty of evaluating the large strain values that are associated with large shear displacement. Based on the methodology developed on thin section analysis, a promising way to improve such measurements would be to develop reliable statistical analysis of cleavage orientations in the field.

Acknowledgements

This project was funded by the European Union's Seventh Framework Programme for research, technological development and demonstration under grant agreement no. 316889 (ITN FlowTrans) and by the Norwegian Research Council grant no. 250661 'HADES' to FR. The authors thank Nathaniel Findling, Valerie Magnin and Valentina Batanova for technical support with sample preparation and XRD and EPMA analytical measurements at ISTerre. The authors thank the editor, Yehuda Ben-Zion, and two reviewers for detailed comments and suggestions that significantly improved the manuscript.

References

Adiyaman, Ö., Chorowicz, J., Nicolas Arnaud, O., Niyazi Gündogdu, M., & Gourgaud, A. (2001). Late Cenozoic tectonics and volcanism along the North Anatolian Fault: new structural and geochemical data. *Tectonophysics*, *338*(2), 135–165.

Akbaş, B., Akdeniz, N., Aksay, A., Altun, İ., Balcı, V., Bilginer, E., Bilgiç, T., Duru, M., Ercan, T., Gedik, İ., Günay, Y., Güven, İ.H., Hakyemez, H. Y., Konak, N., Papak, İ., Pehlivan, Ş., Sevin, M., Şenel, M., Tarhan, N., Turhan, N., Türkecan, A., Ulu, Ü., Uğuz, M. F., & Yurtsever, A, et al. (2016). Turkey Geology Map General Directorate of Mineral Research and Exploration Publications. Ankara, Turkey.

Andreani, M., Boullier, A. M., & Gratier, J. P. (2005). Development of schistosity by dissolution-crystallization in a Californian serpentinite gouge. *Journal of Structural Geology, 27,* 2256–2267.

Armijo, R., Meyer, B., Hubert, A., & Barka, A. (1999). Westward propagation of the North Anatolian fault into the northern Aegean: Timing and kinematics. *Geology, 27,* 267–270. https://doi.org/10.1130/0091-7613.

Armijo, R., Meyer, B., Hubert, A., & Barka, A. (2000). Westward propagation of North Anatolian fault into the northern Aegean: Timing and kinematics: Comment and reply. *Geology, 28*(2), 188. https://doi.org/10.1130/0091-7613.

Ashby, M., & Verrall, R. (1973). Diffusion-accommodated flow and superplasticity. *Acta Metallurgica, 11*(2), 149–163.

Bilham, R., Ozener, H., Mencin, D., Dogru, A., Ergintav, S., Çakir, Z., et al. (2016). Surface creep on the North Anatolian Fault at Ismetpasa, Turkey, 1944–2016. *Journal of Geophysical Research Solid Earth, 121,* 7409–7431.

Bos, B., & Spiers, C. J. (2000). Effect of phyllosilicates on fluid-assisted healing of gouge-bearing faults. *Earth and Planetary Science Letters, 184*(1), 199–210. https://doi.org/10.1016/S0012-821X(00)00304-6.

Bos, B., & Spiers, C. J. (2002). Frictional-viscous flow of phyllosilicate-bearing fault rock: Microphysical model and implications for crustal strength profiles. *Journal of Geophysical Research.* https://doi.org/10.1029/2001jb000301.

Bürgmann, R. (2018). The geophysics, geology and mechanics of slow fault slip. *Earth and Planetary Science Letters, 495,* 112–134.

Çakir, Z., Akoglu, A. M., Belabbes, S., Ergintav, S., & Meghraoui, M. (2005). Creeping along the Ismetpasa section of the North Anatolian fault (Western Turkey): Rate and extent from InSAR. *Earth and Planetary Science Letters, 238,* 225–234. https://doi.org/10.1016/j.epsl.2005.06.044.

Çakir, Z., Ergintav, S., Akoğlu, A. M., Çakmak, R., Tatar, O., & Meghraoui, M. (2014). InSAR velocity field across the North Anatolian Fault (eastern Turkey): Implications for the loading and release of interseismic strain accumulation. *Journal of Geophysical Research Solid Earth, 119,* 7934–7943. https://doi.org/10.1002/2014JB011360.

Carpenter, B. M., Ikari, M. J., & Marone, C. (2016). Laboratory observations of time-dependent frictional strengthening and stress relaxation in natural and synthetic fault gouges. *Journal of Geophysical Research Solid Earth, 121,* 1183–1201. https://doi.org/10.1002/2015JB012136.

Cetin, E., Çakir, Z., Meghraoui, M., Ergintav, S., & Akoglu, A. M. (2014). Extent and distribution of aseismic slip on the Ismetpa?a segment of the North Anatolian Fault (Turkey) from Persistent Scatterer InSAR. *Geochemistry Geophysics Geosystems, 15,* 2883–2894. https://doi.org/10.1002/2014GC005307.

Chen, K. H., & Bürgmann, R. (2017). Creeping faults: Good news, bad news? *Reviews of Geophysics, 55*(2), 282–286. https://doi.org/10.1002/2017RG000565.

Collettini, C., Niemeijer, A., Viti, C., & Marone, C. (2009). Fault zone fabric and fault weakness. *Nature, 462*(7275), 907–910. https://doi.org/10.1038/nature08585.

Emre, Ö., Duman, T. Y., Özalp, S., Elmacı, H., Olgun, Ş., & Şaroğlu, F. (2013). Active Fault Map of Turkey with an Explanatory Text. 1:1,250,000 Scale. General Directorate of Mineral Research and Exploration (MTA), Ankara-Turkey.

Emre, O., Duman, T. Y., Ozalp, S., Saroglu, F., Olgun, S., Elmaci, H., et al. (2016). Active fault database of Turkey. *Bulletin of Earthquake Engineering.* https://doi.org/10.1007/s10518-016-0041-2.

Erslev, E. (1988). Normalized center-to-center strain analysis of packed aggregates. *Journal of Structural Geology, 10,* 201–209. https://doi.org/10.1016/0191-8141(88)90117-4.

Erslev, E., & Ge, H. (1990). Least-squares center-to-center and mean object ellipse fabric analysis. *Journal of Structural Geology, 12,* 1047–1059. https://doi.org/10.1016/0191-8141(90)90100-D.

Fossen, H., & Cavalcante, G. C. G. (2017). Shear zones—a review. *Earth-Science Reviews, 171,* 434–455. https://doi.org/10.1016/j.earscirev.2017.05.002.

Fossen, H., & Tikoff, B. (1993). The deformation matrix for simultaneous simple shearing, pure shearing and volume change, and its application to transpression–transtension tectonics. *Journal of Structural Geology.* https://doi.org/10.1016/0191-8141(93)90137-y.

Fry, N. (1979). Random point distributions and strain measurement in rocks. *Tectonophysics, 60,* 89–105. https://doi.org/10.1016/0040-1951(79)90135-5.

Genier, F., & Epard, J. L. (2007). The Fry method applied to an augen orthogneiss: Problems and results. *Journal of Structural Geology, 29,* 209–224. https://doi.org/10.1016/j.jsg.2006.08.008.

Gratier, J., Dysthe, D. K., & Renard, F. (2013). The role of pressure solution creep in the ductility of the Earth' s upper crust. *Advances in Geophysics, 54,* 47–179. https://doi.org/10.1016/B978-0-12-380940-7.00002-0.

Gratier, J.-P., Noiriel, C., & Renard, F. (2015). Experimental evidence for rock layering development by pressure solution. *Geology, 43*(10), 871–874. https://doi.org/10.1130/G36713.1.

Gratier, J. P., Richard, J., Renard, F., Mittempergher, S., Doan, M. L., Di Toro, G., et al. (2011). Aseismic sliding of active faults by pressure solution creep: Evidence from the San Andreas Fault Observatory at Depth. *Geology, 39,* 1131–1134. https://doi.org/10.1130/G32073.1.

Graymer, R. W., Ponce, D. A., Phelps, G. A., & Wentworth, C. M. (2005). Three-dimensional geologic map of the Hayward fault, northern California: Correlation of rock units with variations in seismicity, creep rate, and fault dip. *Geology, 33,* 521–524. https://doi.org/10.1130/G21435.1.

Harris, R. A. (2017). Large earthquakes and creeping faults. *Reviews of Geophysics, 55,* 169–198. https://doi.org/10.1002/2016RG000539.

Heilbronner, R. (2002). Analysis of bulk fabrics and microstructure variations using tesselations of autocorrelation functions. *Computers and Geosciences, 28,* 447–455. https://doi.org/10.1016/S0098-3004(01)00088-7.

Heilbronner, R., & Barrett, S. (2014). *Image analysis in earth sciences microstructures and textures of earth materials.* Berlin: Springer.

Herece, E. I., & Akay, E. (2003). Atlas of North Anatolian Fault, General Directorate of Mineral Research and Exploration (MTA), Ankara, Turkey.

Hull, J. (1988). Thickness–displacement relationships for deformation zones. *Journal of Structural Geology, 10,* 431–435. https://doi.org/10.1016/0191-8141(88)90020-X.

Hussain, E., Hooper, A., Wright, T. J., Walters, R. J., & Bekaert, D. P. S. (2016). Interseismic strain accumulation across the central North Anatolian Fault from iteratively unwrapped InSAR measurements. *Journal of Geophysical Research Solid Earth, 121,* 9000–9019. https://doi.org/10.1002/2016JB013108.

Imber, J., Holdsworth, R. E., Butler, C. A., & Strachan, R. A. (2001). A reppraisal of the Sibson-Scholz fault model: The nature of the frictional to viscous (brittle-ductile) transition along a long-lived crustal-scale fault, Outer Hebrides, Scotland. *Tectonics, 20,* 601–624.

Janssen, C., Michel, G. W., Bau, M., Lüders, V., & Mühle, K. (1997). The North Anatolian fault zone and the role of fluids in seismogenic deformation. *The Journal of Geology, 105*(3), 387–404. https://doi.org/10.1086/515934.

Jefferies, S. P., Holdsworth, R. E., Wibberley, C. A. J., Shimamoto, T., Spiers, C. J., Niemeijer, A. R., et al. (2006). The nature and importance of phyllonite development in crustal-scale fault cores: An example from the Median Tectonic Line, Japan. *Journal of Structural Geology, 28,* 220–235.

Kaduri, M., Gratier, J.-P., Renard, F., Çakir, Z., & Lasserre, C. (2017). The implications of fault zone transformation on aseismic creep: Example of the North Anatolian Fault, Turkey. *Journal of Geophysical Research Solid Earth., 122,* 4208–4236. https://doi.org/10.1002/2016JB013803.

Kaneko, Y., Fialko, Y., Sandwell, D. T., Tong, X., & Furuya, M. (2013). Interseismic deformation and creep along the central section of the North Anatolian Fault (Turkey): InSAR observations and implications for rate-and-state friction properties. *Journal of Geophysical Research Solid Earth, 118,* 316–331. https://doi.org/10.1029/2012JB009661.

Karmakar, S., Lemaître, A., Lerner, E., & Procaccia, I. (2010). Predicting plastic flow events in athermal shear-strained amorphous solids. *Physical Review Letters, 104,* 215502. https://doi.org/10.1103/PhysRevLett.104.215502.

Lockner, D., Morrow, C., Moore, D., & Hickman, S. (2011). Low strength of deep San Andreas fault gouge from SAFOD core. *Nature, 472,* 82–85. https://doi.org/10.1038/nature09927.

Mavko, G. M. (1981). Mechanics of motion on major faults. *Annual Review of Earth and Planetary Sciences, 9,* 81–111. https://doi.org/10.1146/annurev.ea.09.050181.000501.

Mulchrone, K. F. (2013). Fitting the void: Data boundaries, point distributions and strain analysis. *Journal of Structural Geology, 46,* 22–33. https://doi.org/10.1016/j.jsg.2012.10.011.

Okada, Y. (1985). Surface deformation due to shear and tensile faults in a half-space. *Bulletin of the Seismological Society of America, 75*(4), 1135–1154.

Pennacchioni, G. (2005). Control of the geometry of precursor brittle structures on the type of ductile shear zone in the Adamello tonalites. *Southern Alps (Italy), 27,* 627–644. https://doi.org/10.1016/j.jsg.2004.11.008.

Ramsay, J. G. (1967). *Folding and fracturing of rocks.* New York: McGraw-Hill.

Ramsay, J. G. (1980). Shear zone geometry: A review. *Journal of Structural Geology, 2,* 83–99. https://doi.org/10.1016/0191-8141(80)90038-3.

Ramsay, J. G., & Graham, R. H. (1970). Strain variation in shear belts. *Canadian Journal of Earth Sciences, 7,* 786–813. https://doi.org/10.1139/e70-078.

Reilinger, R., et al. (2006). GPS constraints on continental deformation in the Africa–Arabia–Eurasia continental collision zone and implications for the dynamics of plate interactions. *Journal of Geophysical Research Solid Earth, 111*(5), 1–26. https://doi.org/10.1029/2005JB004051.

Renard, F., Dysthe, D. K., Feder, Bjørlykke, K., & Jamtveit, B. (2001). Enhanced pressure solution creep rates induced by clay particles: Experimental evidence in salt aggregates. *Geophysical Research Letters, 28*(7), 1295–1298. https://doi.org/10.1029/2000GL012394.

Richard, J., Gratier, J.-P., Doan, M., Boullier, A., & Renard, F. (2014). Rock and mineral transformations in a fault zone leading to permanent creep: Interactions between brittle and viscous mechanisms in the San Andreas Fault. *Journal of Geophysical Research Solid Earth, 119,* 8132–8153. https://doi.org/10.1002/2014JB011489.

Rousset, B., Jolivet, R., Simons, M., Lasserre, C., Riel, B., Milillo, P., et al. (2016). An aseismic slip transient on the North Anatolian Fault. *Geophysical Research Letters, 43*(7), 3254–3262. https://doi.org/10.1002/2016GL068250.

Samuelson, J., & Spiers, C. J. (2012). Fault friction and slip stability not affected by CO_2 storage: Evidence from short-term laboratory experiments on North Sea reservoir sandstones and caprocks. *International Journal of Greenhouse Gas Control, 11,* 78–90. https://doi.org/10.1016/j.ijggc.2012.09.018.

Savage, J. C., & Burford, R. O. (1973). Geodetic determination of relative plate motion in central California. *Journal of Geophysical Research, 78,* 832–845. https://doi.org/10.1029/JB078i005p00832.

Scholz, C. H. (2002). *The mechanics of earthquake faulting* (2nd ed.). Boston: Cambridge University Press.

Şengör, A. M. C., Tuysuz, O., Imren, C., Sakinc, M., Eyidogan, H., Gorur, N., et al. (2004). The North Anatolian fault: A new look. *Annual Review of Earth and Planetary Sciences, 33,* 1–75.

Sone, H., Shimamoto, T., & Moore, D. E. (2012). Frictional properties of saponite-rich gouge from a serpentinite-bearing fault zone along the Gokasho-Arashima Tectonic Line, central Japan. *Journal of Structural Geology, 38,* 172–182. https://doi.org/10.1016/j.jsg.2011.09.007.

Stein, R. S., Barka, A., & Dieterich, J. H. (1997). Progressive failure on the North Anatolian fault since 1939 by earthquake stress triggering. *Geophysical Journal International, 128,* 594–604. https://doi.org/10.1111/j.1365-246X.1997.tb05321.x.

Stewart, M. A., Holdsworth, R. E., & Strachan, R. A. (2000). Deformation processes and weakening mechanisms within the frictional-viscous transition zone of major crustal faults: Insights from the Great Glen Fault zone, Scotland. *Journal of Structural Geology, 22,* 543–560.

Thomas, M. Y., Avouac, J.-P., Gratier, J.-P., & Lee, J.-C. (2014). Lithological control on the deformation mechanism and the mode of fault slip on the Longitudinal Valley Fault, Taiwan. *Tectonophysics, 632,* 48–63. https://doi.org/10.1016/j.tecto.2014.05.038.

Toussaint, R., Aharonov, E., Koehn, D., Gratier, J.-P., Ebner, M., Baud, P., et al. (2018). Stylolites: A review. *Journal of Structural Geology, 114,* 163–195. https://doi.org/10.1016/j.jsg.2018.05.003.

van der Pluijm, B. A., & Marshak, S. (2010). *Earth structures: An introduction to structural geology and tectonics* (2nd ed.). New York: W.W. Norton Company.

Vitale, S., & Mazzoli, S. (2008). Heterogeneous shear zone evolution: The role of shear strain hardening/softening. *Journal of Structural Geology, 30,* 1383–1395. https://doi.org/10.1016/j.jsg.2008.07.006.

Weertman, J., & Weertman, J. H. (1964). *Elementary dislocation theory*. New York: Macmilan.

Wilson, M., Tankut, A., & Guleç, N. (1997). Tertiary volcanism of the Galatia province, north-west Central Anatolia, Turkey. *Lithos, 42*(1–2), 105–121.

Zubtsov, S., Renard, F., Gratier, J. P., Guiguet, R., Dysthe, D. K., & Traskine, V. (2004). Experimental pressure solution compaction of synthetic halite/calcite aggregates. *Tectonophysics, 385,* 45–57. https://doi.org/10.1016/j.tecto.2004.04.016.

(Received April 8, 2018, revised October 18, 2018, accepted October 20, 2018, Published online November 13, 2018)

Pure Appl. Geophys. 176 (2019), 1323–1333

https://doi.org/10.1007/s00024-019-02106-6

Pure and Applied Geophysics

Long-Term Evolution of the Deviatoric Background Stress in the Focal Region of the 2011 Tohoku-Oki Earthquake

TIAN-JUE LI[1,2]

Abstract—On 11 March 2011, a giant interplate earthquake of magnitude Mw 9.1 struck Tohoku-Oki area in eastern Japan. The earthquake is acknowledged to have permanently altered the stress field in and around the focal region. To capture the temporal change of the overall stress field, I examined 1251 focal mechanisms, from the NIED MT catalog, that occurred before and after the Tohoku-Oki earthquake from January, 2006 to January, 2017 in the focal region near the subducting slab. The examined observations were grouped into six periods, and based on the selected NIED MT catalog, the stress regime in each period was obtained by using the damped stress tensor inversion method. Based on the temporal evolution of stress rotation, the corresponding deviatoric stress level was estimated using a simplified 2D model. Results of the 10-year seismic stress cycle show that several years before the Tohoku-Oki earthquake, the stress accumulation level seems to have experienced an acceleration process. Studies suggest that this increasingly critically stress state combined with the sufficiently reduced coupling rate off the Tohoku area finally resulted in the unprecedented megathrust event. The coseismic process was violent and released almost all of the deviatoric stress that presented before the main shock. The resultant stress state even reached frictional overshoot. Thus, the postseismic stress pattern in the source region was reshaped significantly, especially for the upper plate and updip portion of the lower plate. After the main shock near the rupture surface, a surprisingly rapid and high-level stress reloading occurred within several postseismic years. To reconcile the classical subduction zone earthquake generation cycle model, the event may be described as an instantaneously decoupled stress state between the upper and inner plates.

Key words: Tohoku-Oki earthquake, stress rotation, stress drop ratio, seismic stress cycle.

Electronic supplementary material The online version of this article (https://doi.org/10.1007/s00024-019-02106-6) contains supplementary material, which is available to authorized users.

[1] State Key Laboratory of Geodesy and Earth's Dynamics, Institute of Geodesy and Geophysics, Chinese Academy of Sciences, Wuhan 430077, China. E-mail: tianjueli@outlook.com

[2] Key Laboratory of Earth and Planetary Physics, Institute of Geology and Geophysics, Chinese Academy of Sciences, Beijing 100029, China.

1. Introduction

Along the Japan trench, the Pacific plate is subducting northwestward beneath the North American plate at a rate of 83.05 mm/year (144°E, 37.5°N) (DeMets et al. 1994). On the date of 11 March 2011, an interplate earthquake of magnitude Mw 9.1 struck the Tohoku-Oki area in eastern Japan and resulted in a huge coseismic slip region spreading approximately 500 km long and 200 km wide, while slipping with an average magnitude of 10–15 m (Hasegawa et al. 2012). This event was so giant that it caused catastrophic consequences to both the local economy and the population. To mitigate such serious earthquake hazards, it is important to understand the physical processes that caused this unpredicted event. The state of the stress field, and changes recorded in the focal region before and after the main shock, could provide some basic information.

Numerous studies about the local stress regime changes off the Pacific coast of Tohoku-Oki at the time of the earthquake have been done, especially by Japanese researchers right after the megathrust event, using the accumulated seismicity data for several months (Asano et al. 2011; Hasegawa et al. 2011, 2012; Chiba et al. 2012; Obana et al. 2012). The combination of precise subducting slab geometry under northeastern Japan from previous studies and accurate location of seismic events from local dense seismic observation arrays (e.g., F-net, Hi-net), allowed the researchers to investigate the variations of stress before and after the mainshock in the upper, inner and lower plates, respectively, yielding several meaningful findings. For example, before the mainshock, a relatively simple and uniform compressive stress along the direction of plate convergence prevailed in the whole source region, which behaved in the same way as many thrust faultings. However, just after the Tohoku-Oki earthquake, the relatively uniform stress regime in the source region changed obviously into a complicated pattern with several different stress regimes in space. In the upper plate (the continental part, or hanging wall) along the plate convergence direction, the compressive stress had been replaced by a tensional stress field, which was verified by the occurrence of many normal faultings. In the lower plate (the subducting slab), stress retained its original compressive pattern along the plate convergence direction in the deep part near the coastal line, but a reversed stress regime occurred in the central area and extended to near-trench portions of the main rupture area. Near the plate interface, seismicity had become significantly weak within the coseismic high-slip zone, and stress drop during the coseismic process estimated from stress rotation seemed to be complete.

Another study by Hardebeck (2012) using the same analysis frame as Hasegawa et al. (2012), but with a longer database from Global Centroid Moment Tensor (GCMT), confirmed that Tohoku-Oki earthquake had released almost all of its deviatoric background stress. What's more, the analysis results also revealed that the plunge of the principle stress had experienced an evident, rapid return rotation in the next 8 months following the main shock. When translating this into the deviatoric background stress, a rapid postseismic reloading emerged, taking the entire rupture zone as a whole. To be quantitative, about 6% of the total deviatoric background stress presented before the Tohoku-Oki earthquake appeared to be reloaded. An independent study (Tormann et al. 2015), estimating *b*-values using the moment tensor (MT) catalog from the National Research Institute for Earth Science and Disaster Prevention (NIED) in the coseismic high-slip zone, showed a similar evolution of the deviatoric background stress. And the postseismic reloading could be as high as 77% of the coseismic stress drop within three years after the main shock, using the hypothesis that *b*-values are negatively correlated with deviatoric background stress linearly. However, understanding the physical process in the background of such a rapid and high-level reloading is quite difficult, considering the possible processes (e.g., afterslip along the megathrust, viscoelastic relaxation in the upper mantle and relocking of the megathrust, Wang et al. 2012; Sun et al. 2014; Hu et al. 2016), as we have learned so far during the postseismic deformation. For example, a stress recovery model (Bürgmann et al. 2016) constrained by geodetic and seismicity data suggested that, although reloading of a megathrust rupture could, indeed, be most rapid during the early period of postseismic relaxation (Hashimoto and Matsu'ura 2002), only a very small fraction (< 5%) of the coseismic stress drop in the high-slip zone could be recovered in the first three postseismic years.

Quantifying the level of such a rapid postseismic reloading event is very important, as it will provide new reliable ingredients for understanding subduction zone processes. This paper pursues the analysis ideas of Hasegawa et al. (2011) and Hardebeck (2012), based on a wider period (from January, 2006 to January, 2017) of the MT catalog from NIED in the focal region of the Tohoku-Oki earthquake (Kubo et al. 2002), to capture the temporal evolution of the deviatoric background stress, taking the entire rupture zone as a whole. The initial motivation of this paper is to see the possible change of deviatoric background stress in the focal region, taken as a whole, from a longer period view. That means a relatively complete picture of the deviatoric background stress change before and after the main shock (or say seismic cycle stress, Hardebeck 2012) is my focus here.

2. Data and Method

Many methods could be used to evaluate the state of a stress field in an area (Zoback 1992). However, the handy and effective way to obtain the regional stress regime, especially for subduction zone, is to rely on inversion based on focal mechanisms (Hasegawa et al. 2011, 2012; Chiba et al. 2012; Hardebeck 2012, 2017). These inversion methods generally begin with two main assumptions (Angelier 1979; Gephart and Forsyth 1984; Michael 1984, 1987): that the fault slips along the tangential traction on the preexisting weak plane, and that the aimed stress tensor keeps uniform for the dataset within the specified spatiotemporal bin. Here, to capture a relatively complete temporal evolution of the stress field, taking the entire rupture zone as a whole before and after the Tohoku-Oki earthquake, I examined 1251 focal mechanisms from NIED Seismic MT Catalogue (Kubo et al. 2002). These focal mechanisms were drawn from the focal region from January, 2006 to January, 2017 within 15 km from the plate interface (subducting slab geometry comes from slab1.0, Hayes et al. 2012), with variance reduction better than 70%, and were obtained using no less than three stations. These were sorted into six period bins (Figs. 1b, c, 3), where each period bin lasted for about 2 years with more than 54 events (the numbers of events selected in the six period bins were 54, 107, 79, 524, 270 and 217, successively). The stress regime in each period bin was obtained by using the damped stress tensor inversion method known as Spatial and Temporal Stress Inversion (SATSI, Hardebeck and Michael 2006), with damping parameters 0, 1.4 and 2.25, respectively (Fig. 2b). By normalizing the tangential traction as an additional assumption, the SATSI method acts efficiently in calculation to obtain the stress tensor through solving linear inversion equations. Here, the confidence region is calculated by the bootstrap method (Efron and Tibshirani 1986; Michael 1987) with 2000 bootstrap resamplings, and the preexisting weak plane is chosen randomly from the two nodal planes each time. Based on the stress inversion results, the average ratio of stress drop to the background deviatoric stress $\Delta\tau/\tau$ (Hardebeck and Hauksson 2001; Hardebeck and Okada 2018) in each period bin was also estimated, using the observed rotation of the maximum principal compressive stress σ_1 axis and the angle from the σ_1 axis to the effective fault geometry (the geometry of the subducting slab, Fig. 2d), where both angles are defined clockwise positive. The NW Pacific plate subducts at a shallow angle of about 10° for the first 20 km depth, and then steepens with a dip angle increasing up to 30° gradually at about 50 km (Hasegawa et al. 2011). Here, the dip angle of the effective fault geometry is selected as 25°. In order to access the deviatoric background stress quantitively in each period bin, I set the deviatoric stress in the first period bin, dating from January, 2006 to January, 2008, to be the normalized one. Then, positive or negative stress drop ratio results which indicating a stress releasing or reloading mechanism (Hardebeck 2012), could be subtracted from or added to the former normalized deviatoric background stress.

Some special result-obtained procedures of which the reader should be aware: the mean misfits of the predicted shear stress to the slip orientation on the fault for all division bins of stress inversion are no more than 45°, which is important to retrieve the homogenous part of the stress regime as suggested by Michael (1991). The stress inversion results used for later stress drop ratio estimation are selected from the median values of the compressive and tensional principal stress axes. They are indicated by plotting as stars, differing from the best fitting stress inversion results (inverted with the all selected events in each period bin with the first nodal plane presented in the NIED MT catalog) which are shown as crosses in Fig. 3. In addition, to access the stability of the results of stress regime change obtained here, several additional processes have also been conducted. Examples include bootstrap resampling with the same number of events in each period bin (SE), using different distances from the plate interface, like 10 km and 20 km to select the interplate events, using different dip angles such as 15°, 20° and 30° as the effective fault geometry, using different timescales for period bin redivision around the occurrence of the main shock, etc.

3. *Result*

By classifying all the selected events (located in the focal region and within 15 km from the plate interface) into normal dip-slip faulting, normal oblique-slip faulting, strike-slip faulting, thrust oblique-slip faulting and thrust dip-slip faulting according to the plunge angle of P, B and T axes from focal mechanisms (Zoback 1992), a notable increase of normal events accompanying the multitude of thrust faultings was observed right after the Tohoku-Oki earthquake (Figs. 3, S6). In space, most of these occurred in the central area and extended to the near-trench portions of the main rupture area. The normal faultings that occurred in the central rupture zone

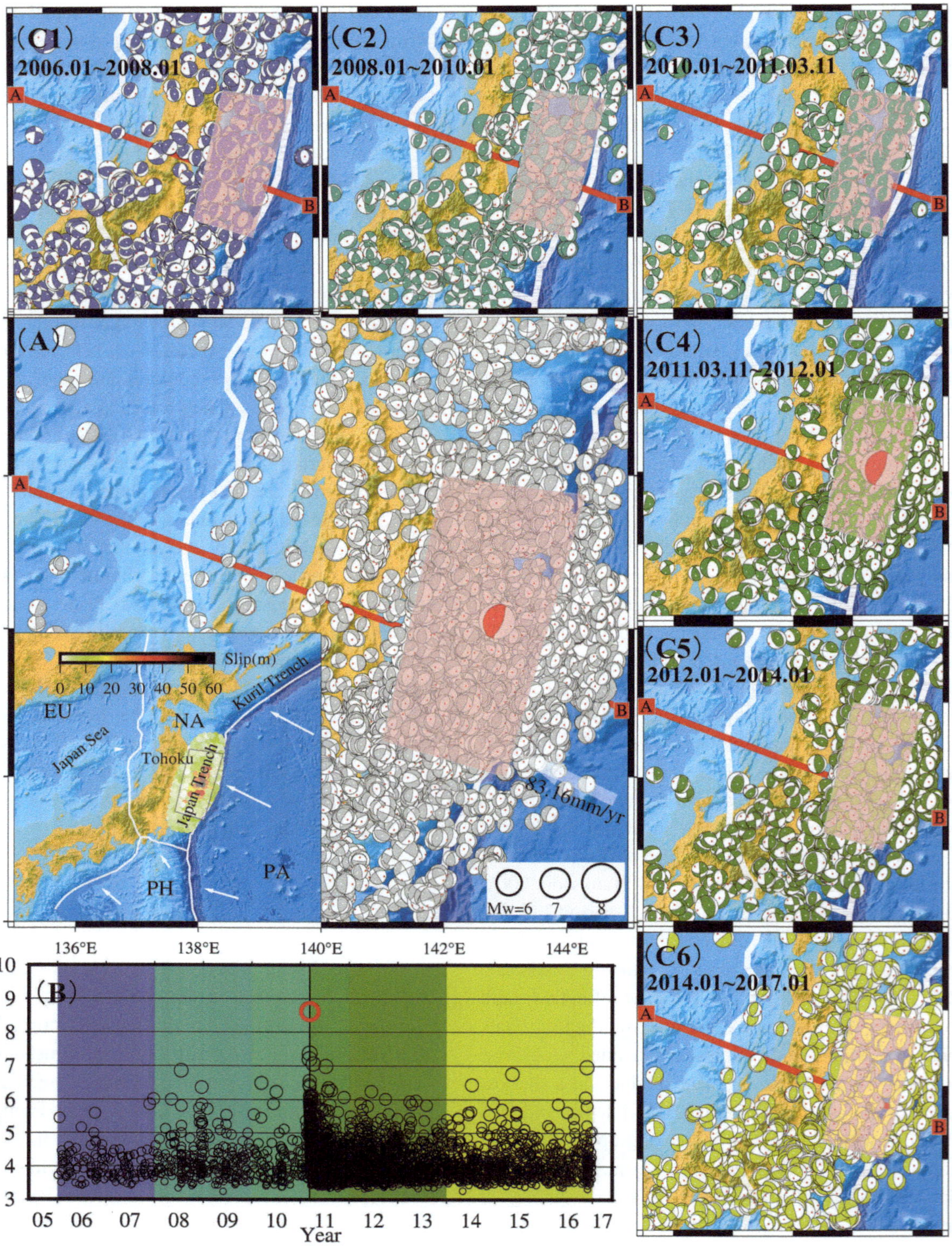
(C1)
2006.01~2008.01
(C2)
2008.01~2010.01
(C3)
2010.01~2011.03.11
(A)
(C4)
2011.03.11~2012.01
(C5)
2012.01~2014.01
(C6)
2014.01~2017.01
(B)
A
B
42°N
40°N
38°N
36°N
34°N
136°E
138°E
140°E
142°E
144°E
Slip(m)
0 10 20 30 40 50 60
EU
NA
Kuril Trench
Japan Sea
Tohoku
Japan Trench
PH
PA
83.16mm/yr
Mw=6 7 8
Magnitude(Mw)
Year
05 06 07 08 09 10 11 12 13 14 15 16 17

◀Figure 1
a Study area is marked in the low-left inset by the solid gray rectangle that covers almost the entire rupture region (the slip distributions come from Simons et al. 2011) of the Tohoku-Oki earthquake. The white solid line depicts the plate boundary and white arrows indicate the relative motion between the two neighboring plates from NUVEL-1A Model (DeMets et al. 1994). PA represents Pacific Plate, PH is Philippine Plate, NA is North America Plate, EU is Eurasia Plate. Outside the inset, the study area is located in the pink shadow, and the red solid line AB is used as the profile to later show events' distribution in depth. All focal mechanisms shown here come from the NIED MT catalog during the period from January, 2006 to January, 2017, and are presented as the lower hemisphere equal-area projections; the beach ball with red compressional quadrant represents the main shock. **b** All events in the focal region have been divided into six period bins according to their occurrence time and are shown as circles with radii in proportion to their magnitudes. The red circle represents the main shock. **c** Events in each period bin have been shown with the corresponding compressional quadrant color shown as the background color of the period bin in **b**

seems to last for about 3 postseismic years, whereas the ones located near the trench last for a longer time, ~ 6 years. The selected fault plane (strike 202°, dip 25°, the geometry approximates the subducting slab) which is plotted as the solid black line in Fig. 3 (fault plane), seems dominant in each period.

Stress inversion results with the damping parameter 0 and using all the selected events in each period bin show a common feature of the stress regime: both the maximum and minimum principal compressive stress are oriented WNW, which coincides well with the direction of the NW Pacific plate convergence. However, for plunge of the principle stress, the steeper tensional regime was replaced by the compressional pattern right after the main shock and lasted for about 10 months from the bin-division timescale used here. When applying a smaller timescale to the period bin redivision right after the main shock, the duration of such an abnormal stress regime lasted for about one and a half months (Fig. S3B). The basic pattern for both the azimuth and plunge of the principal stress seems relatively stable, no matter what distance from the plate interface and damping parameter are chosen (distance 10 km away from the plate interface shown in Fig. S4A, S4B; distance 20 km away from the plate interface shown in Fig. S5A, S5B; different damping parameters shown in Fig. 4).

Evolution of the normalized deviatoric background stress (Fig. 4c) estimated from stress rotation shows that stress accumulation in the rupture zone, taking it as a whole before the main shock, experienced a rise. Although the rise amplitude appears to be damping-dependent and negatively in proportion to the damping parameter, the general rise trend seems rather stable, except for the possible over-damped case (choosing the damping parameter as 2.25). Results using a smaller timescale to redivide the period bins (Fig. S2B) shows the same trend of stress loading process without using the damped constraint. However, the damped inversion results indicate a smooth preseismic process. Just after the main shock, a noticeable deviatoric background stress drop appeared, and the amplitude of this coseismic stress drop ratio could be as high as 85.7% (Fig. S2B; 76.1% shown in Fig. 4c, d) when choosing a smaller timescale to redivide the period bin around the occurrence of the main shock. In the postseismic process, the background deviatoric stress seems to be restored to some degree as it was before the main shock. To my present study result (about 7 years postseismic database used), the deviatoric background stress over the rupture zone, taken as a whole, seems to be totally reloaded to its original state (e.g., the stress level in 2006.1–2008.01). This pattern is also rather stable no matter what damping parameter or dip angle of the effective fault plane is chosen to run the stress inversion and estimate the stress drop ratio. Note that a smaller dip angle as the effective fault plane will bring in some unstable results while estimating the stress drop ratio (Fig. S1B), as the simplified 2D model used here to access stress drop ratio will become invalid when the angle between the principal stress and fault geometry reaches 45° (see Eq. 1 following Hardebeck and Hauksson 2001; Hardebeck and Okada 2018, in the supplementary materials).

4. Discussion

As for the notable increase of normal faultings occurring right after the main shock, the result shown here (Figs. 3, S6) and the high-resolution results (Asano et al. 2011; Hasegawa et al. 2012; Chiba et al.

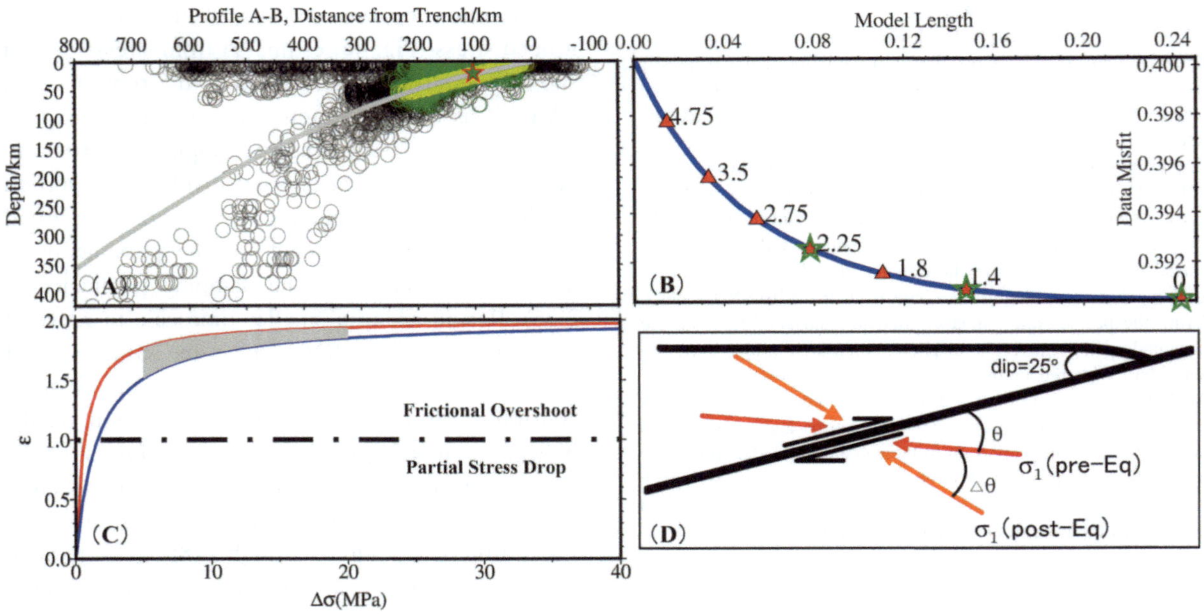

Figure 2

a Events selected from the NIED MT catalog. Events shown in Fig. 1a are plotted as circles here along the profile line AB in depth. Events in the focal region of the Tohoku-Oki earthquake are plotted as green circles, and yellow circles represent the events located within 15 km from the subducting slab interface. The red star indicates the main shock and the gray solid line represents the NW Pacific subducting slab (from slab1.0, Hayes et al. 2012). **b** The damping parameters are selected as 0, 1.4 and 2.25, respectively, in the trade-off curve between model length and data misfit in the inversion process while choosing various damping parameters. **c** An estimate of the amount of partial stress drop or frictional overshoot in the form of the quantity ε (Zúñiga 1993) for the Tohoku-Oki earthquake. The dash-dotted line represents the case in which Orowan's hypothesis works ($\varepsilon = 1$, that means the final stress on the fault stops to be equal to the dynamic frictional stress). An $\varepsilon < 1$ implies a partial stress drop mechanism whereas $\varepsilon > 1$ indicates that frictional overshoot has occurred. The red and blue solid lines represent the theoretical upper and lower limits of ε, respectively, and the estimated ε for the Tohoku-Oki earthquake is located inside the gray shadow. **d** Assumed geometry of stress rotation due to slip on the fault plane. Here, the dip angle of the effective fault geometry is set as 25°. θ is the angle from the fault trend to the maximum principal compressive stress σ_1 axis, clockwise positive. $\Delta\theta$ is the rotation of the stress field due to slip on the fault plane, clockwise positive. Stress drop ($\Delta\tau$) resulting from the main shock is taken to be positive for the sense of slip shown above (thrust event) and negative for the opposite sense of slip (normal event)

2012; Obana et al. 2012), which divided earthquakes into inter- and intraplate events, reveal that the increased normal faultings were mainly concentrated in the upper plate and updip portion of the lower plate near the trench. The occurrence of normal faultings means that the stress state in the hanging wall and updip portion of the lower plate changed totally from compression to tension along the direction of the plate convergence; the wide-spreading trenchward displacements of the upper plate from the postseismic GPS observation right after the main shock (Ozawa et al. 2012) confirmed this stress transformation. At the same time as the events, thrust faultings occurring in almost the total lower plate except for its updip part, continued as usual. This means the stress pattern in the downdip portion of the lower plate seems very stable and remains in a compressional state along the plate convergence direction throughout the whole 10-years pre-, co- and postseismic process.

In the background of the postseismic prevailing normal faultings and trenchward displacements right after the Tohoku-Oki earthquake, is the high-level coseismic stress drop, which could be as high as > 85% of the deviatoric background stress (85.7% in this study shown in Fig.S2B using 25° as the dip angle; 100% in this study when using 20° as the dip angle with the same period division as Fig. S2; > 90% from Hasegawa et al. (2011) and 97% from Hardebeck (2012), both using 20° as the dip angle). The nearly total deviatoric stress release

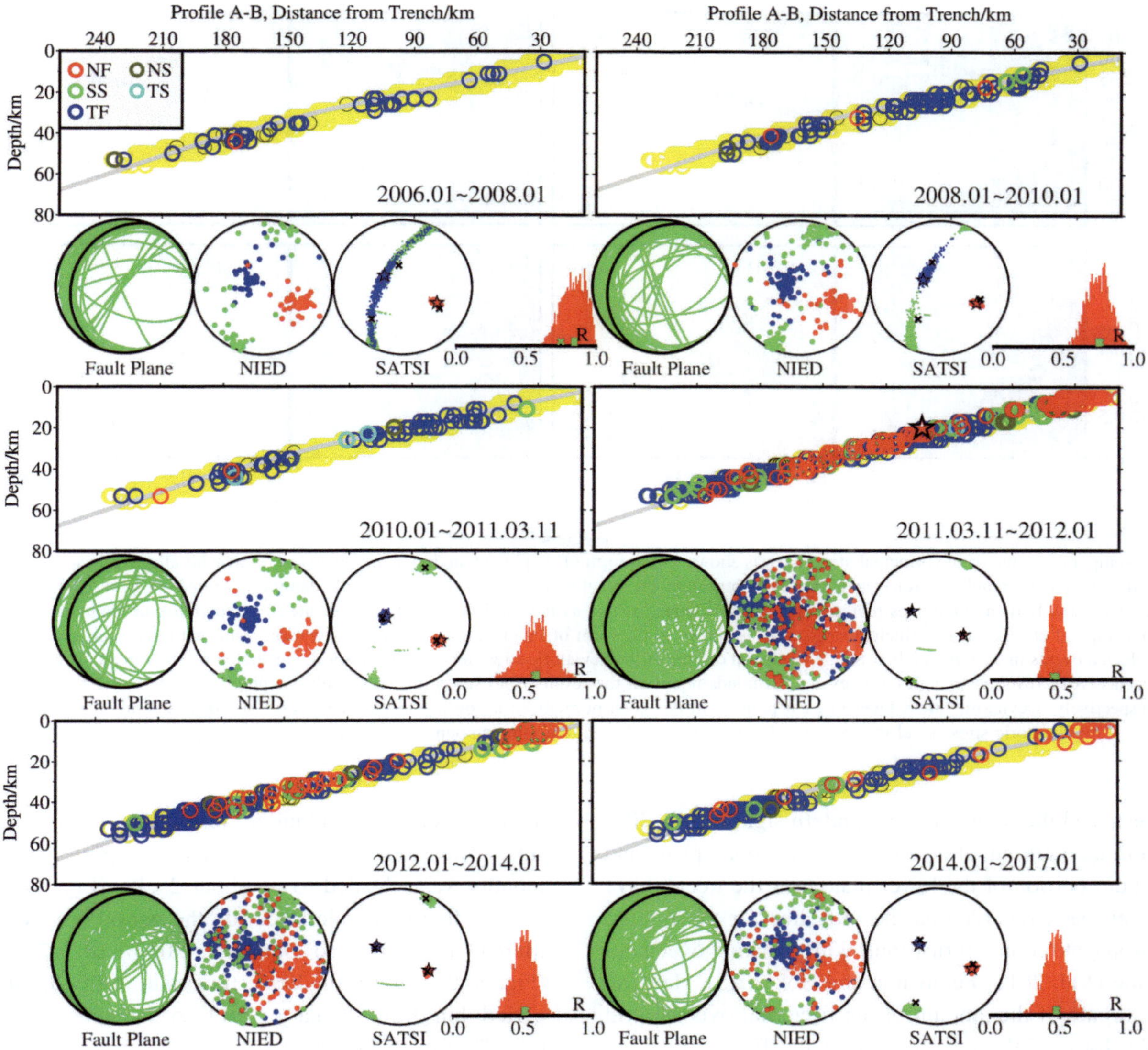

Figure 3

Stress inversion results with damping parameter 0 using all the selected events in each period bin (All). In the picture, the events shown here have been qualified as normal faults (NF), normal faults while with strike-slip component (NS), strike-slip faults (SS), thrust faults while with strike-slip component (TS) and thrust faults (TF) with different-color circles according to the plunge angle of P, B and T axes from the focal mechanisms (Zoback 1992). The gray solid line and black star represent the NW Pacific subducting slab from slab1.0 (Hayes et al. 2012) and the main shock, respectively. Yellow circles indicate all selected events within 15 km from the plate interface over the entire 10-year duration in the focal area. Fault planes shown here are selected as the one nodal plane whose strike angle approximates the rupture direction of the main shock (203°, from USGS, https://earthquake.usgs.gov/earthquakes/eventpage/official20110311054624120_30#executive), and the solid black line in the fault plane result represents the subducting slab with a dip angle of 25°. The NIED MT solutions are plotted with red, green and blue points for P, B, and T axes, and the SATSI inversion results are shown with red, green and blue points for the principal stress σ_1, σ_2 and σ_3 axes, respectively. The cross represents the best fitting principle stress axis, while stars indicate the median results for each principle stress axis. The histogram depicts the stress shape ratio [$(\sigma_2 - \sigma_1)/(\sigma_3 - \sigma_1)$, Gephart and Forsyth 1984] distribution. The green bar indicates the median value of the stress shape ratio, while the green cross indicates the best fitting result. All inversion results are given with 95% confidence level obtained by 2000 bootstrap resamplings

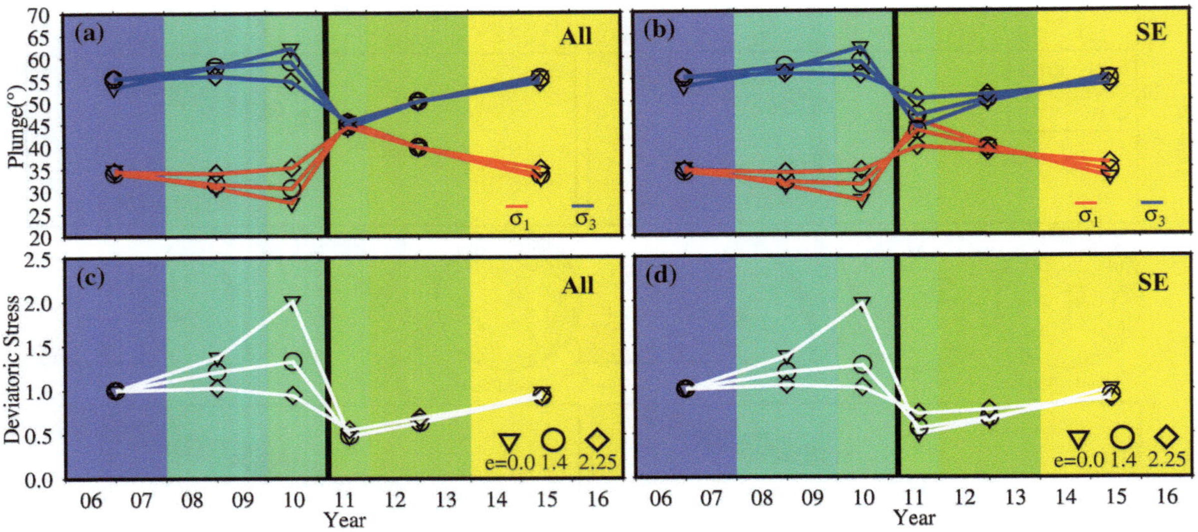

Figure 4

a Temporal change of the principal stress axes is shown in the form of its plunge angle for the inversion result with all the selected events (All). Inverted triangles, circles and diamonds represent the results from stress inversion with damping parameters 0, 1.4 and 2.25, respectively. Red and blue lines represent the principal stress σ_1 and σ_3 in turn. Black solid line indicates the occurrence of the main shock. **b** Temporal change of the principal stress axes is shown in the form of its plunge angle for the inversion result with the same amount of selected events in each period bin (SE). **c** Temporal change of the deviatoric stress level is shown for the inversion result with all the selected events (All). Inverted triangles, circles and diamonds represent the results from stress inversion with damping parameters 0, 1.4 and 2.25, respectively. Deviatoric stress level in each period bin has been normalized according to the first period bin. **d** Temporal change of the deviatoric stress level is shown for the inversion result with the same amount of selected events in each period bin (SE)

reshaped the regional stress pattern significantly. The normal faulting aftershocks in the upper plate and updip portion of the lower plate near the trench have even been regarded as the indicator for the rupture propagation on a thrust fault with dynamic overshoot (Ide et al. 2011; Fukuyama and Hok 2015). Here, an estimate of the amount of the frictional overshoot in the form of the quantity ε (see the Eq. 2 following Zúñiga 1993, in the supplementary materials) for the Tohoku-Oki earthquake (Fig. 2c) confirms that this megathrust event includes a frictional overshoot mechanism ($\varepsilon > 1.52$) when taking the entire rupture zone as a whole. Frictional overshoot means that the final stress reserved on the fault is very low, even lower than the dynamic frictional stress. The quantity ε estimate is done by assuming the radiated seismic energy scaled by seismic scalar moment to be about $0.8\text{–}2.0 \times 10^{-5}$, the stress drop value to be 5–20 MPa, and a uniform rigidity of 40 GPa among the rupture zone (Lay 2018).

After the noticeable coseismic stress releasing process, the stress pattern in the later period shows a trend of stress reloading to restore its preseismic state. The stress level in the first nine postseismic months was reloaded to be about 25.4% (Fig. S3B, using a fine timescale to divide the period bin from March 11, 2011 to January, 2012) or 42.4% (Fig. S2B, using a moderate timescale to divide the period bin from January, 2011 to January, 2012) ($\sim$ 6% from Hardebeck 2012) of the deviatoric background stress presented before the main shock. In the first about three postseismic years, $\sim$ 63.6% (Fig. 4c) ($\sim$ 77% in the greater asperity area (e.g., 10 m-slip contour) using the *b*-values estimate from Tormann et al. 2015) of the preseismic deviatoric stress was reloaded. About 97.7% (Fig. 4c) of the preseismic deviatoric stress was restored in the following six years after the main shock. These results show that the postseismic stress reloading over the rupture zone, taken as a whole, of the Tohoku-Oki earthquake seems fast in time and large in scale.

Understanding the physical process in the background of such a rapid and high-level reloading is quite difficult when considering the possible postseismic processes (e.g., afterslip along the megathrust, viscoelastic relaxation in the upper mantle and relocking of the megathrust, Wang et al. 2012; Sun et al. 2014; Hu et al. 2016). For example, a stress recovery model (Bürgmann et al. 2016) constrained by geodetic and seismicity data suggests that reloading of a megathrust rupture could be indeed most rapid during the early period of postseismic relaxation (Hashimoto and Matsu'ura 2002), but only a very small fraction ($< 5\%$) of the coseismic stress drop in the high-slip zone could be recovered in the first three postseismic years. Besides the difficulties in its physical explanation, there are several paradoxical phenomena (e.g., normal faultings in the upper plate, prevailing trenchward displacements on land) that should vanish if the high-level recovered stress is valid all over the NE Japan area. However, in fact, these are still happening in spite of the fact that the strength of the normal faultings' seismicity and trenchward slip rate has become rather weaker (e.g., slight seismicity of normal faultings till the end of 2016 from this study shown in Fig. S6; low percent of the trenchward slip rate left till about the end of 2015 from Fig. 8a of Ozawa et al. 2012 and Fig. 1 of Tomita et al. 2017) compared to that occurring at the beginning of the postseismic period. To overcome the paradoxes mentioned above, I adhere to the principle that the quantitative stress state estimated from stress rotation need only be valid near the mainshock rupture surface, considering the space this analytical model sampled (Hardebeck 2017; Hardebeck and Hauksson 2001; Hardebeck and Okada 2018). Then, a relatively independent stress state between the upper and inner plates may exist. A possible scenario is that the interplate could be reloaded to a high level a short period after the main shock, while it may take the normal time of a M9 earthquake cycle ($\sim$ 880 years at a coupling rate of 0.5, Uchida and Matsuzawa 2011) to restore the stress state in the upper plate. The weakened lower crust of upper plate and the mantle wedge (or, say, with a low-viscosity structure, Freed et al. 2017) could provide a possible mechanism for the instantaneously decoupled stress state between the upper and inner plates.

Despite the noticeable process of coseismic stress releasing and postseismic stress reloading, it seems that the stress accumulation had experienced a rise several years before the Tohoku-Oki earthquake, although the damping inversion results provide less positive evidence. An interesting thing is that preseismic stress state estimated from other methods (e.g., trend of smaller b-values about 7 years before the main shock like Figs. 2, 3a of Tormann et al. 2015; trend of higher correlation between tides and earthquakes about 10 years before the main shock like Figs. 2, 4a of Tanaka 2012) indicate that the megathrust had been increasingly critically stressed prior to the main shock. Accompanying the increasingly critical stress state, the coupling rate off the Tohoku region had been gradually sufficiently reduced by various forms of strain releasing processes (e.g., partial creep, earthquakes and their afterslip, Ozawa et al. 2012). Thus, former studies (e.g., Ozawa et al. 2012; Sato et al. 2013; Hasegawa and Yoshida 2015) proposed that, based on the favorable preseismic conditions, the source region was critically stressed while sufficiently weakened at the same time, that the unprecedented megathrust earthquake was nucleated and ruptured a huge area finally off the Tohoku-Oki region.

5. *Conclusion*

The results of this study show that several years prior to the Tohoku-Oki earthquake, the stress accumulation in the source region seems to have experienced an acceleration process, and the megathrust event was finally initiated due to the favorable conditions of the critical stress state and sufficiently reduced coupling rate off the Tohoku coast. During the coseismic process of the main shock, stress release was so complete that the rupture zone had even reached frictional overshoot. As the result of the nearly complete coseismic stress drop, the postseismic spatial stress pattern had been reshaped significantly, especially for the upper plate and updip portion of the lower plate, where the stress regime had been changed totally from compression to tension along the plate convergence. After the violent coseismic process, a surprisingly rapid and high-level stress reloading near the rupture surface occurred. The possible scenario is that the relatively independent stress state between the upper and inner plates may favor the classical subduction zone earthquake generation cycle model.

Acknowledgements

I am very grateful to Roland Bürgmann and the two anonymous reviewers for their thoughtful and constructive comments that improved the manuscript. I am also very grateful to the guest editor Yehuda Ben-Zion and the Springer Nature Corrections team for their diligent work to facilitate the publication. In this study, I used the focal mechanisms from the NIED catalog (Kubo et al. 2002). I also used the damped stress tensor inversion method named SATSI (Hardebeck and Michael 2006). The topography data used in Fig. 1 comes from Etopo1 (Amante and Eakins 2009). The figures in this paper were prepared using GMT (Wessel and Smith 1998). This research was supported by the National Natural Science Foundation of China (Grant Nos. 41474041).

References

Amante C., & Eakins, B. W. 2009. *ETOPO1 1 Arc-Minute Global Relief Model: Procedures, data sources and analysis*. National Geophysical Data Center.

Angelier, J. (1979). Determination of the mean principal directions of stresses for a given fault population. *Tectonophysics, 56,* T17–T26.

Asano, Y., Saito, T., Ito, Y., et al. (2011). Spatial distribution and focal mechanisms of aftershocks of the 2011 off the Pacific coast of Tohoku Earthquake. *Earth Planets and Space, 63,* 669–673.

Bürgmann, R., Uchida, N., Hu, Y., et al. (2016). Tohoku rupture reloaded? *Nature Geoscience, 9*(3), 183.

Chiba, K., Iio, Y., & Fukahata, Y. (2012). Detailed stress fields in the focal region of the 2011 off the Pacific coast of Tohoku Earthquake—implication for the distribution of moment release. *Earth Planets and Space, 64,* 1157–1165.

DeMets, C., Gordon, R. G., Argus, D. F., et al. (1994). Effect of recent revisions to the geomagnetic reversal time scale on estimates of current plate motions. *Geophysical Research Letters, 21*(20), 2191–2194.

Efron, B., & Tibshirani, R. (1986). Bootstrap methods for standard errors, confidence intervals, and other measures of statistical accuracy. *Statistical Science, 1*(1), 54–75.

Freed, A. M., Hashima, A., Becker, T. W., et al. (2017). Resolving depth-dependent subduction zone viscosity and afterslip from postseismic displacements following the 2011 Tohoku-oki, Japan earthquake. *Earth and Planetary Science Letters, 459,* 279–290.

Fukuyama, E., & Hok, S. (2015). Dynamic overshoot near trench caused by large asperity break at depth. *Pure and Applied Geophysics, 172*(8), 2157–2165.

Gephart, J. W., & Forsyth, D. W. (1984). An improved method for determining the regional stress tensor using earthquake focal mechanism data: Application to the San Fernando earthquake sequence. *Journal of Geophysical Research, 89,* 9305–9320.

Hardebeck, J. L. (2012). Coseismic and postseismic stress rotations due to great subduction zone earthquakes. *Geophysical Research Letter, 39,* L21313.

Hardebeck, J. L. (2017). The spatial distribution of earthquake stress rotations following large subduction zone earthquakes. *Earth Planets and Space, 69,* 69.

Hardebeck, J. L., & Hauksson, E. (2001). Crustal stress field in southern California and its implications for fault mechanics. *Journal of Geophysical Research, 106*(B10), 21859–21882.

Hardebeck, J. L., & Michael, A. J. (2006). Damped regional-scale stress inversions: Methodology and examples for southern California and the Coalinga aftershock sequence. *Journal of Geophysical Research Solid Earth, 111,* B11310.

Hardebeck, J. L., & Okada, T. (2018). Temporal stress changes caused by earthquakes: A review. *Journal of Geophysical Research Solid Earth, 123,* 1350–1365.

Hasegawa, A., & Yoshida, K. (2015). Preceding seismic activity and slow slip events in the source area of the 2011 Mw 9.0 Tohoku-Oki earthquake: A review. *Geoscience Letters, 2,* 6.

Hasegawa, A., Yoshida, K., Asano, Y., et al. (2012). Change in stress field after the 2011 great Tohoku-Oki earthquake. *Earth and Planetary Science Letters, 355–356,* 231–243.

Hasegawa, A., Yoshida, K., & Okada, T. (2011). Nearly complete stress drop in the 2011 Mw 9.0 off the Pacific coast of Tohoku Earthquake. *Earth Planets and Space, 63,* 703–707.

Hashimoto, C., & Matsu'ura, M. (2002). 3-D simulation of earthquake generation cycles and evolution of fault constitutitve properties. *Pure and Applied Geophysics, 159*(10), 2175–2199.

Hayes, G. P., Wald, D. J., & Johnson, R. L. (2012). Slab1.0: A three-dimensional model of global subduction zone geometries. *Journal of Geophysical Research Solid Earth, 117*(B1), 180–198.

Hu, Y., Bürgmann, R., Uchida, N., et al. (2016). Stress-driven relaxation of heterogeneous upper mantle and time-dependent afterslip following the 2011 Tohoku earthquake. *Journal of Geophysical Research Solid Earth, 121,* 385–411.

Ide, S., Baltay, A., & Beroza, G. C. (2011). Shallow dynamic overshoot and energetic deep rupture in the 2011 Mw 9.0 Tohoku-Oki earthquake. *Science, 332,* 1426–1428.

Kubo, A., Fukuyama, E., Kawai, H., et al. (2002). NIED seismic moment tensor catalogue for regional earthquakes around Japan: Quality test and application. *Tectonophysics, 356,* 23–48.

Lay, T. (2018). A review of the rupture characteristics of the 2011 Tohoku-Oki Mw 9.1 earthquake. *Tectonophysics, 733,* 4–36.

Michael, A. J. (1984). Determination of stress from slip data: Faults and folds. *Journal of Geophysical Research, 89,* 11517–11526.

Michael, A. J. (1987). Use of focal mechanisms to determine stress: A control study. *Journal of Geophysical Research, 92,* 357–368.

Michael, A. J. (1991). Spatial variations in stress within the 1987 Whittier Narrows, California, aftershock sequence: New techniques and results. *Journal of Geophysical Research, 96*(B4), 6303–6319.

Obana, K., Guo, F. J., Takahashi, T., et al. (2012). Normal-faulting earthquakes beneath the outer slope of the Japan Trench after the 2011 Tohoku earthquake: Implications for the stress regime in the incoming Pacific plate. *Geophysical Research Letters, 39,* L00G24.

Ozawa, S., Nishimura, T., Munekane, H., et al. (2012). Preceding, coseismic, and postseismic slips of the 2011 Tohoku earthquake, Japan. *Journal of Geophysical Research, 117,* B07404.

Sato, T., Hiratsuka, S., & Mori, J. (2013). Precursory seismic activity surrounding the high-slip patches of the 2011 Mw 9.0 Tohoku-Oki earthquake. *Bulletin of the Seismological Society of America, 103*(6), 3104–3114.

Simons, M., Minson, S. E., Sladen, A., et al. (2011). The 2011 magnitude 9.0 Tohoku-Oki earthquake: Mosaicking the megathrust from seconds to centuries. *Science, 332,* 1421–1425.

Sun, T., Wang, K. L., Iinuma, T., et al. (2014). Prevalence of viscoelastic relaxation after the 2011 Tohoku-Oki earthquake. *Nature, 514,* 84–87.

Tanaka, S. (2012). Tidal triggering of earthquakes prior to the 2011 Tohoku-Oki earthquake (Mw 9.1). *Geophysical Research Letters, 39,* L00G26.

Tomita, F., Kido, M., Ohta, Y., et al. (2017). Along-trench variation in seafloor displacements after the 2011 Tohoku earthquake. *Science Advances, 3*(7), e1700113.

Tormann, T., Enescu, B., Woessner, J., et al. (2015). Randomness of megathrust earthquakes implied by rapid stress recovery after the Japan earthquake. *Nature Geoscience, 8,* NGEO2343.

Uchida, N., & Matsuzawa, T. (2011). Coupling coefficient, hierarchical structure, and earthquake cycle for the source area of the 2011 off the Pacific coast of Tohoku earthquake inferred from small repeating earthquake data. *Earth Planets and Space, 63,* 675–679.

Wang, K. L., Hu, Y., & He, J. H. (2012). Deformation cycles of subduction earthquakes in a viscoelastic Earth. *Nature, 484,* 327–332.

Wessel, P., & Smith, W. (1998). New, improved version of generic mapping tools released, EOS Transactions. *American Geophysical Union, 79*(47), 579.

Zoback, M. L. (1992). First- and second-order patterns of stress in the lithosphere: The world stress map project. *Journal of Geophysical Research, 97*(B8), 11703–11728.

Zúñiga, F. R. (1993). Frictional overshoot and partial stress drop. Which one? *Bulletin of the Seismological Society of America, 83*(3), 939–944.

(Received April 6, 2018, revised January 6, 2019, accepted January 13, 2019, Published online February 5, 2019)

Pure Appl. Geophys. 176 (2019), 1335–1355

https://doi.org/10.1007/s00024-018-2073-9

Pure and Applied Geophysics

Heterogeneities in Stress and Strength in Tohoku and Its Relationship with Earthquake Sequences Triggered by the 2011 M9 Tohoku-Oki Earthquake

Keisuke Yoshida,[1] Akira Hasegawa,[1] Takeyoshi Yoshida,[2] and Toru Matsuzawa[1]

Abstract—Inland Tohoku has been recognized as under the WNW-ESE compressional stress state before the 2011 M9 Tohoku-Oki earthquake. Earthquakes that occurred there were characterized by reverse faulting with the compressional axis oriented in almost the WNW-ESE direction. The Tohoku-Oki earthquake reduced this WNW-ESE compressional stress and, therefore, should have suppressed the earthquake occurrence. However, several intensive earthquake sequences were triggered in inland Tohoku. In this study, we investigated the triggering mechanism of these remote earthquake sequences in the stress shadow based on the detailed distribution of stress orientations newly determined from pre-mainshock focal mechanism data. The spatial distribution of stress orientations shows that there exist some regions with anomalous stress fields even before the Tohoku-Oki earthquake on the spatial scale of a few tens of kilometers. This spatial heterogeneity in the stress field suggests that the differential stress magnitude in inland Tohoku is low (a few tens of MPa). Locations of the earthquake clusters tend to correspond to regions where the principal stress axis orientations of the pre-mainshock period are similar to those of the static stress change by the Tohoku-Oki earthquake. This observation suggests that these earthquake sequences were triggered by a local increase in differential stress due to the static stress change. However, a few swarm sequences occurred in central Tohoku with delays ranging from a few days to few weeks after the Tohoku-Oki earthquake despite the reduction in differential stress. These sequences have notable characteristics including upward migration of hypocenters. Such features are similar to the fluid-injection induced seismicity. The source regions of these swarms are located near the ancient caldera structures and major geological boundaries. The swarm activities were probably triggered by the upward fluid movement along such pre-existing structures. These observations demonstrate that information about the temporal evolutions of both stress and frictional strength is necessary to understand the triggering mechanism of earthquakes.

Electronic supplementary material The online version of this article (https://doi.org/10.1007/s00024-018-2073-9) contains supplementary material, which is available to authorized users.

[1] Research Center for Prediction of Earthquakes and Volcanic Eruptions, Graduate School of Science, Tohoku University, 6-6 Aza-Aoba, Aramaki, Aoba-ku, Sendai 980-8578, Japan. E-mail: keisuke.yoshida.d7@tohoku.ac.jp

[2] Institute of Mineralogy, Petrology, and Economic Geology, Graduate School of Science, Tohoku University, 6-6 Aza-Aoba, Aramaki, Aoba-ku, Sendai 980-8578, Japan.

Key words: 2011 Tohoku-Oki earthquake, stress shadow, aftershock, stress tensor inversion, Coulomb's stress, pore pressure.

1. Introduction

To obtain a comprehensive view of the earthquake generation process, it is necessary to understand the triggering mechanism of aftershocks or induced earthquakes after a large earthquake. In general, an earthquake occurs when the shear stress acting on a plane exceeds the frictional strength of the plane. We can use the Coulomb failure criterion as a simple approximation of the condition of the earthquake occurrence:

$$\tau = \mu(\sigma_n - P_P) \quad (1)$$

where τ is shear stress, μ is coefficient of friction, σ_n is normal stress, and P_P is pore pressure. Based on this equation, we can consider two causes for the occurrence of an earthquake: increases in shear stress τ and decreases in frictional strength $\mu(\sigma_n - \sigma_P)$ (e.g., Hainzl and Fischer 2002). It has been suggested that both accumulation of stress and reduction in frictional strength due to elevated pore pressure play an important role in earthquake generation (e.g., Hasegawa 2017; Hubbert and Rubey 1959; Miller 2013; Nur and Booker 1972; Rice 1992; Sibson 1992). Thus, the knowledge of the temporal change in stress and pore pressure after a large earthquake is essential to understand the triggering of an earthquake. In particular, the following effects can be considered: coseismic Coulomb stress increase by static deformation and dynamic wave propagation, postseismic Coulomb stress increase by postseismic slip and viscoelastic response, elastic interaction among triggered earthquakes, and the effect of pore pressure change associated with these processes.

The 2011 M9 Tohoku-Oki earthquake caused numerous earthquakes along the plate boundary, within the overriding plate, and within the subducting slab (Asano et al. 2011; Ishibe et al. 2011; Toda et al. 2011; Yukutake et al. 2011; Miyazawa 2011; Okada et al. 2011; Lengliné et al. 2012; Enescu et al. 2012; Kato et al. 2013; Shimojo et al.

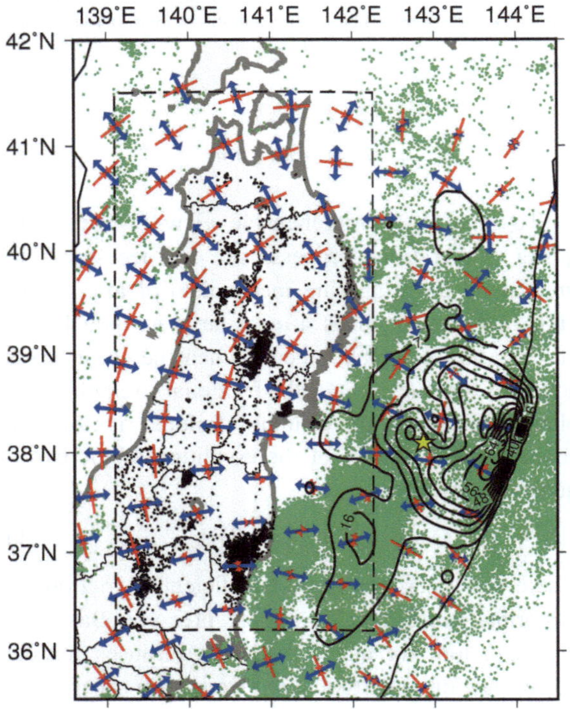

Figure 1
Hypocenter distribution of shallow earthquakes ($z < 40$ km) that occurred after the 2011 Tohoku-Oki earthquake in the overriding plate. Dots show hypocenters of earthquakes listed in the JMA unified catalogue during the 4-year period from March 11, 2011 with the JMA magnitude equal to or greater than 1.5. In this study, we focus on inland earthquakes shown by black colors. Other earthquakes are also shown by green. The rectangle with broken lines indicates the target region of this study. The black contours show the coseismic slip distribution of the Tohoku-Oki earthquake determined by Iinuma et al. (2012). Red and blue arrows show the orientations of the maximum principal stress ($\sigma 1$) axis and the minimum principal stress ($\sigma 3$) axis, respectively, at 10-km depth caused by the static stress change due to the Tohoku-Oki earthquake based on the elastic dislocation model of Okada (1992). The length of arrows varies according to the steepness of the plunge (i.e., shorter arrows are more steeply inclined). *NA* North American plate, *PA* Pacific plate. The arrow indicates the plate convergence direction

2014). A number of earthquakes took place in inland Japan (Fig. 1), which is densely covered by the nationwide seismic network. This provides a unique opportunity to study the mechanism of remote earthquake triggering.

The focal mechanisms of these earthquakes triggered by the Tohoku-Oki earthquake are readily explained by the static stress change by the Tohoku-Oki mainshock (e.g., Asano et al. 2011; Chiba et al. 2013; Hasegawa et al. 2012; Hasegawa and Yoshida 2015; Nakajima et al. 2013; Yoshida et al. 2012). (1) Interplate aftershocks do not occur inside the large slip area of the mainshock rupture. They instead focus on the edge of the large slip area (Asano et al. 2011; Hasegawa et al. 2012; Kato and Igarashi 2012; Nakamura et al. 2016). This feature can be well explained by the redistribution of shear stress inside and around the mainshock slip region. (2) Aftershocks within the subducting Pacific plate intensely occurred in the eastern and the western extensions of the large slip region of the mainshock rupture. *P*-axes of earthquake focal mechanisms were oriented to WNW-ESE in the western extension parallel to the coseismic slip direction, while *T*-axes were oriented to this direction in the eastern extension (Chiba et al. 2013; Hasegawa and Yoshida 2015). This pattern is also well explained by the effect of the static stress change caused by the mainshock slip. (3) Numerous normal fault earthquakes occurred in the overriding plate above the source region of the mainshock (Asano et al. 2011; Hasegawa et al. 2012). Such normal fault earthquakes were scarcely observed before the Tohoku-Oki mainshock, which suggests that the stress field rotated after the Tohoku-Oki earthquake by its static stress change (Hasegawa et al. 2012, 2012; Hardebeck 2012). These normal fault earthquakes seem to almost continuously occur from just above the large slip region to the Fukushima-Ibaraki border region near the Pacific coast. (4) In the Chubu-Kanto district in inland Japan, seismicity rate increased. In this region, the background stress field was characterized by NW–SE compression similar to the static stress change by the mainshock rupture. Yoshida et al. (2012) suggested that the increase in seismicity rate was caused by the increase in differential stress. (5) Even in regions very far from the source region of the Tohoku-Oki mainshock (the static stress change < 0.001 MPa), some earthquake sequences were triggered probably due to the dynamic wave propagation and the associated pore pressure change (Enescu et al. 2012; Kato et al. 2013; Yukutake et al. 2011).

In general, the occurrence of an earthquake not only increases the Coulomb stress, but also reduces it depending on location, referred to as the stress shadow (e.g., Simpson and Reasenberg 1994), even on fault planes with the same specific orientation by the modulation of stress field. Inland Tohoku was recognized as under the WNW-ESE compressional reverse faulting stress regime before the Tohoku-Oki earthquake (e.g., Hasegawa et al. 1994; Terakawa and Matsu'ura 2010; Townend and Zoback 2006; Yoshida et al. 2012). Indeed, all recent large ($M > 6$) earthquakes have reverse faulting focal mechanisms with the P axis oriented to WNW-ESE. This WNW-ESE compressional stress was at least partly formed by the mechanical coupling along the plate boundary (e.g., Suwa et al. 2006). The Tohoku-Oki earthquake reduced this WNW-ESE compressional stress and, therefore, the differential stress (Fig. 1), producing a stress shadow, especially in inland Tohoku, with the typical NNE-SSW striking reverse fault. Therefore, it should have suppressed the earthquake occurrence in inland Tohoku. However, intensive earthquake sequences were activated in inland Tohoku after the Tohoku-Oki event. Thus, the triggering mechanism of earthquake sequences in inland Tohoku is not simple.

In this study, we attempt to improve our understanding of the triggering mechanism of earthquakes in the stress shadow by focusing on the Tohoku-Oki earthquake. In the subsequent section, we review distinctive characteristics of the earthquake sequences triggered by the Tohoku-Oki earthquake in inland Tohoku, such as anomalous focal mechanisms (Sect. 2). We then examine the spatial distribution of the stress orientations in Tohoku to understand why earthquake sequences were triggered in inland Tohoku with anomalous focal mechanisms (Sect. 3). We also discuss the effect of pore pressure change, which probably played an important role in triggering the earthquake in Tohoku (Sect. 4).

2. *Reviews of Notable Features of Inland Earthquake Sequences Triggered by the Tohoku-Oki Earthquake*

2.1. *Hypocenter Distribution and Geological Structure*

Substantial earthquakes occurred in inland Tohoku just after the 2011 Tohoku-Oki earthquake. Hypocenter distributions of earthquakes in inland Tohoku for the period from March 11, 2011 to the end of 2012 are shown in Fig. 2a. Hypocenters of events after the Tohoku-Oki earthquake are concentrated at several locations in clusters, rather than distributed homogeneously throughout the Tohoku region. Hypocenters of earthquakes before the Tohoku-Oki earthquake are plotted over those after the Tohoku-Oki earthquake for comparison and are shown in Fig. 2b. The figure indicates that intense earthquake clusters after the Tohoku-Oki earthquake were mainly located in regions where the pre-mainshock seismicity rate was quite low. This is unlike the cases of the 1992 Landers earthquake and the 1995 Hyogo earthquake, in which post-mainshock seismicity increased in proportion to the level of prior seismicity (Mallman and Zoback 2007). Figure 2c shows the earthquake occurrence time plotted against latitude. Those locations of concentrated seismicity are northern Akita (N1), southern Akita (N2), the Yamagata-Fukushima border (C1), Sendai-Okura (C2), Yamagata (C3), the Fukushima-Ibaraki border (S1), and the Tochigi-Gunma border (S2). Although hypocenters seem scattered, the earthquake occurrence rate increases in Iwate-Prefecture (N3). The major geological boundaries in Tohoku are shown in Fig. 2b by broken curves. These geological boundaries go along the Fukushima-Ibaraki border (S1), Yamagata-Fukushima border (C1), Yamagata (C3), northern Akita (N1) and southern Akita (S1) earthquake clusters. All of the earthquake clusters in the central part of Tohoku (C1–C3) are located near ancient calderas (e.g., Yoshida et al. 2017).

Temporal distributions of the number of earthquakes for the period of 75 days before to 75 days after the Tohoku-Oki earthquake are shown in Fig. 3 for the eight earthquake clusters. The required magnitude was set to 2.0. Although the completeness

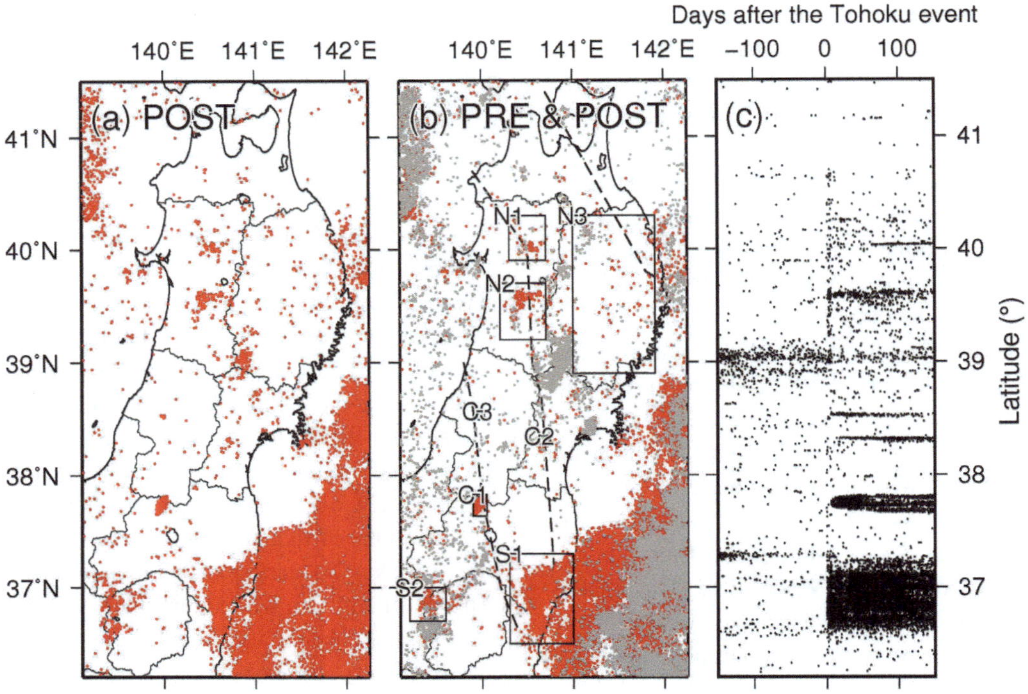

Figure 2
Seismicity before and after the 2011 Tohoku-Oki earthquake. **a** Shallow earthquakes ($z < 40$ km) after the Tohoku-Oki earthquake (2011/3/11–2012). Hypocenters are shown by red circles. **b** Shallow earthquakes before (1997–2011/3/11) the Tohoku-Oki earthquake are plotted over those after (2011/3/11–2012) the earthquake. Hypocenters are shown by gray circles (before) and red circles (after). The rectangles denote focal regions of intense seismicity after the earthquake. Major geological boundaries are shown by broken curves (Yoshida et al. 2014). **c** Space–time plot of earthquakes before and after the Tohoku-Oki earthquake in inland Tohoku. The occurrence time is plotted against latitude by black dots for 300 day periods before and after the Tohoku-Oki earthquake

magnitude was relatively low just after the Tohoku-Oki earthquake (e.g., Kato et al. 2013; Yoshida et al. 2015a), the earthquake numbers still increased abruptly after the Tohoku-Oki earthquake for all these clusters. Seismicity rates were very low before the Tohoku-Oki earthquake in the focal region of these earthquake clusters. However, it was reported that the seismicity rate decreased in the aftershock area of the 2008 M7.2 Iwate-Miyagi Nairiku earthquake, which is located in central Tohoku, in the stress shadow of the Tohoku-Oki earthquake (Suzuki et al. 2014). The initiations of triggered seismic activity of the three clusters C1–C3 in central Tohoku were delayed by a few days to a few weeks after the Tohoku-Oki earthquake, while triggered seismicity of the earthquake clusters N1–N3 and S1–S2 in the northern and southern parts of Tohoku began immediately after the Tohoku-Oki earthquake (Fig. 3).

2.2. *Change in Focal Mechanism*

A notable feature of the earthquake sequences triggered by the Tohoku-Oki earthquake is significant changes in predominant focal mechanisms after the mainshock; earthquakes with anomalous focal mechanisms, such as normal faulting with the *T*-axes oriented to WNW-ESE and strike-slip faulting with *P*-axes oriented to NNE-SSW, started to occur in a wide range of Tohoku from the region beneath the Pacific Ocean in the large slip area of the mainshock rupture (Asano et al. 2011; Hasegawa et al. 2012) to inland Tohoku apart from the source area (Kato et al. 2011; Yoshida et al. 2012). Figure 4 shows focal mechanisms of earthquakes in the overriding plate taken from Yoshida et al. (2012) and Hasegawa et al. (2012). The color shows faulting type based on the classification by Frohlich (1992). The typical focal mechanism in inland Tohoku was known to be

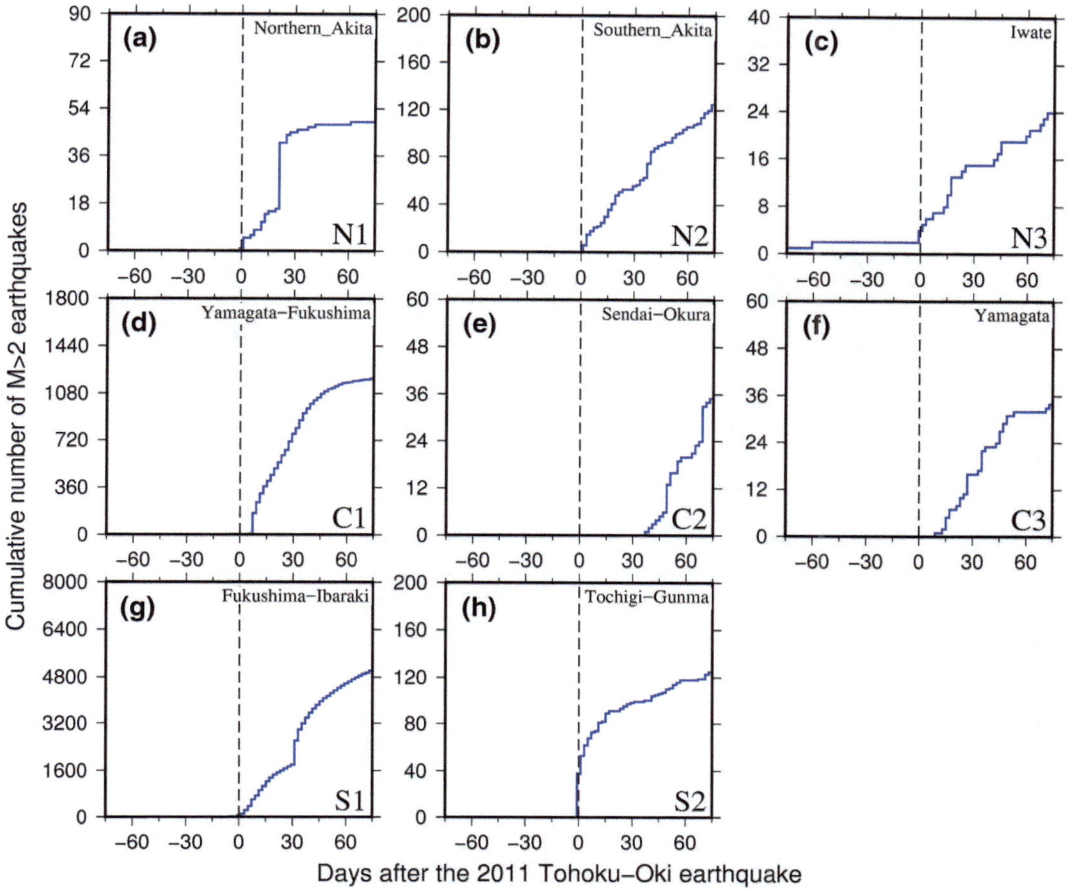

Figure 3
Temporal distributions of earthquake number for 150 day period before and after the Tohoku-Oki earthquake for eight earthquake clusters indicated in Fig. 2b. Cut off magnitude was set at 2.0. The blue curves show the cumulative number of earthquakes

reverse-faulting with WNW-ESE *P*-axis (e.g., Seno 1999; Terakawa et al. 2013). The spatially homogeneous WNW-ESE compressional reverse fault stress regime in Tohoku was supported by geodetically measured principal strain rate axes (e.g., Kato et al. 1998; Miura et al. 2002; Sagiya et al. 2000), geological structures (e.g., Nakamura and Uyeda 1980), earthquake focal mechanisms, and stress tensor inversion analyses (e.g., Hasegawa et al. 1994; Terakawa and Matsu'ura 2010; Townend and Zoback 2006; Yoshida et al. 2012). On the other hand, the earthquake clusters that occurred after the Tohoku-Oki earthquake in northern Tohoku (N1, N2) and those in southern Tohoku (S1) are characterized by strike-slip fault with NNE-SSW *P*-axes and normal fault with E–W ~ NW–SE *T*-axes, respectively. These focal mechanisms cannot be explained by the spatially homogeneous WNW–ESE compressional stress state, which was thought to be dominant before the Tohoku-Oki earthquake.

Three different hypotheses are suggested to explain these anomalous focal mechanisms. We summarize these hypotheses in this section as follows: a possibility of the rotation of the principal stress axes due to the static stress change of the Tohoku-Oki earthquake (Sect. 2.2.1), a possibility of the apparent stress rotation due to the pore pressure change (Sect. 2.2.2), and a possibility of the apparent stress rotation due to the stress heterogeneity in space (Sect. 2.2.3).

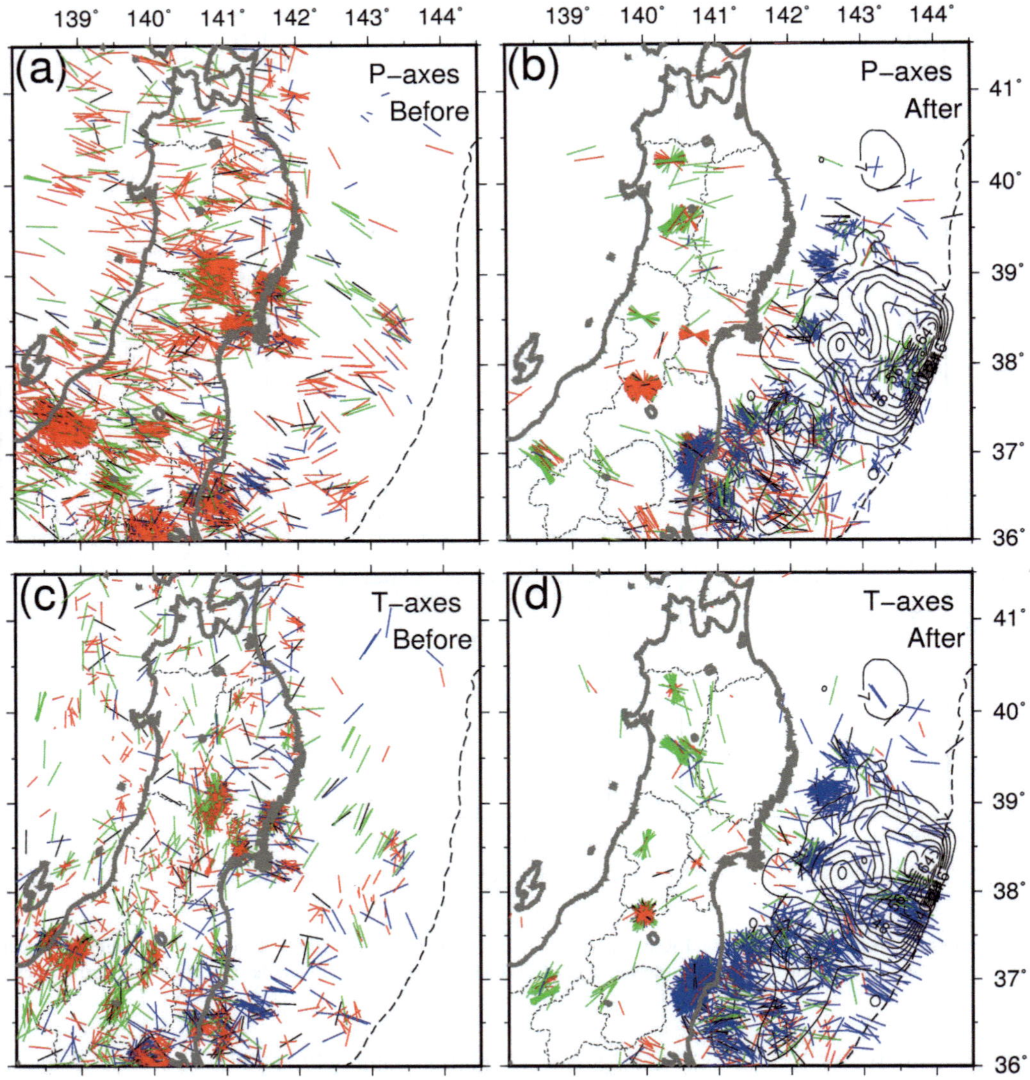

Figure 4
Distributions of *P*- and *T*-axes of focal mechanisms before and after the Tohoku-Oki earthquake. Focal mechanism data are the same as those in Yoshida et al. (2012) and Hasegawa et al. (2012). *P*-axes before and after the earthquake are shown in (**a**) and (**b**), respectively. *T*-axes before and after the earthquake are shown in (**c**) and (**d**), respectively. Red, green, blue, and black colors show reverse faulting, strike-slip faulting, normal faulting and odd type, respectively, based on the classification of Frohlich (1992). Data periods of Yoshida et al. (2012) and Hasegawa et al. (2012) are from 1997 to 20 July 2011 and from 2003 to 30 September 2011, respectively

2.2.1 Possibility of Stress Rotation due to Static Stress Change of the Tohoku-Oki Earthquake

A stress tensor inversion analysis based on earthquake focal mechanisms clearly shows that the principal stress axes in the hanging-wall right above the large slip area of the mainshock rupture was rotated by the Tohoku-Oki earthquake (Hasegawa et al. 2012). Moreover, Yoshida et al. (2012) suggested that the stress axes in a couple of areas in southern Tohoku (S1) and northern Tohoku (N1, N2) also rotated after the Tohoku-Oki earthquake. The stress orientations estimated from the stress tensor inversions in inland Tohoku (Yoshida et al. 2012) along with those in the source region beneath the Pacific Ocean (Hasegawa et al. 2012) are shown in Fig. 5. The figure shows that the orientations of the principal stress axes are significantly different before and after the Tohoku-Oki earthquake not only in the source region just above the large slip area, but also in inland Tohoku. Figure 5 (c) shows the distribution of principal stress axis orientations of the static stress change of the 2011 Tohoku-Oki earthquake computed by Yoshida et al. (2012) and Hasegawa et al. (2012).

Both in the source region beneath the Pacific Ocean and in inland Tohoku, the stress orientations after the Tohoku-Oki earthquake (Fig. 5b) were similar to those of the static stress change (Fig. 5c). This suggests that the stress axes were rotated by the static stress change of the Tohoku-Oki earthquake even in inland Tohoku. If this is the case, since the stress field after the earthquake (Fig. 5b) is considered as the sum of the background stress field (Fig. 5a) and the static stress change (Fig. 5c), magnitudes of deviatoric stress tensor components before the Tohoku-Oki earthquake were smaller than that of the static stress change (Yoshida et al. 2012). Yoshida et al. (2012) quantitatively evaluated deviatoric stress magnitude before the Tohoku-Oki earthquake based on the observed stress rotation by using Wesson and Boyd's (2007) method, and concluded that the differential stress magnitude, as a representative measure of the magnitudes of deviatoric stress tensor components, before the Tohoku-Oki earthquake needs to be less than $\sim$ 1 MPa in northern and southern Tohoku. Differential stress of $\sim$ 1 MPa seems to contradict typically estimated values of stress drop ranging from 1 to 10 MPa (e.g., Allmann and Shearer 2009; Oth et al. 2013). We also consider other possibilities which might explain these anomalous post-earthquake focal mechanisms similar to the static stress change.

2.2.2 Possibility of Apparent Stress Rotation due to Increase in Pore Pessure

Increasing pore pressure can allow unfavorably oriented fault planes, on which the Coulomb stress caused by the regional stress field is small, to slip (e.g., Sibson 1990). A recent fluid injection test confirms that earthquakes occur even on severely misoriented planes in proximity to the injection well during periods of high injection rates (Martinez-Garzon et al. 2016a). If such unfavorably oriented fault planes would be selectively activated by the increase in pore pressure, stress tensor inversion result would be biased. This could lead to apparent stress rotation after the earthquake. By considering the increase in pore pressure after the Tohoku-Oki

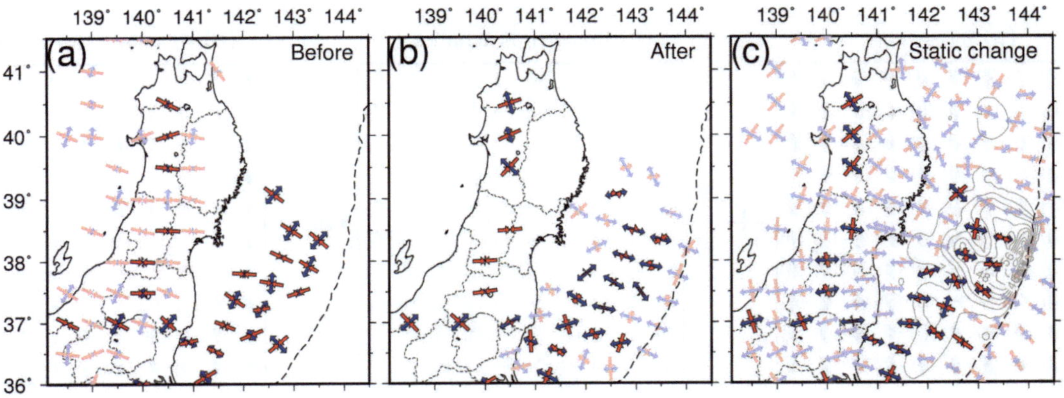

Figure 5

Principal stress axes **a** before and **b** after the Tohoku-Oki earthquake and **c** the static stress change of the earthquake in the overriding plate. Stress tensor inversion results and the static stress change by Yoshida et al. (2012) and Hasegawa et al. (2012) are shown by arrows. Red and blue arrows show $\sigma 1$ and $\sigma 3$ axes, respectively. Dark arrows in (**c**) highlight the results in the region where the stress tensor inversion is performed both before and after the Tohoku-Oki earthquake. The length of arrows varies according to the steepness of the plunge (i.e., shorter arrows are more steeply inclined) earthquake, Terakawa et al. (2013) attempted to explain the change in focal mechanisms after the earthquake. They demonstrated that focal mechanisms in northern Akita (N1) can be explained by the regional WNW–ESE compressional stress states. However, another mechanism is necessary to explain the anomalous focal mechanisms in southern Akita (N2) and southern Tohoku (S1) because their slip directions are largely different from those expected from the regional WNW-ESE compressional stress state. The increase in pore pressure alone does not explain why those anomalous focal mechanisms are similar to the static stress change.

2.2.3 *Possibility of Apparent Stress Rotation due to Heterogeneous Stress Field in the Considered Volume*

Another possibility that might explain the anomalous focal mechanisms is the effect of the spatial heterogeneity of stress orientations (e.g., Smith and Dieterich 2010). As an example, for the 1992 Landers earthquake, Hardebeck and Hauksson (2001) suggested that the stress fields rotated after the earthquake, while Townend and Zoback (2001) found the spatial heterogeneity in the stress field in and around the source region and the stress orientations remained almost stationary in the same locations.

Since the stress tensor inversion method needs to use multiple diverse focal mechanism data to constrain the stress orientation, it assumes the uniform stress orientation in a volume in which focal mechanism data are taken. Violation of this assumption can lead to an apparent rotation of stress field when combined with sample bias effects due to the static stress triggering. Figure 6 illustrates a simple situation in which the considered volume includes subregions where the stress orientations largely differ from those in the other regions (Fig. 6a). Given that the orientation of the static stress change is similar to those in such subregions (Fig. 6b), the differential stress and thus the shear stress on optimally oriented fault locally increases there (Fig. 6c). Earthquakes can be selectively triggered in such subregions with locally anomalous stress axes. This might lead to the observation of the apparent change in focal mechanisms and, therefore, an apparent stress rotation. The spatial extent of the activation depends on the scale of the spatial change.

The spatial heterogeneity in stress orientations in inland Tohoku has been recently found by determining many focal mechanisms using data from the dense seismic network covering this area (Yoshida et al. 2015a). Yoshida et al. (2015a) found that there were regions with stress orientations largely different from the regional WNW-ESE compressional stress state. Fukushima-Ibaraki (S1) is one such region; the

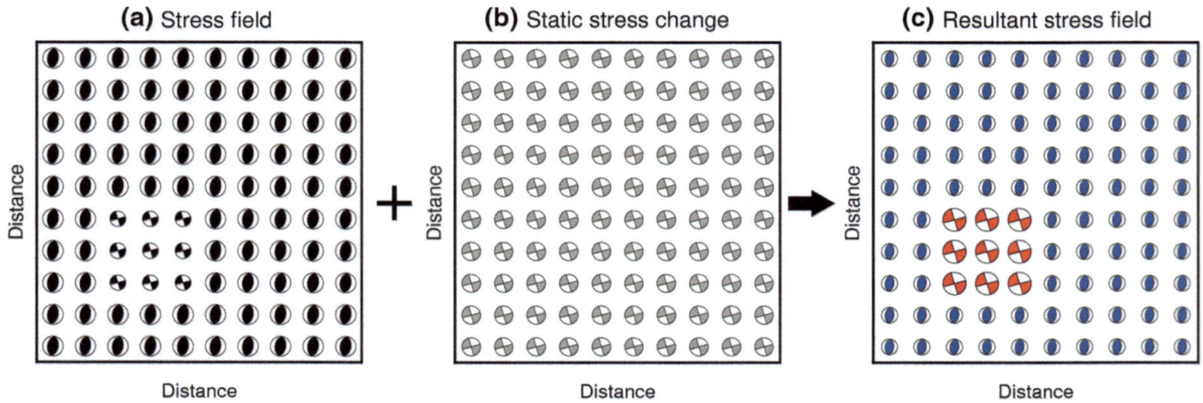

Figure 6
A schematic illustration explaining the drastic change of focal mechanisms after the earthquake by the spatial change in stress field. Orientations of principal stress axis are represented by beach-balls. *P*-, *B*-, and *T*-axes corresponds to $\sigma1$, $\sigma2$, and $\sigma3$ axes, respectively. **a** Stress field before the earthquake. **b** Static stress change caused by the earthquake. **c** Resultant stress field. Red and blue indicate increase and decrease in differential stress, respectively

normal fault stress regime here exists in the shallower part ($z < 12$ km) even before the Tohoku-Oki earthquake, while a reverse fault stress regime exists in the deeper portion. This should have caused the local increase in differential stress in the shallower part of the S1 region by the Tohoku-Oki earthquake because the stress orientation there is similar to the static stress change of the Tohoku-Oki mainshock. The normal fault stress regime in the Fukushima-Ibaraki region was also found previously by Imanishi et al. (2011) who focused their stress inversion study on the earthquake sequence in this region. These observations indicate that the stress rotation in this region after the Tohoku-Oki earthquake were artefact and came from the spatial change in stress fields inside the region in which the stress field is assumed to be uniform.

In northern Tohoku, the WNW-ESE compressional stress state seems to be homogeneously distributed (Fig. 5a). This WNW–ESE compressional stress and, therefore, differential stress should have been reduced by the Tohoku-Oki earthquake. The reason why earthquakes with such focal mechanisms, which are unlikely to occur under the WNW–ESE compressional stress regime, were intensively triggered in northern Tohoku after the Tohoku-Oki earthquake has not been clarified.

2.3. *Summary of Earthquake Sequences Triggered by the Tohoku-Oki Earthquake in Inland Tohoku*

1. The Tohoku-Oki earthquake triggered intensive earthquake sequences even in the stress shadow of inland Tohoku. Hypocenters of these triggered earthquakes were scattered in a wide area concentrated at several locations in clusters rather than distributed homogeneously in space throughout the inland region of Tohoku (Fig. 2).
2. Earthquakes in a few such earthquake clusters (N1, N2, S1) have strikingly different focal mechanisms, such as normal faulting type with *T*-axes oriented to WNW–ESE and strike-slip faulting type with *P*-axes oriented to NNE–SSW, from the typical one in inland Tohoku. Focal mechanisms in the two clusters N2 and S1 cannot be explained by the regional WNW–ESE compressional stress even if pore pressure increases and unfavorably oriented fault planes selectively slip.
3. The orientations of the stress axes that caused earthquake clusters with anomalous focal mechanisms are strikingly similar to those of the static stress change by the Tohoku-Oki earthquake. This observation suggests the two possibilities: (1) the stress axes in areas of those clusters locally rotated by the Tohoku-Oki earthquake and (2) the stress axes orientations before the Tohoku-Oki earthquake have some variations in space. The stress fields have a strong depth variation in the Fukushima-Ibaraki region (S1) and the occurrence of earthquake sequence there with many normal fault earthquakes was due to the local increases in differential stress by the static stress change of the Tohoku-Oki earthquake. In northern Tohoku, the WNW–ESE compressional stress field similar to the regional stress state is homogeneously distributed before the Tohoku-Oki earthquake. The reason why the earthquake clusters with anomalous focal mechanisms, which are unlikely to occur under the WNW-ESE compressional stress regime, are intensively activated in northern Tohoku has not been clarified.

3. *Spatial Heterogeneity of Stress Axes in Tohoku before the Tohoku-Oki Earthquake*

Information about the spatial heterogeneity of the stress orientation is crucial for understanding the triggering mechanism of earthquakes as described in the previous section. In this section, we focus on inland Tohoku and examine in detail the spatial variation of the principal stress axes before the Tohoku-Oki earthquake based on new focal mechanism dataset to understand why earthquake sequences with anomalous focal mechanisms were triggered in Tohoku.

3.1. *Focal Mechanism Data*

Stress orientations are estimated by inverting focal mechanism data using the method of Michael (1987) and Hardebeck and Michael (2006). (1) We use new focal mechanism data of earthquakes for the period 1977 to 2003, which we determined in the present study. (2) Focal mechanism data determined by Yoshida et al. (2015a) for the period 1997 to the occurrence of the 2011 Tohoku-Oki earthquake are used. For the determination of focal mechanisms, we used *P*-wave first motion polarity data manually picked during routine processing at the Tohoku University seismic network. Focal mechanisms were determined in the same way as Yoshida et al. (2015a) by applying the method of Hardebeck and Shearer (2002) to the *P*-wave polarity data. We determined focal mechanisms if the *P*-wave polarity data was larger than 10 and the azimuthal gap was less than 45°. In the method of Hardebeck and Shearer (2002), focal mechanism solutions are evaluated and classified into A–F ranks depending on confidence levels. Only the events with rank A or B were used here. As a result, we could determine 919 focal mechanisms. The number of focal mechanisms with rank A and B are 180 and 739, respectively. The mean numbers of polarity data used for the determination of rank A and B focal mechanisms are 38.7 and 22.7, respectively. The mean RMS values of angular differences of possible nodal planes are 18.7° and 27.5 for rank A and B focal mechanisms, respectively. Examples of focal mechanisms are shown in Fig. S1. Thus, our dataset contains a total of 3,118 focal mechanisms. The lateral distribution of focal mechanisms is shown in Fig. 7.

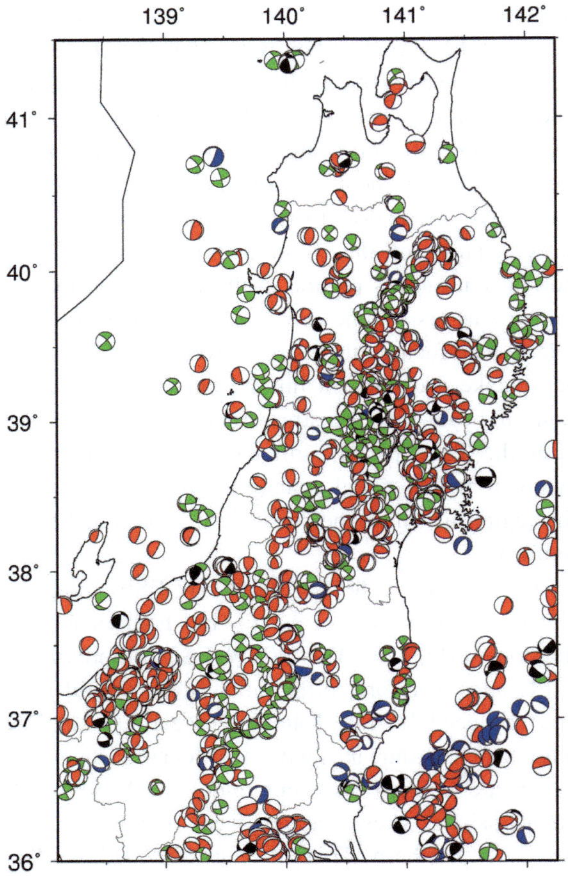

Figure 7
Distribution of focal mechanism data used in this study. Red, green, blue, and black beach-balls show reverse faulting, strike-slip faulting, normal faulting, and odd type, respectively, based on the classification of Frohlich (1992)

3.2. Stress Tensor Inversions

To estimate the stress orientation, we assumed that: (1) earthquakes occurred along pre-existing weak planes having various strikes and dips, (2) slip occurred in the direction of maximum resolved shear stress on those planes, and (3) the stress orientation was uniform in the volume from which the data were taken. By the stress tensor inversion, the orientations of the principal stress axes and stress ratio $R = (\sigma1 - \sigma2)/(\sigma1 - \sigma3)$ are constrained, although their magnitudes cannot be known.

To investigate the lateral variation of the stress orientations, the study area was divided into several subareas. For each subarea, focal mechanisms were inverted assuming a homogeneous stress orientation. However, given that these results may vary depending on how the subareas are defined, we used two different approaches for subdividing the study area following Yoshida et al. (2015a, 2016a).

The first approach is similar to that of Hardebeck and Hauksson (2001). First, we placed a 5-km spaced grid net over the study area. Then, we carried out the stress inversion of Michael (1984, 1987) at each grid node using all events located within 20 km if the number of such events was < 15. Otherwise, we used the 15 events closest to the grid node. If there were < 10 qualifying events, we did not estimate the principal stress orientations at that grid node. In this approach, the spatial resolution of stress fields depends on the density of focal mechanisms and thus on the location. The orientations of σHMAX computed based on the equation by Lund and Townend (2007) are shown in Fig. 8a.

The second approach applies the damped stress inversion method of Hardebeck and Michael (2006) to the focal mechanism data. This involved placing a grid with 0.5° spacing over the study area and assigning each focal mechanism to the nearest grid node. This method avoids the creation of apparent spatial variability, which is actually an artifact due to over-fitting noisy data or non-uniquely fitting data that does not completely constrain the stress tensor. The spatial damping parameter chosen was 0.6 on the basis of a trade-off between model length and data variance. The result of the stress tensor inversions is shown in Fig. 8b. We consider that the second approach yields a more macroscopic and stable view of stress fields than the first approach, while the first approach can provide the higher spatial resolution.

3.3. Spatial Distribution of Stress Orientations before the Tohoku-Oki Earthquake and Static Stress Change

Figure 8a and b shows that the stress analyses performed using the two different approaches basically yield similar results. The inland stress field is characterized by WNW–ESE compression except for the north and south outer arcs in which the σ1 axes are oriented nearly N–S and vertical, respectively. The earthquake sequences N3 and S1 are included in these north and south outer arc regions. Stress orientations are similar to the regional WNW–ESE compressional stress state in regions corresponding to the earthquake sequences N1, N2, C1, C2, and C3 from the result of the second approach (Fig. 8b) which focuses on a macroscopic view (∼ 50 km) of stress fields. On the other hand, we can see some deviations of the stress orientations in regions corresponding to the earthquake sequences N1 and N2 in the first approach (Fig. 8a), which focuses on a spatially high-resolution view of the stress field. The orientations of principal stress axes in N1 and N2 are

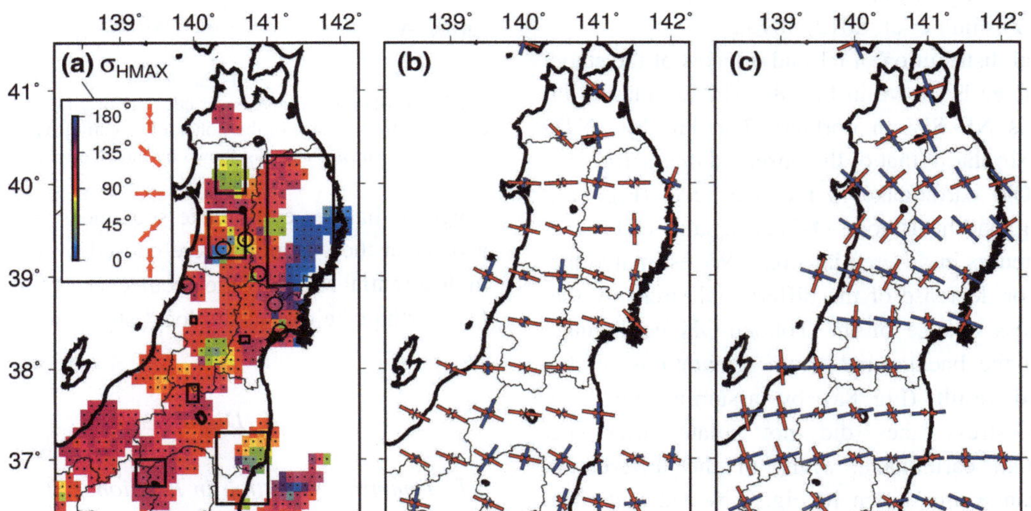

Figure 8
Stress field before the Tohoku-Oki earthquake and the static stress change. **a** Orientations of the observed maximum horizontal compressive stress σHMAX measured in degrees clockwise from north determined based on the first approach described in Sect. 3.2. Orientations of σHMAX are shown by the color scale at the left top. **b** Orientations of the best fit σ1 and σ3 axes projected onto a gridded horizontal plane determined based on the second approach described in Sect. 3.2. **c** Orientations of σ1 and σ3 axes of the static stress change of the Tohoku-Oki earthquake at 20 km depth. In Fig. 8**b** and **c**, σ1 and σ3 axes are indicated by red and blue arrows, respectively, at each grid reference. The length of arrows varies according to the steepness of the plunge (i.e., shorter arrows are more steeply inclined)

NE-SW, which are similar to those observed after the 2011 Tohoku-Oki earthquake (Fig. 5b). This suggests that anomalous stress fields existed there even before the Tohoku-Oki earthquake and that the differential stress magnitude in inland Tohoku can be much higher than 1 MPa as estimated by Yoshida et al. (2012) based on the stress rotation after the earthquake. The difference between the first and second approaches suggests that the stress field is heterogeneous in Tohoku with a scale less than a few tens kilometers.

We estimated the static stress change of the Tohoku-Oki earthquake for comparison with the stress field before the earthquake. We used the coseismic slip distribution determined by Iinuma et al. (2012) by assuming the rigidity of 30 GPa and the Poisson's ratio of 0.25. We used the DC3D code (http://www.bosai.go.jp/study/application/dc3d/DC3Dhtml_E.html) based on the analytical solution for the homogeneous half space summarized by Okada (1992). Although there are various published coseismic slip models (Brown et al. 2015), the results in inland Tohoku, relatively far from the source region, scarcely depend on the difference of the models (Yoshida et al. 2012, 2018a).

Spatial distribution of $\sigma1$ and $\sigma3$ axes of the static stress change is shown in Fig. 8c. The orientation of $\sigma1$ axis is NE–SW in northern Tohoku (N1, N2), which is similar to that of the observed one before the Tohoku-Oki earthquake in the fine scale (Fig. 8a), indicating that the Tohoku-Oki earthquake lead to the local increases in differential stress. We estimated the increase or decrease of the differential stress by the static stress change of the Tohoku-Oki earthquake based on the background stress orientations of the finer grid result (Fig. 8a) by assuming that the principal stress axes did not rotate after the Tohoku-Oki earthquake. Signs of the differential stress change are shown in Fig. 9 by red and blue colors. Figure 9 indicates whether or not differential stress increased by the static stress change at each grid node. Although the computation result is somehow affected by our limited resolution of stress fields, we can see that differential stress increases only locally in some regions. Figure 9 shows the differential stress increased in the focal regions of the earthquake sequences N1, N2, S1, and S2. This suggests that the earthquake sequences N1, N2, S1, and S2 in these regions were caused by the increase in differential stress by the occurrence of the Tohoku-Oki earthquake due to the local stress heterogeneity.

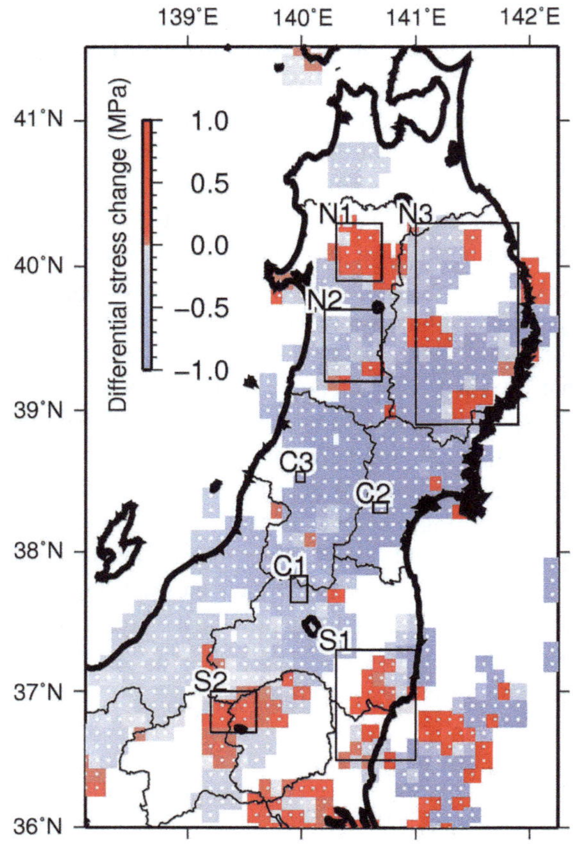

Figure 9
Spatial distribution of increase or decrease of differential stress by the static stress change of the Tohoku-Oki earthquake. Differential stress change is shown by the color scale

4. Discussion

4.1. Temporal Variation in Frictional Strength due to Upward Fluid Movement

The existence of the spatial stress heterogeneities in inland Tohoku can explain why the earthquake sequences in northern and southern Tohoku (N1, N2, N3, S1, and S2) were triggered by the Tohoku-Oki earthquake. However, it is difficult to explain the activation of the earthquake sequences in central Tohoku (C1, C2 and C3) (Fig. 10). Since they have

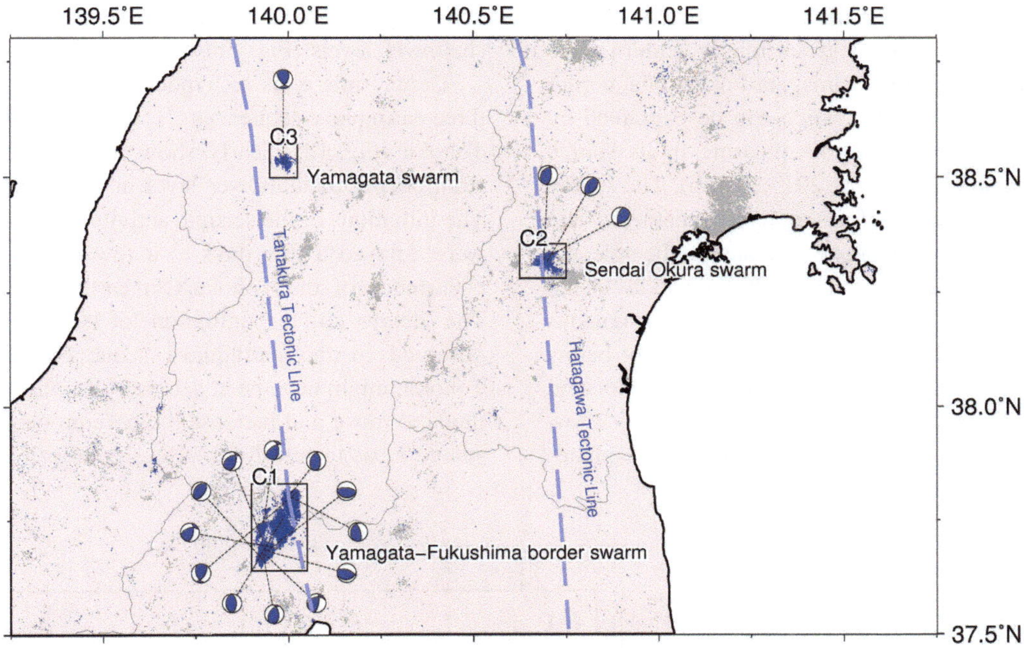

Figure 10
Hypocenter distribution of earthquakes in central Tohoku before and after the Tohoku-Oki earthquake. Gray and blue circles show hypocenters before and after the earthquake, respectively. Beach-balls show focal mechanisms of events in the earthquake sequences C1, C2, and C3 listed in the F-net moment tensor catalogue (Fukuyama et al. 1998, 2003) and the JMA catalogue. Major geological boundaries are shown by broken curves (Yoshida et al. 2014)

reverse fault focal mechanisms with *P*-axes oriented WNW–ESE similarly to the typical focal mechanism in Tohoku, shear stress on the fault planes decreased by the WNW-ESE extension associated with the Tohoku-Oki earthquake (Yoshida et al. 2018a).

Previous studies suggest that those earthquake sequences in central Tohoku were activated in response to the increase in pore pressure due to the upwelling fluids facilitated by the WNW-ESE extension associated with the Tohoku-Oki earthquake (Terakawa et al. 2013; Okada et al. 2011; Yoshida et al. 2016b, 2017, 2018a, b). Those earthquake sequences are characterized by the swarm-like seismicity pattern with a distinct migration behavior of hypocenters, as summarized by Okada et al. (2011). Such migration behaviors of hypocenters are similar to the fluid-injection induced seismicity (e.g., Julian et al. 2010; Rutledge et al. 2004; Shapiro et al. 1997). Differential stress magnitude in inland Tohoku is estimated to be a few tens of MPa or so by recent studies based on the correlation of stress field with topography (Yoshida et al. 2015a) and with the static stress change by the recent large ($\sim$ M7) inland earthquakes (Yoshida et al. 2014, 2015b, 2016c). Given that differential stress magnitude in Tohoku is as small as a few tens of MPa, pore pressure must be much higher than hydrostatic to cause earthquakes under expected effective normal stress at seismogeneic depth and the typical value of coefficient of friction obtained by laboratory experiments (e.g. Sibson 1990). It should be noted that the differential stress magnitude of a few tens of MPa is much higher than 1 MPa estimated based on the stress rotation after the 2011 Tohoku-Oki earthquake in inland Tohoku (Yoshida et al. 2012), which was apparently obtained from ignoring the spatially heterogeneous stress field (Fig. 8a).

A plausible cause for the reduction in the frictional strength is increasing pore pressure (e.g., Hasegawa 2017; Hubbert and Rubey 1959; Miller 2013; Nur and Booker 1972; Rice 1992; Sibson 1992).

In fact, temporal variations in frictional strengths, stress drops, b-values, and seismicity pattern have been detected for the Yamagata-Fukushima border earthquake swarm (C1), which can be explained by the temporal change in pore pressure in its source area (Yoshida et al. 2016b, 2017, 2018b). The source area of this swarm is located just beneath the late Miocene Ohtoge caldera and is related to the volcanic structure (Kanisawa et al. 2006; Yoshida et al. 2016b), which is believed to include shallow igneous bodies with hydrothermal fluids immediately below (Yoshida et al. 2005, 2014a, b). Precisely relocated hypocenters by Yoshida et al. (2018b) clearly show that they are concentrated on several discrete planes and migrate along those planes from deeper to shallower levels (Fig. 11).

Depth–time plots of hypocenters of events in the three earthquake sequences C1, C2, and C3 in central Tohoku (Fig. 12) clearly show hypocenters moved from deeper to shallower level in all the sequences. The initiations of the seismic activity of these swarms were delayed a few days to a few weeks after the occurrence of the Tohoku-Oki earthquake (Fig. 3). The delays of the initiation of seismicity, only observed in the earthquake sequences in central Tohoku, might have been necessary for the upwelling fluids to move and increase the pore pressure in their source areas to fulfill the failure criterion.

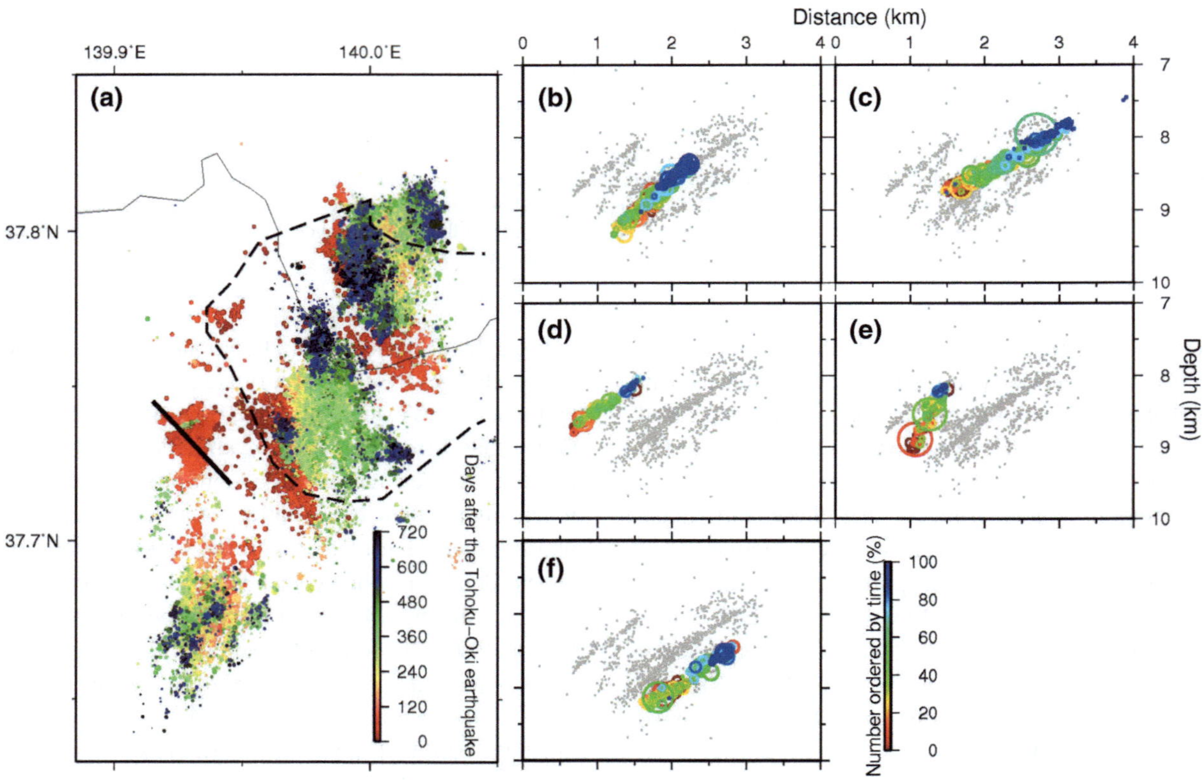

Figure 11

Hypocenter distribution of the earthquake swarm C1 in the Yamagata-Fukushima border. Hypocenters were relocated by Yoshida et al. (2018b). **a** Map view showing hypocenter migration. Dots show hypocenters of earthquakes for the period of 800 days from the beginning of the swarm activity. Elapsed time after the Tohoku-Oki earthquake is shown by the color scale. The thin broken line denotes the border line between Yamagata and Fukushima prefectures. The thick broken line denotes the rim of the Ohtoge caldera (Kanisawa et al. 2006). **b–f** Cross-sectional views showing hypocenter migration along five discrete planes in the western cluster of this earthquake swarm. Hypocenters are shown separately on the five discrete planes plotted on a vertical cross section along the solid line shown in (**a**). Color scale shows the sequence of earthquake occurrence ordered by time. Sizes of circles correspond to fault diameter assuming a stress drop of 10 MPa. Gray circles show hypocenters of other earthquakes

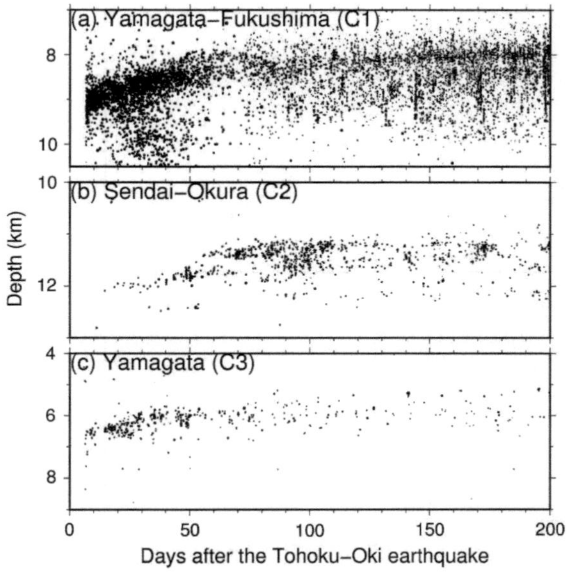

Figure 12
Depth–time plots of the hypocenters of events in the three earthquake sequences in central Tohoku for **a** the Yamagata-Fukushima border swarm C1; **b** the Sendai-Okura swarm C2; **c** the Yamagata swarm C3

All these swarms are located near the ancient caldera structures (Kanisawa et al. 2006; Yoshida et al. 2016b). High b-values were obtained all for these swarms ranging from 1.3 to 1.6 (Fig. 13d, e, f), which are significantly high compared to the other earthquake sequenses (Fig. 13a, b, c, g, h) and the typical estimation value of $\sim$ 0.8 in Tohoku (e.g., Cao and Gao 2002), which might reflect high pore pressure in the source area as suggested from the observations of the fluid-injection induced seismicity (e.g., Wyss 1973; Bachmann et al. 2012). Indeed, Yoshida et al. (2017) reported that b-value changes from 2 to 1 in the source area of the Yamagata-Fukushima border earthquake swarm in association with decreasing pore pressure.

These similarities support that all the three earthquake swarms in central Tohoku were caused by the increase in pore pressure due to upward fluid movements facilitated by the decrease in WNW-ESE compressional stress due to the Tohoku-Oki earthquake. Breaking of low-permeability seals due to the ground shaking might have helped fluid move to the shallower levels. The fluids permeated into several pre-existing planes, reduced the frictional strengths, and satisfied the failure criteria, causing the earthquake swarms despite the reduction in the Coulomb stress. The fluids probably further migrated upward along the planes, which is manifested as the hypocenters migrating along the planes.

Fluid movements after the Tohoku-Oki earthquake affected earthquake occurrences not only in the central part of Tohoku. Kosuga et al. (2014) reported that the earthquake swarm in northern Tohoku (N1) also exhibits a distinct migration behavior of hypocenters. Namely, both the differential stress and the pore pressure increased in this cluster after the Tohoku-Oki earthquake. Delayed triggered swarms with the migration behaviors of hypocenters were also observed after the Mw 7.8 Dusky Sound and the Mw 7.1 Darfield earthquake in New Zealand probably due to the fluid diffusion (Boese et al. 2014). These observations suggest that the pore pressure change after the occurrence of a large earthquake more or less universally plays an important role for subsequent earthquake sequences in association with the stress change.

4.2. A Possible Cause of Heterogeneity of Stress Field

We observed the spatial heterogeneities in stress field in inland Tohoku even before the Tohoku-Oki earthquake (Fig. 8b) based on the stress tensor inversion analyses. The stress tensor inversion methods (e.g., Gephart and Forsyth 1984; Michael 1987) were devised to avoid the effects of the bias in fault planes (McKenzie 1969) by utilizing the diversity of focal mechanisms. It is still difficult, however, to completely distinguish the effect of the fault plane bias from the variations of stress field (Townend 2006). In fact, estimation errors of focal mechanism lead to apparent diversity of focal mechanism. Furthermore, even small earthquakes can perturb the nearby stress field, which violates the assumption of the uniform stress orientation in the considered volume. One way to remove this effect is the use of declustered seismicity catalogue as advocated by Martinez-Garzon et al. (2016b); however, this reduces the number of available focal mechanism data to examine the detailed spatial variation. Therefore, in this study, we have focused on a relatively

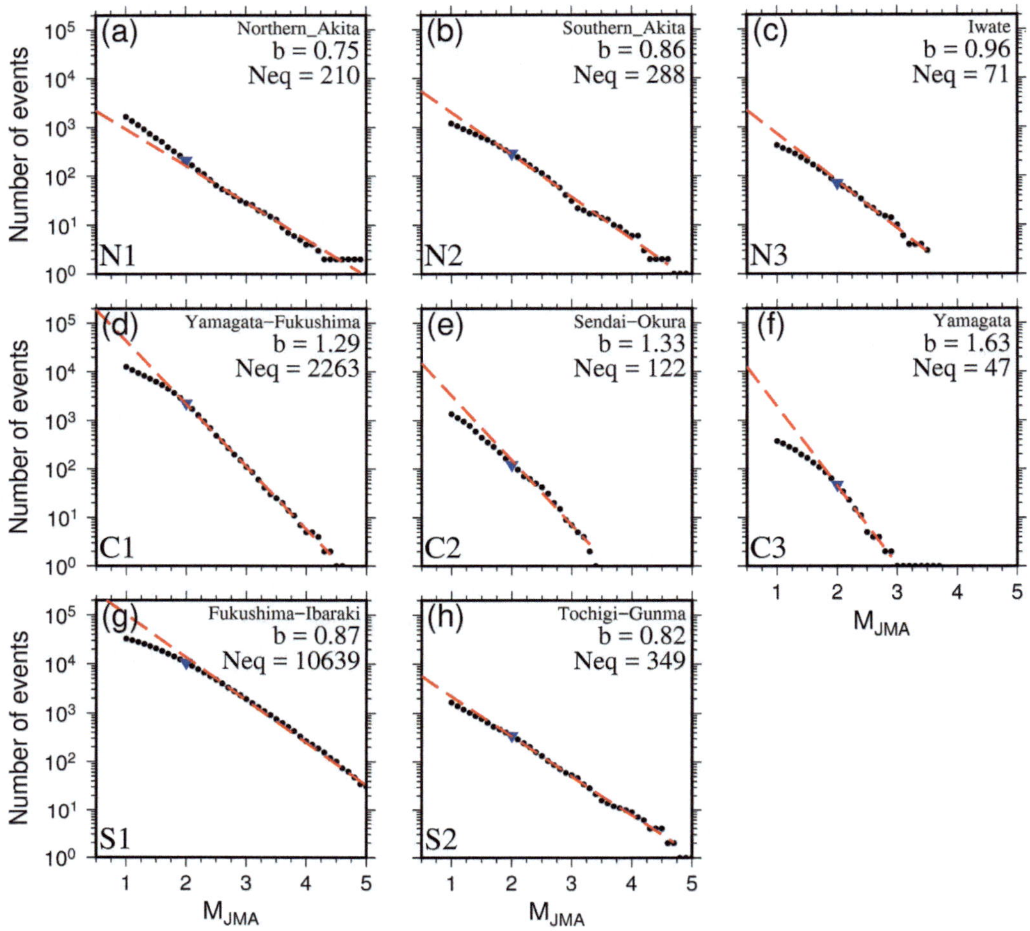

Figure 13
Magnitude-frequency distributions of earthquakes in the earthquake clusters. Black dots indicate the cumulative number of earthquakes. Red broken lines show the best-fit Gutenberg-Richter relation. Blue inverted triangles indicate the cut-off magnitude for the fitting of the Gutenberg-Richter relation

large-scale pattern of stress fields and have discussed the variation in stress fields with such a length scale. Hypocenters of events with anomalous focal mechanisms that occurred after the Tohoku-Oki earthquake are concentrated at several locations in clusters (N1, N2, S1), rather than distributed homogeneously throughout the Tohoku region (Fig. 2a), which suggests that the predominant length scales of the heterogeneity of stress field roughly correspond to those of the clusters (> 10 km).

To confirm the existence of the spatial heterogeneity in stress fields, we used the angle between the direction of the slip and the maximum resolved shear stress computed from the regional stress field (misfit angle). For that, we first performed the stress tensor inversion by assuming that the stress field is uniform in the entire Tohoku, and computed misfit angle of each focal mechanism. The obtained stress field shows the WNW–ESE compressional reverse faulting stress regime (Fig. 14a) with the mean misfit angle of approximately 35° (Fig. 14b). We selected the fault plane from the two nodal planes as having the smaller misfit angle. We then computed the mean values of misfit angles at each grid node based on focal mechanism data used for the stress tensor inversions in Fig. 8a. The spatial distribution of misfit

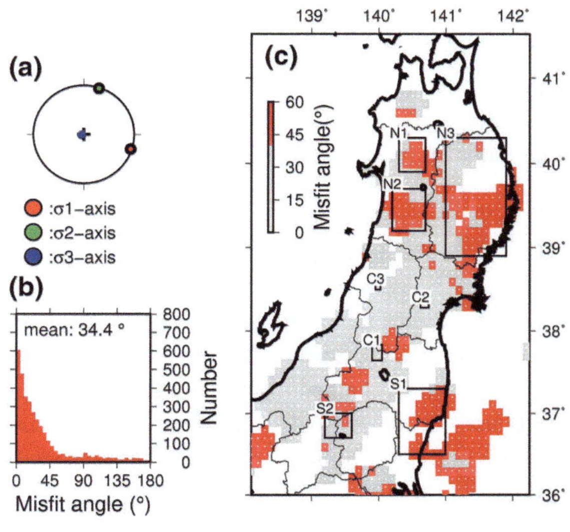

Figure 14
a Principal stress axis orientations determined by the stress tensor inversion based on all the focal mechanism data in Tohoku. Red, green, and blue circles show $\sigma1$, the $\sigma2$, and $\sigma3$-axes, respectively. Circles denote the best-fit solution. **b** Histogram showing the frequency distribution of misfit angles. **c** Spatial distribution of the mean values of misfit angles of focal mechanisms. The mean values are plotted at the same grids used in the stress tensor inversion of Fig. 8a

angles in Fig. 14c indicates a deviation of the stress field from the WNW–ESE compressional reverse faulting stress regime (Fig. 14a). The spatial variation is consistent with the stress tensor inversion results in Fig. 8a. In particular, mean misfit angles are high in the focal regions of earthquakes that occurred after the 2011 Tohoku-Oki earthquake (N1, N2, N3 and S1), which confirms that these earthquakes occurred in regions with anomalous stress field.

One possible explanation of the local stress heterogeneities is the effect of the static stress change of large earthquakes that occurred previously. The static stress change of an earthquake can rotate the principal stress axes in and around the source region (e.g., Hardebeck and Hauksson 2001; Wesson and Boyd 2007), if magnitudes of deviatoric stress tensor components of the background stress field are small compared to those of the static stress change. The differential stress magnitude in inland Tohoku is estimated to be $< \sim$ a few tens of MPa based on the correlation of orientations of the principal stress axes with the topography (Yoshida et al. 2015a). This value of the differential stress is similar to that estimated by Hasegawa et al. (2012) for the source area of the Tohoku-Oki earthquake based on the stress rotation observed after the earthquake. Indeed, spatial heterogeneities in the stress orientations have been detected in the focal regions of the recent three large earthquakes in inland Tohoku: the 2003 M6.3 Northern Miyagi Prefecture earthquake (Yoshida et al. 2016c), the 2008 M7.2 Iwate-Miyagi Nairiku earthquake (Yoshida et al. 2014), and the 2011 M7.0 Fukushima-Hamadori earthquake (Yoshida et al. 2015b). The spatial patterns of the principal stress axes in their source areas after the mainshocks are well explained by the static stress change of those mainshocks, suggesting that the differential stress magnitude in inland Tohoku is very small ($<$ a few ten MPa) and the effects of large earthquake can produce the spatial heterogeneities in stress orientations.

Conversely, if differential stress magnitude in Tohoku is small (less than a few tens of MPa), the principal stress axes should rotate by various effect such as effects of the topography and the static stress change of a large earthquake. The approximate focal regions of large ($M > 6.5$) earthquakes that occurred in Tohoku before 1950 listed by Usami (2003) are shown in Fig. 8a by circles. The focal regions indicated roughly correspond to the regions with anomalous stress orientations. This suggests the possibility that the stress axes locally rotated in these regions after the large earthquakes, and this time the earthquake sequences were triggered by the Tohoku-Oki earthquake because of the local increase in differential stress due to the stress heterogeneities thus produced.

The stress field in northern Tohoku is characterized by the WNW-ESE compression, as well as the N-S compression depending on location (Fig. 8a). This suggests the existence of other causes of the regional stress in northern Tohoku besides the relative movement of the Pacific plate and the overriding plate. Seno (1999) assumed a higher magnitude of N-S compressional stress in central and eastern Japan than in western Japan because of the collision of the Izu Peninsula with central and eastern Japan from south (Matsuda 1971). In fact, the orientation of $\sigma1$-axis is rotated to NNW-SSE direction locally near the Izu Peninsula (Ukawa et al. 1982). The sliver motion of the Kuril fore-arc located just north (e.g., Kimura 1986; Acocella et al. 2008) might not only sustain the collision force but also increase N-S compressional force in northern Tohoku. The gravitational collapse of the mountain range (Wang and He 1999) might reduce the WNW-ESE compressional stress.

5. Conclusions

We examined the spatial variation in stress field in inland Tohoku to understand the triggering mechanisms of earthquake sequences by the 2011 Tohoku-Oki earthquake that occurred in the stress shadow. Focal mechanisms of shallow earthquakes in inland Tohoku are newly determined based on the *P*-wave first motion polarity data, and we inverted them for the stress orientations.

The obtained spatial distribution of the stress orientation shows some variations in inland Tohoku even before the Tohoku-Oki earthquake. Earthquake clusters triggered by the Tohoku-Oki earthquake tend to correspond to the regions in which the orientations of the background stress field are locally similar to those of the static stress change of the earthquake. This observation suggests that those earthquake clusters were triggered by the local increase in shear stress due to the static stress change, which was caused by the spatial heterogeneity of the stress orientation, already existed before the Tohoku-Oki earthquake.

A few earthquake swarms, however, were triggered in central Tohoku where differential stress decreased by the static stress change of the Tohoku-Oki earthquake. All the earthquake swarms have notable characteristics including delays of initiation time of seismic activity by a few days to a few weeks, upward migrations of hypocenters along several thin planes, and high b-values. Such features are similar to the fluid-injection induced seismicity. The source regions of these earthquake swarms are located near the ancient caldera structures and the major geological boundaries (Yoshida et al. 2014). The swarm activities are probably triggered by the upward fluid movement along such pre-existing structures, which was facilitated by the WNW-ESE extension associated with the Tohoku-Oki earthquake. These observations demonstrate that information about the temporal evolutions of stress and frictional strength are necessary to understand the triggering mechanism of earthquakes.

Acknowledgements

We would like to thank the editor Y. Ben-Zion and two anonymous reviewers for their constructive comments which helped improve the manuscript. The figures in the present paper were created using GMT (Wessel and Smith 1998). The present study was partly supported by MEXT KAKENHI (No. 26109002).

References

Acocella, V., Yoshida, T., Yamada, R., & Funiciello, F. (2008). Structural control on late Moicene to Quaternary volcanism in the NE Honshu arc Japan. *Tectonics, 27,* 5008. https://doi.org/10.1029/2008TC002296.

Allmann, B. P., & Shearer, P. M. (2009). Global variations of stress drop for moderate to large earthquakes. *Journal of Geophysical Research: Solid Earth, 114*(1), 1–22. https://doi.org/10.1029/2008JB005821.

Asano, Y., Saito, T., Ito, Y., Shiomi, K., Hirose, H., Matsumoto, T., et al. (2011). Spatial distribution and focal mechanisms of aftershocks of the 2011 off the Pacific Coast of Tohoku Earthquake. *Earth Planets Space, 63,* 669–673. https://doi.org/10.5047/eps.2011.06.016.

Bachmann, C. E., Wiemer, S., Goertz-Allmann, B. P., & Woessner, J. (2012). Influence of pore-pressure on the event-size distribution of induced earthquakes. *Geophysical Research Letters, 39*(9), 1–7. https://doi.org/10.1029/2012GL051480.

Boese, C. M., Jacobs, K. M., Smith, E. G. C., Stern, T. A., & Townend, J. (2014). Background and delayed-triggered swarms in the central Southern Alps, South Island, New Zealand. *Geochemistry, Geophysics, Geosystems, 15*(4), 945–964.

Brown, L., Wang, K., & Sun, T. (2015). Static stress drop in the Mw 9 Tohoku-oki earthquake: Heterogeneous distribution and low average value. *Geophysical Research Letters, 42*(24), 10595–10600. https://doi.org/10.1002/2015GL066361.

Cao, A., & Gao, S. S. (2002). Temporal variation of seismic B-Values beneath Northeastern Japan Island Arc. *Geophysical*

Research Letters, 29(9), 48. https://doi.org/10.1029/2001GL013775/full.

Chiba, K., Iio, Y., & Fukahata, Y. (2013). Detailed stress fields in the focal region of the 2011 off the Pacific coast of Tohoku Earthquake: Implication for the distribution of moment release. *Earth, Planets and Space, 64,* 1157–1165. https://doi.org/10.5047/eps.2012.07.008.

Enescu, B., Aoi, S., Toda, S., Suzuki, W., Obara, K., Shiomi, K., et al. (2012). Stress perturbations and seismic response associated with the 2011 M9.0 Tohoku-oki earthquake in and around the Tokai seismic gap, central Japan. *Geophysical Research Letters, 39,* 10–15. https://doi.org/10.1029/2012GL051839.

Frohlich, C. (1992). Triangle diagrams: Ternary graphs to display similarity and diversity of earthquake focal mechanisms. *Physics of the Earth and Planetary Interiors, 75*(1–3), 193–198. https://doi.org/10.1016/0031-9201(92)90130-N.

Fukuyama, E., & Dreger, D. (2003). Performance test of an automated moment tensor determination system for the future "Tokai" earthquake. *Earth, Planets and Space, 52*(6), 383–392. https://doi.org/10.1186/BF03352250.

Fukuyama, E., Ishida, M., Dreger, D. S., & Kawai, H. (1998). Automated seismic moment tensor determination by using on-line broadband seismic waveforms. *Zisin, 51*(1), 149–156.

Gephart, J. W., & Forsyth, D. W. (1984). An improved method for determining the regional stress tensor using earthquake focal mechanism data: Application to the San Fernando earthquake sequence. *Journal of Geophysical Research, 89,* 9305–9320. https://doi.org/10.1029/JB089iB11p09305.

Hainzl, S., & Fischer, T. (2002). Indications for a successively triggered rupture growth underlying the 2000 earthquake swarm in Vogtland/NW Bohemia. *Journal of Geophysical Research: Solid Earth, 107*(B12), 1–9. https://doi.org/10.1029/2002JB001865.

Hardebeck, J. L. (2012). Coseismic and postseismic stress rotations due to great subduction zone earthquakes. *Geophysical Research Letters, 39,* L21313. https://doi.org/10.1029/2012GL053438.

Hardebeck, J. L., & Hauksson, E. (2001). Crustal stress field in Southern California and its implications for fault mechanics. *Journal of Geophysical Research, 106,* 21859–21882. https://doi.org/10.1029/2001JB000292.

Hardebeck, J. L., & Michael, A. J. (2006). Damped regional-scale stress inversions: Methodology and examples for Southern California and the Coalinga aftershock sequence. *Journal of Geophysical Research, 111,* B11310. https://doi.org/10.1029/2005JB004144.

Hardebeck, J. L., & Shearer, P. M. (2002). A new method for determining first-motion focal mechanisms. *Bulletin of the Seismological Society of America, 92,* 2264–2276. https://doi.org/10.1785/0120010200.

Hasegawa, A. (2017). Role of H_2O in Generating Subduction Zone Earthquakes. *Monographs on Environment, Earth and Planets, 5,* 1–34. https://doi.org/10.5047/meep.2017.00501.0001.

Hasegawa, A., Horiuchi, S., & Umino, N. (1994). Seismic structure of the northeastern Japan convergent margin: A synthesis. *Journal of Geophysical Research: Solid Earth, 99,* 22295–22311. https://doi.org/10.1029/93JB02797.

Hasegawa, A., & Yoshida, K. (2015). Preceding seismic activity and slow slip events in the source area of the 2011 Mw 9.0 Tohoku-Oki earthquake: A review. *Geoscience Letters, 2,* 6. https://doi.org/10.1186/s40562-015-0025-0.

Hasegawa, A., Yoshida, K., Asano, Y., Okada, T., Iinuma, T., & Ito, Y. (2012). Change in stress field after the 2011 great Tohoku-Oki earthquake. *Earth and Planetary Science Letters, 355–356,* 231–243. https://doi.org/10.1016/j.epsl.2012.08.042.

Hino, R., & Kido, M. (2012). Coseismic slip distribution of the 2011 off the Pacific Coast of Tohoku Earthquake (M9.0) refined by means of seafloor geodetic data. *Journal of Geophysical Research: Solid Earth, 117,* B07409. https://doi.org/10.1029/2012JB009186.

Hubbert, M., & Rubey, W. (1959). Role of fluid pressure in mechanics of overthrust faulting I. Mechanics of fluid-filled porous solids and its application to overthrust faulting. *Geological Society of America Bulletin, 70,* 115–166.

Imanishi, K., Ando, R., & Kuwahara, Y. (2011). Unusual shallow normal-faulting earthquake sequence in compressional northeast Japan activated after the 2011 off the Pacific coast of Tohoku earthquake. *Geophysical Research Letters, 39,* L09306. https://doi.org/10.1029/2012GL051491.

Ishibe, T., Shimazaki, K., Satake, K., & Tsuruoka, H. (2011). Change in seismicity beneath the Tokyo metropolitan area due to the 2011 off the Pacific coast of Tohoku Earthquake. *Earth, Planets and Space, 63*(7), 40. https://doi.org/10.5047/eps.2011.06.001.

Julian, B. R., Foulger, G. R., Monastero, F. C., & Bjornstad, S. (2010). Imaging hydraulic fractures in a geothermal reservoir. *Geophysical Research Letters, 37,* 1–5. https://doi.org/10.1029/2009GL040933.

Kanisawa, S., Otsuki, K., Ehiro, M., Yoshida, T., Kazama, M., Kano, K., Takarada, S., Wakita, K., Kyogoku, M., Nakayama, M., Shikama, S., Koyama, T. and Miura, A. (2006). Geology of NE Honshu for construction engineers (with digital geological map of Tohoku district, Japan (1:200,000)) (in Japanese), report, 408 pp., Tohoku Constr. Assoc., Sendai, Japan.

Kato, T., El-Fiky, G. S., Oware, E. N., & Miyazaki, S. (1998). Crustal strains in the Japanese Islands as deduced from dense GPS array. *Geophysical Research Letters, 25*(18), 3445–3448. https://doi.org/10.1029/98GL02693.

Kato, A., Fukuda, J., & Obara, K. (2013). Response of seismicity to static and dynamic stress changes induced by the 2011 M9.0 Tohoku-Oki earthquake. *Geophysical Research Letters, 40,* 3572–3578. https://doi.org/10.1002/grl.50699.

Kato, A., & Igarashi, T. (2012). Regional extent of the large coseismic slip zone of the 2011 Mw 9.0 Tohoku-Oki earthquake delineated by on-fault aftershocks. *Geophysical Research Letters, 39*(15), 2–7. https://doi.org/10.1029/2012GL052220.

Kato, A., Sakai, S., & Obara, K. (2011). A normal-faulting seismic sequence triggered by the 2011 off the Pacific coast of Tohoku earthquake: Wholesale stress regime changes in the upper plate. *Earth, Planets and Space, 63,* 745–748. https://doi.org/10.5047/eps.2011.06.014.

Kosuga, M. (2014). Seismic activity near the Moriyoshi-zan volcano in Akita Prefecture, northeastern Japan: Implications for geofluid migration and a midcrustal geofluid reservoir. *Earth, Planets and Space, 66,* 77. https://doi.org/10.1186/1880-5981-66-77.

Lengliné, O., Enescu, B., Peng, Z., & Shiomi, K. (2012). Decay and expansion of the early aftershock activity following the 2011, Mw 9.0 Tohoku earthquake. *Geophysical Research Letters, 39,* 6–11. https://doi.org/10.1029/2012GL052797.

Lund, B., & Townend, J. (2007). Calculating horizontal stress orientations with full or partial knowledge of the tectonic stress tensor. *Geophysical Journal International*. https://doi.org/10.1111/j.1365-246X.2007.03468.x.

Mallman, E. P., & Zoback, M. D. (2007). Assessing elastic Coulomb stress transfer models using seismicity rates in southern California and southwestern Japan. *Journal of Geophysical Research, 112*(B3), 37.

Martínez-Garzón, P., & Ben-Zion, Y. (2016). A refined methodology for stress inversions of earthquake focal mechanisms. *Journal of Geophysical Research: Solid Earth, 121,* 8666–8687. https://doi.org/10.1002/2016JB013493.

Martínez-Garzón, P., Kwiatek, G., Bohnhoff, M., & Dresen, G. (2016). Impact of fluid injection on fracture reactivation at The Geysers geothermal field. *Journal of Geophysical Research: Solid Earth, 121,* 7432–7449.

Matsuda, T., & Uyeda, S. (1971). On the pacific-type orogeny and its model—extension of the paired belts concept and possible origin of marginal seas. *Tectonophysics, 11*(1), 5–27. https://doi.org/10.1016/0040-1951(71)90076-X/.

McKenzie, D. (1969). The relationship between fault plane solutions for earthquakes and the directions of the principal stresses. *Bulletin of the Seismological Society of America, 59,* 591–601.

Michael, A. J. (1984). Determination of stress from slip data; faults and folds. *Journal of Geophysical Research, 89,* 11517–11526. https://doi.org/10.1029/JB089iB13p11517.

Michael, A. J. (1987). Use of focal mechanisms to determine stress: A control study. *Journal of Geophysical Research, 92,* 357–368. https://doi.org/10.1029/JB092iB01p00357.

Miller, S. A. (2013). The role of fluids in tectonic and earthquake processes. *Advances in Geophysics, 54,* 1–46.

Miura, S., Sato, T., Tachibana, K., Satake, Y., & Hasegawa, A. (2002). Strain accumulation in and around Ou Backbone Range, northeastern Japan as observed by a dense GPS network. *Earth Planets Space, 54,* 1071–1076.

Miyazawa, M. (2011). Propagation of an earthquake triggering front from the 2011 Tohoku-Oki earthquake. *Geophysical Research Letters, 38,* 1–6. https://doi.org/10.1029/2011GL049795.

Nakajima, J., Yoshida, K., & Hasegawa, A. (2013). An intraslab seismic sequence activated by the 2011 Tohoku-oki earthquake: Evidence for fluid-related embrittlement. *Journal of Geophysical Research: Solid Earth, 118,* 3492–3505. https://doi.org/10.1002/jgrb.50246.

Nakamichi, H., Okuda, T., Horikawa, S., Katao, H., Miura, T., Kubo, A., et al. (2015). Hypocenter migration and crustal seismic velocity distribution observed for the inland earthquake swarms induced by the 2011 Tohoku-Oki earthquake in NE Japan: implications for crustal fluid distribution and crustal permeability. *Geofluids, 15,* 293–309. https://doi.org/10.1111/gfl.12112.

Nakamura, W., Uchida, N., & Matsuzawa, T. (2016). Spatial distribution of the faulting types of small earthquakes around the 2011 Tohoku-oki earthquake: A comprehensive search using template events. *Journal of Geophysical Research, 121,* 2591–2607. https://doi.org/10.1002/2015JB012584.

Nakamura, K., & Uyeda, S. (1980). Stress gradient in arc–back arc regions and plate subduction. *Journal of Geophysical Research, 85*(B11), 6419. https://doi.org/10.1029/JB085iB11p06419.

Nur, A., & Booker, J. R. (1972). Aftershocks caused by pore fluid flow? *Science, 175*(4024), 885–887. https://doi.org/10.1126/science.175.4024.885.

Okada, Y. (1992). Internal deformation due to shear and tensile faults in a half-space. *Bulletin of the Seismological Society of America, 82,* 1018–1040.

Okada, T., Yoshida, K., Ueki, S., Nakajima, J., Uchida, N., Matsuzawa, T., et al. (2011). Shallow inland earthquakes in NE Japan possibly triggered by the 2011 off the Pacific coast of Tohoku Earthquake. *Earth, Planets and Space, 63*(7), 749–754. https://doi.org/10.5047/eps.2011.06.027.

Okada, T., Matsuzawa, T., Umino, N., Yoshida, K., Hasegawa, A., Takahashi, H., Yamada, T., Kosuga, M., Takeda, T., Kato, A., Igarashi, T., Obara, K., Sakai, S., Saiga, A., Iidaka, T., Iwasaki, T., Hirata, N., Tsumura, N., Yamanaka, Y., Terakawa, T., Oth, A. (2013). On the characteristics of earthquake stress release variations in Japan. *Earth and Planetary Science Letters, 377–378,* 132–141. https://doi.org/10.1016/j.epsl.2013.06.037.

Rice, J. R. (1992). Fault stress states, pore pressure distributions, and the weakness of the San Andreas Fault. In B. Evans & T. F. Wong (Eds.), *Fault mechanics and transport properties of rocks* (pp. 475–503). New York: Academic Press.

Rutledge, J. T., Phillips, W. S., & Mayerhofer, M. J. (2004). Faulting induced by forced fluid injection and fluid flow forced by faulting: An interpretation of hydraulic-fracture microseismicity, Carthage Cotton Valley gas field, Texas. *Bulletin of the Seismological Society of America, 94,* 1817–1830. https://doi.org/10.1785/012003257.

Sagiya, T., Miyazaki, S., & Tada, T. (2000). Continuous GPS array and present-day crustal deformation of Japan. *Pure and Applied Geophysics, 157,* 2303–2322.

Seno, T. (1999). Syntheses of the regional stress fields of the Japanese islands. *Island Arc, 8*(1), 66–79. https://doi.org/10.1046/j.1440-1738.1999.00225.x.

Shapiro, S. A., Huenges, E., & Borm, G. (1997). Estimating the crust permeability from fluid-injection-induced seismic emission at the KTB site. *Geophysical Journal International, 131,* F15–F18. https://doi.org/10.1111/j.1365-246X.1997.tb01215.x.

Shimojo, K., Enescu, B., Yagi, Y., & Takeda, T. (2014). Fluid-driven seismicity activation in northern Nagano region after the 2011 M9.0 Tohoku-oki earthquake. *Geophysical Research Letters, 41,* 7524–7531. https://doi.org/10.1002/2014GL061763.

Sibson, R. H. (1990). Rupture nucleation on unfavorably oriented faults. *Bulletin of the Seismological Society of America, 80*(6), 1580–1604. https://doi.org/10.1021/jp111520r.

Sibson, R. (1992). Implications of fault-valve behaviour for rupture nucleation and recurrence. *Tectonophysics, 211*(1–4), 283–293. https://doi.org/10.1016/0040-1951(92)90065-E.

Simpson, R. W., and P. A. Reasenberg, Earthquake-induced static-stress changes on central California faults, U.S. Geol. Surv. Prof. Pap., 1550-F, 55-89, 199.

Smith, D. E., & Dieterich, J. H. (2010). Aftershock sequences modeled with 3-D stress heterogeneity and rate-state seismicity equations: Implications for crustal stress estimation. *Pure and Applied Geophysics, 167*(8–9), 1067–1085. https://doi.org/10.1007/s00024-010-0093-1.

Suwa, Y., Miura, S., Hasegawa, A., Sato, T., & Tachibana, K. (2006). Interplate coupling beneath NE Japan inferred from three-dimensional displacement field. *Journal of Geophysical Research: Solid Earth, 111*(4), 1–12. https://doi.org/10.1029/2004JB003203.

Suzuki, Y., S. Toda, K. Yoshida, and T. Okada (2014), Local receiver fault dependency of seismicity shut down in the 2011

Tohoku-oki stress shadow, In AGU Fall Meeting, San Francisco, Abstracts (S23A–4473).

Terakawa, T., Hashimoto, C., & Matsu'ura, M. (2013). Changes in seismic activity following the 2011 Tohoku-oki earthquake: Effects of pore fluid pressure. *Earth and Planetary Science Letters, 365,* 17–24. https://doi.org/10.1016/j.epsl.2013.01.017.

Terakawa, T., & Matsu'ura, M. (2010). The 3-D tectonic stress fields in and around Japan inverted from centroid moment tensor data of seismic events. *Tectonics, 29,* TC6008. https://doi.org/10.1029/2009TC002626.

Toda, S., Stein, R. S., & Lin, J. (2011). Widespread seismicity excitation throughout central Japan following the 2011 M = 9.0 Tohoku earthquake and its interpretation by Coulomb stress transfer. *Geophysical Research Letters, 38,* 1–5. https://doi.org/10.1029/2011GL047834.

Townend, J. (2006). What do faults feel? Observational constraints on the stresses acting on seismogenic faults. In R. Abercrombie, A. McGarr, H. Kanamori, & G. Di Toro (Eds.), *Earthquakes: Radiated Energy and the Physics of Faulting (AGU Monograph Series* (Vol. 170, pp. 313–327). Washington D.C., USA: American Geophysical Union.

Townend, J., & Zoback, M. D. (2001). Implications of earthquake focal mechanisms for the frictional strength of the San Andreas fault system. *Geological Society, London, Special Publications, 186*(1), 13–21.

Townend, J., & Zoback, M. D. (2006). Stress, strain, and mountain building in central Japan. *Journal of Geophysical Research, 111,* B03411. https://doi.org/10.1029/2005JB003759.

Ukawa, M. (1982). Lateral stretching of the philippine sea plate subducting along the nankai-suruga trough. *Tectonics, 1*(6), 543–571. https://doi.org/10.1029/TC001i006p00543.

Usami, T. (2003). *A catalogue of disastrous earthquakes in Japan* (updated ed.). Tokyo: University of Tokyo Press. **(in Japanese)**.

Wang, K., & He, J. (1999). Mechanics of low-stress forearcs: Nankai and Cascadia. *Journal of Geophysical Research, 104*(B7), 15191–15205.

Wessel, P., & Smith, W. H. F. (1998). New, improved version of the Generic Mapping Tools released. *Eos, Transactions AGU, 79,* 579.

Wesson, R. L., & Boyd, O. S. (2007). Stress before and after the 2002 Denali fault earthquake. *Geophysical Research Letters, 34,* L07303. https://doi.org/10.1029/2007GL029189.

Wyss, M. (1973). Towards a physical understanding of the earthquake frequency distribution. *Geophysical Journal of the Royal Astronomical Society, 31*(4), 341–359. https://doi.org/10.1111/j.1365-246X.1973.tb06506.x.

Yoshida, T. (2017). Basic structure of Tohoku District. In T. Yoshida, et al. (Eds.), *Regional geology of Japan 2* (pp. 7–103). Tohoku District: Asakura Publishing Co., Ltd. **(in Japanese)**.

Yoshida, K., & Hasegawa, A. (2018a). Sendai-Okura earthquake swarm induced by the 2011 Tohoku-Oki earthquake in the stress shadow of NE Japan: Detailed fault structure and hypocenter migration. *Tectonophysics, 733,* 132–147.

Yoshida, K., & Hasegawa, A. (2018b). Hypocenter migration and seismicity pattern change in the Yamagata-Fukushima Border, NE Japan, Caused by Fluid Movement and Pore Pressure Variation. *Journal of Geophysical Research [Solid Earth], 95,* 664.

Yoshida, K., Hasegawa, A., & Okada, T. (2015a). Spatial variation of stress orientations in NE Japan revealed by dense seismic observations. *Tectonophysics, 647–648,* 63–72. https://doi.org/10.1016/j.tecto.2015.02.013.

Yoshida, K., Hasegawa, A., & Okada, T. (2015b). Spatially heterogeneous stress field in the source area of the 2011 Mw 6.6 Fukushima-Hamadori earthquake, NE Japan, probably caused by static stress change. *Geophysical Journal International, 201*(2), 1062–1071. https://doi.org/10.1093/gji/ggv068.

Yoshida, K., Hasegawa, A., & Okada, T. (2016a). Heterogeneous stress field in the source area of the 2003 M6.4 Northern Miyagi Prefecture, NE Japan, earthquake. *Geophysical Journal International, 206*(1), 408–419. https://doi.org/10.1093/gji/ggw160.

Yoshida, K., Hasegawa, A., Okada, T., & Iinuma, T. (2014a). Changes in the stress field after the 2008 M7.2 Iwate-Miyagi Nairiku earthquake in northeastern Japan. *Journal of Geophysical Research: Solid Earth, 119*(12), 9016–9030. https://doi.org/10.1002/2014jb011291.

Yoshida, K., Hasegawa, A., Okada, T., Iinuma, T., Ito, Y., & Asano, Y. (2012). Stress before and after the 2011 great Tohoku-oki earthquake and induced earthquakes in inland areas of eastern Japan. *Geophysical Research Letters, 39,* L03302. https://doi.org/10.1029/2011GL049729.

Yoshida, K., Hasegawa, A., & Yoshida, T. (2016b). Temporal variation of frictional strength in an earthquake swarm in NE Japan caused by fluid migration. *Journal of Geophysical Research: Solid Earth.* https://doi.org/10.1002/2015jb012352. received.

Yoshida, T., Kimura, J., Yamada, R., Acocella, V., Sato, H., Zhao, D., et al. (2014b). Evolution of late Cenozoic magmatism and the crust-mantle structure in the NE Japan Arc. *Geological Society, London, Special Publications, 385,* 335–387. https://doi.org/10.1144/SP385.15.

Yoshida, T., Nakajima, J., Hasegawa, A., Sato, H., Nagahashi, Y., Kimura, J., et al. (2005). Evolution of late Cenozoic magmatism in the NE Honshu Arc and its relation to the crust-mantle structures. *Quaternary Research, 44,* 195–216. **[in Japanese with English abstract]**.

Yoshida, K., Pulido, N., & Fukuyama, E. (2016c). Unusual stress rotations within the Philippines possibly caused by slip heterogeneity along the Philippine fault: Stress fields in the Philippines. *Journal of Geophysical Research [Solid Earth], 121*(3), 2020–2036.

Yoshida, K., Saito, T., Urata, Y., Asano, Y., & Hasegawa, A. (2017). Temporal changes in stress drop, frictional strength, and earthquake size distribution in the 2011 Yamagata-Fukushima, NE Japan. Earthquake Swarm, caused by fluid migration: Changes in stress drop and B-value. *Journal of Geophysical Research [Solid Earth], 122*(12), 10379–10397.

Yukutake, Y., Honda, R., Harada, M., Aketagawa, T., Ito, H., & Yoshida, A. (2011). Remotely-triggered seismicity in the Hakone volcano following the 2011 off the Pacific coast of Tohoku Earthquake. *Earth, Planets and Space, 63,* 737–740. https://doi.org/10.5047/eps.2011.05.004.

(Received June 27, 2018, revised November 30, 2018, accepted November 30, 2018, Published online December 7, 2018)

Pure Appl. Geophys. 176 (2019), 1357
© 2018 The Author(s)
https://doi.org/10.1007/s00024-018-2035-2

Pure and Applied Geophysics

Correction

Correction to: Stress Inversion of Regional Seismicity in the Sea of Marmara Region, Turkey

Christopher Wollin,[1] Marco Bohnhoff,[1] Václav Vavryčuk,[2] and Patricia Martínez-Garzón[1]

Correction to: Pure Appl. Geophys. https://doi.org/10.1007/s00024-018-1971-1

The article "Stress Inversion of Regional Seismicity in the Sea of Marmara Region, Turkey", written by Wollin et al., was originally published electronically on the publisher's internet portal (currently SpringerLink) on 03 September 2018 without open access.

With the author(s)' decision to opt for Open Choice the copyright of the article changed on November 2018 to © The Author(s) 2018 and the article is forthwith distributed under the terms of the Creative Commons Attribution 4.0 International License (http://creativecommons.org/licenses/by/4.0/), which permits use, duplication, adaptation, distribution and reproduction in any medium or format, as long as you give appropriate credit to the original author(s) and the source, provide a link to the Creative Commons license and indicate if changes were made.

(Published online November 21, 2018)

The original article can be found online at https://doi.org/10.1007/s00024-018-1971-1.

[1] German Research Center for Geosciences, Section 4.2: "Geomechanics and Rheology", Telegrafenberg, 14473 Potsdam, Germany. E-mail: wollin@gfz-potsdam.de

[2] Institute of Geophysics of the Czech Academy of Science, Bocni II/1401, 14100 Praha 4, Czech Republic.

MIX
Papier aus verantwortungsvollen Quellen
Paper from responsible sources
FSC® C105338

If you have any concerns about our products,
you can contact us on
ProductSafety@springernature.com

In case Publisher is established outside the EU,
the EU authorized representative is:
Springer Nature Customer Service Center GmbH
Europaplatz 3, 69115 Heidelberg, Germany

Printed by Libri Plureos GmbH
in Hamburg, Germany